The Clinical Neurobiology of the Hippocampus

The Clinical Neurobiology of the Hippocampus

An Integrative View

EDITED BY

Thorsten Bartsch

Department of Neurology

University Hospital Schleswig-Holstein

University of Kiel

Kiel

Germany

OXFORD

UNIVERSITY PRESS

OXFORD
UNIVERSITY PRESS

Great Clarendon Street, Oxford, OX2 6DP,
United Kingdom

Oxford University Press is a department of the University of Oxford.
It furthers the University's objective of excellence in research, scholarship,
and education by publishing worldwide. Oxford is a registered trade mark of
Oxford University Press in the UK and in certain other countries

First Edition published in 2012

Impression: 1

British Library Cataloguing in Publication Data

Data available

Library of Congress Cataloging in Publication Data

Library of Congress Control Number: 2012938466

ISBN 978–0–19–959238–8

Printed in Spain by
Grafos S.A., Barcelona

Dedication

To Hilke, and our sons Bela and Jesko

Preface

The hippocampus is one of the most studied structures in the human brain and plays a pivotal role in human memory function. This recognized function is reflected by the presence of an extensive body of neurophysiological, neuropsychological, anatomical, and neurocomputational literature that presents basic mechanisms, theoretical models, and psychological concepts. However, it occurred to me that in the rapidly growing field of hippocampal research, the clinical aspects of the diseases that affect the hippocampus are greatly underrepresented in the literature, and clinical approaches and concepts are scattered throughout various clinical and basic scientific disciplines.

The clinical spectrum of hippocampal dysfunction encompasses a wide range of neurological, behavioural, and psychiatric symptoms and surpasses the ability to encode, store, and retrieve information. The methods of hippocampal imaging and electrophysiology have become more important in recent years. The relevance of hippocampal involvement in clinical diseases goes beyond mere neuropsychological deficits and includes psychopathological states in various conditions, such as acute amnesic syndromes, Alzheimer's disease, temporal lobe epilepsy, sleep, stroke medicine, limbic encephalitis, neurodevelopmental disorders, stress- and trauma-related disorders, depression, rehabilitative medicine, and schizophrenia.

This book aims to integrate the clinical approaches to diseases that affect the hippocampus. This text is not intended to be another neuropsychology or neurophysiology textbook but instead is written for the clinical neurologist, neuroradiologist, neuropaediatrician, psychiatrist, and clinically orientated basic scientist, including interested students. The aim of this project is to integrate and review the common methods, clinical findings, concepts, mechanisms, and, if applicable, therapeutic strategies of these clinical approaches.

The first part of the book covers the basic and integrative features of the hippocampus, such as the anatomy and imaging of this structure, and the basic mechanisms of hippocampal function, including the principles of hippocampus-dependent memory processing in amnesia and sleep, the mechanisms of vulnerability, and adult neurogenesis as well as the effects of stress. The second part of the book covers the various clinical manifestations in which the hippocampus is involved and in which the preceding basic mechanisms are reflected.

With that in mind, it has been a great pleasure to collect a list of specialist authors and experts on the basic and clinical aspects of the hippocampus.

I wish to thank Charlotte Green and Dr Martin Baum from Oxford University Press for their assistance and support throughout the complete creation process. I would like also to thank Professor Günther Deuschl for encouragement and continuous support during the 2 years in which the book was developed. This work was supported by the Deutsche Forschungsgemeinschaft (SFB 654) and the Faculty of Medicine of the University of Kiel.

Lastly, I wish to thank my family for giving me time to devote myself to this book.

If this book contributes to an integrative view of the hippocampus and stimulates interdisciplinary discussions in clinical neuroscience, it might be considered a success.

Contents

List of Contributors xiii
List of Abbreviations xv
Epigraphs xvii

1 Introduction: The Hippocampus in the Clinical Neurosciences 1
Thorsten Bartsch
History of hippocampal research 2
Anatomical explorations 3
Involvement of the hippocampus in neurological and psychiatric diseases 4
Basic principles of the pathophysiology of the hippocampus 5
Functions attributed to the hippocampus 6
Structure of the book 6
Outlook 8

2 Functional Anatomy, Development, and Pathology of the Hippocampus 10
Pierre Lavenex
Introduction 10
Overall organization 11
Postnatal development 27
Neurodevelopmental disorders 32
Conclusion 34

3 Amnesia and the Hippocampal Memory System 39
Howard Eichenbaum
Introduction 39
Case H.M. 40
Spared learning in H.M. and other patients with hippocampal damage 43
What are the fundamental features of declarative and procedural memory? 44
Temporally graded retrograde amnesia and memory consolidation 49
Recent challenges about the defining features of declarative memory and their role in amnesia 51

4 Selective Neuronal Vulnerability in the Hippocampus: Relationship to Neurological Diseases and Mechanisms for Differential Sensitivity of Neurons to Stress 54
Elias K. Michaelis
Importance of the hippocampus in learning, memory, and neurological diseases 55
The concept of selective neuronal vulnerability in the central nervous system 55
Focus on selective neuronal vulnerability in the hippocampus 56
Mechanisms of selective neuronal vulnerability 59
Concluding ideas 70

5 Stress and the Hippocampus 77

E. Ronald de Kloet

Introduction 77
Pulsatility of basal HPA axis activity 80
Stress-induced HPA axis activity 82
MR- and GR-mediated behavioural effects 83
Neuroendocrine regulation 87
Morphology 88
Network function 89
Molecular effects 90
MR and GR polymorphisms 92
Early life experience 93
Implication for diseases of the hippocampus 95

6 Adult Hippocampal Neurogenesis: Functional Implications of a 'New' Type of Plasticity 105

Alexander Garthe and Gerd Kempermann

Introduction: adult neurogenesis in the hippocampus 105
Theories of hippocampal function 107
Experimental evidence for a functional role of adult neurogenesis 112
Conclusion 118

7 Imaging the Hippocampus 122

Bradford C. Dickerson and Jean C. Augustinack

Introduction 122
Hippocampal neuroanatomy for neuroimaging 123
Brief history of hippocampal neuroimaging 123
Imaging and quantifying hippocampal and MTL cortical anatomy 124
Imaging task-related hippocampal and MTL cortical function 128
Imaging hippocampal perfusion and metabolism 132
Imaging hippocampal connectivity 132
Molecular imaging of the hippocampus 134
Imaging hippocampal plasticity 134
Ex vivo hippocampal imaging 135
Technical advances and future directions 137

8 Sleep, Memory, and the Hippocampus 144

Ariane Foret, Laura Mascetti, Caroline Kussé, Mathieu Jaspar, Vincenzo Muto, Anahita Shaffii-Le Bourdiec, Luca Matarazzo, and Pierre Maquet

Introduction 144
Basic facts about sleep 145
Behavioural evidence for a favourable effect of sleep on declarative memories 146
The synaptic downscaling hypothesis 146
The hypothesis of a hippocampo-neocortical dialogue during sleep 147
Hippocampal activity during immediate post-training wakefulness 149

9 The Ageing Hippocampus 152

Lan T. Hoang, James P. Lister, and Carol A. Barnes

Introduction 152
Cross-species cognitive ageing 153
Spatial memory 154
Neuroanatomical changes in the hippocampus 154
Biophysical properties of aged neurons 158
Electrophysiological measures of connectivity 159
Neural plasticity 160
Age-related changes in gene expression 162
Dynamic alterations of Arc transcription—treatment approaches 163
Dynamics of aged neural ensembles 164
Summary 168

10 Neurodevelopmental Disorders 174

Alexander Prehn-Kristensen, Richard Cooke, Lioba Baving, and Cynthia Schumann

Hippocampal size and intelligence 175
Neurodevelopmental outcome in children born preterm and the hippocampus 176
Febrile seizures and hippocampal sclerosis 179
Embryo fetal alcohol syndrome 180
Effects of cocaine on the developing hippocampus 181
Autism spectrum disorder 181
Angelman syndrome 186
Down syndrome 187
Fragile X syndrome 188
Attention-deficit hyperactivity disorder 189
Gilles de la Tourette syndrome 190
Conclusions 191

11 The Hippocampus in Neurological Disease 200

Thorsten Bartsch

The vulnerability of the hippocampus in neurological disease 200
MRI correlates of hippocampal damage 203
Hippocampal ischaemia 203
Anoxic encephalopathy 204
Limbic encephalitis 205
Epilepsy, status epilepticus, and postictal changes 208
Traumatic brain injury 209
Hypoglycaemic encephalopathy 210
Transient global amnesia 210
Multiple sclerosis 215

12 Alzheimer's Disease 223

Maija Pihlajamäki and Hilkka Soininen

Introduction 223
Alzheimer's disease 224
Structural MRI 226

Functional MRI 231
Hippocampal structure and function in AD and MCI 236
Concluding remarks 237

13 Temporal Lobe Epilepsy and the Hippocampus 245
Radwa A. B. Badawy and Graeme D. Jackson
Introduction 245
Mesial temporal sclerosis and epileptogenicity 246
Clinical picture 248
Electroencephalographic findings 250
Neuroimaging in temporal lobe epilepsy 252
Nuclear neuroimaging with positron emission tomography (PET) and single-photon emission computed tomography (SPECT) 254
Treatment 256

14 The Hippocampus and Post-Traumatic Disorders 262
J. Douglas Bremner and Eric Vermetten
Trauma spectrum psychiatric disorders 262
Effects of stress on the hippocampus 263
Hippocampal function in PTSD 264
Glucocorticoids and the hippocampus 266
Hippocampal and other glucocorticoid receptor expression as risk factors for PTSD 267
Functional neuroimaging studies of the hippocampus in PTSD 267
Hippocampus and pharmacotherapy for PTSD 267
Summary and conclusions 268

15 The Hippocampus in Major Depression 273
Glenda M. MacQueen and Thomas Frodl
Introduction 273
Regional morphological variation in the hippocampus in MDD 274
Can hippocampal dysregulation be linked to MDD? 275
Clinical and demographic variables influencing hippocampal volumes in MDD 276
Biological mechanisms that may contribute to small hippocampal volume in MDD 277
The clinical relevance of small hippocampal volumes in MDD 281
Evidence that treatment for MDD may have an impact on hippocampal volumes 282
Summary 283

16 Schizophrenia 288
Peter Falkai, Oliver Gruber, and Andrea Schmitt
Epidemiology and aetiology of schizophrenia 288
Symptoms of schizophrenia with focus on cognitive deficits 290
Hippocampus and cognition in schizophrenia 290
The hippocampus and schizophrenia: findings from MRI studies 290
The hippocampus and schizophrenia: findings from structural postmortem studies 291
From behaviour to the molecular basis of hippocampal alterations in schizophrenia 293
Functional role of schizophrenia risk genes in the hippocampus 295
Influence of antipsychotics on the hippocampus 296

Author Index 303
Subject Index 305

List of Contributors

Jean C. Augustinack Department of Radiology, Athinoula A. Martinos Center for Biomedical Imaging, Massachusetts General Hospital and Harvard Medical School, Boston, MA, USA

Radwa A. B. Badawy Department of Medicine, The University of Melbourne, Brain Research Institute, Department of Neurology, Austin Health, Melbourne, Australia

Lioba Baving Department of Child and Adolescent Psychiatry, University of Kiel, Kiel, Germany

Carol A. Barnes ARL Division of Neural Systems, Memory and Aging Psychology and Neurology Life Sciences, University of Arizona, Tucson, AZ, USA

Thorsten Bartsch Department of Neurology, University Hospital Schleswig-Holstein, University of Kiel, Kiel, Germany

J. Douglas Bremner Department of Psychiatry, Emory University School of Medicine, Atlanta, GA, USA

Richard Cooke University of Liverpool, School of Reproductive and Developmental Medicine, Liverpool, UK

E. Ronald de Kloet Department of Medical Pharmacology, Leiden/Amsterdam Center of Drug Research, Leiden University Medical Center, Leiden, The Netherlands

Bradford C. Dickerson Department of Neurology, MGH Memory Disorders and Frontotemporal Dementias Unit, Massachusetts General Hospital and Harvard Medical School, Boston, MA, USA

Howard Eichenbaum Center for Memory and Brain, Boston University, Boston, MA, USA

Peter Falkai Department of Psychiatry, University of Göttingen, Göttingen, Germany

Ariane Foret Cyclotron Research Centre – Liège University, Liège, Belgium

Thomas Frodl Department of Psychiatry, Trinity College Dublin, Dublin, Ireland

Alexander Garthe Center for Regenerative Therapies, University of Dresden, Dresden, Germany

Oliver Gruber Department of Psychiatry, University of Göttingen, Göttingen, Germany

Lan T. Hoang ARL Division of Neural Systems, Memory and Aging Psychology and Neurology Life Sciences, University of Arizona, Tucson, AZ, USA

Graeme D. Jackson Brain Research Institute, Department of Neurology, Austin Health, Royal Children's Hospital, Melbourne, Australia

Mathieu Jaspar Cyclotron Research Centre – Liège University, Liège, Belgium

Gerd Kempermann Center for Regenerative Therapies, University of Dresden, Dresden, Germany

Caroline Kussé Cyclotron Research Centre – Liège University, Liège, Belgium

Pierre Lavenex Laboratory of Brain and Cognitive Development, Department of Medicine/Physiology Fribourg, Switzerland

James P. Lister ARL Division of Neural Systems, Memory & Aging, University of Arizona, Tucson, AZ, USA

Glenda M. MacQueen Department of Psychiatry, Faculty of Medicine, University of Calgary, Calgary, Canada

Pierre Maquet Cyclotron Research Centre – Liège University, Liège, Belgium

Laura Mascetti Cyclotron Research Centre – Liège University, Liège, Belgium

Luca Matarazzo Cyclotron Research Centre – Liège University, Liège, Belgium

Elias K. Michaelis Higuchi Biosciences Center, Department of Pharmacology & Toxicology, University of Kansas, Lawrence, KS, USA

Vincenzo Muto Cyclotron Research Centre – Liège University, Liège, Belgium

Maija Pihlajamäki Department of Neurology, University of Eastern Finland, Kuopio, Finland

Alexander Prehn-Kristensen Department of Child and Adolescent Psychiatry, University of Kiel, Kiel, Germany

Andrea Schmitt Department of Psychiatry, University of Göttingen, Göttingen, Germany

Cynthia Schumann Psychiatry and Behavioral Sciences. UC Davis MIND Institute, Sacramento, CA, USA

Anahita Shaffii-Le Bourdiec Cyclotron Research Centre – Liège University, Liège, Belgium

Hilkka Soininen Department of Neurology, University of Eastern Finland, Kuopio, Finland

Eric Vermetten Military Mental Health/University Medical Center Utrecht, Utrecht, Netherlands

List of Abbreviations

3D	three-dimensional
5-HT	5-hydroxytryptamine (serotonin)
5-HTT	5-hydroxytryptamine (serotonin) transporter
Ab	amyloid beta
AChA	anterior choroidal artery
ACTH	adrenocorticotropin
AD	Alzheimer's disease
ADC	apparent diffusion coefficient
ADX	adrenalectomized
AED	antiepileptic drug
AHP	after-hyperpolarizing potential
ALS	amyotrophic lateral sclerosis
AMPA	alpha-amino-3-hydroxyl-5-methyl-4-isoxazole-propionate
AS	Angelman syndrome
ATP	adenosine triphosphate
BDNF	brain-derived neurotrophic factor
BOLD	blood-oxygen-level dependent
BPD	borderline personality disorder
CA	cornu ammonis
CBG	corticosteroid binding globulin
Cho	choline
CNS	central nervous system
CORT	cortisol or corticosterone
CREB	cAMP response element binding protein
CRF	corticotropin-releasing factor
CRH	corticotrophin-releasing hormone
CS	conditioning stimulus
CSF	cerebrospinal fluid
CT	computed tomography
DA	dopamine
DBS	deep brain stimulation
DID	dissociative identity disorder
DS	Down syndrome
DTI	diffusion tensor imaging
DTT	diffusion-tensor tracking
DWI	diffusion-weighted imaging
EAE	experimental autoimmune encephalomyelitis
EC	entorhinal cortex
EEG	electroencephalography
EPSP	excitatory postsynaptic potential
ETC	electron transport chain
FA	fractional anisotropy
FAS	fetal alcohol syndrome
FASD	fetal alcohol spectrum disorders
fcMRI	functional connectivity magnetic resonance imaging
FDG	fluorodeoxyglucose
FLAIR	fluid-attenuated inversion recovery
fMRI	functional magnetic resonance imaging
GABA	gamma-aminobutyric acid
GAD	glutamic acid decarboxylase
GR	glucocorticoid receptor
HC	hippocampus
HD	Huntington's disease
HHV	human herpesvirus
HPA	hypothalamic–pituitary–adrenal
HS	hippocampal sclerosis
HSE	Herpes simplex encephalitis
ICA	internal carotidal artery
icv	intracerebroventricular
IEG	immediate early gene
IPI	initial precipitating injury
IQ	intelligence quotient
LCM	laser capture microdissection
LTD	long-term depression
LTLE	lateral temporal lobe epilepsy
LTP	long-term potentiation
MAP2A	microtubule-associated protein 2A
MCI	mild cognitive impairment
MD	mean diffusivity
MDD	major depressive disorder
mEPSC	miniature excitatory postsynaptic current
MPH	methylphenidate
MR	mineralocorticoid
MR	magnetic resonance
MRI	magnetic resonance imaging
MRS	magnetic resonance spectroscopy
MS	multiple sclerosis
MTL	medial temporal lobe
MTLE	mesial temporal lobe epilepsy
MTS	mesial temporal sclerosis
NAA	N-acetylaspartate
NMDA	N-methyl-D-aspartate
NOS	nitric oxide synthase
NPLE	non-paraneoplastic limbic encephalitis
NREM	non-rapid eye movement
O_2^-	superoxide
OS	oxidative stress
PAE	prenatal alcohol exposure
PBMC	peripheral blood mononuclear cell
PCA	posterior cerebral artery
PCR	polymerase chain reaction
PDS	paroxysmal depolarizing shift
PET	positron emission tomography
phMRI	pharmacological magnetic resonance imaging
PLE	paraneoplastic limbic encephalitis
PPF	paired-pulse facilitation
PPI	prepulse inhibition
PTSD	post-traumatic stress disorder
PVN	paraventricular nucleus
RD	radial diffusivity

REM	rapid eye movement
RNS	reactive nitrogen species
ROS	reactive oxygen species
RSC	retrosplenial cortex
sc	subcutaneous
SEM	structural equation modelling
SGV	subgranular zone
SNP	single nucleotide polymorphism
SNRI	serotonin/noradrenaline reuptake inhibitor
SO	stratum oriens
SP	stratum pyramidale
SPECT	single-photon emission computed tomography
SR	stratum radiatum
SSRI	selective serotonin reuptake inhibitor
SVZ	subventricular zone
SWA	slow-wave activity
TBI	traumatic brain injury
Tg	transgenic
TGA	transient global amnesia
TLE	temporal lobe epilepsy
TS	Gilles de la Tourette syndrome
US	unconditioned stimulus
VBM	voxel-based morphometry
VTA	ventral tegmental area
wt	wild type

Epigraphs

Let us suppose that every man has in his mind a block of wax of various qualities, the gift of Memory, the mother of the Muses; and on this he receives the seal or stamp of those sensations and perceptions which he wishes to remember. That which he succeeds in stamping is remembered and known by him as long as the impression lasts; but that, of which the impression is rubbed out or imperfectly made, is forgotten, and not known.

Plato, *Theaetetus* (Translated by Benjamin Jowett)

The stream of thought flows on; but most of its segments fall into the bottomless abyss of oblivion. Of some, no memory survives the instant of their passage. Of others, it is confined to a few moments, hours, or days. Others, again, leave vestiges which are indestructible, and by means of which they may be recalled as long as life endures. Can we explain these differences?

William James, *The Principles of Psychology* (1890, Chapter 16, Memory)

CHAPTER 1
Introduction: The Hippocampus in the Clinical Neurosciences

Thorsten Bartsch

History of hippocampal research 2
Anatomical explorations 3
Involvement of the hippocampus in neurological and psychiatric diseases 4
Basic principles of the pathophysiology of the hippocampus 5
Functions attributed to the hippocampus 6
Structure of the book 6
Outlook 8

ABSTRACT The hippocampus is one of the most studied structures in the human brain and plays a pivotal role in human memory and cognition. The anatomy of this allocortical and phylogenetically old structure was first described over 400 years ago but its function remained enigmatic until the dawn of the modern neurosciences. In neurobiological research, the hippocampus serves as a model for studying the general properties of the nervous system because of the peculiarities in this brain region's anatomy. Eminent functional milestones of brain function, such as the mechanisms of synaptic plasticity as correlates of learning (long-term potentiation), the place–cell system, and the formation of new neuronal cells in the adult hippocampus were discovered in the hippocampus. The function of the hippocampus is affected in various neurological and neuropsychiatric diseases which is a reflection of certain pathophysiological characteristics, such as its vulnerability to ischaemia, metabolic and behavioural stress, as well as its susceptibility to epileptogenic mechanisms. The clinical spectrum of hippocampal dysfunction encompasses clinical conditions such as Alzheimer's disease, temporal lobe epilepsy, ischaemia, limbic encephalitis, neurodevelopmental disorders, depression, and schizophrenia. Functional and structural imaging of the hippocampus has become an important surrogate marker for defining clinical states. The mechanisms of neuronal plasticity in the hippocampus in health and disease have only begun to be unravelled, but may open the door for therapeutic intervention in Alzheimer's disease, neuropsychiatric diseases, and rehabilitative medicine.

KEYWORDS hippocampus, memory, vulnerability, ageing, learning, plasticity

In the Greek mythological underworld of Hades, Lethe (Λήθη), the river of forgetting, sprang from its source under a white cypress, and all who drank from this river experienced complete forgetfulness and oblivion. Illuminated souls, however, were advised to turn to the pool of Mnemosyne (Μνημοσύν), the goddess of memory and remembrance.

As one can observe from the dialogues of Plato and Greek mythology, the ability to remember and forget and the oscillation between these states has fascinated humans since ancient times. Plato assumed that the brain was the topos of remembering, reasoning, and perceptions, but Plato's student Aristotle believed that perceptions induced the emergence of inner images that faded away over time. In his treatise 'On Memory and Reminiscence', Aristotle wrote that remembering required arranging, ordering, and associating perceptions, with the help of reason. This book about the hippocampus is about the wax plate that Plato metaphorically described: the structural underpinning of the ability to remember, and the vulnerability of this brain structure during ageing and disease.

The hippocampus is a peculiar brain structure that is mythological in origin and was virtually unexplored until the 19th century. Although the physiological role of the hippocampus has been elucidated in the last 100 years, this structure's involvement in clinical disorders has only recently been recognized, starting with its pivotal role in amnesia and temporal lobe epilepsy. For example, the impressive two-volume book *The Hippocampus* by Isaacson and Pribram (1975) contained only a brief clinical section on amnesia and the effects of temporal lobectomy on epilepsy, which was in contrast to the extensive reviews of neurophysiological, cellular, and behavioural findings in animals (Isaacson 1975).

Although this book does not present a determination of the function of the hippocampus, it may contribute to a multilevel understanding of the open and controversial question of the 'real' functions of the hippocampus by discussing the dysfunctional states of the hippocampus that are observed in association with neurological and neuropsychiatric diseases. In this respect, this book has a non-reductionist approach. It is evident that the hippocampus is not functionally or anatomically isolated but is embedded in various local and remote functional and structural systems. This finding makes this structure challenging to study, as the integrative nature of the hippocampus has only recently begun to be elucidated. This book is an attempt to provide an integrative view and a multilevel understanding of hippocampal functioning during conditions of health and disease.

History of hippocampal research

The term hippocampus is a Latin word; however, this term also has origins in Greek mythology and was the name of a creature that was depicted as a horse and a sea monster (ἵππος—hippos, horse and κάμπος—kampos, a sea monster, kampe caterpillar)(Lewis 1923). This creature was a horse with two forefeet and a fish's tail or fin and was thought to draw the carriage of the sea god Neptune (Pearce 2001). The earliest English reference to the term hippocampus can be found in the 1606 version of the *Oxford English Dictionary* in a description of Neptune's vehicle (Pearce 2001).

The word hippocampus was coined by the Italian anatomist Julius Caesar Arantius, a student of Vesalius, in 1587 (Arantius 1587; translation in Lewis 1923). However, in his written description, Arantius did not clearly define what he was referring to anatomically as he did not care to draw his observations. This lack of definition subsequently led to confusion for later scholars when they were defining the structure. Soon after Arantius coined the word hippocampus, the term 'cornu ammonis', or 'Ammon's horn', was introduced; the French anatomist René Jacques Croissant de Garengeot referred to this structure in this way in 1742 (Garengeot 1742). The term 'Ammon's horn' refers to a fossil cephalopod shell that was related to the nautilus. These fossils were also called serpent or snake stones because they looked like the horn of Jupiter Ammon (Zeus in Greek mythology). After the Roman conquest of Egypt, Jupiter Ammon was equated with the Egyptian deity Amun ('Amun's Horns') (Lewis 1923). The Swiss anatomist and polymath Albrecht von Haller used the term 'Ammon's horn' in a paper in 1755 (Haller 1774–77). Interestingly, the gyrus dentatus was described by the French anatomist Pierre Tarin in 1750 as the 'dentate fascia' or 'margo denticulatus' (Tarin 1750). The 'pes hippocampi' were added by the Dutch biologist Isbrand van Diemerbroeck (Diemerbroecks 1672). Subsequently, Felix Vicq d'Azur differentiated between the major and minor hippocampus (Vicq d'Azyr 1786) (Figure 1.1). Interestingly, the latter distinction inspired an event in scientific history in 19th-century Victorian England that was known as the famous 'Hippocampus debate'. This evolutionary debate, also known as the Huxley–Owen debate, focused on the question of whether the human brain was unique compared to the

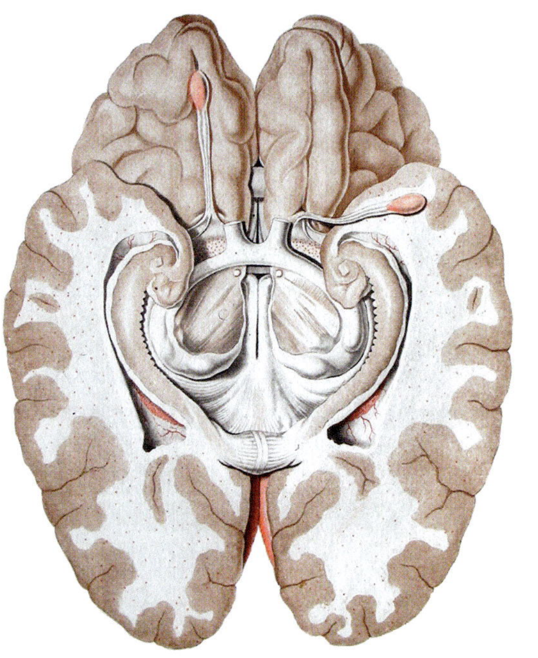

Figure 1.1 Vicq d'Azyr (1748–1794) was a renowned French anatomist who studied the anatomy of the hippocampal region. He is credited with discovering the comparative anatomy of this structure. Reproduction of figure from Vicq d'Azyr, F. *Traité d'anatomie et de physiologie*, © 1786, Didot. Previously reproduced as figure four in André Parent, Anatomy, Medicine, and Revolution, *The Canadian Journal of Neurological Sciences*, 34 (1) pp.30–7. © 2007 The Canadian Journal of Neurological Sciences.

brain of apes, in terms of containing a hippocampus minor (known as the calcar avis). The major players in this debate, Thomas H. Huxley and Richard Owen, were prominent British anatomists, and they acrimoniously debated the anatomical findings that were in favour or against Darwin's theory of evolution (Owen et al. 2009). This debate was later settled in favour of Thomas Huxley and Charles Darwin.

In the late 18th century, Jacques Bénigne Winslow introduced the analogue term 'ram's horn', probably because of the anatomical resemblance of the hippocampus to this shape and the association with the ram's horn of the Egyptian god Amun (Winslow 1732; Olry and Haines 1998). However, today, the hippocampus refers to the following equally accepted and established terms: the 'Hippocampus proper' and 'Ammon's horn' (Federative Committee on Anatomical Terminology 1998). The history of the exploration of the hippocampus, described in detail by Lewis (1923), explains why a certain vagueness still exists regarding the terminology and anatomical description of the hippocampus (Walther 2002).

Anatomical explorations

Bartolomeo Eustachi (lat. Eustachius; ~1500–1574) is credited with creating the first illustration of the hippocampus in 1552; however, this illustration was, unfortunately, not published until 1714 (Eustachius 1714; cited in Lewis, 1923). Interestingly, although he did not label his illustrations, Eustachi invented a metric scaling and coordination system that bears a striking resemblance to the stereotactic atlases of today.

Microstructural and microscopic explorations of the hippocampus by Duval, Golgi, Sala, Kolliker, and Ramon y Cajal (Figures 1.2 and 1.3) were conducted at the end of the 19th century (Duval 1882; Golgi 1886; Sala 1891;

Figure 1.2 Pyramidal cells in the hippocampus. Original preparation (Golgi method) by Ramon y Cajal. Photograph taken by Dr De Carlos. Courtesy of the Cajal Legacy, Instituto Cajal (CSIC), Madrid, Spain.

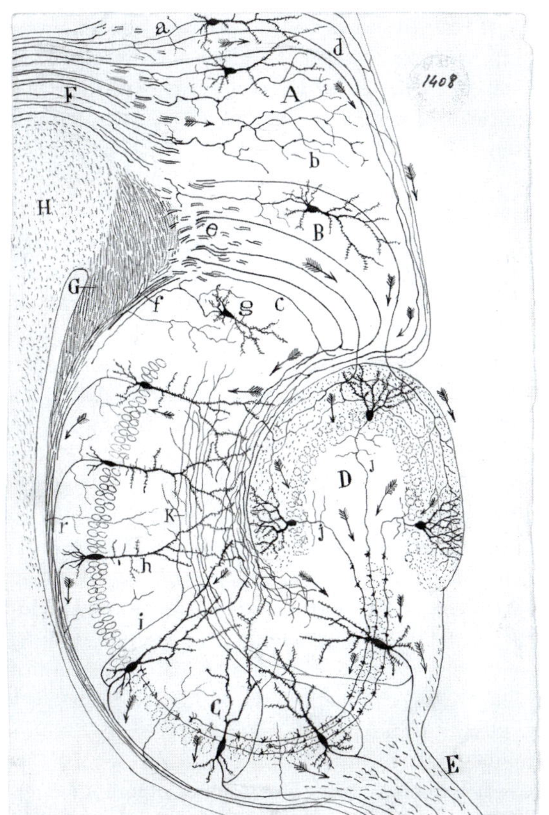

Figure 1.3 Original drawing by Cajal outlining the hippocampal circuitry. Santiago Ramón y Cajal. Courtesy of the Cajal Legacy, Instituto Cajal (CSIC), Madrid, Spain.

Kolliker 1896; Ramón y Cajal 1911). The anatomical exploration of the hippocampus further unfolded (Golgi 2001) in parallel with the revolutionary advances in histological techniques at the turn of the 20th century, such as the silver nitrate impregnation method of Camillo Golgi (1843–1926) (Golgi 1886). Later, Santiago Ramón y Cajal opened the doors to modern neuroscience with a microstructural analysis that supported the Neuron Doctrine, which described the existence of discrete and separate cells as components of the nervous system, the concept of synapses, and the unilateral electrical flow from dendrites to the axon (Ramón y Cajal 1911). Ramon y Cajal provided the foundation for the work of his student Rafael Lorente de Nó (1902–1990). These authors described in detail the cytoarchitecture of the cells and the cellular components in Ammon's horn, including two divisions of this structure according to recurrent circuitry in the pyramidal cell layer. These authors furthermore described the cytoarchitectural properties of small CA1 and large CA3 pyramidal neurons, and Lorente de Nó coined the modern terminology for the sectors of Ammon's horn (CA1–4). To date, a great body of literature exists that describes the anatomy of the hippocampus; for more detailed information, the reader is referred to other fundamental works (Ramón y Cajal 1911; Lorente de Nó 1934; Insausti and Amaral 2004; Duvernoy 2005; Andersen et al. 2006; Nieuwenhuys et al. 2007).

Involvement of the hippocampus in neurological and psychiatric diseases

At the end of the 19th century, neurophysiologists attempted to localize cerebral functions to specific cortical areas of the brain, including the hippocampus (Ferrier 1874). Sir David Ferrier observed a loss of sensation on the contralateral side of the body after thermal ablation of the hippocampus in monkeys, which was presumably due to damage to the adjacent cortical areas (Ferrier 1886). Sir William Macewen in Glasgow, along with Sir Victor Horsley and John Hughlings Jackson at the National Hospital for Nervous Diseases in London, conducted surgery in patients with epilepsy using topological semiological localization. Victor Horsley used intraoperative electrical stimulation to localize epileptic foci. Horsley also studied the impact of lesions on the temporal region, including the hippocampus, in monkeys. John Hughlings Jackson was the first scientist to demonstrate that mesiotemporal lesions that included the hippocampus may cause psychomotor seizures (temporal lobe epilepsy) (Hughlings Jackson and Colman 1898). The first temporal lobe resection, however, was performed by Burckhardt in 1891 on two schizophrenic patients who were suffering from acoustic hallucinations (Manjila et al. 2008). In 1937, Klüver and Bucy observed a remarkable behavioural syndrome after bilateral temporal lobectomy (including the amygdalae) in monkeys that included placidity, changes in food intake, hyperorality, hypersexuality, and visual agnosia (Klüver and Bucy 1937). Similar, behavioural changes such as idiocy, placidity, and bulimia were observed in 1888 after temporal lobe resection (Sanger Brown and Schäfer 1888). The first systematic series of anterior temporal lobe resections for 'non-lesional' psychomotor epilepsy was performed in 1947 by Bailey and Gibbs (1948, 1951). Temporal lobe resections for epilepsy with a selective removal of the amygdala and the hippocampus were further refined for a better localization of the epileptic zone, e.g. in the medial temporal lobe (Niemeyer 1958; Wieser and Yasargil 1982; reviewed in Feindel et al. 2009). Further attempts to localize the epileptic zone and the spread of the epileptic activity in hippocampal seizures were developed at the Montreal Neurological Institute through the use of depth electrodes and electrical stimulation and subsequently with the introduction of electroencephalography (Foerster and Altenburger 1935; Jasper and Kershman 1941; Bailey and Gibbs 1948; Jung 1949; Penfield and Jasper 1954). Wilder Penfield stimulated the uncinate region and elicited automatisms that were recorded using temporal lobe electrodes (Jasper et al. 1951; Penfield and Jasper 1954). In the early 1950s, the Montreal group developed concepts of the causative structure–function relationships of hippocampal epileptogenesis due to tumours or hippocampal sclerosis. In these authors' subsequent studies, they described the functional outcome after temporal lobectomy (Penfield and Milner 1958). In 1949, Richard Jung experimentally demonstrated that the hippocampus exhibited a lower threshold for epileptic activity; he also described the theta rhythm in Ammon's horn (Jung 1949). The neurosurgeon William Scoville performed the now famous bilateral hippocampal resection on patient H.M. as a treatment for epilepsy (Scoville and Milner 1957). Despite his tragic outcome, H.M. opened the doors for defining the role of the hippocampus in memory (Zola-Morgan and Squire 1993; Steinvorth et al. 2005).

In terms of the methodological and technical aspects of functional anatomy of the hippocampus, Talairach, Szikla, and Bancaud in France developed an anatomically orientated approach for their stereotactic technique that was used for the precise placement of depth electrodes in the hippocampus (Brazier 1966; Halgren et al. 1978). Electrical stimulation of the hippocampus induces a great variety of diverse mental phenomena, including amnesia (reviewed in Halgren et al. 1978).

To examine the role of the hippocampus in memory and to monitor postsurgical outcomes, the selective intracarotid sodium amobarbital procedure or 'Wada test' was developed in 1949. Hippocampal function, however, was better assessed using a selective posterior cerebral artery injection of amytal (Wada 1960; Jack et al. 1989). Later, structural and functional neuroimaging were used in conjunction with a diagnostic presurgical assessment of hippocampal function. Recently, deep brain stimulation (DBS) of the hippocampus has been used in patients with intractable temporal lobe seizures; it has been suggested that neuromodulatory stimulation of the hippocampus inhibits local and propagated epileptogenic activity (Velasco et al. 2000). Even more recently, the indirect modulation of the hippocampus (in addition to other components of the Papez circuit) has been elicited using DBS of the anterior thalamic nucleus in patients with intractable epilepsy. Using hippocampal electrodes for presurgical assessment in patients with epilepsy has proven valuable to study the role of the hippocampus in memory formation during wakefulness and sleep (Coleshill et al. 2004; Axmacher et al. 2008; Lacruz et al. 2010).

Basic principles of the pathophysiology of the hippocampus

In mid 20th-century neurobiological research, the hippocampus served as a model for studying the general properties of the nervous system because of the peculiarities in this brain region's structure (Kandel 2009). Functional milestones, including the mechanisms of synaptic plasticity as correlates of (Hebbian) learning such as long-term potentiation (LTP) (Bliss and Lomo 1973), were discovered in the hippocampus in addition to the place-cell system (O'Keefe and Nadel 1978) and the formation of new neuronal cells in the adult hippocampus. Interest in the clinical involvement of the hippocampus increased after the basic mechanisms of vulnerability to ischaemia, degeneration of hippocampal cells in Alzheimer's disease (AD), effects of stress on hippocampal neurons and its endocrine regulation, and the susceptibility to epileptogenic mechanisms were identified in previous studies. The results of these studies suggest that the basic principles and peculiarities of the hippocampus, such as its circuitry, anatomical projections, vulnerability, and plasticity play an important role in clinical states and explain how these mechanisms may be involved in various neuropsychiatric disorders (Small et al. 2011). The exact role of these findings (e.g. reduced hippocampal volumes in psychiatric disorders) with regard to the aetiological mechanisms, effects, and clinical consequences is the subject of intense research. It is unclear if these changes represent a state or a trait marker of neuropsychiatric diseases.

The hippocampus exhibits a pronounced vulnerability to hypoxic, ischaemic, or metabolic noxious events (Sommer 1880). The involvement and vulnerability of the hippocampus in the aetiology of temporal lobe epilepsy was first acknowledged by Wilhelm Sommer in 1880 when he described the pathological findings in the Ammon's horns of epileptic patients (Sommer 1880). In particular, he described a loss of neurons and tissue sclerosis in a sector of the lateral Ammon's horn that appeared to be particularly susceptible to hypoxic-ischaemic neuronal injury; this sector roughly equates to CA1 and has become known as the 'Sommer sector'. Interestingly, Walther Spielmeyer later extended Sommer's findings by describing a neighbouring sector (the 'Spielmeyer sector') that is more resistant to noxious metabolic events and that equates to CA3; however, he noted that the entire hippocampus may be affected in severe conditions (Spielmeyer 1927). Further, a sector of the hippocampus with a medium level of vulnerability was later referred to as the 'Bratz sector' (Bratz 1899). These findings also indicate that it has long been known that the hippocampus is particularly sensitive to epileptogenic mechanisms and that this sensitivity is a key player in temporal lobe epilepsy (TLE). However, gaps in the understanding of the aetiology of TLE still exist in terms of linking neuronal hyperactivity and excitability, structural changes, and neurodevelopmental and genetically driven changes in the hippocampus.

It has been shown that new neuronal cells can be created in the adult human hippocampus; these new cells might play an important role in the lifelong plasticity mechanisms of learning and adaptation (Altman and Das 1965). The plasticity and vulnerability of the hippocampus is tightly connected with the pathophysiological mechanisms and clinical symptoms in neurodegenerative disorders, such as AD (Alzheimer 1907).

Despite the anatomical uniqueness of the hippocampus, the spectrum of hippocampal function is extensive because this structure is functionally and anatomically embedded within the medial temporal lobe system. For a detailed description of the milestones in basic hippocampal research, the reader is referred to *The Hippocampus Book* (Andersen et al. 2007). The hippocampus is so extensively studied that the scientific journal *Hippocampus* is solely devoted to its exploration.

Functions attributed to the hippocampus

In the course of its exploration, various functions and roles have been attributed to the hippocampus. A brief account of these functions indicates that several of these roles are still part of the current clinical models and are thus relevant for understanding clinical disorders. For a detailed account, the reader is directed to *The Hippocampus Book* (Andersen et al. 2007).

Until the beginning of the 20th century, the role of the hippocampus was somewhat limited to a role in olfaction. This role was disputed by Brodal but recently experienced a renaissance in memory research as a result of the use of functional imaging. At the end of the 19th century, Paul Broca described the hippocampus and its surrounding structures and coined the term 'limbic lobe' (Broca 1878). This description was further conceptualized and extended in a model of emotional regulation by Papez ('Papez circuit') (Papez 1937). In this model, the 'limbic structures' were thought to be connected within a reverberating circuitry that extended from the hippocampus to the fornix, mammillary bodies, anterior thalamus, and back to the cortex through the cingulum via the parahippocampal gyrus to the hippocampus. Paul Maclean transformed the 'Papez' model into a model of a 'visceral brain' that included the limbic lobe and its projections to the hypothalamus, amygdala, and septum; he coined the term 'limbic system' and developed a triune concept of the brain and behaviour (Maclean 1952). This model of a triune brain sought to integrate evolutionary aspects, and anatomical components with behavioural phenotypes.

As mentioned earlier, the hippocampus has a reciprocal involvement in the regulation of stress in that this structure influences the hypothalamic–pituitary–adrenal (HPA) axis but is also affected by stress hormones, a mechanism that is widely thought to constitute a conceptual bridge to neuropsychiatric disorders (Moser and Moser 1998; Fanselow and Dong 2010). Although previously suggested by older animal studies, the idea that the hippocampus is involved in attention and novelty has reemerged with the development of new imaging techniques such as functional magnetic resonance imaging (fMRI). The role of the hippocampus in spatial learning and navigation, which began with the discovery of 'place cells' in the hippocampus by O'Keefe and Nadel, has been studied for over 40 years and is outlined in the now classic work by O'Keefe and Nadel (1978), *The Hippocampus as a Cognitive Map*. This text continues to fuel theories on the role of the hippocampus (O'Keefe and Nadel 1978).

The history of hippocampus research reflects the different models of hippocampal function that have been proposed over time and are still relevant today. On the one hand, there is the idea of the 'cognitive hippocampus', which suggests that the hippocampus is a critical relay unit for memory processes. On the other hand, the notion of the 'emotional hippocampus' suggests that this structure is involved in emotional regulation and psychopathology (Gray and McNaughton 1983; Moser and Moser 1998; Fanselow and Dong 2010). For example, it has recently been suggested that the dorsal hippocampus plays a role in the cognitive processes of learning and memory, whereas the ventral hippocampus is involved in the regulation of emotional and motivational behaviour; an intermediate zone of this structure is thought to integrate cognitive and spatial processing into behaviourally relevant actions (Moser and Moser 1998; Fanselow and Dong 2010; Segal et al. 2010). From this epistemological excursion, it is apparent that the hippocampus is not exclusively involved in memory but is likely involved in a variety of functions. It is also apparent that the definitive role of the hippocampus is far from settled.

Structure of the book

As was observed on the short excursion into the history of the exploration of the hippocampus, the anatomy of the hippocampus as an allocortical structure and its nomenclature are complex. The modern view of hippocampal anatomy is reviewed in Chapter 2 of the present work, including the principles of postnatal development that are relevant for neurodevelopmental and neuropsychiatric disorders.

Memory is typically defined as the ability of the brain to encode, store, and retrieve sensory information. Studies of brain-injured patients have provided remarkable insights into the neuroanatomy of memory, how memories are processed in the brain, and the ways in which the hippocampal system is involved in these mechanisms. In 1899, Bechterew described a patient who had memory problems due to bilateral softening of the gyrus uncinatus and Ammon's horn (Bechterew 1900). Over 50 years after Bechterew's study, Grünthal (1947), and Glees and Griffith (1952) described patients with memory disturbances due to the bilateral destruction of Ammon's horn and the cystic degeneration of the hippocampus. As a result, it was recognized that the hippocampus and its surrounding structures were necessary for the maintenance of memory function and normal mental activity. The clinical role of the hippocampus in memory, however, was more strongly supported by the now famous studies of the amnestic deficit of patient H.M. after a hippocampectomy. Chapter 3 presents information about the role of the hippocampus in memory. In this regard, this chapter relates to the clinical chapters that describe the role of the hippocampus in the emergence and maintenance of pathophysiological mechanisms that lead to temporal lobe epilepsy and mesial sclerosis.

A principal component of the hippocampal involvement in clinical neurobiology is this structure's striking vulnerability to insult, stress, and other harmful metabolic events that may be the basis for various neurological and neuropsychiatric disorders. Chapter 4 reviews the basic principles of metabolic vulnerability, including the vulnerability of the hippocampus to the cellular effects of glutamate. Glutamate plays a pivotal role in the plastic synaptic mechanisms of neurotransmission but may also exert cytotoxic effects on hippocampal neurons.

In Chapter 5, the vulnerability of the hippocampus to stress is reviewed systemically within a framework of integrative mechanisms from the cellular to the behavioural level. Because of the discovery of glucocorticoid receptors in the hippocampus, the effects of hippocampal stimulation on the HPA axis and the effects of stress on this structure, it has been suggested that the hippocampus is critically involved in the neuroendocrine regulation of emotion, fear, anxiety, and affection. The deleterious effects of stress and cortisol on the hippocampus are thought to lead to structural changes and functional impairments (Mandell et al. 1963; McEwen et al. 1968; Starkman et al. 1992; Goldstein and McEwen 2002). This chapter conceptualizes the 'emotional' side of the hippocampus and provides a basis for the clinical chapters that describe the involvement of the hippocampus in neuropsychiatric diseases.

In 1965, Altman and Das described the postnatal genesis of granule cells in the dentate gyrus of the rat. This observation was rediscovered in the 1990s when it was shown that new neuronal cells were created in the adult mammalian hippocampus (Altman and Das 1965). Although the functional relevance of neurogenesis is not entirely clear, various experimental results suggest a role for this mechanism in memory, cognition, stress, and emotional regulation. These changes have also been associated with depression and post-traumatic disorders. In Chapter 6, the mechanisms of neurogenesis are reviewed within an integrative framework of hippocampal plasticity.

Traditionally, anatomical studies of the hippocampus have consisted of a postmortem analysis of pathological changes. With the advent of magnetic resonance imaging (MRI), the study of pathological changes was possible *in vivo*. Chapter 7 describes the normal MRI anatomy of the hippocampus and the various imaging techniques that allow for the further study of the function, structure, connectivity, and pharmacological manipulations of the hippocampus.

Sleep has been suggested to play a pivotal role in the consolidation of memories since the beginning of the 20th century (Müller 1900; Heine 1914; Jenkins 1924). Recently, the role of the hippocampus in sleep-dependent memory consolidation has been elucidated in numerous studies (Wilson and McNaughton 1994; Maquet 2001). Chapter 8 reviews the role of the hippocampus in the plasticity of sleep-dependent memory consolidation at a systems level.

The need to study the neurobiological mechanisms that underlie the cognitive changes during healthy ageing has only recently been acknowledged. As lifespans progressively extend, it has become clear that mechanisms involved in ageing are important for understanding age-related diseases. In particular, the mechanisms of neuronal plasticity are worth studying to dissect the physiological and pathophysiological mechanisms. Chapter 9 describes the basic concepts and models of age-related mechanisms in the hippocampus.

With the advent of genetic and imaging methods, it has been shown that the hippocampus is involved in a variety of neurodevelopmental disorders including complex genetically driven syndromes, such as Down syndrome, attention-deficit hyperactivity disorder (ADHD), autism spectrum disorder, and Gilles de la Tourette syndrome (Chapter 10). Additionally, disorders that reflect the vulnerability of the hippocampus, such as preterm birth, febrile seizures, hippocampal sclerosis, and embryo fetal alcohol syndrome, have a severe impact on the neurocognitive outcome of affected children.

The involvement of the hippocampus in neurological disease is further reviewed in Chapter 11 with regard to the vulnerability of the hippocampus to metabolic stress, which has correlates in neurological diseases such as hippocampal stroke, encephalitis, transient global amnesia, and status epilepticus.

Since the clinical description of the patient Auguste Deter in 1907 by Alois Alzheimer, it has been believed that the hippocampus is affected very early in the course of AD and that this has profound clinical consequences. As a result, hippocampal imaging has become an important surrogate marker for defining clinical states. Chapter 12 describes the involvement of the hippocampus in this most common form of dementia and reviews the recent advances in our understanding of the structural and functional MRI correlates and the genetic contribution to risk states to AD.

Temporal lobe seizures are one of the most common forms of epilepsy in adults and are considered to be a form of focal epilepsy that is characterized by seizures that originate in, or primarily involve, the temporal lobe structures including the hippocampus. Chapter 13 describes the clinical phenomenology of TLE, including its histological and electrophysiological correlates.

The final three chapters of this book (Chapter 14–16) describe the important neuropsychiatric disorders that are tightly linked to pathological changes in hippocampal plasticity. Stress-related mechanisms, neuroendocrine dysregulation, neurodegeneration, and changes in neurogenesis are recognized as putative correlates of the final aetiological pathways in depression, schizophrenia, and post-traumatic disorders.

Outlook

The research of the last decades has highlighted the multifaceted structure and function of the hippocampus as an integrative key structure involved in memory and cognitive and behavioural regulation. Recent methodological developments have led to a deeper understanding of hippocampal function and offer the opportunity to study clinically relevant questions such as the detailed structure-function relationship, mechanisms of susceptibility, and vulnerability in health, ageing and disease. From a clinical viewpoint, the unique level of neuroplastic mechanisms in the hippocampus may be further explored with regard to questions of resilience, regeneration, and restoration to metabolic and behavioural stress that could lead to therapeutic interventions on a cellular and systems level in neurological, neurodegenerative, and psychiatric diseases.

Acknowledgements

I would like to thank Dr Stephen Jacyna, Director of the University College London (UCL) Centre for the History of Medicine for his critical comments on the manuscript.

References

Altman, J., and Das, G.D. (1965). Post-natal origin of microneurones in the rat brain. *Nature*, 207, 953–6.

Alzheimer, A. (1907). Über eine eigenartige Erkrankung der Hirnrinde. *Allgemeine Zeitschrift fur Psychiatrie und Psychisch-gerichtliche Medizin*, 64, 146–8.

Andersen, P., Morris, R., Amaral, D., Bliss, T., and O'Keefe, J. (Eds.) (2007). *The Hippocampus Book*. Oxford: Oxford University Press.

Arantius, J. C. (1587). *De Humano Foetu Liber*. Venetiis: Apud Iacobum Brechtanum.

Axmacher, N., Haupt, S., Fernandez, G., Elger, C. E., and Fell, J. (2008). The role of sleep in declarative memory consolidation – direct evidence by intracranial EEG. *Cereb Cortex*, 18, 500–7.

Bailey, P., and Gibbs, F. A. (1948). A preliminary report on the therapeutic results of temporal lobotomy for psychomotor epilepsy. Paper presented at the Central Association of Electroencephalographers. Chicago IL.

Bailey, P., and Gibbs, F. A. (1951). The surgical treatment of psychomotor epilepsy. *J Am Med Assoc*, 145, 365–70.

Bechterew, W. V. von (1900). Demonstration eines Gehirns mit Zerstörung der vorderer und innerer Theile der Hirnrinde in beiden Schlafenlappen. *Neurol Centralbl*, 19, 990–1.

Bliss, T. V., and Lomo, T. (1973). Long-lasting potentiation of synaptic transmission in the dentate area of the anaesthetized rabbit following stimulation of the perforant path. *J Physiol*, 232, 331–56.

Bratz, E. (1899). Ammonshornbefunde bei Epileptikern. *Arch Psychiatr Nervenhlkr*, 32, 820–35.

Brazier, M. (1966). Stimulation of the hippocampus in man using implanted electrodes. In Brazier M.A.B. (Ed.) *RNA and Brain Function, Memory and Learning*, pp. 299–310. Berkeley, CA: University of California Press.

Broca, P. (1878). Anatomie comparee des circonvolutions cerebrales: Le grand lobe limbique et la scissure limbique dans la serie des mammiferes. *Rev Anthrop*, 1, 385. ser. 382.

Coleshill, S. G., Binnie, C. D., Morris, R. G., *et al.* (2004). Material-specific recognition memory deficits elicited by unilateral hippocampal electrical stimulation. *J Neurosci*, 24, 1612–16.

Diemerbroecks, I. van (1672). *Anatome corporis humani, plurimis novis inventis instructa variisque observationibus, et paradoxis, cùm medicis, tùm physiologicis adornata*. Genevae: de Tournes.

Duval, M. (1882). La corne d'Ammon II. *Arch Neurol*, 3, 1–35.

Duvernoy, H. M. (2005). *The Human Hippocampus*. Berlin: Springer Verlag.

Eustachius (1714). *Tabulae anatomicae*. Romae: Bernabò Pagliarini.

Fanselow, M. S., and Dong, H. W. (2010). Are the dorsal and ventral hippocampus functionally distinct structures? *Neuron*, 65, 7–19.

Federative Committee on Anatomical Terminology (FCAT) (1998). *Terminologia Anatomica: International Anatomical Terminology*. Stuttgart: Thieme.

Feindel, W., Leblanc, R., and de Almeida, A. N. (2009). Epilepsy surgery: historical highlights 1909–2009. *Epilepsia*, 50(Suppl 3), 131–51.

Ferrier, D. (1874). On the localisation of the functions of the brain. *Br Med J*, 2, 766–7.

Ferrier, D. (1886). *The Functions of the Brain*. London: Smith Elder & Co.

Foerster, O., and Altenburger, H. (1935). Elektrobiologische Vorgänge an der menschlichen Hirnrinde. *Dtsch Z Nervenheilkd*, 135, 277–88.

Garengeot, R. C. de (1742). *Splanchnologie ou, L'anatomie des viscères; avec des figures originales tirées d'après les cadavres, suivie d'une dissertation sur l'origine de la chirurgie*. Paris: C. Osmont.

Glees, P., and Griffith, H. B. (1952). Bilateral destruction of the hippocampus (cornu ammonis) in a case of dementia. *Monatsschr Psychiatr Neurol*, 123, 193–204.

Goldstein, D. S., and McEwen, B. (2002). Allostasis, homeostats, and the nature of stress. *Stress*, 5, 55–8.

Golgi, C. (1886). *Sulla fina anatomia degli organi centrali di sistema nervoso. V. Sulla fina anatomia di grande piede d'Hippocampo*. Milan: Ulrico Hoepli.

Golgi, C., Bentivoglio, M., and Swanson, L. (2001). On the fine structure of the pes Hippocampi major (with plates XIII–XXIII). 1886. *Brain Res Bull*, 54, 461–83.

Gray, J. A., and McNaughton, N. (1983). Comparison between the behavioural effects of septal and hippocampal lesions: a review. *Neurosci Biobehav Rev*, 7, 119–88.

Grünthal, E. (1947). Über das klinische Bild nach umschriebenem beiderseitigem Ausfall der Ammonshornrinde. *Mschr Psychiat Neurol*, 113, 1–16.

Halgren, E., Walter, R. D., Cherlow, D. G., and Crandall, P. H. (1978). Mental phenomena evoked by electrical stimulation of the human hippocampal formation and amygdala. *Brain*, 101, 83–117.

Haller, A. von (1774–77). *Bibliotheca anatomica*. Tiguri (Zurich): Orell, Gessner, Fuessli, et Socc.

Heine, R. (1914). Über Wiedererkennen und rückwirkende Hemmung. *Z Psychol*, 68, 161–236.

Hughlings Jackson, J., and Colman, W. S. (1898). Case of epilepsy with tasting movements and 'dreamy state' – very small patch of softening in the left uncinate gyrus. *Brain*, 21, 580–90.

Insausti, R., and Amaral, D. G. (2004). Hippocampal formation. In Paxinos, G., and Mai, J.K. (Eds.) *The human nervous system*, pp. 871–914. Amsterdam: Elsevier.

Isaacson, R. L., and Pribram K. H. (1975). *The Hippocampus*. New York: Plenum Press.

Jack, C. R., Jr., Nichols, D. A., Sharbrough, F. W., *et al.* (1989). Selective posterior cerebral artery injection of amytal: new method of preoperative memory testing. *Mayo Clin Proc*, 64, 965–75.

Jasper, H. H., and Kershman, J. (1941). Electroencephalographic classification of the epilepsies. *Arch Neurol Psychiatry*, 45, 903–43.

Jasper, H., Pertuisset, B., and Flanigin, H. (1951). EEG and cortical electrograms in patients with temporal lobe seizures. *Arch Neurol Psychiatry*, 65, 272–90.

Jenkins, J., and Dallenbach, K. M. (1924). Obliviscence during sleep and waking. *Am J Psychol*, 35, 605–12.

Jung, R. (1949). Hirnelektrische Untersuchungen über den Elektrokrampf. Die Erlegungsabläufe in corticalen und subcorticalen Hirnregionen bei Katze und Hund. *Arch Psychiat Nervenkr*, 183, 206–84.

Kandel, E.R. (2009). The biology of memory: a forty-year perspective. *J Neurosci*, 29, 12748–56.

Klüver, H. B., and Bucy, P. C. (1937). 'Psychic blindness' and other symptoms following bilateral temporal lobectomy in Rhesus monkeys. *Am J Physiol*, 119, 352–3.

Kolliker, A. (1896). *Handbuch der Gewebelehre des Menschen*. Leipzig: Engelmann.

Lacruz, M. E., Valentin, A., Seoane J. J., *et al.* (2010). Single pulse electrical stimulation of the hippocampus is sufficient to impair human episodic memory. *Neuroscience*, 170, 623–32.

Lewis, F. (1923). The significance of the term hippocampus. *J Comp Neurol*, 35, 213–30.

Lorente de Nó, R. (1934). Studies on the structure of the cerebral cortex. II. Continuation of the study of the ammonic system. *Journal für Psychologie und Neurologie*, 46, 113–77.

Maclean, P. D. (1952). Some psychiatric implications of physiological studies on frontotemporal portion of limbic system (visceral brain). *Electroencephalogr Clin Neurophysiol*, 4, 407–18.

Mandell, A. J., Chapman, L. F., Rand, R. W., and Walter, R. D. (1963). Plasma corticosteroids: Changes in concentration after stimulation of hippocampus and amygdala. *Science*, 139, 1212.

Manjila, S., Rengachary, S., Xavier, A. R., Parker, B., and Guthikonda, M. (2008). Modern psychosurgery before Egas Moniz: a tribute to Gottlieb Burckhardt. *Neurosurg Focus*, 25, E9.

Maquet, P. (2001). The role of sleep in learning and memory. *Science*, 294, 1048–52.

McEwen, B. S., Weiss, J. M., and Schwartz, L. S. (1968). Selective retention of corticosterone by limbic structures in rat brain. *Nature*, 220, 911–12.

Moser, M. B., and Moser, E. I. (1998). Functional differentiation in the hippocampus. *Hippocampus*, 8, 608–19.

Müller, G. E., and Pilzecker, A. (1900). Experimentelle Beiträge zur Lehre vom Gedächtnis (Experimental Contributions to the Science of Memory). *Z Psychol*, 1, 1–300.

Niemeyer, P. (1958). Transventricular amygdalo-hippocampectomy in temporal lobe epilepsy. In Baldwin, M. B.P.E. (Ed.) *Temporal lobe epilepsy*, pp. 461–82. Springfield, IL: Charles C. Thomas.

Nieuwenhuys, R., Voogd, J., and van Huijzen, C. (2007). *The Human Central Nervous System: A Synopsis and Atlas*. Berlin: Springer.

O'Keefe, J., and Nadel, L. (1978). *The Hippocampus as a Cognitive Map*. Oxford: Oxford University Press.

Olry, R., and Haines, D.E. (1998). Cerebral mythology: a skull stuffed with gods. *Hist Neurosci*, 7, 82–3.

Owen, C. M., Howard, A., and Binder, D. K. (2009). Hippocampus minor, calcar avis, and the Huxley-Owen debate. *Neurosurgery*, 65, 1098–104; discussion 1104–95.

Papez, J. (1937). A proposed mechanism of emotion. *Arch Neurol Psychiatry*, 38, 725–43.

Parent, A. (2007). Félix Vicq d'Azyr: Anatomy, medicine and revolution. *Can J Neurol Sci*, 34, 30–7.

Pearce, J. M. (2001). Ammon's horn and the hippocampus. *J Neurol Neurosurg Psychiatry*, 71, 351.

Penfield, W., and Jasper, H. (1954). *Epilepsy and the Functional Anatomy of the Human Brain*. Boston, MA: Little, Brown.

Penfield, W., and Milner, B. (1958). Memory deficit produced by bilateral lesions in the hippocampal zone. *AMA Arch Neurol Psychiatry*, 79, 475–97.

Ramón y Cajal, S. (1911). *Histologie du Systeme Nerveux de l'Homme et des Vertebretes*. Paris: A. Maloine.

Sala, L. (1891). Zur feinen Anatomie des grossen Seepferdefusses. *Z wiss Zool*, 52, 18–45.

Sanger Brown, S., and Schäfer, E. A. (1888). An investigation into the functions of the occipital and temporal lobes of the monkey's brain. *Phil Trans R Soc Lond B*, 179, 303–27.

Scoville, W. B., and Milner, B. (1957). Loss of recent memory after bilateral hippocampal lesions. *J Neurol Neurosurg Psychiatry*, 20, 11–21.

Segal, M., Richter-Levin, G., and Maggio, N. (2010). Stress-induced dynamic routing of hippocampal connectivity: a hypothesis. *Hippocampus*, 20, 1332–8.

Small, S.A., Schobel, S.A., Buxton, R.B., Witter, M.P., Barnes, C.A. (2011). A pathophysiological framework of hippocampal dysfunction in ageing and disease. *Nat Rev Neurosci*, 12, 585–601.

Sommer, W. (1880). Erkrankungen des Ammonshornes als aetiologisches Prinzip der Epilepsie. *Arch Psychiatr*, 10, 631–75.

Spielmeyer, W. (1927). Die Pathogenese des epileptisches Krampfes. Histopathologischer Teil. *Ztschr Neurol Psychiat*, 109, 501–19.

Starkman, M. N., Gebarski, S. S., Berent, S., and Schteingart, D. E. (1992). Hippocampal formation volume, memory dysfunction, and cortisol levels in patients with Cushing's syndrome. *Biol Psychiatry*, 32, 756–65.

Steinvorth, S., Levine, B., and Corkin, S. (2005). Medial temporal lobe structures are needed to re-experience remote autobiographical memories: evidence from H.M. and W.R. *Neuropsychologia*, 43, 479–96.

Tarin, P. (1750). *Adversaria anatomica*. Paris: Moreau.

Velasco, M., Velasco F., Velasco, A.L. *et al.* (2000). Subacute electrical stimulation of the hippocampus blocks intractable temporal lobe seizures and paroxysmal EEG activities. *Epilepsia*, 41, 158–69.

Vicq d'Azyr, F. (1786). *Traité d'anatomie et de physiologie*. Paris: Didot.

Wada, J., and Rasmussen, T. (1960). Intracarotid injection of sodium amytal for the lateralization of cerebral speech dominance. *Neurosurgery*, 17, 266–82.

Walther, C. (2002). Hippocampal terminology: concepts, misconceptions, origins. *Endeavour*, 26, 41–4.

Wieser, H. G., and Yasargil, M. G. (1982). Selective amygdalohippocampectomy as a surgical treatment of mesiobasal limbic epilepsy. *Surg Neurol*, 17, 445–57.

Wilson, M. A., and McNaughton, B. L. (1994). Reactivation of hippocampal ensemble memories during sleep. *Science*, 265, 676–9.

Winslow, J. (1732). *Exposition anatomique de la structure du corps humain*. Paris: G. Desprez.

Zola-Morgan, S., and Squire L. R. (1993). Neuroanatomy of memory. *Annu Rev Neurosci*, 16, 547–63.

CHAPTER 2
Functional Anatomy, Development, and Pathology of the Hippocampus

Pierre Lavenex

Introduction 10
Overall organization 11
Postnatal development 27
Neurodevelopmental disorders 32
Conclusion 34

ABSTRACT The hippocampal formation is comprised of a group of cortical regions, including the dentate gyrus, hippocampus, subiculum, presubiculum, parasubiculum, and entorhinal cortex. The hippocampal formation plays a central role in the brain network essential for memory function. Paradoxically, the hippocampus is also the brain structure most sensitive to hypoxic-ischaemic episodes. In addition, abnormalities in hippocampal structure and functions have been reported in a number of neuropathological and neurodevelopmental disorders. Here, I provide a basic description of the structural organization of the primate hippocampal formation. I also discuss recent findings regarding its normal postnatal development, which provide an essential framework to consider the aetiologies of different neurodevelopmental disorders affecting hippocampal structure and function. In particular, a developmental decrease in astrocytic processes and functions may be the critical factor underlying the selective vulnerability of CA1 to hypoxic-ischaemic episodes in adulthood. It also provides an explanation for the relative resistance of this brain structure to hypoxia in the perinatal period, especially during the birth process. In contrast, a major benefit that derives from decreased astrocytic coverage is the regulation of synaptic efficacy leading to an increase in synaptic selectivity advantageous for learning. In addition, the relatively high astrocytic coverage of newborn synapses might also play a central role in the generation of febrile seizures. Finally, we identified different periods of postnatal development during which pathogenic factors might impact the structural and functional maturation of distinct hippocampal regions, as is observed in autism, schizophrenia, epilepsy, and Down syndrome.

KEYWORDS hippocampal formation, primate, memory, infantile amnesia, neuron, astrocyte, hypoxia, febrile seizure, epilepsy, schizophrenia, neurodevelopmental disorder.

Introduction

The terms hippocampus and hippocampal formation are often used interchangeably, yet they refer to and include different brain structures depending on the context in which they are used. Here, I do not provide an exhaustive and detailed description of the structural characteristics of the hippocampus, as this was done previously (Insausti and Amaral 2004; Duvernoy 2005; Amaral and Lavenex 2007). Instead, I consider the basic aspects of the structural and functional organization of the primate hippocampal formation, and discuss some of our recent findings in monkeys that provide an essential framework to consider issues regarding the clinical neurobiology of the human hippocampus. The term hippocampus was coined in 1587 by the anatomist Arantius who considered the three-dimensional form

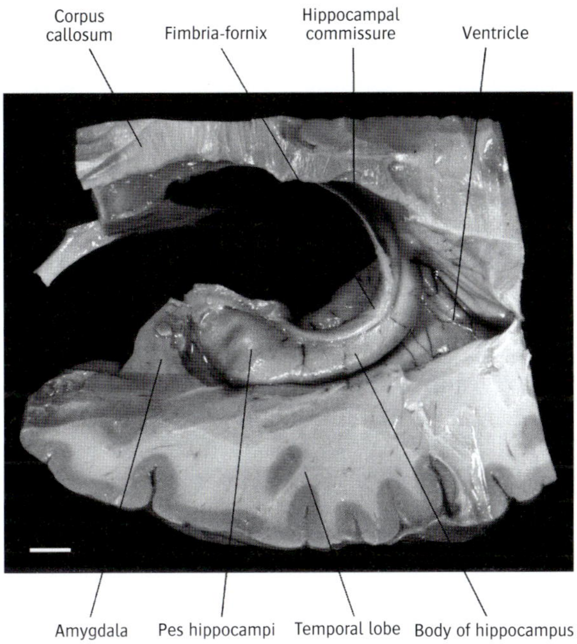

Figure 2.1 Dorsolateral view of the human hippocampus after removing the overlying structures. The lateral ventricle has been opened, exposing the shape of the hippocampus. Scale bar = 1 cm. Reproduced from Andersen, P., Morris, R. G. M., Amaral, D. G., Bliss, T. V., and O'Keefe, J. (Eds). (2007). *The Hippocampus Book*, with permission from Oxford University Press.

Figure 2.2 Human hippocampus dissected free (left) and compared to a specimen of *Hippocampus leria* (right). (Source: courtesy of Professor László Seress, University of Pecs.). Reproduced from Andersen, P., Morris, R. G. M., Amaral, D. G., Bliss, T. V., and O'Keefe, J. (Eds.) (2007). *The Hippocampus Book*, with permission from Oxford University Press.

of the human hippocampus to resemble a sea horse (*hippocampus* in Greek; Figures 2.1 and 2.2) (Arantius 1587). One hundred and fifty-five years later, De Garengeot likened the arched structure of the hippocampus to a ram's horn and named the hippocampus 'Cornu Ammonis' or 'Ammon's horn' after the Egyptian god Amun Kneph, whose symbol was a ram (De Garengeot 1742). The term hippocampus typically refers to the bulge occupying the floor of the temporal horn of the lateral ventricle in the human brain (Figure 2.1). The abbreviations of Cornu Ammonis (CA3, CA2, and CA1) are used to define the different subdivisions of what is sometimes referred to as the *hippocampus proper*. Here, I use the definition of the *hippocampal formation* as a group of cortical regions located in the medial temporal lobe that includes the dentate gyrus, hippocampus (CA3, CA2, CA1), subiculum, presubiculum, parasubiculum, and entorhinal cortex (Amaral and Lavenex 2007). Each of these structures contains a number of different cell types, and different sets of intrinsic connections and interconnections with other brain regions, which exhibit clear and distinct topographical distributions. Altogether, these interconnected structures form a functional brain system essential for memory, which is particularly sensitive to a number of pathologies. As the hippocampal formation is not an isolated brain structure, I will also discuss the organization of its main input and output connections with the rest of the brain, including cortical and subcortical structures. Most of the information regarding the structural organization of functional hippocampal circuits is derived from experimental work performed with animals, in particular rodents but also monkeys. Similarly, animal models are typically used to study the molecular and cellular basis of pathologies affecting human hippocampal structure and functions. Consequently, it is important to be aware of the similarities and differences between species (e.g. rats, monkeys, humans; Figure 2.3), in order to extrapolate from the findings of fundamental research in animals to clinical problems in humans. I will therefore consider this comparative perspective to discuss some issues that might be relevant for questions addressed in other chapters of this book.

Overall organization

The general structural organization of the hippocampal formation is largely conserved across species, despite major species differences in the relative development of brain areas with which it interacts. This is largely due to the fact that very few cortical areas have direct interconnections with the hippocampus. Instead, cortical inputs are relayed through a hierarchy of associational cortices in which significant integration of information takes place before these inputs are forwarded to the next hierarchical level and finally reach the hippocampus (Figure 2.4; Lavenex and Amaral 2000).

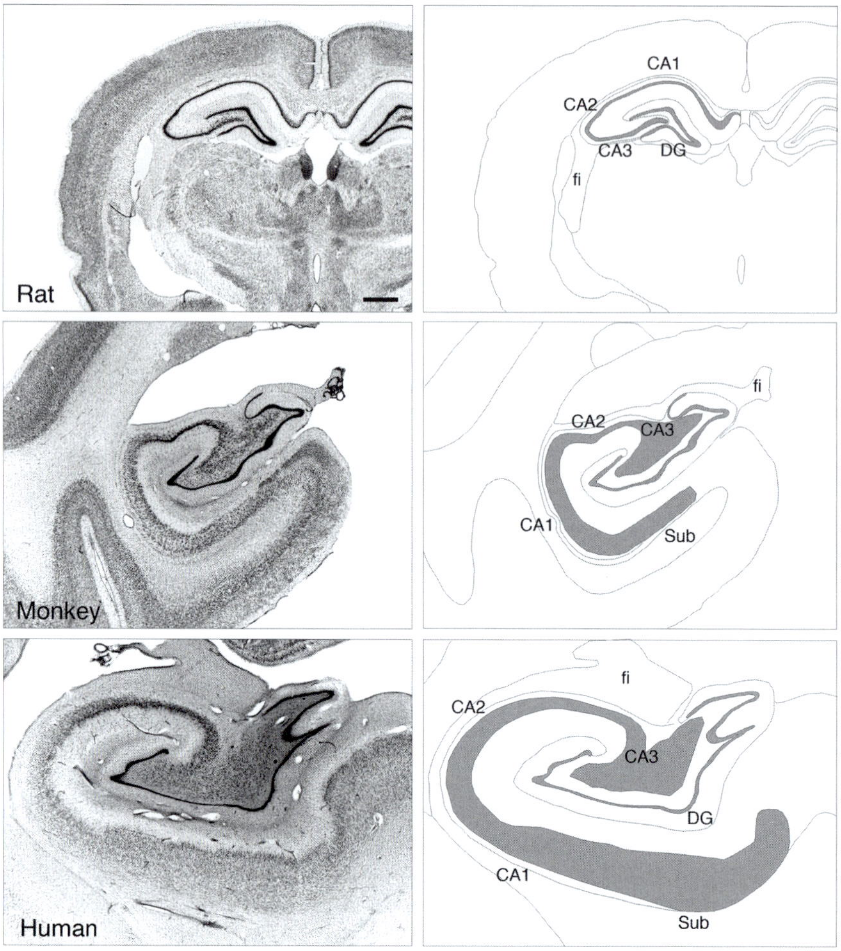

Figure 2.3 Nissl-stained sections and line drawings illustrating the general organization and similarities of the subdivisions of the hippocampal formation in the rat, monkey, and human. Scale bar in A = 1 mm, applies to all panels. Reproduced from Amaral, D. G. & Lavenex, P. (2007) Hippocampal neuroanatomy. In Andersen, P., Morris, R. G. M., Amaral, D. G., Bliss, T. V. & O'Keefe, J. (Eds.) *The Hippocampus Book*. © 2007, with permission from Oxford University Press.

Perirhinal and parahippocampal cortices

In all species, cortical regions send projections converging onto the hippocampal formation via three main structures: the perirhinal cortex, the postrhinal (in rodents) or parahippocampal (in primates) cortex, and the entorhinal cortex. These different regions receive different sets of neocortical inputs and exhibit specific patterns of interconnections with the neocortex (in monkeys: Suzuki and Amaral 1994a; Lavenex et al. 2002; Kobayashi and Amaral 2003; Munoz and Insausti 2005; Insausti and Amaral 2008). The postrhinal cortex of the rat is subdivided into areas PORd and PORv (Burwell 2001). The parahippocampal cortex, its counterpart in monkeys and humans, is subdivided in areas TH, TFm, and TFl (Suzuki and Amaral 2003). The perirhinal cortex is subdivided into five areas in rats (areas 35d, 35v, 36d, 36v, 36p; Burwell 2001). In monkeys, the perirhinal cortex is subdivided into six areas (areas 35, 36d, 36rm, 36rl, 36cm, 36cl; Suzuki and Amaral 2003); whereas the human perirhinal cortex classically refers to areas 35 and 36 (Brodmann 1909; Insausti et al. 1998), which can be further subdivided mediolaterally into areas 35a and 35b, as well as 36a and 36b (Ding and Van Hoesen 2010). In monkeys (Suzuki and Amaral 1994a), the major cortical inputs to the perirhinal cortex arise from unimodal visual areas TE and rostral TEO, and from area TF of the parahippocampal cortex. The perirhinal cortex also receives significant projections from the insular cortex and the orbitofrontal cortex area 13. In contrast, area TF of the parahippocampal cortex receives its strongest input from more caudal visual areas V4, TEO, and caudal TE, and from polymodal association cortices including the retrosplenial cortex and the dorsal bank of the superior temporal sulcus. Area TF also receives projections from areas 7a and LIP of the posterior parietal cortex, insular cortex, and areas 46, 13, 45, and 9 of the frontal cortex. Area TH receives substantial projections from the retrosplenial cortex, moderate projections from the dorsal bank of the superior temporal sulcus, and relatively strong inputs from auditory

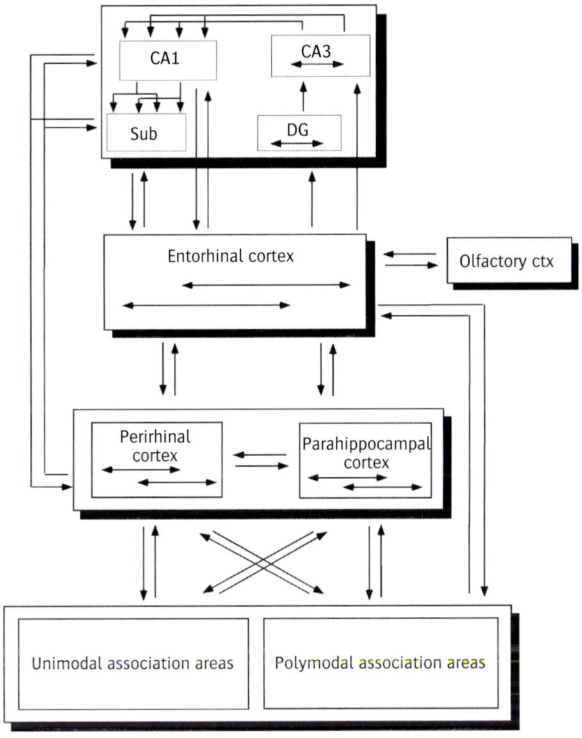

Figure 2.4 Schematic representation of the hierarchical organization of projections within the medial temporal lobe. Reproduced from Lavenex, P. and Amaral, D.G. (2000). Hippocampal-neocortical interaction: A hierarchy of associativity. *Hippocampus*, 10(4), 420–30, © 2000, with permission from John Wiley and Sons.

association areas on the convexity of the superior temporal gyrus. The distribution of cortical inputs to these areas is sometimes oversimplified by stating that the perirhinal cortex receives 'object' visual information, whereas the parahippocampal cortex receives 'spatial' visual information. This is probably due to the fact that most functional studies of these brain regions rely on the use of easily-controlled visual stimuli. What is largely neglected, however, is the fact that other higher-order associational cortical areas project to these two regions (Suzuki and Amaral 1994a), and most importantly perhaps, that the parahippocampal cortex sends significant projections towards the perirhinal cortex (Lavenex et al. 2002, 2004b; Suzuki and Amaral 1994a). Indeed, although perirhinal areas 35, 36rm, 36rl, 36cm, and 36cl are highly interconnected, and parahippocampal areas TH, TFm, and TFl also share an extensive network of associational connections that tend to be heaviest within a given subdivision, area 36c of the perirhinal cortex constitutes an important interface between the perirhinal and parahippocampal cortices. Its heaviest connections are with area 36r and the anterior aspect of area TF. The laminar organization of all these connections is typical of associational projections; the terminals of these projections are distributed through both deep and superficial layers, although heavier projections are directed toward the superficial layers (Lavenex et al. 2004b). However, the laminar distribution of the cells of origin suggests that the projections from caudal areas (36c or TF) to area 36r are of the feedforward type, whereas the projections from areas 36r and 36c to area TF are of the feedback type (Lavenex et al. 2004b). This suggests that the perirhinal cortex is at a higher level than the parahippocampal cortex in the hierarchy of associational cortices. This organization is similar in rats, where the postrhinal cortex projects more heavily in a feedforward manner toward the perirhinal cortex, whereas the perirhinal cortex projects less heavily in a feedback manner toward the postrhinal cortex (Burwell 2001), suggesting that this hierarchical organization is conserved across species and is thus likely similar in humans.

Retrosplenial cortex

The retrosplenial cortex has not been described explicitly in the classical view of the neocortical–hippocampal loop thought to be important for memory consolidation (Lavenex and Amaral 2000). However, there is increasing evidence linking the retrosplenial cortex to amnesia (Aggleton 2010), and its connectivity pattern distinguishes it from other cortical areas (Kobayashi and Amaral 2003, 2007). In both monkeys and humans, it comprises areas 29 and 30, located ventrally relative to the posterior cingulate cortex area 23 (Brodmann 1909; Kobayashi and Amaral 2000). The retrosplenial cortex provides about 20% of the cortical inputs to the entorhinal cortex. It receives about 80% of its cortical afferents from the posterior cingulate cortex areas 23 and 31, while the remainder of its inputs are mainly the efferent projections from the hippocampal formation, including the entorhinal cortex, subiculum, presubiculum, and parasubiculum (Kobayashi and Amaral 2003), as well as from the parahippocampal cortex (Lavenex et al. 2002). Lighter projections arise from prefrontal cortex areas 46, 9, 10, and 11, and occipital cortex area V2. Efferent projections of the retrosplenial cortex largely reciprocate the cortical afferents and include major projections to the frontal cortex areas 46, 9, 10, and 11, as well as substantial projections to the entorhinal cortex, presubiculum, parasubiculum, and areas TH and TF of the parahippocampal cortex. Additional projections reach area V4, the dorsal bank of the superior temporal sulcus, and area 7a of the parietal cortex. There is no systematic investigation of the subcortical connections of the retrosplenial cortex, but its reciprocal connections with the anterior thalamic nuclei are likely to be important for the functions carried out by these brain regions (Aggleton 2010).

Entorhinal cortex

The entorhinal cortex is the next stage in the neocortical–hippocampal loop and the first relay in the hippocampal formation (Figure 2.4; Lavenex and Amaral 2000). In monkeys,

Amaral et al. (1987) defined seven subdivisions for the entorhinal cortex: Eo, Er, Elr, Elc, Ei, Ec, and Ecl. Interestingly, Eo is defined based on the specificity of its afferent connections, as the only portion of the monkey entorhinal cortex receiving olfactory inputs. Eo represents about 15% of the surface area of the entorhinal cortex in monkeys and likely less than 10% in humans (Insausti et al. 2002). In contrast, olfactory inputs reach all subdivisions of the rat entorhinal cortex. Note that the rat entorhinal cortex is typically subdivided into lateral and medial entorhinal areas, although a more elaborate analysis by Insausti and colleagues (Insausti et al. 1997) also recognized seven subdivisions (areas CE and ME that constitute the medial entorhinal cortex; and areas VIE, DIE, DLE, AE that constitute the lateral entorhinal cortex). In humans, Insausti et al. (1995) delineated eight subdivisions for the entorhinal cortex: Eo, Er, Elr, Elc, Ei, Emi, Ec, and Ecl. This is an important point to remember and consider when conducting, for example, postmortem neuropathological studies in humans: the entorhinal cortex is not a uniform structure and its different subdivisions exhibit distinct cytoarchitectonic, neurochemical, and connectional characteristics. Researchers must therefore be able to clearly match sections from control and pathological cases in order to make sound comparisons (Lavenex 2009). In all species, the entorhinal cortex constitutes the main gateway for bidirectional communication between the neocortex and the hippocampal formation; it constitutes the first and last stage for information processing within the hippocampal formation. While some polymodal associational cortical areas, in particular the perirhinal and parahippocampal cortices, project directly to CA1 (Suzuki and Amaral 1990), the entorhinal cortex is the only structure to transmit neocortical inputs to the dentate gyrus, CA3 and CA2.

The perirhinal and parahippocampal cortices provide about two-thirds of the neocortical inputs reaching the entorhinal cortex in monkeys (Suzuki and Amaral 1994b). Other major neocortical projections reaching directly the entorhinal cortex originate in the cingulate and retrosplenial cortices, the insular and orbitofrontal cortices, the superior temporal gyrus, and the olfactory bulb (Insausti et al. 1987a). Distinct cortical regions project to different regions of the entorhinal cortex (Insausti and Amaral 2008; Insausti et al. 1987a). Olfactory inputs are strictly restricted to area Eo, the most rostromedial subdivision of the entorhinal cortex. The perirhinal cortex projects to the rostral two-thirds of the entorhinal cortex, whereas the parahippocampal cortex projects to the caudal two-thirds of the entorhinal cortex. It is interesting to note that the topographical organization of the perirhinal and parahippocampal cortices projections to the entorhinal cortex maintains the same relative segregation of cortical inputs that reach these two cortices. Similarly, the insular cortex (which also projects to the perirhinal cortex) projects to the rostral entorhinal cortex, whereas the retrosplenial cortex (which also projects to the parahippocampal cortex) projects to the caudal entorhinal cortex. In contrast, the orbitofrontal and anterior cingulate cortices project to most subdivisions of the entorhinal cortex. Interestingly, while most neocortical projections toward the medial temporal lobe structures terminate in the superficial layers, the frontal cortex projections reaching the entorhinal cortex are directed toward its deep layers (Insausti and Amaral 2008), and might thus have a different influence on information processing within the medial temporal lobe memory system.

In addition to the cortical inputs just described, the entorhinal cortex receives a number of subcortical inputs (Insausti et al. 1987b; Amaral and Lavenex 2007). There are substantial projections from the claustrum and the amygdaloid complex, mainly from the lateral nucleus (terminating in the deep portion of layer III) and the basal nucleus (terminating diffusely in layers III to V), as well as from the accessory basal, paralaminar nuclei, and periamygdaloid cortex (terminating in layers I and II). The entorhinal cortex sends feedback projections, originating in layer V, which terminate mainly in the basal nucleus. The entorhinal cortex also receives cholinergic innervation from the septum, to which it projects back. The major thalamic inputs to the entorhinal cortex originate in the nucleus reuniens and the nucleus centralis medialis; additional projections originate in the medial pulvinar. Hypothalamic projections to the entorhinal cortex originate mainly in the supramammillary area. Dopaminergic innervation from the ventral tegmental area terminates preferentially in the superficial layers I and II, and is much denser in the rostral subdivisions of the entorhinal cortex (Akil and Lewis 1993, 1994). Serotonergic inputs from the raphe nuclei and noradrenergic inputs from the locus coeruleus are distributed throughout the entorhinal cortex layers, but a detailed description of their topographic distribution is not available (Kobayashi and Amaral 1999).

Intrinsic connections within the entorhinal cortex contribute to a significant integration of information before neocortical inputs are transmitted to other hippocampal regions. In monkeys, Chrobak and Amaral (2007) have shown that the intrinsic entorhinal connections are organized in rostrocaudally oriented bands, which extend for about one-third to one-half of the entire length of the entorhinal cortex. In rats, the rostrocaudally oriented bands extend for the full length of the entorhinal cortex (Dolorfo and Amaral 1998), which suggests a higher segregation of anatomical connections in primates as compared to rodents. Interestingly and similar to what is observed for the interconnections between the parahippocampal and perirhinal cortices, the caudal entorhinal cortex has more extensive associational connections than the rostral entorhinal cortex. The relative segregation of neuroanatomical connections within the entorhinal cortex, which somehow maintains the segregation of the cortical inputs and intrinsic connections of the perirhinal and parahippocampal cortices, is further maintained in the topographical organization of the connections between the entorhinal cortex and the rest of the hippocampal formation (see later in this chapter). But before considering the topographical organization of these projections, I will discuss the laminar organization of the entorhinal projections to the hippocampal formation.

A schematic representation of the main serial and parallel pathways through the monkey hippocampal formation (Figures 2.5 and 2.6), beginning and ending with the entorhinal cortex, helps to characterize the laminar organization of distinct hippocampal regions and functional circuits. Entorhinal cortex layer II neurons give rise to the perforant pathway projections reaching the molecular layer of the dentate gyrus and the stratum lacunosum-moleculare of the distal portion of CA3 and CA2, whereas entorhinal cortex layer III neurons give rise to the perforant and alvear pathways projections reaching CA1 stratum lacunosum-moleculare and the molecular layer of the subiculum. In contrast, entorhinal cortex layer V and VI neurons receive return projections from CA1 and the subiculum, and give rise to the cortical efferent projections of the entorhinal cortex. As described by Insausti and Amaral (2008), layers II and III can be thought of as the input layer of the entorhinal cortex, and thus of the hippocampal formation; whereas layers V and VI can be considered the output layers of the entorhinal cortex, and thus of the hippocampal formation, toward cortical areas. Note, however, that layer V and VI neurons also contribute minor projections reaching the dentate gyrus and hippocampus via the perforant path. The other main output structure of the hippocampal formation is the subiculum, which gives rise to significant projections to a number of subcortical areas (see later in chapter). I will follow the unidirectional flow of information through the serial hippocampal circuits in order to discuss the basic structural characteristics and functional organization of the other hippocampal regions, namely the dentate gyrus, CA3, CA2, CA1, the subiculum, presubiculum, and parasubiculum. I will also consider the organization of the direct parallel pathways from the entorhinal cortex to these structures.

Dentate gyrus

The dentate gyrus is comprised of three layers: the molecular layer, the granule cell layer, and the polymorphic layer (Figure 2.6; Amaral et al. 2007). The molecular layer is a relatively cell-free layer containing the dendrites of the dentate granule cells, a small number of interneurons, the fibres of the perforant path originating in the entorhinal cortex, and a variety of fibres intrinsic to the dentate gyrus (i.e. the local and associational connections) as well as fibres originating in different subcortical areas. The granule cell layer is comprised essentially of densely packed granule cells. An important characteristic of the dentate gyrus is that it is one of only two brain regions in which postnatal neurogenesis has been unequivocally demonstrated (the olfactory bulb being the other). I will come back to this topic when discussing our recent quantitative studies of postnatal neurogenesis and development of the monkey hippocampal formation. A number of other neurons, including in particular the pyramidal basket cells, are located at the border between the granule cell and polymorphic layers. The polymorphic layer, a narrow zone enclosed within the granule cell layer, contains a number

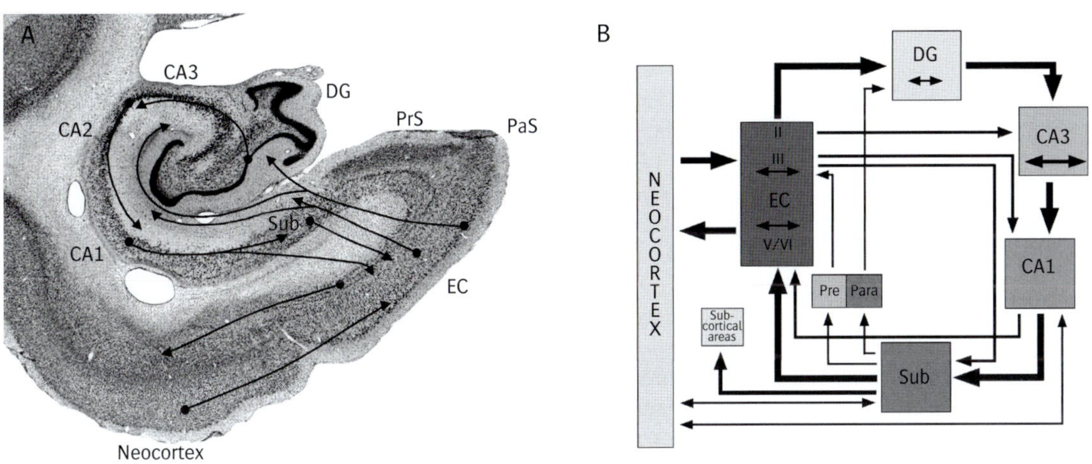

Figure 2.5 Summary diagrams of the serial and parallel pathways through the hippocampal formation. A) Main hippocampal circuits and regions that were analysed quantitatively during early postnatal development in monkeys (Figure 2.17). Reproduced from Jabès, A., Banta Lavenex, P., Amaral, D.G., and Lavenex, P. (2011). Postnatal development of the hippocampal formation: A stereological study in macaque monkeys, *Journal of Comparative Neurology*, 519 (6), 1051–70. © 2011, with permission from John Wiley and Sons. B) Schematic representation including the projections of the presubiculum and parasubiculum. Reproduced from Amaral, D. G. & Lavenex, P. (2007) Hippocampal neuroanatomy. In Andersen, P., Morris, R. G. M., Amaral, D. G., Bliss, T. V. & O'Keefe, J. (Eds.) *The hippocampus book*. © 2007, with permission from Oxford University Press.

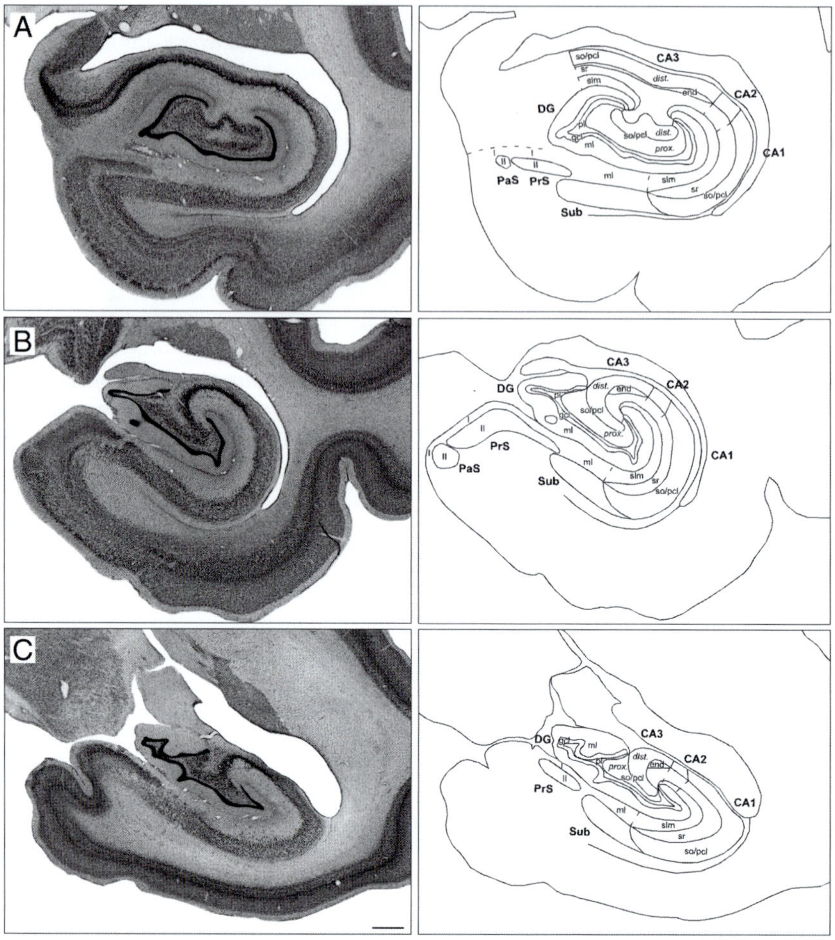

Figure 2.6 A–C) Coronal, Nissl-stained sections and corresponding line drawings at three different rostrocaudal levels of the adult rhesus monkey hippocampal formation. Line drawings delineate the different hippocampal fields and layers. DG, dentate gyrus; ml, molecular layer; gcl, granule cell layer; pl, polymorphic layer; CA3, CA2, CA1, fields of the hippocampus proper; CA3prox, proximal portion of CA3; CA3dist, distal portion of CA3; so, stratum oriens; pcl, pyramidal cell layer; sl, stratum lucidum; end, endbulb; sr, stratum radiatum; slm, stratum lacunosum-moleculare; Sub, subiculum; PrS, presubiculum; PaS, parasubiculum; EC, entorhinal cortex. Scale bar in A = 1mm, applies to all panels. Reproduced from Jabès, A., Banta Lavenex, P., Amaral, D.G., and Lavenex, P. (2011). Postnatal development of the hippocampal formation: A stereological study in macaque monkeys, *Journal of Comparative Neurology*, 519 (6), 1051–70. © 2011, with permission from John Wiley and Sons.

of different cell types, the most prominent of which are the mossy cells (Amaral 1978). The most distinctive feature of the mossy cells is that their proximal dendrites are covered by very large and complex spines called thorny excrescences. These are the distinctive sites of termination of the axons of the granule cells within the dentate gyrus.

Without going into much detail (see Amaral and Lavenex (2007), and references therein for more information), I would like to point out two differences between rodents and primates in the cellular organization of the dentate gyrus, which might be of importance when considering the normal and pathological structure and functions of the dentate gyrus. In adult rats, the granule cells only have apical dendrites extending into the molecular layer, whereas in monkeys and humans many granule cells also have basal dendrites extending into the polymorphic layer (Figure 2.7; Seress and Mrzljak 1987). Interestingly, granule cells with basal dendrites are only observed in rats during development or following epileptic seizures (Spigelman et al. 1998). This suggests that the findings from studies of the cellular correlates of epilepsy in rodents should be considered with caution. Rodent studies clearly provide important information regarding the fundamental mechanisms contributing to the regulation of neuronal excitability in hippocampal circuits. However, the specific characteristics of the cellular organization of the primate dentate gyrus should be considered carefully when extrapolating from the results of studies in rodents to possible clinical applications in humans. Similarly, in rats, the mossy cells located in the polymorphic layer rarely have dendrites entering the molecular layer, whereas in monkeys many mossy cells extend dendrites into the molecular layer (Figure 2.8;

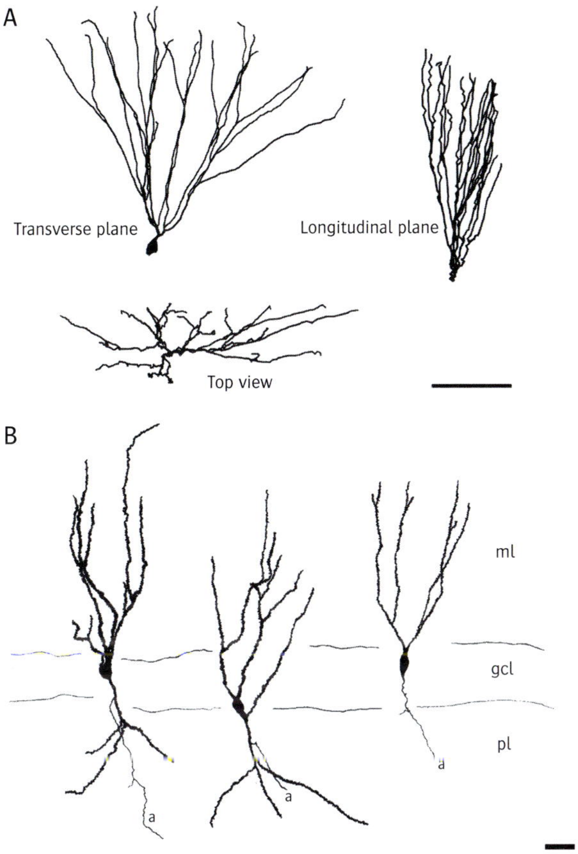

Figure 2.7 Camera lucida drawings of dentate gyrus granule cells. A) Rat granule cell. Scale bar = 100 µm. B) Monkey granule cell. Scale bar = 20 µm. Note the presence of basal dendrites in some monkey granule cells. Reprinted from *Brain Research*, 405, László Seress & Ladislav Mrzljak. Basal dendrites of granule cells are normal features of the fetal and adult dentate gyrus of both monkey and human hippocampal formations. pp. 169–74, Copyright (1987), with permission from Elsevier.

Buckmaster and Amaral 2001). No experimental data is available for humans, but one can reasonably infer that human mossy cells are more likely to be similar to those of monkeys rather than rats. From a functional perspective, primate mossy cells are in a position to receive direct perforant path input in the outer two-thirds of the molecular layer (see later), whereas rodent mossy cells are not in a position to receive direct inputs from entorhinal cortex layer II cells. Such differences could have an important impact on the normal function of the dentate gyrus and on the abnormal regulation of neuronal excitability in neuropathological conditions like epilepsy. Research carried out with non-human primates can provide a necessary perspective by demonstrating, for instance, the structural characteristics and functional mechanisms that are conserved between rats and monkeys, and which are therefore likely to be conserved in humans.

The afferents and intrinsic connections of the dentate gyrus exhibit definite laminar distributions in all species. The perforant path projections terminate exclusively in the outer two-thirds of the molecular layer (Figure 2.9). In addition, perforant path projections originating in the rostral portion of the entorhinal cortex (corresponding to the lateral entorhinal cortex in rodents) terminate in the superficial part of the termination zone, whereas perforant path projections originating in the caudal portion of the entorhinal cortex (corresponding to the medial entorhinal cortex in rodents) terminate more heavily in the middle third of the layer (Witter and Amaral 1991). Interestingly, in monkeys, the medial portion of the entorhinal cortex projects toward the rostral dentate gyrus, whereas the lateral portion of the entorhinal cortex projects more toward the caudal dentate gyrus. In contrast, the associational projection, originating from the mossy cells in the polymorphic layer, terminates in the inner one-third of the molecular layer (Figure 2.10; Kondo et al. 2008). The mossy cells are immunoreactive for glutamate and make asymmetric (excitatory) terminations on the dendrites of the granule cells. The associational projection, though modest at the level of origin, travels both rostrally and caudally from the injection site for as much as 80% of the rostrocaudal extent of the dentate gyrus. The caudally directed projection is typically more extensive and denser than the rostrally directed projection. Another population of cells in the polymorphic layer, some of which colocalize gamma-aminobutyric acid (GABA) and somatostatin, also originate local projections that terminate in the outer two-thirds of the molecular layer, i.e. in the termination zone of perforant path projections. In rats, these cells receive excitatory inputs from the granule cells and send inhibitory projections back to the dendrites of the granule cells (Leranth et al. 1990), thus forming an inhibitory feedback circuit with the granule cells. In monkeys, these projections are densest at the level of the cells of origin but also extend several millimetres rostrocaudally (Kondo et al. 2008). Overall, the topographic organization of the intrinsic connections of the monkey dentate gyrus is largely similar to that of the rat, suggesting that this organization is likely similar in humans. Such extensive longitudinal connections have the potential for integrating information across much of the rostrocaudal extent of the dentate gyrus (Figure 2.11). Note, however, that in contrast to rats, monkeys and likely humans do not have extensive commissural projections from the polymorphic layer to the contralateral molecular layer of the dentate gyrus. Only the most rostral portion of the dentate gyrus demonstrates any commissural projections (Amaral et al. 1984). Interestingly, calretinin-positive cells, located in the polymorphic layer and thought to be associated with the mossy fibre system, are found through the entire rostrocaudal length of the rat dentate gyrus (Gulyas et al. 1992), but only in its most rostral portion in monkeys (Lavenex et al. 2009), where commissural projections are present (Amaral et al. 1984). In addition, the molecular layer of the dentate gyrus is also the target of lighter hippocampal projections originating in the presubiculum and parasubiculum, which

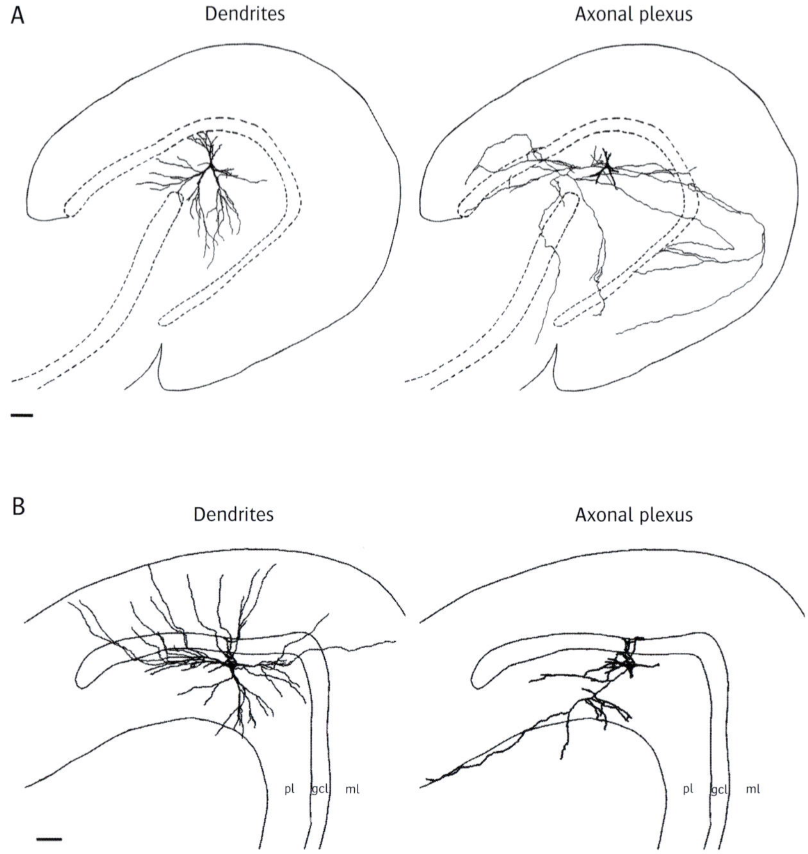

Figure 2.8 Mossy cells of the polymorphic layer of the dentate gyrus. A) Rat mossy cell. Scale bar = 100 μm. Reproduced from Buckmaster, P. S., Strowbridge, B. W., Kunkel, D. D., Schmiege, D. L., and Schwartzkroin, P. A. (1992). Mossy cell axonal projections to the dentate gyrus molecular layer in the rat hippocampal slice. *Hippocampus*, 2, 349–62. © 1992, John Wiley and Sons. B) Monkey mossy cell. Note the presence of apical dendrites entering the molecular layer of the dentate gyrus in monkeys. Scale bar = 100 μm. Reproduced from Buckmaster, P. S., and Amaral, D. G. (2001). Intracellular recording and labeling of mossy cells and proximal CA3 pyramidal cells in macaque monkeys. *Journal of Comparative Neurology*, 430, 264–81. © 2001, with permission from John Wiley and Sons.

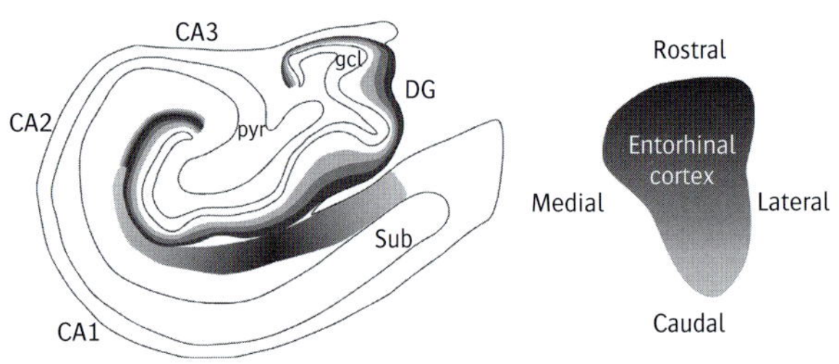

Figure 2.9 Laminar and topographical organization of the entorhinal cortex projection to the dentate gyrus, hippocampus and subiculum. The surface of the entorhinal cortex is represented on the right.

terminate interspersed between the superficial and deeper zones of termination of entorhinal cortex projections. The dentate gyrus receives a number of relatively more modest projections from subcortical areas (Amaral and Lavenex 2007). Projections from the septum terminate in the polymorphic and molecular layers. Interestingly, many of the septal cells projecting to the dentate gyrus are GABAergic, and cholinergic and GABAergic septal projections target different cell types. The heaviest GABAergic septal projections terminate asymmetrically (inhibitory synapses) on interneurons, such as basket cells, in the polymorphic layer; whereas cholinergic septal projections terminate most heavily on the proximal portion of dentate granule cell dendrites in the inner one-third of the molecular layer. Hypothalamic inputs from the supramammillary area terminate most heavily in a narrow zone of the molecular layer just superficial to the granule cell layer, and lightly throughout the molecular layer. Noradrenergic inputs from the locus coeruleus terminate mainly in the polymorphic layer of the dentate gyrus and CA3 stratum lucidum (two zones occupied by the mossy fibre projections of the dentate granule cells). Serotonergic projections from the raphe nuclei terminate most heavily in the polymorphic layer, a zone in which significant postnatal cell division and neurogenesis are observed.

The granule cells are the only neurons to project outside the dentate gyrus, toward the CA3 pyramidal cells

(Figure 2.10; Kondo et al. 2008). However, their axons, the mossy fibres, project first to a restricted portion of the polymorphic layer in which they establish asymmetric (excitatory) contacts with mossy cells located very close to the cells of origin. As described earlier, the mossy cells originate the associational projection terminating in the inner-third of the molecular layer. These associational fibres contact not only the spines of the granule cell dendrites but also the dendritic shafts of the GABAergic basket cells that, in turn, innervate the granule cells. Thus, the associational projection may function both as a feedforward excitatory pathway to distant granule cells and as a disynaptic feedforward inhibitory pathway, via the pyramidal basket cell as intermediary. The mossy fibre projection toward the CA3 pyramidal cells first follows a transverse trajectory with a slight rostral inclination that does not exceed 1 mm, or about 5% of the entire rostrocaudal length of the hippocampus in monkeys (Kondo et al. 2008). After reaching the distal end of CA3, the mossy fibres turn rostrally and travel for about 30% of the entire length of the hippocampus. Consequently, the pyramidal cells located in the proximal and mid portions of CA3 receive mossy fibre input from only a very restricted region of the dentate gyrus located at the same rostrocaudal level. In contrast, the

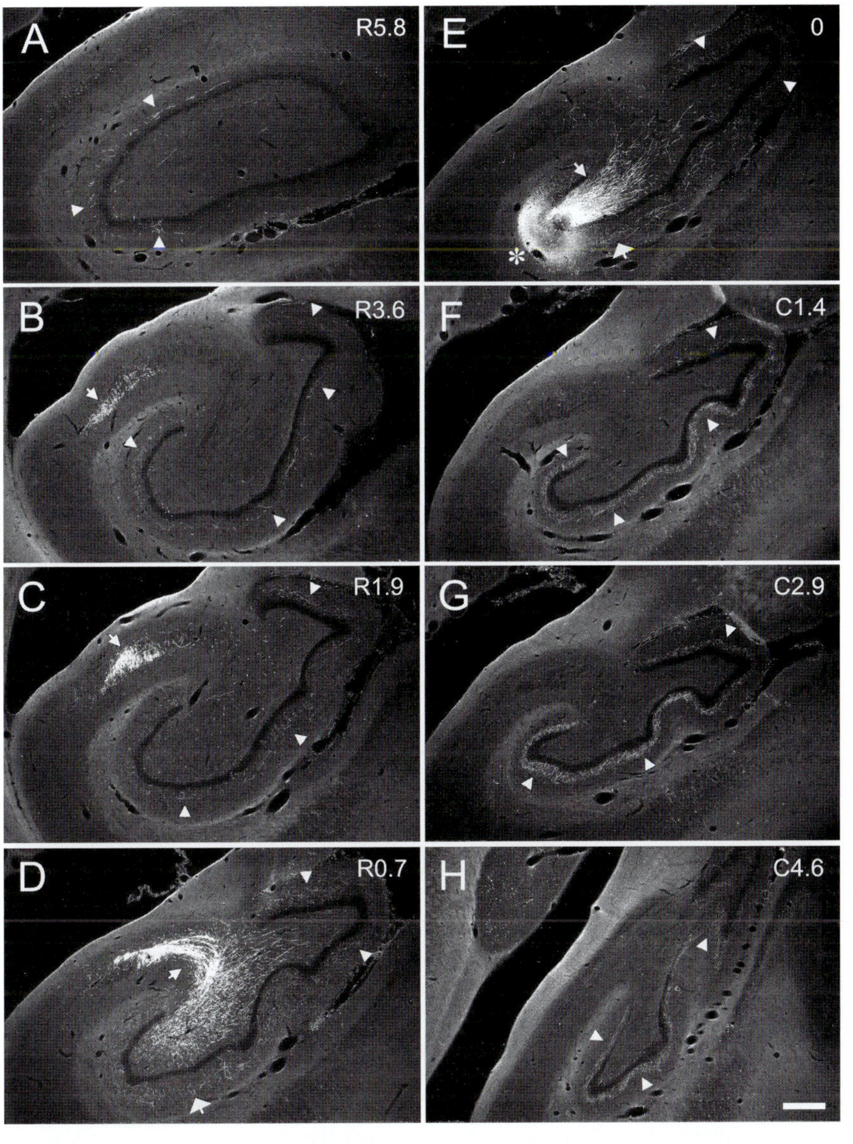

Figure 2.10 A–H) Darkfield photomicrographs of coronal sections of the dentate gyrus arranged from rostral (A) to caudal (H) showing the distribution of PHA-L-labelled fibres and terminals following a tracer injection in the midrostrocaudal one-third of the dentate gyrus (asterisk in E). The numbers to the right of each panel indicate the distance, in millimetres, of the photographed section from the injection site (R and C indicate rostral and caudal to the injection site, respectively). Scale bar = 0.5 mm. Reproduced from Kondo, H., Lavenex, P., and Amaral, D. G. (2008). Intrinsic connections of the macaque monkey hippocampal formation: I. Dentate gyrus. *Journal of Comparative Neurology*, 511, 497–520. © 2008, with permission from John Wiley and Sons.

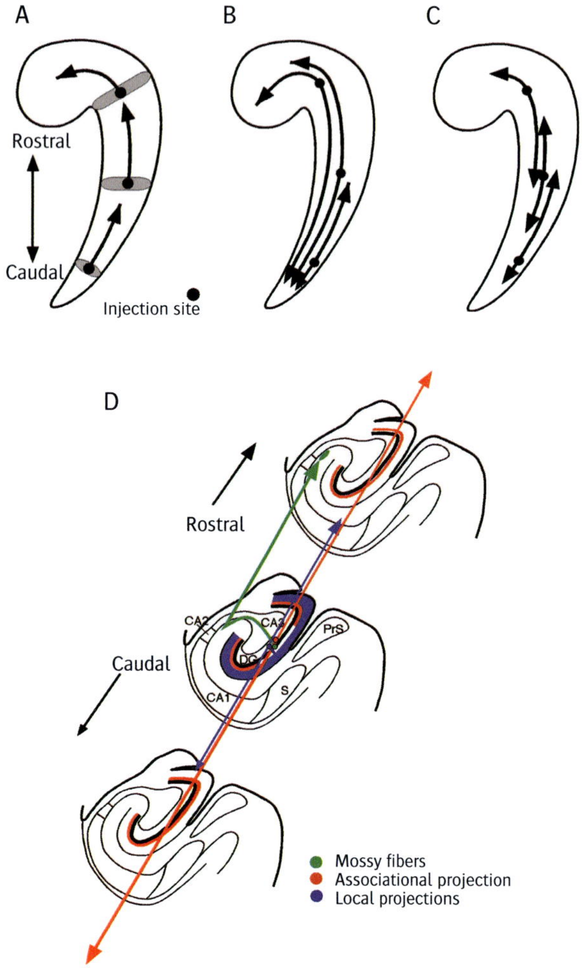

Figure 2.11 Summary diagram showing the topographical organization of the major connections of the dentate gyrus, including the mossy fibres (A; green in D), associational projections (B; red in D), and local projections (C; blue in D). Note that each projection has an extensive distribution along the longitudinal axis. Reproduced from Kondo, H., Lavenex, P., and Amaral, D. G. (2008). Intrinsic connections of the macaque monkey hippocampal formation: I. Dentate gyrus. *Journal of Comparative Neurology*, 511, 497–520. © 2008, with permission from John Wiley and Sons.

pyramidal cells located in the distal portion of CA3 receive mossy fibre inputs from granule cells both from their own rostrocaudal level and from granule cells located as far caudally as one-third of the entire length of the dentate gyrus. In rats, each granule cell communicates with only about 15 CA3 pyramidal cells distributed through the full transverse axis of CA3. Since in monkeys there are about 7.2 million granule cells and 1.3 million CA3 pyramidal cells (Jabès et al. 2011), each pyramidal cell can be expected to receive inputs from about 83 granule cells (72 in rats; Amaral and Lavenex 2007). This pattern of connectivity could be particularly well suited to contribute to one of the fundamental functions assigned to the dentate gyrus: pattern separation, that is the ability to encode similar patterns of activation into distinct, non-overlapping representations (Morris 2007).

Hippocampus

The hippocampus is divided into three distinct fields: CA3, CA2, and CA1 (see Figure 2.6). I describe first the overall structural characteristics of these fields, and then consider the specific organization of their intrinsic and extrinsic connectivity separately. CA3 is characterized by large pyramidal cells (Jabès et al. 2011). The proximal part of CA3 occupies much of the region enclosed within the limbs of the dentate gyrus. The distal part of CA3 appears more compact and darkly stained in Nissl-stained sections than the proximal part. CA2 is differentiated from CA3 by the lack of a mossy fibre input (i.e. projections from the dentate gyrus granule cells; Kondo et al. 2008). CA1 has smaller and more spherical pyramidal cells in its principal cell layer (Jabès et al. 2011), which is substantially thicker and less densely packed than in CA3 and CA2. The widening of the pyramidal cell layer in CA1 therefore marks the border between CA2 and CA1. The hippocampus is also subdivided into a number of laminae that run parallel to the pyramidal cell layer. Subjacent to the pyramidal cell layer is the cell-poor stratum oriens, which in CA1 is less apparent and not clearly identifiable due to the widening and decreased density of the pyramidal cell layer. Stratum oriens can be defined as the infrapyramidal region in which some of the CA3 to CA3 associational connections and the CA3 to CA1 Schaffer collateral connections are located (discussed later). In rats, stratum oriens contains the basal dendrites of the densely packed pyramidal cells, whereas in monkeys and humans the basal dendrites of the pyramidal cells are also largely located in the loosely packed pyramidal cell layer (Altemus et al. 2005). Deep to stratum oriens is the thin, fibre-containing alveus. Superficial to the pyramidal cell layer in CA3 is stratum lucidum, in which some of the mossy fibres travel. The distal part of stratum lucidum, called the end bulb, thickens where the mossy fibres bend rostrally and travel longitudinally for about one-third of the length of the hippocampus (Kondo et al. 2008); the end bulb helps to delineate the border between CA3 and CA2. Unlike in the rat, where the majority of mossy fibres travel in a position just superficial to the pyramidal cell layer, mossy fibres travel both within and above the pyramidal cell layer throughout the entire CA3 in monkeys (Kondo et al. 2008). Stratum radiatum is located superficial to stratum lucidum in CA3, and immediately above the pyramidal cell layer in CA2 and CA1. Stratum radiatum can be defined as the suprapyramidal region in which the CA3 to CA3 associational connections and the CA3 to CA1 Schaffer collateral connections are located (Kondo et al. 2009). The most superficial layer of the hippocampus is called stratum lacunosum-moleculare. It is in this layer that fibres from the entorhinal cortex terminate (Witter and Amaral 1991). The hippocampus also contains a variety of interneurons scattered through its different layers. Although interneurons clearly play an important role in the regulation of network activity in the hippocampal formation (Spruston and McBain 2007), a discussion of the interneurons of the hippocampus is beyond the scope of this chapter. A brief description of these neuronal types can be found in (Amaral and Lavenex 2007); the interested reader can also refer to the detailed synopsis of Freund and Buszaki (1996) for an in-depth description and classification of hippocampal interneurons. An important characteristic of the connectivity of the hippocampus is that most of its synaptic input arises from within its own boundaries. Indeed, CA3 and CA2 are heavily innervated by collaterals of their own axons. CA1, in turn, receives its heaviest inputs from CA3 and CA2. This is not to say that extrinsic inputs are not functionally important, but they account for a relatively smaller number of synapses. Indeed, for example, lighter projections arising from the entorhinal cortex terminate on the distal dendrites of pyramidal neurons and interneurons located in stratum lacunosum-moleculare. And functionally, direct entorhinal cortex inputs play a significant role in the physiological response properties of hippocampal CA1 pyramidal neurons (e.g. (Brun et al. 2002); see (Jabès et al. 2011) for a discussion).

CA3

The dentate gyrus projection to CA3 was described earlier. The axons of the granule cells, the mossy fibres, terminate on the proximal portion of the apical dendrites of CA3 pyramidal cells in stratum pyramidale and stratum lucidum. In contrast, the entorhinal cortex layer II neurons (and to a much lesser extent layer VI cells) projecting to the dentate gyrus also extend collaterals that make synaptic contacts with the distal portion of the apical dendrites of CA3 pyramidal neurons located in stratum lacunosum-moleculare (i.e. in the distal portion of CA3; Witter and Amaral 1991). Although the laminar organization of the entorhinal cortex projections is not as clear in monkeys as in rats or cats, projections from rostral portions of the entorhinal cortex terminate preferentially in the superficial portion of CA3 and CA2 stratum lacunosum-moleculare; whereas projections from caudal portions of the entorhinal cortex terminate preferentially in the deeper portion of stratum lacunosum-moleculare. This topographical organization is similar to that observed for the entorhinal cortex to dentate gyrus projections. CA3 strata oriens, pyramidale, and radiatum are largely occupied by the CA3 to CA3 associational connections, which travel extensively along its longitudinal axis (about 80% of the full hippocampal length, Figure 2.12; Kondo et al. 2009). CA3 to CA3 associational connections are absent from stratum lucidum and the part of the pyramidal cell layer where the mossy fibres travel. Interestingly, the projections from CA3 to CA3 are overall less substantial than those from CA3 to CA1 (Kondo et al. 2009). CA3 associational projections originating in the proximal portion of CA3 (close to the dentate gyrus) are in general weaker and terminate more heavily in the superficial portion of their terminal zone, as compared to projections from the mid and distal portions of CA3 that give rise to heavier and more homogeneous projections throughout their terminal zone. Note that a back projection from CA3 pyramidal cells located in the proximal portion of CA3 to the dentate gyrus has been reported (Buckmaster and Amaral 2001). The structural characterization of this connection and the elucidation of its functional properties have yet to be fully achieved (Myers and Scharfman 2011).

CA3 has no known projections with the neocortex. In contrast, CA3 receives a number of comparatively minor projections from subcortical areas. The septum provides the major subcortical projection to CA3, which terminates most heavily in stratum oriens and to a lesser extent in stratum radiatum. The basal nucleus of the amygdala also originates projections that terminate in strata oriens and radiatum (as shown in rats by (Pitkänen et al. 2000)). Noradrenergic and serotonergic innervation is most densely distributed in stratum lacunosum-moleculare, with much lighter labelling in the other layers. The only sizable projection from CA3 to subcortical structures targets the lateral septal nucleus (Amaral and Lavenex 2007).

CA2

The intrinsic connectivity of CA2 is generally considered to be similar to that of the distal portion of CA3, with the notable exception that CA2 pyramidal cells do not receive any input from the dentate granule cells. Similarly, the entorhinal cortex projections to CA2 are generally considered similar to those reaching distal CA3. One major difference between CA3 and CA2 is that CA2 receives prominent projections from the posterior hypothalamus, in particular from the supramammillary area and the tuberomammillary nucleus. There is no evidence that CA2 returns this projection or projects to any other area outside of the hippocampus.

CA1

There are essentially no CA1 to CA1 associational connections. In contrast, CA1 receives its most prominent projections from CA3 pyramidal neurons, via the so-called Schaffer collaterals, which terminate in strata oriens, pyramidale, and radiatum throughout the entire transverse extent of CA1 (Figure 2.12; Kondo et al. 2009). As mentioned previously, the CA3 to CA1 projections are more substantial than the CA3 to CA3 associational projections. They extend both rostrally and caudally for as much as three-quarters of the longitudinal extent of the hippocampus. Projections from CA3 to CA1 are distributed along a radial gradient, depending on the transverse location of

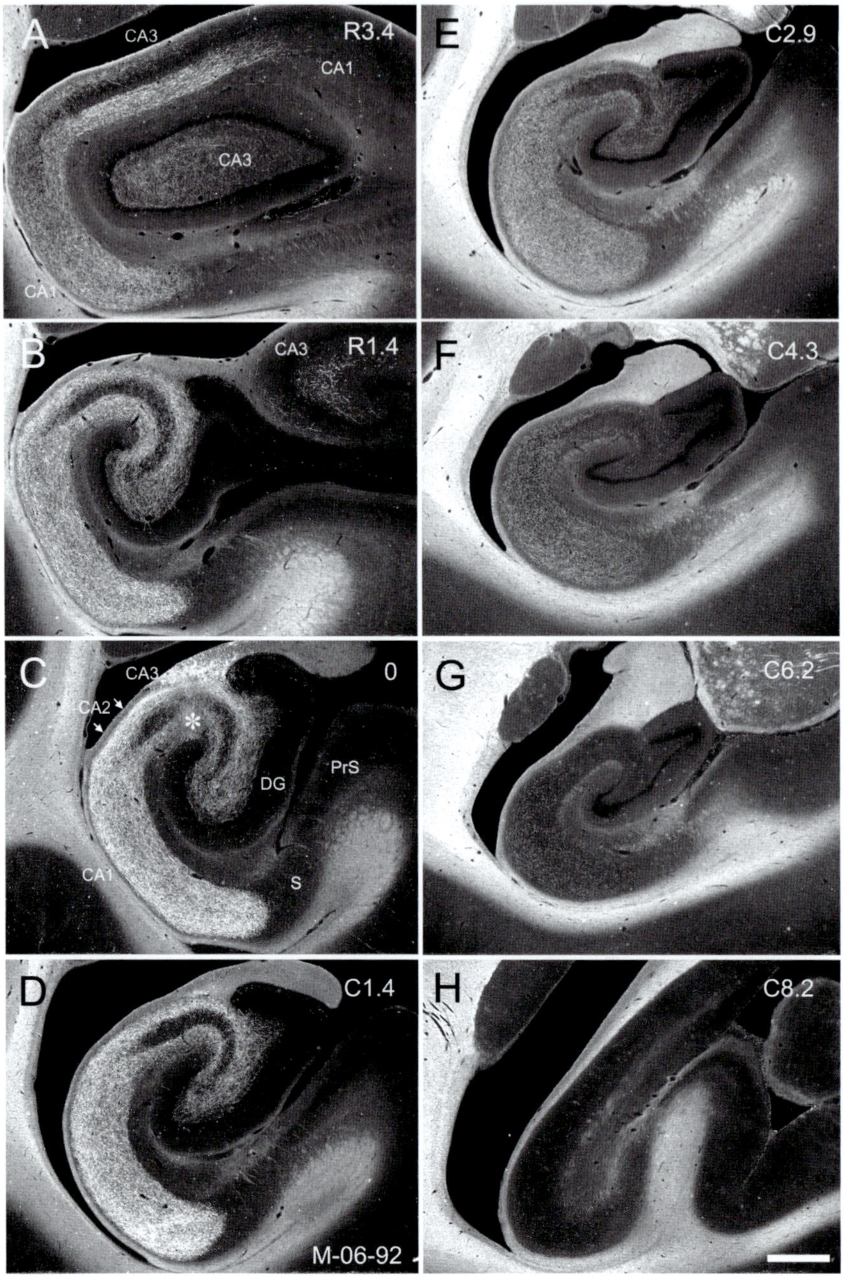

Figure 2.12 Darkfield photomicrographs of sections of the hippocampal formation arranged from rostral (A) to caudal (H) showing the distribution of PHA-L labelled fibres and terminals following a tracer injection in the mid-transverse portion of CA3 at a mid-rostrocaudal level of the hippocampus (asterisk in C, same as (B) in Figure 2.4). The number to the right of each panel indicates the distance, in millimeters, of the photographed section from the injection site. R and C indicate rostral and caudal to the injection site, respectively. Note that the CA3 projections to CA1 and CA3 are distributed very extensively along the rostrocaudal axis. Scale bar = 1 mm. Reproduced from Kondo, H., Lavenex, P., and Amaral, D. G. (2009). Intrinsic connections of the macaque monkey hippocampal formation: II. CA3 connections. *Journal of Comparative Neurology*, 515, 349–77. © 2009, with permission from John Wiley and Sons.

the cells of origin. Proximal CA3 pyramidal neurons (located near the dentate gyrus) generate projections that more densely terminate superficially in the terminal zone of CA1, whereas distal CA3 cells (located closer to CA1) give rise to projections that more heavily terminate deeply in the terminal zone of CA1. CA1 receives a prominent projection from the entorhinal cortex layer III neurons (and to a lesser extent from layer V neurons) that terminates in stratum lacunosum-moleculare. In addition, CA1 receives direct projections from a number of cortical areas, including in particular the perirhinal and parahippocampal cortices (as well as from a few other areas; Suzuki and

Amaral 1990; Rockland and Van Hoesen 1999), which terminate in stratum lacunosum-moleculare. The entorhinal cortex projection to CA1 follows a transverse gradient of distribution, rather than a radial topography as is the case for the dentate gyrus and CA3. Rostral portions of the entorhinal cortex terminate most heavily at the border between CA1 and the subiculum, whereas caudal portions of the entorhinal cortex project toward more proximal portions of CA1 (closer to CA2). In addition, lateral portions of the entorhinal cortex project to caudal levels of CA1, whereas medial portions of the entorhinal cortex project to more rostral levels of CA1.

CA1 is also interconnected with a number of subcortical areas. In monkeys, the rostral two-thirds of the distal portion of CA1 is reciprocally connected with the basal nucleus of the amygdala (Amaral and Insausti 1992). In rats, CA1 receives light projections from the medial septal nucleus and projects back to the lateral septal nucleus. The nucleus reuniens of the thalamus gives rise to a prominent projection to CA1 stratum lacunosum-moleculare, where it overlaps with fibres from the entorhinal cortex. Based on data in rats, CA1 appears to have a similar but lighter distribution of monoaminergic innervation than CA3; noradrenergic and serotonergic innervation is most densely distributed in stratum lacunosum-moleculare.

In contrast to the paucity of CA1 intrinsic connections, there are strong projections from CA1 to the subiculum, which terminate in the pyramidal cell layer and in the deep portion of the molecular layer of the subiculum (Figure 2.13; Kondo et al. unpublished observations). Similar to rats, the proximal portion of CA1 projects preferentially to the distal portion of the subiculum, whereas the distal portion of CA1 projects to the proximal portion of the subiculum. The CA1 projections extend both rostrally and caudally for about half of the length of the hippocampus. In contrast to the dentate gyrus, CA3 and CA2, CA1 also projects back to the deep layers of the entorhinal cortex (layers V and VI). Experiments performed in rats revealed that these projections are in register with the entorhinal cortex inputs to CA1 (Tamamaki and Nojyo 1995; Naber et al. 2001). Although fibres originating from a restricted part of the entorhinal cortex distribute extensively along the longitudinal axis of CA1, only a restricted portion of CA1, receiving the same entorhinal information, gives rise to projections that return to the deep layers of the area of origin of the perforant pathway in the entorhinal

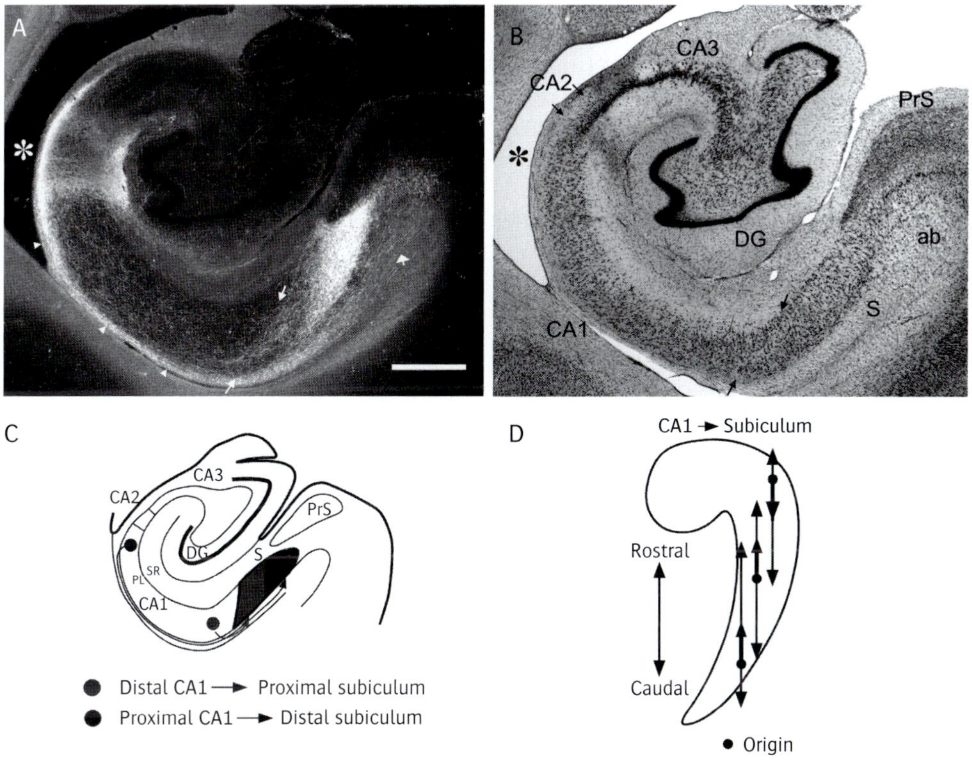

Figure 2.13 A) Darkfield photomicrograph showing PHA-L labelled axons and terminals of the CA1 projections to the subiculum following a tracer injection (asterisk) in the proximal portion of CA1 as well as in CA2 at the mid rostrocaudal level of the hippocampus. B) Adjacent Nissl-stained section. The labelled fibres and terminals are present in the pyramidal cell layer and the deep portion of the molecular layer in the subiculum. Arrows indicate the border of the subiculum with CA1. Note the substantial projections to the distal portion of the subiculum. Scale bar = 1 mm. Summary diagram showing the topographic organization of the CA1 projections to the subiculum in the transverse (C) and rostrocaudal (D) axes.

cortex (Naber et al. 2001). This is also the case for the projections between the subiculum and the entorhinal cortex (see 'Subiculum'). This suggests that in primates, CA1 pyramidal neurons located proximally project toward the caudal portions of the entorhinal cortex, whereas CA1 pyramidal neurons located distally project toward the rostral portions of the entorhinal cortex.

Subiculum

The subiculum is comprised of essentially two layers: the molecular layer, in which the entorhinal cortex projections terminate, is a rather cell-free layer located superficially; and the pyramidal cell layer, in which most of the CA1 projections terminate, is located below. The border of CA1 with the subiculum is oblique; the progressively thinner CA1 pyramidal cell layer actually extends over the initial portion of the subiculum. The pyramidal cell layer of the subiculum is slightly broader than that of CA1. The cell population in the subiculum pyramidal cell layer is more heterogeneous than in CA1 and cell orientation is slightly different. CA1 stratum radiatum ends at the superficial CA1-subiculum border, where it is replaced by the molecular layer of the subiculum. A relatively high level of SMI-32-immunoreactivity, reflecting non-phosphorylated high-molecular-weight neurofilaments, is particularly visible in the proximal segments of apical dendrites and helps to distinguish the subiculum from both CA1 and the presubiculum (Lavenex et al. 2004a, 2009).

The subiculum is interconnected with a number of other brain areas (O'Mara et al. 2001), and constitutes the main output structure of the hippocampal formation toward subcortical brain regions. As described previously, the subiculum receives its main input from CA1, which terminates mostly in the pyramidal cell layer. Entorhinal cortex layer III neurons also send a significant projection that establishes connections with the apical dendrites of subicular neurons located in the molecular layer. Both of these projections exhibit clear topographical organizations. CA1 projections extend both rostrally and caudally for about half of the length of the hippocampus, and proximal CA1 projects to the distal subiculum, whereas distal CA1 projects to the proximal subiculum. The entorhinal cortex projection to the subiculum follows a transverse organization similar to the entorhinal cortex to CA1 projection. Rostral levels of the entorhinal cortex terminate most heavily in the proximal portion of the subiculum (closer to CA1), whereas caudal entorhinal areas terminate more heavily in the distal portion of the subiculum (closer to the presubiculum). In addition, lateral portions of the entorhinal cortex project to the caudal subiculum, whereas medial portions of the entorhinal cortex project to the rostral subiculum. The different regions of CA1 and subiculum that receive similar inputs from the entorhinal cortex are thus preferentially interconnected: rostral entorhinal cortex projects to distal CA1 and proximal subiculum, which are more highly interconnected; and caudal entorhinal cortex projects to proximal CA1 and distal subiculum, which are also more highly interconnected (Figure 2.14). The subiculum projects back to the deep layers, V and VI, of the entorhinal cortex (Figure 2.5). As for the CA1 projections to the entorhinal cortex, the subiculum efferent projections to the deep layers of the entorhinal cortex are in register with the afferents originating in the superficial layers of the entorhinal cortex. In addition to its interconnections with the entorhinal cortex, the rat subiculum receives

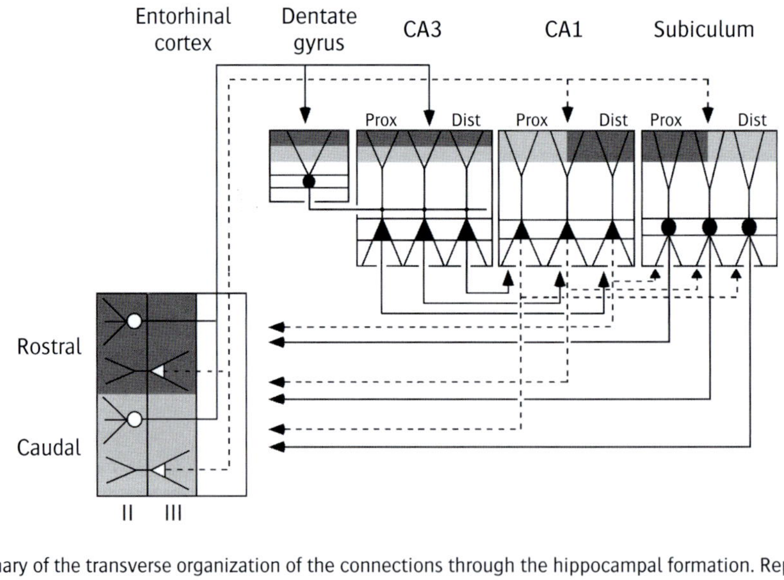

Figure 2.14 Summary of the transverse organization of the connections through the hippocampal formation. Reproduced from *The Hippocampus Book*, Andersen, P., Morris, R. G. M., Amaral, D. G., Bliss, T. V. & O'Keefe, J. (Eds.) © 2007, with permission from Oxford University Press.

direct cortical inputs from the perirhinal and postrhinal cortices. These projections follow the overall organization of the entorhinal cortex projections to the subiculum, so that the perirhinal cortex projection reaches preferentially the proximal portion of the subiculum, whereas the postrhinal cortex projection reaches preferentially the distal portion of the subiculum. The subiculum also projects to two other hippocampal regions: the presubiculum and the parasubiculum. In rats, the subiculum has been shown to provide substantial projections to the superficial layers of the retrosplenial cortex, as well as to the perirhinal cortex, and gives rise to projections to the orbitofrontal, prelimbic, and infralimbic cortices.

The subiculum is the major source of efferent projections from the hippocampal formation toward subcortical structures and receives a number of projections from subcortical areas (see Amaral and Lavenex (2007) for detailed information). The most prominent efferent projections of the subiculum are directed toward the lateral septal nucleus, the nucleus accumbens, and the mammillary nuclei. More minor projections reach the anterior thalamic nuclei (Aggleton et al. 2005). All these projections are organized topographically, but a detailed presentation of their organization is beyond the scope of this chapter. Interestingly, however, the majority of subicular neurons send their axons to only one target structure among the possible targets of the subiculum efferent projections (Naber and Witter 1998). Thus, the subiculum is in a position to provide separate channels to distribute the information processed in the rest of the hippocampal formation to both cortical regions (mostly via the entorhinal cortex) or directly to subcortical regions. The efferent connections of the hippocampal formation toward subcortical structures, via the subiculum, might play an important role in the amnesic syndrome linked to diencephalic pathologies. Indeed, the interconnections between the subiculum, presubiculum, and parasubiculum, and the anterior thalamic nuclei and the retrosplenial cortex are considered by some as critical pathways for memory that could link the classically distinct syndromes of temporal lobe and diencephalic amnesias (Aggleton 2010).

Presubiculum

The presubiculum lies medial to the subiculum. It consists of a cell-free layer I, and a densely packed, superficial layer II composed of small pyramidal cells. The presubiculum receives prominent inputs from the subiculum, as well as a direct projection from the anterior thalamic nuclei. A number of other subcortical areas, including the septum, supramammillary nucleus, raphe nuclei, and locus coeruleus project to the presubiculum. The presubiculum has well-developed intrinsic connections, which have yet to be fully characterized. Its most prominent projection targets the layer III of the entorhinal cortex. Interestingly, in rats, the subicular projections to the entorhinal cortex terminate in a zone corresponding to the medial entorhinal cortex (van Haeften et al. 1997). In monkeys, subicular projections terminate in the caudal portion of the entorhinal cortex (Witter and Amaral unpublished observations), which receives significant projections from the parahippocampal cortex (Suzuki and Amaral 1994b). The presubiculum also gives rise to a major commissural projection to the contralateral caudal entorhinal cortex, but it does not project to the contralateral presubiculum. The presubiculum is also reciprocally interconnected with the retrosplenial cortex. In addition, the presubiculum send back projections to the thalamus as well as minor projections to the molecular layer of the dentate gyrus.

Parasubiculum

The cytoarchitectonic organization of the parasubiculum is quite variable along its rostrocaudal extent and its cell population is heterogeneous. The parasubiculum comprises two layers: a cell-free layer I and a densely packed cell layer II. The parasubiculum receives prominent projections from the subiculum. As the presubiculum, the parasubiculum is reciprocally connected with the anterior thalamic nuclei and the retrosplenial cortex, and receives significant cholinergic innervation from the septum. In parallel to the presubiculum projection directed toward the layer III of the entorhinal cortex, the parasubiculum projects specifically to the layer II of the entorhinal cortex. In rats, the parasubiculum gives rise to a fairly substantial projection to the molecular layer of the dentate gyrus. This projection occupies the superficial two-thirds of the molecular layer, with a preference for the mid portion of the molecular layer, where the projections from the rodent medial entorhinal cortex terminate (corresponding to the caudal entorhinal cortex in primates). Similarly, the parasubiculum projects weakly to stratum lacunosum-moleculare of the hippocampus, the molecular layer of the subiculum, and the presubiculum.

Cortical efferents of the hippocampal formation

So far, I have discussed the general organization of the hippocampal formation and related structures, as well as the subcortical connections and cortical afferents reaching the hippocampal formation via the perirhinal, parahippocampal, retrosplenial and entorhinal cortices. Here, I describe the efferent connections of the hippocampal formation toward cortical regions that are relayed via these cortices. Although the efferent projections from the hippocampal formation to the cortex are generally of the feedback type, their laminar distribution differs in distinct cortical areas, and most importantly, these connections do not necessarily reciprocate the cortical afferents originating in defined brain regions (Figure 2.15).

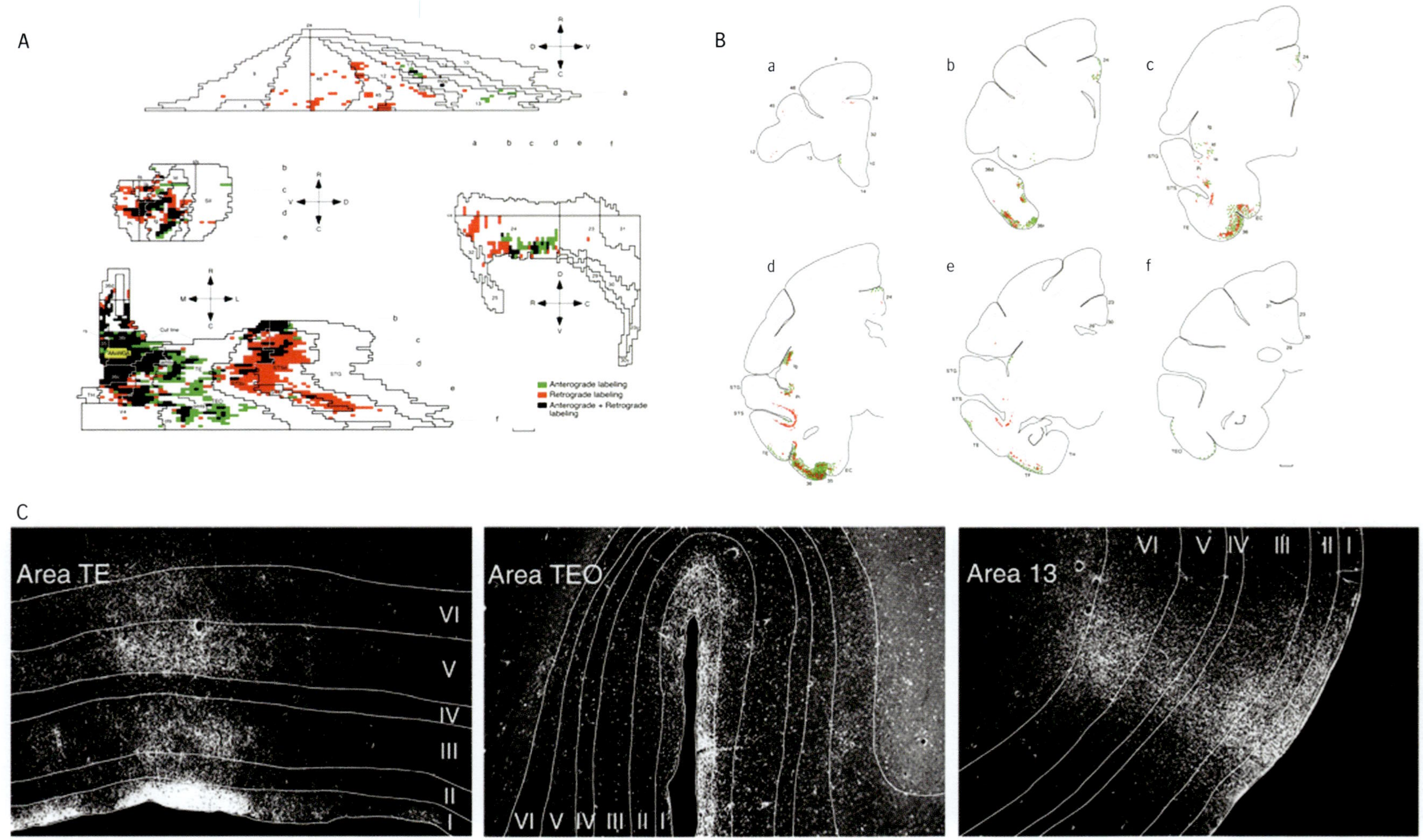

Figure 2.15 Interconnections of the medial temporal lobe and the neocortex. A) Two-dimensional, unfolded maps of the frontal lobe, insular cortex, cingulate and retrosplenial cortices, and temporal lobe, indicating the distribution of anterogradely labelled fibres and terminals and retrogradely labelled cells in the neocortex after combined anterograde and retrograde tracer injection in the medial part of the rostral portion of area 36r (mainly area 36rm). Anterogradely labelled fibres and terminals are represented in green, retrogradely labelled cells in red; columns of cortex that contained both anterograde and retrograde labelling are indicated in black. The letters a–f indicate the levels of the coronal sections illustrated in (B) Scale bar = 2 mm. B. Line drawings of representative coronal sections arranged from rostral (a) to caudal (f), as indicated in panel (A). C) Darkfield photomicrographs illustrating different patterns of fibre and terminal labelling in the neocortex after injections of 3H-AA anterograde tracer in the perirhinal or parahippocampal cortices. Reproduced from Lavenex, P., Suzuki, W. A., and Amaral, D. G. (2002). Perirhinal and parahippocampal cortices of the macaque monkey: Projections to the neocortex. *Journal of Comparative Neurology*, 447, 394–420. © 2002, with permission from John Wiley and Sons.

Neither the dentate gyrus, nor CA3 or CA2 project toward cortical areas. In contrast, CA1, the subiculum, presubiculum, and parasubiculum send modest projections toward a limited number of cortical areas, which bypass the entorhinal cortex (Insausti and Munoz 2001). In particular, the distal portion of CA1 projects to the rostral perirhinal cortex, and the presubiculum and parasubiculum project to the retrosplenial cortex. Very limited projections to the medial and orbital frontal cortex also originate at various levels through these four hippocampal regions. Nevertheless, the vast majority of the efferent projections from the hippocampal formation to the neocortex originate in the deep layers of the entorhinal cortex and are relayed by the perirhinal and parahippocampal cortices (Suzuki and Amaral 1994b; Lavenex et al. 2002; Munoz and Insausti 2005). As described previously, the perirhinal cortex tends to be interconnected with the rostral two-thirds of the entorhinal cortex, while the parahippocampal cortex tends to be interconnected with approximately the caudal two-thirds of the entorhinal cortex. The projections from the perirhinal and parahippocampal cortices to the entorhinal cortex are of the feedforward type, while the projections from the entorhinal cortex to the perirhinal and parahippocampal cortices are of the feedback type (Suzuki and Amaral 1994b). Interestingly, the connections between the parahippocampal and entorhinal cortices are highly reciprocal, whereas the degree of reciprocity of the perirhinal–entorhinal interconnections varies depending on the mediolateral position within the perirhinal cortex; medial portions of the perirhinal cortex exhibit a higher degree of reciprocity with the entorhinal cortex than lateral portions. Considering the next stage of the neocortical–hippocampal loop of information processing (Lavenex and Amaral 2000), although the perirhinal and parahippocampal cortices generally project to the same cortical areas that project to them, there are two particularly important features of these connections to consider: their laminar organization and their degree of reciprocity (Lavenex et al. 2002). As is observed for the parahippocampal–entorhinal cortex interconnections, parahippocampal area TF has largely reciprocal connections with the neocortex, whereas perirhinal area 36 has more asymmetric connections. Area 36, for example, projects to more restricted regions of the frontal cortex and superior temporal sulcus than it receives inputs from. In contrast, it projects to larger portions of areas TE and TEO than it receives inputs from. The efferent projections of areas 36 and TF are primarily directed to the superficial layers of the neocortex, a laminar organization consistent with connections of the feedback type (Lavenex et al. 2002). Projections to unimodal visual areas terminate in large expanses of the cortex, but predominantly in layer I. Projections to other sensory and polymodal areas, in contrast, terminate in a columnar manner predominantly in layers II and III. In all areas receiving heavy projections, the projections extend throughout most cortical layers, largely avoiding layer IV. As is the case for the cortical afferents to the medial temporal lobe, the entorhinal cortex also sends efferent projections toward cortical areas that are not relayed by the perirhinal and parahippocampal cortices (Munoz and Insausti 2005). Interestingly, a number of cortical areas receive convergent projections from the entorhinal cortex and the perirhinal or parahippocampal cortices. With the exception of the medial frontal cortex, efferent projections from the perirhinal and parahippocampal cortices are stronger than those from the entorhinal cortex. In sum, although the medial temporal lobe is primarily organized as a hierarchy of associational cortices (Lavenex and Amaral 2000), there are a number of more limited, alternative routes for information processed in the hippocampal formation to be sent out and reach various cortical brain regions (Insausti and Munoz 2001; Munoz and Insausti 2005).

Finally, the organization of the cortical efferent projections of the perirhinal, parahippocampal, and entorhinal cortices has important implications for theories of memory consolidation. These theories were based on the observation that long-term declarative memories are only initially dependent on the integrity of the medial temporal lobe (Teyler and Discenna 1986; Damasio 1989; Alvarez and Squire 1994; McClelland et al. 1995; Nadel and Moscovitch 1997). The consolidation of long-term memories in cortical areas is thought to depend on reciprocal feedback projections from the medial temporal lobe to the neocortex. The more complex pattern of cortico-cortical connectivity described previously clearly demonstrates that these models are in need of revision. Although all these connections are of the feedback type, they differ and exhibit specific laminar patterns of termination in distinct brain areas; and, most importantly, they do not necessarily reciprocate the cortical afferents to the medial temporal lobe structures. This suggests that different types of information processed by different cortical areas might require different types of feedback projections to enable memory consolidation (Lavenex et al. 2002). This also further demonstrates the need to better take into account the structural organization of the brain in order to build realistic, valid models of the functions that it may subserve.

Postnatal development

Understanding the normal development of the hippocampal formation can provide invaluable information about its functions and its susceptibility to pathologies across the lifespan (Lavenex et al. 2007a). Indeed, there is a large gap in our understanding of the structural development of the different regions of the primate hippocampal formation and the impact of their maturation on the emergence of particular functions (Lavenex et al. 2007a). For example, in the case of developmental amnesia, patients who sustained hippocampal damage early in life exhibit memory impairments affecting preferentially episodic memory (the memory for autobiographical events), whereas semantic memory (the memory for facts about the world) is somehow preserved (Vargha-Khadem et al. 1997). In contrast, lesion of the hippocampus in adults generally impairs both semantic and episodic memory processes (Squire and Zola-Morgan 1996). We have recently shown similar functional plasticity in monkeys that received hippocampal lesions early in life. Hippocampal lesions prevent spatial relational learning in adult-lesioned monkeys (Banta Lavenex et al. 2006), whereas spatial relational learning persist following neonatal lesions (Lavenex et al. 2007b). The possible underlying mechanisms that support recovery of function following early but not late lesions are the subject of ongoing investigations in our laboratory, which may be discussed in future editions of this book. Here, I discuss some of our recent findings regarding the normal postnatal development of the primate hippocampal formation that constitute an essential framework to consider the aetiologies of different neurodevelopmental disorders affecting hippocampal structure and function.

Neurogenesis in the dentate gyrus

The dentate gyrus is one of only two regions of the mammalian brain where substantial neurogenesis occurs postnatally (rats: Altman and Das 1965; monkeys: Rakic and Nowakowski 1981; humans: Eriksson et al. 1998). In a recent study (Jabès et al. 2010), we quantified neurogenesis, cell death, neuron number, and the morphological changes underlying the structural maturation of the monkey dentate gyrus during early postnatal development. We demonstrated that about 40% of the total number of granule cells found in 5–10-year-old monkeys are added to the granule cell layer postnatally, with a peak (about 25%) in the first 3 months after birth. We also found significant levels of cell proliferation, neurogenesis and cell death in the context of an overall stable number of granule cells in mature monkeys. Importantly, we established that the developmental period during which a significant number of neurons are added to the monkey granule cell layer is longer than previously thought. Indeed, although we observed a significant decrease in the rate of neuron addition at about 3 months of age, we calculated that about 3100 neurons are added to the granule cell layer each day between 3 months and 1 year of age in monkeys. These data, which are derived from stereological counts of neurons in Nissl-stained preparations, are supported by immunohistochemical findings on cell proliferation, neurogenesis and cell death (Figure 2.16) (Jabès et al. 2010). Indeed, the number of Ki-67-labelled cells decreased strongly between birth and 3 months of age but remained at an intermediate level, which was ten times higher than that observed in mature monkeys, between 6 months and 1 year of age. The overall distribution of Ki-67-labelled cells that we described in newborn monkeys was similar to that observed in newborn humans (Seress 2001). In humans, cell proliferation also seems to persist until at least the end of the first year after birth, the latest age examined immunohistochemically by Seress et al. (2001). In the absence of strict quantification of older human cases, our findings in monkeys suggest that sustained levels of developmental neurogenesis might continue and impact dentate gyrus structure until about 4 years of age in humans. Importantly, we also estimated the number of new neurons that could potentially be integrated into the granule cell layer of mature monkeys to be about 1300 per day or 0.02% of total neuron number (7.21 million). This daily rate of adult neurogenesis is about ten times lower, in percentage of total neuron number, than that reported for 9–10-week-old rats, where 2250 new neurons are generated per day or 0.2% of total neuron number (1.2 million; Cameron and McKay 2001). However, considering that monkeys live 20–30 years and rats 2–3 years (Havenaar et al. 1993), postnatal neurogenesis has a similar potential in rodents and primates: that is the renewal of the entire population of granule cells during an individual's lifetime. It is therefore likely to be the case also in humans.

Maturation of distinct hippocampal circuits

Information about the postnatal maturation of the primate hippocampal formation is surprisingly sparse, and, until recently, derived mainly from miscellaneous studies using different methodologies and species at various developmental stages. We therefore launched a systematic, quantitative evaluation of the postnatal structural development of individual regions of the monkey hippocampal formation. In a recent study (Jabès et al. 2011), we performed a stereological analysis of neuron number, neuronal soma size, and volume of individual regions and layers, and showed that distinct hippocampal regions and layers exhibit different profiles of structural development during early postnatal life (Figure 2.17). Specifically, we found a protracted period of neuron addition in the dentate gyrus throughout the first postnatal year, and a concomitant late maturation of the granule cell population and individual dentate gyrus layers that extended beyond the first year of life (Jabès et al. 2010, 2011). Together with the late maturation of the granule cells, it has been reported that the mossy cells, the major targets of the granule cell projections in the polymorphic layer, exhibit clear morphological changes in soma and dendritic structure until at least nine months of age in monkeys (qualitative data; Seress and Ribak 1995a) and at least 30 months of age in

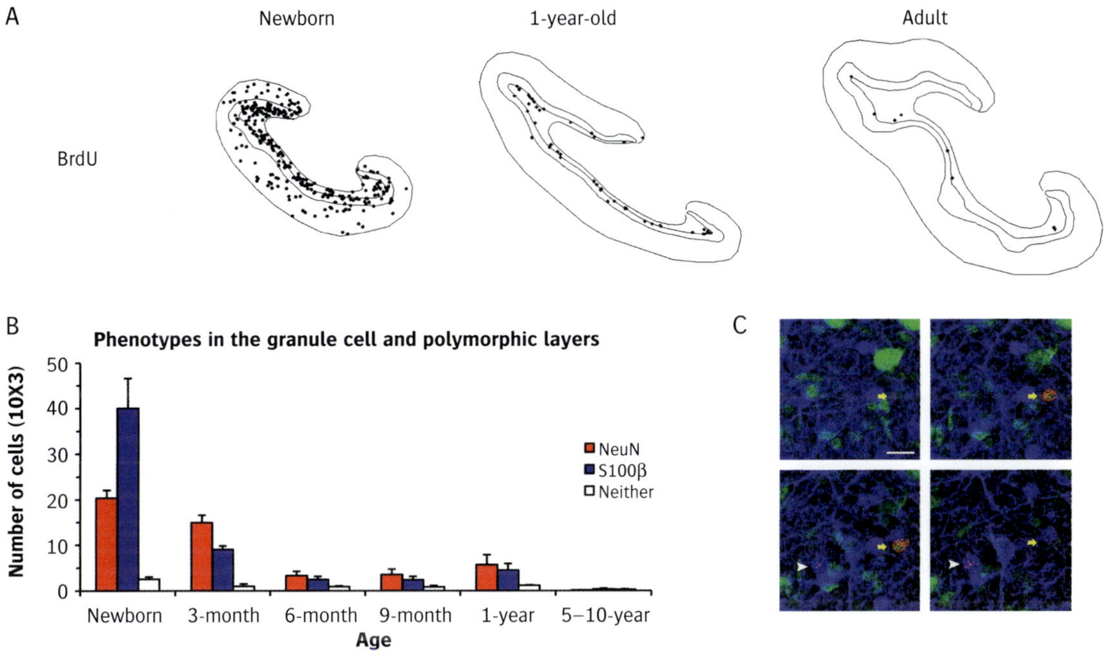

Figure 2.16 Postnatal neurogenesis in the primate dentate gyrus. A) Representative coronal sections through the monkey dentate gyrus illustrating the location of 5′-bromo-2-deoxyuridine (BrdU)-positive cells 4 weeks after BrdU injection, at different ages through postnatal life (newborns; 1-year-olds; 5–10-year-olds). B) Phenotypes of BrdU-labelled cells in the monkey granule cell and polymorphic layers, 4 weeks following BrdU injection (150 mg / kg). C) Confocal images of BrdU/NeuN-positive (yellow arrow) and BrdU/S100beta-positive (white arrowheads) cells at the border between the granule cell and polymorphic layers. Scale bar: 10 µm (applies to all panels). Reproduced from Jabès, A., Banta Lavenex, P., Amaral, D.G., and Lavenex, P. (2010). Quantitative analysis of postnatal neurogenesis and neuron number in the macaque monkey dentate gyrus. *European Journal of Neuroscience*, 31, 273–85. © 2010, with permission from John Wiley and Sons.

humans (qualitative data; Seress and Mrzljak 1992). In addition, the number of spines per 100 µm of dendrite appears to increase by about 10% between 1 year and 4–20 years of age in monkeys (Seress and Mrzljak 1992). Our analyses revealed a 25% increase in volume of the polymorphic layer between 1 year and 5–9 years of age. Although the mossy cells and the axons of granule cells represent a major component of the polymorphic layer, there are a variety of other neuronal types and afferent projections targeting this area (Amaral and Lavenex 2007; Amaral et al. 2007). Thus, although the postnatal development of the polymorphic layer circuits is likely delayed, as compared to that of the dentate gyrus afferents reaching the molecular layer, detailed analyses of the postnatal maturation of the different cell types contained in the dentate gyrus will be necessary to provide a definite answer regarding the functional consequences of this delayed maturation.

Although the development of CA3 generally parallels that of the dentate gyrus, the distal portion of CA3, which receives direct entorhinal cortex projections, matures earlier than the proximal portion of CA3. At the cellular level, we found that the proximal CA3 pyramidal neurons exhibit significant changes in soma size within the first three to six postnatal months. In contrast, the size of distal CA3 pyramidal neurons does not vary during postnatal development. Our quantitative data are thus in agreement with the qualitative report by Seress and Ribak (1995b) showing that the somas and dendrites of distal CA3 pyramidal neurons exhibit adult-like ultrastructural features at birth. There is, unfortunately, no published information on the ultrastructural characteristics of developing proximal CA3 pyramidal neurons.

CA1 matures relatively earlier than the dentate gyrus and CA3, despite the fact that CA3 pyramidal neurons contribute the largest projection to CA1 pyramidal neurons. Interestingly, CA1 stratum lacunosum-moleculare, in which direct entorhinal cortex projections terminate, matures earlier than CA1 strata oriens, pyramidale and radiatum, in which the CA3 projections terminate. Our quantitative measurements are in agreement with recent qualitative reports of a later myelination of fibres in strata pyramidale and radiatum, as compared to stratum lacunosum-moleculare, in CA1 of humans (Abraham et al. 2010).

The subiculum develops earlier than the dentate gyrus, CA3 and CA1, but not CA2. However, similar to CA1, the molecular layer of the subiculum, in which the entorhinal cortex projections terminate, is overall more mature in the first postnatal year, as compared to the stratum pyramidale in which most of the CA1 projections terminate. Unfortunately, there are no other published data on the postnatal maturation of the different layers of the subiculum in other species, in particular humans. Unlike other hippocampal

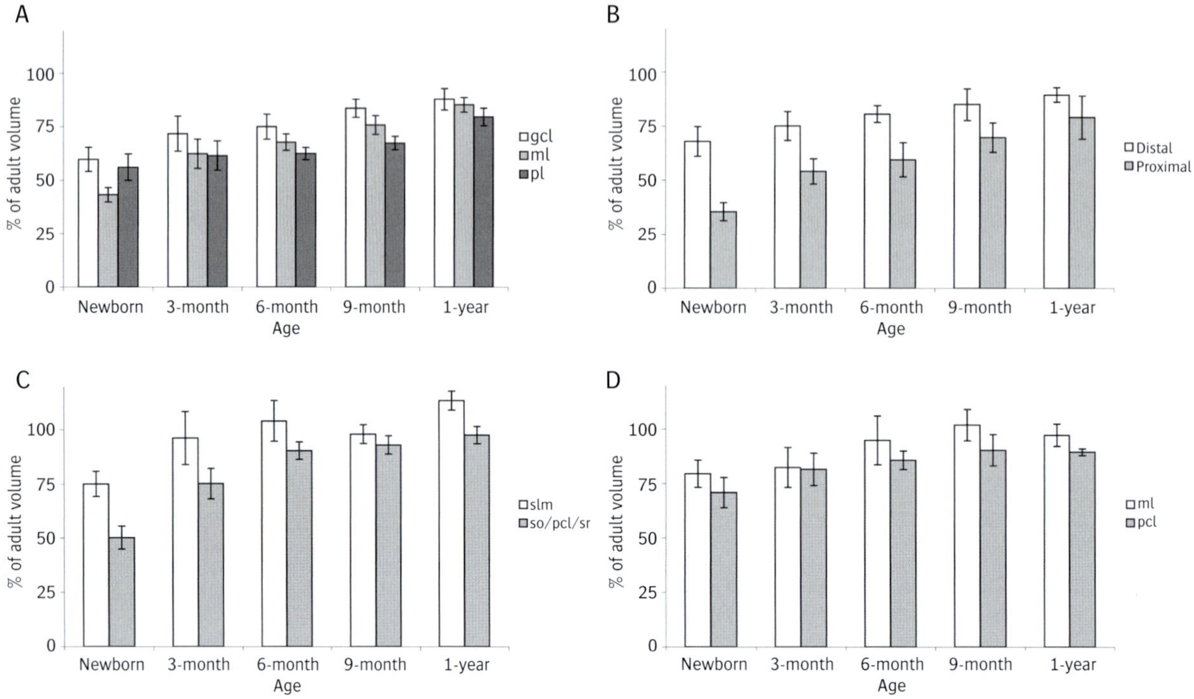

Figure 2.17 Volumes of individual layers/regions of the monkey hippocampal formation at different ages during early postnatal development (expressed as percentage of the volume of the layer/region observed in 5–9-year-old monkeys: average ± standard error of the mean). A) Dentate gyrus: granule cell layer (gcl), molecular layer (ml), polymorphic layer (pl). B) CA3: proximal and distal portions. C) CA1: strata oriens, pyramidale and radiatum (so/pcl/sr), stratum lacunosum-moleculare (slm). D) Subiculum: stratum pyramidale (pcl), stratum moleculare (ml). Reproduced from Jabès, A., Banta Lavenex, P., Amaral, D.G., and Lavenex, P. (2011). Postnatal development of the hippocampal formation: A stereological study in macaque monkeys, *Journal of Comparative Neurology*, 519 (6), 1051–70. © 2011, with permission from John Wiley and Sons.

fields, volumetric measurements suggest regressive events in the structural maturation of presubicular neurons and circuits. Finally, areal and neuron soma size measurements reveal an early maturation of the parasubiculum. Two unique features of these structures, as compared to other hippocampal regions, are their reciprocal connections with the anterior thalamic nuclear complex and their heavy cholinergic innervation. Accordingly, cell circuits in the presubiculum and parasubiculum might contribute to some of the earliest functions subserved by the hippocampal formation. A detailed discussion of the implications of these findings with respect to the role of the hippocampal formation in memory can be found in Jabès et al. (2011).

Molecular analyses

In an effort to understand the molecular basis of the normal development of the hippocampal formation and its susceptibility to a number of different pathologies, we recently launched a series of experiments examining the regulation of gene expression in distinct regions of the monkey hippocampal formation during postnatal development. In a first study, we characterized the molecular signature of individual hippocampal regions in an attempt to understand the paradox that, although the hippocampus plays a central role in the brain network essential for memory function (Banta Lavenex et al. 2006; Morris 2007), it is also the brain structure that is most sensitive to hypoxic-ischaemic episodes (Sommer 1880; Spielmeyer 1925; Zola-Morgan et al. 1986; Nedergaard and Dirnagl 2005; Banta Lavenex et al. 2006; Andersen et al. 2007). We found that the expression of genes associated with glycolysis and glutamate metabolism in astrocytes and the coverage of excitatory synapses by astrocytic processes undergo significant decreases in the CA1 field of the monkey hippocampus during postnatal development (Figure 2.18). Our findings at the gene, protein, and structural levels therefore suggest that the developmental decrease in astrocytic processes and functions may be the critical factor underlying the selective vulnerability of CA1 to hypoxic-ischaemic episodes in adulthood. They also provide an explanation for the relative resistance of this brain structure to hypoxia in the perinatal period, and, in particular, during the birth process. Indeed, hypoxia reduces glycolysis, which in turn leads to a decrease in ATP (Pellerin and Magistretti 1994). As a consequence, the sodium–potassium gradient necessary for the cotransport of glutamate from the synaptic cleft into astrocytes dissipates. This reduces glutamate clearance from the synapse, increasing neuronal depolarization and the potential for neuronal death via

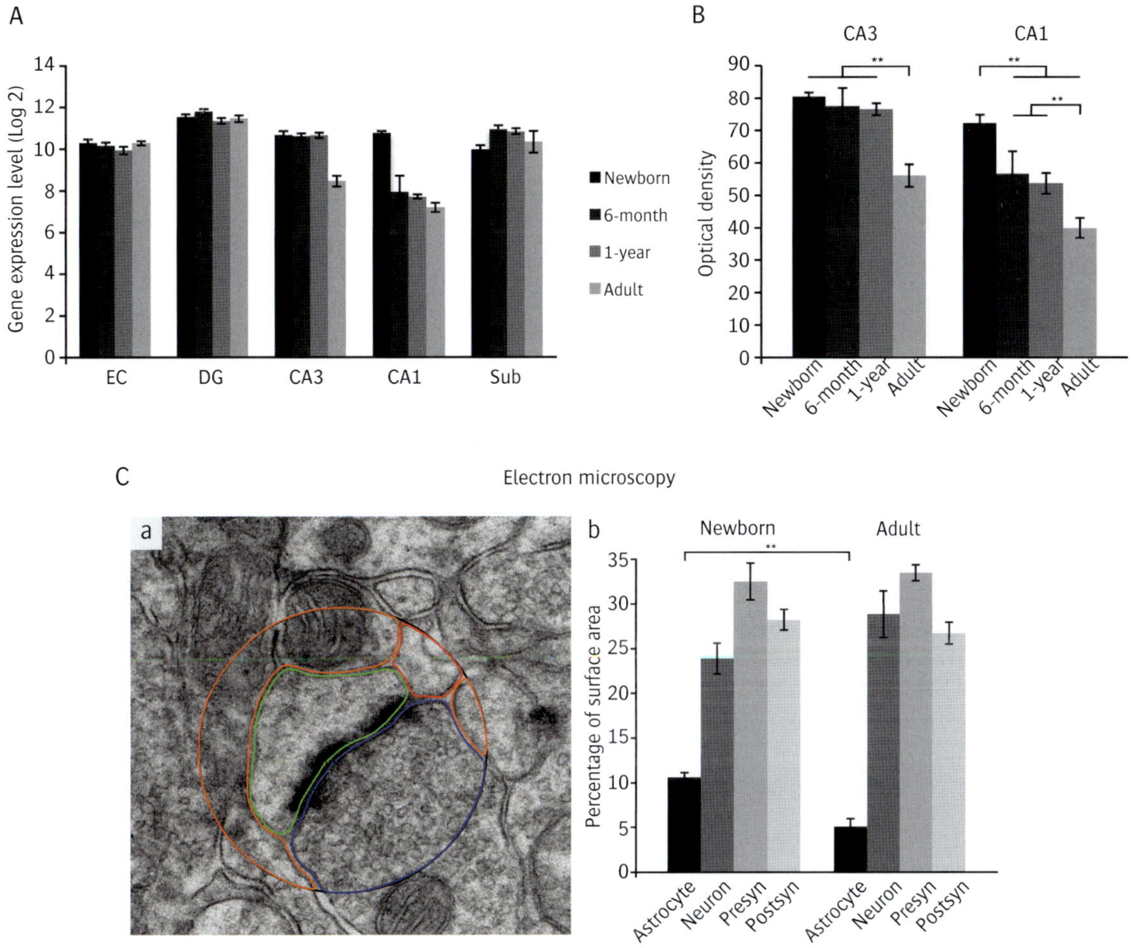

Figure 2.18 A) Microarray analysis of GFAP gene expression in the postnatally developing monkey hippocampal formation. B) Immunohistochemical analysis of GFAP protein expression. C) Electron microscopy evaluation of astrocytic processes around excitatory synapses in the stratum radiatum of CA1. (a) Blue outline: presynaptic neuronal element. Green outline: postsynaptic neuronal element. Orange outline: non-synaptic neuronal elements. Red outline: astrocytic elements. Circle diameter: 1000 nm. (b) The surface area occupied by astrocytic processes decreased between birth and adulthood. Reproduced from Lavenex, P., Sugden, S.G., Davis, R.R., Gregg, J.P., and Banta Lavenex, P. (2011). Developmental regulation of gene expression and astrocytic processes may explain selective hippocampal vulnerability. *Hippocampus*, 21(2), 142–9. © 2011, John Wiley and Sons.

excitotoxicity (Swanson 2005). High-astrocytic coverage, as we have shown in the newborn, likely maintains sufficient glutamate reuptake to limit neuronal depolarization and excitotoxicity during mild to moderate hypoxic-ischaemic events. In the adult, however, a lower expression level of genes associated with glycolysis or glutamate uptake and metabolism (Rao et al. 2001), as well as a lower astrocytic coverage of excitatory synapses in CA1, render the system more vulnerable to a reduction in oxygen concentration.

In contrast, a major benefit that derives from decreased astrocytic coverage is the regulation of synaptic efficacy leading to an increase in synaptic selectivity advantageous for learning (Karlsson and Frank 2008). Indeed, reduction of glutamate clearance associated with a relative decrease in astrocytic processes in the vicinity of synapses can affect transmitter release through modulation of presynaptic metabotropic glutamate receptors (Oliet et al. 2001). Reduced glutamate clearance results in increased glutamate concentration in the extracellular space (Tanaka et al. 1997; Bergles and Jahr 1998), which in turn increases the activation of presynaptic metabotropic glutamate receptors (Scanziani et al. 1997), thus leading to a lower probability of glutamate release by the presynaptic terminal (Oliet et al. 2001). Presynaptic inhibition can be overcome by high-frequency bursts of afferent synaptic potentials (Grover et al. 2009), thus serving as a high-pass filter increasing the signal-to-noise ratio for information transmitted through these synapses (Oliet et al. 2001). Thus, the decreased astrocytic coverage of excitatory synapses in the adult CA1 could serve to ensure that only the most salient information generates synaptic activity in the hippocampal circuits that contribute to learning and memory processes. A developmental decrease of astrocytic processes and functions may therefore contribute to the emergence of adult-like, selective memory function.

Finally, the relatively high astrocytic coverage of the newborn synapses might also play a central role in the generation of febrile seizures. The highest incidence of seizures is in the first year of life in humans and is most often associated with a febrile illness (Hauser 1994). Fever begins with the activation of immune response cells that produce interleukin-1, which in turn increases prostaglandin E2 synthesis (Biddle 2006). Prostaglandins act at the level of the hypothalamus to regulate body temperature and induce fever. Interestingly, prostaglandins also stimulate calcium-dependent glutamate release in astrocytes (Bezzi et al. 1998; Volterra and Meldolesi 2005), which can induce abnormal prolonged depolarization with repetitive spiking in CA1 pyramidal neurons leading to seizures (Kang et al. 2005; Tian et al. 2005). Thus, the relatively high-astrocytic coverage of the dense network of CA1 excitatory synapses in the newborn (twice that of the adult) could explain why infants exhibit a higher incidence of febrile seizures. Conversely, the decrease in the astrocytic coverage of hippocampal excitatory synapses with development might provide the cellular basis for the decreased susceptibility to febrile seizures with age.

Neurodevelopmental disorders

Our neuroanatomical and molecular analyses described in the previous sections identified different periods of postnatal development during which pathogenic factors might impact the structural and functional maturation of distinct regions of the primate hippocampal formation. Here, I review neuropathological findings reported in the literature and discuss the perspective that our findings provide to comprehend the possible aetiologies of human neurodevelopmental disorders affecting hippocampal structure and function.

Autism

Magnetic resonance imaging (MRI) studies have reported abnormal postnatal development of the hippocampal formation in children with autism (Aylward et al. 1999; Saitoh et al. 2001; Sparks et al. 2002; Schumann et al. 2004). Sparks et al. (2002) reported that 2–4-year-old children with autism exhibit bilateral hippocampal enlargement proportional to an overall increase in total cerebral volume. Schumann et al. (2004) found that 7.5–12.5-year-old male children with autism, with and without mental retardation, have a larger right hippocampal volume than typically developing controls, even after controlling for total cerebral volume. Children with autism but without mental retardation have a larger left hippocampal volume relative to controls. In contrast, Saitoh et al. (2001) reported a smaller area (across three contiguous 5-mm MRI coronal sections) of a region including the dentate gyrus and proximal CA3 in a group of children with autism aged 29 months to 42 years (11 subjects aged 29 months to 4 years; 32 subjects aged 5–12 years; 16 subjects aged 13–42 years). Similarly, Aylward et al. (1999) reported a reduction in the volume of the hippocampal complex (including dentate gyrus, CA3, CA2, CA1 and subiculum) in 11–37-year-old subjects with autism, but only after correction for overall differences in brain size. Besides possible differences in diagnostics and inclusion criteria, apparently discrepant reports between studies might derive from the differential, age-dependent maturation of distinct hippocampal regions that we demonstrated (Jabès et al. 2011), and the neuroanatomical definitions of hippocampal regions used in the different studies. These three issues should therefore be considered carefully when interpreting previous results and designing future MRI studies. Nevertheless, cross-sectional studies suggest, altogether, an abnormal pattern of hippocampal development in autism, which persists through adolescence (Amaral et al. 2008).

Postmortem studies describing neuropathological abnormalities in the hippocampal formation of subjects with autism are few and consist largely of preliminary reports by one laboratory (Bauman and Kemper 1985; Raymond et al. 1996; Bauman and Kemper 2005; Lawrence et al. 2010). Nevertheless, the emerging concept is the observation of increased cell density in different hippocampal regions, including the dentate gyrus, CA3 and CA1, and decreased cell size of pyramidal neurons in the proximal portion of CA3. Unfortunately, cell density measurements are insufficient, by themselves, to determine whether these morphological abnormalities are due to differences in neuron number or changes in neuropil volume. Our recent findings (Jabès et al. 2011) indicate that the number of principal neurons in the different hippocampal regions does not vary between birth and young adulthood in monkeys (except, of course, for the dentate gyrus; Jabès et al. 2010). If autism neuropathology develops postnatally, changes in neuron density are likely due to abnormal maturation of the neuropil, suggesting changes in network connectivity (see (Raymond et al. 1996) for qualitative findings in proximal CA3 and CA1). The fact that the size of pyramidal neurons in proximal CA3, but not CA1, is smaller in subjects with autism than in controls, suggest that maturational processes happening during early postnatal development might indeed be impacted in autism spectrum disorders. Such maturational changes, happening when proximal CA3 neurons have not reached adult-like size but CA1 pyramidal neurons have, take place between birth and six month of age in monkeys. This developmental period corresponds to the first 18–24 months of postnatal life in humans; that is, the period immediately preceding the typical appearance of major, detectable autistic symptoms (Amaral et al. 2008).

Schizophrenia

MRI studies consistently report reduced hippocampal volumes in first-episode schizophrenic subjects (Velakoulis et al. 1999; Steen et al. 2006) and asymptomatic, first-degree relatives of subjects with schizophrenia (Sismanlar et al. 2009), indicating that hippocampal pathology is not the result of the illness or treatment but rather contributes to the aetiology of the disorder. Consistent neuropathological findings include alterations in markers of synaptic transmission for both glutamatergic and GABAergic systems (Harrison 2004; Talbot et al. 2004), in particular in proximal CA3 (Harrison et al. 2003). Interestingly, there are no changes in the total number of hippocampal neurons and reports concerning neuron density or cell size differences are rather inconsistent (Harrison 2004). In contrast, there are reliable morphological changes in the dendritic and axonal arborization of dentate granule cells and the synaptic organization of CA3 pyramidal neurons (Lauer et al. 2003; Senitz and Beckmann 2003; Kolomeets et al. 2005; Kolomeets et al. 2007; Kobayashi 2009). Specifically, the frequency of granule cells with basal dendrites is higher (Lauer et al. 2003; Senitz and Beckmann 2003) and the CA3 mossy fibre synapses are both smaller and fewer (Kolomeets et al. 2005, 2007) in schizophrenic patients, as compared to controls. These findings, together with the absence of obvious signs of neurodegeneration (Harrison 2004), suggest that schizophrenia is a neurodevelopmental disorder that might arise following the abnormal maturation of specific hippocampal circuits, namely the dendritic and axonal arborization of dentate granule cells.

Our morphological data indicate that the dentate gyrus matures late, after all the other hippocampal regions. Pathogenic factors, which remained to be determined, might therefore impact the maturation of the dentate granule cells and their projections to CA3 pyramidal neurons during postnatal life. However, the absence of obvious disorganization of the dentate granule cell layer in schizophrenia suggests that pathogenic factors might act after the initial phase of postnatal, developmental neurogenesis and neuron addition to the granule cell layer (i.e. a period corresponding to at least the first postnatal year in monkeys). Pathogenic factors influencing gene expression and the maturation of newly-generated granule cells during late childhood or early adolescence might therefore contribute to the emergence of schizophrenic symptoms during adolescence. Accordingly, genetic predispositions to schizophrenia have been shown (Harrison 2004), and the identification of a number of susceptibility genes involved in postnatal neurogenesis, including DISC1, COMT, NRG1, and NPAS3, provides an additional link between the regulation of postnatal granule cell maturation and schizophrenia (Reif et al. 2007; Le Strat et al. 2009).

Epilepsy

Hippocampal sclerosis is the main neuropathology associated with intractable temporal lobe epilepsy. Structural MRI studies reliably find hippocampal volume reduction ipsilateral to the seizure focus (Bonilha et al. 2003; Seidenberg et al. 2005; Seidenberg et al. 2008). Microscopically, hippocampal sclerosis is characterized by extensive loss of dentate gyrus mossy cells (Blümcke et al. 2000), CA3 and CA1 pyramidal neurons, accompanied by reactive gliosis in these regions (Sloviter 1994). CA2 neurons are generally preserved, but dentate granule cell dispersion is observed in about 40% of temporal lobe epilepsies associated with hippocampal sclerosis (Blümcke et al. 2009). A prominent theory suggests that acquired epilepsy is an immediate network defect caused primarily by initial neuron loss (Sloviter 1994, 2009), most likely a loss of dentate gyrus mossy cells (Sloviter, 1994; Blümcke et al. 2000). Whether hippocampal sclerosis is the cause or the consequence of epileptic seizures has been the focus of intense debate for more than a century (Meencke 2009). However, recent studies suggest that initial precipitating injuries during postnatal development induce hippocampal damage that leads to epileptic seizures (Sloviter 1994; Blümcke et al. 2002; Squier et al. 2003; Sloviter 2009)

Our stereological data suggest a late postnatal maturation of the dentate gyrus, including the polymorphic layer. Mossy cells (Amaral 1978), which besides excitatory contacts with the proximal dendrites of the dentate granule cells provide a net inhibition of granule cell activity via GABAergic basket cells (Sloviter 1994), might be more sensitive to pathogenic insults before they become fully mature. As a result, the dentate gyrus might be particularly at risk during the protracted period of postnatal development of the polymorphic layer circuitry, which extends beyond the period of neuron addition to the granule cell layer (i.e. 1 year of age in monkeys). Interestingly, early seizures before the age of 4 years in humans correlate with granule cell dispersion, whereas seizures after 4 years of age are not associated with disorganization of the granule cell population (Blümcke et al. 2002). Altogether, these findings suggest that the same possible cause (i.e. the loss of mossy cells) might lead to different neuropathologies associated with epilepsy: Precipitating injuries inducing mossy cell loss during early postnatal life might impact the normal pattern of developmental neurogenesis and neuron addition to the granule cell layer, whereas mossy cell loss occurring past this developmental period might not result in major changes in the organization of the dentate granule cell population. In sum, the same pathogenic insults at different developmental ages might have different consequences on the neuropathological correlates of temporal lobe epilepsy, with a consistent loss of mossy cells, but granule cell dispersion in only some of the intractable temporal lobe epilepsy cases associated with hippocampal sclerosis.

Down syndrome

The presence of Alzheimer's disease pathology in the hippocampal formation of Down syndrome (DS) subjects, as early as 8 years of age (Leverenz and Raskind 1998), suggests a disproportionate implication of the hippocampal formation in the cognitive impairments observed in these subjects. Similarly, reduced dendritic spine counts in the CA fields of the hippocampus have been reported in adult DS subjects (Ferrer and Gullotta 1990). Interestingly, structural MRI investigation of the brain reveals smaller hippocampi in 5–25-year-old DS subjects, as compared to normal controls, suggesting that abnormal brain development contributes to the observed pathology (Pinter et al. 2001). Moreover, preliminary behavioural studies reported by Nadel (2003) suggest that, as compared to normally developing children, 26–30-month-old DS children are more severely impaired in a 'hippocampus-dependent', place-learning task than in 'hippocampus-independent', response and cue tasks.

As described earlier, we have recently shown that the developmental regulation of gene expression associated with glycolysis and glutamate metabolism in astrocytes, as well as the structure of astrocytic processes, contribute to the normal maturation of the CA1 and CA3 fields of the hippocampus (Lavenex et al. 2011). Specifically, the astrocytic coverage of excitatory synapses in CA1 decreases during postnatal development and might contribute to an increased selectivity of synaptic transmission beneficial for learning (Lavenex et al. 2011). Interestingly, increased expression of S-100-beta and glial fibrillary acidic protein has been reported in the hippocampus of developing DS subjects (Mito and Becker 1993); these structural abnormalities might reduce the selectivity of synaptic transmission and contribute to cognitive impairments. There is, to our knowledge, no other report of pathological findings in the hippocampal formation of DS subjects under 5 years of age. Our current data therefore provide a developmental timeline regarding the maturation of distinct hippocampal circuits that could be used as a guide to investigate abnormal hippocampal development in DS subjects.

Conclusion

In this chapter, I provided a basic description of the structural organization of the primate hippocampal formation. Experimental work performed with animals, in particular rodents but also monkeys, provides most of the information regarding the structural organization of functional hippocampal circuits. Similarly, animal models are typically used to study the molecular and cellular basis of pathologies affecting human hippocampal structure and functions. It is therefore important to be aware of the similarities and differences between species (e.g. rats, monkeys, humans) in order to extrapolate from the findings of fundamental research in animals to clinical problems in humans. Here, I considered this comparative perspective to introduce some issues that might be relevant for questions addressed in other chapters of this book. I also discussed recent findings from our laboratory regarding the normal postnatal development of the monkey hippocampal formation. I hope that the information presented in this chapter will provide an essential framework to consider the aetiologies of different pathologies affecting hippocampal structure and function discussed in other chapters of the book.

Acknowledgements

This work was supported by grants from the Swiss National Science Foundation (PP00A-106701, PP00P3-124536) and the National Alliance for Research on Schizophrenia and Depression (NARSAD).

References

Abraham, H., Vincze, A., Jewgenow, I., Veszpremi, B., Kravjak, A., Gomori, E., and Seress, L. (2010). Myelination in the human hippocampal formation from midgestation to adulthood. *Int J Dev Neurosci*, 28, 401–10.

Aggleton, J. P. (2010). Understanding retrosplenial amnesia: insights from animal studies. *Neuropsychologia*, 48, 2328–38.

Aggleton, J. P., Vann, S. D., and Saunders, R. C. (2005). Projections from the hippocampal region to the mammillary bodies in macaque monkeys. *Eur J Neurosci*, 22, 2519–30.

Akil, M., and Lewis, D. A. (1993). The dopaminergic innervation of monkey entorhinal cortex. *Cereb Cortex*, 3, 533–50.

Akil, M., and Lewis, D. A. (1994). The distribution of tyrosine hydroxylase-immunoreactive fibers in the human entorhinal cortex. *Neuroscience*, 60, 857–74.

Altemus, K. L., Lavenex, P., Ishizuka, N., and Amaral, D. G. (2005). Morphological characteristics and electrophysiological properties of CA1 pyramidal neurons in macaque monkeys. *Neuroscience,* 136, 741–56.

Altman, J., and Das, G. D. (1965). Autoradiographic and histological evidence of postnatal hippocampal neurogenesis in rats. *J Comp Neurol,* 124, 319–36.

Alvarez, P., and Squire, L. R. (1994). Memory consolidation and the medial temporal lobe: a simple network model. *Proc Natl Acad Sci USA,* 91, 7041–5.

Amaral, D. G. (1978). A Golgi study of cell types in the hilar region of the hippocampus in the rat. *J Comp Neurol,* 182, 851–914.

Amaral, D. G., and Insausti, R. (1992). Retrograde transport of D-[3H]-aspartate injected into the monkey amygdaloid complex. *Exp Brain Res,* 88, 375–88.

Amaral, D. G., and Lavenex, P. (2007). Hippocampal neuroanatomy. In Andersen, P., Morris, R. G. M., Amaral, D. G., Bliss, T. V., and O'Keefe, J. (Eds.) *The Hippocampus Book*, pp. 37–114. Oxford: Oxford University Press.

Amaral, D. G., Insausti, R., and Cowan, W. M. (1984). The commissural connections of the monkey hippocampal formation. *J Comp Neurol,* 224, 307–36.

Amaral, D. G., Insausti, R., and Cowan, W. M. (1987). The entorhinal cortex of the monkey: I. Cytoarchitectonic organization. *J Comp Neurol,* 264, 326–55.

Amaral, D. G., Scharfman, H. E., and Lavenex, P. (2007). The dentate gyrus: fundamental neuroanatomical organization (dentate gyrus for dummies). *Prog Brain Res,* 163, 3–22.

Amaral, D. G., Schumann, C. M., and Nordahl, C. W. (2008). Neuroanatomy of autism. *Trends Neurosci,* 31, 137–45.

Andersen, P., Morris, R. G., Amaral, D. G., Bliss, T., and O'Keefe, J. (2007). Historical perspective: Proposed functions, biological characteristics, and neurobiological models of the hippocampus. In Andersen, P., Morris, R. G., Amaral, D. G., Bliss, T., and O'Keefe, J. (Eds.) *The Hipppocampus Book*, pp. 9–36. Oxford: Oxford University Press.

Arantius, G. (1587). *De humano foetu...Ejusdem anatomicorum observationum liber. Etc.* Vence.

Aylward, E. H., Li, Q., Honeycutt, N. A., *et al.* (1999). MRI volumes of the hippocampus and amygdala in adults with Down's syndrome with and without dementia. *Am J Psychiatry,* 156, 564–8.

Banta Lavenex, P., Amaral, D. G., and Lavenex, P. (2006). Hippocampal lesion prevents spatial relational learning in adult macaque monkeys. *J Neurosci,* 26, 4546–58.

Bauman, M., and Kemper, T. L. (1985). Histoanatomic observations of the brain in early infantile autism. *Neurology,* 35, 866–74.

Bauman, M. L., and Kemper, T. L. (2005). Neuroanatomic observations of the brain in autism: a review and future directions. *Int J Dev Neurosci,* 23, 183–7.

Bergles, D. E., and Jahr, C. E. (1998). Glial contribution to glutamate uptake at Schaffer collateral-commissural synapses in the hippocampus. *J Neurosci,* 18, 7709–16.

Bezzi, P., Carmignoto, G., Pasti, L., *et al.* (1998). Prostaglandins stimulate calcium-dependent glutamate release in astrocytes. *Nature,* 391, 281–5.

Biddle, C. (2006). The neurobiology of the human febrile response. *AANA J,* 74, 145–50.

Blümcke, I., Suter, B., Behle, K., *et al.* (2000). Loss of hilar mossy cells in Ammon's horn sclerosis. *Epilepsia,* 41(Suppl 6), S174–80.

Blümcke, I., Thom, M., and Wiestler, O. D. (2002). Ammon's horn sclerosis: a maldevelopmental disorder associated with temporal lobe epilepsy. *Brain Pathol,* 12, 199–211.

Blümcke, I., Kistner, I., Clusmann, H., *et al.* (2009). Towards a clinico-pathological classification of granule cell dispersion in human mesial temporal lobe epilepsies. *Acta Neuropathol,* 117, 535–44.

Bonilha, L., Kobayashi, E., Rorden, C., Cendes, F., and Li, L. M. (2003). Medial temporal lobe atrophy in patients with refractory temporal lobe epilepsy. *J Neurol Neurosurg Psychiatry,* 74, 1627–30.

Brodmann, K. (1909). *Vergleichende Lokalisationslehre der Grosshirnrinde in ihren Prinzipien dargestellt auf Grund des Zellenbaues*. Leipzig: Johann Ambrosius Barth Verlag.

Brun, V. H., Otnass, M. K., Molden, S., *et al.* (2002). Place cells and place recognition maintained by direct entorhinal-hippocampal circuitry. *Science,* 296, 2243–6.

Buckmaster, P. S., and Amaral, D. G. (2001). Intracellular recording and labeling of mossy cells and proximal CA3 pyramidal cells in macaque monkeys. *J Comp Neurol,* 430, 264–81.

Buckmaster, P. S., Strowbridge, B. W., Kunkel, D. D., Schmiege, D. L., and Schwartzkroin, P. A. (1992). Mossy cell axonal projections to the dentate gyrus molecular layer in the rat hippocampal slice. *Hippocampus,* 2, 349–62.

Burwell, R. D. (2001). Borders and cytoarchitecture of the perirhinal and postrhinal cortices in the rat. *J Comp Neurol,* 437, 17–41.

Cameron, H. A., and Mckay, R. D. G. (2001). Adult neurogenesis produces a large pool of new granule cells in the dentate gyrus. *J Comp Neurol,* 435, 406–17.

Chrobak, J. J., and Amaral, D. G. (2007). Entorhinal cortex of the monkey: VII. intrinsic connections. *J Comp Neurol,* 500, 612–33.

Damasio, A. R. (1989). Time-locked multiregional retroactivation - a systems-level proposal for the neural substrates of recall and recognition. *Cognition,* 33, 25–62.

De Garengeot, D. (1742). *Splanchnologie ou l'anatomie des visceres* (2nd ed.), pp. 250–1. Paris.

Ding, S. L., and Van Hoesen, G. W. (2010). Borders, extent, and topography of human perirhinal cortex as revealed using multiple modern neuroanatomical and pathological markers. *Human Brain Mapping,* 31, 1359–79.

Dolorfo, C. L., and Amaral, D. G. (1998). Entorhinal cortex of the rat: organization of intrinsic connerctions. *J Comp Neurol,* 398, 49–82.

Duvernoy, H. M. (2005). *The human hippocampus*. Berlin: Springer-Verlag.

Eriksson, P. S., Perfilieva, E., Bjork-Eriksson, T., Alborn, A.-M., Nordborg, C., Peterson, D. A., *et al.* (1998). Neurogenesis in the adult human hippocampus. *Nature Med,* 4, 1313–17.

Ferrer, I., and Gullotta, F. (1990). Down's syndrome and Alzheimer's disease: dendritic spine counts in the hippocampus. *Acta Neuropathol,* 79, 680–5.

Freund, T. F., and Buzsaki, G. (1996). Interneurons of the hippocampus. *Hippocampus,* 6, 347–470.

Grover, L. M., Kim, E., Cooke, J. D., and Holmes, W. R. (2009). LTP in hippocampal area CA1 is induced by burst stimulation over a broad frequency range centered around delta. *Learn Mem,* 16, 69–81.

Gulyas, A. I., Miettinen, R., Jacobowitz, D. M., and Freund, T. F. (1992). Calretinin is present in non-pyramidal cells of the rat hippocampus – I. A new type of neuron specifically associated with the mossy fibre system. *Neuroscience,* 48, 1–27.

Harrison, P. J. (2004). The hippocampus in schizophrenia: a review of the neuropathological evidence and its pathophysiological implications. *Psychopharmacology (Berl),* 174, 151–62.

Harrison, P. J., Law, A. J., and Eastwood, S. L. (2003). Glutamate receptors and transporters in the hippocampus in schizophrenia. *Ann N Y Acad Sci,* 1003, 94–101.

Hauser, W. A. (1994). The prevalence and incidence of convulsive disorders in children. *Epilepsia,* 35(Suppl 2), S1–6.

Havenaar, R., Meijer, J. C., Morton, D. B., Ritskes-Hoitinga, J., and Zwart, P. (1993). Biology and husbandry of laboratory animals. In Van Zutphen, L. F. M., Baumans, V., and Beynen, A. C. (Eds.) *Principles of Laboratory Animal Science*. Amsterdam: Elsevier Science.

Insausti, R., and Amaral, D. G. (2004). Hippocampal formation. In Paxinos, G., and Mai, J. K. (Eds.) *The human nervous system*, pp. 871–906. Amsterdam: Elsevier.

Insausti, R., and Amaral, D. G. (2008). Entorhinal cortex of the monkey: IV. Topographical and laminar organization of cortical afferents. *J Comp Neurol*, 509, 608–41.

Insausti, R., and Munoz, M. (2001). Cortical projections of the non-entorhinal hippocampal formation in the cynomolgus monkey (Macaca fascicularis). *Eur J Neurosci*, 14, 435–51.

Insausti, R., Amaral, D. G., and Cowan, W. M. (1987a). The entorhinal cortex of the monkey: II. Cortical afferents. *J Comp Neurol*, 264, 356–95.

Insausti, R., Amaral, D. G., and Cowan, W. M. (1987b). The entorhinal cortex of the monkey: III. Subcortical afferents. *J Comp Neurol*, 264, 396–408.

Insausti, R., Tunon, T., Sobreviela, T., Insausti, A. M., and Gonzalo, L. M. (1995). The human entorhinal cortex: a cytoarchitectonic analysis. *J Comp Neurol*, 355, 171–98.

Insausti, R., Herrero, M. T., and Witter, M. P. (1997). Entorhinal cortex of the rat: Cytoarchitectonic subdivisions and the origin and distribution of cortical efferents. *Hippocampus*, 7, 146–83.

Insausti, R., Juottonen, K., Soininen, H., *et al.* (1998). Mr volumetric analysis of the human entorhinal, perirhinal, and temporopolar cortices. *Am J Neuroradiol*, 19, 659–71.

Insausti, R., Marcos, P., Arroyo-Jimenez, M. M., Blaizot, X., and Martinez-Marcos, A. (2002). Comparative aspects of the olfactory portion of the entorhinal cortex and its projection to the hippocampus in rodents, nonhuman primates, and the human brain. *Brain Res Bull*, 57, 557–60.

Jabès, A., Banta Lavenex, P., Amaral, D. G., and Lavenex, P. (2010). Quantitative analysis of postnatal neurogenesis and neuron number in the macaque monkey dentate gyrus. *Eur J Neurosci*, 31, 273–85.

Jabès, A., Banta Lavenex, P., Amaral, D. G., and Lavenex, P. (2011). Postnatal development of the hippocampal formation: a stereological study in macaque monkeys. *J Comp Neurol*, 519(6), 1051–70.

Kang, N., Xu, J., Xu, Q., Nedergaard, M., and Kang, J. (2005). Astrocytic glutamate release-induced transient depolarization and epileptiform discharges in hippocampal CA1 pyramidal neurons. *J Neurophysiol*, 94, 4121–30.

Karlsson, M. P., and Frank, L. M. (2008). Network dynamics underlying the formation of sparse, informative representations in the hippocampus. *J Neurosci*, 28, 14271–81.

Kobayashi, K. (2009). Targeting the hippocampal mossy fiber synapse for the treatment of psychiatric disorders. *Mol Neurobiol*, 39, 24–36.

Kobayashi, Y., and Amaral, D. G. (1999). Chemical neuroanatomy of the hippocampal formation and the perirhinal and parahippocampal cortices. In Bloom, F., Björklund, A., and Hökfelt, T. (Eds.) *Handbook of Chemical Neuroanatomy* (Vol. 15), pp. 285–337. Amsterdam: Elsevier Science Publishers B.V.

Kobayashi, Y., and Amaral, D. G. (2000). Macaque monkey retrosplenial cortex: I. three-dimensional and cytoarchitectonic organization. *J Comp Neurol*, 426, 339–65.

Kobayashi, Y., and Amaral, D. G. (2003). Macaque monkey retrosplenial cortex: II. Cortical afferents. *J Comp Neurol*, 466, 48–79.

Kobayashi, Y., and Amaral, D. G. (2007). Macaque monkey retrosplenial cortex: III. Cortical efferents. *J Comp Neurol*, 502, 810–33.

Kolomeets, N. S., Orlovskaya, D. D., Rachmanova, V. I., and Uranova, N. A. (2005). Ultrastructural alterations in hippocampal mossy fiber synapses in schizophrenia: a postmortem morphometric study. *Synapse*, 57, 47–55.

Kolomeets, N. S., Orlovskaya, D. D., and Uranova, N. A. (2007). Decreased numerical density of CA3 hippocampal mossy fiber synapses in schizophrenia. *Synapse*, 61, 615–21.

Kondo, H., Lavenex, P., and Amaral, D. G. (2008). Intrinsic connections of the macaque monkey hippocampal formation: I. Dentate gyrus. *J Comp Neurol*, 511, 497–520.

Kondo, H., Lavenex, P., and Amaral, D. G. (2009). Intrinsic connections of the macaque monkey hippocampal formation: II. CA3 connections. *J Comp Neurol*, 515, 349–77.

Lauer, M., Beckmann, H., and Senitz, D. (2003). Increased frequency of dentate granule cells with basal dendrites in the hippocampal formation of schizophrenics. *Psychiatry Res*, 122, 89–97.

Lavenex, P. (2009). Neuroanatomy methods in humans and animals. In Squire, L. R. (Ed.) *Encyclopedia of Neuroscience*, pp. 269–78. Oxford: Academic Press.

Lavenex, P., and Amaral, D. G. (2000). Hippocampal-neocortical interaction: A hierarchy of associativity. *Hippocampus*, 10, 420–30.

Lavenex, P., Suzuki, W. A., and Amaral, D. G. (2002). Perirhinal and parahippocampal cortices of the macaque monkey: Projections to the neocortex. *J Comp Neurol*, 447, 394–420.

Lavenex, P., Banta Lavenex, P., and Amaral, D. G. (2004a). Nonphosphorylated high-molecular-weight neurofilament expression suggests early maturation of the monkey subiculum. *Hippocampus*, 14, 797–801.

Lavenex, P., Suzuki, W. A., and Amaral, D. G. (2004b). Perirhinal and parahippocampal cortices of the macaque monkey: Intrinsic projections and interconnections. *J Comp Neurol*, 472, 371–94.

Lavenex, P., Banta Lavenex, P., and Amaral, D. G. (2007a). Postnatal development of the primate hippocampal formation. *Dev Neurosci*, 29, 179–92.

Lavenex, P., Banta Lavenex, P., and Amaral, D. G. (2007b). Spatial relational learning persists following neonatal hippocampal lesions in macaque monkeys. *Nat Neurosci*, 10, 234–239.

Lavenex, P., Banta Lavenex, P., Bennett, J. L., and Amaral, D. G. (2009). Postmortem changes in the neuroanatomical characteristics of the primate brain: hippocampal formation. *J Comp Neurol*, 512, 27–51.

Lavenex, P., Sugden, S.G., Davis, R.R., Gregg, J.P., and Banta Lavenex, P. (2011). Developmental regulation of gene expression and astrocytic processes may explain selective hippocampal vulnerability. *Hippocampus*, 21(2), 142–9.

Lawrence, Y. A., Kemper, T. L., Bauman, M. L., and Blatt, G. J. (2010). Parvalbumin-, calbindin-, and calretinin-immunoreactive hippocampal interneuron density in autism. *Acta Neurol Scand*, 121, 99–108.

Le Strat, Y., Ramoz, N., and Gorwood, P. (2009). The role of genes involved in neuroplasticity and neurogenesis in the observation of a gene-environment interaction (GxE) in schizophrenia. *Curr Mol Med*, 9, 506–18.

Leranth, C., Malcolm, A. J., and Frotscher, M. (1990). Afferent and efferent synaptic connections of somatostatin-immunoreactive neurons in the rat fascia dentata. *J Comp Neurol*, 295, 111–22.

Leverenz, J. B., and Raskind, M. A. (1998). Early amyloid deposition in the medial temporal lobe of young Down syndrome patients: a regional quantitative analysis. *Exp Neurol*, 150, 296–304.

McClelland, J. L., McNaughton, B. L., and O'Reilly, R. C. (1995). Why there are complementary learning-systems in the hippocampus and neocortex – Insights from the successes and failures of connectionist models of learning and memory. *Psychol Review,* 102, 419–57.

Meencke, H. J. (2009). Clinical neuropathology of the epilepsies in the 100 years of the ILAE (1909–2009). *Epilepsia,* 50(Suppl 3), 8–16.

Mito, T., and Becker, L. E. (1993). Developmental changes of S-100 protein and glial fibrillary acidic protein in the brain in Down syndrome. *Exp Neurol,* 120, 170–6.

Morris, R. G. M. (2007). Theories of hippocampal function. In Andersen, P., Morris, R. G. M., Amaral, D. G., Bliss, T. V., and O'Keefe, J. (Eds.) *The Hippocampus Book,* pp. 581–714. Oxford: Oxford University Press.

Munoz, M., and Insausti, R. (2005). Cortical efferents of the entorhinal cortex and the adjacent parahippocampal region in the monkey (Macaca fascicularis). *Eur J Neurosci,* 22, 1368–88.

Myers, C. E., and Scharfman, H. E. (2011). Pattern separation in the dentate gyrus: A role for the CA3 backprojection. *Hippocampus,* 21(11), 1190–215.

Naber, P. A., and Witter, M. P. (1998). Subicular efferents are organized mostly as parallel projections: a double-labeling, retrograde-tracing study in the rat. *J Comp Neurol,* 393, 284–97.

Naber, P. A., Da Silva, F. H. L., and Witter, M. P. (2001). Reciprocal connections between the entorhinal cortex and hippocampal fields CA1 and the subiculum are in register with the projections from CA1 to the subiculum. *Hippocampus,* 11, 99–104.

Nadel, L. (2003). Down's syndrome: a genetic disorder in biobehavioral perspective. *Genes Brain Behav,* 2, 156–66.

Nadel, L., and Moscovitch, M. (1997). Memory consolidation, retrograde amnesia and the hippocampal complex. *Curr Opin Neurobiol,* 7, 217–27.

Nedergaard, M., and Dirnagl, U. (2005). Role of glial cells in cerebral ischemia. *Glia,* 50, 281–6.

O'Mara, S. M., Commins, S., Anderson, M., and GIGG, J. (2001). The subiculum: a review of form, physiology and function. *Progr Neurobiol,* 64, 129–55.

Oliet, S. H., Piet, R., and Poulain, D. A. (2001). Control of glutamate clearance and synaptic efficacy by glial coverage of neurons. *Science,* 292, 923–6.

Pellerin, L., and Magistretti, P. J. (1994). Glutamate uptake into astrocytes stimulates aerobic glycolysis – a mechanism coupling neuronal-activity to glucose-utilization. *Proc Natl Acad Sci USA,* 91, 10625–9.

Pinter, J. D., Brown, W. E., Eliez, S., Schmitt, J. E., Capone, G. T., and Reiss, A. L. (2001). Amygdala and hippocampal volumes in children with Down syndrome: a high-resolution MRI study. *Neurology,* 56, 972–4.

Pitkänen, A., Pikkarainen, M., Nurminen, N., and Ylinen, A. (2000). Reciprocal connections between the amygdala and the hippocampal formation, perirhinal cortex, and postrhinal cortex in rat. A review. *Ann N Y Acad Sci,* 911, 369–91.

Rakic, P., and Nowakowski, R. S. (1981). The time of origin of neurons in the hippocampal region of the rhesus monkey. *J Comp Neurol,* 196, 99–128.

Rao, V. L., Dogan, A., Todd, K. G., *et al.* (2001). Antisense knockdown of the glial glutamate transporter GLT-1, but not the neuronal glutamate transporter EAAC1, exacerbates transient focal cerebral ischemia-induced neuronal damage in rat brain. *J Neurosci,* 21, 1876–83.

Raymond, G. V., Bauman, M. L., and Kemper, T. L. (1996). Hippocampus in autism – a Golgi analysis. *Acta Neuropathol,* 91, 117–19.

Reif, A., Schmitt, A., Fritzen, S., and Lesch, K. P. (2007). Neurogenesis and schizophrenia: dividing neurons in a divided mind? *Eur Arch Psychiatry Clin Neurosci,* 257, 290–9.

Rockland, K. S., and Van Hoesen, G. W. (1999). Some temporal and parietal cortical connections converge in CA1 of the primate hippocampus. *Cereb Cortex,* 9, 232–7.

Saitoh, O., Karns, C. M., and Courchesne, E. (2001). Development of the hippocampal formation from 2 to 42 years – MRI evidence of smaller area dentata in autism. *Brain,* 124, 1317–24.

Scanziani, M., Salin, P. A., Vogt, K. E., Malenka, R. C., and Nicoll, R. A. (1997). Use-dependent increases in glutamate concentration activate presynaptic metabotropic glutamate receptors. *Nature,* 385, 630–4.

Schumann, C. M., Hamstra, J., Goodlin-Jones, B. L., *et al.* (2004). The amygdala is enlarged in children but not adolescents with autism; the hippocampus is enlarged at all ages. *J Neurosci,* 24, 6392–401.

Seidenberg, M., Kelly, K. G., Parrish, J., *et al.* (2005). Ipsilateral and contralateral MRI volumetric abnormalities in chronic unilateral temporal lobe epilepsy and their clinical correlates. *Epilepsia,* 46, 420–30.

Seidenberg, M., Hermann, B., Pulsipher, D., *et al.* (2008). Thalamic atrophy and cognition in unilateral temporal lobe epilepsy. *J Int Neuropsychol Soc,* 14, 384–93.

Senitz, D., and Beckmann, H. (2003). Granule cells of the dentate gyrus with basal and recurrent dendrites in schizophrenic patients and controls. A comparative Golgi study. *J Neural Transm,* 110, 317–26.

Seress, L. (2001). Morphological changes of the human hippocampal formation from midgestation to early childhood. In Nelson, C. A., and Luciana, M. (Eds.) *Handbook of Developmental Cognitive Neuroscience,* pp. 45–47. Cambridge, MA: The MIT Press.

Seress, L., and Mrzljak, L. (1987). Basal dendrites of granule cells are normal features of the fetal and adult dentate gyrus of both monkey and human hippocampal formations. *Brain Res,* 405, 169–74.

Seress, L., and Mrzljak, L. (1992). Postnatal development of mossy cells in the human dentate gyrus: a light microscopic Golgi study. *Hippocampus,* 2, 127–41.

Seress, L., and Ribak, C. E. (1995a). Postnatal development and synaptic connections of hilar mossy cells in the hippocampal dentate gyrus of rhesus monkeys. *J Comp Neurol,* 355, 93–110.

Seress, L., and Ribak, C. E. (1995b). Postnatal development of CA3 pyramidal neurons and their afferents in the Ammon's horn of rhesus monkeys. *Hippocampus,* 5, 217–31.

Seress, L., Abraham, H., Tornoczky, T., and Kosztolanyi, G. (2001). Cell formation in the human hippocampal formation from mid-gestation to the late postnatal period. *Neuroscience,* 105, 831–43.

Sismanlar, S. G., Anik, Y., Coskun, A., Agaoglu, B., Karakaya, I., and Yavuz, C. I. (2009). The volumetric differences of the fronto-temporal region in young offspring of schizophrenic patients. *Eur Child Adolesc Psychiatry,* 19, 151–7.

Sloviter, R. S. (1994). The functional organization of the hippocampal dentate gyrus and its relevance to the pathogenesis of temporal lobe epilepsy. *Ann Neurol,* 35, 640–54.

Sloviter, R. S. (2009). Experimental status epilepticus in animals: What are we modeling? *Epilepsia,* 50(Suppl 12), 11–13.

Sommer, W. (1880). Erkrankung des Ammonshorns als aetiologisches Moment der Epilepsie. *Arch Psychiatr Nervenkr,* 10, 631–75.

Sparks, B. F., Friedman, S. D., Shaw, D. W., *et al.* (2002). Brain structural abnormalities in young children with autism spectrum disorder. *Neurology,* 59, 184–92.

Spielmeyer, W. (1925). Zur Pathogenese örtlich elektiver Gehirnveränderungen. *Z Ges Neurol Psyhiatr,* 99, 756–76.

Spigelman, I., Yan, X. X., Obenaus, A., Lee, E. Y., Wasterlain, C. G., and Ribak, C. E. (1998). Dentate granule cells form novel basal dendrites in a rat model of temporal lobe epilepsy. *Neuroscience,* 86, 109–20.

Spruston, N., and McBain, C. J. (2007). Structural and functional properties of hippocampal neurons. In Andersen, P., Morris, R. G. M., Amaral, D. G., Bliss, T. V., and O'Keefe, J. (Eds.) *The Hippocampus Book*, pp. 133–202. Oxford: Oxford University Press.

Squire, L. R., and Zola-Morgan, S. (1996). Structure and function of declarative and nondeclarative memory systems. *Proc Natl Acad Sci USA,* 93, 13515–22.

Squier, W., Salisbury, H., and Sisodiya, S. (2003). Stroke in the developing brain and intractable epilepsy: effect of timing on hippocampal sclerosis. *Dev Med Child Neurol,* 45, 580–5.

Steen, R. G., Mull, C., McClure, R., Hamer, R. M., and Lieberman, J. A. (2006). Brain volume in first-episode schizophrenia: systematic review and meta-analysis of magnetic resonance imaging studies. *Br J Psychiatry,* 188, 510–18.

Suzuki, W. A., and Amaral, D. G. (1990). Cortical inputs to the CA1 field of the monkey hippocampus originate from the perirhinal and parahippocampal cortex but not from area TE. *Neuroscience Letters,* 115, 43–8.

Suzuki, W. A., and Amaral, D. G. (1994a). Perirhinal and parahippocampal cortices of the macaque monkey: cortical afferents. *J Comp Neurol,* 350, 497–533.

Suzuki, W. A., and Amaral, D. G. (1994b). Topographic organization of the reciprocal connections between the monkey entorhinal cortex and the perirhinal and parahippocampal cortices. *J Neurosci,* 14, 1856–77.

Suzuki, W. A., and Amaral, D. G. (2003). Where are the perirhinal and parahippocampal cortices? A historical overview of the nomenclature and boundaries applied to the primate medial temporal lobe. *Neuroscience,* 120, 893–906.

Swanson, R. A. (2005). Astrocyte neurotransmitter uptake. In Kettenmann, H., and Ransom, B. R. (Eds.) *Neuroglia*, pp. 346–54. Oxford: Oxford University Press.

Talbot, K., Eidem, W. L., Tinsley, C. L., *et al.* (2004). Dysbindin-1 is reduced in intrinsic, glutamatergic terminals of the hippocampal formation in schizophrenia. *J Clin Invest,* 113, 1353–63.

Tamamaki, N., and Nojyo, Y. (1995). Preservation of topography in the connections between the subiculum, field CA1, and the entorhinal cortex in rats. *J Comp Neurol,* 353, 379–90.

Tanaka, K., Watase, K., Manabe, T., *et al.* (1997). Epilepsy and exacerbation of brain injury in mice lacking the glutamate transporter GLT-1. *Science,* 276, 1699–702.

Teyler, T. J., and Discenna, P. (1986). The hippocampal memory indexing theory. *Behav Neurosci,* 100, 147–54.

Tian, G. F., Azmi, H., Takano, T., *et al.* (2005). An astrocytic basis of epilepsy. *Nat Med,* 11, 973–81.

Van Haeften, T., Wouterlood, F. G., Jorritsma-Byham, B., and Witter, M. P. (1997). GABAergic presubicular projections to the medial entorhinal cortex of the rat. *J Neurosci,* 17, 862–74.

Vargha-Khadem, F., Gadian, D. G., Watkins, K. E., Connelly, A., Van Paesschen, W., and Mishkin, M. (1997). Differential effects of early hippocampal pathology on episodic and semantic memory. *Science,* 277, 376–80.

Velakoulis, D., Pantelis, C., McGorry, P. D., *et al.* (1999). Hippocampal volume in first-episode psychoses and chronic schizophrenia: a high-resolution magnetic resonance imaging study. *Arch Gen Psychiatry,* 56, 133–41.

Volterra, A., and Meldolesi, J. (2005). Astrocytes, from brain glue to communication elements: the revolution continues. *Nat Rev Neurosci,* 6, 626–40.

Witter, M. P., and Amaral, D. G. (1991). Entorhinal cortex of the monkey: V. Projections to the dentate gyrus, hippocampus, and subicular complex. *J Comp Neurol,* 307, 437–59.

Zola-Morgan, S., Squire, L. R., and Amaral, D. G. (1986). Human amnesia and the medial temporal region enduring memory impairment following a bilateral lesion limited to field CA1 of the hippocampus. *J Neurosci,* 6, 2950–67.

CHAPTER 3

Amnesia and the Hippocampal Memory System

Howard Eichenbaum

Introduction 39

Case H.M. 40

Spared learning in H.M. and other patients with hippocampal damage 43

What are the fundamental features of declarative and procedural memory? 44

Temporally graded retrograde amnesia and memory consolidation 49

Recent challenges about the defining features of declarative memory and their role in amnesia 51

ABSTRACT Studies on the amnesic patient H.M. began the modern era of the cognitive neuroscience of memory. These studies, and other case studies, showed that damage limited to the hippocampal region results in an impairment that is selective to memory and spares other perceptual, motor, emotional, or cognitive functions. Furthermore, the amnesic deficit is selective to the permanent establishment of new declarative memories. Thus H.M. and other patients with hippocampal damage have intact short-term and working memory, and can form long-term memories that do not rely on remembering specific past events or on the flexible use of memories to solve new problems. These and other characteristics of amnesia following hippocampal damage indicate that the hippocampal memory system is essential to relational memory, the ability to associate multiple events with one other and with their spatial and temporal context, and the ability to integrate many memories into a network of knowledge.

KEYWORDS declarative memory, procedural memory, priming, skill learning, Pavlovian conditioning, episodic memory, semantic memory, relational memory, retrograde amnesia, memory consolidation, short-term memory, long-term memory

Introduction

The modern era of research on amnesia and the hippocampal system began in 1957 with a case report by Scoville and Milner on the patient Henry Molaison, known only by his initials to protect his anonymity until his death in 2008. H.M. had a medial temporal lobe resection to ameliorate his seizure disorder, and the surgery did reduce the frequency of his seizures considerably and brought them largely under pharmacological control. However, he became severely amnesic as a result of the surgery, and the amnesia was 'pure' in that he showed no other significant neurological deficits. The memory impairment was both severe and selective, such that his case, and similar reports on subsequent patients with hippocampal region damage, changed our fundamental ideas about the organization of memory in the brain.

Prior to the report on H.M., most thought that memories were distributed among specialized areas of the cerebral cortex and the generally held view was that memory and other perceptual and cognitive functions were not anatomically separable. Yet H.M. showed a clear dissociation between fully intact perception and cognition against severely impaired memory. Thus the discovery of a 'pure' memory deficit following selective brain damage also addressed how memory is localized in the brain. In addition, a prominent feature of H.M.'s disorder was the combination of an anterograde and a temporally graded retrograde memory impairment, indicating that the hippocampal region is not the final repository of memories but rather is essential to their consolidation elsewhere in the brain.

H.M.'s amnesia was 'global' in that it appeared to encompass all kinds of memories. Yet, even from the outset, exceptions to global amnesia were identified—he was able to learn a new motor skill in a 'mirror drawing' task (Corkin 1968). In this task the subject views through a mirror two concentric outlines of a star pattern, and the task is to draw a pencil line within the outlines. Normal subjects require several trials before they can successfully draw the line without crossing over the outlines. Similarly, H.M. improved within the initial training session, and showed considerable savings across sessions. His success in learning this sensorimotor skill contrasted with his inability to recall taking the test. In addition, H.M. performed well on the Gollins partial pictures task, a perceptual learning test involves the recognition of fragmented line drawings of common objects (Milner et al. 1968). For each stimulus to be learned, the subject is initially shown just fragments of the drawing of an object, then shown increasingly complete versions of the drawing, and the key measure is the version on which recognition is successful. Like normal subjects, H.M. could recognize the object in earlier versions over repetitions and showed a surprising degree of savings across days, although he could not remember having taken the test from day to day.

A deeper understanding about the exceptions began when Cohen and Squire (1980) examined the performance of amnesic individuals on a reading skill task that involves improvement with practice in the speed of reading of mirror-reversed words. Normal subjects are ordinarily slow in deciphering words that are presented as if seen through a mirror but, with practice, they improve considerably at this perceptual-cognitive skill. Cohen and Squire reported that this kind of perceptual learning was fully normal in a set of amnesic individuals, some of whom had hippocampal region damage. In addition, when normal subjects were presented the same mirror-reversed words a second time, they showed an extra degree of improvement beyond that supported by the general skill, reflecting memory for the particular mirror-reversed words that had been read before. Amnesic individuals were markedly impaired both in facilitation for the familiar words and in recollecting their training experience. This pattern of findings suggested a dissociation between the ability to acquire the general perceptual and cognitive procedures required to read reversed text from the capacity to remember particular words. Cohen and Squire explained this dissociation as a distinction between *declarative memory*, the capacity to consciously recollect specific facts and events supported by the hippocampal region, and *procedural memory*, the ability to tune and modify the brain's networks that underlie general skills supported by other brain systems. A voluminous body of research has supported this dissociation and further characterized the properties of declarative and procedural memory.

The following sections will first provide a more detailed overview of case H.M., in order to provide a closer perspective on the nature of amnesia due to damage in the hippocampal region. Then the distinction between declarative and procedural memory will be explored further, using several examples from the larger experimental literature on amnesia.

Case H.M.

As described in several reports (Scoville & Milner 1957; Milner et al. 1968; Corkin 1984; Corkin 2002), in 1933 at the age of 7 years, H.M. was knocked down by a bicycle and hit his head, and he was unconscious for 5 minutes. Three years later he had a minor seizure, followed by his first major seizure at age 16. Despite continued epileptic attacks, H.M. graduated from high school at 21 in a 'practical' course subsequently worked as a motor winder. However, the seizures became more frequent, on average ten minor attacks each day and a major seizure each week, and he eventually could not perform his job. Attempts to control the seizures with large doses of anticonvulsant drugs were unsuccessful, leading to consideration of a brain operation. There was no evidence of localization from electroencephalography (EEG) studies. Nevertheless, because of the known epileptogenic qualities of the medial temporal lobe, an experimental operation was considered justified as an effort to ameliorate the seizure disorder. In 1953, when H.M. was 27, a bilateral medial temporal lobe resection was performed via a direct approach in which the anterior tip of the temporal lobe was visualized, and the medial part, including the amygdala, anterior two-thirds of the hippocampus, and neighbouring cortex aspirated, and these areas of damage have been confirmed in recent imaging studies (Figure 3.1; Corkin et al. 1997; Salat et al. 2006). The operation reduced the frequency of seizures to a point where they were largely prevented by medication, although minor attacks persisted.

After recovery H.M. returned home where he lived with his parents. He performed household chores, watched TV, and solved crossword puzzles. Following his father's death, H.M. attended a rehabilitation workshop and became a

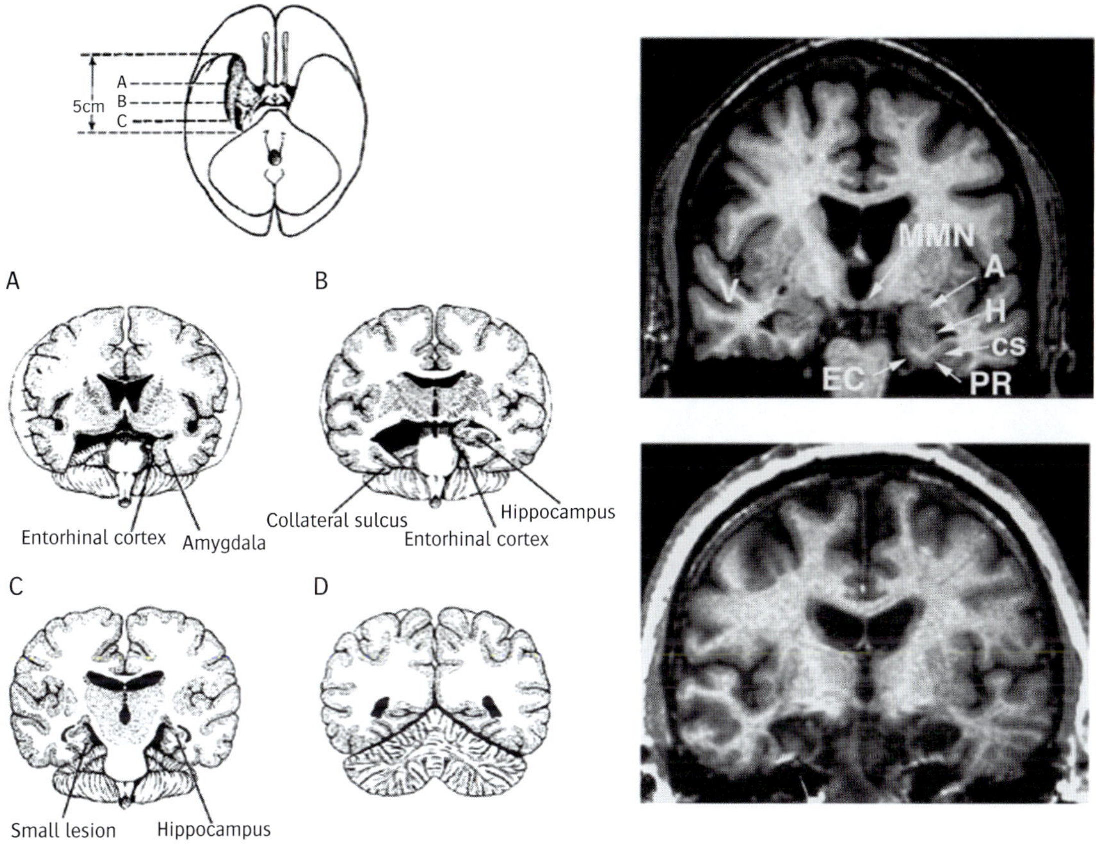

Figure 3.1 Left) Reconstruction of the area of medial temporal removal in patient H.M. The upper panel is a drawing of ventral view of the brain showing area of lesion on right hemisphere and, for comparison, intact appearance of medial temporal lobe on right (the lesion was bilateral). Letters indicate levels of sections drawn on lower panels. Right) T1-weighted images through the temporal lobe of a normal control subject (upper) and H.M. (lower). A, amygdala; CS, collateral sulcus; EC, entorhinal cortex; H, hippocampus; MMN, medial mammillary nucleus; PR, perirhinal cortex. Reproduced from Corkin, S., Amaral, D. G., Gonzalez, R. G., Johnson, K. A., and Hyman. B.T. (1997). H. M.'s medial temporal lobe lesion: Findings from magnetic resonance imaging. *Journal of Neuroscience*, 17, 3964–79. © 1997, Society for Neuroscience, with permission.

handyman, performing simple and repetitive jobs. Eventually his mother and then another relative could no longer care for him, so he was moved to a nursing home where he participated in daily social activities of the home, and continued watching TV and solving difficult crossword puzzles. He was characterized as a highly amiable and cooperative individual. H.M. was considered was very polite, had strong beliefs, values, desires, a sense of humour, and moral views and was altruistic. He rarely complained and had to be quizzed to identify minor problems such as headaches. He never spontaneously asked for food or beverage, or to go to bed, but he readily followed directions for all of his daily activities. His was aware of his memory disorder, but was not consistently concerned about it. When he looked in a mirror, he showed no surprise, concern, or lack of recognition of an older man. Sometimes when given a rather difficult memory question he reminded the examiner, 'You know I have a memory problem'. The severe magnitude of his memory disorder continued unabated since the time of the surgery until his death.

H.M.'s disorder was highly selective, in two important ways. First, as introduced earlier, his impairment was almost entirely selective to memory, as distinguished from other higher order perceptual, motor, and cognitive functions. Second, even within his memory functions, the disorder was selective to particular domains of learning and memory capacity. These two aspects of his preserved and impaired capacities will be discussed next.

H.M.'s impairment is selective to memory

The results of extensive testing of sensory functions showed that his perceptual capacities were entirely normal. He performed within the normal range on tests of visual acuity, adaptation, and other commonly tested visual-perceptual functions, and could recognize and name common objects. He had some loss of touch and fine motor coordination. H.M.'s intelligence was above average in standard IQ tests just before the operation. After the surgery

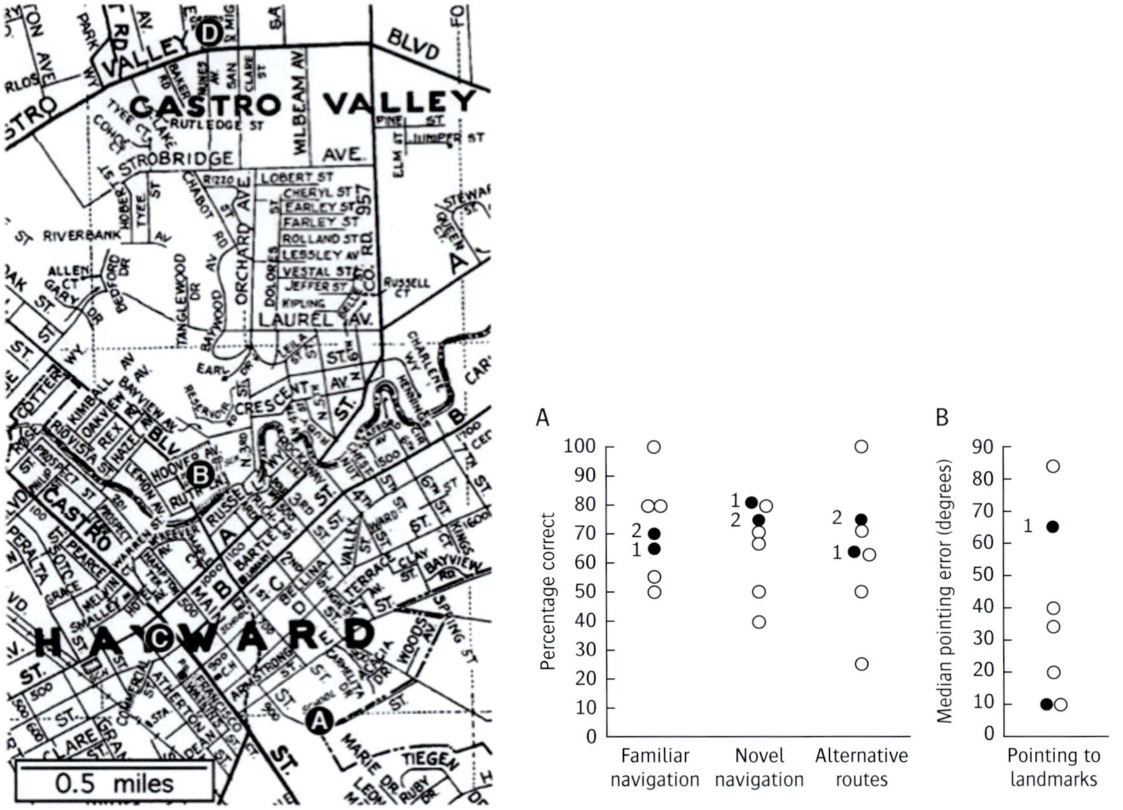

Figure 3.2 Left) Street map of a portion of the Hayward-Castro Valley region from the 1940s (Thomas Bros. Map Co.). The locations of four representative landmarks used in the topographical memory tasks are shown: A: Bret Harte School; B: Hayward Union High School; C: Hayward Theatre; D: Castro Valley Grammar School. Right) Performance on four tasks of topographical memory. Open circles: scores of five control subjects; filled circles: scores of amnesic patient E.P. on two different occasions separated by 9 months. Numbers next to the filled circles indicate scores for E.P.'s first and second test sessions. E.P. scored within the normal range on three verbal navigation tasks that required negotiating either familiar routes, novel routes or alternative routes when the most direct route was blocked, and within the normal range in pointing to particular locations while imagining themselves oriented at other locations in the area. Reprinted by permission from Macmillan Publishers Ltd: Teng, E. and Squire, L. R. (1999). Memory for places learned long ago is intact after hippocampal damage. *Nature*, 400 (6745), 675–7, copyright (1999).

his IQ rose slightly, perhaps because of the alleviation of the seizures. H.M.'s language capacities were largely intact, although he exhibited slight deficits in fluency. He could repeat and manipulate sentence syntax normally, and appreciated puns and linguistic ambiguities. He communicated well and freely. His spatial perceptual capacities that do not depend on memory were mixed. For example, he had some difficulty copying a complex line drawing, and could not use a floor plan to walk a route from one room to another in a testing facility. On the other hand he did well on other complex spatial perceptual tasks, and could draw an accurate floor plan of the house he lived in. Another severely amnesic individual (E.P.) with equally extensive hippocampal region damage can navigate based on geographical knowledge obtained before the onset of amnesia or spatial information that can be held in mind while blindfolded (Figure 3.2; Teng and Squire 1999). Thus, while even modest damage to the hippocampus can cause impairments in new spatial learning, spatial information processing per se is not compromised in amnesia.

H.M.'s anterograde memory impairment is selective to long-term memory

H.M.'s short-term memory was intact. He could immediately reproduce a list of numbers as well as control subjects. Across many tests, his memory span, short-term memory capacity, and working memory capacity were normal. However, the memory deficit became evident as his immediate memory span was exceeded or after a delay with distraction. H.M. also remembered material learned remotely prior to his operation. His memory for the English language was fully intact. He also retained many childhood memories. By contrast, all memory for events for some period preceding the operation were lost.

By contrast to his intact short-term forms of memory and spared remotely acquired memories, H.M. had almost no capacity for new learning, as measured by a large variety of conventional tests. He was not given standard memory tests prior to the operation. After the surgery his scores on standardized scale indicated a severe memory disorder. In particular, he scored zero on tests that assess the persistence of his memory for short stories, lists of words or numbers, pictures, tunes, mazes, public and personal events, or any of a large range of other materials (Corkin 1984).

Spared learning in H.M. and other patients with hippocampal damage

A broad range of learning abilities were spared in H.M. and in other individuals with hippocampal region damage. The observations of Cohen and Squire (1980) showing fully intact perceptual skill learning were properly heralded as a revelation about memory processing functions accomplished by the medial temporal lobe. Subsequently, several laboratories uncovered a variety of examples of spared learning ability in amnesia. Examples from those findings will be presented next, to provide a view of the scope of preserved learning and memory capacities in amnesia. These include another form of perceptual learning called 'priming', skill learning, Pavlovian conditioning, and gradual learning of sensorimotor sequences. Some of these findings were made on H.M. and other findings come from studies on other individuals with damage to the hippocampal region.

Priming

Priming involves initial presentation of a list of words or pictures, then subsequent re-exposure to fragments those items or very brief presentation of the entire item. Learning is measured by increased ability to identify the whole item from a fragment (as in the Gollins partial pictures task described earlier) or by increased speed in making a decision about the item. For example, Graf et al. (1984) used a word stem completion task in which subjects initially study a word list, then are presented with the first three letters of each word (the word 'stem') and asked to complete it. The stimulus words are selected as ones that can be completed more than one way to compose a high frequency word. For example, for the study word 'MOTOR', the stem 'MOT___' can also be completed to form 'MOTEL' or 'MOTHER'. In this experiment priming was measured by the increased likelihood that the subject will complete the stimulus word presented during the study phase. Memory was tested following study of the word list in one of three ways. In the free recall condition, subjects were simply asked to recall the studied words. In the cued recall condition, subjects were asked to use the word stems as cues to remember the study words. In the completion condition, subjects were asked to provide the first word that came to mind after seeing the stem. The amnesic subjects were impaired in recall, but performed well on the completion test and performed no better on cued recall than on completion, suggesting that their performance in cued recall might be entirely supported by priming. By contrast, the normal subjects performed much better on cued recall than on priming, suggesting they used the stems to aid an active search in recalling the words. H.M. also performed normally on stem-completion priming for words that were common before his surgery. However, he was impaired in priming on words that entered the dictionary after he became amnesic, indicating that priming involves the reactivation of word representations that are already established in brain areas outside the hippocampus.

Amnesic individuals also show intact priming for nonverbal information. For example H.M. also shows normal priming in a dot pattern completion task (Gabrieli et al. 1990). In this test, subjects are initially presented with stimuli that consist of dots arranged in a unique pattern and are asked to use the dots to draw a line pattern. Later, the dot pattern again is again presented and normal subjects tend to complete the pattern as they did initially. H.M. showed intact priming for dot patterns, despite his inability to recognize the previously experienced dot patterns.

Skill learning

The mirror drawing test described previously is an example of spared capacity in H.M. and other amnesic individuals for the acquisition of a sensorimotor skill. The capacity to learn skill extends to the acquisition of cognitive rules. For example, in Knowlton et al. (1992) presented subjects with letter strings generated by an 'artificial grammar' that determined general rules for the sequence and length of the letter strings. Following reproduction of several strings, subjects were informed that the sequences followed a complex 'grammar' but were not told the algorithm. Nevertheless, subsequently normal subjects and amnesic individuals were able to correctly classify a set of new letter strings as 'grammatical' or 'non-grammatical' on most trials. By contrast, the amnesic individuals were impaired on recognition of studied letter strings.

Classical Pavlovian conditioning

Modern studies of classical conditioning typically involve repeatedly pairings of a tone or light as the conditioning stimulus (CS) followed by an air puff to the eye as the unconditioned stimulus (US), which produces a reflexive blink. Classical conditioning is measured as the occurrence of eye blinks during the CS period prior to presentation of the US. Amnesic individuals are typically normal in the rate of learning in classical eyelid conditioning in the standard version of the task, known as 'delay-conditioning', in which the CS presentation extends to and overlaps the US presentation. However, in another procedure known as 'trace-conditioning', which involves a brief CS followed by a trace interval during which no stimulus is presented, amnesic individuals are impaired, and this deficit has been related to a lack of conscious awareness of the stimulus contingencies in amnesic individuals (Clark and Squire 1998).

Gradual learning of sensorimotor sequences

Amnesic individuals are impaired in remembering sequential information, for example, in the ability to imitate a sequence of actions that were once demonstrated (McDonough et al. 1995), showing that the deficit is not isolated to memory expression by verbal declaration. However, sequence learning can be accomplished even in amnesia if learning involves the gradual improvement in performance on repetitive stimulus-response sequences. For example, in the serial reaction time test subjects are shown a light at sequences of locations on a computer monitor, and must press keys that corresponded to the light locations (Nissen et al. 1987). As the sequences are repeated, normal subjects as well as amnesic individuals learn to anticipate the upcoming lights by decreasing their reaction times to press the keys. In a control test, subjects do not show improvement when the sequence is random, indicating that the improvement in reaction times on learned sequences does not reflect acquisition of a general sensorimotor ability to coordinate the pressing of keys for appropriate lights.

In addition, multiple experiments have shown that amnesic individuals normally improve reading speed when challenged to re-read aloud the same short story multiple times (Musen et al. 1990). In these studies there is also no general facilitation between stories, indicating that the improvement was not due to acquisition of a general skill. Furthermore, this type of learning extends to pronounceable but meaningless letter strings, showing that this intact capacity does not seem to rely on memory for the content of the story. Also in these tests, the facilitation is specific to the sequence of repeated non-words and is not due to acquisition of a general skill of reading non-words.

What are the fundamental features of declarative and procedural memory?

There have been numerous attempts to identify the common features of learning and memory that are impaired or spared in amnesia. These comparisons have provided insights into the fundamental nature and cognitive mechanisms that underlie declarative memory, as well as properties of the domains of procedural memory.

The earliest findings on H.M. indicated that spared learning in amnesia is limited to motor skills or simple perceptual learning. However, intact learning in amnesia is not limited to general motor skills, perceptual learning, or any other specific category of material. And the preserved learning in amnesia is not limited to any particular category of learning materials, as one can see in the broad range of examples given earlier, including intact learning for words, non-words, common objects, tones, non-verbalizable pictures, motor patterns, and spatial patterns. Rather the scope of spared memory in amnesia includes a variety of forms of highly specific new learning.

Other studies have suggested that spared learning always involves the slow incremental acquisition of habitual routines. However, spared learning in amnesia is not limited to types of learning that involve slow incremental improvement, but instead includes forms of one-trial learning, including numerous repetition priming tests. Priming for single exposures to pictures can be robust and can last at least a week in amnesic subjects. In sum, the domain of intact learning ability is global, and it can be either fast or slow, and includes both general skills and highly specific information content. Also, intact learning in amnesia is not limited to simple forms of perceptual learning or other easy tasks, as clearly shown by the improvements in very difficult tasks such as grammar and sensorimotor sequence learning.

Sufficient conditions for intact memory in amnesia

Having ruled out simple objective parameters such as those just listed, several newer proposals have focused on more complex and higher-order distinctions between declarative and procedural memory, and in particular, the form of memory expression, the extent of conscious access to memories, and the structure of the memory representation. For example, Moscovitch (1984) suggested a list of sufficient conditions for demonstrating preserved memory. He focused on task demands, and argued that amnesic individuals show preserved learning and memory on tests that satisfy three conditions: (1) The task has to be so highly structured that the goal of the task and the means to achieve it are apparent, (2) the means to achieve the goal are available to the subject (i.e. the response strategies are already in the subjects repertoire), and (3) success can be achieved without reference to any particular event or episode. Combining these, Moscovitch suggested that amnesic individuals succeed whenever they simply have to perform a task guided by the conditions and strategies at hand. Intact memory is revealed in changes of task performance itself, typically either a change in the speed of responding or in a bias of choices that are readily available.

As a striking example of learning involving these rules, Cohen and colleagues (Duff et al. 2006) studied the learning new words through *common ground,* the creation of and memory for new terminology that is shared by conversational partners. This kind of learning was investigated in a setting that involved amnesic individuals and their normal communication partners performing a game. At the outset of each test the amnesic subject was given a set of 12 cards, each of which contained a particular shape, in a specified order (Figure 3.3). The normal partner was given an identical set of cards but in random order. The objective of the game was for the partners to collaborate in arranging the randomly organized cards of the normal partner into the same order as those given to the amnesic subject. The partners could not see each other's cards, and could use only conversation between them to convey the correct order. The game was played with the same cards several times, each with a different ordering of the cards. So the speed with which the partners could complete the task increased if they could name the shapes to facilitate their communication. The amnesic individuals were remarkably good at collaborating in the generation of new names, and in remembering the newly created names, often succeeding in completing the task at the same rate as teams composed of two normal individuals (Figure 3.3). The amnesic individuals also remembered the names in succeeding tests with different orders. In contrast, the same amnesic subjects failed to later learn arbitrary names for the same kinds of shapes by explicit memorization. Thus, the successful collaborative learning of new names by amnesic subjects could be attributed to the structured nature of the task, the availability of the conversational means to generate names, and the gradual merging of descriptive names for the shapes without reference to any particular conversational exchange—all properties of spared learning in amnesia described by Moscovitch (1984).

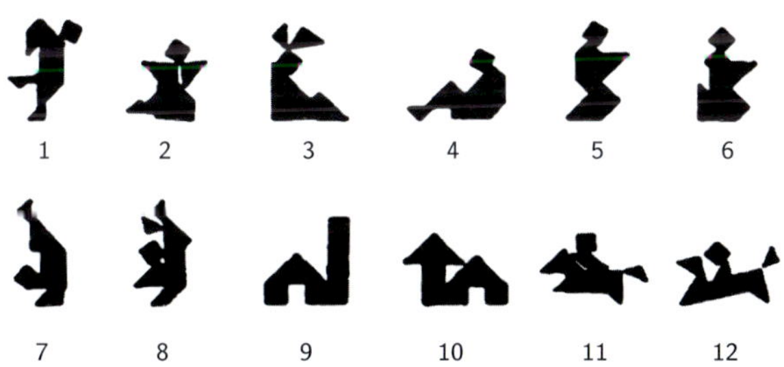

Session number and trial number	Card description
Session 1	
Trial 1	The next one looks almost... the opposite of somebody kind uh slumped down, on the ground, with the same type of sh; head.
Trial 2	The guy that looks like he's sleeping with his knees bent and he's got- **Sitting?**
Trial 3	The guy that's slumped over re- or sleeping or reading a book or something, **Kinda siesta time?**
Trial 4	This would be the siesta guy.
Trial 5	The person that looks like he is siestaing.
Trial 6	The guy taking the siesta.
Session 4	
Trial 1	Siesta man.
Trial 2	Siesta man.
Trial 3	Siesta man.
Trial 4	Siesta man.
Trial 5	Siesta man.
Trial 6	Siesta man.

Bold text is familiar communication partner; '...', a pause in speech of less than 1 s; ':', prolonged sound.

Figure 3.3 Top) The set of 12 tangrams for which participants and their partners collaboratively generated labels. Bottom) Description of card 4 across all 24 trials for an amnesic patient and partner. Reprinted by permission from Macmillan Publishers Ltd: Duff, M. C., Hengst, J., Tranel, D., and Cohen, N. J. (2006). Development of shared information in communication despite hippocampal amnesia. *Nature Neuroscience,* 9(1), 140-6, copyright (2006).

Explicit versus implicit memory

Schacter (1985) characterized memory that is impaired and spared in amnesic by distinguishing 'explicit' and 'implicit' memory. Explicit memory involves conscious recollection generated by direct efforts to access memories. Tests of explicit memory involve direct inquiries that ask the subject to refer to a specific event of learning or a specific fact. Examples include, 'What were the words on the list you studied?' and 'Which of these two items did you see before?' Typical explicit memory tests include direct measures of recall or recognition of word or picture lists, paired associates, story recall, and most of the common tests of memory that are so performed so poorly by amnesic patients. Explicit memory expression also includes most everyday examples of memory, such as recalling what one had for breakfast this morning or the capital of France. In contrast, implicit memory involves unconscious facilitation of performance of a task consequent to a previous experience with the task or material. Implicit tests of memory involve indirect measures, including an increase in performance speed or a biasing of the choices made during performance of a task that can be solved with the information at hand. Examples of implicit memory tests include learning in motor, perceptual, and cognitive skills, habits, conditioning, and repetition priming, the tests described previously at which amnesic patients usually succeed. Notably none of these tests require the subject to be aware of his memory, or to 'remember', a specific event or fact.

Episodic versus semantic memory

Another related characterization is Tulving's (1983) distinction between 'episodic memory' and 'semantic memory'. Episodic memory contains representations of specific personal experiences that occur in a unique spatial and temporal context. Furthermore, episodic memory involves a re-experiencing of the particular event remembered, what Tulving calls 'time travelling'. By contrast, according to Tulving, semantic memory is one's world knowledge, a vast organization of memories not bound to any specific occasion in which the information was acquired. Some investigators have suggested that the pattern of impaired and spared memory capacities in amnesia can be explained as an impaired episodic memory capacity and intact semantic memory. This view readily accounts for the impairment in day-to-day episodic memory ('What did you have for breakfast?'). In addition, episodic memory is reflected in one's ability to recall or recognize a studied list of words as a unique experience distinct from simply knowing the acquired meanings of the words. In contrast, amnesic subjects perform well on semantic memory tasks, because the memory demands avoid reference to the temporal or spatial context in which the information was acquired, and does not require the subject to refer to the learning experience directly. The episodic-semantic distinction is therefore similar to both Moscovitch's and Schacter's characterizations of successful memory performance whenever amnesic individuals do not have to explicitly 'remember' any previous experience or a newly learned fact. Indeed the episodic-semantic and implicit-explicit views are fully compatible, to the extent that implicit memory tests always and only require semantic memory.

In support of this view, Tulving et al. (1988) described a patient, K.C., an individual with normal intelligence and preserved general knowledge, as well as some general knowledge of his past. However, K.C. does not remember any personal experiences from his life before or after his closed head injury, which involves, but is more extensive than, the hippocampal region. K.C. has gradual accumulated semantic knowledge via methods that reduce interference associated with errors. Similarly, H.M.'s amnesia extends far backwards in time specifically for episodic memories. In characterizations of the quality of the memories he had acquired prior to his operation, Steinvorth et al. (2005) found that the remote memories of H.M. and another amnesic patient lacked the rich qualities of episodic memory. For example, when asked for a specific memory of Christmas, H.M. just described how the holiday is differently celebrated in the South, where he grew up, from how it is in the North, where he later moved. These and other memories lacked specific details of a time and place where a particular event occurred, but correctly included factual semantic information.

A problem with this distinction is that the evidence on spared semantic memory in amnesia is mixed. For example, one study directly addressed the issue of semantic learning by training H.M. on new vocabulary words (Gabrieli et al. 1998). H.M. and normal subjects first studied word definitions for eight 8 novel vocabulary words then were given a recognition test asking them to choose a definition for each word. They also studied synonyms for the words, and were tested on a sentence completion task where they were asked to fill in a blank at the end of a sentence with one of the new words. Normal subjects readily learned the new words by these methods but, despite exhaustive testing, H.M. showed virtually no ability to acquire new semantic knowledge.

On the other hand, there are many examples of successful, highly specific learning by amnesic individuals, even for complex materials. For example, one study examined the learning of computer commands and terminology (Glisky et al. 1986). The training involved very gradual procedures by which the commands were introduced in situations where responses were 'error-free' at each stage. Amnesic individuals showed successful learning for new computer terms by this method, but their newly acquired knowledge could be expressed only in replications of the precise training conditions. Glisky and colleagues referred to this characteristic of their successful learning as a 'hyperspecificity' of the preserved memory. In a related study, Bayley and Squire (2002) trained a patient with damage similar to that of H.M. on three-word sentences repeatedly over 3 months. Each training trial involved presentations of the first two words and memory was assessed by the ability to correctly recognize the target word when given a choice between the third word and an alternative word. The amnesic patient was impaired in learning rate, but gradually improved. However, unlike normal subjects, the amnesic subject failed in recognition when target words were replaced with synonyms. Thus, similar to the Glisky findings, the spared memories in amnesia were hyperspecific to the particular features of the words and lacked the semantic content of the sentences.

H.M. also was able to acquire hyperspecific semantic memories during the course of everyday life. This ability was measured in a study where H.M. was cued with the first names of 35 people who had become famous after the onset of his amnesia (O'Kane et al. 2004). H.M. gave the correct last name for 12 of them (Figure 3.4). For example, when cued with the first name 'Ray', he correctly gave the last name 'Charles'; note that Ray Charles became famous in 1960, 7 years after H.M. became amnesic. H.M. remembered many fewer last names than did normal subjects, but he still showed a surprising extent of intact semantic memory. Even more surprising, when H.M. was cued with additional semantic information, such as the field in which the person was famous, his performance nearly doubled and he could recall some additional semantic information about famous individuals for whom he remembered the last name. This finding suggests that H.M.'s memory was not entirely hyperspecific, but was tied to some general information.

Explicit Name-Stem Completion Task* From O'Kane et al. (2004)

First name given as cue	Last name generated by H.M.	Decade became famous	Correct with no semantic cue?
Elvis	Presley	1950	yes
James	Dean	1950	yes
Billie	Graham	1950	yes
Fidel	Castro	1950	yes
Marlon	Brando	1950	no
Yul	Brenner	1950	no
J.F.	Kennedy	1960	yes
Lyndon	Johnson	1960	yes
Mao	Tse-Tung	1960	no
Martin Luther	King	1960	yes
Pablo	Picasso	1960	no
Ray	Charles	1960	yes
Woody	Allen	1960	yes
Ayatollah	Khomeini	1970	no
Billie Jean	King	1970	yes
Jimmy	Carter	1970	no
Liza	Minelli	1970	no
Mother	Teresa	1970	no
Sophia	Loren	1970	yes
Dorothy	Hammil	1980	no
Margaret	Thatcher[a]	1980	no
Princess	Diane[b]	1980	no
Ronald	Reagan	1980	yes

*When given semantic information and first names as cues, H.M. generated all the last names that he had given under implicit instructions (indicated by yes). In addition, he was able to generate 11 last names of individuals who became famous after the onset of his amnesia, whom he did not generate under implicit instructions.
[a]H.M. first said "Thatchinger," then corrected to "Thatcher."
[b]We gave credit for "Diane," although the correct answer is "Diana."

Figure 3.4 Examples of H.M.'s memory for names of people who became famous after he became amnesic. Performance was assessed by the last name generated when cued with the first name. Reproduced from O'Kane, G., Kensinger, E. A., and Corkin, S. (2004). Evidence for semantic learning in profound amnesia: An investigation with patient H.M. *Hippocampus*, 14(4), 417–25, © 2004, with permission from John Wiley and Sons.

H.M. was also tested for his ability to draw a schematic diagram of the house he had lived in 5 years after his operation (Corkin, 2002). His drawing was remarkable accurate both at the time he was still living in the house and 3 years after he had moved out (Figure 3.5). These findings show the existence of a residual capacity for semantic learning in amnesia, albeit only following an enormous extent of exposure to the materials over a prolonged period. Such limited applicability is not characteristic of our typical rapid learning and subsequent flexible use of semantic information in solving everyday problems across a broad range of one's daily challenges. In addition, there are now several cases of childhood brain injury that result in amnesia for everyday life events, but near normal general world knowledge (Vargha-Khadem et al. 1997). Some of these cases have circumscribed damage to the hippocampus, suggesting specific involvement in episodic memory as compared to deficits in the ability to acquire elaborate semantic memories if the damage extends to the cortical areas surrounding the hippocampus, as is the case in H.M.

Finally, a distinction that has been related to episodic and semantic memory is recollection and familiarity (Yonelinas 2001). Thus, recollection refers to the ability to bring to mind the particular circumstances of a previous event, similar to Tulving's characterization of episodic memory. In contrast, familiarity refers to the knowledge that one has experienced a stimulus recently, but without the ability to recall the specific circumstances. Interestingly, this distinction may be reflected in the pattern of memory loss following selective damage within the hippocampal region (Eichenbaum et al. 2007). Thus, selective damage to the hippocampus itself has been reported to result in a specific deficit in recollection, sparing familiarity (Yonelinas et al. 2002; Vann et al. 2009). Conversely, damage to the perirhinal cortex, a cortical component of the hippocampal region, results in selective loss of familiarity, sparing recollection (Bowles et al. 2007).

Relational memory

Another perspective focuses on differences in the types of information processing that may distinguish declarative and non-declarative memory. Thus, the episodic-semantic memory distinction differentiates forms of memory by their contents, with episodic memory containing contextual information and semantic memory being factual and context-free. However, another perspective involves distinguishing declarative memory as involving the processing of relationships among memories, as Tolman (1932) emphasized in his notion of 'cognitive maps' and Bartlett (1932) emphasized in his characterization of 'schemas' as active organizations of multiple past experiences. Similarly, Cohen and Eichenbaum (1993; also Eichenbaum and Cohen, 2001) proposed that the processing of declarative memories is relational because it involves the organization of memories. According to this view, episodic memory involves associating events with each other and with their spatial and temporal context, as well as other associations (e.g. names and faces). And semantic memory is relational because it involves the integrating many memories into a network of knowledge. By contrast, non-declarative memory is characterized by the reinstantiation of representations of isolated prior experiences without interconnecting them, and therefore lacking such a *relational* organization.

Thus new learning involving either episodic or semantic memory should be affected in amnesia, and indeed this is what is observed. The modest sparing of semantic memories observed in H.M. and other amnesic individuals is viewed as reflecting an intact capacity for acquiring individual, isolated memories for specific associations such as those that characterize repeated short sentences or first and last names. As such, the deficits in learning these associations hyperspecific seem to reflect the absence of a relational structure.

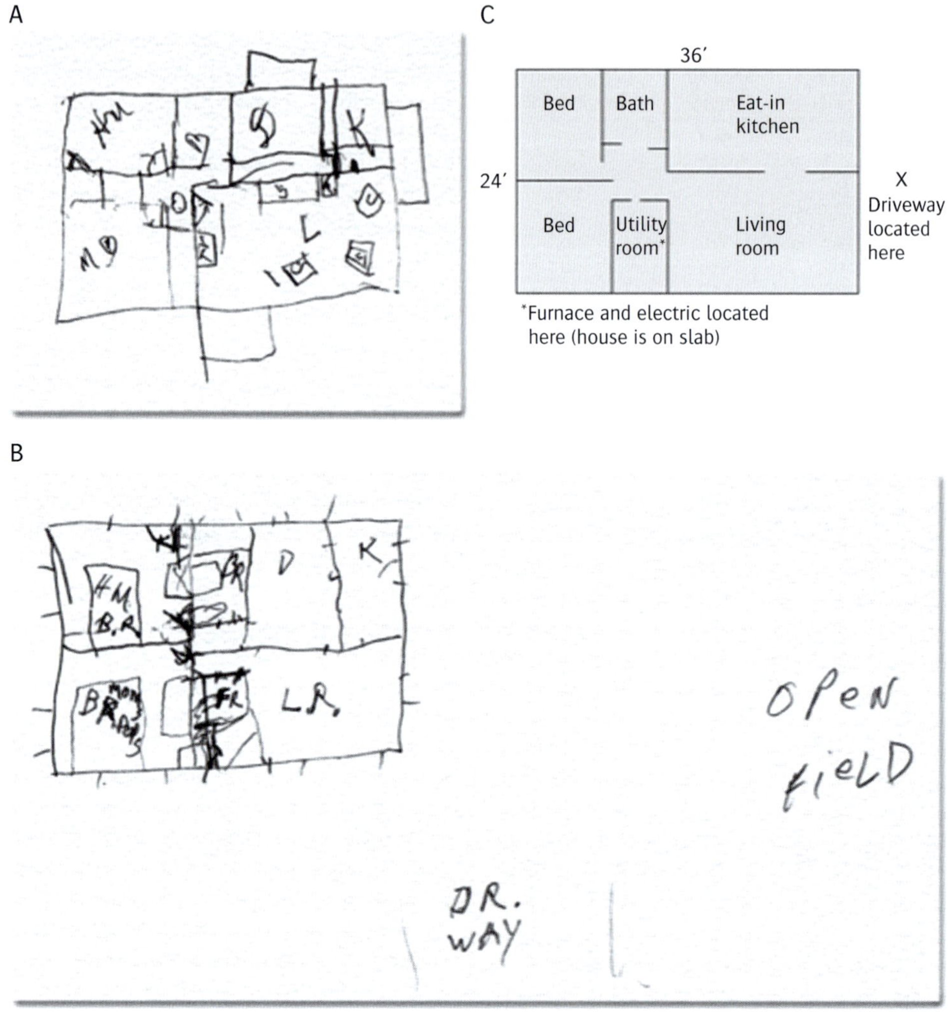

Figure 3.5 Acquisition of a new spatial memory in H.M. A) 1966 drawing of the floor plan of the house where he lived from 1958 (5 years after his operation) to 1974. B) 1977 drawing of the floor plan of the same house, 3 years after he moved to another residence. C) 2002 floor plan of the house. The wall between the kitchen and dining room was taken down by the owner to make an eat-in kitchen. Abbreviations in (A): B, bathroom; C, chair; D, dining room; HM, H.M.'s room; K, kitchen; L, living room; MB, Mom's bedroom; TV, television. Abbreviations in (B): BR, bathroom; D, dining room; FR, furnace room; H.M. B.R., H.M.'s bedroom; K, kitchen; L.R., living room; Mom and Dad's B.R., Mom and Dad's bedroom. Reprinted by permission from Macmillan Publishers Ltd: Corkin, S. (2002). What's new with the amnesic patient H.M.? *Nature Reviews Neuroscience*, 3(2), 153–9, copyright (2002).

The relational memory view is also supported by the studies in which manipulation of cues and task demands require or de-emphasize the need for remembering relations among items in memory. For example, Ryan et al. (2000) exposed subjects to pictures of everyday scenes and monitored the subjects' eye movements as they scanned the scenes. When an exact replica of a scene was re-presented, amnesic individuals spent less time viewing salient objects in the scene, as did normal subjects. This finding provides evidence that a form priming for individual scenes that is spared in amnesia. However, in another condition, the experimenters added, removed, or moved a salient object from where it had appeared in the original scene. Under this condition, normal subjects spent a disproportionate amount of time scanning the area where the scene was changed, whereas amnesic individuals did not (Figure 3.6). These results show that while the normal people represented the spatial relations among objects in the scenes, amnesic subjects encoded the scene as a unified, individual configuration composed of many elements. Subsequent work by Kumaran et al. (2007) has shown that amnesic subjects are impaired in learning both spatial configurations of objects and non-spatial associations between the same objects.

The relational memory view is also supported by studies on verbal memory tasks where associations between

Figure 3.6 Example scene illustrating the relational manipulation effect. Eye movements (black lines) and fixations (white crosses) for two subjects are shown superimposed on the same scene with different viewing histories. A typical-sized critical region is outlined in the box. When the scene was viewed as a repeated scene (left), it was always presented in the same form, and the critical region was always empty of people. Few eye fixations were attracted there. When the scene was viewed as a manipulated scene (right), two girls were in the critical region during study but were removed in the critical block. Although empty in the final block, the critical region attracted many fixations. Reproduced from Ryan, J. D., Althoff, R. R., Whitlow, S., and Cohen, N. J. (2000). Amnesia is a deficit in relational memory. *Psychological Science*, 11, 454-61, copyright © 2000 by Sage Publications. Reprinted by Permission of SAGE Publications.

stimulus elements are required. Giovanello et al. (2003; as well as several other studies; see Eichenbaum, 2004) have reported severe impairment by amnesic patients in learning and remembering associations between arbitrarily paired items, such as pairings of unrelated words, as compared to intact memory for individual words. Notably, in contrast to these results, Manns et al. (2003) have found equal impairments in associative and single item memory, and the situations in which impairment is and is not observed is an active area of investigation. Bird and Burgess (2008) offered a resolution in a study where they reported that amnesic individuals are impaired in memory for common words but not for pictures of faces the subjects had never seen. Perhaps the hippocampal region supports recognition of familiar words by storing memories as related to pre-existing representations of the words in the context of earlier learned associations, whereas familiarity can support recognition of unfamiliar faces as individual representations stored outside the hippocampal region. Taken together, these studies identify a fundamental memory processing function that is lost in amnesia, specifically the ability to represent new events as related to other concurrent or prior events, the kind of relational processing that underlies declarative memory.

Temporally graded retrograde amnesia and memory consolidation

In addition to impairment in long-term storage of new memories, damage to the hippocampal region results in temporally graded retrograde amnesia. Scoville and Milner's (1957) report on H.M. focused on his disorder as a particularly good example of Ribot's law in retrograde amnesia such that his amnesia involves a severe and selective impairment in 'recent memory' in the face of spared remote memory capacities. Interviews with H.M. and his family indicated that the retrograde memory loss reached back 2 years prior to the surgery, with more remotely acquired memories intact. This pattern of temporally graded retrograde impairment has suggested an essential role for the hippocampus in the consolidation of memories. Thus, the temporal gradient is viewed as a combination of the loss of memories that have not had time to consolidate prior to the loss of hippocampal functions and the sparing of memories in which the hippocampus has already completed its role in consolidation.

An extensive formal battery of tests of memory for public and personal events extended the period of impairment back to 11 years prior to the surgery, suggesting the consolidation period is quite prolonged (Milner et al. 1968). Some of these tests evaluated H.M.'s memory for public events, including naming of tunes, verbal recognition of events, or identification of faces that became famous in a particular decade for several years prior to his surgery. For example, a test of recognition for famous events included a series of questions about particularly important public events from the 1940s through the 1970s. H.M., whose surgery was performed in 1953, performed within the normal range of scores for questions about events that occurred in the 1940s, was borderline for events from the 1950s, and was clearly impaired on events from the 1960s onward. Another test focused on H.M.'s memory for personal events (see Steinvorth et al. 2005). In this test, concrete nouns were presented and subjects were asked to relate them to a personal event from any period in their life, and to describe when the event occurred. Normal subjects employed memories from throughout their life span, including predominantly recent memories. By contrast, all of H.M.'s memories dated back to the age of 16 (i.e. 1942) or earlier. Thus, H.M. had no memories of the end of World War II or of his high school graduation (1947), or any other major personal event after that time, consistent with the reports on his retrograde amnesia extending back 11 years prior to his surgery.

Note that this timeframe corresponds with the onset of H.M.'s major seizures, so it is possible that at least some of the retrograde loss might be a result of compromised hippocampal function during the period prior to his surgery. However, other studies on retrograde amnesia have also reported temporally graded memory loss in cases where the deficit is unambiguously a retrograde amnesia. For example, studies on patients receiving electroconvulsive therapy for severe depression examined memory for television programs that were shown in a single viewing season (Cohen and Squire, 1981). These patients were found to have a (temporary) retrograde amnesia extending 1–3 years before the treatments. Other studies have shown temporally graded retrograde amnesia dating 10–20 years backward from an anoxic or ischaemic event (Squire e al. 1989).

In addition, in a recent study of temporally graded retrograde amnesia, Reed and Squire (1998) related the severity of retrograde amnesia to the amount of damage to the medial temporal region. This study involved four patients who had become amnesic without other cognitive impairment following specific brain insult and who had subsequently died. Two patients that developed moderately severe anterograde amnesia following a transient ischaemic event had a selective loss of cells in the CA1 field of the hippocampus and had a very limited retrograde amnesia, extending only 1–2 years. Two other patients had more severe anterograde amnesia and more extensive retrograde amnesia dating 15–25 years. The histopathological findings showed cell loss throughout the hippocampus and in entorhinal cortex. These studies further confirm that damage limited to the hippocampal region can result in temporally limited retrograde amnesia, and that the extent of the temporal gradient of retrograde amnesia might be associated with the anatomical extent of damage within the hippocampal region.

Teng and Squire (1999) examined in detail whether the phenomenon of temporally graded retrograde amnesia extends to spatial memory. This study focused on a patient with extensive hippocampal region damage who had lived in a particular California neighbourhood during the 1930s and 1940s, but had moved away and subsequently returned only occasionally. Spatial memory was evaluated by testing his ability to construct routes between different locations in the community, identified using archival maps. In addition, these tests assessed the ability to plot alternative roundabout routes by asking the subjects to navigate among places when the major route between them was blocked. Teng and Squire also measured accuracy in pointing in the direction of major landmarks from an imagined position in the neighbourhood. On all of these tests, the patient scored at least as well as a group of age-matched control subjects who had lived in the same area during the target period, and who had also subsequently moved away. Control subjects also performed well in solving the same navigational problems based on knowledge about their current neighbourhood, but on these tests of new learning following the brain damage, the patient failed completely. Instead, his pattern of retrograde amnesia for spatial knowledge matched that of temporally-graded non-spatial memory.

Retrograde amnesia without a temporal gradient has been reported in many cases (Moscovitch et al. 2006). These reports involve a variety of aetiologies in which gradients are flat, extending back for to the earliest childhood memories. In addition, other studies have concluded that some retrograde amnesias show a pattern of memory loss characterized less by the age of the memories more by the nature of the memory impairment. These studies suggest that medial temporal lobe damage affects memory for personal episodes more severely than semantic memory, and that this deficit in autobiographical memory extends back many years, and in many cases the entire life span. On the other hand, it has been argued that cases of flat gradients of retrograde amnesia, and particularly those involving autobiographical memory, occur only in cases where there is damage or suspected damage beyond the medial temporal lobe (Bayley et al. 2003). Importantly, amnesia associated with Korsakoff's syndrome, closed head injuries, seizure disorders, and certain other aetiologies often include damage or cell loss in prefrontal cortex. Damage to the prefrontal area is associated with disorders of 'source memory', the ability to recall where and when information was acquired. Thus an inability to identify the circumstances surrounding new learning might be expected to lead to a selective impairment in memory for personal experiences in terms of when and where the events occurred. In these cases, the selectivity of retrograde amnesia to episodic memory would not be related to hippocampal function.

Recent challenges about the defining features of declarative memory and their role in amnesia

Among the hallmark features of amnesia introduced in earlier sections are intact perception, intact short-term memory, and intact cognition outside of memory. Another central feature of declarative memory is the conscious awareness of one's memory. In recent years, each of these features of amnesia has been called into question with regard to involvement of the hippocampal region, and the findings bring new insights into the nature of memory processing by this brain area.

Some studies have suggested that structures in the hippocampal region are involved in perceptual processing of complex objects that include common features. Thus Graham et al. (2010) have shown that amnesic individuals with damage in the hippocampal region are impaired in detecting an 'odd' object among an array of identical objects that differ from the odd one in a single feature. Also, while amnesic subjects perform well in discriminating different faces, there are impaired in discriminating virtual reality scenes composed of slight changes in the contents of the scenes. These findings have led to the proposal that the hippocampal region may be involved in higher-order processing of perceptual configurations.

Hannula et al. (2006) have reported impairments in short-term relational memory for objects that have been manipulated in scenes and relational memory for pairings of faces and scenes. Subsequent work using eye movement analyses showed some reflections of intact memory exist at very short delays (Smith et al. 2006). However, another study using an analysis of eye movement patterns to measure memory for face-scene associations revealed the impairment in amnesic subjects within 500–750 ms (Hannula et al. 2007). A similar impairment appeared within seconds in a test of memory for relations between objects and locations.

Another recent finding is the observation that amnesic individuals are impaired in imagining new experiences. Such a finding seems at odds with the general view that cognition is not impaired in amnesia. However, Hassabis et al. (2007) found that imagining the future was impaired in amnesic individuals. In these studies, subjects were asked to construct new imagined experiences based on simple cues about the nature of those experiences. Whereas normal subjects could readily create rich scenarios for everyday events, amnesic individuals could only produce simple non-specific scenarios. Thus, for example, when asked to imagine lying on a white sandy beach on a beautiful day, normal subjects imagined an elaborate arrangement of objects on the beach, a boat moving by, and personal feelings of the sun and sand. By contrast, an amnesic patient could see only blue sky and white sand without an elaborate array of objects or events contained in the scene.

While amnesia is most typically observed in tests that involve awareness of one's memory, there are now important observations of memory impairment even for memory for which normal subjects are not consciously aware. A particularly striking example is the study by Ryan et al. (2000) where eye movement patterns were used to measure memory. As described earlier, in that experiment subjects study scenes and then were shown the same scenes but with an object added, removed, or moved in the scene. Normal subjects' eye movement patterns reflected relational memory in disproportional viewing of the altered parts of the scenes. Notably, this relational manipulation effect was observed only in the absence of awareness of the change, even though it was this effect that was eliminated in amnesic subjects.

These results seem to call into question key features of episodic and explicit memory. However, each of these findings can be understood from the perspective of the relational processing hypothesis. If the hippocampal region encodes relations among events, then damage to that region would be expected to result in deficits in encoding as well as short-term memory for relations among percepts, as observed in the studies on perception of complex scenes and short term memory for relational materials. Furthermore, to the extent that imagining future scenarios typically involves novel rearrangement of remembered events and their relations from previous experiences, this ability should be affected in amnesia, as observed. Finally, if conscious awareness of the memories is an emergent phenomenon that usually accompanies, but is not essential to relational processing, then it should be possible to observe relational memory deficits in amnesia, even in situations where memories do not reach consciousness. Thus these observations improve our understanding of amnesia as an impairment in relational processing that begins at the outset of encoding, is important to the organization of short-term as well as long-term memory storage, reflects the inherent relational organization of memories even without conscious awareness, and can be called upon in cognitive functions such as imagining future events.

References

Bartlett, F. C. (1932). *Remembering*. London: Cambridge University Press.

Bayley, P. J. and Squire, L. R. (2002). Medial temporal lobe amnesia, Gradual acquisition of factual information by nondeclarative memory. *J Neurosci*, 22, 5741–8.

Bayley, P. J., Hopkins, R. O., and Squire, L. R. (2003). Successful recollection of remote autobiographical memories by amnesic patients with medial temporal lobe lesions. *Neuron*, 38, 135–44.

Bird, C. M. and Burgess, N. (2008). The hippocampus supports recognition memory for familiar words but not unfamiliar faces. *Curr Biol*, 18, 1932–6.

Bowles, B., Crupi, C., Mirsattari, S. M., *et al.* (2007). Impaired familiarity with preserved recollection after anterior temporal-lobe resection that spares the hippocampus. *Proc Natl Acad Sci of the USA*, 104, 16382–7.

Clark, R. E. and Squire, L. R. (1998). Classical conditioning and brain systems: The role of awareness. *Science*, 280, 77–81.

Cohen, N. J. (1984). Preserved learning capacity in amnesia: Evidence for multiple memory systems. In Butters, N. and Squire, L. R. (Eds.) *The Neuropsychology of Memory*, pp. 83–103. New York: Guilford Press.

Cohen, N. J. and Eichenbaum, H. (1993). *Memory, Amnesia, and the Hippocampal System*. Cambridge, MA: MIT Press.

Cohen, N. J. and Squire, L. R. (1980). Preserved learning and retention of a pattern- analyzing skill in amnesia: Dissociation of knowing how and knowing that. *Science*, 210, 207–10.

Cohen, N. J. and Squire, L. R. (1981). Retrograde amnesia and remote memory impairment. *Neuropsychologia*, 19, 337–56.

Corkin, S. (1968). Acquisition of a motor skill after bilateral medial temporal lobe excision. *Neuropsychologia*, 6, 225–65.

Corkin, S. (1984). Lasting consequences of bilateral medial temporal lobectomy: Clinical course and experimental findings in H.M. *Sem Neurol*, 4, 249–59.

Corkin, S. (2002). What's new with the amnesic patient HM. *Nature Rev Neurosci*, 3, 153–9.

Corkin, S., Amaral, D. G., Gonzalez, R. G., Johnson, K. A., and Hyman, B. T. (1997). H. M.'s medial temporal lobe lesion: Findings from magnetic resonance imaging. *J Neurosci*, 17, 3964–3979.

Duff, M. C., Hengs, J., Tranel, D., and Cohen, N. J. (2006). Development of shared information in communication despite hippocampal amnesia. *Nature Neurosci*, 9, 140–6.

Eichenbaum, H. (2004). Hippocampus: Cognitive processes and neural representations that underlie declarative memory. *Neuron*, 44, 109–20.

Eichenbaum, H. & Cohen, N. J. (2001). *From Conditioning to Conscious Recollection: Memory Systems of the Brain*. New York: Oxford University Press.

Eichenbaum, H., Yonelinas, A. R., and Ranganath, C. (2007). The medial temporal lobe and recognition memory. *Annu Rev Neurosci*, 20, 123–52.

Giovanello, K. S., Verfaellie, M., and Keane, M. M. (2003). Disproportionate deficit in associative recognition relative to item recognition in global amnesia. *Cogn Affect Behav Neurosci*, 3, 186–94.

Glisky, E.L., Schacter, D.L., and Tulving, E. (1986). Learning and retention of computer-related vocabulary in memory-impaired patients: Method of vanishing cues. *J Clin Exp Neuropsychol*, 83, 292–312.

Gabrieli, J. D. E., Cohen, N. J., and Corkin, S. (1988). The impaired learning of semantic knowledge following bilateral medial temporal-lobe resection. *Brain Cogn*, 7, 157–77.

Gabrieli, J. D. E., Milberg, W., Keane, M., and Corkin, S. (1990). Intact priming of patterns despite impaired memory. *Neuropsychologia*, 28, 417–27.

Graf, P., Squire, L. R., and Mandler, G. (1984). The information that amnesic patients do not forget. *J Exp Psychol: Learn Mem Cogn*, 10, 164–78.

Graham, K.S., Barense, M. D., and Lee, A. C. H. (2010). Going beyond LTM in the MTL: a synthesis of neuropsychological and neuroimaging findings on the role of the medial temporal lobe in memory and perception. *Neuropsychologia*, 48, 831–53.

Hannula, D. E., Tranel, D., and Cohen, N. J. (2006). The long and the short of it: Relational memory impairments in amnesia, even at short lags. *J Neurosci*, 26, 8352–9.

Hannula, D. E., Ryan, J. D., Tranel, D., and Cohen, N. J. (2007). Rapid onset relational memory effects are evident in eye movement behavior, but not in hippocampal amnesia. *Journal of Cognitive Neuroscience*, 19, 1690–705.

Hassabis, D., Kumaran, D., Vann, S. D., and Maguire, E. A. (2007). Patients with hippocampal amnesia cannot imagine new experiences. *Proc Natl Acad Sci U S A*, 104, 1726–31.

Knowlton, B. J., Ramus, S., and Squire, L. R. (1992). Intact artifical grammar learning in amnesia. *Psychological Sci*, 33, 172–9.

Kumaran, D., Hassabis, D., Spiers, H. J., Van, S. D., Vargha-Khadem, F., and Magurie, E. A. (2007). Impaired spatial and non-spatial configural learning in patients with hippocampal pathology. *Neuropsychologia*, 45, 2699–711.

Manns, J. R., Hopkins, R. O., Reed, J. M., Kitchener, E. G., and Squire, L. R. (2003). Recognition memory and the human hippocampus. *Neuron*, 37, 171–80.

McDonough, L., Mandler, J. M., McKee, R. D., and Squire, L. R. (1995). The deferred imitation task as a nonverbal measure of declarative memory. *Proc Natl Acad Sci U S A*, 92, 7580–4.

Milner, B., Corkin, S., and Teuber, H. L. (1968). Further analysis of the hippocampal amnesic syndrome: 14-year follow-up study of H.M. *Neuropsychologia*, 6, 215–34.

Moscovitch, M. (1984). The sufficient conditions for demonstrating preserved memory in amnesia: A task analysis. In Butters, N. and Squire, L. R. (Eds.) *The Neuropsychology of Memory*, pp. 104–14. New York: Guildford Press.

Moscovitch, M., Nade, L., Winocur, G., Gilboa, A., and Rosenbaum, R. S. (2006). The cognitive neuroscience of remote episodic, semantic and spatial memory. *Curr Opin Neurobiol*, 16, 179–90.

Musen, G., Shimamura, A. P., and Squire, L. R. (1990). Intact text-specific reading skill in amenesia. *J Exp Psychol: Learn Mem Cog*, 6, 1068–76.

Nissen, M. J. and Bullemer, P. (1987). Attentional requirements of learning: evidence from performance measures. *Cog Psychol*, 19, 1–32.

Ogden, J. A. and Corkin, S. (1991). Memories of H.M. In Abraham, W.C., Corballis, M., and White, K.G. (Eds.) *Memory Mechanisms: A Tribute to G.V. Goddard*, pp. 195–215. Hillsdale, NJ: Lawrence Erlbaum.

O'Kane, G., Kensinger, E. A., and Corkin, S. (2004). Evidence for semantic learning in profound amnesia: An investigation with patient HM. *Hippocampus*, 14, 417–25.

Reed, J. M. and Squire, L.R. (1998). Retrograde amnesia for facts and events: Findings from four new cases. *J Neurosci*, 18, 3943–54.

Ryan, J. D., Althoff, R. R., Whitlow, S., and Cohen, N. J. (2000). Amnesia is a deficit in relational memory. *Psychological Science*, 11, 454–61.

Salat, D. H., van der Kouwe, A. J. W., Tuch, D. S., *et al.* (2006). Neuroimaging H.M.: A 10-year follow-up examination. *Hippocampus*, 16, 936–45.

Schacter, D. L. (1985). Multiple forms of memory in humans and animals. In Weinberger, N. M., McGaugh, J. L., and Lynch, G. (Eds.) *Memory Systems of the Brain*, pp. 351–80. New York: Guilford Press.

Scoville, W. B. and Milner, B. (1957). Loss of recent memory after bilateral hippocampal lesions. *J Neurol Neurosurg Psychiat*, 20, 11–12.

Smith, C. N., Hopkins, R. O., and Squire L. R. (2006). Experience-dependent eye movements, awareness, and hippocampus-dependent memory. *J Neurosci*, 26, 11304–12.

Squire, L. R., Haist, F., and Shimamura, A. P. (1989). The neurology of memory: quantitative assessment of retrograde amnesia in two groups of amnesic patients. *J Neurosci*, 9, 828–39.

Squire, L. R., Knowlton, B., and Musen, G. (1993). The structure and organization of memory. *Ann Rev Psychol*, 44, 453–95.

Steinvorth, S., Levine, B., and Corkin, S. (2005). Medial temporal lobe structures are needed to re-experience remote autobiographical memories: evidence from H.M. and W.R. *Neuropsychologia*, 43, 479–96.

Teng, E. and Squire, L. R. (1999). Memory for places learned long ago is intact after hippocampal damage. *Nature*, 400, 675–7.

Tolman, E. C. (1932). *Purposive Behavior in Animals and Men*. Berkeley, CA: University of California Press. (1951 edition).

Tulving, E. (1983). *Elements of Episodic Memory*. New York: Oxford University Press.

Tulving, E., Schacter, D. L., McLachlin, D. R., and Moscovitch, M. (1988). Priming of semantic autobiographcal knowledge:A case study of retrograde amnesia. *Brain and Cognition*, 8, 3–20.

Vann, S. D., Tsivilis, D., Denby, C. E., *et al.* (2009). Impaired recollection but spared familiarity in patients with extended hippocampal system damage: convergence across three methods. *Proc Natl Acad Sci*, 106, 5442–7.

Vargha-Khadem, F., Gadin, D. G., Watkins, K. E., Connelly, A., Van Paesschen, W., and Mishkin, M. (1997). Differential effects of early hippocampal pathology on episodic and semantic memory. *Science*, 277, 376–80.

Yonelinas, A. P. (2001). Components of episodic memory: The contribution of recollection and familiarity. *Philos Trans R Soc Lond B Biol Sci*, 356, 1363–74.

Yonelinas, A. P., Kroll, N. E. A., Quamme, J. R., *et al.* (2002). Effects of extensive temporal lobe damage or mild hypoxia on recollection and familiarity. *Nature Neurosci*, 5, 1236–41.

CHAPTER 4

Selective Neuronal Vulnerability in the Hippocampus: Relationship to Neurological Diseases and Mechanisms for Differential Sensitivity of Neurons to Stress

Elias K. Michaelis

Importance of the hippocampus in learning, memory, and neurological diseases 55
The concept of selective neuronal vulnerability in the central nervous system 55
Focus on selective neuronal vulnerability in the hippocampus 56
Mechanisms of selective neuronal vulnerability 59
Concluding ideas 70

ABSTRACT The phenomenon of selective neuronal vulnerability to neurological diseases or insults, such as ischemia-reperfusion, brain trauma, and ageing-associated neurodegeneration, was described in the hippocampus nearly a century ago. The hippocampus has been the focus for many of the studies designed to identify the stresses to which neurons respond differentially, as well as the molecular, cellular and physiological mechanisms that may be responsible for such differential response patterns. Although we have not yet reached final conclusions on the mechanisms for differential neuronal responses to stress and disease, genomic and proteomic analyses, in conjunction with biochemical and physiological measurements, are beginning to point to endogenous differences between vulnerable and resistant populations of neurons. Collectively, these studies have identified differential patterns of energy metabolism, generation of reactive oxygen species, handling of transient intracellular calcium elevations, and the activity of the neurotransmitter glutamate, as some of the key processes that lead to selective sensitivity to neurological stresses. The genomic and proteomic analyses have revealed that differences in gene expression related to inflammatory and immune responses and responses to oxidative stress represent endogenous processes differentially expressed in neurons selectively vulnerable to stresses. The same is true for genes involved in DNA, RNA, and protein repair. Cells that are more resistant to stresses and disease conditions express genes for energy generation, nervous system development, and synaptic transmission at higher levels than those found in vulnerable neurons. The molecular pathways identified above and described in this chapter may eventually determine the targets for future therapeutic interventions for neurodegenerative diseases.

KEYWORDS CA1/CA3 neurons, ischaemia/reperfusion, ageing, neurodegenerative diseases, energy metabolism, oxidative stress, calcium regulation, glutamate transmission, genomic analyses

Importance of the hippocampus in learning, memory, and neurological diseases

The hippocampus is a relatively small region of the brain yet it has been, and continues to be, the focus of extensive research into the early neuronal development in this structure, the lifelong capacity to produce new neurons, the precise morphology and connectivity of the mature neurons in subregions of this brain structure, and the genetic, biochemical, and physiological characteristics of neurons and glial cells in this region. The reason for such intensive investigation of the hippocampus rests with several facts: it has a precise laminar organization of its key neuronal elements, the connections between or among different populations of neurons can be traced precisely, its genetic and biochemical characteristics can be studied at near microscopic levels, and the physiological properties of the neurons and glial cells in this brain structure can be precisely defined. But, the intrinsic property of the hippocampus that has attracted the detailed scrutiny of neuroscientists is the fact that this structure, in human as well as other animal brains, is involved in contextual temporal-spatial memory formation and retrieval (Squire 1992; Squire et al. 1992) and that damage to this brain region as a result of neurodegenerative diseases or psychiatric disorders, or simply by the ageing process, can lead to deficits in the cognitive functions outlined above (Zola-Morgan et al. 1992; Rajah et al. 2010).

The concept of selective neuronal vulnerability in the central nervous system

Almost a century ago, the neuropathologist Walther Spielmeyer (1931) described the phenomenon of selective neuronal damage in certain regions of the central nervous system (CNS) that were associated with neurological diseases. He stated that 'When we pathologists try to interpret our microscopic pictures, we are confronted with the problem of explaining the process which causes the localization and spread of special pathological lesions ... We see that even in general diffuse spreading there are special localities where the process is most active and others that remain strikingly free'. The term 'selective vulnerability' was used by Meyer (1936) to describe the presence of selective lesions in conditions such as general paralysis, Pick's disease, epidemic encephalitis, Wilson's disease, and senile dementia. At approximately the same period of time, the neuropathologists C. and O. Vogt (Vogt and Vogt 1937) described the localized pattern of neuronal damage observed after ischaemic and hypoxic episodes as 'topistic' neuronal injury, a term that remained in use for some time (Scholz 1953).

Regardless of the term used to describe the lesions that occur in select areas of the brain in neurological diseases, the key feature recognized by the pioneering work of individuals such as Spielmeyer, Meyer, and C. and O. Vogt, was that certain neuronal populations were differentially vulnerable to the injurious effects of certain disease states. Although all of these neuropathologists documented, meticulously, the phenomenon of selective neuronal vulnerability in postmortem brain and drew attention to the concept of selective vulnerability, they differed significantly on what they considered to be the underlying processes responsible for such differential sensitivity to injury among diverse populations of neurons. For example, with respect to the select lesions observed in the hippocampus of cases of epilepsy, the differential damage was attributed to the peculiarities in the vascularization of subregions of the hippocampus, such as the CA1 field (Scholz and Hager 1956). As described by Meyer (1936), 'Spielmeyer believed that at the root of the lesion, there were often vasomotor disturbances'. This was not a far-fetched supposition to explain such differential vulnerabilities in the hippocampus as ischaemia and hypoxia also damage selectively the same region of the hippocampus as does epilepsy. On the other hand, Vogt (1925) proposed the idea of 'pathoclisis', i.e. the concept that selective neuronal vulnerability results from 'physio-chemical' differences in response to a particular disease or stress among certain populations of neurons. The latter concept of the mechanism for selective neuronal vulnerability is much closer to our current ideas about the causation of this phenomenon, but it was a very difficult concept to support with solid experimental evidence at the time that it was proposed.

Selective neuronal vulnerability related to various neurological diseases has been documented repeatedly since the time of the early descriptions of the phenomenon about a century ago. For example, it is now well demonstrated that select populations of neurons in the entorhinal cortex, hippocampus, frontal cortex, and amygdala are the most sensitive to the neurodegeneration associated with Alzheimer's disease (AD) and that this distribution of neuronal damage can be used to stage the disease process (Hyman et al. 1984; Braak and Braak 1991). In Parkinson's disease (PD), neuronal degeneration appears to start in the medulla oblongata and pontine tegmentum, spreads to the midbrain where it affects the dopaminergic neurons of the *substantia nigra* and, in the advanced stages of the disease, it damages neurons in the prefrontal cortex and primary sensory/motor fields (Hirsch et al. 1988; Damier et al. 1999; Braak et al. 2003; Dauer and Przedborski 2003; Braak et al. 2004). The neuronal damage in prefrontal cortex and sensory/motor cortex may account for some of the information processing and visual/spatial difficulties experienced by patients with PD. Similarly, the cognitive difficulties that characterize Huntington's disease (HD) patients, have been attributed to neuronal lesions in the frontal cortex and hippocampus, whereas the abnormal movements manifested by patients with HD are the result of neostriatal neuronal damage (Spargo et al. 1993). And, in amyotrophic lateral sclerosis (ALS), while the motor components of the disease are the result of, primarily, the accumulation of inclusion bodies and hyaline aggregates in spinal motor neurons, resulting in motor neuron death and astrocyte gliosis, the cognitive deterioration seen in some ALS cases is associated with damage in neurons of the prefrontal cortex and dentate gyrus (DG) (Ince et al. 1998; Rowland and Shneider 2001). The very selective nature of neuronal injury in ALS can best be demonstrated by the fact that adjacent neuronal populations of motor neurons in the brainstem have very divergent fates as a result of the disease, with hypoglossal, trigeminal, and facial motor neurons being damaged, while occulomotor, abducens, and trochlear motor neurons having neither intraneuronal accumulations nor suffering any damage (Ince et al. 1998).

The example of nearby populations of neurons with generally similar characteristics and functions, such as the brainstem motor neurons, yet greatly divergent responses to disease-induced cell damage, is a recurrent theme of the process of selective neuronal vulnerability. Other such examples include the differential damage to pyramidal neurons in entorhinal and hippocampal subregions, or those of the different cerebral cortical layers, in diseases such as AD and epilepsy. Selective neuronal vulnerability in the hippocampus and DG has been studied extensively, starting with the early descriptions of the neuropathologists Spielmeyer, Meyer, and Vogt. Such differential sensitivity to disease or stress in the hippocampus can be rather dramatic and easily observed and quantified. Furthermore, selective vulnerability of neurons in the hippocampus where CA1 pyramidal neurons are readily injured or die under conditions that do not affect, for the most part, the adjoining CA3 pyramidal neurons or dentate granule cells, is a phenomenon that has been observed in various neurological diseases. Examples of such subregion selective vulnerability is characteristic of hippocampal sclerosis associated with cerebral ischaemia, neuronal accumulation of neurofibrillary tangles in early stages of AD, and the cell loss that occurs in chronic epileptic seizures, ageing, HD, oxidative stress (OS), anoxia, hypoglycaemia, acquired immunodeficiency syndrome, psychiatric diseases, and traumatic brain injury.

Focus on selective neuronal vulnerability in the hippocampus

The hippocampal formation consists of the histologically identified four subfields of the cornu ammonis (CA), designated as CA1–4, the field of the DG, and the associated hilum. These regions are shown nicely in Figure 4.1 from a drawing of Ramón y Cajal (Cajal 1894). The neuronal populations in these subfields of the hippocampal formation are interconnected and the output of this formation leads to information transfer to other regions of the brain. In terms of input to the hippocampal formation, the entorhinal cortex is of particular importance in both normal functional connectivity to the hippocampus and in the progression of neuronal injury and loss in some neurodegenerative diseases. The differential sensitivity to injury and cell death in the hippocampus will be described in the next section for several of these conditions or disease states. Studies in human brain, as well as those performed with experimental animals, will be summarized in order to demonstrate the generality of the phenomenon of differential vulnerability among hippocampal neurons.

Ischaemia-, hypoxia-, and epilepsy-induced selective brain injury

The cell damage resulting from ischaemia/hypoxia, together with the injury observed in association with certain types of epilepsy, is one of the two earliest examples of selective neuronal vulnerability in the hippocampus. In cases of ischaemia, the CA1 pyramidal and CA4 hilar neurons of the hippocampus are the most sensitive to post-ischaemia neuronal injury, whereas the CA2–3 pyramidal neurons are less sensitive, and the granule neurons of the DG are the least vulnerable (Pulsinelli et al. 1982; Siesjo 1988; Schmidt-Kastner and Freund 1991; Olsson et al. 2003). The differential sensitivity of CA1 and CA4 neurons is best manifested in the early period after a brief episode of focal ischaemia, but it has also been well documented after more prolonged periods of global ischaemia and re-perfusion in both human and experimental animal studies (Diemer and Siemkowicz 1981; Fryd Johansen et al. 1983; Pulsinelli 1985; Winkelmann et al. 2006; DeGracia et al. 2007).

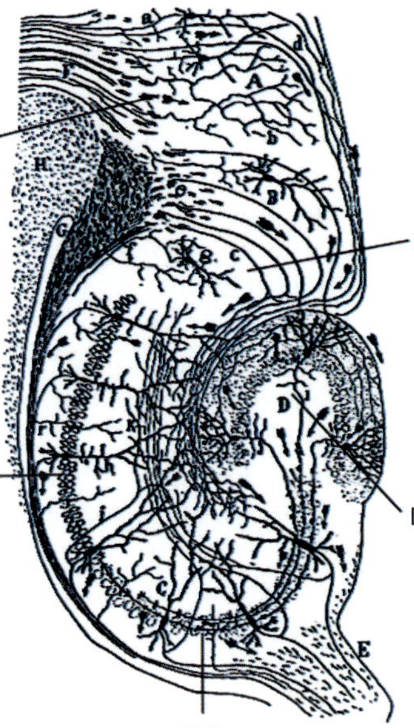

Figure 4.1 Drawing by Cajal (Cajal 1894) of the hippocampus, dentate gyrus (DG), subiculum, and proximal entorhinal cortex. Reproduced from Cajal S.R.Y., *Les nouvelles idées sur la structure du système nerveux chez l'homme et chez les vertébrés* © 1894, C. Reinwald.

Different types of ischaemia have been studied in experimental animals and humans, such as transient occlusion of the carotid or cerebral arteries, human stroke episodes, and cardiac arrest. Cardiac arrest, a common occurrence among people throughout the world, leads to ischaemia and hypoxia of the brain and can produce devastating effects on cognition, memory, and emotional behaviour. The resultant lesions in the brain reveal, once again, the relative sensitivity of the hippocampus, among other brain regions, and in particular of the CA1 region, to the length of exposure to ischaemia and anoxia of the brain (Horn and Schlote 1992; Miclescu et al. 2010; Sharma et al. 2011). Of importance with regard to the effects of cardiac arrest is the observation that following cardiac arrest, there is a period of minimal cell injury and death in the hippocampus and that deterioration of neuronal structure, loss of expression of neuron-specific enolase, and cell death occur after a period of 48–96 h post-cardiac arrest in human (Horn and Schlote 1992) and rat and gerbil hippocampus (Kirino 1982; Kirino et al. 1984). Such delayed neuronal death following episodes of ischaemia has been of fundamental importance in the pursuit of a better understanding of the causes of selective neuronal vulnerability because it offers a time window that might allow for appropriate therapeutic intervention.

Knowing more details about what distinguishes CA1 pyramidal neurons from, for example, CA3 pyramidal neurons in terms of their susceptibility to hypoxia, ischaemia, or other cell injurious conditions, could lead to the identification of appropriate targets for therapeutic intervention. And, delayed injury of even ischaemia-sensitive neurons of the CA1 might work to the advantage of those who attempt to intervene in time to save such neurons from further destruction. As indicated in the preceding section, two broad theories were developed early on to account for differential sensitivities of hippocampal neurons to ischaemia or epilepsy: that of peculiar vascularization and that of 'pathoclisis'. As will be described in subsequent sections, these theories have been put to the test using the newest investigational tools currently available.

The reason for inserting seizure-induced selective neuronal injury within the discussion of ischaemia-associated hippocampal injury is based on the fact that the early studies by Spielmeyer (1927) and Scholz (Scholz and Hager 1956) reported on the analogies between the lesions observed in cases of epilepsy and those of brain ischaemia, especially in the hippocampus. Hippocampal sclerosis has been clearly shown in many subsequent studies and is referred to as 'the most common single lesion to be found post mortem in the brains of chronic epileptics who die a natural death' (Falconer 1974). Hippocampal sclerosis is characterized by the loss of neurons in select subfields of the hippocampus, such as the CA1 and CA4 region, dispersion and cell death of granule cells in the DG, and hippocampal gliosis. These neuropathological characteristics are either causally linked to temporal lobe epilepsy or are consequences of the epileptic seizures (Sagar and Oxbury 1987; Blümcke et al. 2009; Meencke 2009). Given the fact that similar neuronal losses and disruptions, together with proliferation of astrocytes and microglia, are observed in experimental animals following induction of seizures by drugs (Falconer and Taylor 1968; Jorgensen et al. 1993; Mello et al. 1993), it appears that these manifestations of seizure activity are the consequences of abnormal electrical activity. Prevention of anoxia by assisted ventilation of experimental animals does not prevent the most characteristic lesion of seizures in the hippocampus, the loss of neurons in the CA1 region (Meldrum et al. 1973a, 1973b). But, that does mean that there is no connection between tissue blood perfusion and the severity of epileptic seizures or the damage to the hippocampus that may ensue. There is clinical and neuropathological evidence that hippocampal sclerosis resulting from ischaemia or hypoxia can lead to temporal lobe epilepsy (Earle et al. 1953). The observations described here are consistent with the idea that selective neuronal damage in the hippocampus can be both the cause, as well as the result, of epileptic seizures (Falconer and Taylor 1968), and that there is a relationship between ischaemia and seizure diathesis.

Ageing-associated selective neuronal vulnerability in the hippocampus

Ageing is an important component of several neurodegenerative conditions, such as AD and PD, and is, therefore, the biological substrate upon which the selective neuronal vulnerability that is observed in these conditions takes place. Impaired ability to form new memories is a symptom of normal ageing in humans and experimental animals (Barnes 1979; Barnes and McNaughton 1980, 1985; Crook et al. 1986; Uttl and Graf 1993; Barnes et al. 1997; Rosenzweig et al. 1997), and this phenomenon has led investigators to focus their attention to areas that are involved in memory formation and cognitive function, that is the hippocampus and frontal cortex. Neuronal loss in the cortex has been linked to memory deficits in aged primates (Smith et al. 2004) and, although neuronal death in the human cortex is not extensive, large neurons appear to be the most vulnerable to age-related degeneration (Terry et al. 1987; Masliah et al. 1993). An issue that has been addressed in several studies is whether ageing, in the absence of disease, is characterized by selective loss of neurons in different brain regions (Wisniewski and Terry 1973; Terry et al. 1987; Masliah et al. 1993; Simic et al. 1997; Terry and Katzman 2001; La Joie et al. 2010). A general characteristic of the ageing brain in human subjects is the decrease in the volume of the hippocampus as assessed by high resolution magnetic resonance imaging (MRI) (Mueller et al. 2007; La Joie et al. 2010) and, especially, of the CA1 subfield and the subiculum (Mueller et al. 2007; La Joie et al. 2010). The decreases in hippocampal volume and in the volume of the CA1 subfield during the ageing process correlated with statistically significant decreases in the total number of cells in this region of the hippocampal formation (Simic et al. 1997). Although there are some disagreements with regard to the loss of volume in the CA1 region (Mueller et al. 2007; La Joie et al. 2010), there is a general agreement on the overall shrinkage of the hippocampus with advancing age and of the selective loss of neurons and volume in subregions of the hippocampus, such as the subiculum and/or the CA1 subfield.

In addition to the loss of neurons in select brain regions during ageing, there is a well documented loss of dendrites and presynaptic axonal terminals, both of which may contribute to the loss of volume in these regions. Ageing in human and macaque monkey brain leads to a linear, age-dependent decrease in stereologically measured total synapses, dendritic spines, and dendrite length and branching in neurons of the neocortex (Terry et al. 1987; Masliah et al. 1993; Morrison and Hof 1997; Duan et al. 2003). Such changes in synaptic density and dendrite arborizations may correlate well with the observed decrements in memory and cognitive function with advancing age observed in both human subjects and experimental animals.

AD and selective neuronal vulnerability within the hippocampus

The elegant studies of Braak and Braak (1991), focused on staging of AD pathology, demonstrated that the disease as defined by intracellular deposition of neurofibrillary tangles, starts in the transentorhinal layer of the entorhinal cortex in stages I–II and spreads to the CA1 subfield of the hippocampus in stages II–IV. These observations have been confirmed and expanded in various ways in efforts to link neuroanatomical changes to functional and molecular abnormalities associated with the disease. With respect to the hippocampus, there is a disease-related progressive loss of CA1 neurons in AD that exceeds that which is due to the ageing process (Simic et al. 1997). The volume changes detected on the basis of microanatomical studies of the hippocampus have been supported by MRI studies in human subjects with AD and have been correlated with the progression of the disease, including the deterioration of mental function (Jack et al. 2000; Petersen et al. 2000; Xu et al. 2000; Barnes et al. 2009)

Although the deposition of extracellular amyloid beta (Aβ) plaques in the various brain regions was not found to correlate with the progression of the disease (Braak and Braak 1991), the intracellular accumulation in CA1 pyramidal neurons of the primary toxic peptide of AD, Aβ42, is markedly increased in cases of sporadic AD as compared with normal age-matched controls (Aoki et al. 2008). This accumulation of Aβ42 in pyramidal neurons of the CA1 subfield of the hippocampus may be correlated with the early pathological changes in this population of selectively vulnerable neurons in the hippocampus. Similarly, the accumulation of enlarged endosomes and the expression of genes related to early endosome dysfunction in CA1 neurons, are some of the earliest signs of molecular and subcellular pathology in these neurons in early stages of AD (Cataldo et al. 2004; Ginsberg et al. 2010).

But, despite the emphasis on CA1 neurons as being selectively vulnerable to AD as shown by their fate during the early stages of AD, it is quite likely that when more sensitive methods for detecting early molecular and subcellular changes in neuronal populations are applied, neurons in the brain regions that will be ultimately affected by the disease may reveal signs of neuropathology early in AD, even neurons that are considered to be relatively resistant. For example, loss of synapses in the DG outer granule cell layer occurs early in individuals with relatively mild signs of AD, that is those with mild cognitive impairment (Scheff and Price 2003; Scheff et al. 2006). Although the DG is a region that is usually less vulnerable to the disease and suffers neuronal losses at later stages of AD, synaptic loss occurs early in the disease in the DG. This example focuses the issue on how we define selective neuronal vulnerability in the age of sophisticated techniques that can be used to probe the molecular composition and subcellular integrity of neurons.

Selective neuronal vulnerability in hippocampus associated with other neurodegenerative diseases

Although some neurodegenerative diseases have been known to be related to neuronal losses and dysfunction in regions other than the hippocampus (for example, the striatum in HD), nevertheless the hippocampus also appears to be involved in such disease states and may account for some of the cognitive decline observed in these diseases. In the case of HD, neuronal counts from different fields of the hippocampus obtained using the optical dissector method revealed a selective and significant loss of neurons from the CA1 subfield that equalled 35% decreases in neurons of HD patients when compared with normal controls (Spargo et al. 1993). A parallel group of patients with human immunodeficiency virus 1 infection did not show such differential neuronal loss in the hippocampus despite evidence of viral encephalitis associated with abnormal cognitive function. Selective neuronal vulnerability in select brain regions has also been observed in mouse models of human HD, in which the CA1 and CA3 subfields of the hippocampal formation suffered neuronal losses (Reddy et al. 1998). These HD mice over-express the huntingtin gene which contains excess numbers of the disease-causing nucleotide triplet (CAG) repeats. But, despite the widespread expression of this gene, only select neuronal populations appear to suffer from the damage related to the accumulation of polyglutamine-containing huntingtin protein (Reddy et al. 1999).

One of the neurodegenerative disease entities that was described early on as leading to selective neuronal accumulation of neurofibrillary tangles and exhibiting granulovacuolar degeneration of select populations of neurons is Pick's disease. Although this disease affects several regions of the brain, the hippocampus and DG are two regions that contain a high level of accumulation of granules (argyrophilic inclusions) characteristic of the disease (Constantinidis et al. 1974). Specifically, the CA1 subfield, the subiculum, and the entorhinal cortex seem to be the most affected by the disease as determined by the high accumulation of inclusions, the Pick bodies (Ball 1979). And, since the pattern of selective neuronal loss in Pick's disease also appears to be somewhat similar to that of AD, it is not surprising that Pick's disease has some of the same characteristics of intellectual decline as seen in AD.

Selective neuronal vulnerability in hippocampus associated with neuropsychiatric disorders

Despite the fact that a large-scale meta-analysis of 18 studies of the MRI patterns of individuals suffering from schizophrenia indicated that there was significant volume decrease in bilateral hippocampal formations of schizophremics (Nelson et al. 1998), there is still debate about the cytoarchitectural underpinnings of these volume decreases in the hippocampus of schizophrenics (Weinberger 1999). Hippocampal pathology in schizophrenia would fit nicely with the observed loss of memory and other cognitive deficits characteristic of this neuropsychiatric disorder. The most consistent findings of altered cytoarchitecture in the hippocampus have been the observations of significant cell loss in the CA1 subfield of the hippocampus and of reductions in the cell size of neurons in the hippocampal subfields CA1, as well as CA2–4, in some studies (Benes et al. 1991; Arnold et al. 1995). One of the most consistent observations has been the reduction in the numbers of non-pyramidal neurons in the *stratum oriens* and *stratum radiatum* of the CA2–CA3 region of the hippocampus (Benes et al. 1998). The non-pyramidal neurons are inhibitory interneurons that use the neurotransmitter γ-aminobutyric acid (GABA). Deficits in GABA function in the hippocampus of schizophrenics has been repeatedly demonstrated and is thought to result from injury to the interneurons; such injury is brought about by excess activation of glutamate-releasing nerve terminals that synapse upon the affected interneurons (Benes 2007; Benes et al. 2007).

Mechanisms of selective neuronal vulnerability

Vasculature of hippocampal regions—oldest but least likely mechanism

As mentioned in preceding sections, the earliest hypothesis advanced to explain selective neuronal vulnerability to ischaemia or seizures in the hippocampus was that of variations in the vascularization of the most sensitive neurons, the CA1 pyramidal neurons (Spielmeyer 1927; Meyer 1936; Scholz 1953; Pulsinelli 1985; Meencke 2009). However, later studies of ischaemia in experimental animals clearly established the fact that selective lesions of the fimbria/fornix connections to the hippocampus eliminated the sensitivity of CA1 neurons to ischaemic episodes (Buchan and Pulsinelli 1990), observations that would argue against differential patterns of vascularization as being responsible for the selective vulnerability of these neurons. Also, a detailed analysis of the total number, total length, and distribution of capillary diameters in subfields of the human hippocampus and dentate gyrus revealed that it was the dentate gyrus which had fewer capillaries per unit volume than all other subfields of the hippocampus (Lokkegaard et al. 2001). Yet, the neurons of the DG are least sensitive to ischaemia when compared with those of the hippocampal neurons of the CA1, CA4, or CA3 (Pulsinelli 1985). In addition, in hippocampus slices grown in organotypic cultures (i.e. in the absence of any vascular supply), neurons of the CA1 region are the most sensitive to insults such as OS (Wilde et al. 1997; Wang et al. 2005b).

Thus, although some peculiarities in vasculature may contribute to the susceptibility of neurons to ischaemia, hypoxia, seizures, or oxidative stress-induced neuronal injury, they are unlikely to be the primary mechanism for the phenomenon of selective neuronal vulnerability.

High demand for energy metabolism and consequent formation of reactive oxygen species in vulnerable neurons

The most vulnerable neurons in various neurodegenerative diseases, such as AD, are the large pyramidal neurons in the CA1 region of the hippocampus and of the entorhinal cortex layer II (Witter et al. 1989). The large pyramidal neurons represent mainly association neurons that send long projections to other regions of the CNS (Hof et al. 1990; De Lacoste and White 1993). It has been suggested that the vulnerability of these large-sized neurons is related to their high demand for energy and that such high-energy demands may be met by increases in the levels of oxidative phosphorylation.

The energy and respiratory demands in the human brain (approximately 20% of the glucose and O_2 supply of the whole body) reflect those of the neuronal elements, as neurons rely heavily on aerobic metabolism and the oxidative phosphorylation activity of mitochondria. However, it has been reported recently that certain regions of the brain, such as the prefrontal cortex and medial and lateral parietal cortex, maintain high activity of aerobic glycolysis, that is, neurons in these brain regions utilize glucose in excess of that used for oxidative phosphorylation (Vaishnavi et al. 2010; Vlassenko et al. 2010). The high levels of aerobic glycolysis have been linked to enhanced synaptic activity as determined by altered ratios of cerebral metabolic rate for glucose (CMR_{Glu}) and CMR for O_2 during the performance of demanding mental tasks (Madsen et al. 1995). The link between aerobic glycolysis and synaptic activity has received support also from the demonstration of the presence in the post-synaptic region of neuronal dendrites of enzymes associated with the glycolytic pathway (Wu et al. 1997). Interestingly, the hippocampus is a region that has low levels of aerobic glycolysis, suggestive of a greater reliance of hippocampus neurons on oxidative phosphorylation and mitochondrial metabolism than on the glycolytic pathway for the generation of adenosine triphosphate (ATP) (Vaishnavi et al. 2010). Overall, synaptic activity is highly dependent on the energy provided by both aerobic glycolysis and by mitochondrial metabolic activity. Thus, abnormalities in energy production would lead to alterations in synaptic transmission, changes that would be expected to affect neuronal viability, physiological activity, and the control of behavioural states, such as cognition and memory (Gibson et al. 1998; Du et al. 2008). Based on this concept, it has been proposed that synaptic degeneration in AD is the result of mitochondrial dysfunction (Reddy and Beal 2008).

Mitochondria use 90% or more of a cell's O_2 to produce ATP. Oxidative phosphorylation is directly linked to the activity of the electron transport chain (ETC) enzymes. Because of electron leakage in the reactions of the ETC, 1–2% of the electrons react with O_2 to generate reactive oxygen species (ROS) (Cadenas and Davies 2000). Thus, the hyperactivation of these enzymes in mitochondria can lead to OS and cause the oxidative modification of proteins, DNA, and lipids in neurons (i.e. molecular changes that are injurious to cells). Excessive formation of ROS has been invoked as a mechanism for neurodegeneration associated with insults to CNS, such as hypoxia and hypoglycaemia (Halliwell et al. 1992; Friberg et al. 2002), neurodegenerative diseases, such as AD (Zhu et al. 2004a), PD (Fahn and Cohen 1992; Jenner 2003; van Muiswinkel et al. 2004), ALS (Carri et al. 2003), and neuropsychiatric disorders, such as schizophrenia and bipolar disorders (Benes 2007).

This process may be particularly relevant to aging-associated differential vulnerability of certain populations of neurons. In an elegant study of gene expression changes in human frontal cortex during ageing, genes in the ontogenetic categories related to synaptic transmission, vesicle transport, Ca^{2+} homeostasis, and mitochondrial function, that is categories of genes important to both learning and memory as well as energy metabolism, are down-regulated during the ageing process and such down-regulation is, apparently, through selective vulnerability of the promoter regions of the DNAs for these genes to oxidative modification of DNA bases (Lu et al. 2004). One of the key genes whose promoter region was susceptible to oxidative modification during ageing is that of the ATP synthase of mitochondria, the enzyme that synthesizes ATP. Thus, oxidative modification of the gene for a key enzyme in the synthesis of mitochondrial ATP, if it occurs in select neuronal populations, would lead to decreases in ATP formation in these neurons and this would affect processes that are dependent on ATP, such as the plasma and endoplasmic reticulum ATPases that maintain trans-membrane ionic gradients and low levels of intracellular calcium (Ca^{2+}), the ATP-dependent chaperones involved in protein folding, the process of axoplasmic transport of organelles, mRNA and proteins, and the myriad intracellular signalling events that require activation of ATP-dependent protein kinases.

If mitochondrial ROS levels exceed the capacity of intramitochondrial antioxidant defences (i.e. those of Mn superoxide dismutase and glutathione peroxidase), then cell proteins, lipids, and DNA become targets of oxidative molecular modification (Wang et al. 2005a). Oxidative modification of DNA, proteins and lipids may be highest in mitochondria because of the proximity of these macromolecules to the site of generation of ROS. Furthermore, mitochondria are deficient in mechanisms involved in base excision and repair of DNA molecules (Bensch et al. 2009) and, since most of the mitochondrial DNA is composed of coding sequences, oxidative modification of mitochondrial DNA bases could lead to a higher rate of mutations in mitochondrial genes than nuclear genes. This may account for the observation that mitochondrial DNA in AD brain, for example, contains oxidized bases at levels up to 10-fold higher than nuclear DNA (Mecocci et al. 1994; Wang et al. 2005a). And, as part of the free radical hypothesis of ageing, it has been proposed that mitochondrial DNA mutations accumulate during the ageing process and lead to abnormalities in enzymes of oxidative phosphorylation, which lead to greater ROS generation, and more mitochondrial DNA mutations (Harman 1956).

Thus, OS may lead to dysfunctional mitochondria and dysfunctional mitochondria will produce less ATP, generate more ROS, and lead to a further exacerbation of OS. Neurons which are vulnerable to ageing, hypoxia, seizures, or neurodegenerative diseases, might be expected to show, in the long run, opposite trends in terms of mitochondrial function and mitochondrial DNA oxidative modification, i.e. decreases in mitochondrial function and increases in oxidative modification of DNA or RNA. Such trends have been observed during the ageing process (Kujoth et al. 2005) and in neurodegenerative diseases, such as AD (Gibson et al. 1998; Castellani et al. 2002; Grazina et al. 2006). Such mitochondrial dysfunction is best exhibited by the decreases in the activities of enzymes in the tricarboxylic acid cycle (α-ketoglutarate dehydrogenase and pyruvate dehydrogenase) and in the ETC (cytochrome oxidase) (Parker et al. 1990, 1994a, 1994b; Gibson et al. 1998; Cottrell et al. 2001). These mitochondrial changes in neurons are associated with structural damage to neurons, including the presence of lysosomal particles that contain fragmented and dense mitochondria, as observed in AD (Zhu et al. 2004b).

Even though the general mechanism of neuronal injury that results from altered metabolic states and the formation of ROS may appear to be a fairly general mechanism that can be used to explain neuronal susceptibility to the various abnormal states described in preceding sections, it is nevertheless true that many brain neurons do not suffer from the damaging effects of altered metabolism and the formation of ROS. As a matter of fact, OS-induced neuronal death is very circumscribed. For example, counter to the idea that very similar types of neurons might be equally susceptible to hypoxia, ageing, and oxidative stress, the large pyramidal neurons in two adjacent subregions of the hippocampus, CA1 versus CA2/CA3, are markedly different in their sensitivity to these stresses, in particular, OS (Wilde et al. 1997; Wang et al. 2005b, 2007, 2009). Whereas approximately 80% of CA1 pyramidal neurons die upon exposure to agents that generate ROS, only roughly 8% of CA3 neurons suffer likewise (Wang et al. 2005b, 2009) (Figure 4.2). The longer the period of exposure of CA1 neurons to a superoxide-generating system, the more neurons die but the same length of exposure to the same ROS-generating chemical agent produced very little change in cell death of neurons in the CA3 region (Figure 4.2). What these observations point to is the fact that OS may be a common underlying factor in the various adverse

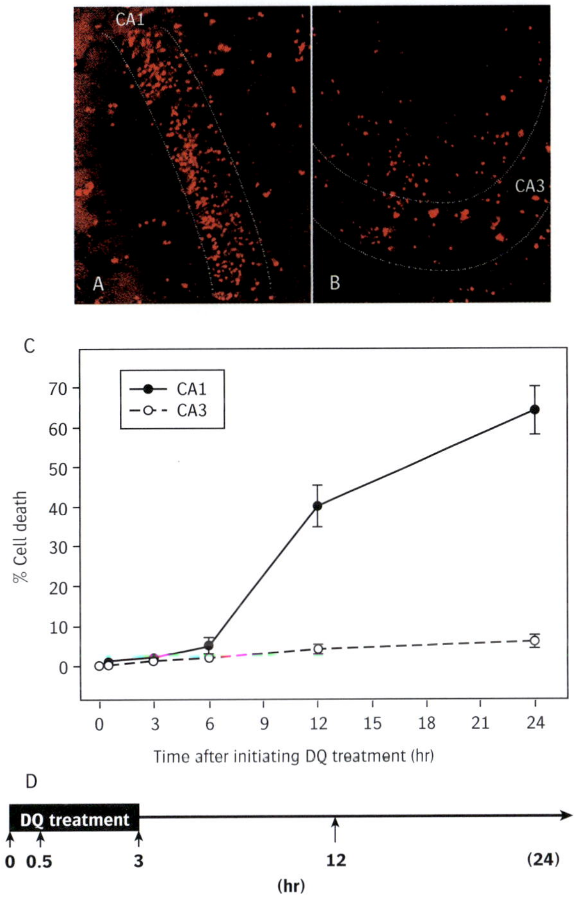

Figure 4.2 Image of selective vulnerability of neurons in the pyramidal layer of the hippocampal CA1 as compared with the CA3 region (A, B), and quantification of the differential, time-dependent neuronal death in these two regions (C) following oxidative stress induced by a 3-h exposure to 100 μM of the superoxide-generating compound, duroquinone (DQ) (experimental paradigm shown in D). Cell death was detected through the use of the membrane-impermeable DNA-binding dye, propidium iodide, which labels cells whose plasma membranes have disintegrated. Note the high levels of pyramidal cell death in CA1 and the comparatively modest neuronal loss in the adjacent CA3 region of the same organotypic slice culture treated with duroquinone. (A, B): Reprinted from *Molecular Brain Research*, 40(1–2), Wang, X., Pal, R., Chen, X-W., Limpeanchob, N., Kumar, K.N., Michaelis, E.K. High intrinsic oxidative stress may underlie selective vulnerability of the hippocampal CA1 region, pp. 120–6, Copyright (2005), with permission from Elsevier. (C, D) Reprinted from *Genomics*, 90 (2), Wang, X., Pal, R., Chen, X-W., Kumar, K.N., Kim, O-J., and Michaelis, E.K., Genome-wide transcriptome profiling of region-specific vulnerability to oxidative stress in the hippocampus, pp. 201–12, Copyright (2007), with permission from Elsevier.

conditions that are characterized by the appearance of selective neuronal vulnerability (Sayre et al. 2001; Valko et al. 2007) and yet the response to OS may determine the differential resistance to injury or cell death exhibited by select populations of neurons in the CNS. The key question is what are the molecular and cell biological mechanisms that determine such differential sensitivity to altered metabolic states and OS.

Selective neuronal vulnerability to glutamate-induced and Ca^{2+}-mediated neurotoxicity

Not all neurons that are injured following episodes of ischaemia, hypoxia, or brain trauma, nor all of those that are damaged during ageing or by the processes involved in neurodegenerative diseases and neuropsychiatric disorders, are large association neurons in the CNS. For example, it was mentioned in a preceding section that hippocampal neurons that are most consistently lost in the brains of schizophrenics, as well as the neurons in the CA4 region of the hippocampus that are lost in individuals with chronic epilepsy, are interneurons (non-pyramidal neurons) of relatively small size and with considerably shorter axons than the projection neurons of the cortex (Houser et al. 1986; Ribak et al. 1986; Benes et al. 1998; Meencke 2009). The mechanisms involved in such neuronal shrinkage or loss may not include the same processes as those invoked to explain how large neurons may be selectively vulnerable to certain stress conditions. Vulnerability of large neurons has been attributed to the high demand for metabolic energy, the relatively large surface area of the association neurons (which increases their exposure to extracellular toxic agents), the high demand for transport of materials along great distances, and the dependence on the expression of high levels of neurofilament proteins (i.e. proteins that are susceptible to aggregation and disruption of neuronal function and structure) (Mattson and Magnus 2006).

Neurons of the CNS differ not only in the size of the soma and the length of axonal processes, but also in the number and complexity of the dendrites, the number of synaptic connections, and the types of neurotransmitters that they use for neuron to neuron communication. Among the various transmitters, L-glutamic acid has been associated most frequently with neuronal injury and loss. Glutamate is the most prevalent excitatory neurotransmitter in the CNS and produces both short-term and long term effects upon postsynaptic neurons (Michaelis 1998). Glutamate activates both ion-channel-forming (ionotropic), and G-protein coupled (metabotropic) receptors (Gasic and Hollmann 1992). Activation of glutamate receptors is important in the formation of long-term potentiation and long-term depression of synaptic activity (Cotman et al. 1988), synapse formation, neuronal migration, and survival of some neurons during the development of the CNS (Balazs et al. 1988). However, excessive activation of glutamate receptors can lead to neuronal injury and cell death (Choi and Rothman 1990), i.e. the phenomenon known as excitotoxicity. Most of the studies of glutamate-induced neurotoxicity have been performed using neuronal cultures, either primary neuronal cultures or slice (organotypic) cultures. Nevertheless, the validity of the observed toxicity produced by glutamate has not been doubted as the injection of glutamate analogues into the brain of experimental animals produced similar neuronal injury and death as that observed in cell cultures. Furthermore, the recent development of null mutant mice that accumulate large quantities of extracellular glutamate and suffer from extensive neuronal damage and loss, is a strong indication that excessive glutamate accumulation in the extracellular space can lead to extensive neuronal damage.

There are three gene knockout animal models for which inadequate glutamate transport into glial cells leads to excessive accumulation of glutamate in the extracellular space. The mice that have large and sustained excess accumulation of extracellular glutamate in the CNS are the null mutant mice for the glutamate transporters Slc1a2 (Eaat2 or Glt-1) and Slc1a3 (Eaat1 or Glast), and the null mutant mouse for the tuberous sclerosis complex-1 gene (Tsc1) in glial cells of the CNS (Rothstein et al. 1996; Tanaka et al. 1997; Matsugami et al. 2006; Zeng et al. 2007). The first two gene knockout mice obviously suffer from loss of the activity of the glial cell transporters EAAT1 and EAAT2, the major pathways for the clearance of synaptically released glutamate. The null mutant mice for Tsc1 with the mutation selectively introduced in glial cells, also exhibit much reduced glutamate transport activity in glial cells and the same extensive neuronal damage throughout the brain as that observed for the null mutant mice for *Slc1a2* and *Slc1a3*. These animal models, although suffering from excessive glutamate accumulation and widespread neuronal damage rather than selective neuronal injury or death, may still be relevant models for conditions or diseases in which glutamate transport is diminished. For example, a gradual diminution of glutamate clearance from synapses has been observed in the mammalian brain during ageing (Zoia et al. 2004, 2005; Nickell et al. 2005) and in some neurodegenerative diseases such as AD (Li et al. 1997; Scott et al. 2002; Zoia et al. 2004, 2005) and ALS (Rothstein et al. 1990, 1993). Unfortunately, the large accumulation of extracellular glutamate causes such extensive damage in brain that the animals do not survive to an old age, thus becoming a model of widespread excitotoxicity but not of selective vulnerability and age-associated modest neuronal losses.

A fourth mouse model is that of excessive glutamate synthesis and synaptic release, the transgenic mouse in which the enzyme Glud1 (glutamate dehydrogenase 1) is overexpressed in neurons only (Bao et al. 2009). These mice suffer neuronal damage and cell loss in select brain regions, such as the hippocampus CA1 region, but not the CA3, the somatosensory cortex, the striatum, and the anterior horn of the spinal cord (Michaelis et al. 2011). The pyramidal cell layer in the hippocampus of 10–12 month-old Glud1 transgenic (Tg) mice is populated by shrunken cell bodies and fewer neurons as compared with the same region of littermate wild type (wt) mice (Figure 4.3). Neuronal damage in the Tg mice is most easily observed in terms of the loss of dendrite and nerve ending labelling of hippocampal, cortical, and striatal neurons but, even in the hippocampus, such neuronal injury is quite prominent in CA1 but not in the CA3 region. An example of the loss in

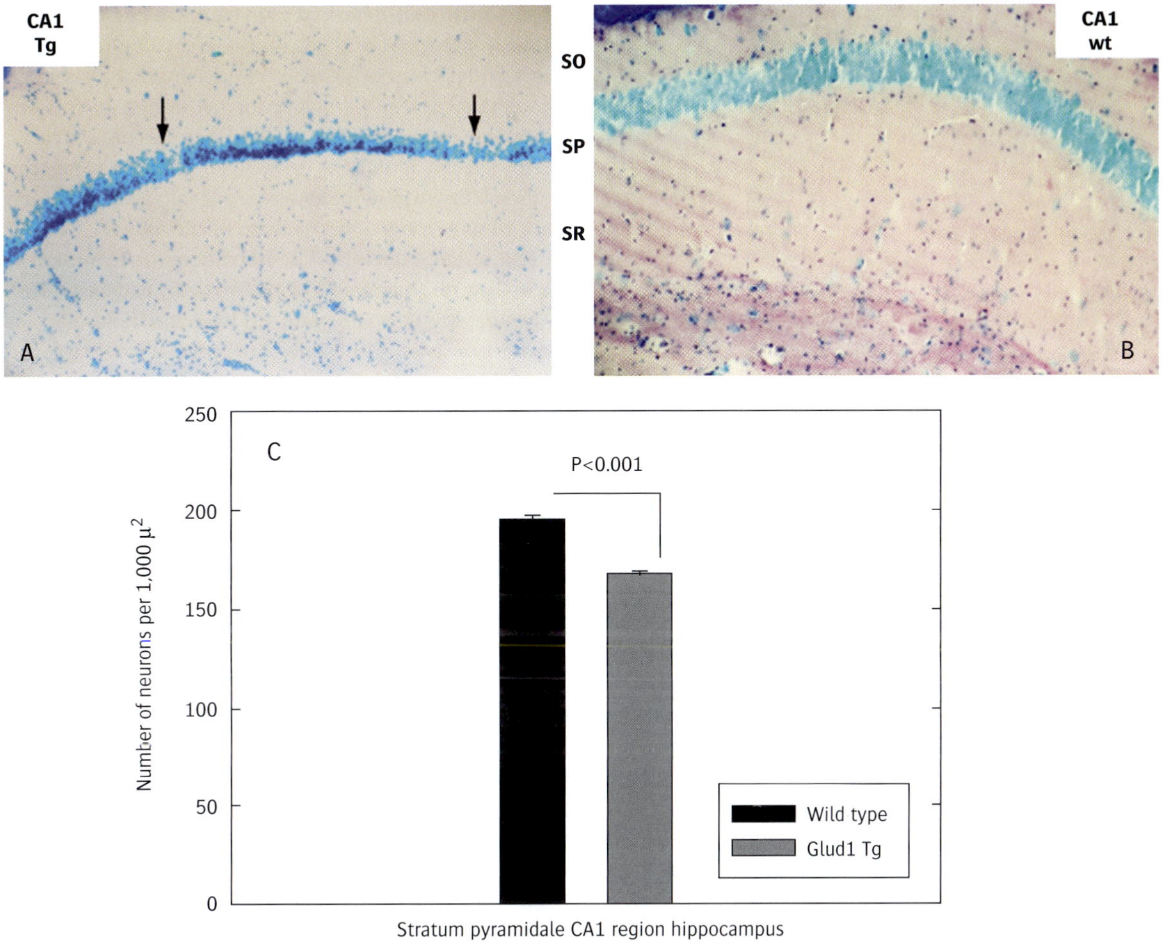

Figure 4.3 Neuronal volume loss and thinning of the pyramidal cell layer in the CA1 region of the hippocampus of Glud1 Tg mice (A) as compared with the same region of littermate wt mice (B), and quantification of overall neuronal density in the CA1 of Tg versus wt mice (C). The neurons in brain sections from 16-month-old mice were stained using methylene blue and the images were obtained at the same magnification. Arrows in (A) point to areas of marked thinning of the CA1 pyramidal cell layer. Neuronal cell bodies in a set area of the pyramidal layer (1100 μ^2) were counted and the cell counts in C are the mean (±SEM) from 4 wt and 4 Tg age-matched mice.

labelling of hippocampal CA1 neurons of Tg versus wt mice is shown in Figure 4.4. The dendritic and synaptic changes described earlier are associated with moderate, age-related cell losses in select brain regions, such as the hippocampus CA1 (Bao et al. 2009; Michaelis et al. 2011). These observations have led to the suggestion that a moderate excess release of glutamate in the absence of high extracellular accumulation of glutamate in the brain of these mice, leads to the appearance of the phenomenon of selective neuronal vulnerability of hippocampal, somatosensory cortical, striatal, and spinal neurons (Michaelis 1998; Bao et al. 2009; Hascup et al. 2011). Important to the theme of this discussion is the observation that whereas CA1 pyramidal neurons suffer progressive dendritic and, eventually, whole cell loss during the life span of the Glud1 transgenic mice, the CA3 neurons do not exhibit any such losses despite the fact that equal levels of the enzyme GLUD1 are expressed in both regions of the hippocampus. Therefore, the neuronal injury in these transgenic mice would support the notion that glutamate-induced injury, if not excessive, might result in the appearance of the phenomenon of selective neuronal vulnerability.

With regard to other clinically relevant conditions and the possible contribution of glutamate hyperactivity to neuronal loss, it is well known that an excess of glutamate released in the penumbra of brain regions that have suffered an episode of focal ischaemia leads to neuronal death through the process of excitotoxicity initiated through activation of the *N*-methyl-D-aspartate (NMDA) ionotropic receptors (Szatkowski and Attwell 1994). Besides the excitotoxicity observed in association with ischaemia, glutamate excitotoxicity also occurs in other pathologic states in the nervous system, such as seizures, hypoglycaemia, and inflammation associated with viral infections (Schwarcz and Meldrum 1985; Lipton et al. 1991; Dawson and Dawson 1998). Importantly, most of these pathological states are also ones that have been associated with the appearance of selective neuronal injury and loss.

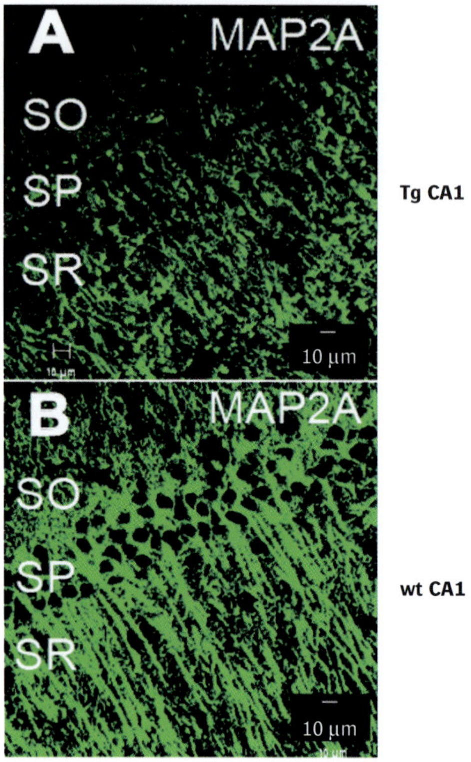

Figure 4.4 Changes in dendrite structure of pyramidal neurons in the CA1 region of 12-month-old Glud1 Tg mice (A) as compared with same age wt mice (B). Neurons were immunolabelled with anti-MAP2A antibodies (specifically labelling cell bodies and dendrites. SO, stratum oriens; SP, stratum pyramidale; SR, stratum radiatum. Bars: 10µm. Reproduced from Wang X, Bao X., Pal R., Agbas A., Michaelis E.K., Transcriptomic responses in mouse brain exposed to chronic excess of the neurotransmitter glutamate, *BMC Genomics* 11, p. 360 © 2010, Wang et al, with permission.

Excitotoxicity can result from the activation of all three forms of glutamate ionotropic receptors, NMDA, kainate, and alpha-amino-3-hydroxy-5-methyl-4-isoxazole propionic acid (AMPA) receptors, and it is closely linked to both Na^+ and Ca^{2+} entry and accumulation in neurons (Choi 1992; Michaelis 1998; Greenwood et al. 2007b; Hoskison et al. 2007; Kintner et al. 2010). Hyper-stimulation of glutamate receptors causes swelling of the neuronal cell body, localized swelling or 'beading' in dendrites, destruction of intracellular organelles, and, ultimately, cell death (Choi 1992; Hoskison and Shuttleworth 2006; Greenwood et al. 2007a; Hoskison et al. 2007; Kintner et al. 2010). The lesions in dendrites following activation of the inotropic receptors have been traced to increased accumulation of intracellular Na^+ as a result of the activation of the Na^+/H^+ exchange carriers, the reversal of the normal function of the Na^+/Ca^{2+} exchange carriers, the resulting intracellular and intramitochondrial accumulation of Ca^{2+}, the depolarization of mitochondria, the decreases in ATP levels, and the disruption of both plasma membrane potentials and microtubule-dependent transport processes (Greenwood et al. 2007a; Kintner et al. 2010). Although these events of dendritic and neuronal damage have been studied in primary neuronal and organotypic slice cultures, rather than under conditions of *in vivo* excitotoxicity, they are thought to be the pathways for neurotoxicity following glutamate receptor activation either *in vivo* or *in vitro*.

However, it is difficult to explain differential neuronal vulnerability to glutamate-related pathological states on the basis of such general mechanisms of excitotoxicity. The key questions are, therefore, whether differential neuronal vulnerability to glutamate-induced toxicity can be observed either *in vivo* or *in vitro*, and whether the molecular pathways for resistance to glutamate-induced toxicity in neurons can be identified. As indicated for the Glud1 Tg mice, selective neuronal vulnerability, presumably due to glutamate overactivity at synapses, has been demonstrated *in vivo*. Such differential vulnerability to glutamate-induced neuronal damage in the Glud1 Tg mouse hippocampus, for example, is not a surprising finding because even from the earliest descriptions of glutamate as a neurotoxic agent for neurons in culture, it was noted that some cells exhibited marked resistance to glutamate-induced cell death. This was true for striatal and cortical neurons that contain the enzyme NADPH diaphorase (identical to nitric oxide synthase) (Koh et al. 1986; Behrens et al. 1996). This is a particularly relevant observation with respect to CNS disease states as the NADPH diaphorase-positive neurons in the striatum have also been shown to be the ones that survive in patients with HD (Ferrante et al. 1985). But, the neurons that contain elevated levels of nitric oxide synthase (NOS) represent a relatively small number in striatum, cortex, and hippocampus. In the cortex, they have been described as medium to large size aspiny neurons, they are GABA-containing neurons, i.e. inhibitory interneurons, are more abundant in the entorhinal cortex and in layers II and III of the somatosensory cortex, and constitute approximately 0.5–2% of the total population of neurons (Dawson et al. 1991; Valtschanoff et al. 1993b). In the hippocampus and DG, these neurons represent, once again, a small percentage, they are localized in the subiculum and CA3 region, and are GABA-containing inhibitory interneurons (Valtschanoff et al. 1993a). Thus, this population of NOS-containing neurons may not account for all neuronal populations that are resistant to glutamate-induced toxicity. Yet, it may be clinically relevant that cortical and striatal neurons containing NOS are less susceptible to ischaemia; unfortunately, the same has not been reported for hippocampal neurons that express high levels of NOS (Ferriero et al. 1988). Nevertheless, relative sparing of NOS-containing neurons in the hippocampus of cases with AD has been described (Hyman et al. 1992), thus NOS expression in neurons may have an overall protective effect against glutamate-induced toxicity.

It should be pointed out that a different mechanism, and one that is not related to nitric oxide, has been described to explain the differential neuronal vulnerability to ischaemia of CA1 and CA4 hippocampal neurons. This differential sensitivity has been attributed to the vulnerability of the hypoxia-sensitive neurons to the zinc (Zn) that is released from glutamatergic nerve terminals and that is taken up by the postsynaptic pyramidal cells of the CA1 and CA4 neurons (Koh et al. 1996). The Zn-induced neuronal death in the CA1 region and the control of such sensitivity by the concurrent activation of NMDA receptors, is thought to correlate with the high sensitivity of these neurons to ischaemia-induced neuronal injury and death. Thus, either the post-receptor activation of NOS or the pre-receptor release of Zn may be positive or negative determinants of selective neuronal vulnerability to glutamate-induced neurotoxicity. However, the role of Zn in ischaemia-induced selective vulnerability has been questioned in other studies in which chelation of intra- and extracellular Zn had no effect on the ischaemia-induced selective accumulation of Ca^{2+} in sensitive CA1 neurons (Stanika et al. 2010). The neuronal and, specifically, the intramitochondrial accumulation of Ca^{2+} correlated best with the differential sensitivity of CA1 neurons to ischaemia or glutamate-induced cell injury. The differences between sensitive and resistant neurons, the CA3 neurons being the resistant ones, were shown to be linked to the differential levels of NMDA receptor-activated Ca^{2+} in neuronal mitochondria (Stanika et al. 2010).

The three possible mechanisms of differential sensitivity to glutamate excitotoxicity described above may not be the only such processes that determine neuronal injury or survival. Other mechanisms for such differential sensitivity to glutamate-induced toxicity may include factors such as the levels of expression of the endoplasmic reticulum molecular chaperone (ORP150); the expression of a stress protein that is important in the maintenance of metabolic homeostasis (Kitao et al. 2001); the expression and activity of the pre-mRNA base-editing enzyme ADAR2 which determines the ratio of glutamate receptor 2 (GluR2) subunits that contain glutamine 607 to those that contain arginine 607, a critical ratio in terms of Ca^{2+} permeability of these glutamate receptors (Michaelis 1998; Peng et al. 2006); or the production and local release of agents such as tumor necrosis factor α (TNFα) and brain-derived neurotrophic factor (BDNF) (Glazner and Mattson 2000). It is important to state, however, that despite the fact that the levels of expression of all of these macromolecules have been shown to correlate with or induce relative degrees of sensitivity or resistance to glutamate-dependent excitotoxicity, they have not been directly linked to the differential vulnerability of select populations of neurons to glutamate-related toxicity.

Neurons that are vulnerable to increases in extracellular glutamate are frequently the same population of neurons as those that use ROS, such as superoxide (O_2^-) and hydrogen peroxide (H_2O_2), as intra- and transcellular signalling molecules (Garthwaite 1991; Thiels et al. 2000; Finkel 2003; Kamsler and Segal 2003; Esposito et al. 2004). This correlation led to the hypothesis that the pro-oxidant state of these populations of neurons may determine their vulnerability to excess activation of glutamate receptors (Wang et al. 2005b). The oxidative stress burden on these neurons probably results from altered metabolic states in mitochondria and may be related to the uncoupling of electron transfer reactions and the formation of excess O_2^- (Turrens 1997; Lenaz 2001). Excess glutamate receptor activation leads to elevations in intracellular Ca^{2+} concentrations ($[Ca^{2+}]$), increases in intramitochondrial $[Ca^{2+}]$, enhanced activation of the 2-oxoglutarate dehydrogenase complex and inactivation of complex I of the ETC, and both of these events can lead to high levels of O_2^- formation and further oxidative stress (McCormack et al. 1990; Rizzuto et al. 2000; Sadek et al. 2004). In support of this idea are the observations that elevated levels of O_2^- and H_2O_2 are responsible for the mitochondrial damage and neuronal injury observed in *Drosophila* whose glial glutamate transporters were knocked down (Rival et al. 2004). Thus, neurons that live 'at the edge' in terms of their production and use of ROS for signalling purposes may suffer greater damage than other populations of neurons, i.e. they would exhibit differential vulnerability to glutamate.

Differential gene expression patterns in selectively vulnerable neurons

All of the preceding descriptions of potential mechanisms related to selective neuronal vulnerability have focused on either a differential physiological process that might explain such vulnerability, e.g. vascular differences in subregions of the hippocampus, or on a differential biochemical activity, such as mitochondrial energetics or ROS formation. However, it is quite likely that multiple molecular processes or pathways may be differentially activated or suppressed in select populations of neurons and that these pathways may determine the response of vulnerable or resistant neurons to exogenous or endogenous stress stimuli. The study of single molecular events, one at a time, is not only time-consuming but it may not offer a comprehensive view of what makes a vulnerable neuron sensitive to stress. To begin to address the likely multitude of contributions at the molecular and cellular level to the phenomenon of selective neuronal vulnerability to stress and disease, the methods of large-scale, whole-genome gene expression have been used recently. Examples of this approach can be gleaned from studies of gene expression in select brain regions during ageing, AD, hypoxia, and OS.

An analysis of the patterns of gene expression in different brain regions, both susceptible and resistant to aging- or AD-associated damage, have been summarized nicely in a previous publication (Wang et al. 2010b). As pointed out in that review of the literature, brain regions that respond differentially to the stresses associated with increasing age, e.g. cerebellar cortex versus cerebral cortex, exhibit distinct patterns of gene expression. Whereas two regions of the cerebral cortex have similar levels of expression of most genes, the two cortical regions differ significantly in terms of gene expression patterns from the cerebellar cortex

(Evans et al. 2003). What is more significant is that the ontological categories of differentially expressed genes are those of *signal transduction, neurogenesis, synaptic transmission,* and *transcription factor regulation*. Some of the same gene ontological categories are also significantly different in terms of expression in several brain regions that are more vulnerable to ageing-related neuronal loss (dorsolateral prefrontal cortex, anterior cingulate cortex, primary visual cortex, and Broca's area) as compared with the expression patterns in regions more resistant to ageing-related effects (cerebellum and caudate nucleus) (Khaitovich et al. 2004). The gene ontology categories that differed most in terms of expression between the four cortical regions and either the caudate nucleus or the cerebellum were those of *synaptic transmission, signal transduction, neurogenesis, neuronal development*, and *Ca^{2+} regulation*. It should be noted that these gene ontology categories are similar to those described in another study of frontal cortex changes in gene expression during the ageing process (cited previously), changes that correlate with oxidative modification of the DNA in the promoter regions of several of the genes that fall into these ontological categories (Lu et al. 2004). In that study, the expression of genes in the functional categories of *synaptic transmission, Ca^{2+} homeostasis, vesicular transport, cell signalling,* and *mitochondrial metabolism* were significantly downregulated with advancing age, whereas those of *antioxidant response, DNA repair, metal ion homeostasis,* and *stress response* (*protein misfolding/correction*) were significantly upregulated.

The gene microarray studies described previously have brought forth many more gene and gene-related pathways for consideration as potentially important molecular and cellular parameters that may determine the fate of neurons under conditions of stress than the studies of single molecular entities or single pathways did. It is important to note, however, that many of the gene array studies have been performed using samples of whole tissue from a given brain region, samples that contain not only neurons, but also glia and possibly cells from the vascular endothelium and blood. Therefore, some of the observed differential gene patterns may represent non-neuronal cell differences in certain brain regions. To ensure that the pool of cells sampled for such microarray gene expression studies are highly enriched in neurons, the technique of laser capture microdissection (LCM) has been used to capture neurons under the microscope from either vulnerable or resistant brain regions. Gene expression in neurons captured by LCM from the hippocampus and entorhinal cortex identified 102 genes that are expressed at lower levels than neurons from the posterior cingulate and superior frontal gyrus. The molecular pathways to which these genes belong include *Ca^{2+} signalling, CREB pathway, cytoplasmic/mitochondrial transport of proapoptotic proteins*, and *genes involved in cell cycle transition* (Liang et al. 2007). Under-expression of genes in these pathways in neurons that are susceptible to the effects of advancing age would indicate abnormal Ca^{2+} handling, deficient memory formation (CREB pathway), and abnormal regulation of cell cycle, which may lead to apoptotic cell death. Therefore, the down-regulation of gene expression during the ageing process of genes involved in Ca^{2+} homeostasis, synaptic transmission, and energy metabolism identified in this and previous studies, point to some common molecular themes for neurons that are susceptible to stress stimuli. Furthermore, the observations made using human brain samples are clearly generalizable by the fact that the results from human brain studies are similar to those obtained using brain from rodents at various ages. For example, gene expression analyses of the hippocampus and cerebral cortex in mouse brain during the ageing process shows age-associated decreases in the expression of genes related to mitochondrial function (Xu et al. 2007).

Of greater relevance to the issue of selective vulnerability of neurons in the hippocampus, are studies using the approach of whole genome transcription differences in subregions of the hippocampus, the CA1 and CA3 fields, as well as the DG (Lein et al. 2004; Torres-Munoz et al. 2004; Ginsberg and Che 2005; Wang et al. 2005b; Newrzella et al. 2007; Wang et al. 2007, 2009; Wang and Michaelis 2010; Zeier et al. 2011). Differential gene expression patterns in neurons, captured by LCM or by aspiration of the cell contents from human CA1 and CA3 subfields, revealed some differences between these two subregions of the hippocampus, but did not point to a particular gene pattern that might account for the differential vulnerability to stress of neurons in these two regions (Torres-Munoz et al. 2004; Ginsberg and Che 2005). On the other hand, studies in rodent hippocampus have shown that gene expression in neurons of the CA1 region across several gene ontology categories differs significantly from the pattern of gene expression in CA3 neurons (Wang et al. 2005b, 2007, 2009; Zeier et al. 2011). Differentially expressed genes between these two regions indicate that cells in the CA1 region have higher transcriptional activity for genes that fall in the biological function and molecular pathways related to *stress/inflammatory responses, transition metal transport, ferroxidase activity*, and *presynaptic signalling activity*, while cells of the CA3 region have higher expression of genes in the *GABA-signalling, postsynaptic activity*, and *calcium and potassium channel activity* pathways or molecular functions (Wang et al. 2007). With respect to the issue of differential vulnerability of CA1 neurons to oxidative stress, it is important to note that cells of the CA1 region have significantly higher O_2^- levels and express several genes that are involved in the control of the formation of ROS, as well as the defence of cells against OS (Wang et al. 2005b). And, although most of the emphasis in these studies has been on the differential levels of expression of genes related to neuronal function, the genes associated with stress and inflammatory responses, such as those involved in cytokine and chemokine formation, are genes related to glial function and the response of these cells to stress. Most inflammatory responses in glial cells of the CNS, in particular microglia, have an initial protective effect in that they are involved in the removal of damaged neuronal elements; but if such responses are excessively activated, then they may lead to enhanced generation of ROS and further damage to neurons (Wang and Michaelis 2010). In ischaemia, for example, microglia in brain regions that contain neurons susceptible to ischaemic damage, such as the CA1 field, are rapidly activated whereas microglia in regions more resistant to ischaemia, such as layers II, III, IV, and V of the cortex, microglia activation occurred at a much later point in time (Bernaudin et al. 1998). Thus, with regard to the pattern of gene expression the cells of the CA1 region appear to be poised at the tipping point between normal generation and use of ROS as signalling molecules and excessive production of ROS leading to injury to neurons (Wang et al. 2005b). Furthermore, these neurons may be exposed to excess ROS generated by microglia in the CA1 region through an exaggerated inflammatory response.

What is particularly significant is that the gene transcription differences between CA1 and CA3 neurons may represent a more general pattern of molecular pathways that characterize neurons vulnerable to various forms of stress and those that are resistant to such stresses. This idea is based on the observations that CA1 neurons that are vulnerable to OS exhibit similar patterns of differential gene expression as cerebellar granule neurons which are also relatively sensitive to oxidative stress (Wang et al. 2009). And, what is true for neurons vulnerable to OS is also true for those that are resistant to such stress, i.e. the CA3 neurons that are relatively resistant to OS have similar patterns of gene expression as other resistant cerebral cortical neurons. All of these analyses were performed on neurons captured by LCM, therefore, they represent neuronal patterns of gene expression in different populations of neurons in different brain regions. An examination of the gene expression patterns of vulnerable neurons as compared with those of resistant neurons shows that the vulnerable neurons have higher expression levels of genes related to cell stress, such as *repair/response to DNA and RNA damage, and immune response*, while they have lower expression levels than resistant neurons for genes related to *energy generation, nervous system development, neurotransmission,* and *signal transduction* (Wang et al. 2009). Of particular importance is the recurring theme of cellular stress in neurons sensitive to OS and of diminished energy metabolism and neurotransmission. The deficits in energy metabolism identified on the basis of gene expression studies were correlated at the biochemical level to significantly lower ATP levels in the neurons that are sensitive to OS as compared with those that are resistant (Wang et al. 2009).

In other studies, it was shown that CA1 cells are more susceptible to age-related stress and exhibit differential levels of gene expression during ageing as compared with CA3 or DG cells (Zeier et al. 2011). Based on those studies, the categories of genes that are most highly expressed in CA1 cells are those related to *immune response* and *apoptosis*, both categories of a state of generalized cell stress. Thus, differences between CA1 and CA3 neurons include increased expression in CA1 of genes related to *cell stress, inflammatory response,* and *immune response*, but decreased expression of *energy generation, cell signalling,* and *neurogenesis*. It is becoming apparent that these patterns of gene expression are emerging as pathways of differential molecular and cellular processes that characterize vulnerable as compared with resistant hippocampal neurons or other populations of CNS neurons.

Most of the studies of differential gene expression between CA1 and CA3 or other hippocampal regions were designed so that baseline differences in gene expression patterns might be discerned and possibly linked to differential susceptibility of neurons in these two regions to stresses such as OS or ischaemia. An alternative and just as informative an approach would be to study both the baseline differences in gene expression, as well as the differences in response to stress on these neuronal populations induced either *in vivo* or *in vitro*. There are three such studies that have been performed and which can be used to glean differences not only in baseline gene expression patterns but also in the type of response that neurons mount to endogenous or exogenous stressors. When organotypic slice cultures of the hippocampus are exposed to an agent that generates OS in cells, duroquinone, both CA1 and CA3 cells upregulate or downregulate a subpopulation of the same genes in response to OS (0.5% of the total genes; Figure 4.5) (Wang et al. 2007). The functional categories of *signal transduction, stress response, apoptosis, neurogenesis,* and *calcium signalling* pathway represent the common gene expression responses in CA1 and CA3 following the application of OS across time (0–12 h). It is noteworthy that a few genes associated with immune responses, such as *chemokine motif ligand* genes, were actually downregulated in both CA1 and CA3 during the period of exposure to OS. What is interesting about the genes that were up- or downregulated during exposure to OS is the fact that cell injury and cell death differed dramatically between CA1 and CA3 neurons, yet the gene expression patterns for this small population of genes appeared to be very similar in cells from these two hippocampal regions. Thus, in addition to defining the gene expression patterns that distinguish the responses of CA1 versus those of CA3 neurons to OS, some interesting insights about the cellular responses to stress and the compensatory molecular and cellular events that occur over the period of exposure to stress, may be gleaned from the common responses to OS among these two neuronal populations.

The fact that CA1 and CA3 populations of pyramidal cells may have similar patterns of upregulation of genes following stress is not a characteristic of only OS but appears to be the case also for the effects of ischaemia on gene expression (Figure 4.5) (Newrzella et al. 2007). Ischaemia induction *in vivo* leads to altered gene expression of over 5000 genes in each region, CA1 and CA3, yet only 97 differ in their expression between CA1 and CA3 following

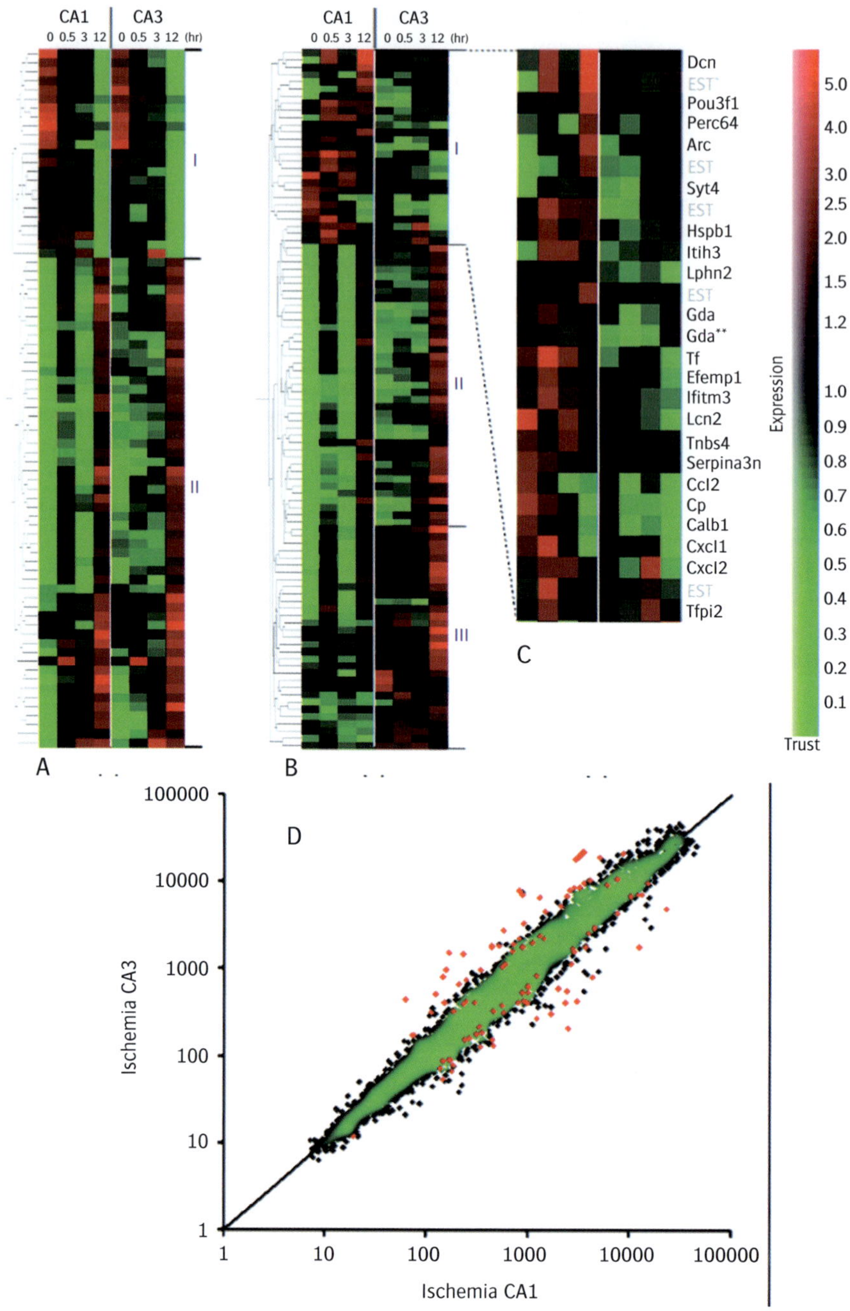

Figure 4.5 Patterns of gene expression (gene clustering) of pyramidal cells of the CA1 and CA3 regions measured following exposure of organotypic slice cultures to 100 μM duroquinone (A–C), and of gene expression (scatterplot) in CA1 versus CA3 following induction of ischemia in rat brain. Shown in (A–C) are clusterings of genes with similar (A) or divergent (B, C) transcriptional responses to duroquinone in CA1 versus CA3 across four time periods of exposure to duroquinone. The average relative expression levels are depicted by different colours and the colour bar on the right indicates the relative levels of expression. Genes with similar patterns of response in CA1 and CA3 are shown in (A) and are subgrouped based on the patterns of response across time. In (B), are shown the gene clusters that exhibited differential expression patterns between CA1 and CA3. The genes with consistently higher expression in CA1 than CA3 were labelled as type 'I' and the specific genes in this subcategory are shown in the enlarged view in (C). The scatterplot in (D) shows that the majority of genes in CA1 covaried with those in CA3 following a period of ischaemia through ligation of the common carotid. Only 97 genes (red circles) are differentially expressed in CA1 versus CA3 at a significant level after a period of ischaemia (see text for additional information). (A–C) Reprinted from *Genomics*, 90 (2), Wang, X., Pal, R., Chen, X-W., Kumar, K.N., Kim, O-J., and Michaelis, E.K. Genome-wide transcriptome profiling of region-specific vulnerability to oxidative stress in the hippocampus, pp. 201–12, Copyright (2007), with permission from Elsevier. (D) Reproduced from Newrzella D., et al,The functional genome of CA1 and CA3 neurons under native conditions and in response to ischemia, *BMC Genomics* 8, p.370 © Newrzella et al, with permission.

induction of ischaemia. In other words, cells of the CA1 and CA3 region respond to the stress of ischaemia in a very similar manner with regard to gene expression, except for some unique genes which are differentially expressed in CA1 versus CA3 under these conditions. The genes with differential patterns of expression between non-ischaemia and ischaemia conditions in CA1 and CA3 belong to some classes of genes that exhibit differential levels of expression in these two regions of the hippocampus even in the absence of a stress stimulus.

The studies of genes that may be up- or downregulated in their expression following the application of a stress stimulus such as OS or ischaemia, were performed over a relatively short period of time after the application of the stress stimulus. This was necessary in order to avoid the confounding variable in data gathering and analysis due to extensive cell death in one region as compared with another. Nevertheless, it would be informative to have more information about the changes in gene expression in different hippocampal or other brain regions following various types of stress that do not bring about rapid neuronal death. Such patterns of gene expression might represent adaptive or compensatory responses of neurons to the continuous presence of a stress condition. Such a study was undertaken in examining the gene expression patterns in the hippocampus of nine-month old Glud1 Tg mice and comparing them with those of their wt littermates (Wang et al. 2010a). These analyses of gene expression in the hippocampus have provided insights into possible adaptive or compensatory changes at the molecular level in mice whose brain neurons have been exposed to the stresses induced by high levels of synaptic glutamate release throughout their development and early adult life. Exposure to elevated levels of synaptic glutamate caused mostly up-regulation of the expression of genes in the categories of *synaptic neurotransmission, cognition and behaviour, ion transport and homeostasis, nervous system development, neurotrophin guidance signalling, cytoskeleton and focal adhesion,* and *intracellular signal transduction* (Figure 4.6). These patterns of gene expression were considered to be strongly indicative of compensatory responses to the neuronal, dendritic, and synaptic damage or loss suffered by the neurons of 9-month-old Glud1 Tg mice as compared with their wt littermates. Included in this pattern of potentially compensatory gene expression responses to the stress of a hyper-glutamatergic state are the up-regulation of genes in pathways of inflammatory and oxidative stress responses, such as, *NRF2-mediated oxidative stress response,*

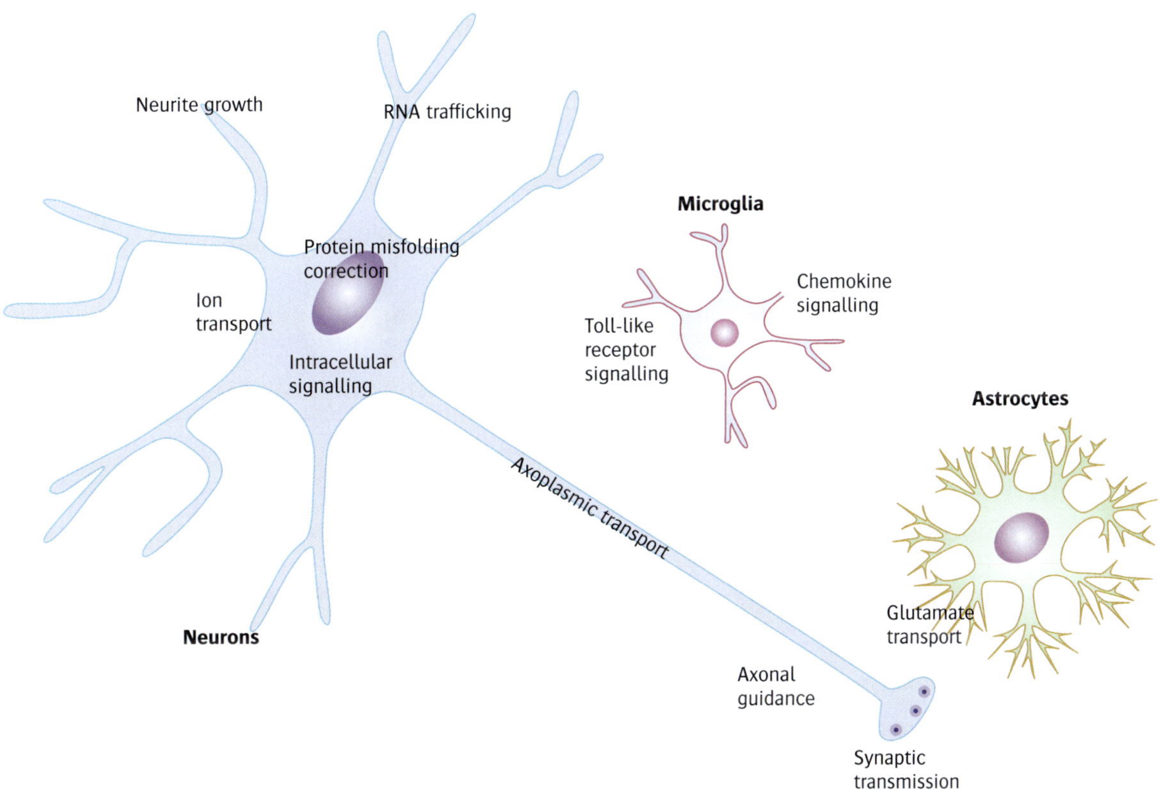

Figure 4.6 A schematic representation of the molecular and cellular functions and pathways that are altered in the hippocampus of Glud1 Tg as compared with those of wt mice. The molecular functions and pathways depicted are based on the patterns of differential gene expression, including differential protein expression for some of the genes, between Glud1 and wt mice. The gene categories are described in the text. Reprinted from Neurochemistry International, 59 (4), E.K. Michaelis, X. Wang, R. Pal, X. Bao, K.N. Hascup,Y. Wang, W.-T. Wang, D. Hui, A. Agbas, I.-Y. Choi, A. Belousov, and G.A. Gerhardt, NeuronalGlud1(glutamate dehydrogenase 1) over-expressing mi Increased glutamate formation and synaptic release, loss of synaptic activity, and adaptive changes in genomic expression, pp. 473–81, Copyright (2011), with permission from Elsevier.

protein misfolding and correction, chemokine signalling, and *toll-like receptor signalling*. In terms of the stress response genes, the expression patterns in the hippocampal cells of the Tg mice resembled those identified previously for human brain neurons susceptible to aging-associated stress. And, with respect to the pattern of expression of genes that represent adaptive or compensatory responses, such as those related to *synaptic neurotransmission, cognition and behaviour, nervous system development, neurotrophin guidance signalling,* and *intracellular signal transduction*, the hippocampal cells of the Glud1 Tg mice were very similar to cerebral cortical cells subjected to transient ischaemia and 24 h of reperfusion (Rickhag et al. 2006). The conclusion that may be drawn from these gene expression analyses is that future studies of neurons which are differentially vulnerable to stress stimuli should focus not only on endogenous differences prior to the application of a stress stimulus, but also to the differences in gene expression following a period of partial recovery from or adaptation to stressful conditions. Neurons susceptible to such stresses may differ from those that are resistant to stress or disease not only in terms of their baseline gene expression characteristics but also in terms of the short-term responses to injury or stress and the long-term adaptive or compensatory responses to these conditions.

Concluding ideas

In the preceding sections, the evidence of the generality of the phenomenon of selective neuronal vulnerability in the hippocampus and other brain regions was elaborated and our current concepts of the molecular, cellular, and physiological parameters that contribute to such differential vulnerabilities examined in some detail. Although we have not yet reached a conclusion about the mechanisms for differential neuronal responses to stress and disease in a region such as the hippocampus CA1 subfield, the collective body of work that has addressed this issue has

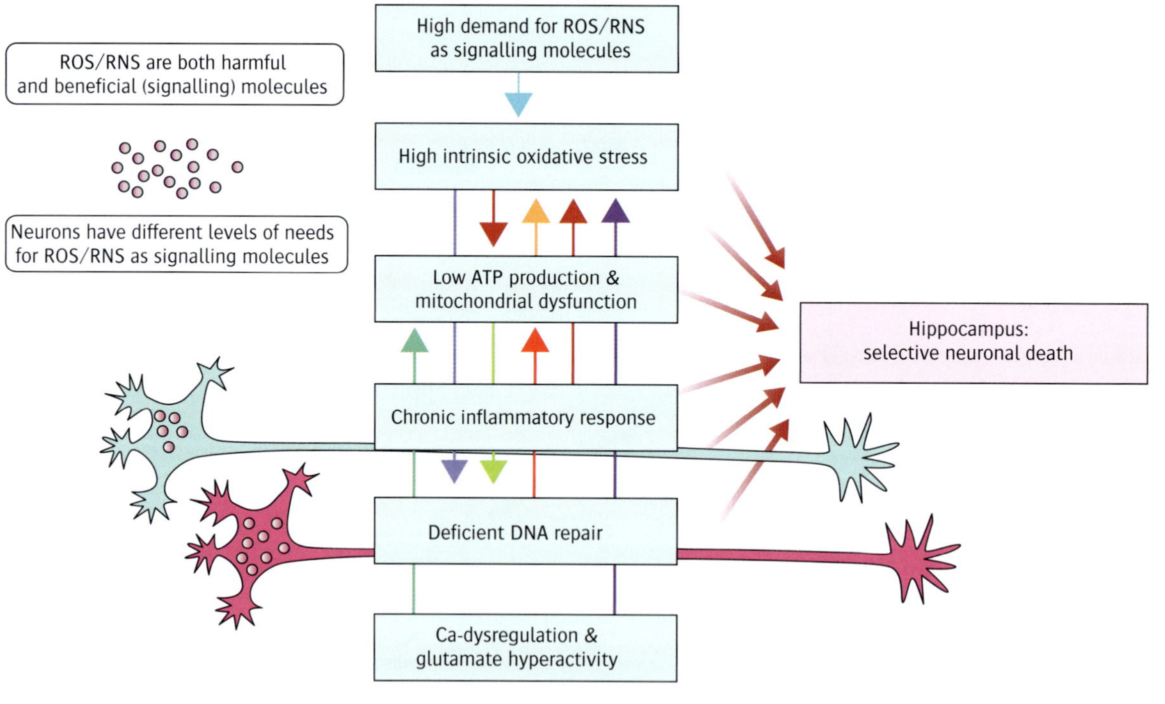

Figure 4.7 Schematic representation of the molecular and cellular factors that contribute to selective vulnerability of neurons to oxidative stress. As indicated in the text, neurons with enhanced sensitivity to OS are intrinsically generating ROS at a higher rate and are characterized by mitochondrial dysfunction, low ATP production, low homeostatic regulation of intracellular and intramitochondrial Ca^{2+}, and deficient DNA repair mechanisms and, therefore, increased DNA oxidative damage. Genes that are upregulated in these OS-vulnerable neurons suggest the activation of inflammatory responses. The combination of these processes, especially the low Ca^{2+}-buffering capacity and likely exposure to glutamate hyperactivity, would enhance OS for these neurons and tip the scales towards cell injury or cell death. The coloured arrows that link the factors shown in the figure indicate direct relationships between the various factors. Reproduced from Wang X., Michaelis E.K., Selective neuronal vulnerability to oxidative stress in the brain. *Front Aging Neurosci* 2 (12) © 2010, Wang X, Michaelis E.K.

provided ample support for the concept proposed nearly one hundred years ago, that of 'pathoclisis' (Vogt 1925). At the time that concept was advanced, the methods for individual cell characterization at the molecular and genetic level were simply not available, thus the idea of the 'physio-chemical' differences described as the basis of neuronal 'pathoclisis' in response to disease or stress could not have been explored. The combination of sophisticated genomic and proteomic analyses, in conjunction with biochemical and physiological measurements, are beginning to point to the endogenous differences between a vulnerable population of neurons and one that is resistant to the same sets of stresses. For example, inflammatory response and immune response have been described repeatedly as being associated with ageing and various neurodegenerative diseases. The same is true for differential regulation of Ca^{2+} in neurons or the generation of ROS (Figure 4.7). The definition of certain molecular pathways that may determine differential sensitivity of neurons to stress and disease should eventually lead to better therapeutic interventions for several neurodegenerative diseases.

Acknowledgements

This chapter was prepared with the support of the Higuchi Biosciences Center, University of Kansas, and grants from the National Institutes of Health, National Institute on Aging, AG12993 and AG035982.

References

Aoki, M., Volkmann, I., Tjernberg, L. O., Winblad, B., and Bogdanovic, N. (2008). Amyloid beta-peptide levels in laser capture microdissected cornu ammonis 1 pyramidal neurons of Alzheimer's brain. *Neuroreport*, 19, 1085–9.

Arnold, S. E., Franz, B. R., Gur, R. C., *et al.* (1995). Smaller neuron size in schizophrenia in hippocampal subfields that mediate cortical-hippocampal interactions. *Am J Psychiatry*, 152, 738–48.

Balazs, R., Jorgensen, O. S., and Hack, N. (1988). N-methyl-D-aspartate promotes the survival of cerebellar granule cells in culture. *Neuroscience*, 27, 437–51.

Ball, M. J. (1979). Topography of Pick inclusion bodies in hippocampi of demented patients. A quantitative study. *J Neuropathol Exp Neurol*, 38, 614–20.

Bao, X., Pal, R., Hascup, K. N., *et al.* (2009). Transgenic expression of Glud1 (glutamate dehydrogenase 1) in neurons: in vivo model of enhanced glutamate release, altered synaptic plasticity, and selective neuronal vulnerability. *J Neurosci*, 29, 13929–44.

Barnes, C. A. (1979). Memory deficits associated with senescence: a neurophysiological and behavioral study in the rat. *J Comp Physiol Psychol*, 93, 74–104.

Barnes, C. A., and McNaughton, B. L. (1980). Physiological compensation for loss of afferent synapses in rat hippocampal granule cells during senescence. *J Physiol*, 309, 473–85.

Barnes, C. A., and McNaughton, B. L. (1985). An age comparison of the rates of acquisition and forgetting of spatial information in relation to long-term enhancement of hippocampal synapses. *Behav Neurosci*, 99, 1040–8.

Barnes, C. A., Suster, M. S., Shen, J., and McNaughton, B. L. (1997). Multistability of cognitive maps in the hippocampus of old rats. *Nature*, 388, 272–5.

Barnes, J., Bartlett, J. W., van de Pol, L. A., *et al.* (2009). A meta-analysis of hippocampal atrophy rates in Alzheimer's disease. *Neurobiol Aging*, 30, 1711–23.

Behrens, M. I., Koh, J. Y., Muller, M. C., and Choi, D. W. (1996). NADPH diaphorase-containing striatal or cortical neurons are resistant to apoptosis. *Neurobiol Dis*, 3, 72–5.

Benes, F. M. (2007). Searching for unique endophenotypes for schizophrenia and bipolar disorder within neural circuits and their molecular regulatory mechanisms. *Schizophr Bull*, 33, 932–6.

Benes, F. M., Sorensen, I., and Bird, E. D. (1991). Reduced neuronal size in posterior hippocampus of schizophrenic patients. *Schizophr Bull*, 17, 597–608.

Benes, F. M., Kwok, E. W., Vincent, S. L., and Todtenkopf, M. S. (1998). A reduction of nonpyramidal cells in sector CA2 of schizophrenics and manic depressives. *Biol Psychiatry*, 44, 88–97.

Benes, F. M., Lim, B., Matzilevich, D., Walsh, J. P., Subburaju, S., and Minns, M. (2007). Regulation of the GABA cell phenotype in hippocampus of schizophrenics and bipolars. *Proc Natl Acad Sci U S A*, 104, 10164–9.

Bensch, K. G., Mott, J. L., Chang, S. W., *et al.* (2009). Selective mtDNA mutation accumulation results in beta-cell apoptosis and diabetes development. *Am J Physiol Endocrinol Metab*, 296, E672–80.

Bernaudin, M., Nouvelot, A., MacKenzie, E. T., and Petit, E. (1998). Selective neuronal vulnerability and specific glial reactions in hippocampal and neocortical organotypic cultures submitted to ischemia. *Exp Neurol*, 150, 30–9.

Blümcke, I., Kistner, I., Clusmann, H., *et al.* (2009). Towards a clinico-pathological classification of granule cell dispersion in human mesial temporal lobe epilepsies. *Acta Neuropathol*, 117, 535–44.

Braak, H., and Braak, E. (1991). Neuropathological stageing of Alzheimer-related changes. *Acta Neuropathol*, 82, 239–59.

Braak, H., Del Tredici, K., Rub, U., de Vos, R. A., Jansen Steur, E. N., and Braak, E. (2003). Staging of brain pathology related to sporadic Parkinson's disease. *Neurobiol Aging*, 24, 197–211.

Braak, H., Ghebremedhin, E., Rub, U., Bratzke, H., and Del Tredici, K. (2004). Stages in the development of Parkinson's disease-related pathology. *Cell Tissue Res*, 318, 121–34.

Buchan, A. M., and Pulsinelli, W. A. (1990). Septo-hippocampal deafferentation protects CA1 neurons against ischemic injury. *Brain Res*, 512, 7–14.

Cadenas, E., and Davies, K. J. (2000). Mitochondrial free radical generation, oxidative stress, and aging. *Free Radic Biol Med*, 29, 222–30.

Cajal, S. R. Y. (1894). *Les nouvelles idées sur la structure du système nerveux chez l'homme et chez les vertébrés*. Paris: C. Reinwald.

Carri, M. T., Ferri, A., Cozzolino, M., Calabrese, L., and Rotilio, G. (2003). Neurodegeneration in amyotrophic lateral sclerosis: the role of oxidative stress and altered homeostasis of metals. *Brain Res Bull*, 61, 365–74.

Castellani, R., Hirai, K., Aliev, G., *et al.* (2002). Role of mitochondrial dysfunction in Alzheimer's disease. *J Neurosci Res*, 70, 357–60.

Cataldo, A. M., Petanceska, S., Terio, N. B., *et al.* (2004). Abeta localization in abnormal endosomes: association with earliest Abeta elevations in AD and Down syndrome. *Neurobiol Aging*, 25, 1263–72.

Choi, D. W. (1992). Excitotoxic cell death. *J Neurobiol*, 23, 1261–76.

Choi, D. W., and Rothman, S. M. (1990). The role of glutamate neurotoxicity in hypoxic-ischemic neuronal death. *Annu Rev Neurosci*, 13, 171–82.

Constantinidis, J., Richard, J., and Tissot, R. (1974). Pick's disease. Histological and clinical correlations. *Eur Neurol*, 11, 208–17.

Cotman, C. W., Monaghan, D. T., and Ganong, A. H. (1988). Excitatory amino acid neurotransmission: NMDA receptors and Hebb-type synaptic plasticity. *Annu Rev Neurosci*, 11, 61–80.

Cottrell, D. A., Blakely, E. L., Johnson, M. A., Ince, P. G., and Turnbull, D. M. (2001). Mitochondrial enzyme-deficient hippocampal neurons and choroidal cells in AD. *Neurology*, 57, 260–4.

Crook, T., Bartus, R. T., Ferris, S., Whitehouse, P., Cohen, G., and Gershon, S. (1986). Age-associated memory impairment: Proposed diagnostic criteria and measures of clinical change - Report of a National Institue of Mental Health Work Group. *Dev Neuropsychiatry*, 2, 261–76.

Damier, P., Hirsch, E. C., Agid, Y., and Graybiel, A. M. (1999). The substantia nigra of the human brain. II. Patterns of loss of dopamine-containing neurons in Parkinson's disease. *Brain*, 122(Pt 8), 1437–48.

Dauer, W., and Przedborski, S. (2003). Parkinson's disease: mechanisms and models. *Neuron*, 39, 889–909.

Dawson, V. L., and Dawson, T. M. (1998). Nitric oxide in neurodegeneration. *Prog Brain Res*, 118, 215–29.

Dawson, V. L., Dawson, T. M., London, E. D., Bredt, D. S., and Snyder, S. H. (1991). Nitric oxide mediates glutamate neurotoxicity in primary cortical cultures. *Proc Natl Acad Sci U S A*, 88, 6368–71.

De Lacoste, M. C., and White, C. L., 3rd (1993). The role of cortical connectivity in Alzheimer's disease pathogenesis: a review and model system. *Neurobiol Aging*, 14, 1–16.

DeGracia, D. J., Rudolph, J., Roberts, G. G., Rafols, J. A., and Wang, J. (2007). Convergence of stress granules and protein aggregates in hippocampal cornu ammonis 1 at later reperfusion following global brain ischemia. *Neuroscience*, 146, 562–72.

Diemer, N.H., and Siemkowicz, E. (1981). Regional neurone damage after cerebral ischaemia in the normo- and hypoglycaemic rat. *Neuropathol Appl Neurobiol*, 7, 217–27.

Du, H., Guo, L., Fang F., *et al.* (2008). Cyclophilin D deficiency attenuates mitochondrial and neuronal perturbation and ameliorates learning and memory in Alzheimer's disease. *Nat Med*, 14, 1097–105.

Duan, H., Wearne, S. L., Rocher, A. B., Macedo, A., Morrison, J. H., and Hof, P. R. (2003). Age-related dendritic and spine changes in corticocortically projecting neurons in macaque monkeys. *Cereb Cortex*, 13, 950–61.

Earle, K. M., Baldwin, M., and Penfield, W. (1953). Incisural sclerosis and temporal lobe seizures produced by hippocampal herniation at birth. *AMA Arch Neurol Psychiatry*, 69, 27–42.

Esposito, F., Ammendola, R., Faraonio, R., Russo, T., and Cimino, F. (2004). Redox control of signal transduction, gene expression and cellular senescence. *Neurochem Res*, 29, 617–28.

Evans, S. J., Choudary, P. V., Vawter, M. P., *et al.* (2003). DNA microarray analysis of functionally discrete human brain regions reveals divergent transcriptional profiles. *Neurobiol Dis*, 14, 240–50.

Fahn, S., and Cohen, G. (1992). The oxidant stress hypothesis in Parkinson's disease: evidence supporting it. *Ann Neurol*, 32, 804–12.

Falconer, M. A. (1974). Mesial temporal (Ammon's horn) sclerosis as a common cause of epilepsy. Aetiology, treatment, and prevention. *Lancet*, 2, 767–70.

Falconer, M. A., and Taylor, D. C. (1968). Surgical treatment of drug-resistant epilepsy due to mesial temporal sclerosis. Etiology and significance. *Arch Neurol*, 19, 353–61.

Ferrante, R. J., Kowall, N. W., Beal, M. F., Richardson, E. P., Jr., Bird, E. D., and Martin, J. B. (1985). Selective sparing of a class of striatal neurons in Huntington's disease. *Science*, 230, 561–3.

Ferriero, D. M., Arcavi, L. J., Sagar, S. M., McIntosh, T. K., and Simon, R. P. (1988). Selective sparing of NADPH-diaphorase neurons in neonatal hypoxia-ischemia. *Ann Neurol*, 24, 670–6.

Finkel, T. (2003). Oxidant signals and oxidative stress. *Curr Opin Cell Biol*, 15, 247–54.

Friberg, H., Wieloch, T., and Castilho, R. F. (2002). Mitochondrial oxidative stress after global brain ischemia in rats. *Neurosci Lett*, 334, 111–14.

Fryd Johansen, F., Balslev Jorgensen, M., and Diemer, N. H. (1983). Resistance of hippocampal CA-1 interneurons to 20 min of transient cerebral ischemia in the rat. *Acta Neuropathol*, 61, 135–40.

Garthwaite, J. (1991). Glutamate, nitric oxide and cell-cell signalling in the nervous system. *Trends Neurosci*, 14, 60–7.

Gasic, G. P., and Hollmann, M. (1992). Molecular neurobiology of glutamate receptors. *Annu Rev Physiol*, 54, 507–36.

Gibson, G. E., Sheu, K. F., and Blass, J. P. (1998). Abnormalities of mitochondrial enzymes in Alzheimer disease. *J Neural Transm*, 105, 855–70.

Ginsberg, S. D., and Che, S. (2005). Expression profile analysis within the human hippocampus: comparison of CA1 and CA3 pyramidal neurons. *J Comp Neurol*, 487, 107–18.

Ginsberg, S. D., Alldred, M. J., Counts, S. E., *et al.* (2010). Microarray analysis of hippocampal CA1 neurons implicates early endosomal dysfunction during Alzheimer's disease progression. *Biol Psychiatry*, 68, 885–93.

Glazner, G. W., and Mattson, M. P. (2000). Differential effects of BDNF, ADNF9, and TNF[alpha] on levels of NMDA receptor subunits, calcium homeostasis, and neuronal vulnerability to excitotoxicity. *Exp Neurol*, 161, 442–52.

Grazina, M., Pratas, J., Silva, F., Oliveira, S., Santana, I., and Oliveira, C. (2006). Genetic basis of Alzheimer's dementia: role of mtDNA mutations. *Genes Brain Behav*, 5(Suppl 2), 92–107.

Greenwood, S. M., Mizielinska, S. M., Frenguelli, B. G., Harvey, J., and Connolly, C. N. (2007a). Mitochondrial dysfunction and dendritic beading during neuronal toxicity. *J Biol Chem*, 282, 26235–44.

Greenwood, S. M., Mizielinska, S. M., Frenguelli, B. G., Harvey, J., and Connolly, C. N. (2007b). Mitochondrial dysfunction and dendritic beading during neuronal toxicity. *J Biol Chem*, 282, 26235–44.

Halliwell, B., Gutteridge, J. M., and Cross, C. E. (1992). Free radicals, antioxidants, and human disease: where are we now? *J Lab Clin Med*, 119, 598–620.

Harman, D. (1956). Aging: a theory based on free radical and radiation chemistry. *J Gerontol*, 11, 298–300.

Hascup, K. N., Bao, X., Hascup, E. R., *et al.* (2011). Differential levels of glutamate dehydrogenase 1 (GLUD1) in BALB/c and C57BL/6 mice and the effects of over-expression of Glud1 gene on glutamate release in striatum. *ASN Neuro*, 3(2), e00057.

Hirsch, E., Graybiel, A. M., and Agid, Y. A. (1988). Melanized dopaminergic neurons are differentially susceptible to degeneration in Parkinson's disease. *Nature*, 334, 345–8.

Hof, P. R., Cox, K., and Morrison, J. H. (1990). Quantitative analysis of a vulnerable subset of pyramidal neurons in Alzheimer's disease: I. Superior frontal and inferior temporal cortex. *J Comp Neurol*, 301, 44–54.

Horn, M., and Schlote, W. (1992). Delayed neuronal death and delayed neuronal recovery in the human brain following global ischemia. *Acta Neuropathol*, 85, 79–87.

Hoskison, M. M., and Shuttleworth, C. W. (2006). Microtubule disruption, not calpain-dependent loss of MAP2, contributes to enduring NMDA-induced dendritic dysfunction in acute hippocampal slices. *Exp Neurol*, 202, 302–12.

Hoskison, M. M., Yanagawa, Y., Obata, K., and Shuttleworth, C. W. (2007). Calcium-dependent NMDA-induced dendritic injury and MAP2 loss in acute hippocampal slices. *Neuroscience*, 145, 66–79.

Houser, C. R., Harris, A. B., and Vaughn, J. E. (1986). Time course of the reduction of GABA terminals in a model of focal epilepsy: a glutamic acid decarboxylase immunocytochemical study. *Brain Res*, 383, 129–45.

Hyman, B. T., Van Horsen, G. W., Damasio, A. R., and Barnes, C. L. (1984). Alzheimer's disease: cell-specific pathology isolates the hippocampal formation. *Science*, 225, 1168–70.

Hyman, B. T., Marzloff, K., Wenniger, J. J., Dawson, T. M., Bredt, D. S., and Snyder, S. H. (1992). Relative sparing of nitric oxide synthase–containing neurons in the hippocampal formation in Alzheimer's disease. *Ann Neurol*, 32, 818–20.

Ince, P. G., Lowe, J., and Shaw, P. J. (1998). Amyotrophic lateral sclerosis: current issues in classification, pathogenesis and molecular pathology. *Neuropathol Appl Neurobiol*, 24, 104–17.

Jack, C. R., Jr., Petersen, R. C., Xu, Y., *et al.* (2000). Rates of hippocampal atrophy correlate with change in clinical status in aging and AD. *Neurology*, 55, 484–9.

Jenner, P. (2003). Oxidative stress in Parkinson's disease. *Ann Neurol*, 53(Suppl 3), S26–36; discussion S36–8.

Jorgensen, M. B., Finsen, B. R., Jensen, M. B., Castellano, B., Diemer, N. H., and Zimmer, J. (1993). Microglial and astroglial reactions to ischemic and kainic acid-induced lesions of the adult rat hippocampus. *Exp Neurol*, 120, 70–88.

Kamsler, A., and Segal, M. (2003). Paradoxical actions of hydrogen peroxide on long-term potentiation in transgenic superoxide dismutase-1 mice. *J Neurosci*, 23, 10359–67.

Khaitovich, P., Muetzel, B., She, X., *et al.* (2004). Regional patterns of gene expression in human and chimpanzee brains. *Genome Res*, 14, 1462–73.

Kintner, D. B., Chen, X., Currie, J., *et al.* (2010). Excessive Na^+/H^+ exchange in disruption of dendritic Na^+ and Ca^{2+} homeostasis and mitochondrial dysfunction following in vitro ischemia. *J Biol Chem*, 285, 35155–68.

Kirino, T. (1982). Delayed neuronal death in the gerbil hippocampus following ischemia. *Brain Res*, 239, 57–69.

Kirino, T., Tamura, A., and Sano, K. (1984). Delayed neuronal death in the rat hippocampus following transient forebrain ischemia. *Acta Neuropathol*, 64, 139–47.

Kitao, Y., Ozawa, K., Miyazaki, M., *et al.* (2001). Expression of the endoplasmic reticulum molecular chaperone (ORP150) rescues hippocampal neurons from glutamate toxicity. *J Clin Invest*, 108, 1439–50.

Koh, J. Y., Peters, S., and Choi, D. W. (1986). Neurons containing NADPH-diaphorase are selectively resistant to quinolinate toxicity. *Science*, 234, 73–6.

Koh, J. Y., Suh, S. W., Gwag, B. J., He, Y. Y., Hsu, C. Y., and Choi, D. W. (1996). The role of zinc in selective neuronal death after transient global cerebral ischemia. *Science*, 272, 1013–16.

Kujoth, G. C., Hiona, A., Pugh, T. D., *et al.* (2005). Mitochondrial DNA mutations, oxidative stress, and apoptosis in mammalian aging. *Science*, 309, 481–4.

La Joie, R., Fouquet, M., Mezenge, F., *et al.* (2010). Differential effect of age on hippocampal subfields assessed using a new high-resolution 3T MR sequence. *Neuroimage*, 53, 506–14.

Lein, E. S., Zhao, X., and Gage, F. H. (2004). Defining a molecular atlas of the hippocampus using DNA microarrays and high-throughput in situ hybridization. *J Neurosci*, 24, 3879–89.

Lenaz, G. (2001). The mitochondrial production of reactive oxygen species: mechanisms and implications in human pathology. *IUBMB Life*, 52, 159–64.

Li, S., Mallory, M., Alford, M., Tanaka, S., and Masliah, E. (1997). Glutamate transporter alterations in Alzheimer disease are possibly associated with abnormal APP expression. *J Neuropathol Exp Neurol*, 56, 901–11.

Liang, W. S., Dunckley, T., Beach, T. G., *et al.* (2007). Gene expression profiles in anatomically and functionally distinct regions of the normal aged human brain. *Physiol Genomics*, 28, 311–22.

Lipton, S. A., Sucher, N. J., Kaiser, P. K., and Dreyer, E. B. (1991). Synergistic effects of HIV coat protein and NMDA receptor-mediated neurotoxicity. *Neuron*, 7, 111–18.

Lokkegaard, A., Nyengaard, J. R., and West, M. J. (2001). Stereological estimates of number and length of capillaries in subdivisions of the human hippocampal region. *Hippocampus*, 11, 726–40.

Lu, T., Pan, Y., Kao, S. Y., *et al.* (2004). Gene regulation and DNA damage in the ageing human brain. *Nature*, 429, 883–91.

Madsen, P. L., Hasselbalch, S. G., Hagemann, L. P., *et al.* (1995). Persistent resetting of the cerebral oxygen/glucose uptake ratio by brain activation: evidence obtained with the Kety-Schmidt technique. *J Cereb Blood Flow Metab*, 15, 485–91.

Masliah, E., Mallory, M., Hansen, L., DeTeresa, R., and Terry, R. D. (1993). Quantitative synaptic alterations in the human neocortex during normal aging. *Neurology*, 43, 192–7.

Matsugami, T. R., Tanemura, K., Mieda, M., *et al.* (2006). From the Cover: Indispensability of the glutamate transporters GLAST and GLT1 to brain development. *Proc Natl Acad Sci U S A*, 103, 12161–6.

Mattson, M. P., and Magnus, T. (2006). Ageing and neuronal vulnerability. *Nat Rev Neurosci*, 7, 278–94.

McCormack, J. G., Halestrap, A. P., and Denton, R. M. (1990). Role of calcium ions in regulation of mammalian intramitochondrial metabolism. *Physiol Rev*, 70, 391–425.

Mecocci, P., MacGarvey, U., and Beal, M. F. (1994). Oxidative damage to mitochondrial DNA is increased in Alzheimer's disease. *Ann Neurol*, 36, 747–51.

Meencke, H. J. (2009). Clinical neuropathology of the epilepsies in the 100 years of the ILAE (1909–2009). *Epilepsia*, 50(Suppl 3), 8–16.

Meldrum, B. S., Vigouroux, R. A., and Brierley, J. B. (1973a). Systemic factors and epileptic brain damage. Prolonged seizures in paralyzed, artificially ventilated baboons. *Arch Neurol*, 29, 82–7.

Meldrum, B. S., Vigouroux, R. A., Rage, P., and Brierley, J. B. (1973b). Hippocampal lesions produced by prolonged seizures in paralyzed artificially ventilated baboons. *Experientia*, 29, 561–3.

Mello, L. E., Cavalheiro, E. A., Tan, A. M., *et al.* (1993). Circuit mechanisms of seizures in the pilocarpine model of chronic epilepsy: cell loss and mossy fiber sprouting. *Epilepsia*, 34, 985–95.

Meyer, A. (1936). The selective regional vulnerability of the brain and its relation to psychiatric problems. *Proc R Soc Med*, 29, 1175–81.

Michaelis, E. K. (1998). Molecular biology of glutamate receptors in the central nervous system and their role in excitotoxicity, oxidative stress and aging. *Prog Neurobiol*, 54, 369–415.

Michaelis, E. K., Wang, X., Pal, R., *et al.* (2011). Neuronal Glud1 (glutamate dehydrogenase 1) over-expressing mice: Increased glutamate formation and synaptic release, loss of synaptic activity, and adaptive changes in genomic expression. *Neurochem Int*, 59, 473–81.

Miclescu, A., Sharma, H. S., Martijn, C., and Wiklund, L. (2010). Methylene blue protects the cortical blood-brain barrier against ischemia/reperfusion-induced disruptions. *Crit Care Med*, 38, 2199–206.

Morrison, J. H., and Hof, P. R. (1997). Life and death of neurons in the aging brain. *Science*, 278, 412–19.

Mueller, S. G., Stables, L., Du, A. T., *et al.* (2007). Measurement of hippocampal subfields and age-related changes with high resolution MRI at 4T. *Neurobiol Aging*, 28, 719–26.

Nelson, M. D., Saykin, A. J., Flashman, L. A., and Riordan, H.J. (1998). Hippocampal volume reduction in schizophrenia as assessed by magnetic resonance imaging: a meta-analytic study. *Arch Gen Psychiatry*, 55, 433–40.

Newrzella, D., Pahlavan, P. S., Kruger, C., *et al.* (2007). The functional genome of CA1 and CA3 neurons under native conditions and in response to ischemia. *BMC Genomics*, 8, 370.

Nickell, J., Pomerleau, F., Allen, J., and Gerhardt, G. A. (2005). Age-related changes in the dynamics of potassium-evoked L-glutamate release in the striatum of Fischer 344 rats. *J Neural Transm*, 112, 87–96.

Olsson, T., Wieloch, T., and Smith, M. L. (2003). Brain damage in a mouse model of global cerebral ischemia. Effect of NMDA receptor blockade. *Brain Res*, 982, 260–9.

Parker, W. D., Jr., Filley, C. M., and Parks, J. K. (1990). Cytochrome oxidase deficiency in Alzheimer's disease. *Neurology*, 40, 1302–3.

Parker, W. D., Jr., Mahr, N. J., *et al.* (1994a). Reduced platelet cytochrome c oxidase activity in Alzheimer's disease. *Neurology*, 44, 1086–90.

Parker, W. D., Jr., Parks, J., Filley, C. M., and Kleinschmidt-DeMasters, B. K. (1994b). Electron transport chain defects in Alzheimer's disease brain. *Neurology*, 44, 1090–6.

Peng, P. L., Zhong, X., Tu, W., *et al.* (2006). ADAR2-dependent RNA editing of AMPA receptor subunit GluR2 determines vulnerability of neurons in forebrain ischemia. *Neuron*, 49, 719–33.

Petersen, R. C., Jack, C. R., Jr., Xu, Y. C., *et al.* (2000). Memory and MRI-based hippocampal volumes in aging and AD. *Neurology*, 54, 581–7.

Pulsinelli, W. A. (1985). Selective neuronal vulnerability: morphological and molecular characteristics. *Prog Brain Res*, 63, 29–37.

Pulsinelli, W. A., Brierley, J. B., and Plum, F. (1982). Temporal profile of neuronal damage in a model of transient forebrain ischemia. *Ann Neurol*, 11, 491–8.

Rajah, M. N., Kromas, M., Han, J. E., and Pruessner, J. C. (2010). Group differences in anterior hippocampal volume and in the retrieval of spatial and temporal context memory in healthy young versus older adults. *Neuropsychologia*, 48, 4020–30.

Reddy, P. H., and Beal, M. F. (2008). Amyloid beta, mitochondrial dysfunction and synaptic damage: implications for cognitive decline in aging and Alzheimer's disease. *Trends Mol Med*, 14, 45–53.

Reddy, P. H., Williams, M., Charles, V., *et al.* (1998). Behavioural abnormalities and selective neuronal loss in HD transgenic mice expressing mutated full-length HD cDNA. *Nat Genet*, 20, 198–202.

Reddy, P. H., Charles, V., Williams, M., Miller, G., Whetsell, W. O., Jr., and Tagle, D. A. (1999). Transgenic mice expressing mutated full-length HD cDNA: a paradigm for locomotor changes and selective neuronal loss in Huntington's disease. *Philos Trans R Soc Lond B Biol Sci*, 354, 1035–45.

Ribak, C. E., Hunt, C. A., Bakay, R. A., and Oertel, W. H. (1986). A decrease in the number of GABAergic somata is associated with the preferential loss of GABAergic terminals at epileptic foci. *Brain Res*, 363, 78–90.

Rickhag, M., Wieloch, T., Gido, G., *et al.* (2006). Comprehensive regional and temporal gene expression profiling of the rat brain during the first 24 h after experimental stroke identifies dynamic ischemia-induced gene expression patterns, and reveals a biphasic activation of genes in surviving tissue. *J Neurochem*, 96, 14–29.

Rival, T., Soustelle, L., Strambi, C., Besson, M. T., Iche, M., and Birman, S. (2004). Decreasing glutamate buffering capacity triggers oxidative stress and neuropil degeneration in the Drosophila brain. *Curr Biol*, 14, 599–605.

Rizzuto, R., Bernardi, P., and Pozzan, T. (2000). Mitochondria as all-round players of the calcium game. *J Physiol*, 529(Pt 1), 37–47.

Rosenzweig, E. S., Rao, G., McNaughton, B. L., and Barnes, C. A. (1997). Role of temporal summation in age-related long-term potentiation-induction deficits. *Hippocampus*, 7, 549–58.

Rothstein, J. D., Tsai, G., Kuncl, R. W., *et al.* (1990). Abnormal excitatory amino acid metabolism in amyotrophic lateral sclerosis. *Ann Neurol*, 28, 18–25.

Rothstein, J. D., Jin, L., Dykes-Hoberg, M., and Kuncl, R. W. (1993). Chronic inhibition of glutamate uptake produces a model of slow neurotoxicity. *Proc Natl Acad Sci U S A*, 90, 6591–5.

Rothstein, J. D., Dykes-Hoberg, M., Pardo, C. A., *et al.* (1996). Knockout of glutamate transporters reveals a major role for astroglial transport in excitotoxicity and clearance of glutamate. *Neuron*, 16, 675–86.

Rowland, L. P., and Shneider, N. A. (2001). Amyotrophic lateral sclerosis. *N Engl J Med*, 344, 1688–700.

Sadek, H. A., Szweda, P.A., and Szweda, L. I. (2004). Modulation of mitochondrial complex I activity by reversible Ca^{2+} and NADH mediated superoxide anion dependent inhibition. *Biochemistry*, 43, 8494–502.

Sagar, H. J., and Oxbury, J. M. (1987). Hippocampal neuron loss in temporal lobe epilepsy: correlation with early childhood convulsions. *Ann Neurol*, 22, 334–40.

Sayre, L. M., Smith, M. A., and Perry, G. (2001). Chemistry and biochemistry of oxidative stress in neurodegenerative disease. *Curr Med Chem*, 8, 721–38.

Scheff, S. W., and Price, D. A. (2003). Synaptic pathology in Alzheimer's disease: a review of ultrastructural studies. *Neurobiol Aging*, 24, 1029–46.

Scheff, S. W., Price, D. A., Schmitt, F. A., and Mufson, E. J. (2006). Hippocampal synaptic loss in early Alzheimer's disease and mild cognitive impairment. *Neurobiol Aging*, 27, 1372–84.

Schmidt-Kastner, R., and Freund, T. F. (1991). Selective vulnerability of the hippocampus in brain ischemia. *Neuroscience*, 40, 599–636.

Scholz, W. (1953). Selective neuronal necrosis and its topistic patterns in hypoxemia and olegemia. *J Neuropathol Exp Neurol*, 12, 249–61.

Scholz, W., and Hager, H. (Eds.) (1956). *Epilepsie*. Berlin: Springer.

Schwarcz, R., and Meldrum, B. (1985). Excitatory aminoacid antagonists provide a therapeutic approach to neurological disorders. *Lancet*, 2, 140–3.

Scott, H. L., Pow, D. V., Tannenberg, A. E., and Dodd, P. R. (2002). Aberrant expression of the glutamate transporter excitatory amino acid transporter 1 (EAAT1) in Alzheimer's disease. *J Neurosci*, 22, RC206.

Sharma, H. S., Miclescu, A., and Wiklund, L. (2011). Cardiac arrest-induced regional blood-brain barrier breakdown, edema formation and brain pathology: a light and electron microscopic study on a new model for neurodegeneration and neuroprotection in porcine brain. *J Neural Transm*, 118(1), 87–114.

Siesjo, B. K. (1988). Mechanisms of ischemic brain damage. *Crit Care Med*, 16, 954–63.

Simic, G., Kostovic, I., Winblad, B., and Bogdanovic, N. (1997). Volume and number of neurons of the human hippocampal formation in normal aging and Alzheimer's disease. *J Comp Neurol*, 379, 482–94.

Smith, D. E., Rapp, P. R., McKay, H. M., Roberts, J. A., and Tuszynski, M. H. (2004). Memory impairment in aged primates is associated with focal death of cortical neurons and atrophy of subcortical neurons. *J Neurosci*, 24, 4373–81.

Spargo, E., Everall, I. P., and Lantos, P. L. (1993). Neuronal loss in the hippocampus in Huntington's disease: a comparison with HIV infection. *J Neurol Neurosurg Psychiatry*, 56, 487–91.

Spielmeyer, W. (1927). Die Pathogenese des epileptisches Krampfes. Histopathologischer Teil. (Pathogenesis of the epileptic seizure. Histopathological part). *Ztschr Neurol Psychiat*, 109, 501–19.

Spielmeyer, W. (1931). The significance of local factors for electivity in central nervous system disease processes. *Medicine*, 10, 243–56.

Squire, L. R. (1992). Memory and the hippocampus: a synthesis from findings with rats, monkeys, and humans. *Psychol Rev*, 99, 195–231.

Squire, L. R., Ojemann, J. G., Miezin, F. M., Petersen, S. E., Videen, T. O., and Raichle, M. E. (1992). Activation of the hippocampus in normal humans: a functional anatomical study of memory. *Proc Natl Acad Sci U S A*, 89, 1837–41.

Stanika, R. I., Winters, C. A., Pivovarova, N. B., and Andrews, S. B. (2010). Differential NMDA receptor-dependent calcium loading and mitochondrial dysfunction in CA1 vs. CA3 hippocampal neurons. *Neurobiol Dis*, 37, 403–11.

Szatkowski, M., and Attwell, D. (1994). Triggering and execution of neuronal death in brain ischaemia: two phases of glutamate release by different mechanisms. *Trends Neurosci*, 17, 359–65.

Tanaka, K., Watase, K., Manabe, T., *et al.* (1997). Epilepsy and exacerbation of brain injury in mice lacking the glutamate transporter GLT-1. *Science*, 276, 1699–702.

Terry, R. D., and Katzman, R. (2001). Life span and synapses: will there be a primary senile dementia? *Neurobiol Aging*, 22, 347–8; discussion 353–44.

Terry, R. D., DeTeresa, R., and Hansen, L. A. (1987). Neocortical cell counts in normal human adult aging. *Ann Neurol*, 21, 530–9.

Thiels, E., Urban, N. N., Gonzalez-Burgos, G. R., *et al.* (2000). Impairment of long-term potentiation and associative memory in mice that overexpress extracellular superoxide dismutase. *J Neurosci*, 20, 7631–9.

Torres-Munoz, J. E., Van Waveren, C., Keegan, M. G., Bookman, R. J., and Petito, C. K. (2004). Gene expression profiles in microdissected neurons from human hippocampal subregions. *Brain Res, Mol Brain Res*, 127, 105–14.

Turrens, J. F. (1997). Superoxide production by the mitochondrial respiratory chain. *Biosci Rep*, 17, 3–8.

Uttl, B., and Graf, P. (1993). Episodic spatial memory in adulthood. *Psychol Aging*, 8, 257–73.

Vaishnavi, S. N., Vlassenko, A. G., Rundle, M. M., Snyder, A. Z., Mintun, M. A., and Raichle, M. E. (2010). Regional aerobic glycolysis in the human brain. *Proc Natl Acad Sci U S A*, 107, 17757–62.

Valko, M., Leibfritz, D., Moncol, J., Cronin, M. T., Mazur, M., and Telser, J. (2007). Free radicals and antioxidants in normal physiological functions and human disease. *Int J Biochem Cell Biol*, 39, 44–84.

Valtschanoff, J. G., Weinberg, R. J., Kharazia, V. N., Nakane, M., and Schmidt, H. H. (1993a). Neurons in rat hippocampus that synthesize nitric oxide. *J Comp Neurol*, 331, 111–21.

Valtschanoff, J. G., Weinberg, R. J., Kharazia, V. N., Schmidt, H. H., Nakane, M., and Rustioni, A. (1993b). Neurons in rat cerebral cortex that synthesize nitric oxide: NADPH diaphorase histochemistry, NOS immunocytochemistry, and colocalization with GABA. *Neurosci Lett*, 157, 157–61.

van Muiswinkel, F. L., de Vos, R. A., Bol, J. G., *et al.* (2004). Expression of NAD(P)H:quinone oxidoreductase in the normal and Parkinsonian substantia nigra. *Neurobiol Aging*, 25, 1253–62.

Vlassenko, A. G., Vaishnavi, S. N., Couture, L., *et al.* (2010). Spatial correlation between brain aerobic glycolysis and amyloid-beta (Abeta) deposition. *Proc Natl Acad Sci U S A*, 107, 17763–7.

Vogt, C., and Vogt, O. (1937). Sitz und Wesen der Krankheiten im Lichte der topistischen Hirnforschung und des Variierens der Tiere . *J Psychol Neurol*, 47, 237–457.

Vogt, O. (1925). Der Begriff der Pathoklise. *J Psychol Neurol*, 31, 245–55.

Wang, X., and Michaelis, E. K. (2010). Selective neuronal vulnerability to oxidative stress in the brain. *Front Aging Neurosci*, 2, 12.

Wang, J., Xiong, S., Xie, C., Markesbery, W. R., and Lovell, M. A. (2005a). Increased oxidative damage in nuclear and mitochondrial DNA in Alzheimer's disease. *J Neurochem*, 93, 953–62.

Wang, X., Pal, R., Chen, X. W., Limpeanchob, N., Kumar, K. N., and Michaelis, E. K. (2005b). High intrinsic oxidative stress may underlie selective vulnerability of the hippocampal CA1 region. *Brain Res Mol Brain Res*, 140, 120–6.

Wang, X., Pal, R., Chen, X. W., Kumar, K. N., Kim, O. J., and Michaelis, E. K. (2007). Genome-wide transcriptome profiling of region-specific vulnerability to oxidative stress in the hippocampus. *Genomics*, 90, 201–12.

Wang, X., Zaidi, A., Pal, R., *et al.* (2009). Genomic and biochemical approaches in the discovery of mechanisms for selective neuronal vulnerability to oxidative stress. *BMC Neurosci*, 10, 12.

Wang, X., Bao, X., Pal, R., Agbas, A., and Michaelis, E. K. (2010a). Transcriptomic responses in mouse brain exposed to chronic excess of the neurotransmitter glutamate. *BMC Genomics*, 11, 360.

Wang, X., Michaelis, M. L., and Michaelis, E. K. (2010b). Functional genomics of brain aging and alzheimer's disease, focus on selective neuronal vulnerability. *Curr Genom*, 11, 618–33.

Weinberger, D. R. (1999). Cell biology of the hippocampal formation in schizophrenia. *Biol Psychiatry*, 45, 395–402.

Wilde, G. J., Pringle, A. K., Wright, P., and Iannotti, F. (1997). Differential vulnerability of the CA1 and CA3 subfields of the hippocampus to superoxide and hydroxyl radicals in vitro. *J Neurochem*, 69, 883–6.

Winkelmann, E. R., Charcansky, A., Faccioni-Heuser, M. C., Netto, C. A., and Achaval, M. (2006). An ultrastructural analysis of cellular death in the CA1 field in the rat hippocampus after transient forebrain ischemia followed by 2, 4 and 10 days of reperfusion. *Anat Embryol (Berl)*, 211, 423–34.

Wisniewski, H. M., and Terry, R. D. (1973). Morphology of the aging brain, human and animal. *Prog Brain Res*, 40, 167–86.

Witter, M. P., Groenewegen, H. J., Lopes da Silva, F. H., and Lohman, A. H. (1989). Functional organization of the extrinsic and intrinsic circuitry of the parahippocampal region. *Prog Neurobiol*, 33, 161–253.

Wu, K., Aoki, C., Elste, A., Rogalski-Wilk, A. A., and Siekevitz, P. (1997). The synthesis of ATP by glycolytic enzymes in the postsynaptic density and the effect of endogenously generated nitric oxide. *Proc Natl Acad Sci U S A*, 94, 13273–8.

Xu, X., Zhan, M., Duan, W., *et al.* (2007). Gene expression atlas of the mouse central nervous system: impact and interactions of age, energy intake and gender. *Genome Biol*, 8, R234.

Xu, Y., Jack, C. R., Jr., O'Brien, P. C., *et al.* (2000). Usefulness of MRI measures of entorhinal cortex versus hippocampus in AD. *Neurology*, 54, 1760–7.

Zeier, Z., Madorsky, I., Xu, Y., Ogle, W. O., Notterpek, L., and Foster, T. C. (2011). Gene expression in the hippocampus: regionally specific effects of aging and caloric restriction. *Mech Ageing Dev*, 132, 8–19.

Zeng, L. H., Ouyang, Y., Gazit, V., *et al.* (2007). Abnormal glutamate homeostasis and impaired synaptic plasticity and learning in a mouse model of tuberous sclerosis complex. *Neurobiol Dis*, 28, 184–96.

Zhu, X., Raina, A. K., Lee, H. G., Casadesus, G., Smith, M. A., and Perry, G. (2004a). Oxidative stress signalling in Alzheimer's disease. *Brain Res*, 1000, 32–9.

Zhu, X., Smith, M. A., Perry, G., and Aliev, G. (2004b). Mitochondrial failures in Alzheimer's disease. *Am J Alzheimers Dis Other Demen*, 19, 345–52.

Zoia, C., Cogliati, T., Tagliabue, E., *et al.* (2004). Glutamate transporters in platelets: EAAT1 decrease in aging and in Alzheimer's disease. *Neurobiol Aging*, 25, 149–57.

Zoia, C. P., Tagliabue, E., Isella, V., *et al.* (2005). Fibroblast glutamate transport in aging and in AD: correlations with disease severity. *Neurobiol Aging*, 26, 825–32.

Zola-Morgan, S., Squire, L. R., Rempel, N. L., Clower, R. P., and Amaral, D. G. (1992). Enduring memory impairment in monkeys after ischemic damage to the hippocampus. *J Neurosci*, 12, 2582–96.

CHAPTER 5

Stress and the Hippocampus

E. Ronald de Kloet

Introduction 77

Pulsatility of basal HPA axis activity 80

Stress-induced HPA axis activity 82

MR- and GR-mediated behavioural effects 83

Neuroendocrine regulation 87

Morphology 88

Network function 89

Molecular effects 90

MR and GR polymorphisms 92

Early life experience 93

Implication for diseases of the hippocampus 95

ABSTRACT Environmental stimuli perceived as stressors and processed in higher brain circuits including the hippocampus activate the release of several neuropeptides eventually leading to the secretion of adrenal cortisol which feeds back on the brain. To exert this feedback, the action of cortisol is mediated by nuclear receptors that operate as gene transcription factors but can also directly affect neurotransmission. There are two types of receptors, the mineralocorticoid (MR) and glucocorticoid receptors (GR). Both serve complementary as master switch in the control of neural network responses that facilitate the defence and recovery of homeostasis underlying behavioural adaptation. Imbalance in MR:GR driven pathways, caused either by genetic receptor variants or by experince-related factors, compromises processing of stressful information. Therapies are therefore envisioned to *rebalance* the stress system for protection, resilience or repair from damaging signalling pathways that can introduce a bias towards stress-related brain disease. In this chapter, the basal pulsatility and the stress-induced modes of operation of the hypothalamic–pituitary–adrenal axis are described. Further, central highlights of cortisol action are presented on different levels of biological complexity from emotions and cognitive performance to their molecular underpinnings. Next is a discussion on genetic receptor variants and the programming of brain and behaviour by early experience. A brief synthesis of cortisol actions for clinical neurobiology of the hippocampus and some thoughts on potential treatment strategies conclude the chapter.

KEYWORDS hippocampus, stress, brain, development, behaviour, hormones, glucocorticoids, receptor polymorphisms, gene transcription, depression, ageing

Introduction

The late Seymour Levine defined 'stress' as a composite, multidimensional construct, in which three components interact: (1) the input, when the stressor is perceived and appraised, (2) the processing of stressful information and (3) the output or stress response. All three components interact via complex self-regulating feedback loops with the goal to restore homeostasis through behavioural and physiological adaptations (Levine 2005).

Selye (1936, 1950), who coined the stress concept in the 1930s called this effort of the organism to adapt to noxious stimuli the 'general adaptation syndrome' and distinguished during the course of stress exposure an initial phase of *alarm* with a subsequent phase of *resistance* over days and weeks in which the individual seemingly coped with the chronic stressor and finally leading to an *exhaustion* phase. The latter phase of exhaustion refers to breakdown of adaptation. This negative chain of events initiated by failure to cope was termed *distress* by Selye. *Eustress*, positive stress, refers to activation of the stress system in a positive context such as exercise or anticipation of rewards, a condition that is actually beneficial to the organism.

While Selye focused mainly on the (patho-)physiology of the stressor and the stress response, the following research of Mason, Weiss, Levine, and others emphasized that it is the *processing of information* resulting in the experience of the stressor and that the ability to cope and to adapt is the quintessence of the understanding of stress. Thus, the most severe stressor is a psychological condition characterized by lack of information to predict upcoming events. The sense of control is lost with an uncertain anxious feeling of threat, either real or imagined.

The ability to cope with such a psychological stressor is dependent on experience- and gene-related factors, and is affected by cognitive and environmental inputs. Coping resources rely also on the context associated with the stressful experience. This context can be determined by psychosocial factors such as, for example, social position, social support, or attachment to a care giver. If any of these factors is disrupted—e.g. loss of control in a social environment, expulsion from social support, homelessness, or deprivation of (maternal) care—the challenge may exceed the coping resources and produce strong emotional reactions, which ultimately may lead to a condition of *chronic stress* (Selye's exhaustion phase) and enhanced vulnerability to mental disease (Lazarus 2006).

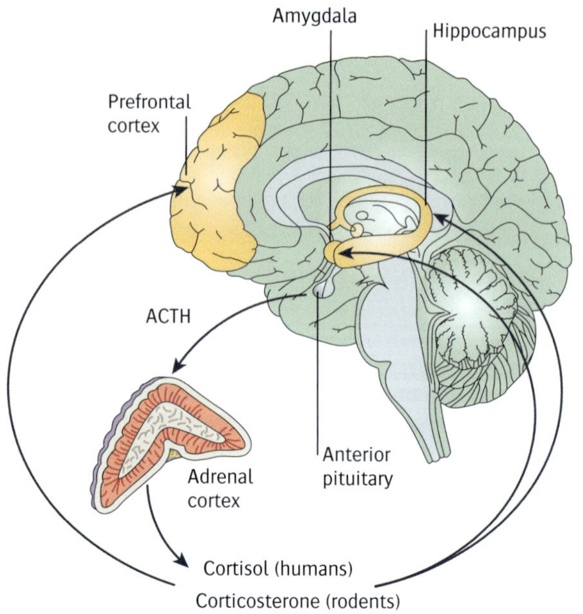

Figure 5.1 Schematic representation of the hypothalamic–pituitary–adrenal (HPA) axis. The scheme demonstrates parts of the limbic system, i.e. amygdala, hippocampus, prefrontal cortex, the anterior pituitary, and the adrenal cortex in the context of the intrinsic feedback connectivity. Reprinted by permission from Macmillan Publishers Ltd: Krugers, H. J., Hoogenraad, C. C., and Groc, L. (2010). Stress hormones and AMPA receptor trafficking in synaptic plasticity and memory. *Nature Reviews Neuroscience*, 11(10), 675–81, © 2010.

The neuroendocrine substrate of stress

It is commonly thought that the enhanced vulnerability to stress is the result of the interaction of experience-related factors with genetic predisposition. To understand the stress-diathesis concept, it is essential to study how the integrative communication systems operate that process stressful information and coordinate the reaction to a stressor. These systems include the two branches of the autonomic nervous system, the parasympathetic and sympathetic nervous system which will not further be discussed here. The responses of the sympathetic nervous system and the hypothalamic–pituitary–adrenal (HPA) axis are orchestrated by corticotrophin-releasing hormone (CRH) and vasopressin secreted by the parvocellular neurons of the paraventricular nucleus (PVN) in the hypothalamus (Watts 2005) (Figure 5.1).

CRH and its co-secretagogue vasopressin drive the HPA axis by enhancing the synthesis and release of pro-opiomelanocortin (POMC) peptides of the anterior pituitary corticotrophs. One of these peptides, adrenocorticotropin (ACTH), stimulates the adrenocortical secretion of glucocorticoids via melanocortin (MC-2) receptors. Subsequently, the glucorticoids secreted by the adrenals feedback on every tissue including the brain. The glucocorticoids also exert negative feedback action on the core of the HPA axis, the PVN, and the pituitary corticotrophs to facilitate recovery of baseline activity after stress-induced activation.

The hippocampus and amygdala are major targets for the glucocorticoid hormones, cortisol and/or corticosterone secreted after stress. In these tissues, the hormone regulates information processing by affecting synaptic plasticity, neurotransmission, and energy metabolism (McEwen 2007). Moreover, neurogenesis in a very discrete part of the hippocampus, the dentate gyrus, is also regulated by the glucocorticoid hormone. The hormone operates in concert with the other stress signals of the HPA axis as well as the autonomic nervous system to promote the plasticity of genes, circuits, and behaviour while processing (stressful) information. This dynamic change in brain function and structure has led to the concept of allostasis. Bruce McEwen defined 'allostasis' as 'acquiring stability through change' and the cost to achieve these changes 'allostatic load' (McEwen and Wingfield 2010). In this concept allostatic mediators of brain plasticity underlying adaptation (such as, e.g. glucocorticoids) are distinguished from those mediators that do *not* facilitate change but rather maintain stable homeostatic setpoints essential for life such as ion concentrations and pH.

The hippocampus *must have* a crucial role in the allostasis concept underlying the diathesis–stress model of mental disease. It *must have* because it is an important target for glucocorticoids (McEwen et al. 1968) and the hippocampus, together with the amygdala, are part of the limbic circuitry involved in the processing of novel (stressful) information. Emotional reactions generated from the amygdala are burned into memory. The hippocampus is for the declarative facts: it is thought that the 'emotional input' from the amygdala receives hippocampal labelling in time, space, and context during information processing and storage in memory. This multidimensional encoding, if needed later, facilitates the retrieval of a proper coping response (Akirav et al. 2001).

Receptors

Here, the effect of stress on the hippocampus refers to the effects of the main glucocorticoids cortisol and corticosterone (the latter for rodents only), collectively called CORT. These actions exerted by CORT are 'conditional'—CORT by itself does not evoke brain activity—meaning that they operate only in the context of information processing. The effects of CORT in the CNS are mediated by two types of nuclear receptors: mineralocorticoid (MR) and glucocorticoid (GR) receptors, which act in the cell nucleus as gene transcription factors (Reul and de Kloet 1985; Evans and Arriza 1989). The MR is abundantly expressed in limbic structures such as the hippocampus, amygdala, septum, and prefrontal cortical regions. The GR is ubiquitous with highest abundance in the typical stress centres of the brain: PVN, hippocampus, amygdala, ascending aminergic neurons (Herman et al. 2003).

Pivotal for the understanding of integrative effects is that MR displays a tenfold higher affinity for CORT than GR (Reul and de Kloet 1985). This has far reaching consequences since for this reason GR is only activated by stress levels of CORT. Recently, however, also rapid non-genomic actions at the level of the cell membrane were firmly established for these very same receptor proteins (Karst et al. 2005, 2010; Joëls et al. 2008). The membrane MR and GR have a lower affinity for CORT making these rapid non-genomic actions also eligible for regulation of stress-related events (Karst et al. 2005) (Figure 5.2).

The brain MR binds both the mineralocorticoid aldosterone as well as the glucocorticoid CORT with high affinity, but is structurally identical to the MR present in the epithelial cells of kidney, colon, and sweat glands where aldosterone is the preferred ligand. This is because in these epithelial cells CORT is inactivated by the enzyme 11βHSD

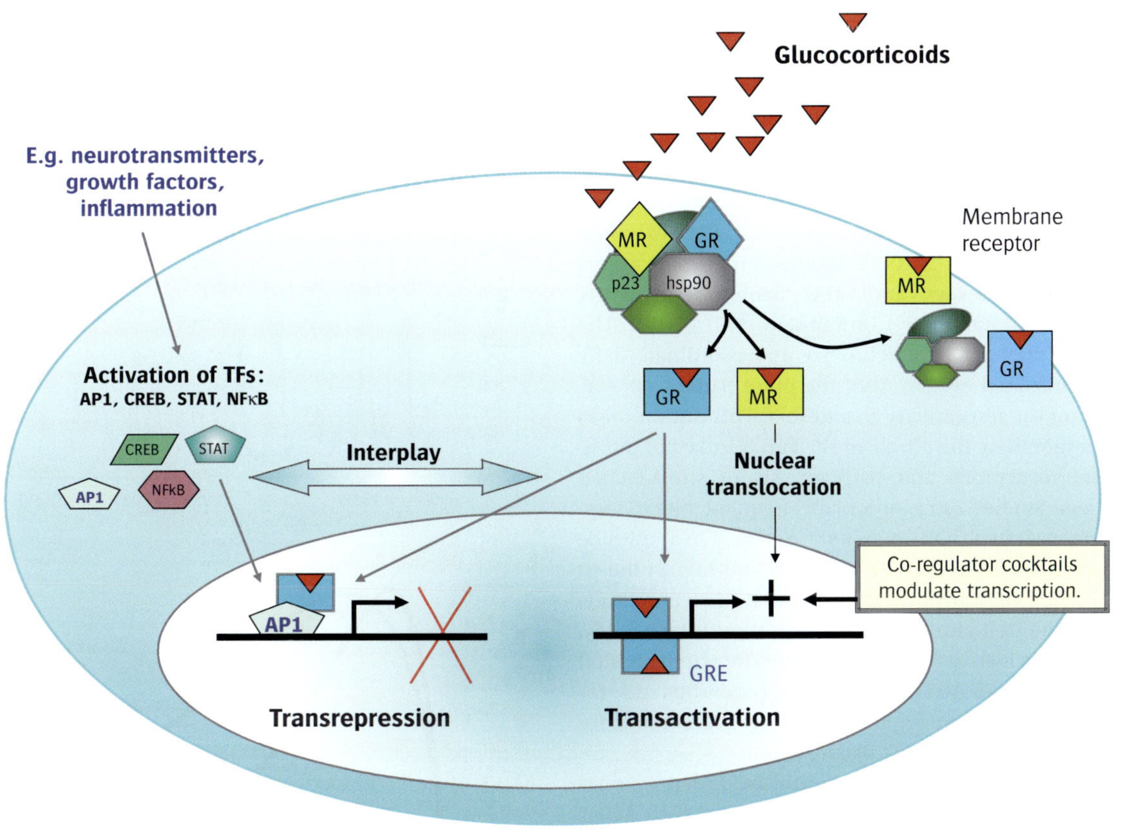

Figure 5.2 Mechanism of glucocorticoid action. The receptors as part of a multimeric protein bind the glucocorticoids, which then dissociate and translocate to the cell nucleus. MR, mineralocorticoid receptors; GR, glucocorticoid receptors. MR and GR are transcription factors that regulate as dimers gene transcription; receptor function is also regulated by a cocktail of co-regulators. GR also interacts as a monomer with transcription factors (CREB, AP1, NFκB). MR and GR also function as membrane receptors modulating neurotransmission directly. Image courtesy of Dr N. Datson, Medical Pharmacology, University of Leiden, The Netherlands.

type 2 rendering aldosterone as the certified mineralocorticoid controlling Na retention. In the brain, 11βHSD type 2 is only expressed in the circumventricular organs that mediate effects of aldosterone on salt appetite (Holmes and Seckl 2006). Instead, the 11β-HSD type 1 isoform is present in non-epithelial cells of heart, vascular wall, and brain, and there the enzyme acts as a reductase capable to regenerate bioactive cortisol from its inactive precursor (Funder 2010). Both aldosterone and CORT therefore have access to the brain including hippocampal MR, but because CORT circulates in much higher concentrations than aldosterone the brain MR *sees* predominantly CORT (Figure 5.3).

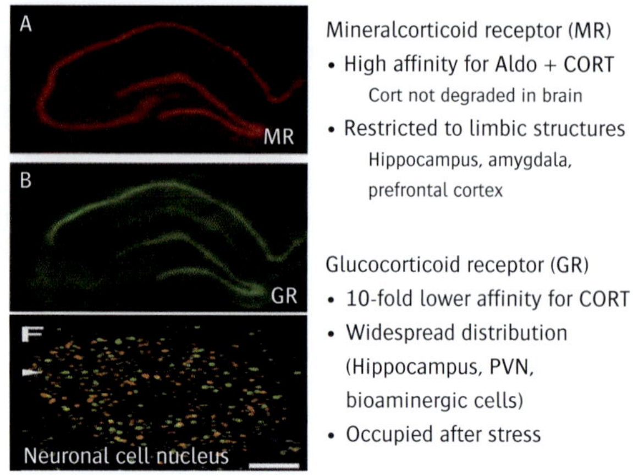

Figure 5.3 Colocalization of MR and GR in the hippocampus. The properties and localization of MR and GR are also described.

Goal

In this chapter, I will focus on CORT and its receptors MR and GR in the hippocampus with respect to their functioning under basal and stressful conditions from gene to behaviour and how the function of these receptors is affected by gene polymorphisms and (early life) experience. The leading theme in the chapter is the notion that the balance in MR- and GR-mediated actions is important for stable neuroendocrine regulations and effective behavioural performance (de Kloet 1991; de Kloet et al. 1998, 2005). The clinical implications are that a balanced MR:GR mediated action aims to promote adaptation to stress and resilience. In contrast, a MR:GR imbalance enhances vulnerability to stress-related mental disorders.

Pulsatility of basal HPA axis activity

CORT is secreted under basal conditions in hourly pulses, which are produced within minutes and usually last about 20 min followed by reciprocal period until the next pulse arrives (Walker et al. 2010). The ultradian rhythm occurs in humans, monkeys, and other mammals including rodents (Lightman and Conway-Campbell 2010). The pulses are thought to synchronize and coordinate daily and sleep-related events. Also, the pulse pattern appears crucial for the responsivity to stressors. This notion raises the question how the ultradian rhythm is linked to physiological regulations and pathogenesis of stress-related disorders. Studies using automatic frequent blood sampling showed that CORT pulses are increased in amplitude towards the activity period, which is the night-time in rodents, and day-time in man. This variation in amplitude conveys the well-known circadian rhythm. Interestingly, the profound increase in CORT towards the activity period is uncoupled from ACTH, suggesting an additional role for the splanchnic nerve input sensitizing the adrenal for ACTH at the circadian peak (Figure 5.4).

Thus, the hypercortisolaemia during severe depression is most likely a result of increased amplitudes of both ACTH and CORT pulses (Holsboer 2000; Carroll et al. 2007), particularly at the nadir of the rhythm. This finding would explain the flattening of the circadian rhythm characteristic for depression. Inflammatory disorders are characterized by increased frequency rather than amplitude of the

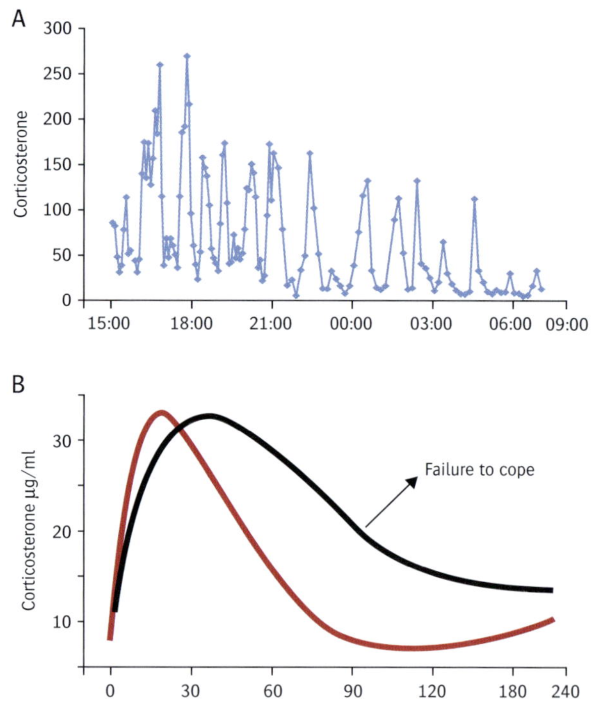

Figure 5.4 Pulsatile (4A) and stress-induced CORT (4B) secretion. The latter figure shows that a prolonged secretion of CORT develops under conditions of failure to cope with stress.

ultradian rhythm (Windle et al. 2001). During the process of ageing, the ACTH–CORT pulsatile patterns become disordered which is reflected in the loss of circadian patterns in daily activity and sleep (Veldhuis et al. 2000; Young et al. 2004). In exercised rats, the increase in amplitude towards the activity period is more pronounced. The pattern of pulsatility therefore varies over physiological and pathological conditions. Accordingly, frequency encoding is an important modus operandi of the HPA axis.

The ultradian pattern of CORT release is reflected in the free CORT concentration in hippocampal extracellular fluid as measured with microdialysis (Droste et al. 2008). The fraction of free CORT not bound to corticosteroid binding globulin (CBG) varies as a function of the concentration, genetic polymorphisms, and in particular of temperature changes, such as those occurring during fever (Cabrera et al. 2000). Interestingly, the stress-induced rise in CORT lags in the brain with a 20-min delay before maximal concentrations are reached after which the ultradian rhythm is rapidly resumed. The reason for this delay in brain uptake of CORT during stress is not known. It may be due to rapidly changing properties of CBG in the blood to which the majority of circulating CORT is bound (only a fraction is free hormone under resting conditions) or the blood–brain barrier may change in properties during stress (Qian et al. 2011). How the access of the steroids to the brain is regulated, is not precisely known. Corticosterone is not a substrate for multidrug resistance P-glycoprotein (Pgp), while exogenous steroids for the rat such as cortisol and the synthetic glucocorticoids are exported by the Pgp transporter (Meijer et al. 1998; Karssen et al. 2001).

The consequence of ultradian, circadian, and stress-induced rises in CORT for receptor translocation to the nucleus of hippocampal neurons has been extensively studied. MR and GR are abundantly co-expressed in CA1 and CA2 pyramidal neurons and dentate gyrus granular neurons, while GR expression (but not MR) is low in CA3 neurons. As pointed out previously, the tenfold higher affinity of CORT for MR than GR has profound consequences for the occupancy of the receptors during the ultradian rhythm. Using immunocytochemistry and confocal microscopy, it was found that the pulses in CORT are sufficient to maintain occupancy of MR and retention of the steroid-MR complex in the nucleus over the hourly interval. In contrast, the immunocytochemical staining also demonstrated that GR translocates in and out the nucleus in parallel with the ultradian rhythm (Conway-Campbell et al. 2007; Sarabdjitsingh et al. 2010b).

The concomitant GR activation and DNA binding was demonstrated by glucocorticoid response element oligonucleotide binding, and verified for the Clock gene Period 1(Per 1) promoter region by chromatin immunoprecipitation assays. Strikingly, each CORT pulse induced a 'burst' of transcription of Per 1 measured by heterogeneous nuclear RNA quantitative polymerase chain reaction. The net effect of pulsatile CORT exposure on accumulation of the mature transcript was also assessed, revealing a plateau of mRNA levels throughout the time course of pulsatile exposure, indicating that the pulse timing works optimally for steady state Per1 expression. The plateau dropped to baseline within 120 min of the final pulse, further indicating a relatively short half-life for hippocampal Per1. The significance of this strict temporal control is that any perturbation to the pulse frequency or duration would have rapid quantitative effects on the levels of Per1. This in turn could affect hippocampal function, especially with regard to circadian-related memory and learning processes (Conway-Campbell et al. 2010).

The question then is how responsiveness of hippocampal target (plasticity) genes may change under different regimes of pulsatility. This question was examined by comparing the response of the target genes Gilz and Sgk-1. Rats were implanted subcutaneously (sc) with vehicle or 40% or 100% CORT pellets known to flatten ultradian and circadian rhythmicity while maintaining daily average levels or mimicking pathologically high CORT levels. The findings showed that the stable (non-pulsatile) patterns of CORT following sc pellets dose-dependently downregulated GR and attenuated MR and GR translocation to the acute challenge. Interestingly, whereas target gene Gilz expression to the challenge was already attenuated by stable levels around the daily average (40%), Sgk-1 was affected only after constant high CORT exposure (100%), indicating altered receptor responsiveness due to treatment. The data suggest that sustained CORT levels, disturbing pulsatility, attenuate GR signalling, and target gene responsiveness to an acute challenge in the rat hippocampus (Sarabdjitsingh et al. 2010b). Interestingly, this lack of tissue responsiveness (resistance or lack of resilience) also occurs during psychotic and depressive states characterized by flattened circadian rhythmicity and resistance to endogenous CORT.

CORT secretory bursts can be triggered by a stressor any time and are superimposed on the ultradian and circadian rhythms. Interestingly, stress-induced ACTH and CORT release is differentially regulated, depending on the phase of the ultradian pulse (Windle et al. 1998). Post hoc analysis revealed that responsiveness to stressors is enhanced in the ascending phase, and attenuated if experienced during the descending slope of the pulse. In a recent study, we (Sarabdjitsingh et al. 2010a, 2010c) have modelled ultradian variations in stress responsiveness by artificially creating different patterns of CORT in adrenalectomized (ADX) rats. As was the case in the post hoc analysis pulsatile administration of CORT facilitated a brisker response, which was markedly greater in the rising than in the falling phase of a CORT pulse. This differential phase-dependent effect was also seen in emotional reactivity and the behavioural response to noise, which was much greater in the rising phase. This finding raises the possibility that stress responsivity and coping responses show hourly changes, a notion that has not been investigated yet.

In summary, the HPA axis displays an hourly pulsatility, which is prominent at the adrenal level where the ACTH signal can be amplified by a non-ACTH mechanism. CORT pulsatility coordinates brain and body function under basal conditions. The pulse generator itself is not known, but any system with a delay in feedback action has to oscillate. Frequency, amplitude, and pattern of the pulses may change with profound consequences for responsiveness and resilience of the stress system and its targets.

Stress-induced HPA axis activity

Severely depressed individuals suffering from psychotic episodes generally display periods of *hypercortisolaemia* because of CORT feedback resistance, adrenal growth, and adrenocortical hyper-responsivity. In contrast, *hypocortisolaemia* due to enhanced feedback action and impaired adrenocortical responsivity was proposed to be characteristic for individuals suffering from anxiety disorders such as post-traumatic stress disorder (PTSD) (Yehuda 2002). The link of hypocortisolaemia with PTSD is, however, not that strong since a meta-analysis of all available studies found controversial results, as PTSD is often accompanied by depression (Meewisse et al. 2007). In fact, recent evidence links PTSD as an anxiety disorder with the sympathetic nervous system and adrenergic receptors (Rasch et al. 2009). In a discussion of some distinct controversies in stress research, Fink (2011) quoted Mason et al. (2002) stating that 'rather than a simplistic static *hypo- or hyper*-cortisolemia linked to mental disorders, the lability in CORT changes pointed to a more complex regulatory dysfunction of the HPA axis involving a propensity for relatively large swings in CORT levels, probably in both downward as upward directions'. Indeed, in response to a stressor a rapid onset of CORT secretion is a feature of a healthy organism as long as termination of the CORT response also occurs efficiently. If a CORT response starts sluggish, and/or is either excessive, prolonged, or inadequate, this aberrant stress response pattern is usually indicative for (proneness to) a stress-related disorder.

These dysregulative patterns in the HPA axis can be in part identified by endocrine challenge tests such as the dexamethasone suppression test with or without combination with CRH (Dex-CRH). These tests are primarily designed for identification of pituitary function; in mental disorders they provide some additional insight in the balance between central drive and hormonal feedback indicative of impaired processing of stressful information in higher brain regions. Note that during severe depression and psychosis, the Dex-CRH test reveals glucocorticoid resistance implying that CORT action is actually inadequate to contain the stress or CRH-induced activations (Heuser et al. 1994; Holsboer and Ising 2010).

The action exerted by CORT in the brain on processing of information underlying psychological stress reactions follows the guiding principle in any stress reaction as was explained in detail by Allan Munck (Munck et al. 1984; Sapolsky et al. 2000) and Tausk (1952). CORT feeds back precisely on those processes that initially have activated the HPA axis. This can be an inflammatory response, a metabolic disturbance, a reduction in blood volume or—as is the topic in this chapter—a limbic molecular and cellular reaction to a psychological stressor.

The action of CORT has two modes of operation. On the one hand, CORT promotes the immediate stress reaction and, on the other hand, subsequently prevents these initial reactions from overshooting and becoming damaging themselves. In initial stress reactions, CORT operates in a *pro-active* mode as is obvious from its pro-inflammatory action. The later action preventing these initial stress reactions from overshooting is *reactive* and was metaphorically defined by Marius Tausk (1952) as: 'glucocorticoids protect against the water damage caused by the fire brigade', and represents the well-known anti-inflammatory action of synthetic glucocorticoids such as prednisone. In this context, Selye distinguished the pro-inflammatory mineralocorticoid actions from anti-inflammatory glucocorticoid actions in his pendulum hypothesis. As will be pointed out later in his chapter, these opposing actions of mineralocorticoids and glucocorticoids observed by Selye are exerted in the limbic brain (hippocampus) by one single hormone: CORT. We discovered that in the limbic brain the action of CORT on processing of stressful information is mediated by MR and GR, operating in a complementary fashion (Reul and de Kloet 1985; de Kloet and Reul 1987).

Our research has demonstrated that during a stressful experience CORT first facilitates appraisal processes conveyed by the limbic circuitry with the goal to retrieve and organize an immediate behavioural response to deal with the stressor, an action mediated by MR. Subsequently, CORT via GR promotes encoding of the information, and stimulates consolidation of the experiences in the hippocampal circuits for future use in case a similar challenge occurs (de Kloet et al. 2005). These actions exerted by CORT are linked to sympathetic nervous activity. On the one hand, CORT promotes these immediate actions of the sympathetic nervous as displayed behaviourally by either the fight, flight, or fright modes of the coping response with immediate danger, whereas later these actions are suppressed by CORT during recovery and behavioural adaptation.

The limbic amygdala, hippocampus, and prefrontal cortical regions express MR and GR in abundance. Inputs from the amygdala and ascending aminergic pathways convey emotional and stressful information that activates the limbic circuitry and the PVN, producing CRH and vasopressin, in the latter nucleus for activation of the sympathetic nervous system and the HPA axis. The hippocampus and prefrontal cortex in turn feed back to the amygdala and communicate in trans-synaptic fashion with the PVN core of the HPA axis. Thus, excitatory inputs from the hippocampus activate a local inhibitory GABAergic network innervating the parvocellular neurons of the PVN providing an inhibitory influence on the HPA axis (Herman et al. 2003). Accordingly, via MR and GR, CORT modulates neuronal input from the limbic structures towards the PVN on multiple sites. The HPA core in parvocellular CRH neurons and the pituitary corticotrophs predominantly express GR. Considerable MR and GR coexpression is found in hippocampal pyramidal cell fields (except the adult CA3 with minimal GR expression) and dentate gyrus, the amygdaloid and lateral septal nuclei as well as some cortical areas. MR and GR colocalization is found in the hippocampus of all species including humans and new world primates.

In summary, in concert with the other stress mediators, the action of CORT in the hippocampus is a double-edged sword. Via MR, CORT promotes the initial reaction to novel information by activating processes underlying appraisal and selection of coping responses. Subsequently, via GR the experience is stored in memory for future use. CORT provides the energy for these costly processes underlying behavioural adaptation. The resulting neuroendocrine response as regulated in the HPA axis operates with CORT as a closed feedback loop, modulated by information from higher brain regions reaching the PVN via trans-synaptic communication.

MR- and GR-mediated behavioural effects

After these general aspects of CORT action on endocrine regulation and behaviour, a number of specific examples will be discussed in which the role of the hippocampus is highlighted in distinct MR- and GR-mediated actions regarding: (1) timing, (2) emotional input, (3) findings with transgenesis, (4) switching between memory systems, (5) the role of other stress mediators, (6) the implications of receptor imbalance, and (7) its translational aspects.

Timing

Complex learning paradigms involving the hippocampus such as spatial orientation are affected by CORT in a coordinate manner through MR and GR. This can be demonstrated using the Morris water maze by the intracerebroventricular (icv) administration of GR antagonists which, if administered immediately after the learning phase, impaired retrieval measured 24 h later (Oitzl and de Kloet 1992). Since the GR antagonist was not effective immediately before the retrieval test, this finding indicated that the antagonist interferes with consolidation rather than retrieval processes. Further support came from a study employing GR dim/dim mouse mutants which carry a point mutation preventing GR dimerization and DNA binding: these mice were also impaired in memory storage since CORT, released during the learning task, could not exert its effect (Oitzl et al. 2001). Clearly, GR activation in the hippocampus seems a prerequisite for storage of relevant information to be used in the future (Figure 5.5).

Conversely, MR blockade did not interfere with memory storage and the latency to locate the platform when given after the learning trial, but was effective when given before the retrieval session 24 h later. To reveal the effect of MR blockade the platform was removed. Control rats swim directly to the former platform location and remain there. MR blockade reduced latency in the former platform quadrant and the rats displayed an altered search pattern. This emphasizes the role of MR in the rapid appraisal of environmental stimuli and subsequent choice of behavioural strategy to escape from the water maze (Oitzl and de Kloet 1992).

While task-related elevations of CORT thus seem essential for consolidation (Oitzl and de Kloet 1992; Quirarte et al. 1997; Sandi et al. 1997), GR activation prior to retrieval can apparently interfere with earlier acquired information. For instance, stress prior to the retrieval test in a Morris maze also reduced latency spent in the former platform quadrant (de Quervain et al. 1998). This could be interpreted as a retrieval deficit entirely due to a GR-mediated action. The data, however, can also be regarded as behavioural extinction through active reconsolidation of new, more appropriate information implying, for example, that the previous emotion-provoking situation is no longer relevant—a view similar to the extinction of Pavlovian conditioning (de Kloet et al. 1999; Myers and Davis 2002).

Extinction and retrieval are closely related memory processes. Apparently, incoming information not relevant to the situation is suppressed by CORT, as demonstrated by the plasticity of hippocampal neurons (Wiegert et al, 2006; Pu et al. 2007). These actions are predominantly GR dependent, but require the concomitant action of CORT action via MR and of other stress signals such as, for example, CRH, vasopressin, and the catecholamines.

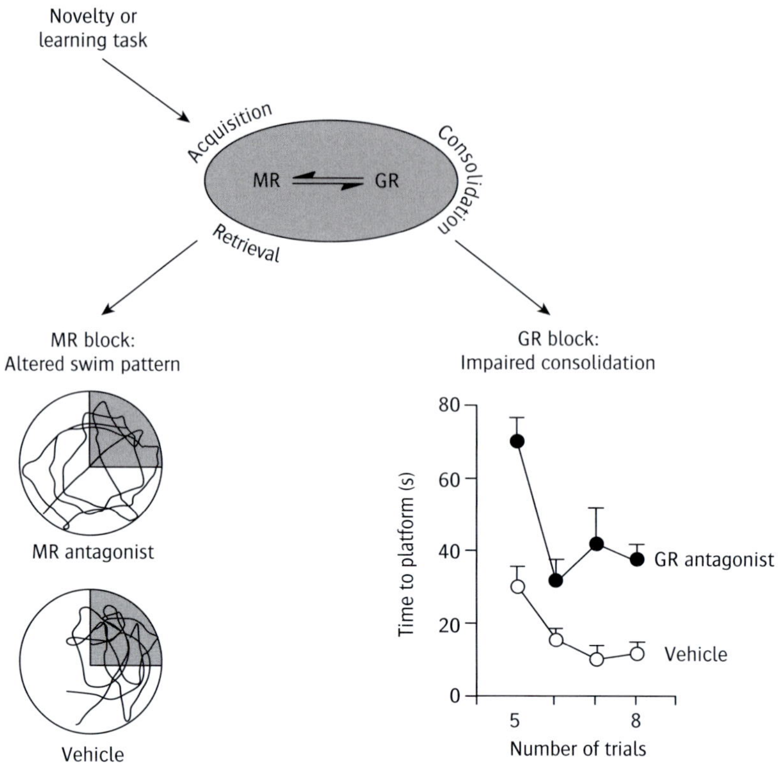

Figure 5.5 Information processing modulated by CORT action via MR and GR in the Morris water maze. GR blockade with a glucocorticoid antagonist administered immediately after the learning session blocks consolidation of the learned swimming pattern measured 24 h later at the retrieval session. MR blockade only works immediately before the retrieval session and results in an altered search pattern for the platform. Reproduced from Oitzl, M. S. and de Kloet, E. R. (1992). Selective corticosteroid antagonists modulate specific aspects of spatial orientation learning. *Behavioural Neuroscience*, 106(1), 62–71, © 1992.

Emotions

Arousing emotions are not only relevant to strengthen memory, but, if occurring in another context than previously experienced, can also facilitate its extinction (Bohus and de Kloet 1981; Oitzl and de Kloet 1992; Brinks et al. 2009b). This feature of CORT action had been demonstrated already about three decades ago in a forced extinction paradigm. In a one-step-through inhibitory (passive) avoidance response paradigm animals are placed in the shock compartment 3 h after exposure to the mild electric shock, and this is sufficient to extinguish the passive avoidance response. This so-called 'forced extinction' was eliminated if 1 h prior to the extinction session the adrenals were removed. It was found that the extinction of the fear-related response depended selectively on CORT replacement at the time of ADX as steroids such as dexamethasone and deoxycorticosterone were not effective (Bohus and de Kloet 1981). CORT was also effective icv and the action could be blocked by anti-mineralocorticoids indicating a crucial role for the brain MR in this rapid appraisal during the forced extinction paradigm that the shock compartment had become a safe place.

Administration of the GR-antagonist RU38486 in rats prior to as well as briefly after training in a fear conditioning paradigm impaired contextual memory when tested 24 h later. The antagonist was not effective, however, if retention was tested 3 h after training. This supports the role of GR in the consolidation of memory. Interestingly, pre- (but not post-) training administration of MR antagonist spironolactone impaired contextual memory, both at 3 and 24 h after training. Spironolactone also impaired tone-cue memory, but only at 4 h after training. These results reveal that—in addition to GR—MR also play a critical role in establishing fear memories, particularly in the early phase of memory formation (Zhou et al. 2010).

The combined contribution of MR and GR is also illustrated for another fear-motivated hippocampus-dependent behaviour in mice, the hole board task, which is a task that allows simultaneous registration of emotional and memory parameters (Brinks et al. 2007a). In contrast to rats, mice that undergo ADX continue to secrete low concentrations of CORT from residual adrenal cells (Hummel 1958). The concentrations of circulating CORT are too low for GR activation, but provide an excellent model for predominant MR activation. Different degrees of continuous GR activation were achieved via CORT releasing sc pellet implants. By using 'principal component analysis statistics', the contribution of fear-induced emotion and motivation to behavioural performance could be determined. The data revealed a U-shaped dose response curve for cognitive performance. Predominant MR activation facilitated the precipitation of an emotional state characterized by low anxiety, high motivation, goal-directed exploration, and enhanced behavioural reactivity, but learning was not promoted. At the other end, the excessive GR activation by high CORT concentrations precipitates high anxiety with impaired learning. This pattern of inappropriate emotions linked to decline of cognitive functions also occurs during chronic stress, while the opposite is observed with chronic anti-glucocorticoid administration icv, which actually improves cognitive performance (Oitzl et al. 1998). *Optimal cognitive performance with low anxiety and high motivation* was observed in mice having in addition to MR a moderate GR activation, produced by the average level of circulating CORT under physiological conditions (Brinks et al. 2007b).

Acute stress can facilitate territorial aggression which persists for weeks (Wood et al. 2003), and CORT rising in anticipation of the violent encounter is critical for this escalated aggression (Kruk et al. 2004). Also, CORT facilitates the transition of playfight into serious fighting (Wommack and Delville 2007). By contrast, blocking CORT synthesis with metyrapone (Mikics et al. 2004), or blocking the MR with spironolactone briefly prior to the first encounter strongly inhibits territorial aggression (Haller et al. 1998). Further studies showed that the CORT response to stress has an enduring MR dependent effect on the future propensity for violent aggression in individuals initially inexperienced with such serious conflicts (Kruk and de Kloet unpublished).

Transgenesis

To specifically examine the role of GR in the brain, Tronche et al. (1999) generated mice with a brain-specific deletion of GR ($GR^{Nes/Cre}$) that showed an impaired behavioural response to stress and in spite of increased CRH expression, as expected reduced anxiety-related behaviour. Gass and coworkers have generated mouse strains that underexpress or overexpress GR, but maintain the regulatory genetic context controlling the GR gene (Ridder et al. 2005). To achieve this goal, they used GR-heterozygous mutant mice ($GR^{+/-}$) with a 50% GR gene dose reduction having a phenotype of increased helplessness and anxiety. They also generated mutants overexpressing GR by a yeast artificial chromosome resulting in a twofold gene dose elevation (YGR mice) showing the opposite, less helplessness, and less anxiety. However, different results are reported when transgenic mice overexpress GR specifically in forebrain (Wei et al. 2004), namely increased anxiety- and depression-like behaviours in line with the effect of high CORT concentration. Hence, it seems that cell specificity in GR manipulation contributes to the overall behavioural phenotype.

MR overexpression in the forebrain decreases anxiety-like behaviour and modifies GR and serotonin 5HT-1a receptor expression in the hippocampus (Rozeboom et al. 2007; Lai et al 2007). In contrast, forebrain $MR^{-CaMK-Cre}$ knockout in mice that have the MR inactivated using the Cre/loxP-recombination system enhances emotional arousal after stress. Acquisition of contextual fear memory is also enhanced. These mutant mice show impaired learning of the water-maze task and deficits in measures of working memory on the radial maze due to behavioural perseverance and stereotypy (Berger et al. 2006). The study of the role of the MR is complex because the (loss of) function of the receptor becomes particularly apparent upon stress exposure a few minutes before testing and the effects vary over different mouse strains (Oitzl et al. 1994; Brinks et al. 2009a). This is in line with the finding that the MR antagonists only act prior to the acquisition and encoding of context dependent fear motivated responses (Zhou et al. 2010). The behavioural changes are associated with abnormalities of the mossy fibre projection and an upregulation of GR expression in the hippocampus (Berger et al. 2006).

Viral transfer to site-specific overexpression or knockdown of genes, also rapidly modulates MR and/or GR gene expression (Ferguson and Sapolsky 2008). A transient increase in MR levels in dentate gyrus granule cells enhanced the acquisition (a consolidation) of non-spatial memory and protected the rats from the disruptive effects of high CORT on the retrieval of non-spatial memory. On the other hand, knockdown of GR by siRNA bilaterally injected into the dentate gyrus of male BALB/c mice resulted 3–4 weeks later in an impaired consolidation of fear memories, while the initial anxiety-related behaviour during conditioning were comparable to control injected mice. This GR-mediated effect in the progenitor cells appears related to the positioning of the newborn neurons in the hippocampal circuitry (Fitzsimons et al. submitted).

Switching

Switching between memory systems was examined in a series of pioneering studies by Melly Oitzl and Lars Schwabe (Schwabe et al. 2010a, 2010b). To examine whether stress and CORT are capable of facilitating the switch between multiple memory systems in mice, a task was designed that allowed measurement of nucleus caudate-based *stimulus-response* and hippocampus-based *spatial learning* strategies. Naïve mice used spatial strategies to locate an exit hole on a circular hole board at a fixed location flagged by a proximal stimulus. When the mice were either stressed or received CORT before the task, 30–50% of the mice switched to a stimulus-response strategy. This switch between learning strategies was accompanied by a rescue of performance, whereas performance declined in the stressed mice that kept using the spatial strategy. Pretreatment with an MR antagonist prevented the switch toward the stimulus-response strategy but led to a deterioration of hippocampus-dependent performance in these stressed mice (Schwabe et al. 2010a). These findings show that CORT promotes the transition from spatial to stimulus-response memory systems in an MR-dependent fashion. In humans, stress prior to learning facilitated simple stimulus-response learning strategies at the expense of a more cognitive learning strategy (Schwabe et al. 2010b).

Other stress mediators

Memories are stronger when a situation is particularly arousing (Joëls et al. 2006) and convergence of adrenaline, neuropeptides such as e.g. vasopressin and CRH, and CORT actions is extremely important for this purpose. If the signals do not converge, prior CORT attenuates adrenaline action as was demonstrated by Borrell et al (1983, 1984), who showed that CORT replacement of ADX rats attenuated the facilitatory effect of adrenalin on the retention of an inhibitory avoidance response. In the prior presence of CORT the dose of epinephrine administered post-training had to be increased ten thousand fold in order to enhance retention of behaviour in this fear conditioning test. For an adequate consolidation of memory, the concomittant pharmacological manipulation of aminergic and CORT input into limbic structures is required (McGaugh 2000; Roozendaal et al. 2004, 2006, 2009).

The relevance of convergence in signalling is reflected at the cellular hippocampal level: CORT regulates β-adrenergic modulation of synaptic plasticity in CA1 and dentate gyrus in opposite directions, depending on the timing of hormone application (Joëls and de Kloet 1989; Pu et al. 2007). Pretreatment with CORT attenuates the adrenergic signal, but the steroid amplifies if administered at the time of adrenergic stimulation.

Stress models

Experimental procedures for chronic stress models differ over studies, but the consensus is that the most powerful designs are based on lack of predictability and control. Without doubt, long-lasting exposure to high amounts of glucocorticoids and its impact on energy resources threatens the adaptive capacity of the organism. Further, cognitive processes are endangered. Studies usually report deficits in hippocampus-dependent spatial learning and memory tasks (Lupien et al. 2009). However, chronic stress enhanced fear conditioning to the context, thus strengthening memory for the adverse event, most probably as a result of stress-induced sensitisation of amygdala function (Sandi et al. 2001; Sandi 2004). Thus, chronic stress impairs hippocampal spatial memory, but sensitizes for amygdaloid fear conditioning and striatal habit learning; the opposite responses in hippocampus and amygdala demonstrate the loss of adequate behavioural responses.

The effects exerted by chronically elevated CORT and chronic stress depend on genotype (Grootendorst et al. 2001; Brinks et al. 2009b). For example, apolipoprotein E (apoE) knockout mice are impaired in spatial learning, which is ameliorated by exposure to the stress of chronic rat exposure or additional exogenous CORT. Non-mutant littermates and C57BL/6J mice (the background strain) had the opposite response: the chronic rat stressor or high exogenous CORT impaired their cognitive functions (Grootendorst et al. 2001).

Translational aspects

Chronic stress procedures and prolonged elevated CORT levels resulted in depression-associated behaviour and cognitive impairment in rats which was restored after a few days treatment with the GR antagonist mifepristone (RU38486) (Wu et al. 2007). At the same time, the 4-day GR antagonist treatment blocked the suppression of neurogenesis and reversed many other morphological and neurochemical markers of hippocampal atrophy (Oomen et al. 2007). Already displayed in naïve Wistar rats, chronic icv infusion of RU38486 improved spatial learning and memory of rats in the water maze (Oitzl et al. 1998). Studies in humans demonstrated that the age-related increase in CORT and cognitive decline can be partially overcome by blocking the synthesis of endogenous CORT with metyrapone (Lupien et al. 2002). These chronic conditions are distinct from acute interference of the antiglucocorticoid with the consolidation of spatial learning and fear-conditioning.

In recent studies, hippocampal morphology and function was studied in a mouse model for type 1 diabetes (T1D) generated by treatment of mice with streptozotocin, which destroys the beta cells in the pancreas and hence circulating insulin levels drop. The lack of insulin causes hyperglycaemia and cellular starvation, a condition known as metabolic stress. One of the concomitant hormonal changes to cope with this metabolic stressor is a profound activation of the adrenal CORT secretion even occurring without ACTH (Revsin et al. 2008). As a result of hypercorticism cells including the neurons in the central nervous system are further deprived of glucose further triggering a sustained elevation of CORT in these mice.

In this model, remodelling of hippocampal circuits is accelerated and its vulnerability to cognitive dysfunctions is enhanced during diabetes in a process that can be blocked by antiglucocorticoid therapy or even removal of the adrenals and simultaneous replacement with normal CORT concentrations (Stranahan et al. 2008; Revsin et al. 2009). Previously, hyperglycaemia and insulin deficiency were thought to underlie the hippocampal deficits (Biessels et al. 2008), but these factors can be ruled out by the experiments of Revsin et al. (2009) and Stranahan et al. (2008), although it cannot be excluded that synergism of these conditions with the deleterious effect of hypercortisolaemia plays a role. A rationale behind this observation is that the blockade of the GR would allow a more prominent function of neuroprotective MR-mediated actions

(Joëls et al. 2008). Hence this would predict that during GR blockade in the face of high circulating glucocorticoids the maintenance of the integrity of the hippocampus is a necessary condition for hippocampal-dependent behavioural performance in the spatial learning task.

In summary, MR- and GR-mediated effects of CORT modulate the integration of emotional arousal and cognitive performance. Hence, MR- and GR-mediated actions interact, proceed in a coordinated manner, and are linked in time to particular stages of information processing. These integrative actions are set in contexts where different cocktails of stress mediators participate. MR amplifies attention, vigilance, and appraisal, facilitates behavioural reactivity, and the organization of subsequent coping responses. In this context, in fear-motivated or other stressful paradigms, MR activation prior to learning tasks facilitates acquisition and hence storage of the experience in memory. At retrieval, MR facilitates perseverance of learned behaviour and promotes under stressful conditions the switch towards stimulus response (caudate) strategies rather than spatial (hippocampus).

GR primarily is engaged in the management of later adaptive phases. GR promotes the consolidation of learned information and facilitates the extinction or elimination of behaviours that are of no more relevance. Preventing excessive and prolonged GR activation by antiglucocorticoids, CORT reduction, or knockdown of GR in specific limbic regions protects from the cognitive impairing influences of high CORT exposure. The underlying mechanism seems to be the recovery of the MR:GR balance since enhanced MR mediated action is one of the consequences of chronic GR antagonist treatment.

Neuroendocrine regulation

Networks in the limbic brain regions also project to the PVN (Herman et al. 2003). As pointed out in the previous section, manipulation of MR and GR in the hippocampus therefore can be expected to have neuroendocrine effects as well. These effects are conveyed in trans-synaptic fashion from the excitatory hippocampal projection to the inhibitory GABAergic network of the PVN. From classical studies involving electrical stimulation of the hippocampus an inhibitory influence of the hippocampal output on the HPA axis was inferred. However, the situation is more complex. The hippocampal outflow modulating the core of the HPA axis depends on behavioural adaptation also affected by CORT and its receptors. Various studies have been performed by blocking the receptors with selective antagonists or by using a genetic approach ablating or overexpressing the different receptor types.

In experiments using an automatic blood injection and sampling system allowing sampling every 5 min, an intravenous bolus of the MR antagonist spironolactone under basal resting conditions enhanced the hourly pulses of ACTH and CORT, while a rapid suppressive effect was observed by the agonist aldosterone (Atkinson et al. 2008). Administration of 100 ng MR antagonist RU28318 icv or of 10 ng in the dorsal hippocampus also caused a rise in basal circulating CORT levels and facilitates the CORT response to a novelty stressor (Ratka et al. 1989; van Haarst et al. 1997; Spencer et al. 1998). This initial rise occurring after chronic MR antagonist icv (100 ng/h) administration via an Alzet minipump readily disappeared, resulting in an altered setpoint level characterized by enhanced adrenal sensitivity to ACTH (van Haarst et al. 1996; Reul et al. 1997; Yau et al. 1999). Conditional forebrain MR knockout animals have normal basal CORT levels and an enhanced CORT secretion and emotional arousal during fear conditioning (Berger et al. 2006; Brinks et al. 2009a), while forebrain MR overexpression mice showed reduced emotional arousal and an attenuated CORT response to stressors.

An acute administration of 100ng mifepristone icv enhanced the release of ACTH and CORT if administered at the circadian peak at the start of dark period. If administered in the hippocampus in a dose of 10ng the GR antagonist blocked the release of ACTH, however, which is opposite to the effect of the MR antagonist. Chronic mifepristone administration through Alzet minipumps (100 ng/h) also triggered an enhanced afternoon rise in both ACTH and CORT already on the first day. In the following days, the HPA response including CRH mRNA expression was not different from controls, but at day 4 the enhanced circadian rise in CORT reappeared at a time when adrenal weight and sensitivity to ACTH were markedly enhanced (van Haarst et al. 1996). An altered pattern of CORT secretion was also observed using frequent blood sampling. After a treatment for 5 days twice a day with an antiglucocorticoid, the pulse amplitude at the circadian peak was increased (Spiga et al. 2007). This finding is of particular interest since it demonstrates that by chronic blockade of the GR the negative feedback is impaired and as a consequence the oscillations in HPA axis activity and CORT secretion are expected to be enhanced.

Mice carrying a targeted mutation of the GR all have in common that the HPA axis is disinhibited. This includes the antisense GR knockdown mice (Barden et al. 1997), the GR knockout animals developed by Cole et al. (1995), that still had a truncated GR variant expressing minimal amounts of GR, rescuing about 10% of the animals and finally, complete GR knockout heterozygotes showing impaired feedback in the Dex suppression test (Ridder et al. 2005). The $GR^{dim/dim}$ mice, in which dimerization of the GR is prevented, had elevated basal ACTH and CORT demonstrating direct transcriptional control at the pituitary level, while stress-activated signalling pathways are suppressed by transrepression that still operates in these mutants (Oitzl et al. 1998; Reichardt et al. 1998). The GR brain knockout mice (GR^{NesCre}) of Tronche et al (1999) also showed increased HPA axis activity, as did the forebrain specific GR mutants (Boyle et al. 2005). GR overexpression transgenic mice showed no differences in circadian changes of HPA axis activity, but enhanced suppression in the Dex suppression test (Wei et al. 2004; Ridder et al. 2005). None of these models were investigated for amplitudes of ultradian or circadian rhythmicity.

In summary, as in rodents, in man CORT action via MR also controls basal ultradian pulsatility and the onset of the stress reaction, which is terminated via GR (Kellner et al. 2002; Young et al. 2003). Blockade of GR disinhibits the stress-induced HPA activation and, if GR blockade is extended, enhances circadian oscillations of CORT as one of the consequences. The primary site of CORT feedback in the disinhibition of the HPA axis is in the PVN–pituitary core of the HPA axis. Secondary feedback actions modulate the HPA axis via limbic influences in a state dependent manner either in pulsatility or as output during behavioural adaptation to stress. The chronic blockade of GR function exerted by GR antagonists in combination with enhanced MR function could well be fundamental for its therapeutic efficacy in stress-related mental disorders as the resulting increased oscillations also enhance responsiveness (resilience) of the target tissue.

Morphology

Patients treated with synthetic glucocorticoids, or patients suffering from Cushing's syndrome have a reduced volume of the hippocampus (Starkman et al. 1992). The principal relationship between hippocampal volume and function, however, has been controversially debated (Czéh and Lucassen 2007). In fact Swaab et al. (2005), using postmortem measures of hippocampal volume, did not find a significant effect of glucocorticoids on hippocampal size. A similar unclear picture refers to the association of hippocampal volume and depression. Hence, Fink (2011) noted that with regard to the relationship between depression, hippocampal volume, and hypercortisolaemia, this phenomenon is state dependent and reversible and does not show a disease specificity. Further, it seems to be individually different already from early life (Lupien et al. 2007). Moreover, it is not clear whether hippocampal volume changes are the cause or the consequence of diseases and hypercortisolaemia. Therefore, it can be concluded that the current data are inconsistent.

Nevertheless, from animal studies it is obvious that CORT has an important influence on the integrity and stability of the hippocampal network. Removal of CORT by ADX results in apoptosis in the dentate gyrus (Sloviter et al. 1989), which can be reversed by activation of the MR with low amounts of CORT or aldosterone (Nair et al. 2004). In mouse mutants with a site-specific knock out of MR, viability of neurons is also compromised (Chourbaji et al. 2008). ADX or excess CORT also results in extensive dendritic remodelling, which is, however, also reversible.

As support for the 'glucocorticoid neurotoxicity' hypothesis neuronal death in the CA3 and CA1 subregions has been emphasized as cause of the volume reduction (Sapolsky 2000). However, this cannot be the sole explanation since subsequent studies showed that the stress-induced volume changes are reversible and disappear after recovery from the stressor (Heine et al. 2004c), or when elevated CORT levels are back to baseline (Czéh and Lucassen 2007). Cell loss in the hippocampal neuronal layers is therefore not responsible for the hippocampal volume changes observed after stress. Other candidate cellular mechanisms may be related to somatodendritic components, neurogenesis and glial changes, while factors like shifts in fluid balance cannot be excluded either (Czéh and Lucassen 2007; Lucassen et al. 2011).

Stress-induced dendritic reorganization has been observed together with alterations in postsynaptic densities suggesting altered neuronal connectivity. Chronic stress or experimentally increased CORT concentrations induce retraction of the apical dendrites of the CA3 and to a lesser extent of CA1 pyramidal cells and dentate granule cells (Stewart et al. 2005: Fuchs et al. 2006). Recent studies show that such dendritic alterations in hippocampal neurons can already occur within a very short period of time, particularly in relatively immature cells (Alfarez et al. 2009). This synaptic remodelling may also extend to cortical areas, where changes in cell adhesion molecule expression, attention, spatial memory and fear conditioning often occur in parallel (Sandi et al. 2001; Sandi 2004). The changes in neuronal morphology are therefore likely to contribute to various cognitive deficits occurring as a result of chronic stress exposure.

Neurogenesis still occurs in the adult brain of most species including humans. In young adult rats, thousands of new granule neurons are generated every day in the subventricular zone (SVZ) (Kempermann and Gage 2002; Weber et al. 2011). The process of neurogenesis is regulated by stress and declines with age (Heine et al. 2004a). Adult neurogenesis starts with the proliferation of progenitor cells, followed by their morphological and physiological maturation and migration towards the outer layers of the dentate gyrus. This ends with a fully functional neuron that is integrated into the pre-existing hippocampal network (Balu and Lucki 2009). As the newborn neurons mature, they extend axons and dendrites, followed by the formation of spines and functional synapses. CORT and stress affect the process of maturation, migration, and integration, as it seems by determining the positioning and organization of the newborn neurons (Fitszimons et al. submitted).

New neurons display characteristic functional properties such as lower threshold for induction of long-term potentiation (LTP), while subsequent survival of the newly generated neurons is regulated by their input-dependent activity (Tashiro et al. 2006). There is significant overproduction of newborn cells and many of them are rapidly eliminated by apoptotic cell death. A significant turnover of granule cells thus occurs in the dentate gyrus of young rodents. In monkeys, this turnover rate is significantly lower while no quantitative data are yet available for humans (Kornack and Rakic 1999).

Stressors inhibit one or more phases of the neurogenesis process and inhibit both, proliferation and survival (Oomen et al. 2007). Stress-induced reductions in proliferation could result, for example, from apoptosis of progenitor cells, or from cell cycle arrest (Heine et al. 2004b). Stress reduces the expression of several growth factors that all can influence neurogenesis (Schmidt and Duman 2007). The proximity of the precursors to blood vessels further suggests a strong interaction with the vasculature and it is this population that is particularly sensitive to stress (Heine et al. 2005). Many of the symptoms of stress or HPA hyperactivity can typically be reversed with tricyclic antidepressant treatment in both humans and animal models. The observation that antidepressants with distinct mechanisms of action can restore adult hippocampal neurogenesis supports the possibility that stimulating neurogenesis is a common pathway through which antidepressants exert their slow therapeutic effects (Sahay and Hen 2007).

In summary, there is no conclusive evidence for hippocampal volume changes as a function of mental disease and circulating cortisol levels. Yet, variations in CORT in various stressful contexts can alter neurogenesis, apoptosis, and dendritic reorganization in the hippocampus, reversibly, and in regionally distinct fashion. There is evidence suggesting that newborn neurons are integrated into hippocampal circuitry involved in spatial learning and memory processes. How these aspects of stress and CORT-induced hippocampal plasticity contribute to cognitive performance in health and disease requires further in-depth research.

Network function

Basal CORT levels, occupying mostly MR, are associated with very effective induction of LTP in the hippocampus (Diamond et al. 1992). In contrast, stress or high levels of CORT impair electrophysiological correlates of memory formation in CA1 hippocampal neurons such as LTP and paired-pulse facilitation (PPF), probably due to a downscaling by a stress-induced saturation of glutamatergic transmission. Long-term depression (LTD) is, however, facilitated (Pavlides et al. 1996; Xu et al. 1998; Kim and Diamond 2002). The underlying mechanism of these stress-induced effects is currently not known, although NMDA receptors appear to be involved as these effects can be blocked by blockade of both, NMDA and GR in CA1 neurons (Kim et al. 1996). Clearly, GR or stress-mediated changes in glutamate transmission and calcium influx will affect the potential for synaptic plasticity, though not necessarily in the direction of impaired LTP. The crucial issue seems to be the *timing* of events. If stress hormones reach limbic areas, they may alter network function—e.g. through rapid effects of CRH (Blank et al. 2002) or even CORT itself (Korz and Frey 2003)—such that subsequent patterned input unrelated to the stress (either by electrical stimulation or because of learning-induced situations) is less effective in strengthening the synapses (Diamond et al. 2004). However, if stress hormones are released in conjunction with the learning situation, causing hormone release and network activation to concur, they may very well prime synapses such that synaptic potentiation is facilitated.

Electrophysiological studies have revealed some common principles of the effects of stress hormones on electrical properties of hippocampal cells (Joëls 1997). *Firstly*, corticosteroid hormones, particularly when acting via GR, usually work in a conditional manner, i.e. they have little effect on cellular properties under resting conditions. Their effects become evident when cells have been driven from their resting potential. In this manner GR activation can for instance reverse norepinephrine-induced excitation (Joëls and de Kloet 1989). *Secondly*, CORT action has an enormous diversity and can be considered a pleiotropic hormone, as expected from hormones that regulate transcription of multiple genes. Functions of diverse proteins involved, for example, in voltage gated ion channel function but also in G-protein coupled receptor signalling are strongly affected by CORT and stress. Of the latter category, particularly the 5-HT1A receptor was studied in detail (Joëls et al. 1991). Inhibitory responses mediated through the 5-HT1A receptors were found to be small under basal conditions but markedly increased with subsequent GR activation. In the absence of the hormone, responses were very similar to those seen with high and concurrent activation of the receptors, yielding a U-shaped dose-dependency. This turned out to be a typical feature of nearly all cellular properties affected by CORT in the CA1 hippocampal area (Joëls 1997). Still, steroids—though pleiotropic in their effects—certainly do not change all cellular properties. For instance, L-type calcium channel function is markedly affected (Kerr et al. 1992; Karst et al. 2000), while function of channels for potassium or sodium seems to be less prominently altered.

Effects of stress and CORT on ionotropic receptors generally are quite rapid and independent of the slow gene-mediated pathways. This certainly pertains to effects on glutamate release in hippocampus (Venero and Borrell 1999), but also in hypothalamus, involving endocannabinoids. Di et al. (2003) discovered that in parvocellular PVN neurons CORT rapidly reduces the frequency of miniature excitatory postsynaptic currents (mEPSCs) representing the spontaneous release of a glutamate-containing vesicle by a process requiring G-proteins. The effects were mimicked by a membrane-impermeable conjugate of CORT with bovine serum albumin. Blockade of endocannabinoid signalling prevented the CORT effects. The data suggest that CORT, by binding to postsynaptically located G-protein coupled GRs inserted into the membrane, caused enhanced release of retrogradely transported endocannabinoids, which in turn suppress the probability that glutamate-containing vesicles fuse with the presynaptic membrane.

Experiments in the hippocampus showed that CORT conjugated to bovine serum albumin rapidly and reversibly enhanced the frequency of mEPSCs and the release probability of glutamate containing vesicles in CA1 pyramidal neurons (Karst et al. 2005). The underlying signalling pathway of the rapid CORT actions in the hippocampus differs from that described for PVN neurons because the effect appeared mediated by membrane MR rather than GR. Blockers of retrograde signals like endocannabinoids or nitric oxide were ineffective (Olijslagers et al. 2008). It was concluded that the rapid CORT effects in the hippocampus involve MR localized on the presynaptic membrane where they activate the ERK1/2 signalling pathway, causing more vesicles containing glutamate to be spontaneously released. CORT previously had been found to enhance the release of glutamate in hippocampus. Interestingly, MR enhanced mEPSC frequency and probability also in the amygdala, but the effect was much more prolonged and cooperated with a genomic GR-mediated action (Karst et al. 2010).

CORT was found recently to regulate in cultured hippocampal neurons α-amino-3-hydroxyl-5-methyl-4-isoxazole-propionate (AMPA) receptor trafficking, which contributes to synaptic plasticity (Krugers et al. 2010). These effects exerted by CORT involve time-dependent increases in GluR2-AMPAR surface mobility and synaptic surface GluR2 contents induced by a chemical LTP stimulus (Groc et al. 2008). The enhanced lateral diffusion properties may underlie both the facilitating and suppressive effects of CORT on synaptic plasticity. GluR2 trafficking also appeared involved in the facilitation of water maze learning by stress and fear motivated behaviour (Zhou et al. 2009; Conboy and Sandi 2010). The facilitated learning by stress was accompanied by enhanced synaptic expression of GluR2's that was not observed in mice trained under less stressful conditions. Both the memory facilitation and the enhanced GluR2 trafficking induced by stressful learning can be blocked by metyrapone, a blocker of CORT synthesis. Moreover, blockade of GluR2 synaptic trafficking impaired immediate performance at learning as well as long-term memory retrieval, supporting a causal role for GluR2 trafficking in stress-induced facilitation of spatial learning and memory (Figure 5.5).

Overall, on the cellular level MR effects serve to maintain a high excitatory tone and stability of networks. Conversely, GR (in addition to MR) activation leads to a delayed suppression or normalization of stress-induced network activity. At the same time, by allowing more calcium influx into the cell upon depolarization, GR activation may enable retention of information, in a process that involves AMPA receptor trafficking. The actions exerted by CORT traditionally were considered to be genomic, but recently rapid actions mediated by membrane MR and GR were discovered which regulate cellular responses in coordination with the gene-mediated receptor actions.

Molecular effects

Targeted approach

GR targets include enzymes and receptors for biogenic amines, neuropeptides (Sabban and Kvetnansky 2001), growth factors (Schaaf et al. 1997; Hansson et al. 2000), and cell adhesion factors (Sandi 2004). GR is abundantly expressed in the ascending mono-aminergic neurons and the rate-limiting enzymes tryptophan hydroxylase of serotonin (5HT) and tyrosine hydroxylase of dopamine (DA), norepinephrine (NE) are under control of CORT. Epinephrine (E) synthesis is also regulated by CORT but only in the adrenal medulla and not in the brain stem.

Biochemical studies showed that 5-HT turnover in hippocampus is enhanced by CORT through stimulation of tryptophan hydroxylase activity (de Kloet et al. 1983; Chaouloff 1993), but reduced after sustained CORT activation (de Souza and van Loon 1986). The interaction of 5-HT and stress-induced HPA axis activity occurs at various levels. 5HT can stimulate the release of CRH in the PVN. In the hippocampus acute stress enhances 5-HT turnover in the synaptic terminals, but 5-HT released from the projecting neurons in the raphe nuclei cell body attenuates stress-induced activation by an auto-inhibitory feedback loop (Price et al. 1998). The postsynaptic 5-HT1A receptors in the hippocampus are downregulated by stress and CORT, particularly in the dentate gyrus (Meijer and de Kloet 1998) by an MR mediated process. As noted electrophysiologically, the hyperpolarization response of 5HT1A receptors is suppressed via MR and enhanced via GR. Moreover, the 5-HT2 receptors are desensitized by acute stress, but under conditions of chronic stress 5-HT2 receptor binding is increased (Fernandes et al. 1997), particularly of the 5HT2A and 5HT2C receptors (Rënyi et al. 2001), potentially contributing to heightened anxiety. The interaction is bi-directional. For instance, enhancing the serotonergic input to hippocampal cells pharmacologically using 5-HT reuptake blockers increases the synthesis of MR and GR (Seckl and Fink 1992; Reul et al. 1993).

CORT also enhanced tyrosine hydroxylase and hence DA turnover and release, particularly in the mesolimbic system, but DA turnover is reduced after prolonged stress (Puglisi-Allegra et al. 1991). These effects of stress on the DA system are regionally different because the nigrostriatal area is not affected while DA turnover in the hypothalamus is suppressed after acute stress, leading to enhanced prolactin release. In subsequent studies using either GR antagonists or site-specific GR knockout mice, Piazza, Tronche and coworkers demonstrated that CORT increases DA release in the nucleus accumbens mesolimbic terminal fields, enhancing locomotion and the propensity to self-administer cocaine (Deroche-Gamonet et al. 2003; Ambroggi et al. 2009). Sustained (uncontrollable) stress ultimately attenuates dopaminergic activity. Finally, sustained stress- and hypercortisolaemia-induced changes in the NE system which are comparable with the changes observed during depression. Particularly, the reverberating positive feedback loop between locus coeruleus and amygdala has received much attention, a feedback loop that is activated in synergy by CRH and CORT (van Praag et al. 2004).

MAPK signalling pathway and the downstream transcription factor Egr-1 were found to be activated by CORT. Using *in vitro* and *in vivo* approaches Piazza's group identified the CORT induced synapsin-Ia/Ib synthesis and phosphorylation as one of the effectors in stress-induced memory processes. Their data provide a complete molecular pathway (GR/Egr-1/MAPK/Syn-Ia/Ib) through which stress and CORT enhance the memory of stress-related events and highlight the function of synapsin-Ia/Ib as molecular effector of the behavioural effects of stress (Revest et al. 2010).

Unbiased approach

Large-scale gene expression profiling methods, like serial analysis of gene expression (SAGE) and microarrays, have been applied to identify responsive genes, using a design that mimics differential MR and GR occupation. In the hippocampus, MR and GR activation each led to altered expression of some 70–100 genes, of which roughly 50% was upregulated and one-third responsive to both receptor types (Datson et al. 2001). The hormone was found to coordinate expression of high and medium abundant genes underlying aspects of energy metabolism, cell adhesion, synaptic transmission, oxidative stress, neurogenesis, cell cycle, and structural plasticity. There are large differences in overall gene expression patterns between hippocampal subregions (Datson et al. 2004, Lein et al. 2007), which has consequences for the repertoire of available coactivators and corepressors as well as the availability of transcription factors that can interact with GR.

CORT-responsive gene networks respond in a time- and context-dependent fashion. The dynamics of the transcriptional response to GR activation appears to be conserved between different neuronal substrates. Both in hippocampal slices used for electrophysiology as well as in neuronally differentiated PC12 cells, a rapid wave of exclusively downregulated genes was observed 1 h after activating GR, followed by a second later wave of upregulated genes (Morsink et al. 2006a, 2006b). This suggests that transrepression involving interactions with other transcription factors may be the predominant mode of CORT action immediately after GR-activation, followed by a later wave of transactivation. Since transrepression is an important route through which CORT receptors can interact with other activated signalling pathways, this fast downregulation of genes may be the preferred route dampening initial stress reactions as part of the recovery phase from exposure to the stressor.

In the dentate gyrus, chronic stress exposure affected the expression of 90 genes ($P < 0.01$), with an overrepresentation of genes involved in brain development and morphogenesis and synaptic transmission. RU486 treatment of stressed animals affected expression of 107 genes; however, mostly different genes than those responding to stress. Interestingly, we found CREB-BP to be normalized by RU486 treatment to levels observed in control animals, suggesting that CREB-signalling may play a central role in mediating the chronic stress effects on neurogenesis, LTP and calcium currents. The identified genomic pathways provide insight into the stress-induced adaptive plasticity including neurogenesis of the hippocampal dentate gyrus (Datson et al. 2012).

While large-scale expression profiling also pointed to structural proteins as endpoints of CORT action, this approach is not informative on proteins involved in electrical and chemical communication; the transcripts of most genes involved in such communication are too low-abundant to be detected reliably with the present methodologies.

The neurochemical studies on stress, CORT and monoamines seem to suggest that chronic stress reduces 5HT metabolism, downregulates 5HT1A receptors, and upregulates 5HT2 receptors. The effects on the NE system are variable, while reduction of mesolimbic DA activity is observed under conditions reminiscent to some forms of depression characterized by elevated CORT. A CORT-dependent pathway was identified driving synapsin 1a/1b as a key molecule in stress and memory processes. The unbiased approaches have identified a number of tentative pathways beyond the scope of this review and point to an enormous diversity in CORT signalling: at least 6000 genes were identified that respond to CORT in different contexts.

MR and GR polymorphisms

CORT signalling is mediated by MR and GR, each encoded by one gene. MR and GR occur, however, in dozens of genetic variants and protein isoforms (Revollo and Cidlowski 2009). Moreover, receptor function is influenced by at least 50 proteins such as heatshock proteins, coregulators, and transcription factors, that are also modulated by environmental signals, all together providing an enormous potential in signalling diversity.

Genetic variation is widely present in the human genome, ranging from 1 basepair to loss of chromosome parts. Single nucleotide polymorphisms (SNPs) are probably the most abundant type of variation in the human genome. Besides direct effects on the primary amino acid sequence, as often seen in overt 'mutations', SNPs can influence gene expression by affecting promoter activity, transcription efficiency, gene-splicing, or mRNA stability, while effects on translation efficacy and ligand binding have also been described (Kimchi-Sarfaty et al. 2007). Moreover, these SNPs often appear in combinations which are designated haplotypes.

Several genetic variants have been described in both the MR and GR genes showing associations with different levels of HPA axis reactivity (DeRijk and de Kloet 2008). The MR (chromosome 4q31.1) and GR (chromosome 5q31-32) gene structure is comparable, with the coding region of the genes formed by exons 2–9, and the 5′ promoter region, with several exon 1's. Considering the importance of the CORT receptors in regulation of limbic circuitry and the HPA axis, it is expected that common functional genetic variants will modulate HPA axis reactivity associated with cognitive performance and emotional arousal.

Two MR-gene variants have been studied in relation to HPA axis reactivity. At position −2, two nucleotides before the first ATG start codon, a G/C SNP (rs2070951, frequency approx 50%) changes *in vitro* transactivational activity (van Leeuwen et al. 2010) resulting in increased expression of the MR-protein. The −2 C variant is associated with lower basal cortisol levels (Kuningas et al. 2007) and also modulates morning cortisol levels following 0.5 mg Dex administration the previous day (van Leeuwen et al. 2010). Furthermore, in exon 2 of the human MR-gene, at codon 180, the GTT to ATT variation results in an isoleucine to valine change (rs5522, frequency approx 11%). This MR 180V shows *in vitro* loss of function using cortisol as a ligand (DeRijk et al. 2006). Using a psychosocial challenge, the so-called Trier Social Stress Test (TSST), both these MR-gene variants were found to associate with changes in plasma ACTH, plasma cortisol and saliva cortisol levels (DeRijk et al. 2006). In addition, heart beat frequency was modulated by these two SNPs, indicative of effects on the autonomic nervous system (Figure 5.6).

Three common haplotypes based on the functional MR −2 G/C and I180V SNPs (hap 1: −2G/180I; hap 2: −2C/180I; hap 3: −2C/180V) are in linkage disequilibrium (LD) with SNPs in the gene's promoter region. A promoter region of 4 kilobases containing haplotype 2 resulted in 1.4 or 2.2 times higher gene transcription after transfection in human neuroblastoma cells in comparison to respectively haplotype 1 or 3 (Van Leeuwen et al. 2010). Together with previous work the data show that haplotype 2 results in the highest gene transcription, translation, and transactivation of target genes. Genetic association analysis showed that haplotype 2 (freq. 0.36) was associated with heightened dispositional optimism ($p = 0.001$) in one study and with less hopelessness ($p < 0.05$) and rumination ($p < 0.001$) in a follow-up study. In both studies effects were restricted to women (Klok et al. 2011b). We propose that the common and functional MR haplotypes might relate to significant variability in MR expression in the brain, conferring inter-individual variability in resilience to psychopathology in women.

Regarding the GR-gene, several SNPs have been tested *in vitro* for functionality with increased or decreased activity in different cell types or systems. Just as in the MR-gene, genetic variants of the GR-gene are expected to modulate several levels of HPA axis control. With respect to 'basal' non-modulated levels of CORT, the GR TthIIII was found to associate with higher basal cortisol secretion in men (Rosmond et al. 2000). The GR N363S and the BclI were found to associate with lower morning CORT levels following 0.25 mg Dex (Huizenga et al. 1998; Stevens et al. 2004). This is in line with the *in vitro* data suggesting increased CORT sensitivity induced by this N363S GR-variant. In contrast, the NR3C1-1 (in linkage with the GR A3669G),

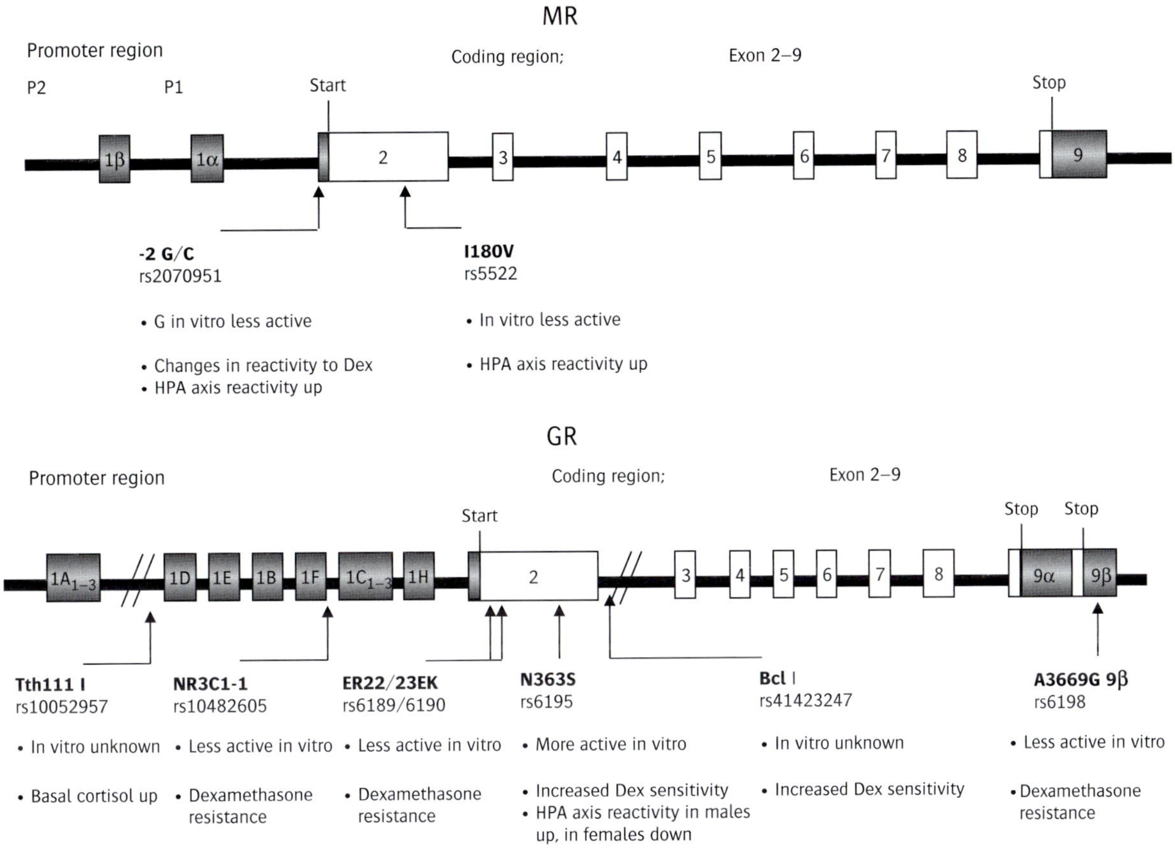

Figure 5.6 An overview of the common genetic variants of the MR and GR and their effects on HPA axis reactivity. Data are from Van Leeuwen et al. (2010).

the ER22/23EK and the A3669G displayed decreased sensitivity to Dex in males resulting in higher post-Dex plasma ACTH and CORT levels. The decreased *in vitro* activity of these GR-gene variants fits with these findings, suggesting a relative glucocorticoid resistance *in vivo*.

In the TSST, highest levels of ACTH were observed in carriers of the NR3C1-1 and the A3669G (in LD), but not for CORT. Heterozygotes of the BclI showed increased responses of CORT, while unexpectedly, homozygotes displayed lower levels. Male N363S carriers especially have high levels of cortisol. Importantly, in a recent follow-up study including both, males and females (Kumsta et al. 2007), males showed typically similar effects with N363S carriers the highest responses. Of note, none of the GR-gene variants affected heart-beat responses, suggesting little effect on autonomic regulation during the TSST.

The data show that MR and GR-gene variants affect different aspects of HPA axis regulation with strong gender effects appearing. Moreover, the MR I180V also modifies autonomic function during stress, with effects evolving within minutes. Moreover, the strong gender dependent effects indicate the involvement of other (sex) hormones, which might interact with MR and GR variants in their own specific manner.

The MR variants generally affect the initial response to the stressor, which is then affected in termination by GR and its variants. These SNPs or haplotypes affect receptor synthesis, affinity for CORT, binding of other partners or are important for the stability of the receptor. Studies are beginning to identify the functional implications of these variants with promising perspectives as biomarkers for vulnerability and resilience to psychopathology (van Rossum et al. 2006).

Early life experience

Thanks to the pioneering studies of the late Seymour Levine (1925–2007), it is common knowledge that early life experience has profound influences on stress-related neuroendocrine regulations and adaptive behaviour. In his seminal studies rat pups were subjected to a daily brief separation from the dam—a procedure called handling. It was found that these handled pups display reduced emotional and neuroendocrine responses to common stressors at adulthood (Levine 2005). These lasting effects correlate with the extent of maternal care the pups had received upon reunion with the mother. In the studies of Michael Meaney and colleagues, it was observed that the offspring from mothers showing endogenous 'high licking and grooming' also display attenuated stress-induced HPA activity, improved cognitive performance in a spatial learning test, and reduced anxiety-like behaviour at adulthood as compared with the offspring that had received relatively little care. If rat pups received little maternal care, a high stress-responsiveness was observed in adulthood, but this outcome could be reversed by cross-fostering the pups with high licking and grooming mothers (Liu et al. 1997).

These very challenging studies have shown that early-life experiences have profound consequences, probably through epigenetic modification of key molecules of the HPA axis, e.g. vasopressin and CRH (Murgatroyd et al. 2010) and the CORT receptors, changing their expression (Szyf et al. 2005). In support of a role for CORT during brain development, the hormone appeared crucial for maturation of neuronal circuitry involved in processing of odour and fearful information. Rat pups readily learn maternal odour to support attachment behaviour during the first 2 weeks of postnatal life at a time that CORT concentrations normally are stable and very low, i.e. the so-called stress hyporesponsive period (SHRP). If exposed to CORT, the infant can switch towards avoidance behaviour by a mechanism involving the amygdala (Moriceau and Sullivan 2006).

Early life experience paradigms have been used to program stress-related behaviour. One commonly used laboratory model for neglect is prolonged separation of mother and pup. In our experiments, pups exposed to a 24-h maternal separation display a slow rise in CORT level and enhanced adrenal (both medulla and cortex) responsiveness to stressors and exogenous ACTH. The immediate effects of maternal deprivation could be obliterated when maternal behaviour was mimicked by feeding and stroking the pups, albeit stroking only reinstated the central aspects and feeding was needed to restore adrenal function (van Oers et al. 1998). When adult, the maternally-deprived animals showed enhanced CORT and sympathetic responses to a common stressor and spatial learning was impaired (Oitzl et al. 2000). A recent study in primates actually suggests that maternal care or the early experience of stressful events can 'inoculate' for the ability to cope with stress in later life (Parker et al. 2006).

Another frequently used model is repeated maternal separations. A large variation of protocols have been applied varying in daily dam-pup separation for 1, 3, 6, or 8 h. Maternal separation slowly activates the HPA axis with CORT levels increasing from the first separation over a period of several hours (Schmidt et al. 2004). The rodent, however, readily adapts to this procedure, because the next day absence of the dam does not produce this rise in CORT anymore. Apparently, the pup readily learns to predict return of the mother. In spite of this fast adaptation to maternal absence, the pup's stress system stays on alert and continues to respond to a novelty stressor (Enthoven et al. 2008). Hence it appears to be the combination of maternal absence exposure to stressors and CORT exposure that is capable to prime and activate the amygdala fear pathway (Daskalakis et al. submitted).

Developmental pathways therefore may be shaped for life by gene–environment interactions during early life in a process involving maternal care and the dual brain CORT receptor system. Such pathways could be biased towards processing of fearful information if previously exposed to an adverse early life experience. Hence, the animal model to test the three hit hypothesis should have an assessment of the quality of mother–pup interaction and a defined predisposing genetic background. The outcome of the third hit at adulthood is regionally different in the brain: in response to a chronic unpredictable stressor the hippocampus and prefrontal cortex are usually atrophied (Magariños et al. 2006; Czéh et al. 2001), while hypertrophy occurs in the basolateral amygdala (Vyas et al. 2002; Roozendaal et al. 2009).

A striking example how context may affect the third hit has been observed with the offspring of low and high caring mothers. Hippocampal neurons of rats exposed to a high caring mother have a much larger dendritic length in hippocampal CA1 neurons, while an excellent LTP response could be evoked, much in line with their reduced HPA responsiveness and enhanced cognitive performance. However, this phenotype deteriorated if exposed to CORT or a stressor, while the offspring of the low care mother thrived under these conditions (Champagne et al. 2008). In a stressful fear conditioning paradigm, the low care offspring performed much better in the contextual freezing response and had an improved LTP response. Likewise, 24 h of maternal deprivation resulted at adulthood in reduced neurogenesis and an altered dendritic morphology, poor LTP and impaired Morris water maze performance. If exposed to CORT and stress, a clear LTP response evolved and contextual fear conditioning was dramatically improved. Hence, these are striking examples of the 'predictive adaptive plasticity hypothesis' (Champagne et al. 2009). In other words, early life experience prepares for life ahead (Figure 5.7).

However, the outcome of an early life experience cannot be generalized. In fact a dichotomy was observed in ageing animals exposed as pups to 24 h of maternal deprivation (Oitzl et al. 2000). Around midlife, these animals experience a period of enhanced stress responsiveness and subsequently in the deprived animals cognitive performance was driven to either excellent or poor, at the expense of the

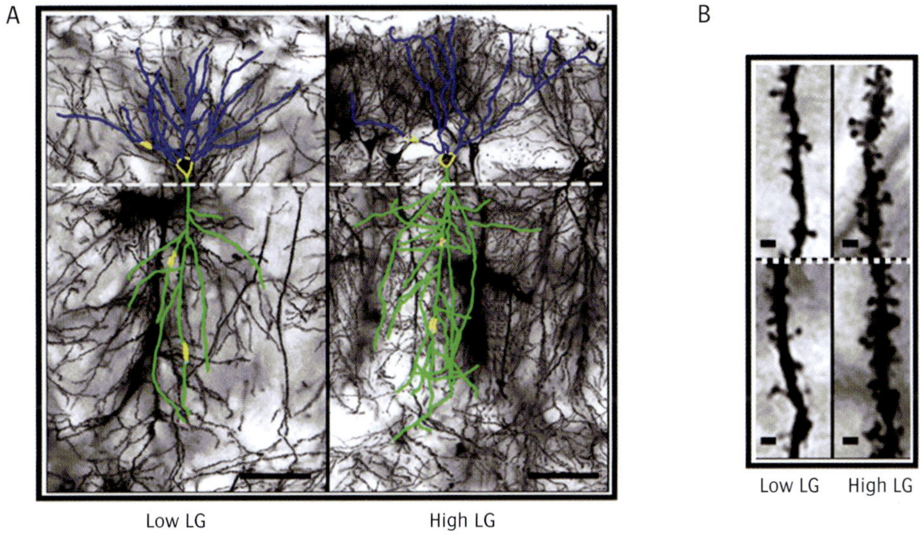

Figure 5.7 Effect of the quality and quantity of maternal care on dendrite structure of Golgi-stained hippocampal CA1 neurons of the 2-month-old offspring. LG, maternal licking and grooming Reproduced from Champagne, D. L., Bagot, R. C., van Hasselt, F., et al. (2008). Maternal care and hippocampal plasticity: Evidence for experience-dependent structural plasticity, altered synaptic functioning and differential responsiveness to glucocorticoids and stress. *Journal of Neuroscience*, 28, 6037–45 © 2008, Society for Neuroscience, with permission.

average observed in the controls. BDNF expression in the hippocampus correlated with this dichotomy in cognitive performance in the maternally deprived animals (Schaaf et al. 2001). Whether this phenomenon is another example of 'inoculation by stress' in early life (Parker et al. 2006) or if it demonstrates the importance of 'context' in the third hit is not clear at present. It does, however, demonstrate that adverse early life conditions do not inexorably or inevitably lead to enhanced vulnerability, but rather seem to equip the individual in the proper context with adequate coping abilities.

Early life experiences programme emotional and stress reactivity in later life. In rodents, variations in sensory stimulation and feeding of the pup during the SHRP have lasting consequences for cognitive and emotional functions involving the hippocampus. If pups are deprived from care and feeding, the SHRP is disrupted and stressors can evoke a CORT response from the otherwise quiescent adrenal. Such inappropriate early life exposure to high CORT levels facilitates programming the amygdala fear pathway at the expense of olfactory/hippocampal pathways and the genes involved 'learn' from this by epigenesis. Then in later life, the expression of the programming seems to depend on the match between the expected conditions and the real circumstances in later life, i.e. the predictive adaptive plasticity hypothesis.

Implication for diseases of the hippocampus

The MR:GR balance hypothesis

This chapter focuses on the action of CORT in the hippocampus mediated by two distinct receptor types: MR and GR. Both operate as master switch in control of plasticity, resilience, and adaptation to stress. Thus, MR modulates appraisal processes involved in the onset of psychological stress reactions while GR is more involved in management of later adaptations. The MR:GR balance hypothesis predicts that, upon imbalance, threats to the dynamic equilibrium (allostasis) are less well communicated. At a certain threshold, this may lead to a condition of neuroendocrine dysregulation and impaired behavioural adaptation, which potentially can promote susceptibility to age-related neurodegenerations and mental disorders (Figure 5.8).

If the organism suffers from a condition of chronic stress these MR:GR responses are altered. Such a condition is achieved by for instance loss of social support, loss of a dominant position, or loss of attachment to a loved one, conditions that are mostly characterized by an impaired ability to predict or control upcoming events. Remarkably, then the stress system is sensitised to a brief additional exposure to a stressor and an inadequate recovery of homeostasis. Thus alterations of dendrite and spine morphology in region specific manner (increased arborizations in amygdala, decreased arborizations in hippocampus and

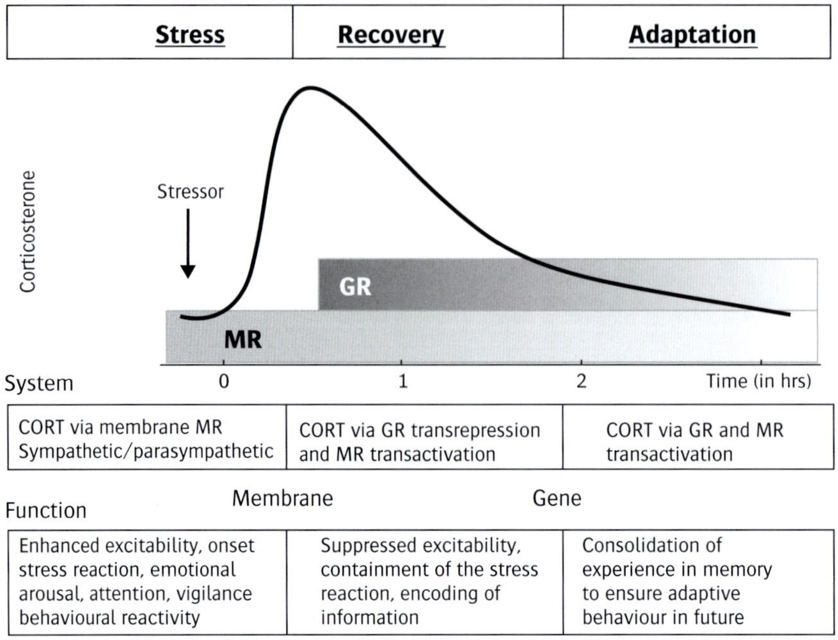

Figure 5.8 Coordinate MR and GR mediated effects of CORT. The graph shows the phases of stress (the initial stress reaction) recovery and adaptation, during which the listed molecular, cellular and behavioural effects, exerted by CORT, occurs. Reprinted from *Neuroscience & Biobehavioral Reviews*, 34(6), Oitzl, M. S., Champagne, D. L., van der Veen, R., and de Kloet, E. R. (2010). Brain development under stress: Hypotheses of glucocorticoid actions revisited 853–66, © 2010, with permission from Elsevier.

prefrontal cortex) and suppressed neurogenesis occur and profound gene and cellular responses to an acute stressor are observed (Joëls et al. 2007; Datson et al. 2012). The quest is what this altered responsiveness of the chronically stressed individual to acute challenges means. Is it representative for Selye's resistance phase, or today's allostatic load, and what does it tell about the pathogenesis of stress-related mental disease?

Highlights

The MR:GR hypothesis was first formulated in 1987 (de Kloet and Reul 1987) when the properties, localization and function of MR and GR became apparent (Reul and de Kloet 1985; Arriza et al. 1987). Electrophysiological studies outlined the conditional nature of MR and GR mediated signalling (Joëls and de Kloet 1992; Joëls et al. 2009), the crucial importance of convergence of signalling in time and space for optimal information processing (Joëls and Baram 2009). Moreover, it was discovered that the very same receptors that operate as gene transcription factors also function as membrane receptors rapidly modulating excitatory neurotransmission (Di et al 2003; Karst et al. 2005).

Classical textbook knowledge depicting CORT secretion in smooth circadian patterns became obsolete; CORT is secreted in hourly pulses which are a significant determinant for stress responsiveness and resilience (Lightman and Conway-Campbell 2010; Sarabdjitsingh et al. 2010a). Finally, a mechanistic solution is dawning to resolve the glucocorticoid paradox, i.e. *distress* and *eustress* both cause lasting elevations in CORT but with either a damaging or a protective outcome respectively (Dupret et al. 2007; Oomen et al. 2007). The molecular solution to resolve this paradox seems to be in the molecular aggregate of MR and GR recruiting co-regulators at the gene transcription level (Van der Laan and Meijer, 2008; Lachize et al. 2009) (Figure 5.2).

One line of support for the MR:GR balance hypothesis stems from gain of function or loss of function SNPs and haplotypes. Thus, the ER22/23EK polymorphism (frequency about 9%), a finding from Steven Lambert's group, confers a relative resistance (or loss of function) to the GRs, which was found to have a significantly higher prevalence in the older population (van Rossum et al. 2004). Likewise, in animal studies gain of function mutations in the MR are characteristic for the Brown Norway rat strain, which is known for its healthy cognitive and metabolic longevity that appeared linked to its low circulating CORT and efficient neuroendocrine stress regulation (Oitzl et al. 2001; Workel et al 2001; de Kloet et al. 2008). The gain of function haplotype 2 of the MR is another example; the haplotype (frequency 30%) is linked to dispositional optimism and protects against depression (Klok et al. 2011b).

Another line of support stems from the epigenetic down-regulation of hippocampal MR and GR expression as a hallmark in programming enhanced emotional and stress reactivity biasing the amygdalean emotional fear pathway. However, the expression of this phenotype to the benefit of coping and adaptation depends on the context and conditions the individual is facing in later life. The evidence so far supports the predictive adaptive plasticity hypothesis stating that early experience programmes for life ahead (Gluckman et al. 2007).

Hippocampal disease

The hippocampus is damaged during, for example, hypoxia, chronic epileptic seizures, and diabetes, while it is the first region affected by Alzheimer's disease (AD). High CORT does not produce AD pathology (Swaab et al. 2005) nor epileptic seizures. Clinical data suggest, however, that the onset and progression of AD is enhanced under stressful and high CORT conditions. The atrophy and death of hippocampal cells affects information processing and impairs the feedback action of CORT thus compromising the coordinating and integrating role of the hormone. As a consequence, the HPA axis becomes dysregulated resulting in elevated CORT levels for prolonged periods of time which will further aggravate the neurodegenerative cascades (Herman and Seroogy, 2006). Patients with a high level of distress are 2.7 times more likely to develop AD (Green and DeKosky, 2006).

Epilepsy patients report stress among the most frequent precipitant of seizures and stress mediators such as CRH and CORT could contribute to this phenomenon. In a recent review, Marian Joëls has summarized the evidence for stress effects on hippocampus as follows:

> In hippocampus, CORT rapidly enhances excitatory transmission and subsequently increases calcium influx several via a gene-mediated actions. This later effect serves to normalize activity but also imposes a risk for neuronal injury if and when neurons are concurrently strongly depolarized, for example, during epileptic activity. Under conditions of repetitive stress (e.g., caused from experiencing repetitive and unpredictable seizures) enhanced calcium influx and increased excitation become more important. In agreement, elevated CORT levels accelerate epileptogenesis and lower seizure threshold in various animal models for epilepsy.

Diabetes has peripheral and autonomous neuropathies as well-known and devastating complications. Studies on diabetic patients demonstrated mild to moderate slowing of mental speed and diminished mental flexibility. Although the alterations in cognitive functions under normal conditions are not severe, mild cognitive defects can influence everyday activities in more demanding situations (Biessels et al, 2008). Sandeep et al. (2004) reported that hypercortisolism in diabetic patients may contribute to their hippocampal dysfunction. We have used the streptozotocin (STZ) treated mouse, which spontaneously develops Type 1 diabetes and hypercorticism. Using a brief treatment with the GR-antagonist mifepristone we found normalization of hippocampal dysfunction and cognitive performance (Revsin et al. 2009) supporting other evidence for detrimental effects of high CORT at the early stages of diabetes (Stranahan et al. 2008).

Several lines of evidence have linked the function of the hippocampus with depression, schizophrenia, and PTSD (see Holsboer, 2000; Yehuda 2002; Phillips et al. 2006).

Therapeutic opportunities

Pathways responsive to stress system mediators such as CRH (Holsboer and Ising 2010) and CORT in specific limbic-cortical areas provide an opportunity to rebalance the stress system. For this purpose the following options are to be considered.

First, if treatment requires correction of aberrant CORT, action it is desirable to mimic its circadian and ultradian secretion patterns. For instance, if Cushing patients are treated or if an ADX is performed patients often still complain of symptoms (Tiemensma et al. 2011). This could be due to the 'imperfect' replacement therapy. Even though average hormone level over 24 h is achieved, there is no hourly pulsatile pattern reinstated. Alternatively, transient overexposure to CORT may have irreversibly altered brain circuitry as one of the consequences (Ratka et al. 1988).

Secondly, understanding how the timing of the cortisol and other signalling pathways converge with the psychological context and the environmental demands. Time and context are important because during consolidation and retrieval, memory traces are unstable and can be modulated by pharmacological treatments. Exogenous CORT and/or beta-blockers for instance can facilitate the extinction of behaviour that is no longer relevant and thus allow reconsolidation of relevant novel information (de Kloet et al. 1999; Joëls et al. 2006). This reasoning calls for an approach where the psychological context is created that allows CORT to promote the storage of corrective experiences and to suppress retrieval of unwanted memories. The success of this approach was apparent recently in human studies in which CORT reduced phobic fear and symptoms of PTSD (Kindt et al. 2009; Hauer et al. 2011).

Thirdly, mifepristone also was applied as therapeutic agent in man. Several years ago, mifepristone appeared to ameliorate the depressive and psychotic symptoms in Cushing patients in a fast and effective manner (van der Lely et al. 1991). More recently, patients suffering from psychotic depression also rapidly improved (Belanoff et al. 2001; De Battista and Belanoff 2006). In these patients, mifepristone enhanced the amplitude of the flattened CORT rhythm (Keller et al. 2006), a result reminiscent to the effect of chronic treatment of the antagonist in rats (van Haarst et al. 1996), as mentioned in the previous paragraph. Other researchers (Wolkowitz 2001) also reported positive results of reduction in glucocorticoid signalling in affect and cognitive performance. Although mifepristone is a mixed glucocorticoid and progesterone antagonist, its effect on cognition, mood, and HPA regulation in animals and man are due to blockade of the GR. Cortisol can also be reduced by blockade of the CORT generating enzyme

11-β HSD2 with beneficial consequences for cognitive performance during the ageing process (Sooy et al. 2010).

Fourthly, MR ligands. The genomic MR actions are important for maintaining stability and neuronal integrity of hippocampal circuits. Situations with a predominant MR activation and few GR are associated with small calcium currents, reduced spike frequency accommodation, stable responses to repeated stimulation of glutamatergic pathways, and relatively small responses to biogenic amines (Joëls et al. 2008, 2009) guaranteeing a stable background of neuronal firing. The genomic MR actions are therefore important for maintaining stability and neuronal integrity of hippocampal circuits. Then, activation of membrane MR rapidly boosts ongoing excitatory transmission.

Recent studies have shown reduced hippocampal MR in depression (Klok et al. 2011a). Also MR variants subject to less efficient translation associate with reduced dispositional optimism, enhanced rumination, and increased feelings of depression in females, but not in males (Klok et al. 2011b). Antidepressants and neurotrophic peptides enhance MR expression in the hippocampus (Reul et al. 1988, 1993). Moreover, fludrocortisone as add on promote the efficacy of antidepressants, while antimineralocorticoids impair their beneficial action (Holsboer 1999; Otte et al. 2010). Activation of brain MR is considered neuroprotective (Kang et al. 2009).

Collectively, these findings call for the identification of MR:GR-based biomarkers to monitor the functioning of susceptibility pathways for disease the individual is genetically predisposed (de Kloet et al. 2008). In fact, the very same CORT-induced signalling pathways (Datson et al. 2008) that may cause deterioration and chronification of mental disorders to chronic distress also hold the key towards novel therapies enhancing plasticity and resilience still present in the diseased brain.

Acknowledgements

The support by the Royal Netherlands Academy of Sciences, Top Institute Pharma #5-209, ESF-Eurostress, EU Lifespan & HFSP is gratefully acknowledged. I also would like to express my gratitude to the colleagues of the Department of Medical Pharmacology, Leiden University, to Professor Marian Joëls and her research groups at the University of Amsterdam and Utrecht, and my dear colleagues in The Netherlands and abroad for sharing experimental data and stimulating discussions.

References

Akirav, I., Sandi, C., and Richter-Levin, G. (2001). Differential activation of hippocampus and amygdala following spatial learning under stress. *Eur J Neurosci*, 14, 719–25.

Alfarez, D. N., De Simoni, A., Velzing, E. H., *et al.* (2009). Corticosterone reduces dendritic complexity in developing hippocampal CA1 neurons. *Hippocampus*, 19, 828–36.

Ambroggi, F., Turiault, M., Milet, A., *et al.* (2009). Stress and addiction: glucocorticoid receptor in dopaminoceptive neurons facilitates cocaine seeking. *Nat Neurosci*, 12, 247–9.

Atkinson, H. C., Wood, S. A., Castrique, E. S., Kershaw, Y. M., Wiles, C. C., and Lightman, S. L. (2008). Corticosteroids mediate fast feedback of the rat hypothalamic-pituitary-adrenal axis via the mineralocorticoid receptor. *Am J Physiol Endocrinol Metab*, 294, E1011–22.

Balu, D. T., and Lucki, I. (2009). Adult hippocampal neurogenesis: Regulation, functional implications, and contribution to disease pathology. *Neurosci Biobehav Rev*, 33, 232–52.

Barden, N., Stec, I. S., Montkowski, A., Holsboer, F., and Reul, J. M. (1997). Endocrine profile and neuroendocrine challenge tests in transgenic mice expressing antisense RNA against the glucocorticoid receptor. *Neuroendocrinology*, 66, 212–20.

Belanoff, J. K., Flores, B. H., Kalezhan, M., Sund, B., and Schatzberg, A. F. (2001). Rapid reversal of psychotic depression using mifepristone. *J Clin Psychopharmacol*, 21, 516–21.

Berger, S., Wolfer, D. P., Selbach, O., *et al.* (2006). Loss of the limbic mineralocorticoid receptor impairs behavioral plasticity. *Proc Natl Acad Sci USA*, 103, 195–200.

Biessels, G. J., Deary, I. J., and Ryan, C. M. (2008). Cognition and diabetes: a lifespan perspective. *Lancet Neurol*, 7, 184–90.

Blank, T., Nijholt, I., Eckart, K., and Spiess, J. (2002). Priming of long-term potentiation in mouse hippocampus by corticotropi-releasing factor and acute stress: Implications for hippocampus-dependent learning. *J Neurosci*, 22, 3788–94.

Bohus, B., and de Kloet, E. R. (1981). Adrenal steroids and extinction behavior: antagonism by progesterone, deoxycorticosterone and dexamethasone of a specific effect of corticosterone. *Life Sci*, 28, 433–40.

Borrell, J., de Kloet, E. R., Versteeg, D. H., and Bohus, B. (1983). Inhibitory avoidance deficit following short-term Adrenalectomy in rats: The role of adrenal catecholamines. *Behav Neural Biol*, 39, 241–58.

Borrell, J., de Kloet, E. R., and Bohus, B. (1984). Corticosterone decreases the efficacy of adrenaline to affect passive avoidance retention of adrenalectomized rats. *Life Sci*, 34, 99–104.

Boyle, M. P., Brewer, J. A., Funatsu, M., *et al.* (2005). Acquired deficit of forebrain glucocorticoid receptor produces depression-like changes in adrenal axis regulation and behaviour. *Proc Natl Acad Sci USA*, 102, 473–8.

Brinks, V., de Kloet, E. R., and Oitzl, M. S. (2007a). Emotion and cognition in high and low stress sensitive mouse strains: a combined neuroendocrine and behavioral study in BALB/c and C57BL/6J mice. *Front Behav Neurosci*, 1, 8.

Brinks, V., van der Mark, M. H., de Kloet, E. R., and Oitzl, M. S. (2007b). Differential MR/GR activation in mice results in emotional states beneficial or impairing for cognition. *Neural Plast*, Article ID 90163.

Brinks, V., Berger, S., Gass, P., de Kloet, E. R., and Oitzl, M. S. (2009a). Mineralocorticoid receptors in control of emotional arousal and fear memory. *Horm Behav*, 56, 232–8.

Brinks, V., de Kloet, E. R., and Oitzl, M. S. (2009b). Corticosterone facilitates extinction of fear memory in BALB/c mice but strengthens cue related fear in C57BL/6 mice. *Exp Neurol*, 216, 375–82.

Cabrera, R., Korte, S. M., and Lentjes, E. G. W. M. (2000). The amount of free corticosterone is increased during lipopolysaccharide-induced fever. *Life Sci*, 40, 553–62.

Carroll, B. J., Cassidy, F., Naftolowitz, D., *et al.* (2007). Pathophysiology of hypercortisolism in depression. *Acta Psychiatr Scand. Suppl*, 433, 90–103.

Champagne, D. L., Bagot, R. C., van Hasselt, F., *et al.* (2008). Maternal care and hippocampal plasticity: evidence for experience-dependent structural plasticity, altered synaptic functioning and differential responsiveness to glucocorticoids and stress. *J Neurosci*, 28, 6037–45.

Champagne, D. L., de Kloet, E. R., and Joëls, M. (2009). Fundamental aspects of the impact of glucocorticoids on the (immature) brain. *Semin Fetal Neonatal Med*, 14, 136–42.

Chaouloff F. (1993). Physiopharmacological interactions between stress hormones and central serotonergic systems. *Brain Res Rev*, 18, 1–32.

Chourbaji, S., Vogt, M. A., and Gass, P. (2008). Mice that under- or overexpress glucocorticoid receptors as models for depression or posttraumatic stress disorder. *Prog Brain Res*, 167, 65–77.

Cole, T. J., Blendy, J. A., Monaghan, A. P., Schmid, W., Aguzzi, A., and Schütz, G. (1995). Molecular genetic analysis of glucocorticoid signaling during mouse development. *Steroids*, 60, 93–6.

Conboy, L., and Sandi, C. (2010). Stress at learning facilitates memory formation by regulating AMPA receptor trafficking through a glucocorticoid action. *Neuropsychopharmacology*, 35, 674–85.

Conway-Campbell, B. L., McKenna, M. A., Wiles, C. C., Atkinson, H. C., de Kloet, E. R., and Lightman, S. L. (2007). Proteasome-dependent down-regulation of activated nuclear hippocampal glucocorticoid receptors determines dynamic responses to corticosterone. *Endocrinology*, 148, 5470–7.

Conway-Campbell, B. L., Sarabdjitsingh, R. A., and McKenna, M. A. (2010). Glucocorticoid ultradian rhythmicity directs cyclical gene pulsing of the clock gene period 1 in rat hippocampus. *J Neuroendocrinol*, 22, 1093–100.

Czéh, B., and Lucassen, P. J. (2007). What causes the hippocampal volume decrease in depression? Are neurogenesis, glial changes and apoptosis implicated? *Eur Arch Psychiatry Clin Neurosci*, 257, 250–60.

Czéh, B., Michaelis, T., Watanabe, T., *et al.* (2001). Stress-induced changes in cerebral metabolites, hippocampal volume, and cell proliferation are prevented by antidepressant treatment with tianeptine. *Proc Natl Acad Sci USA*, 98, 12796–801.

Daskalakis, N. P., Remmers, E., Fung, W. L. (submitted). Early life stress induces context dependent amygdala activation that predicts adult neuroendocrine and behavioral fear response: limited role of neonatal handling and maternal care.

Datson, N. A., van der Perk, J., de Kloet, E. R., and Vreugdenhil, E. (2001). Identification of corticosteroid-responsive genes in rat hippocampus using serial analysis of gene expression. *Eur J Neurosci*, 14, 675–89.

Datson, N. A., Meijer, L., Steenbergen, P. J., *et al.* (2004). Expression profiling in laser-microdissected hippocampal subregions in rat brain reveals large subregion-specific differences in expression. *Eur J Neurosci*, 20, 2541–54.

Datson, N. A., Morsink, M. C., Meijer, O. C., and de Kloet, E. R. (2008). Central corticosteroid actions: Search for gene targets. *Euro J Pharmacol*, 583, 272–89.

Datson, N. A., Speksnijder, N., and Mayer, J. L. (2012). The transcriptional response to chronic stress and glucocorticoid receptor blockade in the hippocampal dentate gyrus. *Hippocampus*, 22(2), 359–71.

DeBattista, C., and Belanoff, J. (2006). The use of mifepristone in the treatment of neuropsychiatric disorders. *Trends Endocrinol Metab*, 17, 117–21.

de Kloet, E. R. (1991). Brain corticosteroid balance and homeostatic control. *Front Neuroendocrinol*, 12, 95–164.

de Kloet, E. R., Versteeg, D. H. G., and Kovacs, G. L. (1983). Aldosterone blocks theresponse to corticosterone in the raphe-hippocampal serotonin system. *Brain Res*, 264, 323–7.

de Kloet, E. R. and Reul, J. M. H. M. (1987). Feedback action and tonic influence of corticosteroids on brain function: a concept arising from the heterogenity of brain receptor system. *Pyschoneuroendocrinology*, 12, 83–105.

de Kloet, E. R., Vreugdenhil, E., Oitzl, M. S., and Joëls, M. (1998). Brain corticosteroid receptor balance in health and disease. *Endocr Rev*, 19, 269–301.

de Kloet, E. R., Oitzl, M. S., and Joëls, M. (1999). Stress and cognition: are corticosteroids good or bad guys? *Trends Neurosci*, 22, 422–6.

de Kloet, E. R., Joëls, M., and Holsboer, F. (2005). Stress and the brain: from adaptation to disease. *Nat Rev Neurosci*, 6, 463–75.

de Kloet, E. R., DeRijk, R. H., and Meijer, O. C. (2008). Therapy Insight: is there an imbalanced response of mineralocorticoid and glucocorticoid receptors in depression? *Nat Clin Pract Endocrinol Metab*, 3, 168–79.

DeRijk, R. H. and de Kloet, E. R. (2008). Corticosteroid receptor polymorphisms: determinants of vulnerability and resilience. *Euro J Pharmacol*, 583, 303–11.

DeRijk, R. H., Wüst, S., Meijer, O. C., *et al.* (2006). A common polymorphism in the mineralocorticoid receptor modulates stress responsiveness. *J Clin Endocrinol Metab*, 91, 5083–9.

De Quervain, D. J., Roozendaal, B., and McGaugh, J. L. (1998). Stress and glucocorticoids impair retrieval of long-term spatial memory. *Nature*, 394, 787–90.

Deroche-Gamonet, V., Sillaber, I., Aouizerate, B., *et al.* (2003). The glucocorticoid receptor as a potential target to reduce cocaine abuse. *J Neurosci*, 23, 4785–90.

De Souza, E. B. and Van Loon, G. R. (1986). Brain serotonin and catecholamine responses to repeated stress in rats. *Brain Res*, 367, 77–86.

Di, S., Malcher-Lopes, R., Halmos, K. C., and Tasker, J. G. (2003). Nongenomic glucocorticoid inhibition via endocannabinoid release in the hypothalamus: a fast feedback mechanism. *J Neurosci*, 23, 4850–57.

Diamond, D. M., Bennett, M. C., Fleshner, M., and Rose, G. M. (1992). Inverted-U relationship between the level of peripheral corticosterone and the magnitude of hippocampal primed burst potentiation. *Hippocampus*, 2, 421–30.

Diamond, D. M., Park, C. R., and Woodson, J. C. (2004). Stress generates emotional memories and retrograde amnesia by inducing an endogenous form of hippocampal LTP. *Hippocampus*, 14, 281–91.

Droste, S.K., de Groote, L., Atkinson, H. C., Lightman, S. L., Reul, J. M., and Linthorst, A. C. (2008). Corticosterone levels in the brain show a distinct ultradian rhythm but a delayed response to forced swim stress. *Endocrinology*, 149, 3244–53.

Dupret, D., Fabre, A., Döbrössy, M. D., *et al.* (2007). Spatial learning depends on both the addition and removal of new hippocampal neurons. *PLoS Biolology*, 5, e214.

Enthoven, L., Oitzl, M. S., Koning, N., van der Mark, M., and de Kloet, E. R. (2008). HPA axis activity of newborn mice rapidly desensitizes to repeated maternal absence, but becomes highly responsive to novelty. *Endocrinology*, 149, 6366–77.

Evans, R. M., and Arriza, J. L. (1989). A molecular framework for the actions of glucocorticoid hormones in the nervous system. *Neuron*, 2, 1105–12.

Ferguson, D., and Sapolsky, R. M. (2008). Overexpression of mineralocorticoid and transdominant glucocorticoid receptor blocks the impairing effects of glucocorticoids on memory. *Hippocampus*, 18, 1103–11.

Fernandes, C., McKittrick, C. R., File, S. E., and McEwen, B. S. (1997). Decreased 5-HT1a and increased 5-HT2a receptor binding after chronic corticosterone associated with a behavioural indication of depression but not anxiety. *Psychoneuroendocrinology*, 22, 477–91.

Fink, G. (2011). Stress controversies: Post traumatic stress disorder, hippocampal volume, gastroduodenal ulceration. *J Neuroendocrinol*, 23, 107–17.

Fitzsimons, C. P., Van Hooijdonk, L., Brinks, V. (submitted) Knockdown of the glucocorticoid receptor accelerates functional integration of newborn neurons in the adult hippocampus and impairs fear-motivated behaviour.

Fuchs, E., Flügge, G., and Czéh, B. (2006). Remodeling of neuronal networks by stress. *Front Biosci*, 11, 2746–58.

Funder, J. W. (2010). Minireview: Aldosterone and mineralocorticoid receptors: past, present, and future. *Endocrinology*, 151, 5098–102.

Gluckman, P. D., Hanson, M. A., and Beedle, A. S. (2007). Early life events and their consequences for later disease: a life history and evolutionary perspective. *Am J Hum Biol*, 19, 1–19.

Green, R. C., and DeKosky, S. T. (2006). Primary prevention trials in Alzheimer disease. *Neurology*, 14, 67, S2–5.

Groc, L., Choquet, D., and Chaouloff, F. (2008). The stress hormone corticosterone conditions AMPAR surface trafficking and synaptic potentiation. *Nat Neurosci*, 11, 868–70.

Grootendorst, J., de Kloet, E. R., Dalm, S., and Oitzl, M. S. (2001). Reversal of cognitive deficit of apolipoprotein E knockout mice after repeated exposure to a common environmental experience. *Neuroscience*, 108, 237–47.

Haller, J., Millar, S., and Kruk, M. R. (1998). Mineralocorticoid receptor blockade inhibits aggressive behaviour in male rats. *Stress*, 2, 201–7.

Hansson, A. C., Cintra, A., Belluardo, N., *et al.* (2000). Gluco- and mineralocorticoid receptor-mediated regulation of neurotrophic factor gene expression in the dorsal hippocampus and the neocortex of the rat. *Eur J Neurosci*, 12, 2918–34.

Hauer, D., Weis, F., Papassotiropoulos, A., and Schmoeckel, M. (2011). Relationship of a common polymorphism of the glucocorticoid receptor gene to traumatic memories and posttraumatic stress disorder in patients after intensive care therapy. *Crit Care Med*, 39(4), 643–50.

Heine, V. M., Maslam, S., Joëls, M., and Lucassen, P. J. (2004a). Prominent decline of newborn cell proliferation, differentiation, and apoptosis in the aging dentate gyrus, in absence of an age-related hypothalamus-pituitary-adrenal axis activation. *Neurobiol Aging*, 25, 361–75.

Heine, V. M., Maslam, S., Joëls, M., and Lucassen, P. J. (2004b). Increased P27KIP1 protein expression in the dentate gyrus of chronically stressed rats indicates G1 arrest involvement. *Neuroscience*, 129, 593–601.

Heine, V. M., Maslam, S., Zareno, J., Joëls, M., and Lucassen, P. J. (2004c). Suppressed proliferation and apoptotic changes in the rat dentate gyrus after acute and chronic stress are reversible. *Eur J Neurosci*, 19, 131–44.

Heine, V. M., Zareno, J., Maslam, S., Joëls, M., and Lucassen, P. J. (2005). Chronic stress in the adult dentate gyrus reduces cell proliferation near the vasculature and VEGF and Flk-1 protein expression. *Eur J Neurosci*, 21, 1304–14.

Herman, J. P., and Seroogy, K. (2006). Hypothalamic-pituitary-adrenal axis, glucocorticoids, and neurologic disease. *Neurol Clin*, 24, 461–81.

Herman, J. P., Figueiredo, H., Mueller, N. K., *et al.* (2003). Central mechanisms of stress integration: hierarchical circuitry controlling hypothalamo-pituitary-adrenocortical responsiveness. *Front Neuroendocrinol*, 24, 151–80.

Heuser, I., Yassouridis, A., and Holsboer, F. (1994). The combined dexamethasone/CRH test: a refined laboratory test for psychiatric disorders. *J Psychiatry Res*, 28, 341–56.

Holmes, M. C., and Seckl, J. R. (2006). The role of 11β-hydroxysteroid dehydrogenases in the brain. *Mol Cell Endocrinol*, 248, 9–14.

Holsboer, F. (1999). The rationale for corticotropin releasing hormone receptor (CRH-R) antagonists to treat depression and anxiety. *J Psychiatry Res*, 33, 181–214.

Holsboer, F. (2000). The corticosteroid receptor hypothesis of depression. *Neuropsychopharmacology*, 23, 477–501.

Holsboer, F., and Ising, M. (2010). Stress hormone regulation: biological role and translation into therapy. *Ann Rev Psychol*, 61, 81–109, C1–11.

Huizenga, N. A., Koper, J. W., De Lange, P., *et al.* (1998). A polymorphism in the glucocorticoid receptor gene may be associated with and increased sensitivity to glucocorticoids in vivo. *J Clin Endocrinol Metab*, 83, 144–51.

Hummel K. P. (1958). Accessory adrenal cortical nodules in the mouse. *Anatomy Recueil*, 132, 281–95.

Joëls, M. (1997). Steroid hormones and excitability in the mammalian brain. *Front Neuroendocrinol*, 18, 2–48.

Joëls, M., and Baram, T. Z. (2009). The neuro-symphony of stress. *Nat Rev Neurosci*, 10, 459–66.

Joëls, M., and de Kloet, E. R. (1989). Effects of glucocorticoids and norepinephrine on the excitability in the hippocampus. *Science*, 245, 1502–5.

Joëls, M., and de Kloet, E. R. (1992). Control of neuronal excitability by corticosteroid hormones. *Trends Neurosci*, 15, 25–30.

Joëls, M., Hesen, W., and de Kloet, E. R. (1991). Mineralocorticoid hormones suppress serotonin-induced hyperpolarization of rat hippocampal CA1 neurons. *J Neurosci*, 11, 2288–94.

Joëls, M., Pu, Z., Wiegert, O., Oitzl, M. S., and Krugers, H. J. (2006). Learning under stress: How does it work? *Trends Cogn Sci*, 10, 152–8.

Joëls, M., Karst, H., Krugers, H. J., and Lucassen, P. J. (2007). Chronic stress: implications for neuronal morphology, function and neurogenesis. *Front Neuroendocrinol*, 28, 72–96.

Joëls, M., Karst, H., DeRijk, R., and de Kloet, E. R. (2008). The coming out of the brain mineralocorticoid receptor. *Trends Neurosci*, 31, 1–7.

Joëls, M., Krugers, H. J., Lucassen, P. J., and Karst, H. (2009). Corticosteroid effects on cellular physiology of limbic cells. *Brain Res*, 1293, 91–100.

Kang, P., Rogalska, J., Walker, C.A., *et al.* (2009). Injury-induced mineralocorticoid receptor expression involves differential promoter usage: a novel role for the rat MRbeta variant. *Mol Cell Endocrinol*, 305, 56–62.

Karssen, A. M., Meijer, O. C., van der Sandt, I. C., *et al.* (2001). Multidrug resistance P-glycoprotein hampers the access of cortisol but not of corticosterone to mouse and human brain. *Endocrinology*, 142, 2686–94.

Karst, H., Karten, Y. J., Reichardt, H. M., de Kloet, E. R., Schütz, G., and Joëls, M. (2000). Corticosteroid actions in hippocampus require DNA binding of glucocorticoid receptor homodimers. *Nat Neurosci*, 3, 977–8.

Karst, H., Berger, S., Turiault, M., Tronche, F., Schütz, G., and Joëls, M. (2005). Mineralocorticoid receptors are indispensable for nongenomic modulation of hippocampal glutamate transmission by corticosterone. *Proc Natl Acad Sci USA*, 102, 19204–7.

Karst, H., Berger, S., Erdmann, G., Schütz, G., and Joëls, M. (2010). Metaplasticity of amygdalar responses to the stress hormone corticosterone. *Proc Natl Acad Sci USA*, 107, 14449–54.

Kempermann, G., and Gage, F. H. (2002). Genetic determinants of adult hippocampal neurogenesis correlate with acquisition, but not probe trial performance, in the water maze task. *Eur J Neurosci*, 16, 129–36.

Keller, J., Flores, B., Gomez, R.G., *et al.* (2006). Cortisol circadian rhythm alterations in psychotic major depression. *Biol Psychiatry*, 60, 275–81.

Kellner, M., Baker, D. G., Yassouridis, A., *et al.* (2002). Mineralocorticoid receptor function in patients with posttraumatic stress disorder. *Am J Psychiatry*, 159, 1938–40.

Kerr, D. S., Campbell, L. W., Thibault, O., and Landfield, P. W. (1992). Hippocampal glucocorticoid receptor activation enhances voltage-dependent Ca^{2+} conductances: relevance to brain aging. *Proc Natl Acad Sci U S A*, 89, 8527–31.

Kim, J. J., and Diamond, D. (2002). The stressed hippocampus, synaptic plasticity and lost memories. *Nat Rev Neurosci*, 3, 453–62.

Kim, J. J., Foy, M. R., and Thompson, R. F. (1996). Behavioral stress modifies hippocampal plasticity through N-methyl-D-aspartate receptor activation. *Proc Natl Acad Sci U S A*, 93, 4750–53.

Kimchi-Sarfaty, C., Marple, A.H., Shinar, S. *et al.* (2007). Ethnicity-related polymorphisms and haplotypes in the human ABCB1 gene. *Pharmacogenomics*, 8, 29–39.

Kindt, M., Soeter, M., and Vervliet, B. (2009). Beyond extinction: erasing human fear responses and preventing the return of fear. *Nat Neurosci*, 12, 256–8.

Klok, M. D., Alt, S. R., Lafitte, I. A. J., *et al.* (2011a). Decreased expression of mineralocorticoid receptor mRNA and its splice variants in postmortem brain regions of patients with major depressive disorder. *J Psychiatry Res*, 45(7), 871–8.

Klok, M. D., Giltay, E. J., Van der Does, W., *et al.* (2011b). A common and functional mineralocorticoid receptor haplotype enhances optimism and protects against depression. *Transl. Psychiatry*, 1, e62.

Kornack, D. R., and Rakic, P. (1999). Continuation of neurogenesis in the hippocampus of the adult macaque monkey. *Proc Natl Acad Sci USA*, 96, 5768–73.

Korz, V., and Frey, J. U. (2003). Stress-related modulation of hippocampal long-term potentiation in rats: Involvement of adrenal steroid receptors. *J Neurosci*, 23, 7281–87.

Krugers, H. J., Hoogenraad, C. C., and Groc, L. (2010). Stress hormones and AMPA receptor trafficking in synaptic plasticity and memory. *Nat Rev Neurosci*, 11, 675–81.

Kruk, M. R., Halasz, J., Meelis, W., and Haller, J. (2004). Fast positive feedback between the adrenocortical stress response and a brain mechanism involved in aggressive behavior. *Behav Neurosci*, 118, 1062–70.

Kumsta, R., Entringer, S., Koper, J. W., van Rossum, E. F., Hellhammer, D. H., and Wüst, S. (2007). Sex specific associations between common glucocorticoid receptor gene variants and hypothalamus-pituitary-adrenal axis responses to psychosocial stress. *Biol Psychiatry*, 62, 863–9.

Kuningas, M., de Rijk, R. H., Westendorp, R. G., Jolles, J., Slagboom, P. E., and van Heemst, D. (2007). Mental performance in old age dependent on cortisol and genetic variance in the mineralocorticoid and glucocorticoid receptors. *Neuropsychopharmacology*, 32, 1295–301.

Lachize, S., Apostolakis, E. M., van der Laan, S., *et al.* (2009). Steroid receptor coactivator-1 is necessary for regulation of corticotropin-releasing hormone by chronic stress and glucocorticoids. *Proc Natl Acad Sci USA*, 106, 8038–42.

Lai, M., Horsburgh, K., Bae, S. E., *et al.* (2007). Forebrain mineralocorticoid receptor overexpression enhances memory, reduces anxiety and attenuates neuronal loss in cerebral ischaemia. *Eur J Neurosci*, 25, 1832–42.

Lazarus, R. S. (2006). Emotions and interpersonal relationships: toward a person-centered conceptualization of emotions and coping. *J Pers*, 74, 9–46.

Lein, E. S., Hawrylycz, M. J., and Ao, N. (2007). Genome-wide atlas of gene expression in the adult mouse brain. *Nature*, 445, 168–76.

Levine, S. (2005). Developmental determinants of sensitivity and resistance to stress. *Psychoneuroendocrinology*, 30, 939–94.

Lightman, S. L., and Conway-Campbell, B. L. (2010). The crucial role of pulsatile activity of the HPA axis for continuous dynamic equilibration. *Nat Rev Neurosci*, 11, 710–8.

Liu, D., Diorio, J., and Tannenbaum, B. (1997). Maternal care, hippocampal glucocorticoid receptors, and the hypothalamic-pituitary-adrenal responses to stress. *Science*, 277, 1659–62.

Lucassen, P. J., Fitzsimons, C. P., Vreugdenhil, E., *et al.* (2011). Regulation of structural plasticity and neurogenesis during stress and diabetes; protective effects of anti-glucocorticoids. In Gravanis, A.G., and Mellon, S. H. (Eds.) *Hormones in Neurodegenration, Neuroprotection and Neurogenesis*, pp. 103–20. New York: Wiley.

Lupien, S. J., Wilkinson, C. W., Brière, S., *et al.* (2002). The modulatory effects of corticosteroids on cognition: studies in young human populations. *Psychoneuroendocrinology*, 27, 401–16.

Lupien, S. J., Evans, A., Lord, C., *et al.* (2007). Hippocampal volume is as variable in young as in older adults: implications for the notion of hippocampal atrophy in humans. *Neuroimage*, 34, 479–85.

Lupien, S. J., McEwen, B. S., Gunnar, M. R., and Heim, C. (2009). Effects of stress throughout the lifespan on the brain, behaviour and cognition. *Nat Rev Neurosci*, 10, 434–45.

Magariños, A. M., Verdugo, J. M., and McEwen, B. S. (2006). Chronic stress alters synaptic terminal structure in hippocampus. *Proc Natl Acad Sci USA*, 94, 14002–8.

Mason, J. W., Wang, S., Yehuda, R., *et al.* (2002). Marked lability in urinary cortisol levels in subgroups of combat veterans with posttraumatic stress disorder during an intensive exposure treatment program. *Psychosom Med*, 64, 238–46.

McEwen, B. S. (2007). Physiology and neurobiology of stress and adaptation: central role of the brain. *Physiol Rev*, 87, 873–904.

McEwen, B. S., and Wingfield, J. C. (2010). What is in a name? Integrating homeostasis, allostasis and stress. *Horm Behav*, 57, 105–11.

McEwen, B. S., Weiss, J. M., and Schwartz, L. S. (1968). Selective retention of corticosterone by limbic structures in rat brain. *Nature*, 220, 911–12.

McGaugh, J. L. (2000). Memory – a century of consolidation. *Science*, 287, 248–51.

Meijer, O. C., and de Kloet, E. R. (1998). Corticosterone and serotonergic neurotransmission in the hippocampus: functional implications of central corticosteroid receptor diversity. *Crit Rev Neurobiol*, 12, 1–20.

Meijer, O. C., de Lange, E. C., Breimer, D. D., de Boer, A. G., Workel, J. O., and de Kloet, E. R. (1998). Penetration of dexamethasone into brain glucocorticoid targets is enhanced in mdr1A P-glycoprotein knockout mice. *Endocrinology*, 139, 1789–93.

Meewisse, M. L., Reitsma, J. B., de Vries, G. J., Gersons, B. P., and Olff, M. (2007). Cortisol and post-traumatic stress disorder in adults: systematic review and meta-analysis. *B J Psychiatry*, 191, 387–92.

Mikics, E., Kruk, M. R., and Haller, J. (2004). Genomic and non-genomic effects of glucocorticoids on aggressive behavior in male rats. *Psychoneuroendocrinology*, 29, 618–35.

Moriceau, S., and Sullivan, R. M. (2006). Maternal presence serves as a switch between learning fear and attraction in infancy. *Nat Neurosci*, 9, 1004–6.

Morsink, M. C., Joëls, M., Sarabdjitsingh, R. A., Meijer, O. C., De Kloet, E. R., and Datson, N. A. (2006a). The dynamic pattern of glucocorticoid receptor-mediated transcriptional responses in neuronal PC12 cells. *J Neurochem*, 99, 1282–98.

Morsink, M. C., Steenbergen, P. J., Vos, J. B., *et al.* (2006b). Acute activation of hippocampal glucocorticoid receptors results in different waves of gene expression throughout time. *J Neuroendocrinol*, 18, 239–52.

Munck, A., Guyre, P. M., and Holbrook, N. J. (1984). Physiological functions of glucocorticoids in stress and their relation to pharmacological actions. *Endocr Rev*, 5, 25–44.

Murgatroyd, C., Wu, Y., Bockmühl, Y., and Spengler, D. (2010). Genes learn from stress: How infantile trauma programs us for depression. *Epigenetics*, 31, 5(3).

Myers, K. M., and Davis, M. (2002). Behavioral and neural analysis of extinction. *Neuron*, 36, 567–84.

Nair, S. M., Karst, H., Dumas, T., *et al.* (2004). Gene expression profiles associated with survival of individual rat dentate cells after endogenous corticosteroid deprivation. *Eur J Neurosci*, 20, 3233–43.

Oitzl, M. S., and de Kloet, E. R. (1992). Selective corticosteroid antagonists modulate specific aspects of spatial orientation learning. *Behav Neurosci*, 106, 62–71.

Oitzl, M. S., Fluttert, M., and de Kloet, E. R. (1994). The effect of corticosterone on reactivity to spatial novelty is mediated by central mineralocorticosteroid receptors. *Eur J Neurosci*, 6, 1072–9.

Oitzl, M. S., Fluttert, M., Sutanto, W., and de Kloet, E. R. (1998). Continuous blockade of brain glucocorticoid receptors facilitates spatial learning and memory in rats. *Eur J Neurosci*, 10, 3759–6.

Oitzl, M. S., Workel, J. O., Fluttert, M., Frösch, F., and de Kloet, E. R. (2000). Maternal deprivation affects behaviour from youth to senescence: amplification of individual differences in spatial learning and memory in senescent Brown Norway rats. *Eur J Neurosci*, 12, 3771–80.

Oitzl, M. S., Reichardt, H. M., Joëls, M., and de Kloet, E. R. (2001). Point mutation in the mouse glucocorticoid receptor preventing DNA binding impairs spatial memory. *Proc Natl Acad Sci USA*, 98, 12790–5.

Oitzl, M. S., Champagne, D. L., van der Veen, R., and de Kloet, E. R. (2010). Brain development under stress: hypotheses of glucocorticoid actions revisited. *Neurosci Biobehav Rev*, 34, 853–66.

Olijslagers, J. E., de Kloet, E. R., Elgersma, Y., van Woerden, G. M., Joëls, M., and Karst, H. (2008). Rapid changes in hippocampal CA1 pyramidal cell function via pre- as well as postsynaptic membrane mineralocorticoid receptors. *Eur J Neurosci*, 27, 2542–50.

Oomen, C. A., Mayer, J., de Kloet, E. R., Joëls, M., and Lucassen, P. J. (2007). Brief treatment with the glucocorticoid receptor antagonist mifepristone normalizes the reduction in neurogenesis after chronic stress. *Eur J Neurosci*, 26, 3395–401.

Otte, C., Hinkelmann, K., Moritz, S., *et al.* (2010). Modulation of the mineralocorticoid receptor as add-on treatment in depression: a randomized, double-blind, placebo-controlled proof-of-concept study. *J Psychiatry Res*, 44, 339–46.

Parker, K. J., Buckmaster, C. L., Sundlass, K., Schatzberg, A. F., and Lyons, D. M. (2006). Maternal mediation, stress inoculation, and the development of neuroendocrine stress resistance in primates. *Proc Natl Acad Sci USA*, 103, 3000–5.

Pavlides, C., Ogawa, S., Kimura, A., and McEwen, B. S. (1996). Role of adrenal steroid mineralocorticoid and glucocorticoid receptors in long-term potentiation in the CA1 field of hippocampal slices. *Brain Res*, 738, 229–35.

Phillips, L. J., McGorry, P. D., Garner, B., *et al.* (2006). Stress, the hippocampus and the hypothalamic-pituitary-adrenal axis: implications for the development of psychotic disorders. *Aust N Z J Psychiatry*, 40, 725–41.

Price, M. L., Curtis, A. L., Kirby, L. G., Valentino, R. J., and Lucki, I. (1998). Effects of corticotropin-releasing factor on brain serotonergic activity. *Neuropsychopharmacology*, 18, 492–502.

Pu, Z., Krugers, H. J., and Joëls, M. (2007). Corticosterone time-dependently modulates beta-adrenergic effects on long-term potentiation in the hippocampal dentate gyrus. *Learn Mem*, 14, 359–67.

Puglisi-Allegra, S., Imperato, A., Angelucci, L., and Cabib, S. (1991). Acute stress induces time-dependent responses in dopamine mesolimbic system. *Brain Res*, 554, 217–22.

Qian, X., Droste, S.K., Gutièrrez-Mecinas, M., *et al.* (2011). A rapid release of corticosteroid-binding globulin from the liver restrains the glucocorticoid hormone response to acute stress. *Endocrinology*, 152, 3738–48.

Quirarte, G. L., Roozendaal, B., and McGaugh, J. L. (1997). Glucocorticoid enhancement of memory storage involves noradrenergic activation in the basolateral amygdala. *Proc Natl Acad Sci USA*, 94, 14048–53.

Rasch, B., Spalek, K., Buholzer, S., *et al.* (2009). A genetic variation of the noradrenergic system is related to differential amygdala activation during encoding of emotional memories. *Proc Natl Acad Sci U S A*, 106, 19191–6.

Ratka, A., Sutanto, W., Bloemers, M., and de Kloet, E. R. (1989). On the role of brain mineralocorticoid (type I) and glucocorticoid (type II) receptors in neuroendocrine regulation. *Neuroendocrinology*, 50, 117–23.

Ratka, A., Sutanto, W., and De Kloet, E. R. (1988). Long-lasting glucocorticoid suppression of opioid-induced antinociception. *Neuroendocrinology*, 48, 439–44.

Reichardt, H. M., Kaestner, K. H., Tuckermann, J., *et al.* (1998). DNA binding of the glucocorticoid receptor is not essential for survival. *Cell*, 93, 531–41.

Rënyi, L., Evenden, J. L., and Fowler, C. J. (2001). The pharmacological profile of (R)-3,4-Dihydro-N-isopropyl-3-(N-isopropyl-N-propylamino)-2H-1-benzopyran-5-carboxamide, a selective 5- hydroxytryptamine1A receptor agonist. *J Pharmacol Exp Ther*, 299, 883–93.

Reul, J. M., and de Kloet, E. R. (1985). Two receptor systems for corticosterone in rat brain: microdistribution and differential occupation. *Endocrinology*, 117, 2505–11.

Reul, J. M., Tonnaer, J. A., and De Kloet, E. R. (1988). Neurotrophic ACTH analogue promotes plasticity of type I corticosteroid receptor in brain of senescent male rats. *Neurobiol Aging*, 9, 253–60.

Reul, J. M., Stec, I., Söder, M., and Holsboer, F. (1993). Chronic treatment of rats with the antidepressant amitriptyline attenuates the activity of the hypothalamic-pituitary-adrenocortical system. *Endocrinology*, 133, 312–21.

Reul, J. M., Probst, J. C., Skutella, T., *et al.* (1997). Increased stress-induced adrenocorticotropin response after long-term intracerebroventricular treatment of rats with antisense mineralocorticoid receptor oligodeoxynucleotides. *Neuroendocrinology*, 65, 189–99.

Revest, J. M., Kaouane, N., Mondin, M., *et al.* (2010). The enhancement of stress-related memory by glucocorticoids depends on synapsin-Ia/Ib. *Mol Psychiatry*, 15, 1140–51.

Revollo, J. R., and Cidlowski, J. A. (2009). Mechanisms generating diversity in glucocorticoid receptor signaling. *Ann N Y Acad Sci*, 1179, 167–78.

Revsin, Y., Rekers, N. V., Louwe, M. C., *et al.* (2009). Glucocorticoid receptor blockade normalizes hippocampal alterations and cognitive impairment in streptozotocin-induced type 1 diabetes mice. *Neuropsychopharmacology*, 34, 747–58.

Revsin, Y., van Wijk, D., Saravia, F. E., Oitzl, M. S., De Nicola, A. F., and de Kloet, E. R. (2008). Adrenal hypersensitivity precedes chronic hypercorticism in Streptozotocin-induced diabetes mice. *Endocrinology*, 149, 3531–9.

Ridder, S., Chourbaji, S., Hellweg, R., *et al.* (2005). Mice with genetically altered glucocorticoid receptor expression show altered sensitivity for stress-induced depressive reactions. *J Neurosci*, 25, 6243–50.

Rosmond, R., Chagnon, Y. C., Chagnon, M., Pérusse, L., Bouchard, C., and Björntorp, P. (2000). A polymorphism of the 5'-flanking region of the glucocorticoid receptor gene locus is associated with basal cortisol secretion in men. *Metabolism*, 49, 1197–9.

Rozeboom, A. M., Akil, H., and Seasholtz, A. F. (2007). Mineralocorticoid receptor overexpression in forebrain decreases anxiety-like behavior and alters the stress response in mice. *Proc Natl Acad Sci USA*, 104, 4688–93.

Roozendaal, B., Hahn, E. L., Nathan, S. V., de Quervain, D. J. F., and McGaugh, J. L. (2004). Glucocorticoid effects on memory retrieval require concurrent noradrenergic activity in the hippocampus and basolateral amygdala. *J Neurosci*, 24, 8161–69.

Roozendaal, B., Hui, G. K., Hui, I. R., Berlau, D. J., McGaugh, J. L., and Weinberger, N. M. (2006). Basolateral amygdala noradrenergic activity mediates corticosterone-induced enhancement of auditory fear conditioning. *Neurobiol Learn Mem*, 86, 249–55.

Roozendaal, B., McEwen, B. S., and Chattarji, S. (2009). Stress, memory and the amygdala. *Nat Rev Neurosci*, 10, 423–33.

Sabban, E. L., and Kvetnansky, R. (2001). Stress-triggered activation of gene expression in catecholaminergic systems: dynamics of transcriptional events. *Trends Neurosci*, 24, 91–8.

Sahay, A., and Hen, R. (2007). Adult hippocampal neurogenesis in depression. *Nat Neurosci*, 10, 1110–5.

Sandeep, T. C., Yau, J. L., MacLullich, A. M., *et al.* (2004). 11Beta-hydroxysteroid dehydrogenase inhibition improves cognitive function in healthy elderly men and type 2 diabetics. *Proc Natl Acad Sci USA*, 101, 6734–9.

Sandi, C. (2004). Stress, cognitive impairment and cell adhesion molecules. *Nat Rev Neurosci*, 5, 917–30.

Sandi, C., Loscertales, M., and Guaza, C. (1997). Experience-dependent facilitating effect of corticosterone on spatial memory formation in the water maze. *Eur J Neurosci*, 9, 637–42.

Sandi, C., Merino, J. J., Cordero, M. I., Touyarot, K., and Venero, C. (2001). Effects of chronic stress on contextual fear conditioning and the hippocampal expression of the neural cell adhesion molecule, its polysialylation, and L1. *Neuroscience*, 102, 329–39.

Sapolsky, R. M. (2000). Glucocorticoids and hippocampal atrophy in neuropsychiatric disorders. *Arch Gen Psychiatry*, 57, 925–35.

Sapolsky, R. M., Romero, L. M., and Munck, A. U. (2000). How do glucocorticoids influence stress responses? Integrating permissive, suppressive, stimulatory, and preparative actions. *Endocr Rev*, 21, 55–89.

Sarabdjitsingh, R. A., Conway-Campbell, B. L., Leggett, J. D., *et al.* (2010a). Stress responsiveness varies over the ultradian glucocorticoid cycle in a brain-region-specific manner. *Endocrinology*, 151, 5369–79.

Sarabdjitsingh, R. A., Isenia, S., and Polman, A. (2010b). Disrupted corticosterone pulsatile patterns attenuate responsiveness to glucocorticoid signaling in rat brain. *Endocrinology*, 151, 1177–86.

Sarabdjitsingh, R. A., Spiga, F., Oitzl, M., *et al.* (2010c). Recovery from disrupted ultradian glucocorticoid rhythmicity reveals a dissociation between hormonal and behavioral stress responsiveness. *J Neuroendocrinol*, 22, 862–71.

Schaaf, M. J., Hoetelmans, R. W., de Kloet, E. R., and Vreugdenhil, E. (1997). Corticosterone regulates expression of BDNF and trkB but not NT-3 and trkC mRNA in the rat hippocampus. *J Neurosci Res*, 48, 334–41.

Schaaf, M. J., Workel, J. O., Lesscher, H. M., Vreugdenhil, E., Oitzl, M. S., and de Kloet, E. R. (2001). Correlation between hippocampal BDNF mRNA expression and memory performance in senescent rats. *Brain Res*, 915, 227–33.

Schmidt, H. D., and Duman, R. S. (2007). The role of neurotrophic factors in adult hippocampal neurogenesis, antidepressant treatments and animal models of depressive-like behavior. *Behav Pharmacol*, 18, 391–418.

Schmidt, M., Enthoven, L., van Woezik, J. H., Levine, S., de Kloet, E. R., and Oitzl, M. S. (2004). The dynamics of the hypothalamic-pituitary-adrenal axis during maternal deprivation. *J Neuroendocrinol*, 16, 52–7.

Schwabe, L., Schächinger, H., de Kloet, E. R., and Oitzl, M. S. (2010a). Stress impairs spatial but not early stimulus-response learning. *Behav Brain Res*, 213, 50–5.

Schwabe, L., Schächinger, H., de Kloet, E. R., and Oitzl M. S. (2010b). Corticosteroids operate as a switch between memory systems. *J Cogn Neurosci*, 22, 1362–72.

Seckl, J. R., and Fink, G. (1992). Antidepressants increase glucocorticoid and mineralocorticoid receptor mRNA expression in rat hippocampus in vivo. *Neuroendocrinology*, 55, 621–9.

Selye, H. (1950). Stress and the general adaptation syndrome. *Br Med J*, 17, 1383–92.

Selye, H. (1936). A syndrome produced by diverse nocuous agents. *Nature*, 138, 132.

Sloviter, R. S., Valiquette, G., Abrams, G. M., *et al.* (1989). Selective loss of hippocampal granule cells in the mature rat brain after adrenalectomy. *Science*, 243, 535–8.

Sooy, K., Webster, S. P., Noble, J., *et al.* (2010). Partial deficiency or short-term inhibition of 11beta-hydroxysteroid dehydrogenase type 1 improves cognitive function in aging mice. *J Neurosci*, 30, 13867–72.

Spencer, R. L., Kim, P. J., Kalman, B. A., and Cole, M. A. (1998). Evidence for mineralocorticoid receptor facilitation of glucocorticoid receptor-dependent regulation of hypothalamic-pituitary-adrenal axis activity. *Endocrinology*, 139, 2718–26.

Spiga, F., Harrison, L. R., Wood, S. A., *et al.* (2007). Effect of the glucocorticoid receptor antagonist Org 34850 on basal and stress-induced corticosterone secretion. *J Neuroendocrinol*, 19, 891–900.

Starkman M. N., Gebarski, S. S., Berent, S., and Schteingart, D. E. (1992). Hippocampal formation volume, memory dysfunction, and cortisol levels in patients with Cushing's syndrome. *Biol Psychiatry*, 32, 756–65.

Stevens, A., Ray, D. W., and Zeggini, E. (2004). Glucocorticoid sensitivity is determined by a specific glucocorticoid receptor haplotype. *J Clin Endocrinol Metab*, 89, 892–7.

Stewart, M. G., Davies, H. A., Sandi, C., *et al.* (2005). Stress suppresses and learning induces plasticity in CA3 of rat hippocampus: a three-dimensional ultrastructural study of thorny excrescences and their postsynaptic densities. *Neuroscience*, 131, 43–54.

Stranahan, A. M., Arumugam, T. V., Cutler, R. G., Lee, K., Egan, J. M., and Mattson, M. P. (2008). Diabetes impairs hippocampal function through glucocorticoid-mediated effects on new and mature neurons. *Nat Neurosci*, 11, 309–17.

Swaab, D. F., Bao, A. M., and Lucassen, P. J. (2005). The stress system in the human brainin depression and neurodegeneration. *Ageing Res Rev*, 4, 141–94.

Szyf, M., Weaver, I. C., Champagne, F. A., Diorio, J., and Meaney, M. J. (2005). Maternal programming of steroid receptor expression and phenotype through DNA methylation in the rat. *Front Neuroendocrinol*, 26, 139–62.

Tashiro, A., Sandler, V. M., Toni, N., Zhao, C., and Gage, F. H. (2006). NMDA-receptor mediated, cell-specific integration of new neurons in adult dentate gyrus. *Nature*, 442, 929–33.

Tausk, M. (1952). Hat die Nebennieren Rinde wirklich eine Verteidigungs Funktion? *Das Hormon*, 3.

Tiemensma, J., Kaptein, A. A., Pereira, A. M., Smit, J. W., Romijn, J. A., and Biermasz, N. R. (2011). Coping strategies in patients after treatment for functioning or nonfunctioning pituitary adenomas. *J Clin Endocrinol Metab*, 96(4), 964–71.

Tronche, F., Kellendonk, C., Kretz, O., *et al.* (1999). Disruption of the glucocorticoid receptor gene in the nervous system results in reduced anxiety. *Nature Genet*, 23, 99–103.

van der Laan, S., and Meijer, O. C. (2008). Pharmacology of glucocorticoids: beyond receptors. *Eur J Pharmacol*, 585, 483–91.

van der Lely, A. J., Foeken, K., van der Mast, R. C., and Lamberts, S. W. (1991). Rapid reversal of acute psychosis in the Cushing syndrome with the cortisol-receptor antagonist mifepristone (RU 486). *Ann Inter Med*, 114, 143–4.

van Haarst, A. D., Oitzl, M. S., Workel, J. O., and de Kloet, E. R. (1996). Chronic brain glucocorticoid receptor blockade enhances the rise in circadian and stress-induced pituitary-adrenal activity. *Endocrinology*, 137, 4935–43.

van Haarst, A. D., Oitzl, M. S., and de Kloet, E. R. (1997). Facilitation of feedback inhibition through blockade of glucocorticoid receptors in the hippocampus. *Neurochem Res*, 22, 1323–8.

van Leeuwen, N., Kumsta, R., Entringer, S., *et al.* (2010). Functional mineralocorticoid receptor (MR) gene variation influences the cortisol awakening response after dexamethasone. *Psychoneuroendocrinology*, 35, 339–49.

van Oers, H. J., de Kloet, E. R., Whelan, T., and Levine, S. (1998). Maternal deprivation effect on the infant's neural stress markers is reversed by tactile stimulation and feeding but not by suppressing corticosterone. *J Neurosci*, 18, 10171–9.

van Praag, H. M., de Kloet, E. R., and Van Os, J. (2004). *Stress, the Brain and Depression*. Cambridge: Cambridge University Press.

van Rossum, E. F., Binder, E. B., Majer, M., *et al.* (2006). Polymorphisms of the glucocorticoid receptor gene and major depression. *Biol Psychiatry*, 59, 681–8.

van Rossum, E. F., Voorhoeve, P. G., te Velde, S. J., et al. (2004). The ER22/23EK polymorphism in the glucocorticoid receptor gene is associated with a beneficial body composition and muscle strength in young adults. *J Clin Endocrinol Metab*, 89, 4004–9.

Veldhuis, J. D., Iranmanesh, A., Godschalk, M., and Mulligan, T. (2000). Older men manifest multifold synchrony disruption of reproductive neurohormone outflow. *J Clin Endocrinol Metab*, 85, 1477–86.

Venero, C., and Borrell, J. (1999). Rapid glucocorticoid effects on excitatory amino acid levels in the hippocampus: a microdialysis study in freely moving rats. *Eur J Neurosci*, 11, 2465–73.

Vyas, A., Mitra, R., Shankaranarayana Rao, B. S., and Chattarji, S. (2002). Chronic stress induces contrasting patterns of dendritic remodeling in hippocampal and amygdaloid neurons. *J Neurosci*, 22, 6810–18.

Walker, J. J., Terry, J. J., and Lightman, S. L. (2010). Origin of ultradian pulsatility in the hypothalamic-pituitary-adrenal axis. *Proc Biol Sci*, 277, 1627–33.

Watts, A. G. (2005). Glucocorticoid regulation of peptide genes in neuroendocrine CRH neurons: a complexity beyond negative feedback. *Front Neuroendocrinol*, 26, 109–30.

Weber, T., Baier, V., Pauly, R., *et al.* (2011). Inducible gene expression in GFAP+ progenitor cells of the SGZ and the dorsal wall of the SVZ-A novel tool to manipulate and trace adult neurogenesis. *Glia*, 59, 615–26.

Wei, Q., Lu, X. Y., Liu, L., *et al.* (2004). Glucocorticoid receptor overexpression in forebrain: a mouse model of increased emotional lability. *Proc Natl Acad Sci USA*, 101, 11851–6.

Wiegert, O., Joëls, M., and Krugers, H. (2006). Timing is essential for rapid effects of corticosteroids on synaptic potentiation in the mouse hippocampus. *Learn Mem*, 13, 110–13.

Windle, R. J., Wood, S. A., Shanks, N., Lightman, S. L., and Ingram, C. D. (1998). Ultradian rhythm of basal corticosterone release in the female rat: dynamic interaction with the response to acute stress. *Endocrinology*, 139, 443–50.

Windle, R. J., Wood, S. A., Kershaw, Y., Lightman, S. L., Ingram, C. D., and Harbuz, M. S. (2001). Increased corticosterone pulse frequency during adjuvant-induced arthritis and its relationship to alterations in stress responsiveness. *J Neuroendocrinol*, 13, 905–11.

Wolkowitz, O. M., Epel, E. S., and Reus, V. I. (2001). Stress hormone-related psychopathology: pathophysiological and treatment implications. *World J Biol Psychiatry*, 2, 115–43.

Wommack, J. C., and Delville, Y. (2007). Cortisol controls the pubertal development of agonistic behavior in male golden hamsters via type II corticosteroid receptors. *Horm Behav*, 51, 306–12.

Wood, G. E., Young, L. T., Reagan, L. P., and McEwen, B. S. (2003). Acute and chronic restraint stress alter the incidence of social conflict in male rats. *Horm Behav*, 43, 205–13.

Workel, J. O., Oitzl, M. S., Fluttert, M., Lesscher, H., Karssen, A., and de Kloet, E. R. (2001). Differential and age-dependent effects of maternal deprivation on the hypothalamic-pituitary-adrenal axis of brown norway rats from youth to senescence. *J Neuroendocrinol*, 13, 569–80.

Wu, L. M., Han, H., Wang, Q. N., *et al.* (2007). Mifepristone repairs region-dependent alteration of synapsin I in hippocampus in rat model of depression. *Neuropsychopharmacology*, 32, 2500–10.

Xu, L., Anwyl, R., and Rowan, M. J. (1998). Spatial exploration induces a persistent reversal of long-term potentiation in rat hippocampus. *Nature*, 394, 891–4.

Yau, J. L., Noble, J., and Seckl, J. R. (1999). Continuous blockade of brain mineralocorticoid receptors impairs spatial learning in rats. *Neurosci Lett*, 277, 45–8.

Yehuda, R. (2002). Post-traumatic stress disorder. *N Engl J Med*, 346, 108–14.

Yehuda, R. (2006). Advances in understanding neuroendocrine alterations in PTSD and their therapeutic implications. *Ann N Y Acad Sci*, 1071, 137–66.

Young, E. A., Abelson, J., and Lightman, S. L. (2004). Cortisol pulsatility and its role in stress regulation and health. *Front Neuroendocrinol*, 25, 69–76.

Young, E. A., Lopez, J. F., Murphy-Weinberg, V., Watson, S. J., and Akil, H. (2003). Mineralocorticoid receptor function in major depression. *Arch Gen Psychiatry*, 60, 24–8.

Zhou, M., Bakker, E. H., Velzing, E. H., *et al.* (2010). Both mineralocorticoid and glucocorticoid receptors regulate emotional memory in mice. *Neurobiol Learn Mem*, 94, 530–7.

Zhou, M., Conboy, L., Sandi, C., Joëls, M., and Krugers, H. J. (2009). Fear conditioning enhances spontaneous AMPA receptor-mediated synaptic transmission in mouse hippocampal CA1 area. *Eur J Neurosci*, 30, 1559–64.

CHAPTER 6

Adult Hippocampal Neurogenesis: Functional Implications of a 'New' Type of Plasticity

Alexander Garthe and Gerd Kempermann

Introduction: adult neurogenesis in the hippocampus 105

Theories of hippocampal function 107

Experimental evidence for a functional role of adult neurogenesis 112

Conclusion 118

ABSTRACT Adult hippocampal neurogenesis has changed our view on the adult mammalian brain by adding a new kind of neuroplasticity with strong implications for cognition and behaviour of older mammals. Because the new neurons are found in the hippocampus, a structure well known to be the gateway to memory, adult neurogenesis appears to be of especially high relevance for humans. The integration of new neurons into existing and well-working networks depends on an intimate regulation of the divisions the putative stem cells undergo in the subgranular zone, the expansion of the precursor cell pool, the processes of maturation and differentiation, and, finally, the appropriate functional integration into the network of the dentate gyrus. While much progress has been made in the last decade on the genetic and molecular underpinnings of adult neurogenesis, as well as how these processes are regulated in the context of different 'lifestyles', its functional contribution remains unclear. From a clinical perspective, however, all the attention adult neurogenesis deserves depends on the specific functional contribution of the new neurons to hippocampal function. Only recently, a coherent picture is about to begin to emerge what the new neurons might be good for. Importantly, results from studies testing rodents in spatial learning paradigms suggest a similar role for adult neurogenesis in human spatial navigation. This chapter reviews the findings on the functional relevance of adult neurogenesis along with a recently published attempt to integrate the data into a common theoretical framework. Beside experimental data, results and implications from studies based on theoretical models regarding the function of the new neurons are considered.

KEYWORDS adult neurogenesis, hippocampus, function, theoretical models, pattern separation, catastrophic interference, spatial learning, spatial navigation, regulation, enriched environment, running

Introduction: adult neurogenesis in the hippocampus

In 1965 Altman and Das published the first report on new-born neurons in the adult brain (Altman and Das 1965). Although their study should have initiated a fundamental change in our view on the plasticity in the mature mammalian brain the phenomenon of adult neurogenesis had to be rediscovered in the 1990s to attract the attention it deserves. Since then, thousands of studies have yielded a wealth of information about where the new neurons are coming from, how they mature, what kind of stimuli regulate their development, and, last but not least, what the new neurons might be good for.

There are two well-defined neurogenic regions in the brain of most adult mammals: the subventricular zone (SVZ) of the wall of the lateral ventricles and the subgranular zone (SGZ) in the dentate gyrus within the hippocampal formation. New-born cells in the SVZ migrate towards the olfactory bulb where they differentiate into a range of, to date, seven known types of interneurons. In the hippocampus the new neurons migrate a much shorter distance and end up only as excitatory granule cells, the principal cell type of the dentate gyrus (Figure 6.1). The putative stem cells of the SGZ (type 1 cells) have low proliferative activity and produce highly proliferative intermediate progenitor cells (type 2/3). Expansion mainly takes place at the stage of type 2 cells thereby expanding the pool of new cells available for cell cycle exit and further neuronal differentiation and maturation. Most of the new cells, however, undergo apoptosis early after cell cycle exit, but 10–50% survive, extend neurites, and become functionally integrated into the existing networks. Once integrated, they are likely to persist for months and probably lifelong. Intriguingly, proliferation and survival of the precursor cells and their progeny are differentially regulated by physical and cognitive training. These regulatory effects are additive: physical exercise increases the pool of progenitor cells that can—potentially—be selected for survival (van Praag et al. 1999; Fabel et al. 2009). Cognitive stimuli such as learning or exposure to a challenging environment increase survival and integration (Kempermann et al. 1997).

At the functional level new hippocampal neurons have been implicated in learning processes across multiple paradigms, although their exact functional contribution remains difficult to establish. For example, spatial learning performance in the Morris water maze, probably the best-validated learning task for hippocampal function, appeared to be correlated with new neuron numbers in some studies whereas other authors did not find such a correlation (Leuner et al. 2006; Deng et al. 2010). The experimental setups and protocols that were used turned out to be relevant for the exact interpretation, however. Without drawing a clear positive picture as yet, the discrepancies suggest a high specificity of the functional role of adult neurogenesis. Studies directly assessing the role of the dentate gyrus in pattern separation could less ambiguously demonstrate a role of new-born neurons for that type of information processing (Clelland et al. 2009; Creer et al. 2010).

Adult hippocampal neurogenesis has also been described for humans (Eriksson et al. 1998). Because the hippocampal formation is known to be a key structure in the formation and retrieval of new memories, the question is whether adult neurogenesis contributes to these processes and how it could be related to human hippocampal pathologies such as Alzheimer disease and other types of dementia, major depression, and schizophrenia, as well as aspects of so-called normal cognitive and mental ageing.

The facts that adult hippocampal neurogenesis is found in the human brain, that it contributes to hippocampal function, and that it is regulated by physical and cognitive

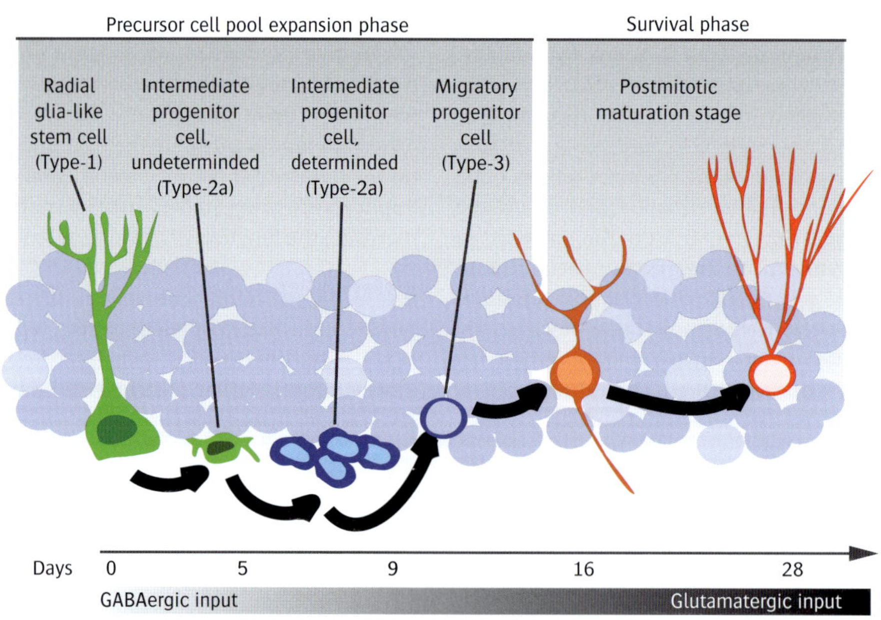

Figure 6.1 Developmental stages in adult hippocampal neurogenesis. Each stage can be identified by a unique combination of morphological, immunohistochemical and electrophysiological features. Reprinted from *Neuroscience*, 154(2), Steiner, B., Zurborg, S., Hörster, H., Fabel, K., and Kempermann, G. (2008). Differential 24 h responsiveness of Prox1-expressing precursor cells in adult hippocampal neurogenesis to physical activity, environmental enrichment, and kainic acid-induced seizures. 521–9, Copyright (2008), with permission from Elsevier.

training in rather specific ways make the new neurons interesting and relevant for medicine, even though many concrete links remain speculative. There are still many white spots on the map and gaps in the arguments. And, to date, too much of our knowledge is still based on rodent data only. Nevertheless, the new neurons have started to fundamentally change our view on (and hence our therapy and prevention of) essentially all brain disorders affecting the hippocampus. Hippocampal function presumably cannot be seen and reasonably understood without adult neurogenesis.

Because the new neurons gain all their relevance for clinical medicine from their contributions to brain function in health and disease and the structural and molecular underpinnings are rarely an issue for clinical neurobiology, this chapter focuses on the relevance of adult neurogenesis for learning and memory processes. For broader overviews on adult neurogenesis we refer to a number of larger review articles, which, however, largely lack the emphasis that we have placed (Lledo et al. 2006; Gould 2007; Zhao et al. 2008; Suh et al. 2009). Important findings from physiological, molecular, and genetic studies are emphasized where needed.

Theories of hippocampal function

To understand how new neurons contribute to hippocampal and brain function a clear concept is necessary of how the hippocampus works. Because of its key involvement in learning and memory processes, the hippocampus is one of the best-studied brain structures in mammals (Amaral and Lavenex 2007). However, the concept of 'memory' involves different types of information being processed and represented in different brain systems. Two classical theoretical frameworks still represent the scaffold of most modern ideas of hippocampal function. In humans, the type of learning classically assumed to be most dependent on the hippocampus is to form memories for declarative information (Squire 2004; Moscovitch et al. 2005). Declarative memory is the knowledge of facts (semantic memory) and events (episodic memory) that can be consciously recalled and expressed in words. The nature and importance of this kind of memory were explored and determined by the case of patient H.M. who showed an almost complete anterograde amnesia after his hippocampi had been removed to alleviate his intractable epileptic seizures (Scoville and Milner 1957). However, the hippocampal formation is not assumed to be the long-term storage site for the memories. The hippocampus rather serves as a 'gateway to memory' and with time the learned information becomes successively more independent from the allocortex as it gets transferred to neocortical areas. These suppositions are summarized and referred to as the declarative memory theory and over time some influencing spin-offs have been derived such as the multiple-trace theory or the idea that the hippocampus might be involved in general relational processing (Nadel and Moscovitch 1997; Eichenbaum and Cohen 2001). Also because human memories for word lists, groups of items, or events separated in time can be easily assessed using interviews or query sheets, declarative memory theories are still predominant in research assessing hippocampal function in humans.

An altogether different perspective has been taken by the cognitive map theory. Brought up in 1978, upon discovery of hippocampal 'place cells' in rats, it essentially limits the information that is processed by the hippocampus to the spatial domain, such as in orientation and navigation (Nadel and O'Keefe 1978). As in the declarative memory theory the hippocampus is proposed to encode and consolidate information it receives by its various afferents from higher sensory cortical areas. The difference to the declarative memory theory lies in the fact that the hippocampus is supposed to represent a specialized locale system allowing navigation through the environment by linking events to places. Perceived stimuli get organized and represented relative to a cognitive map of the environment. The reference points of such an 'allocentric' cognitive map are not identical to the individual ('egocentric') and thus have achieved a certain degree of independence. Landmarks such as distant visible cues encountered during explorative activity serve as the basic constituent elements of the hippocampal cognitive map. An effective spatial navigation through an environment, where the organism has to find food, mates, and shelter is believed to be the main purpose of the representations formed. Albeit being exclusively spatial, episodic memories are considered to be formed in the context of the cognitive maps, too. Navigation can be seen as a sequence of events that requires an ordered memory to allow, for example, tracing back one's path. Because most studies related to the cognitive map theory have been done using rats or mice this theory appears to be somewhat biased towards a rodent's perspective. However, several studies have shown that the general principles underlying hippocampal information processing identified in rodents can also be found in humans (Kesner and Hopkins 2006).

There are specific aspects of information processing uniquely found in the hippocampal formation and its subdivisions. Despite being fundamental for how information is being processed in the hippocampus and how the representations formed contribute to brain function, pattern separation, pattern completion, and pattern association were recognized in detail by just a few authors (Rolls 1996, 2010; Hunsaker and Kesner 2008; Hunsaker et al. 2008).

Besides assessing the functional properties of specific subregions, the relationship of the hippocampus to other brain systems in the context of learning is an important direction for future research. Lisman and Grace, for example, proposed a role for the hippocampus together with the ventral tegmental area (VTA) in providing teaching signals to control the entry of information into long-term memory (Lisman and Grace 2005).

Network theories and information processing in the hippocampal subdivisions

The hippocampus has an appealingly clear backbone structure, the so called trisynaptic circuit (Amaral and Lavenex 2007). Inputs from virtually all higher sensory cortical areas as well as from other brain regions including the parietal, temporal, and frontal lobes converge in the entorhinal cortex (EC) and, via the perforant path, are conveyed to the first relay station of the circuit, the dentate gyrus. From here the signal is send to CA3 (second relay station), further relayed to CA1 (third station) and via the subiculum transmitted back to EC. While in general the flow of information through the hippocampus is unidirectional, there are shortcut-like direct connections between the EC and the hippocampal subdivisions appearing later in the trisynaptic sequence. Usually, when considering hippocampal function these shortcuts are ignored, but they might provide new, yet unknown twists to information processing, especially in the context of different phases of the learning process. The hippocampus also receives inputs from the amygdala and the cholinergic and other regulatory systems.

The hippocampus represents a well-conserved part of the brain (allocortex) as it consists of three layers compared to the six layers found in the cortex (Amaral and Witter 1995). Given that a huge amount of sensory and other information converges in the EC and subsequently floods into hippocampal circuitry it is clear that the hippocampus as such represents a bottleneck structure. Compared to the amount of information arriving from the EC, only an extremely sparse pattern of activation is observed in the dentate gyrus. Thus, the dentate gyrus appears to act as a competitive learning network implementing a strong compression of data intended to be stored in CA3 (Rolls 1996).

The highly recurrent network structure of area CA3 strongly implies that is serves as a temporary autoassociative memory (Marr 1971; Rolls 2010). Among the hippocampal substructures in area CA3 the lowest number of principal neurons is found. Judging from the dendritic contacts each CA3 pyramidal cell receives, the highest number of inputs comes from recurrent connections with other pyramidal cells (about 12,000 in the rat), followed by direct input from the EC via the perforant path (about 4000 in the rat). Importantly, there are just a few synapses emanating from axons of the dentate granule cells (about 50 in the rat), the mossy fibres (Rolls 1996). Therefore for new contents conveyed through the dentate gyrus, CA3 represents a bottleneck within the bottleneck and thus, the main function generally assigned to the dentate gyrus is the compression and orthogonalization of activity patterns to make them suitable for storage (Marr 1971; Treves et al. 2008). Adult hippocampal neurogenesis provides new neurons exclusively to the dentate gyrus and thus produces axons that are part of the mossy fibre connection between the DG and CA3.

Because the mossy fibres are the strongest excitatory afferents of the pyramidal cells in area CA3 the firing of granule cells in the dentate gyrus dominate the representations formed. This is an effective way to solve a problem found in all autoassociative networks: Because of an extremely high degree of recurrent connections between the pyramidal cells activation patterns previously stored tend to reverberate in the network hampering new representations to be instantiated (Treves and Rolls 1992). Providing a strong and sparse input to the CA3 principal cells, the mossy fibres simply force the formation of efficient representational activity patterns. Thus, every memory to be stored initiates its own arbitrary and small subset of CA3 pyramidal cells to be active. Given that the dentate gyrus makes information suitable for effective, non-overlapping encoding in area CA3, sparsification of activity patterns ensures highly orthogonalized representations to be formed and—in more common sense—guarantees the successful storage and use of memories. Usually, such an orthogonalization of representations is referred to as pattern separation. Without pattern separation the representations stored in CA3 would suffer from catastrophic interference, that is new memories would interfere with older, already encoded ones, thereby preventing the effective storage and retrieval of highly similar but nonetheless different representations. Because adult hippocampal neurogenesis adds new neurons to the dentate gyrus, it is likely that the function of the new adult-born granule cells is to prevent 'catastrophic interference' and ensure pattern separation while experiencing a high variety of different but often similar environments—that is: input-patterns—over a lifetime. That pattern separation is indeed a key function of the dentate gyrus was elegantly demonstrated by (Kesner et al. 2004). As discussed later, we have presented a computational model of the function of adult hippocampal neurogenesis based on the central assumption that new neurons are an effective means of avoiding catastrophic interference in CA3. Aimone and colleagues have focused on the related issue of pattern separation.

Hippocampal networks show at least two different modes of operation assigned to the encoding of new and retrieval of already stored memories (Hasselmo 2005). Upon exposure to a new situation a large amount of new information enters the hippocampus. In order to improve an organism's behaviour in the future all relevant contents have to be temporally stored in CA3 followed by a consolidation period where hippocampal representations are transferred to long-term memory. The consolidation process involves the repeated retrieval of contents previously stored in the recurrent CA3 network. It is still not clear to which degree information represented in CA3 remains permanently in the hippocampus or is further passed to cortical regions due to consolidation. Given the amount of memories to be stored across the life period and the modest size of CA3 the latter seems more plausible. To reinitiate an activation pattern already memorized, even exposure to a small subset of the known cues is sufficient. This ability strongly depends on the autoassociative network structure in CA3 and is referred to as pattern completion.

Because area CA3 is involved in both memory storage and retrieval these processes constantly need to be separated. A key role in setting up hippocampal networks for storage has been assigned to the activity of hippocampal neurons rhythmically modulated by afferent inputs from cholinergic and gamma-aminobutyric acid (GABA)ergic neurons of the septum (Düzel et al. 2010). The rhythm found (4–8Hz) is referred to as theta-rhythm and it allows the encoding of new memories to become temporally separated apart from retrieval processes. In rodents a strong theta-activity is found during exploration of an unknown environment, suggesting theta as being an obligatory correlate for the formation of new representations in CA3 (O'Keefe 2007). Theta rhythms can be recorded in humans as well where they appear to be involved in spatial navigation (Cornwell et al. 2008; Staudigl et al. 2010). Oscillatory neuronal activity is thought to play an important role in the regulation of adult hippocampal neurogenesis by promoting differentiation and survival but to date no specific involvement of theta rhythms has been shown (Babu et al. 2009).

The third and final relay station of the tri-synaptic circuit is area CA1, which is thought to be a decoding network. In a general sense, its function is to reintegrate the highly separated activity patterns stored in CA3 into real episodes and to re-expand the compressed information after passing the CA3 bottleneck before it is conveyed back to cortical areas via the EC (Rolls 2010).

Another interesting idea is to see CA1 as a comparator, providing an error signal by evaluating how well a representation previously stored in CA3 fits with reality (Lisman and Grace 2005). This model is appealing because it couples the formation and retrieval of hippocampal memories to the VTA as a key component in the dopaminergic modulation of reward-related aspects of learning. Since adult hippocampal neurogenesis is known to be regulated by different neuromodulators, physiological conditions, and behavioural states, a tight integration of the new neurons into concepts of general learning rather than just hippocampal processing is not only highly attractive but presumably also necessary. Such consideration will probably underscore the relevance of this 'new' form of plasticity far beyond the hippocampal formation.

Nevertheless, despite the wealth of anatomical and neurophysiological data as well as a number of theoretical models providing insight into the principles of information processing in the hippocampus, it is not yet really understood how the abstract concepts relate to concrete elements of actual behaviour. Obviously, this also hampers the interpretation of data concerning the functional contribution of adult-generated granule cells. The discovery of place cells in the hippocampus almost three decades ago can serve as a good starting point: place cells and their corresponding place fields can be characterized in relation to different experimental situations, some of them stressing supposedly adult neurogenesis-related aspects of information processing. Separating patterns or places located closely to each other is central for successful navigation and other types of hippocampal learning. Place cells have been shown to encode not only space but also abstract, task-related information (Hok et al. 2007). Thus, separation of hippocampal activity patterns becomes especially relevant when the actual goal becomes updated while the general task and its context remain unchanged. Thereby, a theoretical concept such as 'pattern separation' can be principally linked to concrete behaviours.

Theoretical models of adult neurogenesis—chances and limitations

Based on the knowledge about hippocampal circuitry, physiological investigations at the synaptic and cellular levels, and specific ideas about the functional contributions the hippocampal subregions might provide to the overall functionality a number of theoretical models for the contribution of adult neurogenesis to hippocampal function have been developed (Meltzer et al. 2005; Aimone et al. 2006; Wiskott et al. 2006; Becker and Wojtowicz 2007; Appleby et al. 2011). Such models allow testing the impact of neurogenesis on specific aspects of the performance of neuronal networks. The so-called stability-plasticity dilemma refers to a fundamental problem common to all neuronal networks that are supposed to learn and refers to the trade-off between the necessity of a stable network to preserve learned information and plasticity of the same network to acquire new information. Adult neurogenesis poses a stability-plasticity dilemma of its own in that adding new easily excitable neurons to a mature, stable and functional network might seem a risky strategy. Although computational modelling is sometimes criticized for supposedly allowing to design a model for each intended outcome, modelling distinct aspects of neuronal networks in a highly reductionistic way here opens the view on specific adult neurogenesis-related functional details that are difficult to assess in classical behavioural or electrophysiological studies.

In general, models concerning the functional relevance of adult neurogenesis can be assigned to one of two groups defined by the main focus the model. First, because adult hippocampal neurogenesis is exclusively found in the dentate gyrus, which as a structure is implied to be important for pattern separation, most models have asked for a role of the new neurons in pattern separation and preventing catastrophic interference of representations in area CA3. The main implications of such models for our understanding of what adult neurogenesis might be good for in the context of real-world spatial navigation will be discussed further below. Models relating to spatial or temporal pattern separation can make predictions that are testable in animal experiments or possibly human subjects. For instance, Wiskott and colleagues have proposed a highly reductionistic computational model that implied a role for adult neurogenesis in avoiding catastrophic interference when the CA3 network has to encode a new environment similar to the one previously memorized (Wiskott et al. 2006). That implication was later confirmed in a study using mice with suppressed adult neurogenesis (Garthe et al. 2009). After successful acquisition of the water maze task, animals with low numbers of new neurons were shown to be significantly impaired in re-learning the goal position when the hidden platform was moved to the opposite side of the pool. As another example, a prediction based on a model introduced by Aimone et al. assessing pattern separation in both spatial and temporal domains was later experimentally confirmed by Clelland et al. (Aimone et al. 2006, 2010; Clelland et al. 2009).

The second group of theoretical models primarily deals with a role of network homeostasis in adult neurogenesis. Here, the focus lies on the potentially detrimental effects of adding new, highly excitable new neurons to existing circuits and how the hippocampus might compensate for those side effects (Deisseroth et al. 2004). Possible negative effects of adult neurogenesis on hippocampal function encompass epileptiform processes by mossy fibre sprouting as well as impaired retrieval of already encoded memories due to overall increased excitation levels. One possible strategy to reduce the risks of adding new neurons is a turnover homeostasis where older granule cells were removed by cell death. Indeed, new neurons do turnover. The immature neurons go through a period of increased synaptic plasticity and thus provide a population of highly plastic cells. Importantly, not all young new-born neurons survive as most of them undergo apoptosis before entering the postmitotic phase. There might thus be two contributions of adult neurogenesis: one transient one based on a turnover of new cells, one based on long-term changes of the network. Interestingly in this context, according to one as yet unconfirmed study most neurons in the dentate gyrus appear to be 'retired', in that they are not longer responsive even for inputs those cells were originally tuned for earlier in life (Alme et al. 2010). A parsimonious interpretation of the available body of evidence is that the lifelong addition of new neurons might be necessary to uphold a certain level of a very specific kind of plasticity in the dentate gyrus. This plasticity is necessary only here, under the particular network conditions ruling here and seems dispensable elsewhere in the brain. Adult neurogenesis continues to add new neurons to the network of older, less active ('retired') granule cells that obviously still seem to be useful and have thus not yet been removed from the network.

There are now indeed theoretical models that point towards different functions of new born neurons depending on their respective maturation stages (Aimone et al. 2006, 2009). Adult-born granule cells are derived from precursor cells they are not simply integrated as new 'ready-to-use neurons'. Instead, there is a maturation process encompassing multiple stages each characterized by a unique pattern of morphology, marker expression, responsiveness to electrical stimulation, and sensitivity to regulation by physical or cognitive challenges (Figure 6.1). There is, thus, a feedback link from functionality (and behaviour) into the process of adult neurogenesis itself. Development is adjusted to the functional needs. In some sense there is thus not just a single type of neurons available for functional integration which can be used or not. Rather the distinct maturational stages with their different properties and different responsiveness to regulatory stimuli provide a considerable variety of neuronal cells that might fulfil different functional roles in different contexts. Activity-dependent regulation lies at the heart of adult neurogenesis and the entire process that adds to plasticity is plastic and malleable itself.

Ablation of adult neurogenesis and the causality problem

A growing number of studies have used ablation of adult neurogenesis as a reductionistic approach to reveal the functional relevance of the new granule cells. Blocking adult hippocampal neurogenesis by irradiation, cytostatic drugs or genetic manipulations is thought to cause specific functional deficits that can in turn be attributed to the lack of new adult born neurons. There is, however, always the risk of unspecific and unwanted side effects. For instance, irradiation can cause a strong inflammatory response in the exposed tissues, making it difficult to attribute learning deficits observed specifically to the lack of new hippocampal neurons. Systemic treatments using cytostatic drugs such as MAM or temozolomide bear the risk of systemic side effects. Genetic approaches can provide a much higher specificity here and the conditional ablation of certain types of stem- or progenitor-cell populations in the SGZ and are thought to represent the state-of-the-art method for this kind of studies. Here the problem is that the genetic approach leads to highly complex breeding schemes involving numerous constructs, all of which in theory would call for appropriate controls. Genetic ablation can be only as specific as the underlying genes are and as specific the strategies are to manipulate them.

In the end no model can as yet solve the causality problem. This is because the functional role of the hippocampal formation in different learning tasks and the contribution of the dentate gyrus to this information processing are far from being fully understood. Adult neurogenesis might rather contribute to elucidating these functions than being explorable before a backdrop of a finished picture. Although some of the more recent theoretical models take into account a considerable amount of anatomical and neurophysiological data, many aspects and details must remain unmentioned. It is also not clear, what would have to be included to make a model and hypothesis 'good', nor what a 'good' model or hypothesis actually is. Thus, as any functional change observed after ablating new-born cells in the dentate gyrus should be interpreted in the specific context of the learning task itself and given that our knowledge how the hippocampus is involved in that task appears to be limited, a straightforward causal relationship between the adult-generated neurons and certain aspects of learning is difficult to make. This is illustrated by the seemingly contradictory results obtained using a wide range of ablation methods and testing for hippocampal function with the Morris water maze task (Leuner et al. 2006; Deng et al. 2010). Despite being by far the most frequently used and best-validated hippocampus-dependent learning task, studies using the water maze to address the functional role of adult neurogenesis in spatial learning largely failed to demonstrate a clear and consistent functional contribution of the new cells. This turned out to be largely a matter of the specific test setup and protocol used (and hence the underlying assumptions for specific subfunctions to be preferentially tested) and the analysis tools employed.

Recognizing our lack of understanding what the hippocampus really does in a certain task using a certain setup and test protocol may turn out to be more critical for a successful interpretation than aiming for ever greater specificity in the ablation methods.

Modulation of adult neurogenesis by different housing conditions

Adult neurogenesis is regulated by activity and certain types of behaviours increase the presence of new neurons and hence presumably the functionality of the hippocampus. Because of this link, clinical interest in adult neurogenesis is considerable, because adult neurogenesis might provide a concrete biological explanation, why certain paradigms in prevention and neurorehabilitation (as far as the underlying problem affects the hippocampus) are so successful. Our overall hypothesis is that adult neurogenesis provides a means of lifelong adaptability of the dentate gyrus to particular functional needs. This is an optimization process: the addition of new neurons is triggered by the behaviour itself. And activity maintains the potential for more neurogenesis in the future. We have summarized these observations in the 'neurogenic reserve hypothesis', stating that leading an active life maintains the hippocampus in a state of plasticity that is best suited to cope with the level of cognitive challenges that were encountered by the individual in the past (Kempermann 2008). The system thus only seems to afford the level of cellular plasticity that it 'knows' it will use.

Rodents are nocturnal animals and in the wild they run several kilometres per night. Access to a running wheel in their home cage allows mice kept under laboratory conditions to express this innate behaviour. Running as a presumably largely non-specific stimulus significantly enhances proliferation of hippocampal precursor cells in the dentate gyrus and thereby increases neurogenesis (van Praag et al. 1999; Kronenberg et al. 2003; Snyder et al. 2009).

Similarly, exposure to an enriched environment also increases adult neurogenesis but does so by what seems to be more specific means. Enriched environment housing consists of a larger cage, more social contacts, as well as an availability of multiple toys, tubes, and shelters. As a consequence animals can display a considerably broader range of social and spatial behaviours such as establishment of social hierarchies, territorial behaviour, foraging or increased exploratory activity. In most variants the topology of the enriched environment changes at least weekly, challenging the animals' ability to cope with changes in the spatial layout of the world they are living in. In contrast to the primarily pro-proliferative effects of voluntary wheel running, exposure to an enriched environment increases the survival of new born neurons, presumably at later stages of development after cell cycle exit when the new-born cells show more neuronal properties (Kempermann et al. 1997; van Praag et al. 2000; Steiner et al. 2008).

An argument occasionally raised against voluntary wheel running and living in an enriched environment as being natural counterparts to the ablation of adult neurogenesis is that both stimuli are to some extent unspecific as well. Running might elicit its pro-proliferative effects either as a consequence of increased levels of growth factors such as FGF2, IGF1 or VEGF or due to a local induction of theta-rhythmic activity, both of which might have effects on learning that are independent of new neurons. Further, exposure to an enriched environment does not simply provide more opportunities to learn, it allows a much broader range of behaviours to be expressed (and thus trained), so that an animal is likely to have a greater repertoire of behavioural strategies to master a given task not all of which explicitly fall into the category of 'learning'. Whenever learning appears to be increased after running or living in an enriched environment, any observed gain of function cannot be exclusively explained by the presence of more new neurons alone. Again, the lack of knowledge on what the hippocampus exactly does represents a considerable limitation when interpreting the data gathered in gain-of-function studies using running or enrichment paradigms.

Since the effects of running and enrichment were found to be additive, both modulatory effects might be also functionally related on the behavioural level. In this view locomotion significantly increases the likelihood to encounter situations where something has to be learned. Because physical activity also increases proliferation of precursor cells in the SGZ more young new neurons are available to become activated in the dentate gyrus. This provides an effective mechanism to adjust adult neurogenesis to an animal's overall behaviour and prepares the system for future challenges (Kempermann et al. 2010).

Modulation of adult neurogenesis by learning

Whereas environmental enrichment might provide some sort of seminaturalistic cognitive challenge, regulation of adult neurogenesis in response to specific learning events has also been studied. Döbrössy et al. have reported that proliferation of precursor cells was induced during the asymptotic phase of water maze learning when the animals' improvements relative to the previous trial get increasingly smaller (Döbrössy et al. 2003). This might represent a feedback mechanism by which the size of the available pool of precursor cells is adjusted to the stimuli that call for more neurogenesis. A later different study, however, failed to confirm this link between proliferation and learning (Mohapel et al. 2006).

Using the trace eye-blink conditioning paradigm Elizabeth Gould and colleagues showed that learning a hippocampus-dependent task enhances survival of new-born granule cells (Gould et al. 1999). Importantly, it was later found that the actual stimulus responsible for an increased survival is the formation of memories underlying the learning process as opposed to the mere exposure to training (Leuner et al. 2004). For spatial learning paradigms a survival promoting effect was also found using the water maze, and a more recent study even described a correlation between survival and overall performance in learning the task (Gould et al. 1999; Hairston et al. 2005).

Interestingly, new neurons seem to get involved in spatial learning in a subregion-specific manner. Rather than activating all new neurons everywhere in the dentate gyrus learning the water maze task preferentially affects new granule cells in the ventral part of the dentate gyrus whereas the activation of granule cells as a whole population is higher in the dorsal part (Snyder et al. 2009).

Effects of learning on hippocampal structure have also been assessed in humans. For example, the hippocampi of taxi drivers appear to be larger than those of navigational less experienced subjects (Maguire et al. 2000). Furthermore, extensive learning of abstract, declarative information also results in structural grey matter changes in the hippocampus (Draganski et al. 2006). These studies did not describe changes specifically related to adult neurogenesis, however. But because both navigating a cab or passing a medical exam explicitly stress functions provided by hippocampal networks, the results fit well into the idea that the increase in grey matter is—at least in part—due to adult neurogenesis.

Experimental evidence for a functional role of adult neurogenesis

As for most histological, molecular, or genetic data regarding adult neurogenesis, the functional contribution of the new neurons has been assessed preferentially using laboratory rats and mice. In rodents the hippocampus is primarily implicated in spatial learning processes, while in humans the functional contribution of the hippocampus likely goes well beyond the spatial domain. Nevertheless, spatial learning in humans is dependent on the hippocampus as it is in rodents. For that reason, the functional relevance of adult generated hippocampal granule cells is discussed primarily in the context of spatial pattern discrimination and spatial navigation, even though it is acknowledged that this view is still selective.

From an evolutionary perspective and given that the new neurons become increasingly indistinguishable from their older siblings during maturation it is unlikely that adult neurogenesis enables the dentate gyrus to perform new kinds of information processing. Thus, the contribution of adult neurogenesis to hippocampal function might primarily lie in a rather quantitative domain, i.e. preventing a decline in processing efficacy in situations demanding the network to deal with a high load of information. Given that pattern separation is a key function of the dentate gyrus, a straightforward approach to identify a functional role of adult born neurons is to look for links between adult neurogenesis and pattern separation.

A role for new-born neurons in pattern separation and pattern integration

To directly assess pattern separation in rodents, Clelland and colleagues devised a new task using a touch screen device (Clelland et al. 2009). In this task, mice had to differentiate between visual patterns presented either distant in space or close to each other. Ablation of neurogenesis impaired the animals' ability to differentiate between the two spatially adjacent patterns, whereas performance appeared to be normal for both groups, when the patterns were well separated in space. Another task in the same study showed similar results. Here, mice with suppressed adult neurogenesis had to remember a previously baited arm of a radial arm maze and to choose another arm on the subsequent trial. The authors reported deficiencies only when the sample and the tested arm were adjacent to each other in space. These findings were further supported by a study testing the impact of voluntary wheel running on pattern separation (Creer et al. 2010). After being trained in an easy differentiation task using the touch screen paradigm mentioned before and then tested on a more difficult condition in the same task mice previously exposed to a running wheel outperformed the not-running controls. As running increases proliferation of precursor cells in the SGZ the authors assigned the gain-of-function observed in pattern discrimination to the presence of more young new-born neurons in the dentate gyrus. Together, these results suggest that new hippocampal neurons significantly improve pattern separation in rodents, albeit still following a rather simple 'the-more-the-better' rule.

Results obtained in other studies, however, appear to be at odd with such an interpretation. Also using a spatial working memory protocol in the radial arm maze, Saxe et al. found an improvement after ablation of adult neurogenesis (Saxe et al. 2007). As proposed by Aimone et al., a possible explanation for this paradoxical finding might be the ability to encode the underlying rules when learning a task (Aimone et al. 2006, 2010). According to that hypothesis, immature new neurons are more excitable, thus forming broader less differentiated representations for events closely related in time. This allows the generalization of 'rules' encountered repetitively and over multiple trials. In the studies by Clelland and Saxe et al. animals were tested in the same paradigm but the respective protocols used were considerably different. Saxe et al. were placing a higher demand on differentiating between visits in closely adjacent arms and those closely related in time. Animals lacking new-born neurons might thus show a reduced ability to generalize. Each visit would be treated as a distinct event. In that special case, reduced plasticity would allow solving the task more efficiently than for animals with normal levels of adult neurogenesis. From this perspective, improving pattern separation by adding new neurons to the dentate gyrus might not be an optimal way to generally improve the performance for all kinds of learning. The fact that a better ability to separate events does not necessarily lead to better learning again points towards the problem that the demands that different testing paradigms pose to the animals simply do not match the range of challenges that might be encountered under more naturalistic conditions. Therefore, the mentioned studies not only provide good evidence for a role of new-born neurons in improving pattern separation, they also implicate that the functional relevance of adult neurogenesis strongly depends on the specific problem that needs to be dealt with.

The classical pioneering study on the functional relevance of adult generated hippocampal neurons by Tracey Shors and Elizabeth Gould in 2001 also fits well into the framework proposed by Aimone et al. Using a pharmacological approach to ablate neurogenesis it was found that only animals lacking new neurons were impaired in learning a hippocampus-dependent trace conditioning task while no difference between groups could be found in the hippocampus-independent version of the same paradigm (Shors et al. 2001). Trace conditioning is a classical conditioning task and expects the tested animals to associate an unconditioned stimulus with a conditioned one. Both stimuli can be separated in time (500 ms) and the hippocampus acts to keep up a trace of the first, conditioned stimulus, allowing both stimuli to become linked together when the second, unconditioned stimulus occurs. Although this task is usually used to assess cerebellar function and the exact contribution of the hippocampal component is not fully understood, the results are supporting the notion that young new-born neurons in the dentate gyrus might facilitate the binding of episodes occurring closely related in time. Another example of conditioning paradigms that has been extensively investigated in the context of adult neurogenesis is the contextual fear conditioning task (Phillips and LeDoux 1992). Here, animals have to associate a conditioned neutral stimulus and its respective context with an aversive unconditioned stimulus, usually a mild electric footshock. Encoding of the context serves as the hippocampal component here. Learning is assessed by measuring the 'freezing' time (the time of fearful immobility), when the trained animal is re-exposed to the conditioned context. Longer freezing times here indicate successful learning. Several studies have found an impaired performance in contextual fear conditioning after ablation of adult neurogenesis in the hippocampus (Shors et al. 2002; Saxe et al. 2006; Winocur et al. 2006; Wojtowicz et al. 2008; Hernández-Rabaza et al. 2009; Drew et al. 2010; Stone et al. 2011). Again, the new neurons' greater excitability might be important for integrating the sensory context with the unconditioned, aversive footshock. Adult neurogenesis is thus not only increasing pattern separation. It might also even allow for a transient reduction in pattern separation, thus helping to find a trade-off between encoding the characteristic details differentiating distinct episodes and integrating all the information belonging to a certain episode spanning multiple events separated in time.

At this point it should be mentioned that even slightly disturbances of the intimate processes regulating proliferation, maturation, and integration of the new neurons might lead to the formation of false-positive correlations. The presence of such unwanted associations has been implicated in the pathogenesis of human major depression (Kempermann et al. 2008).

Functional role of adult born neurons in spatial navigation

Among the different cognitive skills spatial navigation is one of the most accessible domains. The hippocampal formation including the dentate gyrus is critically involved in this particular kind of learning. Because the hippocampus is part of larger networks with roles not only in learning and memory, but also for example affective behaviour and attention, hippocampal data processing affects a much broader range of behaviours than those considered strictly 'hippocampal'. Spatial navigation is thus an imperfect proxy for broader-sense hippocampal function.

During evolution the ability to successfully navigate an environment in order to find food, mates, and shelter is fundamental for survival. The fittest navigate well. We have proposed that this navigation is invariably associated with locomotion and hence physical activity, thereby explaining why a non-specific stimulus such as exercise might have become so critically involved in regulating adult neurogenesis (Kempermann et al. 2010). Navigating through space relies on the formation of internal representations of events and places (as well as a temporal order of this information). Such cognitive maps of views, places, events, etc. are closely linked to the hippocampal formation (Nadel and O'Keefe 1978). But although the hippocampus is an essential prerequisite for spatial navigation, to make use of a cognitive map in order to guide situation-specific behaviours contributions from many other brain regions are mandatory.

The water maze is probably the best-validated paradigm to assess hippocampus-dependent spatial learning in rodents (Morris 1984). While the experimental setup and protocol usually seem quite simple, a water maze test can provide a vast amount of possible read-out parameters. The apparatus consists of a circular pool filled with opaque water and a small hidden escape platform, located just beneath the surface of the water. Being placed in the water is an aversive stimulus for mice and finding the platform is the only way to escape. Among the two principal variants of the paradigm, the cued version provides a visible platform usually marked by a flag whereas in the hidden version the only way to find the goal is to use distant visual cues outside the pool to guide navigation. Changes in latency times to reach the platform or length of the paths swum gives a learning curve as animals acquire the task over multiple trials.

Spatial navigation in the hidden version of the water maze depends on the successful integration of place-specific and temporal information over a range of stereotypic behavioural patterns, i.e. exploration and approaching behaviour. Learning the task consists of different phases, characterized by contributions of different brain systems and a clear progression towards the use of hippocampal, truly spatial search strategies (Nadel and O'Keefe 1978; Redish 1999). The strategies that are employed by the animals to navigate to the platform can be distinguished by analysing the swim patterns (Wolfer et al. 2001; Janus 2004; Garthe et al. 2009). Assessing the qualitative aspects of spatial learning to some degree allows distinguishing the hippocampal components of behaviour from contributions of other brain systems. For example, even after hippocampal lesions rats can learn the water maze task in the sense that latencies to reach the platform become significantly shorter with training (Morris et al. 1982). However, analysing the search strategies revealed that in lesioned animals true spatial and hippocampus-dependent strategies are absent. The animals had simply found out to swim in a circle at the correct distance from the wall until they unfailingly hit the platform. This is a clever strategy but does not involve the hippocampus and thus also no adult neurogenesis. The example demonstrates that latency and path length alone are not sensitive enough measures to reliably dissociate hippocampal and non-hippocampal aspects of behaviour. In the case of adult neurogenesis, identifying the specific contributions of the dentate gyrus to water maze learning appears to be vital in order to characterize the new neurons' functional relevance.

The water maze paradigm allows the assessment of different forms of spatial memory. Usually, learning about cues, places and directions within one single trial is referred to as spatial short-term or working memory, whereas incremental learning over multiple trials is interpreted as a long-term or spatial reference memory (Morris 2007). As a further variation the starting point for the animals tested can be varied across trials or is held constant at least within days. The same applies for the goal position in protocols including platform reversals.

At a first glance, studies trying to show a functional contribution of adult generated hippocampal neurons to spatial learning in the water maze yielded contradicting results (for a comprehensive review see (Deng et al. 2010)). However, combining theoretical models with the data obtained using other paradigms allows sketching a hypothesis that can integrate the seemingly inconsistent findings and provide testable predictions (Aimone et al. 2010). As a first step it can be asked whether pattern separation (or integration) is also involved in the more complex water maze task and which parameters of the task would actually describe pattern separation. In most cases, once introduced into the pool mice or rats begin to explore the arena almost immediately. During that initial phase sampling of the available distal visual cues takes place and as shown for exploration of dry-land environments a cognitive map is formed as indicated by the appearance of place cells with their respective stable place fields (Moser et al. 2008). In subsequent trials the cognitive map is used in an increasingly flexible way to find the hidden goal reliably from all possible starting points (Figure 6.2). However, it is important to note here that learning within a single trial represents not only a trajectory leading to the goal but also a

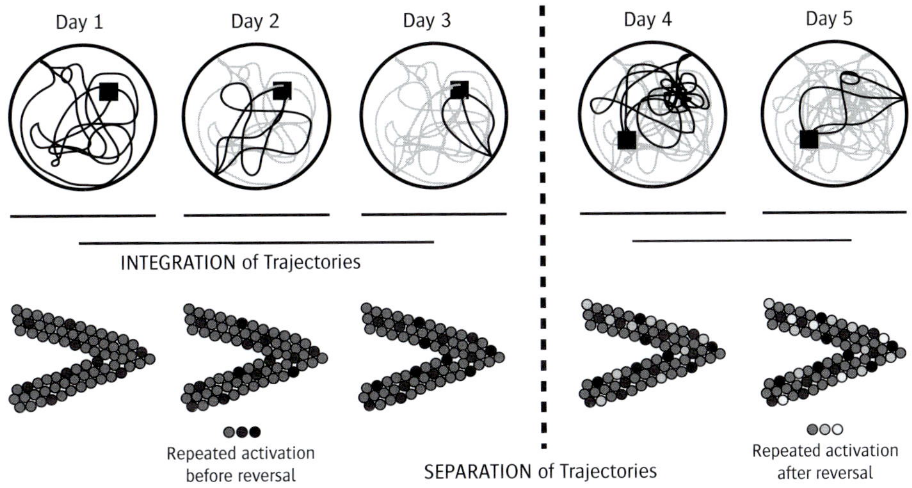

Figure 6.2 Morris water maze experiment using the Lipp–Wolfer protocol. Animals were allowed to search for the hidden platform using fixed starting points repetitively. After three days of acquisition the goal is moved to the opposite side of the pool. Re-experiencing parts of trajectories already taken results in the activation of overlapping sets of dentate granule cells.

sequence of events experienced within a short time window at certain places each defined by a unique configuration of distal cues. This set of cues in time is the context. Because of the overlapping activity of new granule cells during subsequent situations the experiences from within a single trial (e.g. approaching a certain region from different directions) or even from across trials can be integrated in order to allow an increasingly more efficient navigation. Thus, referring to the findings on pattern separation mentioned above to keep the trajectories from different starting points well separated would actually prevent the general knowledge contained within all those experiences to become part of a single, integrated representation of some subsequent trials. More new adult born neurons would allow for such integration and consequently adult hippocampal neurogenesis should enhance the acquisition in the Morris water maze task.

Indeed, some studies described a positive correlation between the numbers of new neurons in the hippocampus and specific learning parameters of the water maze. Using inbred mice from the BXD panel, Kempermann et al. found a weak but significant correlation between the number of new neurons and the animals' acquisition performance in the water maze task (Kempermann and Gage 2002). As the baseline of adult neurogenesis in those mice is solely determined by the genotype this study also emphasized the importance of genetic factors in determining spatial learning abilities. The complementary result was provided by studies using different methods to ablate adult neurogenesis: a lack of young new neurons resulted in an impairment in learning the water maze task: for example, Dupret et al. explicitly assessed the ability of mice to integrate experiences across multiple trials after conditioned genetically ablation of adult neurogenesis (Dupret et al. 2008). A transgenic mouse model that allows the selective removal of young, still immature new neurons significant impairments 'in forming robust, long-term spatial memory and displayed impaired performance in extinction tasks' (Deng et al. 2009). The functional deficits found in that study provided further evidence that the integration of subsequent events into more general representations might be facilitated by the highly excitable adult-born neurons. Interestingly, in some other studies impairments of water maze learning were absent (or at least not recognized). Attempting to interpret results from water maze studies with respect to pattern separation requires taking into account the specifics of the experimental setup and protocol (see also Deng et al. 2010). To analyse the behavioural strategies of the mice to find the hidden goal further increases sensitivity. Using an algorithm-based approach to classify the search strategies shown by mice in the water maze, we have found that animals with suppressed adult neurogenesis were specifically impaired in using spatially precise (hippocampus-dependent) strategies in a flexible way (Garthe et al. 2009). Because those strategies principally rely on integrated, generalized representations, the finding that suppressed adult neurogenesis effectively prevents the development of the most advanced search patterns further supports the hypothesis proposed of Aimone and colleagues.

Finally there is good evidence that the integrative function of adult neurogenesis depends on new-born cells at an early stage of their maturation process. The increased excitability and synaptic plasticity of the immature neurons likely corresponds to a particular functionality during this period. Indeed, the specific ablation of still immature granule cells resulted in deficits in spatial learning (Jessberger et al. 2009). Another study showed that learning the water maze task especially calls for new neurons older than 4–6 weeks (Kee et al. 2007). Compared to earlier (2–4 weeks) and later (> 8 weeks) stages, synaptic plasticity (long-term potentiation, LTP) is significantly increased in that time window. Thus, after having begun to participate in information processing at an age of 2–4 weeks, the new neurons can now become part of a memory in a Hebbian selection process. This indicates that beginning at 4 weeks of age the new-born cells might serve a particular function related to the long-term storage of hippocampal memories.

Since the 1990s virtual water maze paradigms allow to assess spatial learning in human subjects using the setups and protocols already known from rodent studies (Rizzo and Buckwalter 1997). During virtual spatial navigation the same brain regions are activated in humans as found for rats and mice, and theta-rhythmic activity can be recorded (Barrash et al. 2000; Skelton et al. 2006; Mormann et al. 2008). It is well known that older human subjects show specific impairments in spatial learning when tested in a virtual water maze task (Driscoll et al. 2005). Interestingly, spatial navigation abilities appear to be significantly impaired (Cornwell et al. 2010). Besides measuring the latencies to reach the goal, human participants can be asked which strategy they chose, which cues they used, and whether there were any obvious difficulties experienced. Even the 'swim patterns' can be analysed using classification schemata similar to those applied for mice (Schoenfeld et al. 2010). However, given that clear experimental evidence for the functional contribution of adult neurogenesis in animal studies is only beginning to emerge the function of new neurons has not been assessed in humans, so far.

Adult neurogenesis and the capacity problem

Because the storage capacity of the CA3 network is limited, the amount and overlap of hippocampal representations become critical once multiple, highly similar activity patterns have to be differentiated. Theoretical models suggest a role for new-born neurons in avoiding catastrophic interference of representations in the CA3 network (Wiskott et al. 2006). To effectively orthogonalize new activity patterns arriving from the EC, adding new neurons to the dentate gyrus significantly improves storage and retrieval of new representations that are highly similar but relevantly different to those already stored. Without adult neurogenesis the new patterns would interfere with the old ones, resulting in overall impaired encoding and reduced retrieval. The problem of catastrophic interference becomes especially relevant on longer timescales. Because a selected subset of new neurons persists for months (or years), the adult-generated cells probably improve pattern separation beyond the first encoding process (Kempermann et al. 2004; Leuner et al. 2004). Thereby adult neurogenesis might affect the consolidation or retrieval stages of hippocampal memory processing.

The experimental evidence for such long-term functional relevance still is, however, rather sketchy. As measured by probe trial performance (i.e. the initial perseverance of the animals to search for the platform location in the hidden platform version of the water maze, when the platform has in fact been removed) several weeks after task acquisition, ablation of adult neurogenesis impaired long-term memory in rodents (Snyder et al. 2005). This finding is in-line with the hypothesis that adult neurogenesis in the dentate gyrus prevents catastrophic interference in CA3. Interestingly, as opposed to the results mentioned above the animals tested showed no differences during acquisition (Snyder et al. 2005; Imayoshi et al. 2008; Deng et al. 2009; Jessberger et al. 2009;). Differences in the protocols and setups used likely serve as an explanation here. In contrast, we not only found impairment during task acquisition, but also after platform reversal within the same experiment (Garthe et al. 2009). In our study, groups with and without suppressed adult neurogenesis successfully learned to find the platform, but when the goal was moved to a new position, mice lacking new neurons perseverated in searching at the platform's old position (i.e. they had an excellent probe trial performance!). Animals were given multiple trials per day and a constant starting position for each day was used. The platform was moved to the opposite side of the pool after 3 days. Using that protocol, for the first 3 days similar trajectories have to be used repeatedly to reach the platform and a similar activation pattern has to be stored in CA3 for each trajectory. This might have caused the dentate gyrus reaching its maximum capacity to deal with highly similar input patterns. Because the goal is always at the same position, however, interference of activation patterns stored in CA3 does not seriously impair finding the platform. Upon goal reversal the need to separate the representations becomes essential in order to cope effectively with the new situation. An effective orthogonalization is prevented when the new neurons are absent and as the new representations interfere with those previously stored the animal's behaviour becomes inflexible.

Another line of evidence that supports the idea that adult neurogenesis improves the capacity of the dentate gyrus network to orthogonalize input patterns derives from studies, in which the activation of granule cells after exposure to different environments was analysed (Tashiro et al. 2007). After animals were housed in an enriched environment and re-exposed to either the same environment or trained in a water maze, the authors assessed the expression of immediate early genes (IEGs) in the new neurons. It was found that the fraction of new-born cells expressing IEGs was significantly greater in mice re-experiencing the enriched environment compared to water maze learners. The implications of this study are far-reaching, because it showed that the highly excitable adult born neurons not only respond to an acute set of stimuli, but also become 'tagged' by that experience making the new cells less likely to respond to later occurring, different stimuli. Therefore, encoding input patterns encountered the first time might be strongly improved by the presence of adult born new neurons, which become dedicated to that particular kind of input. Without adult neurogenesis the encoding needs to be provided by a smaller fraction of older dentate gyrus cells already tagged by stimuli that were experienced earlier in life. The resulting representation in CA3 is less effectively separated from already stored ones. Thus, continuously adding new neurons by adult neurogenesis provides the dentate gyrus with a reserve of young, untagged new granule cells.

Capacity might also be defined in terms of the time needed for memories to become consolidated and independent from the hippocampus and transferred to the cortex. The earlier and shorter this transfer time period, the sooner the dentate gyrus should be available for new input patterns. Indeed, a study based on contextual fear conditioning showed that the presence of more new neurons might accelerate the consolidation process (Kitamura et al. 2009). However, the exact mechanism of such shortening of the hippocampus-dependent period by adult neurogenesis is not known.

A unified framework for multiple functions of adult hippocampal neurogenesis

Combining the conclusions from the studies inspired by hypotheses regarding pattern separation or catastrophic interference points towards a dual role of adult hippocampal neurogenesis. When they are still young and immature, adult-born neurons in the hippocampus seem to contribute to the integration of events closely related in time. This contribution appears to be transient only: once the new cells have reached a maturation level associated with increases levels of LTP, adult neurogenesis facilitates long-term memory by preventing catastrophic interference and allowing a more effective consolidation process. It remains to be shown how these functions differentially affect more complex behaviours beyond the hippocampus and on longer timescales.

All stages of information processing in the hippocampal formation are dependent on the dentate gyrus working properly. Therefore, any functional change at that level also affects the representations formed in the subsequent structures. A lack of new neurons in the dentate gyrus is thus likely to alter many processes of learning and the resulting behaviours, potentially even those not explicitly linked to the hippocampus. In humans the consequences of overlapping, not effectively orthogonalized representations in CA3 caused by impaired adult neurogenesis are likely to include many subtle aspects of behaviour. Thus, putting the bare concept of pattern separation into a broader, more differentiated behavioural context such as the hidden water maze task, allows: (1) to assess the specific conditions when adult neurogenesis becomes functionally relevant within a learning process encompassing multiple brain systems and related to this (2) to answer the question whether the new neurons play different roles at different levels of network function.

Adult hippocampal neurogenesis and affective behaviour

Despite the progress in elucidating the functional relevance of adult neurogenesis with respect to the key role of the hippocampus in learning and memory, the hypothesis that has stirred most clinical interest in this context so far is the idea that failing adult hippocampal neurogenesis is involved in the pathogenesis and course of major depression and schizophrenia. The idea had initially been largely based on structural evidence (i.e. hippocampal volume changes in affective disorders), the observation that all known antidepressants regulate adult neurogenesis, as well as theoretical considerations. The core hypotheses have been reviewed elsewhere: in a simplistic way they are almost certainly not correct. But a relevant link between adult neurogenesis and affective disorders is plausible. Two main options exist: first, disturbed adult neurogenesis might relate only to the cognitive symptoms of affective disorders, largely subsumed under the functions so far described in this chapter. Second, adult hippocampal neurogenesis might also serve distinct functions related to affective behaviour itself. Two studies suggest so. Santarelli et al. suppressed adult neurogenesis by irradiation and found that in the novelty-suppressed feeding task, in which the mice have to overcome their inborn fear of open plains to get to the food, the anxiety could not be reversed by antidepressant treatment, when the new neurons were present (Santarelli et al. 2003). Revest and colleagues chose a more direct route and found a significant increase in anxiety-related behaviours after suppression of adult neurogenesis in a transgenic mouse model (Revest et al. 2009). The study showed that adult neurogenesis is involved in the regulation of affective behaviour providing a strong link to human anxiety disorders such as depression and addiction.

Generally, the hippocampal formation does take part in affective behaviour and is intricately linked to the limbic system. In fact, limbic input is critically involved also cognitive performance in that emotions help to prioritize and 'colour' learning events. A partial separation of functions takes place in that the dorsal (or medial or anterior) portions of the dentate gyrus and hippocampus are primarily active in cognitive behaviour, whereas the ventral (or temporal or posterior) portion plays a greater role in affective behaviour. Adult neurogenesis generally takes place in both parts of the hippocampus, but in a few instances a differential regulation or relationship to different types of behaviours has been reported (Snyder et al. 2009). The picture is, however, still far from clear. While there can be little doubt that adult neurogenesis is somehow involved in affective behaviour as well, it is not known, whether this involvement is direct or indirect and what the presumed relationship to roles in cognitive behaviour is.

Two sides of the coin: risks and potentials

Given the far-reaching functional implications of adding new neurons to a brain structure central to learning and memory, a sophisticated regulation of adult neurogenesis is required. Disruption or dysfunction in the responsible regulatory systems might cause severe functional impairments or even damages to the underlying neuronal networks. As an example, during medial temporal lobe epilepsy seizures lead to a massive increase in cell proliferation in the dentate gyrus. The new neurons generated following seizures migrate ectopically into the hilus and display hilar basal dendrites thereby forming recurrent excitatory networks (Parent et al. 1997; Ribak et al. 2000; Overstreet-Wadiche et al. 2006). Interestingly, the functional consequences of new neurons generated in the context of epileptic seizures at the same time might also have beneficial effects for the network (Jakubs et al. 2006). The example illustrates how disturbed network activity can influence the cellular and molecular mechanisms regulating proliferation, migration, and differentiation of the newborn neurons and how the system might try to compensate for the resulting impairments in network function. The potential risk of adding new neurons to a working brain structure must be balanced against its benefits as a particular source of plasticity, well after development has ceased.

Caveats

Studies in laboratory rats or mice might lead to inconsistent or even contradictory results simply due to the impact of strain differences. The influence of the genetic background on adult neurogenesis gains even more importance when the results from highly inbred laboratory animals are to be translated to humans. In ablation studies, which lie at the heart of present-day research on what adult neurogenesis is good for, the details of the behavioural paradigm and the specific protocol turn out to be probably more relevant to determine a certain phenotype than the ablation method itself. As an example, in most studies the main read-outs from the Morris water maze are the latencies and times in quadrants. Such parameters only allow an indirect assessment of dentate gyrus specific aspects of hippocampal learning. However, since many brain regions outside the hippocampus are involved in learning the comparatively complex water maze task the regulation of adult neurogenesis will be affected by several modulatory systems arising from the involvement of multiple brain regions.

Conclusion

Although the canonical statement that the hippocampus is involved in spatial learning has been validated by many studies its exact contribution to more complex forms of learning like spatial navigation is still not fully understood. Further, how pattern separation, the key function assigned to the dentate gyrus, relates to learning and especially outside the realm of spatial navigation is far from clear. We have here summarized what is known about the potential role of adult hippocampal neurogenesis in these functions. Given the many open questions to precisely identify this function is a challenging task. Based on data from the literature and own research we propose that adult-generated neurons contribute to both information processing and long-term network adaptations. Both serve improving generalization within episodes closely related in time and sharing the same behavioural endpoints as well as reducing interference between episodes and thus increasing flexibility, thereby allowing a lasting optimization of the mossy fibre connections to levels of cognitive (and possibly affective) challenges, realistically encountered by the organism. We have also proposed the 'neurogenic reserve hypothesis', which tries to explain how the activity-dependent regulation of adult neurogenesis contributes to lifelong plasticity as compensatory mechanisms in the face of 'physiological' cognitive ageing and neurodegenerative disease.

Because the basic forms of information processing in the dentate gyrus appear to be the same in the murine as in the human brain the new neurons likely affect the same aspects of behaviour in all species showing adult neurogenesis. A deeper understanding of how the dentate gyrus and pattern separation contribute to spatial learning will help to bridge the conceptual gaps in our understanding of navigation between rodents and humans. Using virtual water maze paradigms for human subjects allows a direct comparison with studies in rodents. Such a translation of insights into the functional relevance of the new neurons from animal studies to humans will clarify how adult neurogenesis can help to keep hippocampal learning in humans effective and flexible far beyond spatial navigation.

References

Aimone, J. B., Wiles, J., and Gage, F. H. (2006). Potential role for adult neurogenesis in the encoding of time in new memories. *Nature Neurosci*, 9(6), 723–7.

Aimone, J. B., Wiles, J., and Gage, F. H. (2009). Computational influence of adult neurogenesis on memory encoding. *Neuron*, 61(2), 187–202.

Aimone, J. B., Deng, W., and Gage, F. H. (2010). Adult neurogenesis: integrating theories and separating functions. *Trends Cogn Sci*, 14(7), 325–37.

Alme, C.B., Buzzetti, R.A., Marrone, D.F., *et al.* (2010). Hippocampal granule cells opt for early retirement. *Hippocampus*, 20(10), 1109–23.

Altman, J., and Das, G. (1965). Post-natal origin of microneurones in the rat brain. *Nature*, 207, 953–6.

Amaral, D.G., and Witter, M.P. (1995). Hippocampal formation. In Paxinos, G. (Ed.) *The Rat Nervous System*, pp. 443–493. San Diego, CA: Academic Press Inc.

Amaral, D. G., and Lavenex, P. (2007). Hippocampal neuroanatomy. In Andersen, P., Morris, R. G. M., Amaral, D. G., Bliss, T. V., and O'Keefe, J. (Eds.) *The Hippocampus Book*, pp. 37–114. Oxford: Oxford University Press.

Appleby, P.A., Kempermann, G., and Wiskott, L. (2011). The role of additive neurogenesis and synaptic plasticity in a hippocampal memory model with grid-cell like input. *PLoS Comput Biol*, 7, e1001063.

Babu, H., Ramirez-Rodriguez, G., Fabel, K., Bischofberger, J., and empermann, G. (2009). Synaptic network activity induces neuronal differentiation of adult hippocampal precursor cells through BDNF signaling. *Front Neurosci*, 3, 49.

Barrash, J., Damasio, H., Adolphs, R., and Tranel, D. (2000). The neuroanatomical correlates of route learning impairment. *Neuropsychologia*, 38(6), 820–36.

Becker, S., and Wojtowicz, J. (2007). A model of hippocampal neurogenesis in memory and mood disorders. *Trends Cogn Sci*, 11(2), 70–6.

Clelland, C.D., Choi, M., Romberg, C., *et al.* (2009). A functional role for adult hippocampal neurogenesis in spatial pattern separation. *Science*, 325(5937), 210–13.

Cornwell, B.R., Johnson, L. L., Holroyd, T., Carver, F. W., and Grillon, C. (2008). Human hippocampal and parahippocampal theta during goal-directed spatial navigation predicts performance on a virtual Morris water maze. *J Neurosci*, 28(23), 5983–90.

Cornwell, B.R., Salvadore, G., Colon-Rosario, V., *et al.* (2010). Abnormal hippocampal functioning and impaired spatial navigation in depressed individuals: evidence from whole-head magnetoencephalography. *Am J Psychiatry*, 167(7), 836–44.

Creer, D.J., Romberg, C., Saksida, L. M., van Praag, H., and Bussey, T. J. (2010). Running enhances spatial pattern separation in mice. *Proc Natl Acad Sci USA*, 107(5), 2367–72.

Deisseroth, K., Singla, S., Toda, H., Monje, M., Palmer, T. D., and Malenka, R. C. (2004). Excitation-neurogenesis coupling in adult neural stem/progenitor cells. *Neuron*, 42(4), 535–52.

Deng, W., Saxe, M. D., Gallina, I. S., and Gage, F. H. (2009). Adult-born hippocampal dentate granule cells undergoing maturation modulate learning and memory in the brain. *J Neurosci*, 29(43), 13532–42.

Deng, W., Aimone, J.B., and Gage, F. H. (2010). New neurons and new memories: how does adult hippocampal neurogenesis affect learning and memory? *Nature Rev Neurosci*, 11(5), 339–50.

Draganski, B., Gaser, C., Kempermann, G., *et al.* (2006). Temporal and spatial dynamics of brain structure changes during extensive learning. *J Neurosci*, 26(23), 6314–17.

Drew, M.R., Denny, C.A., and Hen, R. (2010). Arrest of adult hippocampal neurogenesis in mice impairs single- but not multiple-trial contextual fear conditioning. *Behav Neurosci*, 124(4), 446–54.

Driscoll, I., Hamilton, D. A., Yeo, R. A., Brooks, W.M., and Sutherland, R. J. (2005). Virtual navigation in humans: the impact of age, sex, and hormones on place learning. *Horm Behav*, 47(3), 326–35.

Dupret, D. *et al.* (2008). Spatial relational memory requires hippocampal adult neurogenesis. *PloS One*, 3(4), e1959.

Döbrössy, M.D., Drapeau, E., Aurousseau, C., Le Moal, M., Piazza, P. V., and Abrous, D. N. (2003). Differential effects of learning on neurogenesis: learning increases or decreases the number of newly born cells depending on their birth date. *Mol Psychiatry*, 8(12), 974–82.

Düzel, E., Penny, W. D., and Burgess, N. (2010). Brain oscillations and memory. *Curr Opin Neurobiol*, 20(2), 143–9.

Eichenbaum, H., and Cohen, N. (2001). *From Conditioning to Conscious Recollection: Memory Systems of the Brain*. New York: Oxford University Press.

Eriksson, P.S., Perfilieva, E., Björk-Eriksson, T., *et al.* (1998). Neurogenesis in the adult human hippocampus. *Nat Med*, 4(11), 1313–7.

Fabel, K., Wolf, S. A., Ehninger, D., Babu, H., Leal-Galicia, P., and Kempermann, G. (2009). Additive effects of physical exercise and environmental enrichment on adult hippocampal neurogenesis in mice. *Front Neurosci*, 3, 50.

Garthe, A., Behr, J., and Kempermann, G., (2009). Adult-generated hippocampal neurons allow the flexible use of spatially precise learning strategies. *PloS One*, 4(5), e5464.

Gould, E., Beylin, A., Tanapat, P., Reeves, A., and Shors, T. J. (1999). Learning enhances adult neurogenesis in the hippocampal formation. *Nat Neurosci*, 2(3), 260–5.

Gould, E. (2007). How widespread is adult neurogenesis in mammals? *Nature Rev Neurosci*, 8(6), 481–8.

Hairston, I.S., Little, M. T., Scanlon, M. D., *et al.* (2005). Sleep restriction suppresses neurogenesis induced by hippocampus-dependent learning. *J Neurophysiol*, 94(6), 4224–33.

Hasselmo, M.E. (2005). What is the function of hippocampal theta rhythm? Linking behavioral data to phasic properties of field potential and unit recording data. *Hippocampus*, 15(7), 936–49.

Hernández-Rabaza, V., Llorens-Martín, M., Velázquez-Sánchez, C., *et al.* (2009). Inhibition of adult hippocampal neurogenesis disrupts contextual learning but spares spatial working memory, long-term conditional rule retention and spatial reversal. *Neuroscience*, 159(1), 59–68.

Hok, V., Lenck-Santini, P. P., Roux, S., Save, E., Muller, R. U., and Poucet, B. (2007). Goal-related activity in hippocampal place cells. *J Neurosci*, 27(3), 472–82.

Hunsaker, M. R., and Kesner, R. P. (2008). Evaluating the differential roles of the dorsal dentate gyrus, dorsal CA3, and dorsal CA1 during a temporal ordering for spatial locations task. *Hippocampus*, 18(9), 955–64.

Hunsaker, M. R., Rosenberg, J. S., and Kesner, R. P. (2008). The role of the dentate gyrus, CA3a,b, and CA3c for detecting spatial and environmental novelty. *Hippocampus*, 18(10), 1064–73.

Imayoshi, I., Sakamoto, M., Ohtsuka, T., *et al.* (2008). Roles of continuous neurogenesis in the structural and functional integrity of the adult forebrain. *Nature Neurosci*, 11(10), 1153–61.

Jakubs, K., Nanobashvili, A., Bonde, S., *et al.* (2006). Environment matters: synaptic properties of neurons born in the epileptic adult brain develop to reduce excitability. *Neuron*, 52(6), 1047–59.

Janus, C. (2004). Search strategies used by APP transgenic mice during navigation in the Morris water maze. *Learn Mem*, 11(3), 337–46.

Jessberger, S., Clark, R. E., Broadbent, N. J., *et al.* (2009). Dentate gyrus-specific knockdown of adult neurogenesis impairs spatial and object recognition memory in adult rats. *Learn Mem*, 16(2), 147–54.

Kee, N., Teixeira, C. M., Wang, A.H., and Frankland, P. W. (2007). Preferential incorporation of adult-generated granule cells into spatial memory networks in the dentate gyrus. *Nat Neurosci*, 10(3), 355–62.

Kempermann, G. (2008). The neurogenic reserve hypothesis: what is adult hippocampal neurogenesis good for? *Trends Neurosci*, 31(4), 163–9.

Kempermann, G., Kuhn, H.-G., and Gage, F. H. (1997). More hippocampal neurons in adult mice living in an enriched environment. *Nature*, 386(6624), 493–5.

Kempermann, G. and Gage, F. H. (2002). Genetic determinants of adult hippocampal neurogenesis correlate with acquisition, but not probe trial performance, in the water maze task. *Eur J Neurosci*, 16(1), 129–36.

Kempermann, G., *et al.* (2004). Milestones of neuronal development in the adult hippocampus. *Trends Neurosci*, 27(8), 447–52.

Kempermann, G., Krebs, J., and Fabel, K. (2008). The contribution of failing adult hippocampal neurogenesis to psychiatric disorders. *Curr Opin Psychiatry*, 21(3), 290–5.

Kempermann, G., Fabel, K., Ehninger, D., *et al.* (2010). Why and how physical activity promotes experience-induced brain plasticity. *Front Neurosci*, 4, 189.

Kesner, R.P., and Hopkins, R.O. (2006). Mnemonic functions of the hippocampus: a comparison between animals and humans. *Biol Psychol*, 73(1), 3–18.

Kesner, R.P., Lee, I., and Gilbert, P. (2004). A behavioral assessment of hippocampal function based on a subregional analysis. *Rev Neurosci*, 15(5), 333–51.

Kitamura, T., Saitoh, Y., Takashima N., *et al.* (2009). Adult neurogenesis modulates the hippocampus-dependent period of associative fear memory. *Cell*, 139(4), 814–27.

Kronenberg, G., Reuter, K., Steiner, B., *et al.* (2003). Subpopulations of proliferating cells of the adult hippocampus respond differently to physiologic neurogenic stimuli. *J Comp Neurol*, 467(4), 455–63.

Leuner, B., Mendolia-Loffredo, S., Kozorovitskiy, Y., Samburg, D., Gould, E., and Shors, T. J. (2004). Learning enhances the survival of new neurons beyond the time when the hippocampus is required for memory. *J Neurosci*, 24(34), 7477–81.

Leuner, B., Gould, E., and Shors, T. J. (2006). Is there a link between adult neurogenesis and learning. *Hippocampus*, 16, 216–24.

Lisman, J. E., and Grace, A. A. (2005). The hippocampal-VTA loop: controlling the entry of information into long-term memory. *Neuron*, 46(5), 703–13.

Lledo, P.-M., Alonso, M., and Grubb, M. S. (2006). Adult neurogenesis and functional plasticity in neuronal circuits. *Nature Rev Neurosci*, 7(3), 179–93.

Maguire, E.A., Gadian, D. G., Johnsrude, I. S., *et al.* (2000). Navigation-related structural change in the hippocampi of taxi drivers. *Proc Natl Acad Sci USA*, 97(8), 4398–403.

Marr, D. (1971). Simple memory: a theory for archicortex. *Philos Trans R Soc Lond B Biol Sci*, 262(841), 23–81.

Meltzer, L.A., Yabaluri, R., and Deisseroth, K. (2005). A role for circuit homeostasis in adult neurogenesis. *Trends Neurosci*, 28(12), 653–60.

Mohapel, P., Mundt-Petersen, K., Brundin, P., and Frielingsdorf, H. (2006). Working memory training decreases hippocampal neurogenesis. *Neuroscience*, 142(3), 609–13.

Mormann, F., Osterhage, H., Andrzejak, R. G., *et al.* (2008). Independent delta/theta rhythms in the human hippocampus and entorhinal cortex. *Front Hum Neurosci*, 2, 3.

Morris, R. (1984). Developments of a water-maze procedure for studying spatial learning in the rat. *J Neurosci Methods*, 11(1), 47–60.

Morris R. (2007). Theories of hippocampal function. In Andersen, P., Morris, R. G. M., Amaral, D. G., Bliss, T. V., and O'Keefe, J. (Eds.) *The Hippocampus Book*, pp. 581–714. Oxford: Oxford University Press.

Morris, R. G., Garrud, P., Rawlins, J. N., and O'Keefe, J. (1982). Place navigation impaired in rats with hippocampal lesions. *Nature*, 297(5868), 681–3.

Moscovitch, M., Rosenbaum, R. S., Gilboa, A., *et al.* (2005). Functional neuroanatomy of remote episodic, semantic and spatial memory: a unified account based on multiple trace theory. *J Anat*, 207(1), 35–66.

Moser, E. I., Kropff, E., and Moser, M.-B. (2008). Place cells, grid cells, and the brain's spatial representation system. *Annu Rev Neurosci*, 31, 69–89.

Nadel, L. and Moscovitch, M. (1997). Memory consolidation, retrograde amnesia and the hippocampal complex. *Curr Opin Neurobiol*, 7(2), 217–27.

Nadel, L. and O'Keefe, J. (1978). *The Hippocampus as a Cognitive Map*. Oxford: Oxford University Press.

O'Keefe, J. (2007). Hippocampal neurophysiology in the behaving animal. In Andersen, P., Morris, R. G. M., Amaral, D. G., Bliss, T. V., and O'Keefe, J. (Eds.) *The Hippocampus Book*, pp. 475–548. Oxford: Oxford University Press.

Overstreet-Wadiche, L.S., Bromberg, D. A., Bensen, A. L., and Westbrook, G. L. (2006). Seizures accelerate functional integration of adult-generated granule cells. *J Neurosci*, 26(15), 4095–103.

Parent, J. M., Yu, T. W., Leibowitz, R. T., *et al.* (1997). Dentate granule cell neurogenesis is increased by seizures and contributes to aberrant network reorganization in the adult rat hippocampus. *J Neurosci*, 17(10), 3727–38.

Phillips, R. G., and LeDoux, J. E. (1992). Differential contribution of amygdala and hippocampus to cued and contextual fear conditioning. *Behav Neurosci*, 106(2), 274–85.

Praag, H. van, Kempermann, G., and Gage, F. H. (1999). Running increases cell proliferation and neurogenesis in the adult mouse dentate gyrus. *Nat Neurosci*, 2(3), 266–70.

Praag, H. van, Kempermann, G., and Gage, F. H. (2000). Neural consequences of environmental enrichment. *Nat Rev Neurosci*, 1(3), 191–8.

Redish, A. D. (1999). *Beyond the Cognitive Map*. London: MIT Press.

Revest, J.-M., Dupret, D., Koehl, M., *et al.* (2009). Adult hippocampal neurogenesis is involved in anxiety-related behaviors. *Mol Psychiatry*, 14(10), 959–67.

Ribak, C. E., Tran, P. H., Spigelman, I., Okazaki, M. M., and Nadler, J. V. (2000). Status epilepticus-induced hilar basal dendrites on rodent granule cells contribute to recurrent excitatory circuitry. *J Comp Neurol*, 428(2), 240–53.

Rizzo, A. A. and Buckwalter, J. G. (1997). The status of virtual reality for the cognitive rehabilitation of persons with neurological disorders and acquired brain injury. *Stud Health Technol Inform*, 39, 22–33.

Rolls, E. T. (1996). A theory of hippocampal function in memory. *Hippocampus*, 6(6), 601–20.

Rolls, T. (2010). A computational theory of episodic memory formation in the hippocampus. *Behav Brain Res*, 215(2), 180–96.

Santarelli, L., Saxe, M., Gross, C., *et al.* (2003). Requirement of hippocampal neurogenesis for the behavioral effects of antidepressants. *Science*, 301(5634), 805–9.

Saxe, M.D., Battaglia, F., Wang, J. W., *et al.* (2006). Ablation of hippocampal neurogenesis impairs contextual fear conditioning and synaptic plasticity in the dentate gyrus. *Proc Natl Acad Sci USA*, 103(46), 17501–6.

Saxe, M. D., Malleret, G., Vronskaya, S., *et al.* (2007). Paradoxical influence of hippocampal neurogenesis on working memory. *Proc Natl Acad Sci USA*, 104(11), 4642–6.

Schoenfeld, R., Moenich, N., Mueller, F. J., Lehmann, W., and Leplow, B. (2010). Search strategies in a human water maze analogue analyzed with automatic classification methods. *Behav Brain Res*, 208(1), 169–77.

Scoville, W.B., and Milner, B. (1957). Loss of recent memory after bilateral hippocampal lesions. *J Neurol, Neurosurg, Psychiatry*, 20(1), 11–21.

Shors, T. J., Miesegaes, G., Beylin, A., Zhao, M., Rydel, T., and Gould, E. (2001). Neurogenesis in the adult is involved in the formation of trace memories. *Nature*, 410(6826), 372–6.

Shors, T. J., Townsend, D. A., Zhao, M., Kozorovitskiy, Y., and Gould, E. (2002). Neurogenesis may relate to some but not all types of hippocampal-dependent learning. *Hippocampus*, 12(5), 578–84.

Skelton, R. W., Ross, S. P., Nerad, L., and Livingstone, S. A. (2006). Human spatial navigation deficits after traumatic brain injury shown in the arena maze, a virtual Morris water maze. *Brain Inj*, 20(2), 189–203.

Snyder, J. S., Hong, N. S., McDonald, R. J., andWojtowicz J. M. (2005). A role for adult neurogenesis in spatial long-term memory. *Neuroscience*, 130(4), 843–52.

Snyder, J., Glover, L. R., Sanzone, K. M., Kamhi, J.F., and Cameron, H. A. (2009a). The effects of exercise and stress on the survival and maturation of adult-generated granule cells. *Hippocampus*, 19(10), 898–906.

Snyder, J. S., Radik, R., Wojtowicz, J. M., and Cameron, H. A. (2009b). Anatomical gradients of adult neurogenesis and activity: young neurons in the ventral dentate gyrus are activated by water maze training. *Hippocampus*, 19(4), 360–70.

Squire, L. R. (2004). Memory systems of the brain: a brief history and current perspective. *Neurobiol Lear Mem*, 82(3), 171–7.

Staudigl, T., Hanslmayr, S., and Bauml, K.-H.T. (2010). Theta oscillations reflect the dynamics of interference in episodic memory retrieval. *J Neurosci*, 30(34), 11356–62.

Steiner, B., Zurborg, S., Hörster, H., Fabel, K., and Kempermann, G. (2008). Differential 24 h responsiveness of Prox1-expressing precursor cells in adult hippocampal neurogenesis to physical activity, environmental enrichment, and kainic acid-induced seizures. *Neuroscience*, 154(2), 521–9.

Stone, S. S. D., Teixeira, C. M., Zaslavsky, K., *et al*. (2011). Functional convergence of developmentally and adult-generated granule cells in dentate gyrus circuits supporting hippocampus-dependent memory. *Hippocampus*, 21(12), 1348–62.

Suh, H., Deng, W., and Gage, F. H. (2009). Signaling in adult neurogenesis. *Annu Rev Cell Dev Biol*, 25, 253–75.

Tashiro, A., Makino, H., and Gage, F. H. (2007). Experience-specific functional modification of the dentate gyrus through adult neurogenesis: a critical period during an immature stage. *J Neurosci*, 27(12), 3252–9.

Treves, A., and Rolls, E. T. (1992). Computational constraints suggest the need for two distinct input systems to the hippocampal CA3 network. *Hippocampus*, 2(2), 189–99.

Treves, A., Tashiro, A., Witter, M. E., and Moser, E. I. (2008). What is the mammalian dentate gyrus good for? *Neuroscience*, 154(4), 1155–72.

Winocur, G., Wojtowicz, J. M., Sekeres, M., Snyder, J. S., and Wang, S. (2006). Inhibition of neurogenesis interferes with hippocampus-dependent memory function. *Hippocampus*, 16(3), 296–304.

Wiskott, L., Rasch, M. J., and Kempermann, G. (2006). A functional hypothesis for adult hippocampal neurogenesis: avoidance of catastrophic interference in the dentate gyrus. *Hippocampus*, 16(3), 329–43.

Wojtowicz, J. M., Askew, M. L., and Winocur, G. (2008). The effects of running and of inhibiting adult neurogenesis on learning and memory in rats. *Eur J Neurosci*, 27(6), 1494–502.

Wolfer, D. P., Madani, R., Valenti, P., and Lipp, H. P. (2001). Extended analysis of path data from mutant mice using the public domain software Wintrack. *Physiol Behav*, 73(5), 745–53.

Zhao, C., Deng, W., and Gage, F. H. (2008). Mechanisms and functional implications of adult neurogenesis. *Cell*, 132(4), 645–60.

CHAPTER 7

Imaging the Hippocampus

Bradford C. Dickerson and Jean C. Augustinack

Introduction 122
Hippocampal neuroanatomy for neuroimaging 123
Brief history of hippocampal neuroimaging 123
Imaging and quantifying hippocampal and MTL cortical anatomy 124
Imaging task-related hippocampal and MTL cortical function 128
Imaging hippocampal perfusion and metabolism 132
Imaging hippocampal connectivity 132
Molecular imaging of the hippocampus 134
Imaging hippocampal plasticity 134
Ex vivo hippocampal imaging 135
Technical advances and future directions 137

ABSTRACT An extensive body of literature has accrued on the imaging of the human hippocampus since it was first visualized clearly in a living person using magnetic resonance imaging (MRI) in the late 1980s. Clinicians now routinely perform high-resolution MRI to inspect the structure of the hippocampus and surrounding cortices in detail in patients suspected of having Alzheimer's disease (AD) and related disorders, temporal lobe epilepsy and related conditions, or limbic encephalitis, transient global amnesia, or other forms of amnesia. Researchers have developed protocols for measuring the size of the hippocampus and are now working toward visualization and measurement of hippocampus subfields *in vivo*. Investigators are also using advanced imaging techniques to measure hippocampal activity during memory and other tasks as well as its perfusion, metabolism, and molecular constituents. New technical advances in imaging continue to be applied to the investigation of the hippocampus, medial temporal cortices, and connected brain regions, and will certainly lead to novel clinical insights that should improve patient care.

KEYWORDS magnetic resonance imaging, functional MRI, volumetric analysis, neuroanatomy

Introduction

Brain imaging scientists have long been fascinated with the hippocampus. Nearly every new neuroimaging technique is quickly applied to the study of its structure, function, molecular composition, or other features. These investigations have provided fundamental insights into hippocampal neuroanatomy and physiology, the role of the hippocampus in memory and other abilities, and disease-related abnormalities. A number of these insights have been applied clinically, particularly in the fields of Alzheimer's disease (AD) and epilepsy.

Here we will survey the field of hippocampal neuroimaging, providing an overview of contemporary approaches to imaging its structure and function, and will touch on both scientific and clinical applications of these methods. We will also discuss emerging technical advances that promise to extend the frontiers of hippocampal neuroimaging, both *in vivo* and *ex vivo*. But first, anatomical concepts of specific relevance to imaging and the history of hippocampal neuroimaging will be briefly covered.

Hippocampal neuroanatomy for neuroimaging

The hippocampal and parahippocampal cortex are often considered together, as they make up the medial temporal lobe (MTL) memory system (Squire and Zola-Morgan 1991). These cortices are located in the ventromedial temporal lobe, and are visible grossly as the uncus and ventral portion of the limbic lobe. The parahippocampal gyrus encompasses cortex that includes entorhinal cortex (rostromedially), perirhinal cortex (rostrolaterally), and posterior parahippocampal cortex (caudally). The collateral sulcus is the major sulcal landmark associated with these structures. The term perirhinal cortex is often used today in a manner inclusive of two cytoarchitecturally distinct regions, Brodmann's area 35 (Braak's transentorhinal cortex), located medially in the collateral sulcus, and Brodmann's area 36 (the ectorhinal cortex), located laterally in the collateral sulcus. The entorhinal cortex (Brodmann's areas 28 and 34) lies medially along the crown of the anterior parahippocampal gyrus. The amygdala and hippocampal formation lie deep to the parahippocampal cortical regions, and are visible in dissected preparations of medial temporal lobe (Rosene and Van Hoesen 1987; Duvernoy 2005). The subicular cortices (including parasubiculum, presubiculum, and subiculum proper) lie on the ventral bank of the hippocampal sulcus, and on the opposite bank dorsally the hippocampus proper and dentate gyrus have folded together during development to form two interlocking structures anatomically. The *hippocampus proper* comprises cornu Ammonis (CA) regions CA1, CA2, CA3, CA4 (Lorente de No 1934) only whereas the subicular cortices plus the hippocampus proper and the dentate gyrus together make up the *hippocampal formation* (Rosene and Van Hoesen 1987). See Figure 7.3 later in this chapter for illustrations of these regions.

The rostrocaudal extent of the structures of the MTL memory system in adult humans is approximately 5 cm, with the perirhinal and entorhinal cortices appearing most rostrally, first visible about 1–2 cm caudal to the temporal pole. Moving caudally, the amygdala appears next, followed by the subiculum and then hippocampus proper. The entorhinal cortex extends for approximately 2.5 cm, ending at about the level of the lateral geniculate nucleus of the thalamus. Perirhinal cortex ends slightly more caudally, encircling entorhinal cortex. Continuing caudally, the posterior parahippocampal cortex makes up the caudal MTL cortical tissue occupying the crown of the parahippocampal gyrus. Posterior parahippocampal cortex does not have a Brodmann number but instead is referred to as TH and TF. As the MTL gradually follows the bend of the lateral ventricle caudally, the hippocampal and parahippocampal subdivisions become less well-differentiated as the fornix appears and carries the major output fibres dorsally, ventral to the corpus callosum. The reader should refer to Chapter 2 (Lavenex, this volume) for more detail on anatomy and connectivity.

Brief history of hippocampal neuroimaging

The hippocampal formation and neighbouring MTL cortical structures play critical roles in episodic memory abilities (Squire et al. 2004; Eichenbaum et al. 2007). Much of our knowledge about these brain regions has accrued from studies of patients with focal lesions, most prominently the late H. M. (Squire 2009), and investigations of rodents, primates, and other animals. In the late 1980s, the first magnetic resonance imaging (MRI) data began appearing that enabled the clear visualization of the human hippocampal formation *in vivo* (Naidich et al. 1987a, 1987b; Seab et al. 1988; Jack et al. 1989). Initial efforts at quantification provided data indicating that manual tracings by trained operators were valid and reliable (Jack et al. 1988c, 1989). These new methods were immediately applied to visualize and quantify atrophy in patients with AD (Jack et al. 1987, 1992; Seab et al. 1988), temporal lobe epilepsy (Jack et al. 1988a, 1988b; 1990), and schizophrenia (Suddath et al. 1989). Remarkably for that time, they were also applied at high resolution (0.625 × 0.625 × 5 mm) to visualize and manually quantify anatomic abnormalities of the hippocampal formation in three patients with chronic amnesic syndromes (Press et al. 1989) (Figure 7.1). As stated by Jack (Jack et al. 1992), these new methods were quickly seen as having utility in the diagnostic evaluation of AD:

> The rationale for quantitative MRI of MTL atrophy in the diagnosis of AD is: (1) a memory impairment is usually the earliest and most severe clinical manifestation of AD, (2) MTL limbic structures are central to the integrity of declarative memory function, (3) MTL limbic structures are involved earliest and most extensively in the pathology of AD, and (4) several principal MTL limbic structures are amenable to accurate volumetric quantitation by MRI—the hippocampal formation, amygdala, and parahippocampal gyrus (PHG).

During the 1990s, the scope of quantitative structural MRI efforts expanded to include more sophisticated studies of larger samples of patients with AD, epilepsy, and other neuropsychiatric disorders, in addition to investigations of normal development and ageing. Those efforts presaged huge developments in the past decade on the use of MRI measures of the hippocampal formation and MTL

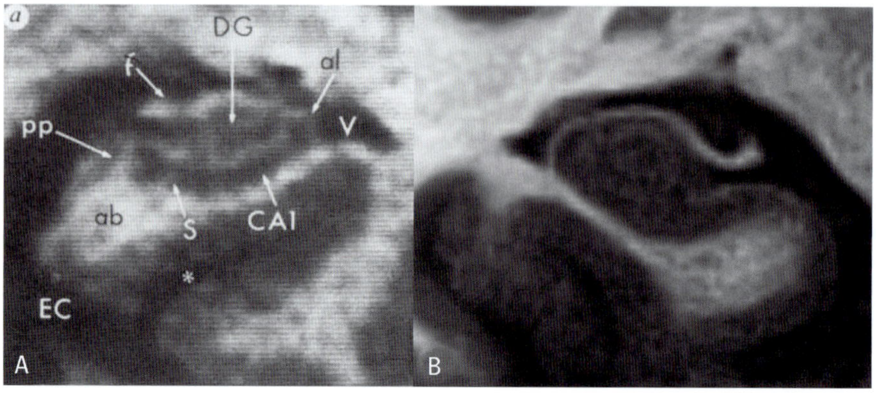

Figure 7.1 Coronal views of MRI images of the human hippocampal formation *in vivo* in (A) 1989 (0.625-mm in-plane resolution Reprinted by permission from Macmillan Publishers Ltd: Press, G. A., Amaral, D. G., and Squire, L.R. (1989). Hippocampal abnormalities in amnesic patients revealed by high-resolution magnetic resonance imaging. *Nature*, 341, 54–7 © 1989, and in (B) 2009 (0.380-mm in-plane resolution).

in AD for the purposes of early diagnosis and monitoring (Jack et al. 2008).

The advent of functional MRI technology in the 1990s opened the door to rapidly proliferating studies of hippocampal and MTL function. In the 1990s and first decade of this century, other neuroimaging techniques also began to be applied to investigations of hippocampal perfusion, molecular composition, and connectivity.

Imaging and quantifying hippocampal and MTL cortical anatomy

Because of its position and shape, the hippocampal formation and MTL cortex are typically best visualized in the coronal plane, ideally with a slightly obliquely oriented plane of acquisition. Previously, the MRI data acquisition methodology for obtaining these types of images existed primarily in specialized research centres. It is now commonplace to obtain three-dimensional (3D; which can be reformatted in multiple planes, as opposed to the early scans that were essentially only viewable in one orientation) MRI data with high signal-to-noise properties at resolutions of 1 mm^3 or better. Using current widely available MRI technology, radiologists can set up protocols (e.g. 'Epilepsy protocol' or 'Dementia protocol') in which relatively high-resolution 3D coronal sequences are obtained in addition to routine diagnostic images. Particularly with the use of new parallel acquisition techniques that enable faster data collection, such high-resolution sequences can be acquired in 5–8 mins as part of a comprehensive multi-sequence MRI protocol (i.e. total time typically on the order of about half an hour). We routinely obtain both T1-weighted and T2-weighted scans, since T1-weighted sequences typically show many anatomic details very clearly and T2-weighted sequences demonstrate signal hyperintensities associated with cerebrovascular, gliotic/sclerotic, inflammatory, neoplastic, or other types of pathology. These images can be visually inspected for the purpose of identifying the variety of pathologic changes, including atrophy, that are associated with diseases affecting the MTL (Figure 7.2). A number of efficient visual rating protocols have been described with the goal of enabling relatively standardized descriptions of the magnitude of hippocampal/MTL atrophy (Vermersch et al. 1994; Galton et al. 2001). With these procedures now being available, it is relatively straightforward for a specialized team to be assembled for the purposes of evaluating these disorders in a comprehensive fashion using contemporary clinical and radiological methods (typically including: a radiologist or neuroradiologist, and a neurologist, psychiatrist, and/or neuropsychologist).

Hippocampal volumetric analysis

An extensive body of literature exists on quantitative volumetric analysis of the hippocampus and nearby structures in healthy adults across the lifespan and in patients with a variety of neurological, psychiatric, and medical disorders. A detailed review of these studies is beyond the scope of this chapter and the interested reader is referred to several excellent recent reviews on this topic (Van Petten 2004; Geuze et al. 2005a; Peper et al. 2007; Barnes et al. 2009b). Instead here we will summarize some methodological issues and highlight a few observations of interest.

Given the large body of literature on the topic, it would be useful to try to assemble a database of reported values of

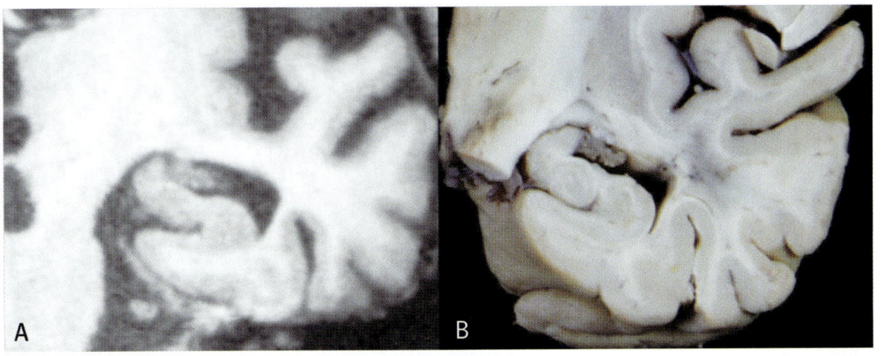

Figure 7.2 Coronal views of the human hippocampal formation and medial temporal lobe from the same patient with frontotemporal dementia, showing *in vivo* MRI (A) 1.5 years prior to death and postmortem brain section (B). Efforts are underway to more directly link *in vivo* imaging measures and postmortem neuropathological measures.

hippocampal and other MTL volumes from the many studies that have been done. This has been done in part by Dr David Kennedy and colleagues as part of the Internet Brain Volume Database (IBVD) (http://www.cma.mgh.harvard.edu/ibvd/). An important problem for such efforts is the variability in measurement protocols. Although there are many similarities between protocols, investigators differ in the rostrocaudal extent of measurement, with some protocols aiming to measure every bit of visible hippocampal tissue, while others opt for the higher level of reliability that can often be obtained by starting and stopping at locations that do not cover the full extent of the hippocampus (e.g. the slice on which the crus of the fornix is first clearly visible). In addition, alveus/fimbria/fornix white matter may or may not be included. The amygdalo-hippocampal transition area in the caudal uncus is another challenging area that has led to differences in tracing protocols. Although these differences may seem relatively minor, they contribute to between-laboratory variability in results. More importantly, laboratories differ in how they deal with between-subject differences in head size/intracranial volume (Barnes et al. 2010), a general factor that influences the volume of the hippocampus or any other brain structure, and also whether they report raw hippocampal volume measures, volume measures that have been normalized to estimates of intracranial volume, or both. Several recently published studies have examined these issues in detail (Geuze et al. 2005b; Konrad et al. 2009).

Since it typically takes an experienced operator approximately 20–30 mins to trace both hippocampi in a single subject, manual tracing is usually considered too inefficient for routine clinical use, although some groups incorporate it into specialized evaluation procedures (e.g. suspected temporal lobe epilepsy; mild cognitive impairment/early AD). Fortunately, accompanying the advances in data acquisition mentioned previously, there has been an explosion in the development of analytic techniques drawn from a variety of branches of computer science (Ashburner et al. 2003). In part because of its unique topographic properties, a number of these techniques have been applied to analyses of the human hippocampal formation. One such approach is the use of automated computational techniques for measuring hippocampal volume (so-called automated segmentation techniques). Although many of these methods are still computationally intensive, often requiring hours of computer processing time, several of them are nearly fully or entirely automated, enabling the hippocampal formation to be labelled and volume to be derived (Figure 7.3). With most of these methods, though, there is still a non-negligible failure rate, particularly for patients with anatomical abnormalities, so a trained individual is typically required to check the results to ensure a reasonable level of accuracy. With these caveats in mind, however, hippocampal volume can be obtained from a large sample of subjects with a reasonably high level of efficiency. Methodological advances in automated hippocampal volumetry have thus enabled multiple laboratories to study sample sizes on the order of hundreds of subjects to address important questions in development, ageing, AD, and other neuropsychiatric conditions. To date, however, relatively little effort has been devoted to comparisons of methodologies for the detection of abnormalities of hippocampal structure (Bergouignan et al. 2009). In addition to hippocampal volume, automated and semi-automated cortical surface-based methods can be used to measure the volume, thickness, and surface area of multiple cortical regions including medial temporal lobe cortical regions (Feczko et al. 2009) (Figure 7.3). Preliminary observations suggest that volumetric decline of MTL cortical regions can result from reduced surface area, which is seen prominently in normal ageing, as well as reduced thickness, which is seen most prominently in AD and other neurodegenerative diseases (Dickerson et al. 2009).

In addition to fundamental scientific studies aimed at advancing knowledge of hippocampal abnormalities in ageing and various neuropsychiatric conditions, measures of hippocampal volume are now being proposed as surrogate markers for use in clinical trials of putative disease-modifying

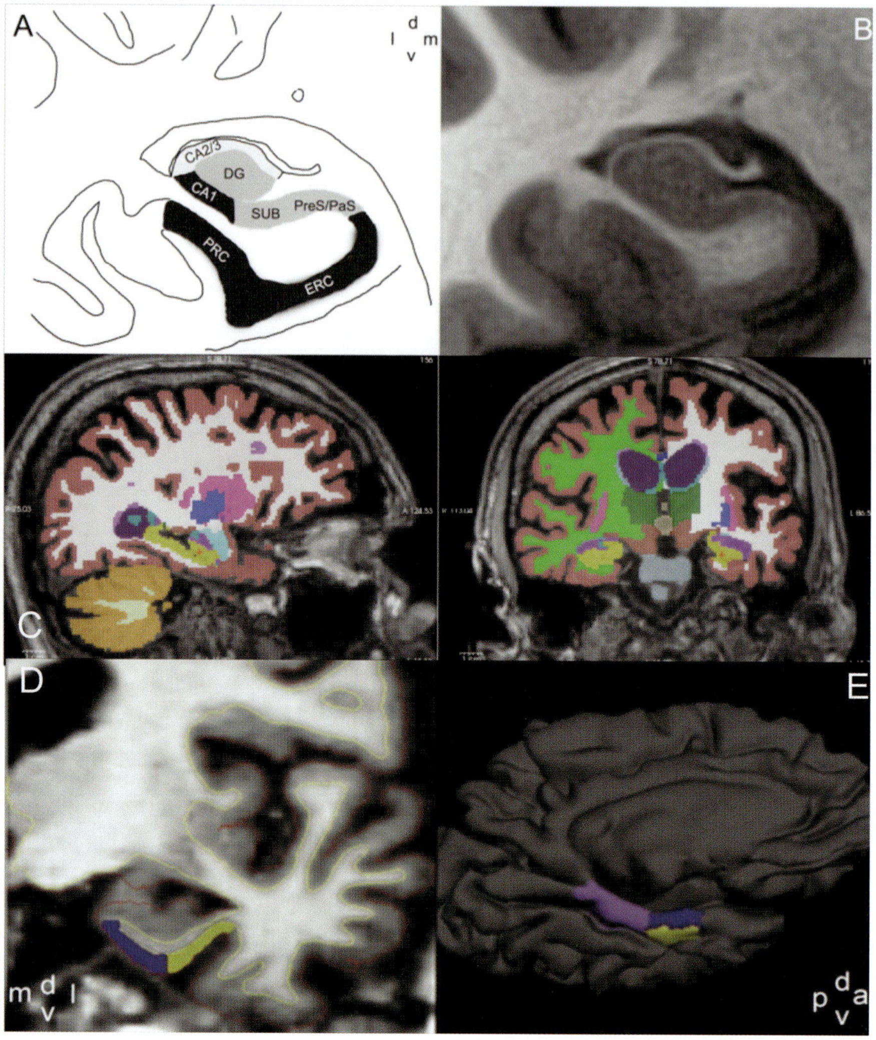

Figure 7.3 A) Drawing of hippocampal subregions based on (B) *in vivo* MRI scan. C) Fully automated segmentation of patient's brain from Figure 7.2, showing hippocampus in yellow in both coronal and sagittal views (crosshair indicates intersection between planes). Semi-automated and automated algorithms have now made hippocampal volumetric measurements much more efficient, although inaccuracies are still sometimes present. D) Coronal MRI with grey–white surface (yellow line) and grey–CSF surface (red line) overlaid from automated reconstruction; locations of perirhinal (yellow) and entorhinal (dark purple) cortices were drawn in manually. E) Manual tracings from the volume were projected onto surface rendering to show locations of perirhinal, entorhinal, and posterior parahippocampal (light purple) cortices. Letters indicate dorsal (d), ventral (v), lateral (l), medial (m), anterior (a), and posterior (p) orientations for reference.

therapies for AD (Jack et al. 2003; Barnes et al. 2009a; Frisoni et al. 2010). Some of the major challenges in translating fundamental scientific observations regarding hippocampal atrophy into quantitative imaging biomarkers for use in clinical trials is the need to define generalizable links between the imaging measures and pathologic markers as well as, probably more importantly, clinical outcomes (Hampel et al. 2010). Substantial progress has been made on both of these fronts. However, new questions have arisen as a result of some of the initial disease-modifying clinical trials in AD—for example, patients who received an antibody aimed at clearing amyloid from the brain demonstrated greater hippocampal atrophy than the control group (Fox et al. 2005), a counterintuitive finding that underscored the need for further research.

Finally, a few comments should be made about voxel-based morphometry (VBM) of the hippocampal formation. VBM has been used extensively over the past decade in studies of the size of different components of brain tissue (i.e. grey matter or white matter) in relation to normal individual differences as well as neuropsychiatric disorders. A technical challenge of particular relevance to the hippocampus is that VBM is heavily dependent on registration of each subject's brain into a common voxel template space. Several investigations have demonstrated that this procedure may result in suboptimal alignment of the hippocampus between individuals (Vandenbroucke et al. 2004), and more sophisticated registration techniques can improve inter-subject alignment (Miller et al. 2005; Miller et al. 2009). This is a critical issue and deserves further study, although some investigations have demonstrated that major findings are detectable whether measured by VBM or by manual ROI tracings (Maguire et al. 2000).

Hippocampal shape and subregional analysis

Although hippocampal volumetry has provided many scientific insights and is undoubtedly a clinically useful tool for certain conditions, the entire volume of a structure as complex as the hippocampus provides a relatively crude measure. Therefore, a number of teams have devoted effort to the analysis of the shape of the surface of the hippocampal formation. The scientific techniques involved draw from fields of geometry and topology and in their most straightforward implementation essentially determine where along the entire surface of the hippocampus there are deformations. For example, a localized inward deformation of the hippocampal surface may or may not be detectable as a reduction in overall volume but may nevertheless be an indicator of significant abnormalities in particular regions of the hippocampus. The location of such a deformation can then be used to identify, at least in probabilistic terms, the subregion(s) where the abnormality is most likely localized. Several groups have used this

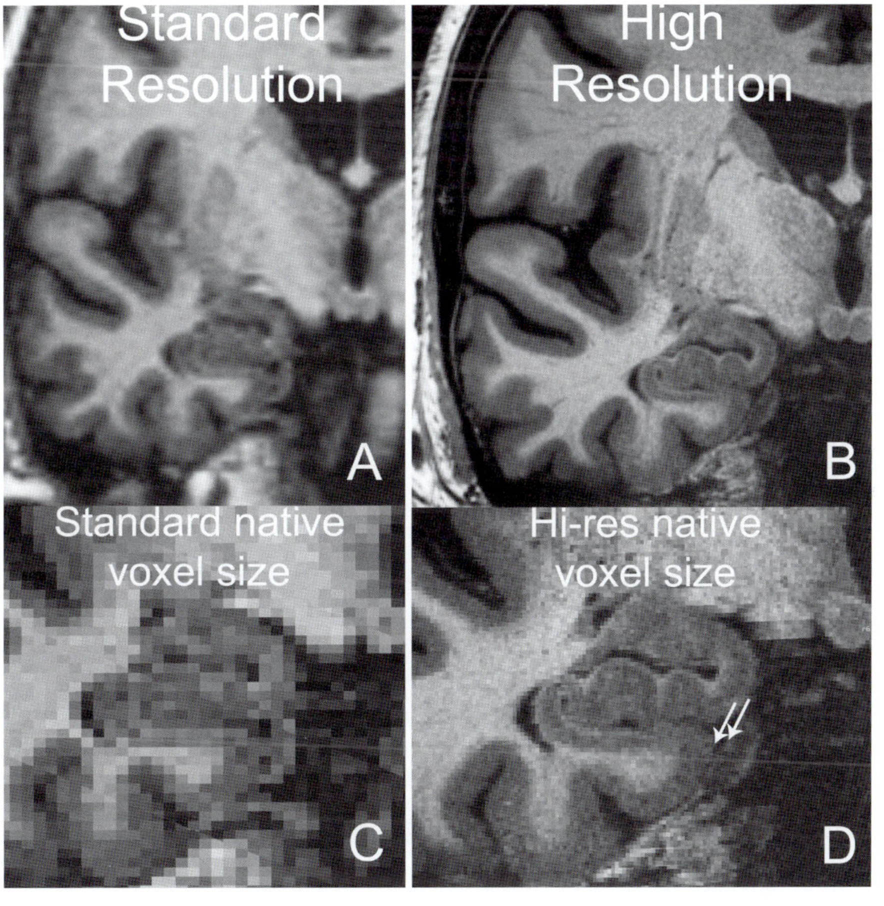

Figure 7.4 Technical advances are leading to much higher resolution MRI data acquisition. This image shows a standard 1-mm isotropic 3D T1-weighted MRI scan from a healthy young adult (A and C) and an ultra-high resolution 3D T1-weighted MRI scan for comparison (B and D). Both sets of images were acquired using a custom-built 32-channel head coil on a Siemens 3.0T TIM Trio scanner. The ultra-high resolution data are 380um in-plane resolution (plane is oblique coronal) with 0.8mm thick slices; five 6-min acquisitions were averaged together to produce this volume. The images in (C) and (D) show native voxel elements; the images in (A) and (B) show data that has been smoothed for better visualization. Arrows in (D) indicate the separation between entorhinal cortical surface and dura mater that can be seen in ultra-high resolution images, which is not clear in standard resolution images.

approach to identify localized hippocampal shape abnormalities in AD, epilepsy, schizophrenia, depression, autism spectrum disorders, among other neuropsychiatric conditions (Csernansky et al. 2000, 2002; Posener et al. 2003; Tepest et al. 2003; Hogan et al. 2004; Nicolson et al. 2006; L. Wang et al. 2006; Dager et al. 2007; Morra et al. 2008; Miller et al. 2009; Boccardi et al. 2010; Cole et al. 2010).

In addition to inferring the hippocampal subregion-specific localization of abnormalities in the shape of the hippocampal surface, several groups have begun developing and applying protocols for the measurement of the volumes of hippocampal subregions (Mueller et al. 2007, 2008; Yushkevich et al. 2008, 2009, 2010; Mueller and Weiner, 2009; Van Leemput et al. 2009; Kerchner et al. 2010; Neylan et al. 2010; Z. Wang et al. 2010;). The development of new methods for higher resolution MRI data acquisition is fuelling many of these efforts, along with new computational analytic methods (Figure 7.4). Some of these methods can be used to visualize and measure detailed aspects of MTL atrophy in AD and other neurodegenerative diseases (Figure 7.5).

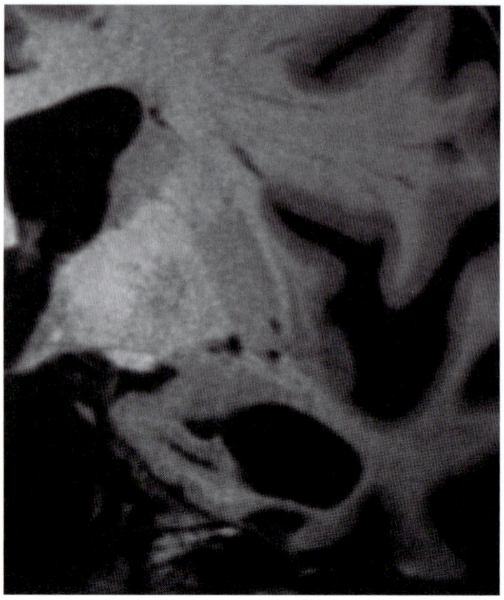

Figure 7.5 Ultra-high resolution view of neurodegenerative atrophy in the hippocampus and medial and lateral temporal cortex in a patient with Alzheimer's disease.

Imaging task-related hippocampal and MTL cortical function

Since functional neuroimaging tools assess inherently dynamic processes that may change over short time intervals in relation to a host of factors, these measures have unique characteristics that may offer both strengths and weaknesses as potential biomarkers of neurological disease. Functional neuroimaging measures may be affected by transient brain and body states at the time of imaging, such as arousal, attention, sleep deprivation, sensory processing of irrelevant stimuli, or the effects of substances with pharmacologic central nervous system activity. Imaging measures of brain function may also be more sensitive than structural measures to constitutional or chronic differences between individuals, such as genetics, intelligence or educational level, learning, mood, or medication use. While these may be effects of interest in certain experimental settings, they need to be controlled when the focus is on disease-related changes and differences between subject groups or within individuals over time.

Among functional neuroimaging techniques, functional MRI (fMRI) has many potential advantages in studying MTL activity, as it is a non-invasive imaging technique that does not require the injection of contrast agent. It can be repeated and thus lends itself well to longitudinal studies. It has relatively high spatial and temporal resolution, and the use of event-related designs enables the haemodynamic correlates of specific behavioural events to be measured. Thus, although positron emission tomography (PET) was used in numerous early experiments aiming to identify changes in the functional activity of the hippocampus during learning or retrieval, the vast majority of functional neuroimaging experiments on this topic over the past decade have been conducted with fMRI.

There are, however, significant challenges to performing fMRI studies in patients with memory or other cognitive impairments. The technique is particularly sensitive to even small amounts of head motion. Differences in task performance between patient and control groups complicate data interpretation (Price and Friston 1999). Finally, it is critical to complete further reliability experiments if fMRI is to be used in longitudinal or pharmacological studies. Although there are now a few studies of fMRI test-retest reliability in young subjects (Machielsen et al. 2000; Manoach et al. 2001; Sperling et al. 2002), reproducibility studies are only beginning to be performed in older adults and patient populations.

Functional neuroimaging of the hippocampus in memory

One of the early applications of fMRI aimed to determine whether the hippocampal formation and nearby MTL structures were engaged during learning or retrieval of new information. Although several initial studies did not demonstrate increased hippocampal/MTL activity during the acquisition or retrieval of episodic verbal information (Squire et al. 1992; Shallice et al. 1994; Buckner et al. 1995), other early studies demonstrated increased blood-oxygen-level dependent (BOLD) fMRI or glucose metabolic activity during the encoding of novel information compared to the presentation of repeated, highly familiarized information (Tulving et al. 1994; Grady et al. 1995), and also during the verbal recall of words (Schacter et al. 1996). It is important to point out that the activation of a brain region as detected with functional neuroimaging methods is relative, since functional neuroimaging measures obtained during task performance are typically compared to measures obtained during a control condition, thus identifying relatively greater activity in one condition than another. Furthermore, similarly to VBM, co-registration techniques that are usually required to align data from multiple subjects to each other are imperfect but continue to be improved (Miller et al. 2005). From a conceptual standpoint, given the variability of MTL subregion size and shape between subjects, some investigators prefer to localize fMRI data at the individual subject level first and then perform statistical analyses, rather than trying to localize activity at the group-level after aligning subjects to a template. This continues to be a major issue and creative computational techniques are being developed to coregister higher resolution structural MRI data with relatively lower resolution fMRI data to better localize activations measured using this method.

An important technical and experimental design development in the late 1990s, 'event-related' fMRI (drawn from event-related potential electroencephalographic methodology), enabled the detection of relatively greater hippocampal and other MTL activation during the encoding of stimuli that were subsequently remembered as compared to those that were not remembered (Burock et al. 1998). In the decade since the first papers demonstrating relatively greater MTL activity during subsequently remembered items (Brewer et al. 1998; Wagner et al. 1998), many clever experiments have built on this approach to elucidate specific functional neuroanatomical properties of MTL and other brain regions (Figure 7.6).

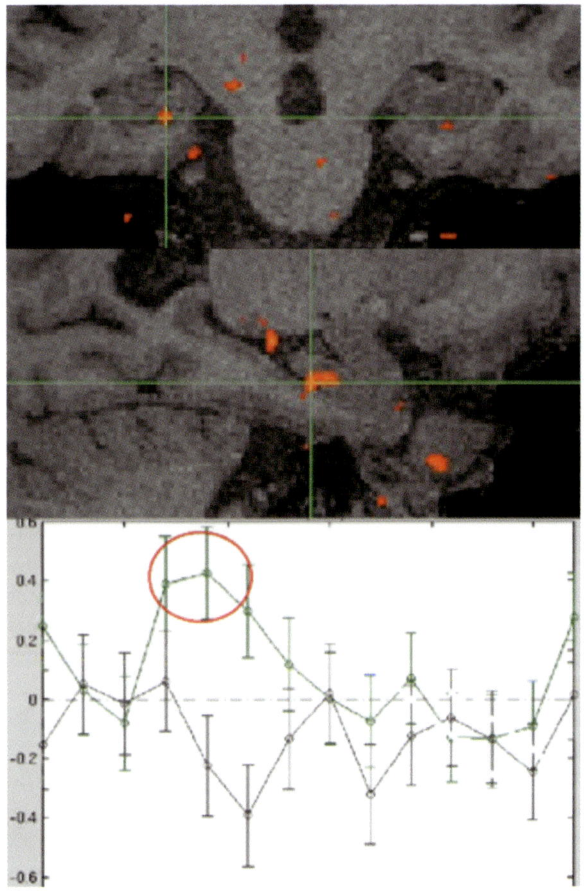

Figure 7.6 Event-related functional MRI scan of memory encoding in a single 73-year-old female, showing statistical map of hippocampal activation indicated with crosshair in coronal (top) and sagittal (middle) views, and timecourse (bottom) of signal showing relatively greater activation for items that were subsequently remembered (green line) versus items that were not subsequently remembered (blue line).

One issue that has been investigated since the early days of functional neuroimaging is the differential activation of the MTL during encoding as opposed to retrieval. Partly because of the large number of variables that often differ between experiments, there is still not consensus regarding whether there is a specific topographic organization in the MTL that differentiates the functional neuroanatomy of encoding or retrieval. Some investigators find that the rostral hippocampal formation is relatively more active during encoding while the caudal hippocampus is relatively more active during retrieval (Zeineh et al. 2003), but the opposite findings have also been observed (Gabrieli et al. 1997). Within the canonical MTL subregions, some investigators have found encoding related activity in hippocampus-proper CA fields with retrieval-related activity in subiculum (Zeineh et al. 2003), while others have found a pattern that differs (Pihlajamaki et al. 2003).

A major challenge when trying to connect human with animal data regarding the MTL is the gap in resolution. The human functional neuroimaging technique with the highest spatial resolution, fMRI, is still striving to obtain resolution on the order of 1mm, with most studies at advanced centres still working in the 1.5–2-mm range (and most routine hippocampal fMRI being performed with 3–4-mm voxels). Furthermore, the signal-to-noise ratio of BOLD fMRI in the MTL is notoriously poor. Although a few groups are performing investigations of hippocampal function at higher resolution (Carr et al. 2010), technical advances are still badly needed for further advances that would bring human MTL imaging closer to animal data.

Questions related to the specificity of MTL regions for memory processes have been the subject of a number of studies, including associative versus item-based memory and the dual process model of episodic memory (recollection versus familiarity). Some evidence supports the concept that the hippocampal formation is selectively activated for recollection or for associative memories, and the perirhinal or entorhinal cortex is activated for familiarity or for item-based non-associative memories (Davachi et al. 2003; Squire et al. 2004; Eichenbaum et al. 2007). However, other studies of these same topics have produced conflicting results (Squire et al. 2007; Brickman et al. 2011).

The specificity of MTL regions for the particular content of episodic memory has received some attention. Successful memory for faces tends to be associated with activity in perirhinal and rostral hippocampal regions, while objects tend to be associated with somewhat more caudal activity (Preston et al. 2010). Memory for indoor or outdoor scenes or other spatial context information is fairly reliably associated with posterior parahippocampal activity (Stern et al. 1996; Maguire 1997). Surprisingly little work has been performed on object-in-context associative memory performance, although one study distinguished activation of the hippocampus and posterior parahippocampal region associated with memory for objects in imagined contextual scenes from activation of the perirhinal cortex associated with memory for the objects alone (Davachi et al. 2003).

Neuroimaging investigations have recently turned to the remarkably fine-grained ability of components of the MTL memory system to form associations even in the absence of complete information, while at the same time distinguishing related but distinct objects or contexts from each other—pattern completion versus pattern separation. Similarly to rodent studies, initial experiments suggest that CA3 or dentate gyrus play an important role in pattern separation within the hippocampal formation (Bakker et al. 2008).

Recent neuroimaging research has begun to provide evidence convergent with animal data that the MTL may play specific roles in the flexible use of episodic memory trace information for processes that probably subserve inferential reasoning and generalization. Transitive inference, the ability to use previously learned associative relationships between items to draw inferences about probable associations that have never been specifically learned, appears to depend on the hippocampal formation (Dusek and Eichenbaum 1997). Functional neuroimaging evidence also supports the specific role of the hippocampal formation in transitive inference in humans (Heckers et al. 2004).

FMRI of the hippocampus in ageing and AD

A number of fMRI studies in patients with clinically diagnosed AD, using a variety of visually presented stimuli, have identified decreased activation in hippocampal and parahippocampal regions compared to control subjects during episodic encoding tasks (Small et al. 1999; Rombouts et al. 2000; Kato et al. 2001; Machulda et al. 2003; Sperling et al. 2003b). As for mild cognitive impairment (MCI), the prodromal stage of AD, studies have demonstrated conflicting results with some findings of hippocampal hypoactivation (Machulda et al. 2003; Johnson et al. 2006) similar to that seen in AD dementia and some findings of hyperactivation (Dickerson et al. 2005; Hämäläinen et al. 2007; Kircher et al. 2007). Despite the inconsistencies, which we believe can be explained by a number of factors (Dickerson and Sperling 2008), there is replicated evidence to support the hypothesis that there is a phase of increased MTL activation early in the course of prodromal AD, prior to clinical dementia (Figure 7.7). This increase may represent, at least in part, an attempted compensatory response to AD neuropathology; given that greater MTL activation is associated with both better memory performance and smaller hippocampal volume (Dickerson et al. 2004; Hämäläinen et al. 2007). Nevertheless, it may also be a marker of cascading neuropathology, with evidence that greater hippocampal activation is predictive of more rapid clinical decline (Miller et al. 2008). Additional studies employing event-related fMRI paradigms (Sperling et al. 2003a; Kircher et al. 2007) will be very helpful in determining whether increased MTL activation in MCI patients is specifically associated with successful memory, as opposed to a general effect that is present regardless of success (possibly indicating increased effort).

We believe that the accumulating evidence indicates that task-related regional brain hyperactivation may be a universal neural response to insult, as it occurs in a variety of neuropsychiatric disorders and conditions, including AD/MCI, Huntington's disease (Rosas et al. 2004), Parkinson's disease (Monchi et al. 2004), cerebrovascular disease (Carey et al. 2002; Johansen-Berg et al. 2002), multiple sclerosis (Reddy et al. 2000; Morgen et al. 2004), traumatic brain injury (McAllister et al. 1999), HIV (Ernst et al. 2002), alcoholism (Desmond et al. 2003), schizophrenia (Callicott et al. 2003), sleep deprivation (Drummond et al. 2000), and ageing (Cabeza et al. 2002). The evidence discussed also indicates that increased MTL activation can be seen in the absence of significant MTL atrophy, which provides *in vivo* support for laboratory and animal data suggesting that physiological alterations may precede significant structural abnormalities in very early AD (Selkoe 2002; Walsh and Selkoe 2004) and may represent inefficient neural circuit function (Stern et al. 2004). Thus, fMRI may provide a means to detect changes in human memory circuit function that underlie the earliest symptoms of AD, and may be useful in identifying groups of subjects at high risk for future cognitive decline prior to a diagnosis of AD.

Recent studies are extending these initial insights using longitudinal and multimodal approaches. We recently showed that MCI subjects who declined over 2 years of clinical follow-up demonstrated an initial hippocampal hyperactivation during encoding but after 2 years a reduced level of activity, consistent with the 'inverse U-shaped curve' model (O'Brien et al. 2010). Evidence of the earliest

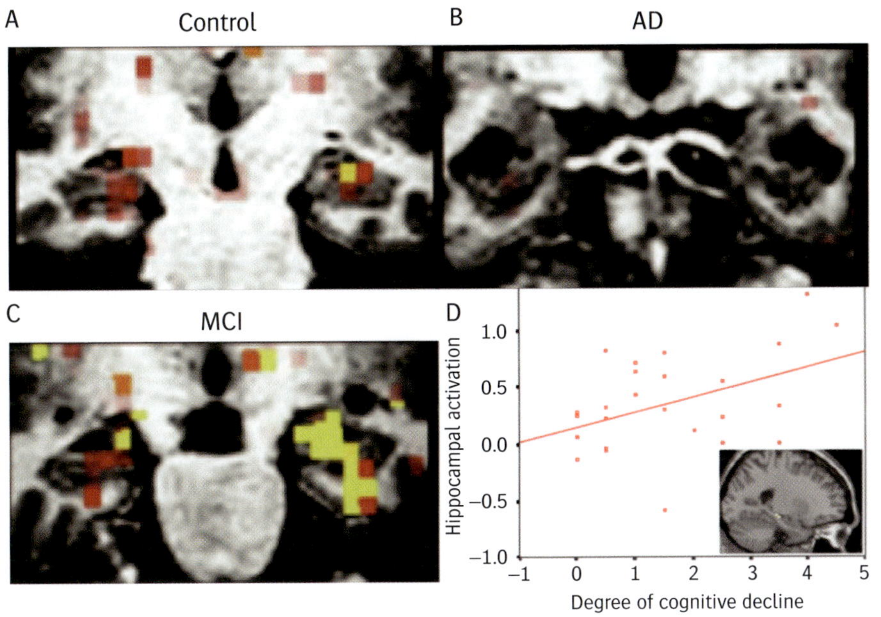

Figure 7.7 The hippocampal formation and medial temporal cortex activates in (A) healthy elderly individuals during memory encoding and is not only (B) atrophic but also hypoactive in AD patients. Relative to controls, the hippocampus and medial temporal cortex is (C) hyperactivated in MCI patients, and (D) the magnitude of activation in MCI predicts rate of clinical decline over the coming years, with those individuals with the highest activation declining the most rapidly to dementia.

stages of this process can be seen in cognitively intact older adults with brain amyloid, who hyperactivate the hippocampus during memory encoding relative to the activation of similarly aged adults without brain amyloid (Sperling et al. 2009). Extending these approaches to genetic risk, in cognitively intact older adults (Bookheimer et al. 2000; Bondi et al. 2005; Fleisher et al. 2005; Xu et al. 2009) at elevated genetic risk for AD, hippocampal activation during memory tasks is greater than in those without such elevated risk.

Pharmacological fMRI

Pharmacological fMRI (phMRI) is being used increasingly to study the effects of neurochemical perturbations on brain function at a regionally-specific or systems level (Leslie and James, 2000; Honey and Bullmore, 2004). Besides the cholinergic system, phMRI has been applied to studies of dopaminergic modulation of working memory (Mattay et al. 2000), GABAergic modulation of repetition priming (Thiel et al. 2001), and noradrenergic effects on attention/orienting tasks (Coull et al. 2001). In addition, fMRI has been used to study the effects of intrinsic or extrinsic steroid hormonal modulation on cognitive tasks (Goekoop et al. 2006; Goldstein et al. 2005). PhMRI has begun to reveal regional brain activation responses associated with individual differences in behavioural response to pharmacologic agents (Gibbs and D'Esposito 2005). In clinical populations, phMRI has been applied to investigations of cocaine dependence (Breiter et al. 1997), depression (Kalin et al. 1997), schizophrenia (Honey et al. 1999), and Parkinson's disease (Mattay et al. 2002). A recently emerging, exciting application of phMRI is in the prediction of clinical response to pharmacologic treatment—e.g. patients with depression who are treated with antidepressant medications demonstrate changes in activation of affective circuitry that are associated with clinical response to pharmacologic treatment (Fu et al. 2004) and may enable prediction of response (Davidson et al. 2003). The addition of genetic factors to these investigations holds fascinating and powerful promise to illuminate fundamental pharmacogenetic mechanisms of central nervous system drug effects—e.g. in schizophrenic patients, there is a catechol-o-methyl-transferase polymorphism-dependent effect of atypical antipsychotic therapy on prefrontal activation during working memory task performance (Bertolino et al. 2004).

With respect to the cholinergic system, phMRI has been used to study effects of cholinergic antagonists or agonists on brain function and behaviour in healthy young individuals (Thiel et al. 2001, 2002a, 2002b, 2002c; Sperling et al. 2002; Bentley et al. 2003; Schon et al. 2005) and in ageing (Wink et al. 2006). Initial studies of cholinergic enhancement in AD showed increased fusiform activation during episodic encoding (Rombouts et al. 2002), and cholinergic inhibition in controls was associated with reduced hippocampal, fusiform, and prefrontal activation during episodic encoding (Sperling et al. 2002).

Imaging hippocampal perfusion and metabolism

Another measure related to brain function is perfusion. Perfusion refers to the delivery of blood, oxygen, and nutrients to tissue, and is tightly coupled with metabolism. Thus, although measures of perfusion are of direct relevance for clinical disorders of blood flow, such as stroke, these measures are also of value as surrogates of regional brain metabolism and function. The 'gold standard' method for quantifying brain perfusion *in vivo* is PET scanning with ^{15}O-labeled water, and that for metabolism is ^{18}fluorodeoxyglucose (FDG). With regard to perfusion, water in arterial blood can also be 'labelled' via MRI using a technique called arterial spin labelling (ASL). This method has been developed over the last two decades but has received increasing attention in the past few years in part because it is non-invasive and relatively easy to obtain as part of a standard MRI protocol on widely available scanners.

In part because of the relatively low resolution of PET scanning technology, there has been relatively little investigation of hippocampal perfusion and metabolism. Although some studies suggest that hippocampal perfusion may be reduced in normal older adults compared with younger adults (Heo et al. 2010) while some indicate that it is preserved (Rusinek et al. 2011), analysis of hippocampal metabolism in relation to global grey matter metabolism has shown that it is relatively spared (much more so than many frontal and parietal regions) or even increased in relation to global metabolism in normal ageing (Willis et al. 2002; Kalpouzos et al. 2009). One longitudinal study of older adults demonstrated that, in individuals with preserved memory over an 8-year interval, hippocampal blood flow was also preserved (relatively increased compared to global blood flow) (Beason-Held et al. 2008). More detailed high-resolution studies suggest that particular hippocampal subregions, especially the dentate gyrus, are vulnerable to metabolic decline in normal ageing while other subregions may be resilient (Small et al. 2004).

Other investigations have highlighted the differential genetic influences on hippocampal metabolism in ageing. In adults aged 50–63 with a family history of AD, APOE-ε4 carriers exhibited longitudinal decline in hippocampal metabolism over 2 years compared to non-carriers, despite no differences in memory performance (Reiman et al. 2001). Similar findings have been reported for individuals with a family history of AD regardless of APOE genotype (Mosconi et al. 2007). Thus, although data are still sparse, focused investigations reveal a complex picture of age-related hippocampal perfusion and metabolism with familial/genetic factors likely playing an important role in the degree to which these measures change or remain stable.

Complicating matters further, several recent reports suggest that the hippocampus and nearby MTL cortex may be hyper-perfused in AD compared to controls (Scarmeas et al. 2004; Alsop et al. 2008). A recent investigation of the relationship between hippocampal volume and perfusion in MCI and AD provided further support for these observations, indicating that MTL perfusion and atrophy were not correlated (Luckhaus et al. 2010). As expected based on prior literature, this study found that hippocampal volume was smaller in APOE ε4 carriers than in non-carriers, but that MTL perfusion was not different based on APOE genotype. In temporal lobe epilepsy, as another clinical example, there is typically ictal MTL hyperperfusion (Tae et al. 2005) and interictal MTL hypoperfusion (Wolf et al. 2001).

Imaging hippocampal connectivity

Lesion and neuroimaging data are often focused on the contributions of one or more specific regions to task performance, although it is widely acknowledged that most abilities depend on large-scaled distributed brain networks (Mesulam 1998). Prefrontal-temporal interactions have been a particularly attractive starting point for theoretical work on episodic memory networks (Simons and Spiers 2003). Although lesion analysis studies in monkeys and humans and other data have been used to develop hypotheses about large-scale distributed episodic memory networks, human neuroimaging data inherently provide fruitful data for analysis at the systems level. Besides the coactivation of brain regions during the performance of episodic encoding and retrieval tasks, as described earlier, new systems-level analytic approaches to neuroimaging data have begun to reveal the interactions between brain regions that make up the large-scale episodic memory network, which contains the hippocampus and MTL cortex at its epicentre.

So-called functional connectivity neuroimaging data analytic approaches have existed for nearly a decade, and have been applied in a few studies to questions related to the interaction between brain regions engaged in memory tasks. These approaches enable the detection of correlated activity increases between spatially distributed brain regions, and causal modelling using activity increases in one region to predict activity increases in another. These types of analyses have provided evidence for prefrontal-medial temporal-ventrolateral temporal interactions in episodic encoding that leads to successful subsequent free recall, for example (Dickerson et al. 2007).

A prominent relatively recent line of functional neuroimaging research has identified a set of brain regions that are relatively more active during undirected epochs of time during experiments than during epochs involving the directed performance of a variety of tasks, including sensorimotor, attentional, language, and other types of tasks. These brain regions include posterior cingulate/precuneus, lateral inferior parietal lobule, lateral and medial temporal, and medial prefrontal cortices, and have been called the 'default mode' network (Raichle et al. 2001; Buckner et al. 2008). Many of these brain regions are activated during various processes involved in specific memory encoding or retrieval tasks, as described previously using data from task-related functional neuroimaging experiments. This observation has led to the hypothesis that one activity that may be prominently occurring during these 'rest' epochs is thinking about the past or future, or 'mental time travel,' which engages episodic memory systems of the brain.

The convergence of thinking about functional brain systems and observations derived from studies of undirected 'rest' epochs of experiments has led to an explosion of studies employing 'resting-state' functional connectivity MRI (fcMRI), which is an exciting new neuroimaging technique for investigating large-scale functional-anatomic systems. Spontaneous fluctuations in BOLD signal can be readily detected in subjects whether or not a task is being performed. The degree to which these spontaneous fluctuations are correlated between different brain regions can be interrogated using a systems perspective. The initial experiments in this field have validated findings using this technique against findings using the same technique in anesthetized monkeys, and have demonstrated that functional brain networks identified using fcMRI correspond to networks identified using traditional tract tracing methods (Vincent et al. 2007). These techniques have already been used to demonstrate the presence of a hippocampal-parietal network (Vincent et al. 2006), and the distinction between subnetworks within the MTL memory system, including the rostral hippocampal–perirhinal network versus the caudal hippocampal–parahippocampal–retrosplenial network (Kahn et al. 2004). This approach will clearly provide valuable novel contributions to knowledge about human memory networks in the brain, and will likely help provide further links between human and animal data. We have recently shown that the magnitude of hippocampal connectivity with other regions of the large-scale episodic memory network predicts memory performance in young (L. Wang et al. 2010a) and older (L. Wang et al. 2010b) healthy adults. Figure 7.8 illustrates the spatial localization of the brain regions in the large-scale episodic memory network.

A recent combined BOLD fMRI and perfusion MRI investigation demonstrated the presence of these resting state networks in both types of signal, suggesting that BOLD fluctuations are similar to fluctuations in perfusion, which the authors interpreted as reflecting underlying metabolism

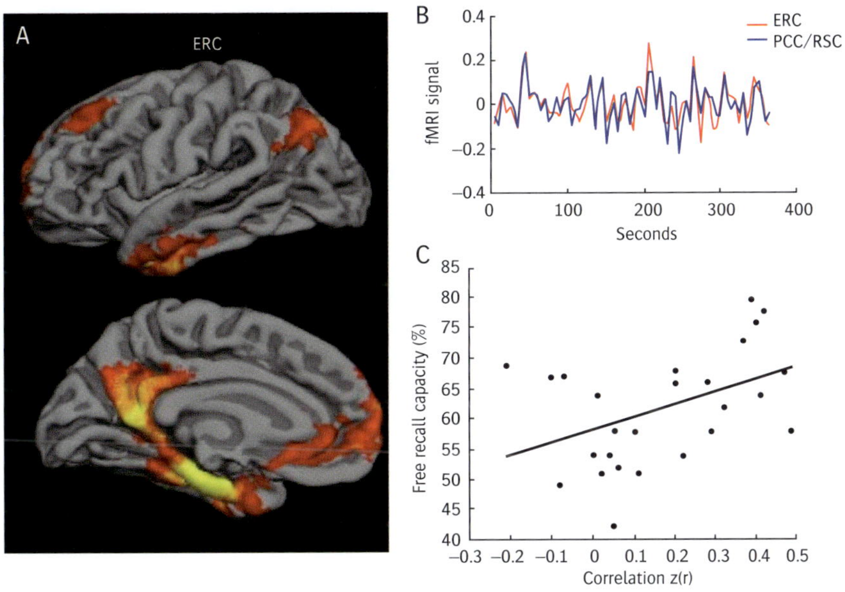

Figure 7.8 Resting state functional connectivity MRI (fcMRI) is illuminating the large-scale network properties of the hippocampus and medial temporal lobe cortex in relation to other brain regions with which they are connected. A) Using the entorhinal cortex as a 'seed region' the large-scale cortical network of connected regions can be mapped to illustrate PCC/RSC, angular gyrus, ventrolateral anterior temporal cortex, and regions in the prefrontal cortex. Low-frequency oscillations in fMRI signal are correlated between brain regions that share connectivity such as entorhinal cortex (red) and posterior cingulate (PCC)/retrosplenial cortex (RSC) (blue). B) Low-frequency oscillations in fMRI signal are correlated between brain regions that share connectivity such as entorhinal cortex (red) and posterior cingulate (PCC)/retrosplenial cortex (RSC) (blue). Using the entorhinal cortex as a 'seed region' the large-scale cortical network of connected regions can be mapped to illustrate PCC/RSC, angular gyrus, ventrolateral anterior temporal cortex, and regions in the prefrontal cortex. C) Within young adults, the strength of entorhinal-PCC/RSC connectivity (correlation values) predicts performance on a challenging memory task (free recall capacity).

and neuronal activity (Wu et al. 2009). Although there has been an explosion of studies of regional brain connectivity using diffusion tensor imaging and tractography, relatively little investigation of the connectivity of the hippocampus and MTL cortex has been performed using these techniques (Powell et al. 2004).

Molecular imaging of the hippocampus

Proton magnetic resonance spectroscopic imaging (MRS) is a technique that enables the quantitation of particular neurochemical constituents of brain tissue. Metabolic components that are commonly measured include *N*-acetyl aspartate (NAA), an indirect marker of neuronal and synaptic integrity and particularly indicative of glutamate levels (Petroff et al. 2002); myoinositol (mI), thought to represent the density of glial cells; and choline (Cho), a marker of cell membrane turnover.

Although MR spectroscopy of the MTL has been performed for more than 10 years, there are still many technical challenges involved in using this technique to study the hippocampus in part due to its small size; a relatively recent reliability study demonstrated a 14% coefficient of variation in measures of hippocampal NAA, more than 50% greater variability than the anterior cingulate cortex (Venkatraman et al. 2006).

Nevertheless, a number of studies have demonstrated reductions of NAA in the MTL in ageing (Schuff et al. 1999), AD (Schuff et al. 1997; Jessen et al. 2000; Kantarci et al. 2000), schizophrenia (Maier et al. 2000), temporal lobe epilepsy (Maier et al. 2000), depression, and other neuropsychiatric disorders. Since NAA measured by MRS may be a marker of synaptic glutamate concentrations, it has been hypothesized to be a dynamic indicator of hippocampal neurochemical integrity. Indeed, several studies have demonstrated that abnormally low levels of NAA in patients with depression increase after treatment in a manner related to clinical improvement (Block et al. 2009; Huang et al. 2010). Additional recent work with MRS, summarized next, has begun to provide tantalizing data suggestive of *in vivo* evidence of hippocampal plasticity in healthy humans.

Imaging hippocampal plasticity

The hippocampal formation is one of the most plastic structures in the mammalian brain (Klintsova and Greenough 1999). As a result, it has been the subject of a number of intriguing imaging studies, although the need for a high degree of test–retest reliability has presented technical challenges. Many of the techniques described earlier have been applied to the investigation of hippocampal plasticity.

In 2000, Eleanor Maguire and colleagues reported that the hippocampal formation of London taxi cab drivers is larger than that of similarly aged control individuals (Maguire et al. 2000). This finding was interpreted as evidence for hippocampal plasticity as a result of acquiring the extensive spatial knowledge of the city required for licensure. Although until prospective longitudinal studies are performed of this kind of training it is still possible that individuals who choose to become taxi drivers possess larger hippocampi and superior spatial abilities prior to undergoing such training, a number of findings in the original study (Maguire et al. 2000) and subsequent study (Maguire et al. 2006) support the hypothesis that hippocampal enlargement may occur as a result of the acquisition of spatial experience. Probably the most compelling evidence is that hippocampal grey matter correlates with experience, such that more seasoned taxi drivers have the largest hippocampi. Very recently, a new study demonstrates that intense spatial training was associated with an increase in hippocampal NAA as measured with magnetic resonance spectroscopy, which then normalized several months after cessation of training; furthermore, this effect was modulated by the specific genotype of brain-derived neurotrophic factor (BDNF) (Lovden et al.), thought to be a critical growth factor for hippocampal synaptic plasticity and previously shown to influence hippocampal NAA and memory task-related fMRI activation in humans (Egan et al. 2003).

Exercise-induced cerebral plasticity has been the subject of a great deal of recent investigation (Churchill et al. 2002), and a 2010 study provided impressive data regarding exercise-induced hippocampal plasticity (Pajonk et al. 2010). After 3 months of daily aerobic bicycle exercise, schizophrenic patients showed a 16% increase in hippocampal volume and an increase in hippocampal NAA, compared with a group that played table football. Surprisingly, the control group of neuropsychiatrically healthy adults showed a 12% increase in hippocampal volume.

In addition to these exciting developments, several new techniques are being applied to the fascinating topic of hippocampal neurogenesis, which appears to occur relatively specifically within the dentate gyrus (Deng et al. 2010). While many technical challenges remain to provide convincing validation of the imaging findings, both MR spectroscopic and cerebral blood volume measures have been reported as potential surrogate markers of hippocampal neurogenesis in humans (Manganas et al. 2007; Pereira et al. 2007).

Ex vivo hippocampal imaging

In the last section of this chapter, we will briefly review *ex vivo* imaging of the hippocampal formation. *Ex vivo* imaging is a relatively new field of study in which autopsied brain tissue scanned with magnetic resonance produce high quality images. Of course, the parameters typically used for *in vivo* imaging do not yield the same attributes and must be adjusted for fixation effects. In *ex vivo* or formalin fixed specimens, several lines of research have shown that longitudinal relaxation times T1 decreases in grey matter and white matter but that the decrease in grey matter is greater than in white matter (Nagara et al. 1987; Tovi and Ericsson 1992; Blamire et al. 1999; Schumann et al. 2001; Pfefferbaum 2004; Shepherd et al. 2009). The transverse relaxation T2 also decreases but to a lesser degree than the changes observed in T1 (Tovi and Ericsson 1992) and T2 decreases to a point but then increases slightly and plateaus (Dawe et al. 2009). Thus, *ex vivo* images predominantly exhibit a proton density weighting and have little T1 weighting. The physical changes of fixation for imaging occur within the first week of fixation (Tovi and Ericsson 1992) and stabilize after several weeks. Fixation increases contrast to noise (CNR) and creates a sharper grey matter–white matter boundary and cortical lamina (Augustinack et al. 2010).

Several advantages stand out when considering *ex vivo* imaging. First, autopsy specimens can undergo very long scan sessions with essentially no motion, a thus a greater number of acquisitions can be obtained. As a result, *ex vivo* MRI produces high resolution and high signal-to-noise images (Fatterpekar et al. 2002; Pfefferbaum 2004; Augustinack et al. 2005; Chakeres et al. 2005; Hinds et al. 2008). Second, *ex vivo* images have been validated histologically. With concomitant *ex vivo* and histological stained sections, registration between the two modalities is possible and authenticates the 'new' magnetic resonance microscopy image with the 'old' histological stain (e.g. cell density for Nissl stain). Histologically validated high-resolution MRI is thus creating the ability to examine cortical architecture without sectioning. Examples of neuroimaging and histological validation have been carried out in several cortical areas, including MTL (Augustinack et al. 2005), temporal pole (Blaizot et al. 2010), orbitofrontal cortex (Uylings et al. 2010), parietal cortex (Caspers et al. 2006) visual cortex (Fatterpekar et al. 2002; Eickhoff et al. 2005) and several rodent models (Poduslo et al. 2002; Wadghiri 2003; Jack et al. 2004; Zhang et al. 2004; Hogan et al. 2009). Moreover, sophisticated quantitative methods have been developed to extract signals indicative of neuroanatomical laminae (Schleicher et al. 2005; Eickhoff et al. 2006), improve MRI with contrast enhancement (Huang et al. 2009), and validate longitudinal *in vivo* neuroimaging data with *ex vivo* data (Pieperhoff et al. 2008).

Third, the correlation of *ex vivo* and *in vivo* imaging data permits lamina, cellular input projections, and other data below the resolution of typical *in vivo* scans to be extrapolated from *ex vivo* to *in vivo* data. Utilizing the confirmed histological data and applying it to *in vivo* cases or brain averages enables location of specific cytoarchitectural cortical regions to be predicted based on topology (Fischl et al. 2008). In addition to using the information contained in *in vivo* brain volume images and surface-based topologic reconstructions, *ex vivo* imaging data can draw on coregistered histological data such as cell density, fibre density, specific protein and receptor profiles to determine boundaries (Amunts et al. 2005, 2007; Scheperjans et al. 2008). Given that cerebral sulci and brain volumetrics contain variation among subjects, capitalizing on additional components intrinsic to the tissue improves the accuracy of boundary definitions in macroscopic imaging data. Probablistic atlases based on cytoarchitecture and surface based analysis accurately predict locations of cytoarchitecturally defined areas (Amunts et al. 2005, 2007; Fischl et al. 2008, 2009) and have been used to compare human cortex to primate cortex (Schenker et al. 2010). Finally, *ex vivo* MRI data contributes a valuable resource to neuroanatomical analyses because it easily allows for re-slicing in multiple planes, reconstruction for 3D modelling, and for viewing at multiple sites.

Neuropathological assessment has begun to utilize *ex vivo* imaging in several diseases such as stroke and oedema (Boyko et al. 1994), multiple sclerosis (Bo et al. 2004), Alzheimer's disease (Bobinski et al. 2000; Jagust et al. 2008), AIDS (Grafe et al. 1990; Everall et al. 1997), autism (Schumann et al. 2001) and white matter pathology (Braffman et al. 1988; Grafton et al. 1991; Bronge et al. 2002). In addition to high-resolution structural *ex vivo* MRI, investigators have utilized *ex vivo* diffusion imaging (for methods review, Dyrby et al. 2011) to corroborate microstructure such as fibre distributions in myeloarchitecture and probabilistic tractography findings in the hippocampus. High-resolution diffusion imaging in the hippocampus proper (Shepherd et al. 2007) and associated pathways (Augustinack et al. 2010) hint at the enormous value *ex vivo* imaging may provide for the investigation of hippocampal systems as well as biomarkers of diseases such as AD.

The basic pipeline for *ex vivo* imaging entails: (1) formalin or paraformaldehyde tissue fixation, (2) scanning whole brain or brain hemisphere at low resolution and usually low field strength (i.e. 1.5 T) to make surface models, 3) scanning a smaller tissue block at high resolution and at high field strength (i.e. 7 T), 4) sectioning tissue and photographing a blockface image, 5) mounting and histologically staining tissue, and finally, 6) registration of the two modalities. In Figure 7.9, a sample case shows a coronal view for a medial temporal block at the anterior-posterior level of the hippocampal body at 1 mm^3 resolution (Figure 9A), at 120 μm resolution[3] (Figure 7.9B), and Nissl stained sections (Figure 7.9C).

It is obvious that the level of detail increases from the 1-mm resolution *ex vivo* MRI to the Nissl stained section. The low-resolution images were acquired at 1 mm^3 at 1.5 T on a Siemens scanner (TR = 20 ms, TE = 1.80 ms, matrix 256 × 256 × 256, acquisition time per run = 6 min) and show the hippocampal body where the fimbria begins (Figure 7.9A). The overall shape and the ability to distinguish grey from white matter, especially the alveus, is clear in low-resolution scans but delineation of other substructures is difficult. High resolution *ex vivo* MRI fast low angle shot (FLASH) scans were also acquired on a Siemens platform (TR = 20 ms, TE = 10 ms, flip angle = 20, matrix 512 × 384 × 320) with an acquisition time of 1.42 h per run, yielding 120um^3 resolution. In *ex vivo* images at 120 μm^3 (Figure 7.9B), distinct cortical lamina emerge and as a result the capability to delineate different cortical areas, entorhinal, parasubiculum and presubiculum, subiculum (Augustinack et al. 2005). The high cell density layers result in light/white contrast in the FLASH images. For example, the granule cells of the dentate gyrus, the hippocampal pyramidal layer, the principal lamina externa (layer II) of the presubiculum and the entorhinal layer II islands show light contrast in Figure 7.9B. The granule cells of the dentate gyrus reveal the brightest contrast in the MTL likely due to the tight packing density of neurons in this layer. In this image, the dentate granule cells form a sharp white 'C' in the middle panel. The pyramidal layer of hippocampus displays a more homogeneous light contrast where the pyramidal layer width narrows as the hippocampus proper transitions from subiculum/CA1 to CA3. The alternating dark and light bands of contrast in presubiculum and entorhinal cortex illustrate the cell dense collections of the principal lamina externa of presubiculum and the entorhinal layer II islands (Brodmann's area 28) respectively in these temporal mesocortical areas. While the contrast of closely packed cells hippocampus and associated cortical structures are represented as light contrast in *ex vivo* MRI, they are represented as darkly stained contrast in histological section (Figure 7.9C). The 120-μm^3 *ex vivo* MRI mirrors that in the histological section (original section thickness, 50 μm) (Figures 7.9B and 7.9C) but the cell dense structures denote dark contrast in the Nissl substance. In addition to the cortical lamina, high resolution *ex vivo* MRI highlights the sublamina of the hippocampus including the alveus, stratum pyramidal, and stratum radiatum/stratum lacunosum/stratum moleculare (i.e. the molecular zone). Although at this MRI resolution, distinguishing the alveus from the stratum oriens or the stratum radiatum from the stratum lacunosum/moleculare is impractical and lacks the level of detail needed to evaluate fibre bundle differences. Precise registration with histological specimens is necessary for this level of anatomical specification. Further histological assessment is needed to validate the functional CA fields of the hippocampus. However, in the dentate gyrus, the shape and simplified trilamina structure (polymorphic layer, granule cells of dentate and molecular layer) help discriminate the organization of the dentate gyrus in MRI.

Ultimately, the value of *ex vivo* MRI will be vastly increased by linking it with *in vivo* MRI (Figure 7.2). A very recent example that illustrates the value of *in vivo–ex vivo* datasets has been demonstrated with the famous amnesic

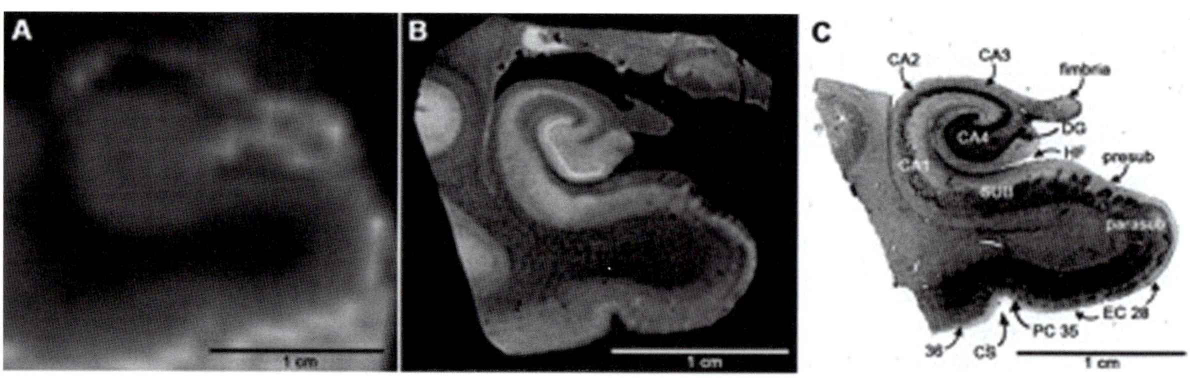

Figure 7.9 *Ex vivo* MRI and histology of the human hippocampus. *Ex vivo* MRI acquired at 1.5 T (A) and 7.0 T (B) with fast low angle shot sequences and corresponding Nissl preparation for this case (C). Note bright dentate gyrus granule cells in (B) and the same lamina is dark in (C).

H.M. Several *in vivo* MRI datasets were acquired with H.M. that clarified and updated our existing knowledge regarding the lesion size, remaining structures, and changes due to ageing (Corkin et al. 1997; Salat et al. 2006). Soon after H.M.'s death in 2008, his brain was scanned *in situ* and *ex vivo* at very high resolution at Massachusetts General Hospital (Van Der Kouwe et al. 2010). After MRI scanning was concluded, H.M.'s brain was sectioned at the Brain Observatory in San Diego, where histological evaluation for H.M. is underway (Annese, 2010). Correlation between *in vivo* and *ex vivo* imaging or *ex vivo* imaging and histochemistry will yield opportunities to connect diverse fields of neuroscience between the macroscopic structural and microscopic cytoarchitectural features and the vast behavioural data that exists on H.M.

As the postmortem hippocampal formation begins to be imaged at ultra-high field strengths (Fatterpekar et al. 2002; Augustinack et al. 2005; Chakeres et al. 2005; Yushkevich et al. 2009), the large gaps that currently exist between *in vivo* and *ex vivo* data will be reduced. This will undoubtedly lead to increased clarity in our interpretations of the cellular underpinnings of *in vivo* measurements of hippocampal structure, function, connectivity, and molecular composition, as well as pathology.

Technical advances and future directions

In the past 20 years, technology has advanced substantially in MRI data acquisition and analysis. It is now commonplace to obtain 3D MRI data with high signal-to-noise properties at resolutions of 1 mm^3 or better. Specialized techniques using higher field strengths and other advances in data acquisition can obtain data at much higher resolution; several recent studies have reported on the visualization of fine structure within the hippocampus that may enable subregions to be relatively directly measured (Mueller et al. 2007; Yushkevich et al. 2009; Kerchner et al. 2010; Prudent et al. 2010). Accompanying these advances in data acquisition, there has been an explosion in the development of analytic techniques drawn from a variety of branches of computer science (Ashburner et al. 2003; Dickerson 2009). Because of its unique topographic properties, a number of these techniques have been applied to analyses of the human hippocampal formation, and have begun to provide valuable insights regarding, among other topics, normal anatomy and function of the human hippocampal formation and MTL, and the more specific identification and quantification of disease effects on this system so critical for episodic memory and other functions. It is exciting to imagine the new discoveries that will emerge as our capacities continue to become more sophisticated in neuroimaging of the hippocampus and related brain structures.

References

Alsop, D. C., Casement, M., de Bazelaire, C., Fong, T., and Press, D. Z. (2008). Hippocampal hyperperfusion in Alzheimer's disease. *Neuroimage*, 42, 1267–74.

Amunts, K., Schleicher, A., and Zilles, K. (2007). Cytoarchitecture of the cerebral cortex – more than localization. *Neuroimage*, 37, 1061–5; discussion 1066–8.

Amunts, K., Kedo, O., Kindler, M., *et al.* (2005). Cytoarchitectonic mapping of the human amygdala, hippocampal region and entorhinal cortex: intersubject variability and probability maps. *Anat Embryol (Berl)*, 210, 343–52.

Annese, J. (2010). Deconstructing Henry. The neuroanatomy and neuroinformatics of the brain of the amnesic patient H.M. In *Society for Neuroscience Abstract*. San Diego, CA: SFN.

Ashburner, J., Csernansky, J. G., Davatzikos, C., Fox, N. C., Frisoni, G. B., and Thompson, P. M. (2003). Computer-assisted imaging to assess brain structure in healthy and diseased brains. *Lancet Neurol*, 2, 79–88.

Augustinack, J. C., Helmer, K., Huber, K. E., Kakunoori, S., Zollei, L., and Fischl, B. (2010). Direct visualization of the perforant pathway in the human brain with ex vivo diffusion tensor imaging. *Front Hum Neurosci*, 4, 42.

Augustinack, J. C., van der Kouwe, A. J., Blackwell, M. L., *et al.* (2005). Detection of entorhinal layer II using 7Tesla [corrected] magnetic resonance imaging. *Ann Neurol*, 57, 489–94.

Bakker, A., Kirwan C. B., Miller, M., and Stark, C. E. (2008). Pattern separation in the human hippocampal CA3 and dentate gyrus. *Science*, 319, 1640–2.

Barnes, J., Ourselin, S., Fox, N. C. (2009a). Clinical application of measurement of hippocampal atrophy in degenerative dementias. *Hippocampus*, 19, 510–16.

Barnes, J., Bartlett, J. W., van de Pol, L. A., *et al.* (2009b). A meta-analysis of hippocampal atrophy rates in Alzheimer's disease. *Neurobiol Aging*, 30, 1711–23.

Barnes, J., Ridgway, G. R., Bartlett, J., *et al.* (2010). Head size, age and gender adjustment in MRI studies: a necessary nuisance? *Neuroimage*, 53, 1244–55.

Beason-Held, L. L., Kraut, M. A., and Resnick, S. M. (2008). I. Longitudinal changes in aging brain function. *Neurobiol Aging*, 29, 483–96.

Bentley, P., Vuilleumier, P., Thiel, C. M., Driver, J., and Dolan, R. J. (2003). Effects of attention and emotion on repetition priming and their modulation by cholinergic enhancement. *J Neurophysiol*, 90, 1171–81.

Bergouignan, L., Chupin, M., Czechowska, Y., *et al.* (2009). Can voxel based morphometry, manual segmentation and automated segmentation equally detect hippocampal volume differences in acute depression? *Neuroimage*, 45, 29–37.

Bertolino, A., Caforio, G., Blasi, G., *et al.* (2004). Interaction of COMT (Val(108/158)Met) genotype and olanzapine treatment on prefrontal cortical function in patients with schizophrenia. *Am J Psychiatry*, 161, 1798–805.

Blaizot, X., Mansilla, F., Insausti, A. M., *et al.* (2010). The human parahippocampal region: I. temporal pole cytoarchitectonic and MRI correlation. *Cereb Cortex*, 20, 2198–212.

Blamire, A. M., Rowe J. G., Styles, P., and McDonald, B. (1999). Optimising imaging parameters for post mortem MR imaging of the human brain. *Acta Radiol*, 40, 593–7.

Block, W., Traber, F., von Widdern, O., *et al.* (2009). Proton MR spectroscopy of the hippocampus at 3 T in patients with unipolar major depressive disorder: correlates and predictors of treatment response. Int J *Neuropsychopharmacol*, 12, 415–22.

Bo, L., Geurts, J. J., Ravid, R., and Barkhof, F. (2004). Magnetic resonance imaging as a tool to examine the neuropathology of multiple sclerosis. *Neuropathol Appl Neurobiol*, 30, 106–17.

Bobinski, M., de Leon, M. J., Wegiel, J., *et al.* (2000). The histological validation of post mortem magnetic resonance imaging-determined hippocampal volume in Alzheimer's disease. *Neuroscience*, 95, 721–5.

Boccardi, M., Ganzola, R., Rossi, R., *et al.* (2010). Abnormal hippocampal shape in offenders with psychopathy. *Hum Brain Mapp*, 31, 438–47.

Bondi, M. W., Houston, W. S., Eyler, L. T., and Brown, G. G. (2005). fMRI evidence of compensatory mechanisms in older adults at genetic risk for Alzheimer disease. *Neurology*, 64, 501–8.

Bookheimer, S. Y., Strojwas, M. H., Cohen, M. S., *et al.* (2000). Patterns of brain activation in people at risk for Alzheimer's disease. *N Engl J Med*, 343, 450–6.

Boyko, O. B., Alston, S. R., Fuller, G. N., Hulette, C. M., Johnson, G. A., and Burger, P. C. (1994). Utility of postmortem magnetic resonance imaging in clinical neuropathology. *Arch Pathol Lab Med*, 118, 219–25.

Braffman, B. H., Zimmerman, R. A., Trojanowski, J. Q., Gonatas, N. K., Hickey, W.F., and Schlaepfer, W. W. (1988). Brain MR: pathologic correlation with gross and histopathology. 2. Hyperintense white-matter foci in the elderly. *AJ Am J Roentgenol*, 151, 559–66.

Breiter, H. C., Gollub, R. L., Weisskoff, R. M., *et al.* (1997). Acute effects of cocaine on human brain activity and emotion. *Neuron*, 19, 591–611.

Brewer, J.B., Zhao, Z., Desmond, J.E., Glover, G.H., and Gabrieli, J.D. (1998). Making memories: brain activity that predicts how well. visual experience will be remembered. *Science*, 281, 1185–7.

Brickman, A. M., Stern, Y., and Small, S. A. (2011). Hippocampal subregions differentially associate with standardized memory tests. *Hippocampus*, 21, 923–8.

Bronge, L., Bogdanovic, N., and Wahlund, L. O. (2002). Postmortem MRI and histopathology of white matter changes in Alzheimer brains. A quantitative, comparative study. *Dement Geriatr Cogn Disord*, 13, 205–12.

Buckner, R. L., Petersen, S. E., Ojemann, J. G., Miezin, F. M., Squire, L. R., and Raichle, M. E. (1995). Functional anatomical studies of explicit and implicit memory retrieval tasks. *J Neurosci*, 15, 12–29.

Buckner, R.L., Andrews-Hanna, J.R., and Schacter, D.L. (2008). The brain's default network: anatomy, function, and relevance to disease. *Ann N Y Acad Sci*, 1124, 1–38.

Burock, M.A., Buckner, R.L., Woldorff, M.G., Rosen, B.R., and Dale, A.M. (1998). Randomized event-related experimental designs allow for extremely rapid presentation rates using functional MRI. *Neuroreport*, 9, 3735–9.

Cabeza, R., Anderson, N. D., Locantore, J. K., and McIntosh, A. R. (2002). Aging gracefully: compensatory brain activity in high-performing older adults. *Neuroimage*, 17, 1394–402.

Callicott, J. H., Mattay, V. S., Verchinski, B. A., Marenco, S., Egan, M. F., and Weinberger, D. R. (2003). Complexity of prefrontal cortical dysfunction in schizophrenia: more than up or down. *Am J Psychiatry*, 160, 2209–15.

Carey, J. R., Kimberley, T. J., Lewis, S. M., *et al.* (2002). Analysis of fMRI and finger tracking training in subjects with chronic stroke. *Brain*, 125, 773–88.

Carr, V. A., Rissman, J., and Wagner, A. D. (2010). Imaging the human medial temporal lobe with high-resolution fMRI. *Neuron*, 65, 298–308.

Caspers, S., Geyer, S., Schleicher, A., Mohlberg, H., Amunts, K., and Zilles, K. (2006). The human inferior parietal cortex: cytoarchitectonic parcellation and interindividual variability. *Neuroimage*, 33, 430–48.

Chakeres, D. W., Whitaker, C. D., Dashner, R. A., *et al.* (2005). High-resolution 8 Tesla imaging of the formalin-fixed normal human hippocampus. *Clin Anat*, 18, 88–91.

Churchill, J. D., Galvez, R., Colcombe, S., Swain, R. A., Kramer, A. F., and Greenough, W. T. (2002). Exercise, experience and the aging brain. *Neurobiol Aging*, 23, 941–55.

Cole, J., Toga, A. W., and Hojatkashani, C. (2010). Subregional hippocampal deformations in major depressive disorder. *J Affect Disord*, 126, 272–7.

Corkin, S., Amaral, D. G., Gonzalez, R. G., Johnson, K. A., and Hyman, B. T. (1997). H. M.'s medial temporal lobe lesion: findings from magnetic resonance imaging. *J Neurosci*, 17, 3964–79.

Coull, J. T., Nobre, A. C., and Frith, C. D. (2001). The noradrenergic alpha2 agonist clonidine modulates behavioural and neuroanatomical correlates of human attentional orienting and alerting. *Cereb Cortex*, 11, 73–84.

Csernansky, J.G., Wang, L., Jones, D., *et al.* (2002). Hippocampal deformities in schizophrenia characterized by high dimensional brain mapping. *Am J Psychiatry*, 159, 2000–6.

Csernansky, J. G., Wang, L., Joshi, S., *et al.* (2000). Early DAT is distinguished from aging by high-dimensional mapping of the hippocampus. Dementia of the Alzheimer type. *Neurology*, 55, 1636–43.

Dager, S. R., Wang, L., Friedman, S. D., *et al.* (2007). Shape mapping of the hippocampus in young children with autism spectrum disorder. *AJNR Am J Neuroradiol*, 28, 672–7.

Davachi, L., Mitchell, J.P., and Wagner, A.D. (2003). Multiple routes to memory: distinct medial temporal lobe processes build item and source memories. *Proc Natl Acad Sci U S A*, 100, 2157–62.

Davidson R. J., Irwin, W., Anderle, M. J., and Kalin, N. H. (2003). The neural substrates of affective processing in depressed patients treated with venlafaxine. *Am J Psychiatry*, 160, 64–75.

Dawe, R. J., Bennett, D. A., Schneider, J. A., Vasireddi, S. K., and Arfanakis, K. (2009). Postmortem MRI of human brain hemispheres: T2 relaxation times during formaldehyde fixation. *Magn Reson Med*, 61, 810–18.

Deng, W., Aimone, J. B., and Gage, F. H. (2010). New neurons and new memories: how does adult hippocampal neurogenesis affect learning and memory? *Nat Rev Neurosci*, 11, 339–50.

Desmond, J. E., Chen, S. H., DeRosa, E., Pryor, M. R., Pfefferbaum, A., and Sullivan, E. V. (2003). Increased frontocerebellar activation in alcoholics during verbal working memory: an fMRI study. *Neuroimage*, 19, 1510–20.

Dickerson, B. C. (2009). New frontiers in computational analysis of human hippocampal anatomy. *Hippocampus*, 19, 507–9.

Dickerson, B. C., and Sperling, R. A. (2008). Functional abnormalities of the medial temporal lobe memory system in mild cognitive impairment and Alzheimer's disease: insights from functional MRI studies. *Neuropsychologia*, 46, 1624–35.

Dickerson, B. C., Salat, D. H., Bates, J. F., *et al.* (2004). Medial temporal lobe function and structure in mild cognitive impairment. *Ann Neurol*, 56, 27–35.

Dickerson, B. C., Salat, D. H., Greve, D. N., *et al.* (2005). Increased hippocampal activation in mild cognitive impairment compared to normal aging and AD. *Neurology*, 65, 404–11.

Dickerson, B.C., Miller, S.L., Greve, D.N., Dale, A.M., Albert, M.S., Schacter, D.L., *et al.* (2007). Prefrontal-hippocampal-fusiform activity during encoding predicts intraindividual differences in free recall ability: an event-related functional-anatomic MRI study. *Hippocampus*, 17, 1060–70.

Dickerson, B. C., Feczko, E., Augustinack, J. C., *et al.* (2009). Differential effects of aging and Alzheimer's disease on medial temporal lobe cortical thickness and surface area. *Neurobiol Aging*, 30, 432–40.

Drummond, S. P., Brown, G. G., Gillin, J. C., Stricker, J. L., Wong, E. C., and Buxton, R. B. (2000). Altered brain response to verbal learning following sleep deprivation. *Nature*, 403, 655–7.

Duvernoy, H. M. (2005). *The human hippocampus* (Third edn.). Berlin: Springer-Verlag.

Dusek, J.A. and Eichenbaum, H. (1997). The hippocampus and memory for orderly stimulus relations. *Proc Natl Acad Sci U S A*, 94, 7109–14.

Dyrby, T. B., Baare, W. F., Alexander, D. C., Jelsing, J., Garde, E., and Sogaard, L. V. (2011). An ex vivo imaging pipeline for producing high-quality and high-resolution diffusion-weighted imaging datasets. *Hum Brain Mapp*, 32(4), 544–63.

Egan, M. F., Kojima, M., Callicott, J. H., *et al.* (2003). The BDNF val66met polymorphism affects activity-dependent secretion of BDNF and human memory and hippocampal function. *Cell*, 112, 257–69.

Eichenbaum, H., Yonelinas, A. P., and Ranganath, C. (2007). The medial temporal lobe and recognition memory. *Annu Rev Neurosci*, 30, 123–52.

Eickhoff, S., Walters, N. B., Schleicher, A., *et al.* (2005). High-resolution MRI reflects myeloarchitecture and cytoarchitecture of human cerebral cortex. *Hum Brain Mapp*, 24, 206–15.

Eickhoff, S. B., Schleicher, A., Zilles, K., and Amunts, K. (2006). The human parietal operculum. I. Cytoarchitectonic mapping of subdivisions. *Cereb Cortex*, 16, 254–67.

Ernst, T., Chang, L., Jovicich, J., Ames, N., and Arnold, S. (2002). Abnormal brain activation on functional MRI in cognitively asymptomatic HIV patients. *Neurology*, 59, 1343–9.

Everall, I. P., Chong, W. K., Wilkinson, I. D., *et al.* (1997). Correlation of MRI and neuropathology in AIDS. *J Neurol Neurosurg Psychiatry*, 62, 92–5.

Fatterpekar, G. M., Naidich, T. P., Delman, B. N., *et al.* (2002). Cytoarchitecture of the human cerebral cortex: MR microscopy of excised specimens at 9.4 Tesla. *AJNR Am J Neuroradiol*, 23, 1313–21.

Feczko, E., Augustinack, J. C., Fischl, B., and Dickerson, B. C. (2009). An MRI-based method for measuring volume, thickness and surface area of entorhinal, perirhinal, and posterior parahippocampal cortex. *Neurobiol Aging*, 30, 420–31.

Fischl, B., Rajendran, N., Busa, E., *et al.* (2008). Cortical folding patterns and predicting cytoarchitecture. *Cereb Cortex*, 18, 1973–80.

Fischl, B., Stevens, A. A., Rajendran, N., *et al.* (2009). Predicting the location of entorhinal cortex from MRI. *Neuroimage*, 47, 8–17.

Fleisher, A. S., Houston, W. S., Eyler, L. T., *et al.* (2005). Identification of Alzheimer disease risk by functional magnetic resonance imaging. *Arch Neurol*, 62, 1881–8.

Fox, N. C., Black, R. S., Gilman, S., *et al.* (2005). Effects of Abeta immunization (AN1792) on MRI measures of cerebral volume in Alzheimer disease. *Neurology*, 64, 1563–72.

Frisoni, G. B., Fox, N. C., Jack, C. R., Jr., Scheltens, P., and Thompson, P. M. (2010). The clinical use of structural MRI in Alzheimer disease. *Nat Rev Neurol*, 6, 67–77.

Friston, K. J. (1999). Scanning patients with tasks they can perform. *Hum Brain Mapp*, 8, 102–8.

Fu, C. H., Williams, S. C., Cleare, A. J., *et al.* (2004). Attenuation of the neural response to sad faces in major depression by antidepressant treatment: a prospective, event-related functional magnetic resonance imaging study. *Arch Gen Psychiatry*, 61, 877–89.

Gabrieli, J.D., Brewer, J.B., Desmond, J.E., and Glover, G.H. (1997). Separate neural bases of two fundamental memory processes in the human medial temporal lobe. *Science*, 276, 264–6.

Galton, C. J., Gomez-Anson, B., Antoun, N., *et al.* (2001). Temporal lobe rating scale: application to Alzheimer's disease and frontotemporal dementia. *J Neurol Neurosurg Psychiatry*, 70, 165–73.

Geuze, E., Vermetten, E., and Bremner, J. D. (2005a). MR-based in vivo hippocampal volumetrics: 2. Findings in neuropsychiatric disorders. *Mol Psychiatry*, 10, 160–84.

Geuze, E., Vermetten, E., and Bremner, J. D. (2005b). MR-based in vivo hippocampal volumetrics: 1. Review of methodologies currently employed. *Mol Psychiatry*, 10, 147–59.

Gibbs, S. E., and D'Esposito, M. (2005). Individual capacity differences predict working memory performance and prefrontal activity following dopamine receptor stimulation. *Cogn Affect Behav Neurosci*, 5, 212–21.

Goekoop, R., Barkhof, F., Duschek, E. J., *et al.* (2006). Raloxifene treatment enhances brain activation during recognition of familiar items: a pharmacological fMRI study in healthy elderly males. *Neuropsychopharmacology*, 31, 1508–18.

Goldstein, J. M., Jerram, M., Poldrack, R., *et al.* (2005). Hormonal cycle modulates arousal circuitry in women using functional magnetic resonance imaging. *J Neurosci*, 25, 9309–16.

Grady, C. L., McIntosh, A. R., Horwitz, B., *et al.* (1995). Age-related reductions in human recognition memory due to impaired encoding. *Science*, 269, 218–21.

Grafe, M. R., Press, G. A., Berthoty, D. P., Hesselink, J. R., and Wiley, C. A. (1990). Abnormalities of the brain in AIDS patients: correlation of postmortem MR findings with neuropathology. *AJNR Am J Neuroradiol*, 11, 905–11; discussion 912–13.

Grafton, S. T., Sumi, S. M., Stimac, G. K., Alvord, E. C., Jr., Shaw, C. M., and Nochlin, D. (1991). Comparison of postmortem magnetic resonance imaging and neuropathologic findings in the cerebral white matter. *Arch Neurol*, 48, 293–8.

Hämäläinen, A., Pihlajamaki, M., Tanila, H., *et al.* (2007). Increased fMRI responses during encoding in mild cognitive impairment. *Neurobiol Aging*, 28, 1889–903.

Hampel, H., Frank, R., Broich, K., *et al.* (2010). Biomarkers for Alzheimer's disease: academic, industry and regulatory perspectives. *Nat Rev Drug Discov*, 9, 560–74.

Heckers, S., Zalesak, M., Weiss, A.P., Ditman, T., and Titone, D. (2004). Hippocampal activation during transitive inference in humans. *Hippocampus*, 14, 153–62.

Heo, S., Prakash, R. S., Voss, M. W., *et al.* (2010). Resting hippocampal blood flow, spatial memory and aging. *Brain Res*, 1315, 119–27.

Hinds, O. P., Rajendran, N., Polimeni, J. R., *et al.* (2008). Accurate prediction of V1 location from cortical folds in a surface coordinate system. *Neuroimage*, 39, 1585–99.

Hogan R. E., Wang L., Bertrand M. E., *et al.* (2004). MRI-based high-dimensional hippocampal mapping in mesial temporal lobe epilepsy. *Brain*, 127, 1731–40.

Hogan, R. E., Bouilleret, V., Liu, Y. R., *et al.* (2009). MRI-based large deformation high dimensional mapping of the hippocampus in rats: development and validation of the technique. *J Magn Reson Imaging*, 29, 1027–34.

Honey, G., and Bullmore, E. (2004). Human pharmacological MRI. *Trends Pharmacol Sci*, 25, 366–74.

Honey, G. D., Bullmore, E. T., Soni, W., Varatheesan, M., Williams, S. C., and Sharma, T. (1999). Differences in frontal cortical activation by a working memory task after substitution of risperidone for typical antipsychotic drugs in patients with schizophrenia. *Proc Natl Acad Sci U S A*, 96, 13432–7.

Huang, S., Liu, C., Dai, G., Kim, Y. R., and Rosen, B. R. (2009). Manipulation of tissue contrast using contrast agents for enhanced MR microscopy in ex vivo mouse brain. *Neuroimage*, 46, 589–99.

Huang, Y., Chen, W., Li, Y., Wu, X., Shi, X., and Geng, D. (2010). Effects of antidepressant treatment on N-acetyl aspartate and choline levels in the hippocampus and thalami of post-stroke depression patients: a study using (1)H magnetic resonance spectroscopy. *Psychiatry Res*, 182, 48–52.

Jack, C. R., Jr., Mokri, B., Laws, E. R., Jr., Houser, O. W., Baker, H. L., Jr., and Petersen, R. C. (1987). MR findings in normal-pressure hydrocephalus: significance and comparison with other forms of dementia. *J Comput Assist Tomogr*, 11, 923–31.

Jack, C. R., Jr., Sharbrough, F. W., and Marsh, W. R. (1988a). Use of MR imaging for quantitative evaluation of resection for temporal lobe epilepsy. *Radiology*, 169, 463–8.

Jack, C. R., Jr., Nichols, D. A., Sharbrough, F. W., Marsh, W. R., and Petersen, R. C. (1988b). Selective posterior cerebral artery Amytal test for evaluating memory function before surgery for temporal lobe seizure. *Radiology*, 168, 787–93.

Jack, C. R., Jr., Gehring, D. G., Sharbrough, F. W., *et al.* (1988c). Temporal lobe volume measurement from MR images: accuracy and left-right asymmetry in normal persons. *J Comput Assist Tomogr*, 12, 21–9.

Jack, C. R., Jr., Twomey, C. K., Zinsmeister, A. R., Sharbrough, F. W., Petersen, R. C., and Cascino, G. D. (1989). Anterior temporal lobes and hippocampal formations: normative volumetric measurements from MR images in young adults. *Radiology*, 172, 549–54.

Jack, C. R., Jr., Sharbrough, F. W., Twomey, C. K., *et al.* (1990). Temporal lobe seizures: lateralization with MR volume measurements of the hippocampal formation. *Radiology*, 175, 423–9.

Jack, C. R., Jr., Petersen, R. C., O'Brien, P. C., and Tangalos, E. G. (1992). MR-based hippocampal volumetry in the diagnosis of Alzheimer's disease. *Neurology*, 42, 183–8.

Jack, C. R., Jr., Slomkowski, M., Gracon, S., *et al.* (2003). MRI as a biomarker of disease progression in a therapeutic trial of milameline for AD. *Neurology*, 60, 253–60.

Jack, C. R., Jr., Garwood, M., Wengenack, T. M., *et al.* (2004). In vivo visualization of Alzheimer's amyloid plaques by magnetic resonance imaging in transgenic mice without a contrast agent. *Magn Reson Med*, 52, 1263–71.

Jack, C. R., Jr., Bernstein, M. A., Fox, N. C., *et al.* (2008). The Alzheimer's Disease Neuroimaging Initiative (ADNI): MRI methods. *J Magn Reson Imaging*, 27, 685–91.

Jagust, W. J., Zheng, L., Harvey, D. J., *et al.* (2008). Neuropathological basis of magnetic resonance images in aging and dementia. *Ann Neurol*, 63, 72–80.

Jessen, F., Block, W., Traber, F., *et al.* (2000). Proton MR spectroscopy detects a relative decrease of N-acetylaspartate in the medial temporal lobe of patients with AD. *Neurology*, 55, 684–8.

Johansen-Berg, H., Dawes, H., Guy, C., Smith, S. M., Wade, D. T., and Matthews, P. M. (2002). Correlation between motor improvements and altered fMRI activity after rehabilitative therapy. *Brain*, 125, 2731–42.

Johnson, S. C., Schmitz, T. W., Moritz, C. H., *et al.* (2006). Activation of brain regions vulnerable to Alzheimer's disease: the effect of mild cognitive impairment. *Neurobiol Aging*, 27, 1604–12.

Kahn, I., Davachi, L., and Wagner, A.D., (2004). Functional-neuroanatomic correlates of recollection: implications for models of recognition memory. *J Neurosci*, 24, 4172–80.

Kalin, N. H., Davidson, R. J., Irwin, W., *et al.* (1997). Functional magnetic resonance imaging studies of emotional processing in normal and depressed patients: effects of venlafaxine. *J Clin Psychiatry*, 58(Suppl 16), 32–9.

Kalpouzos, G., Chetelat, G., Baron, J. C., *et al.* (2009). Voxel-based mapping of brain gray matter volume and glucose metabolism profiles in normal aging. *Neurobiol Aging*, 30, 112–24.

Kantarci, K., Jack, C. R., Jr., Xu, Y. C., *et al.* (2000). Regional metabolic patterns in mild cognitive impairment and Alzheimer's disease: A 1H MRS study. *Neurology*, 55, 210–17.

Kato, T., Knopman, D., and Liu, H. (2001). Dissociation of regional activation in mild AD during visual encoding: a functional MRI study. *Neurology*, 57, 812–16.

Kerchner, G. A., Hess, C. P., Hammond-Rosenbluth, K. E., *et al.* (2010). Hippocampal CA1 apical neuropil atrophy in mild Alzheimer disease visualized with 7-T MRI. *Neurology*, 75, 1381–7.

Kircher, T., Weis, S., Freymann, K., *et al.* (2007). Hippocampal activation in MCI patients is necessary for successful memory encoding. *J Neurol Neurosurg Psychiatry*, 78, 812–18.

Klintsova, A. Y., and Greenough, W. T. (1999). Synaptic plasticity in cortical systems. *Curr Opin Neurobiol*, 9, 203–8.

Konrad, C., Ukas, T., Nebel, C., Arolt, V., Toga, A. W., and Narr, K.L. (2009). Defining the human hippocampus in cerebral magnetic resonance images—an overview of current segmentation protocols. *Neuroimage*, 47, 1185–95.

Leslie, R. A., and James, M. F. (2000). Pharmacological magnetic resonance imaging: a new application for functional MRI. *Trends Pharmacol Sci*, 21, 314–18.

Lorente de No, R. (1934). Studies on the structure of the cerebral cortex II Continuation of the study of ammonic system. *Journal Fur Psychologie and Neurologie*, 46, 113–77.

Lovden, M., Schaefer, S., Noack, H., *et al.* (2011). Performance-related increases in hippocampal N-acetylaspartate (NAA) induced by spatial navigation training are restricted to BDNF Val homozygotes. *Cereb Cortex*, 21, 1435–42.

Luckhaus, C., Cohnen, M., Flüss, M. O., *et al.* (2010). The relation of regional cerebral perfusion and atrophy in mild cognitive impairment (MCI) and early Alzheimer's dementia. *Psychiatry Res*, 183(1), 44–51.

Machielsen, W. C., Rombouts, S. A., Barkhof, F., Scheltens, P., and Witter, M. P. (2000). FMRI of visual encoding: reproducibility of activation. *Hum Brain Mapp*, 9, 156–64.

Machulda, M. M., Ward, H. A., Borowski, B., *et al.* (2003). Comparison of memory fMRI response among normal, MCI, and Alzheimer's patients. *Neurology*, 61, 500–6.

Maguire, E. A., Gadian, D. G., Johnsrude, I. S., *et al.* (2000). Navigation-related structural change in the hippocampi of taxi drivers. *Proc Natl Acad Sci U S A*, 97, 4398–403.

Maguire, E. A., Woollett, K., Spiers, H. J. (2006). London taxi drivers and bus drivers: a structural MRI and neuropsychological analysis. *Hippocampus*, 16, 1091–101.

Maier, M., Mellers, J., Toone, B., Trimble, M., Mattay, V. S., Callicott, J. H., Bertolino, A., *et al.* (2000). Effects of dextroamphetamine on cognitive performance and cortical activation. *Neuroimage*, 12, 268–75.

Maier, M., Mellers, J., Toone, B., Trimble, M., and Ron, M. A. (2000). Schizophrenia, temporal lobe epilepsy and psychosis: an in vivo magnetic resonance spectroscopy and imaging study of the hippocampus/amygdala complex. *Psychol Med*, 30, 571–81.

Mesulam, M.M. (1998). From sensation to cognition. *Brain*, 121 (Pt 6), 1013–52.

Maguire, E.A. (1997). Hippocampal involvement in human topographical memory: evidence from functional imaging. *Philos Trans R Soc Lond B Biol Sci*, 352, 1475–80.

Manganas, L. N., Zhang, X., Li, Y., *et al.* (2007). Magnetic resonance spectroscopy identifies neural progenitor cells in the live human brain. *Science*, 318, 980–5.

Manoach, D. S., Halpern, E. F., Kramer, T. S., *et al.* (2001). Test-retest reliability of a functional MRI working memory paradigm in normal and schizophrenic subjects. *Am J Psychiatry*, 158, 955–8.

Mattay, V. S., Callicott, J. H., Bertolino, A., *et al.* (2000). Effects of dextroamphetamine on cognitive performance and cortical activation. *Neuroimage*, 12, 268–75.

Mattay, V. S., Tessitore, A., Callicott, J. H., *et al.* (2002). Dopaminergic modulation of cortical function in patients with Parkinson's disease. *Ann Neurol*, 51, 156–64.

McAllister, T. W., Saykin, A. J., Flashman, L. A., *et al.* (1999). Brain activation during working memory 1 month after mild traumatic brain injury: a functional MRI study. *Neurology*, 53, 1300–8.

Miller, M. I., Beg, M. F., Ceritoglu, C., and Stark, C. (2005). Increasing the power of functional maps of the medial temporal lobe by using large deformation diffeomorphic metric mapping. *Proc Natl Acad Sci U S A*, 102, 9685–90.

Miller, M. I., Priebe, C. E., Qiu, A., *et al.* (2009). Collaborative computational anatomy: an MRI morphometry study of the human brain via diffeomorphic metric mapping. *Hum Brain Mapp*, 30, 2132–41.

Miller, S. L., Fenstermacher, E., Bates, J., Blacker, D., Sperling, R. A., and Dickerson, B. C. (2008). Hippocampal activation in adults with mild cognitive impairment predicts subsequent cognitive decline. *J Neurol Neurosurg Psychiatry*, 79, 630–5.

Monchi, O., Petrides, M., Doyon, J., Postuma, R. B., Worsley, K., and Dagher, A. (2004). Neural bases of set-shifting deficits in Parkinson's disease. *J Neurosci*, 24, 702–10.

Morgen, K., Kadom, N., Sawaki, L., *et al.* (2004). Training-dependent plasticity in patients with multiple sclerosis. *Brain*, 127, 2506–17.

Morra, J. H., Tu, Z., Apostolova, L. G., *et al.* (2008). Validation of a fully automated 3D hippocampal segmentation method using subjects with Alzheimer's disease mild cognitive impairment, and elderly controls. *Neuroimage*, 43, 59–68.

Mosconi, L., Brys, M., Switalski, R., *et al.* (2007). Maternal family history of Alzheimer's disease predisposes to reduced brain glucose metabolism. *Proc Natl Acad Sci U S A*, 104, 19067–72.

Mueller, S. G., and Weiner, M. W. (2009). Selective effect of age, Apo e4, and Alzheimer's disease on hippocampal subfields. *Hippocampus*, 19, 558–64.

Mueller, S. G., Schuff, N., Raptentsetsang, S., Elman, J., and Weiner, M. W. (2008). Selective effect of Apo e4 on CA3 and dentate in normal aging and Alzheimer's disease using high resolution MRI at 4 T. *Neuroimage*, 42, 42–8.

Mueller, S. G., Stables, L., Du, A. T., *et al.* (2007). Measurement of hippocampal subfields and age-related changes with high resolution MRI at 4T. *Neurobiol Aging*, 28, 719–26.

Nagara, H., Inoue, T., Koga, T., Kitaguchi, T., Tateishi, J., and Goto, I. (1987). Formalin fixed brains are useful for magnetic resonance imaging (MRI). study. *J Neurol Sci*, 81, 67–77.

Naidich, T. P., Daniels, D. L., Haughton, V. M., Williams, A., Pojunas, K., and Palacios, E. (1987a). Hippocampal formation and related structures of the limbic lobe: anatomic-MR correlation. Part I. Surface features and coronal sections. *Radiology*, 162, 747–54.

Naidich, T. P., Daniels, D. L., Haughton, V. M., *et al.* (1987b). Hippocampal formation and related structures of the limbic lobe: anatomic-MR correlation. Part II. Sagittal sections. *Radiology*, 162, 755–61.

Neylan, T. C., Mueller, S. G., Wang, Z., *et al.* (2010). Insomnia severity is associated with a decreased volume of the CA3/dentate gyrus hippocampal subfield. *Biol Psychiatry*, 68, 494–6.

Nicolson, R., DeVito, T. J., Vidal, C. N., *et al.* (2006). Detection and mapping of hippocampal abnormalities in autism. *Psychiatry Res*, 148, 11–21.

O'Brien, J. L., O'Keefe, K. M., LaViolette, P. S., *et al.* (2010). Longitudinal fMRI in elderly reveals loss of hippocampal activation with clinical decline. *Neurology*, 74, 1969–76.

Pajonk, F. G., Wobrock, T., Gruber, O., *et al.* (2010). Hippocampal plasticity in response to exercise in schizophrenia. *Arch Gen Psychiatry*, 67, 133–43.

Peper, J. S., Brouwer, R. M., Boomsma, D. I., Kahn, R. S., and Hulshoff Pol, H. E. (2007). Genetic influences on human brain structure: a review of brain imaging studies in twins. *Hum Brain Mapp*, 28, 464–73.

Pereira, A. C., Huddleston, D. E., Brickman, A. M., *et al.* (2007). An in vivo correlate of exercise-induced neurogenesis in the adult dentate gyrus. *Proc Natl Acad Sci U S A*, 104, 5638–43.

Petroff, O. A., Errante, L. D., Rothman, D. L., Kim, J. H., and Spencer, D. D. (2002). Neuronal and glial metabolite content of the epileptogenic human hippocampus. *Ann Neurol*, 52, 635–42.

Pfefferbaum, A., Sullivan, E. V., Adalsteinsson, E., Garrick, T., and Harper, C. (2004). Postmortem MR imaging of formalin-fixed human brain. *Neuroimage*, 21, 1585–95.

Pieperhoff, P., Sudmeyer, M., Homke, L., Zilles, K., Schnitzler, A., and Amunts, K. (2008). Detection of structural changes of the human brain in longitudinally acquired MR images by deformation field morphometry: methodological analysis, validation and application. *Neuroimage*, 43, 269–87.

Pihlajamaki, M., Tanila, H., Hanninen, T., Kononen, M., Mikkonen, M., Jalkanen, V., *et al.* (2003). Encoding of novel picture pairs activates the perirhinal cortex: an fMRI study. *Hippocampus*, 13, 67–80.

Poduslo, J. F., Wengenack, T. M., Curran, G. L., *et al.* (2002). Molecular targeting of Alzheimer's amyloid plaques for contrast-enhanced magnetic resonance imaging. *Neurobiol Dis*, 11, 315–29.

Posener, J. A., Wang, L., Price, J. L., *et al.* (2003). High-dimensional mapping of the hippocampus in depression. *Am J Psychiatry*, 160, 83–9.

Powell, H. W., Guye, M., Parker, G. J., *et al.* (2004). Noninvasive in vivo demonstration of the connections of the human parahippocampal gyrus. *Neuroimage*, 22, 740–7.

Press G. A., Amaral, D. G., and Squire, L. R. (1989). Hippocampal abnormalities in amnesic patients revealed by high-resolution magnetic resonance imaging. *Nature*, 341, 54–7.

Preston, A.R., Bornstein, A.M., Hutchinson, J.B., Gaare, M.E., Glover, G.H., and Wagner, A.D. (2010). High-resolution fMRI of content-sensitive subsequent memory responses in human medial temporal lobe. *J Cogn Neurosci*, 22, 156–73.

Price, C.J., Friston, K.J., (1999). Scanning patients with tasks they can perform. *Hum Brain Mapp*, 8, 102–8.

Price, C. J., Rombouts, S. A., Barkhof, F., Veltman, D. J., *et al.* (2000). Functional MR imaging in Alzheimer's disease during memory encoding. *AJNR Am J Neuroradiol*, 21, 1869–75.

Prudent, V., Kumar, A., Liu, S., Wiggins, G., Malaspina, D., and Gonen, O. (2010). Human hippocampal subfields in young adults at 7.0 T: feasibility of imaging. *Radiology*, 254, 900–6.

Raichle, M.E., MacLeod, A.M., Snyder, A.Z., Powers, W.J., Gusnard, D.A., and Shulman, G.L. (2001). A default mode of brain function. *Proc Natl Acad Sci U S A*, 98, 676–82.

Reddy, H., Narayanan, S., Arnoutelis, R., *et al.* (2000). Evidence for adaptive functional changes in the cerebral cortex with axonal injury from multiple sclerosis. *Brain*, 123(Pt 11), 2314–20.

Reiman, E. M., Caselli, R. J., Chen, K., Alexander, G. E., Bandy, D., and Frost, J. (2001). Declining brain activity in cognitively normal apolipoprotein E epsilon 4 heterozygotes: A foundation for using positron emission tomography to efficiently test treatments to prevent Alzheimer's disease. *Proc Natl Acad Sci U S A*, 98, 3334–9.

Rombouts, S. A., Barkhof, F., Veltman, D. J., *et al.* (2000). Functional MR imaging in Alzheimer's disease during memory encoding. *AJNR Am J Neuroradiol*, 21, 1869–75.

Rombouts, S. A., Barkhof, F., Van Meel, C. S., and Scheltens, P. (2002). Alterations in brain activation during cholinergic enhancement with rivastigmine in Alzheimer's disease. *J Neurol Neurosurg Psychiatry*, 73, 665–71.

Rosas, H. D., Feigin, A. S., and Hersch, S. M. (2004). Using advances in neuroimaging to detect, understand, and monitor disease progression in Huntington's disease. *NeuroRx*, 1, 263–72.

Rosene, D. L., and Van Hoesen, G. W. (1987). The hippocampal formation of the primate brain, a review of some comparative aspects of cytoarchitecture and connections. In Jones, E.G., and Peters, A. (Eds.) *Cerebral Cortex. Further Aspects of Cortical Function, Including Hippocampus*, pp. 345–456. New York: Plenum Press.

Rusinek, H., Brys, M., Glodzik, L., *et al.* (2011). Hippocampal blood flow in normal aging measured with arterial spin labeling at 3T. *Magn Reson Med*, 65, 128–37.

Salat, D. H., van der Kouwe, A. J., Tuch, D. S., *et al.* (2006). Neuroimaging H.M.: a 10-year follow-up examination. *Hippocampus*, 16, 936–45.

Scarmeas, N., Habeck, C. G., Zarahn, E., *et al.* (2004). Covariance PET patterns in early Alzheimer's disease and subjects with cognitive impairment but no dementia: utility in group discrimination and correlations with functional performance. *Neuroimage*, 23, 35–45.

Schacter, D. L., Alpert, N. M., Savage, C. R., Rauch, S. L., and Albert, M. S. (1996). Conscious recollection and the human hippocampal formation: evidence from positron emission tomography. *Proc Natl Acad Sci U S A*, 93, 321–5.

Schenker, N. M., Hopkins, W. D., Spocter, M. A., *et al.* (2010). Broca's area homologue in chimpanzees (Pan troglodytes): probabilistic mapping, asymmetry, and comparison to humans. *Cereb Cortex*, 20, 730–42.

Scheperjans, F., Hermann, K., Eickhoff, S. B., Amunts, K., Schleicher, A., and Zilles, K. (2008). Observer-independent cytoarchitectonic mapping of the human superior parietal cortex. *Cereb Cortex*, 18, 846–67.

Schleicher, A., Palomero-Gallagher, N., Morosan, P., *et al.* (2005). Quantitative architectural analysis: a new approach to cortical mapping. *Anat Embryol (Berl)*, 210, 373–86.

Schon, K., Atri, A., Hasselmo, M. E., Tricarico, M. D., LoPresti, M. L., and Stern, C. E. (2005). Scopolamine reduces persistent activity related to long-term encoding in the parahippocampal gyrus during delayed matching in humans. *J Neurosci*, 25, 9112–23.

Schuff, N., Amend, D. L., Knowlton, R., Norman, D., Fein, G., and Weiner, M. W. (1999). Age-related metabolite changes and volume loss in the hippocampus by magnetic resonance spectroscopy and imaging. *Neurobiol Aging*, 20, 279–85.

Schuff, N., Amend, D., Ezekiel, F., *et al.* (1997). Changes of hippocampal N-acetyl aspartate and volume in Alzheimer's disease. A proton MR spectroscopic imaging and MRI study. *Neurology*, 49, 1513–21.

Schumann, C. M., Buonocore, M. H., and Amaral, D. G. (2001). Magnetic resonance imaging of the post-mortem autistic brain. *J Autism Dev Disord*, 31, 561–8.

Seab, J. P., Jagust, W. J., Wong, S. T., Roos, M. S., Reed, B. R., and Budinger, T. F. (1988). Quantitative NMR measurements of hippocampal atrophy in Alzheimer's disease. *Magn Reson Med*, 8, 200–8.

Selkoe, D. J. (2002). Alzheimer's disease is a synaptic failure. *Science*, 298, 789–91.

Shallice, T., Fletcher, P., Frith, C. D., Grasby, P., Frackowiak, R. S., and Dolan, R. J. (1994). Brain regions associated with acquisition and retrieval of verbal episodic memory. *Nature*, 368, 633–5.

Shepherd, T. M., Ozarslan, E., Yachnis, A. T., King, M. A., and Blackband, S. J. (2007). Diffusion tensor microscopy indicates the cytoarchitectural basis for diffusion anisotropy in the human hippocampus. *AJNR Am J Neuroradiol*, 28, 958–64.

Shepherd, T. M., Flint, J. J., Thelwall, P. E., *et al.* (2009). Postmortem interval alters the water relaxation and diffusion properties of rat nervous tissue – implications for MRI studies of human autopsy samples. *Neuroimage*, 44, 820–6.

Simons, J.S., and Spiers, H.J. (2003). Prefrontal and medial temporal lobe interactions in long-term memory. *Nat Rev Neurosci*, 44, 637–48.

Small, S. A., Perera, G. M., DeLaPaz, R., Mayeux, R., and Stern, Y. (1999). Differential regional dysfunction of the hippocampal formation among elderly with memory decline and Alzheimer's disease. *Ann Neurol*, 45, 466–72.

Small, S. A., Chawla, M. K., Buonocore, M., Rapp, P. R., and Barnes, C. A. (2004). Imaging correlates of brain function in monkeys and rats isolates a hippocampal subregion differentially vulnerable to aging. *Proc Natl Acad Sci U S A*, 101, 7181–6.

Sperling, R., Greve, D., Dale, A., *et al.* (2002). Functional MRI detection of pharmacologically induced memory impairment. *Proc Natl Acad Sci U S A*, 99, 455–60.

Sperling, R., Chua, E., Cocchiarella, A., *et al.* (2003a). Putting names to faces: successful encoding of associative memories activates the anterior hippocampal formation. *Neuroimage*, 20, 1400–10.

Sperling, R. A., Bates, J. F., Chua, E. F., *et al.* (2003b). fMRI studies of associative encoding in young and elderly controls and mild Alzheimer's disease. *J Neurol Neurosurg Psychiatry*, 74, 44–50.

Sperling, R. A., Laviolette, P. S., O'Keefe, K., *et al.* (2009). Amyloid deposition is associated with impaired default network function in older persons without dementia. *Neuron*, 63, 178–88.

Squire, L. R. (2009). The legacy of patient H.M. for neuroscience. *Neuron*, 61, 6–9.

Squire, L. R., and Zola-Morgan, S. (1991). The medial temporal lobe memory system. *Science*, 253, 1380–6.

Squire, L. R., Ojemann, J. G., Miezin, F. M., Petersen, S. E., Videen, T. O., and Raichle, M. E. (1992). Activation of the hippocampus in normal humans: a functional anatomical study of memory. *Proc Natl Acad Sci U S A*, 89, 1837–41.

Squire, L. R., Stark, C. E., and Clark, R. E. (2004). The medial temporal lobe. *Annu Rev Neurosci*, 27, 279–306.

Squire, L.R., Wixted, J.T., and Clark, R.E. (2007). Recognition memory and the medial temporal lobe: a new perspective. *Nat Rev Neurosci*, 8, 872–83.

Stern, C.E., Corkin, S., Gonzalez, R.G., Guimaraes, A.R., Baker, J.R., Jennings, P.J., *et al.* (1996). The hippocampal formation participates in novel picture encoding: evidence from functional magnetic resonance imaging. *Proc Natl Acad Sci U S A*, 93, 8660–5.

Stern, E. A., Bacskai, B. J., Hickey, G. A., Attenello, F. J., Lombardo, J. A., and Hyman, B. T. (2004). Cortical synaptic integration in vivo is disrupted by amyloid-beta plaques. *J Neurosci*, 24, 4535–40.

Suddath, R. L., Casanova, M. F., Goldberg, T. E., Daniel, D. G., Kelsoe, J. R., Jr., and Weinberger, D. R. (1989). Temporal lobe pathology in schizophrenia: a quantitative magnetic resonance imaging study. *Am J Psychiatry*, 146, 464–72.

Tae, W. S., Joo, E. Y., Kim, J. H., *et al.* (2005). Cerebral perfusion changes in mesial temporal lobe epilepsy: SPM analysis of ictal and interictal SPECT. *Neuroimage*, 24, 101–10.

Tepest, R., Wang, L., Miller, M. I., Falkai, P., and Csernansky, J. G. (2003). Hippocampal deformities in the unaffected siblings of schizophrenia subjects. *Biol Psychiatry*, 54, 1234–40.

Thiel, C. M., Henson, R. N., Morris, J. S., Friston, K. J., and Dolan, R. J. (2001). Pharmacological modulation of behavioral and neuronal correlates of repetition priming. *J Neurosci*, 21, 6846–52.

Thiel, C. M., Bentley, P., and Dolan, R. J. (2002a). Effects of cholinergic enhancement on conditioning-related responses in human auditory cortex. *Eur J Neurosci*, 16, 2199–206.

Thiel, C. M., Friston, K. J., and Dolan, R. J. (2002b). Cholinergic modulation of experience-dependent plasticity in human auditory cortex. *Neuron*, 35, 567–74.

Thiel, C. M., Henson, R. N., and Dolan, R. J. (2002c). Scopolamine but not lorazepam modulates face repetition priming: a psychopharmacological fMRI study. *Neuropsychopharmacology*, 27, 282–92.

Tovi, M., and Ericsson, A. (1992). Measurements of T1 and T2 over time in formalin-fixed human whole-brain specimens. *Acta Radiol*, 33, 400–4.

Tulving, E., Markowitsch, H. J., Kapur, S., Habib, R., and Houle, S. (1994). Novelty encoding networks in the human brain: positron emission tomography data. *Neuroreport*, 5, 2525–8.

Uylings, H. B., Sanz-Arigita, E. J., de Vos, K., Pool, C. W., Evers, P., and Rajkowska, G. (2010). 3-D cytoarchitectonic parcellation of human orbitofrontal cortex correlation with postmortem MRI. *Psychiatry Res*, 183, 1–20.

Van Der Kouwe, A. J. W., Salat, D. H., Benner, T., *et al.* (2010). Technical description of post-mortem magnetic resonance neuroimaging of H.M. In *Society for Neuroscience Abstract*. San Diego, CA: SFN.

Van Leemput, K., Bakkour, A., Benner, T., *et al.* (2009). Automated segmentation of hippocampal subfields from ultra-high resolution in vivo MRI. *Hippocampus*, 19, 549–57.

Van Petten, C. (2004). Relationship between hippocampal volume and memory ability in healthy individuals across the lifespan: review and meta-analysis. *Neuropsychologia*, 42, 1394–413.

Vandenbroucke, M. W., Goekoop, R., Duschek, E. J., *et al.* (2004). Interindividual differences of medial temporal lobe activation during encoding in an elderly population studied by fMRI. *Neuroimage*, 21, 173–80.

Venkatraman, T. N., Hamer, R. M., Perkins, D. O., Song, A. W., Lieberman, J. A., and Steen, R. G. (2006). Single-voxel 1H PRESS at 4.0 T: precision and variability of measurements in anterior cingulate and hippocampus. *NMR Biomed*, 19, 484–91.

Vermersch, P., Leys, D., Scheltens, P., and Barkhof, F. (1994). Visual rating of hippocampal atrophy: correlation with volumetry. *J Neurol Neurosurg Psychiatry*, 57, 1015.

Wadghiri, Y. Z., Sigurdsson, E. M., Sadowski, M., *et al.* (2003). Detection of Alzheimer's amyloid in transgenic mice using magnetic resonance microimaging. *Magn Reson Med*, 50, 293–302.

Wagner, A.D., Schacter, D.L., Rotte, M., Koutstaal, W., Maril, A., Dale, A.M., *et al.* (1998). Building memories: remembering and forgetting of verbal experiences as predicted by brain activity. *Science*, 281, 1188–91.

Walsh, D. M., and Selkoe, D. J. (2004). Deciphering the molecular basis of memory failure in Alzheimer's disease. *Neuron*, 44, 181–93.

Wang, L., Miller, J. P., Gado, M. H., *et al.* (2006). Abnormalities of hippocampal surface structure in very mild dementia of the Alzheimer type. *Neuroimage*, 30, 52–60.

Wang, L., Negreira, A., LaViolette, P., Bakkour, A., Sperling, R.A., and Dickerson, B.C. (2010a). Intrinsic interhemispheric hippocampal functional connectivity predicts individual differences in memory performance ability. *Hippocampus*, 20, 345–51.

Wang, L., Laviolette, P., O'Keefe, K., *et al.* (2010b). Intrinsic connectivity between the hippocampus and posteromedial cortex predicts memory performance in cognitively intact older individuals. *Neuroimage*, 51, 910–17.

Wang, Z., Neylan, T. C., Mueller, S. G., *et al.* (2010). Magnetic resonance imaging of hippocampal subfields in posttraumatic stress disorder. *Arch Gen Psychiatry*, 67, 296–303.

Willis, M. W., Ketter, T. A., Kimbrell, T. A., *et al.* (2002). Age, sex and laterality effects on cerebral glucose metabolism in healthy adults. *Psychiatry Res*, 114, 23–37.

Wink, A. M., Bernard, F., Salvador, R., Bullmore, E., and Suckling, J. (2006). Age and cholinergic effects on hemodynamics and functional coherence of human hippocampus. *Neurobiol Aging*, 27, 1395–404.

Wolf, R. L., Alsop, D. C., Levy-Reis, I., *et al.* (2001). Detection of mesial temporal lobe hypoperfusion in patients with temporal lobe epilepsy by use of arterial spin labeled perfusion MR imaging. *AJNR Am J Neuroradiol*, 22, 1334–41.

Wu, C. W., Gu, H., Lu, H., Stein, E. A., Chen, J. H., and Yang, Y. (2009). Mapping functional connectivity based on synchronized CMRO2 fluctuations during the resting state. *Neuroimage*, 45, 694–701.

Xu, G., McLaren, D. G., Ries, M. L., *et al.* (2009). The influence of parental history of Alzheimer's disease and apolipoprotein E epsilon4 on the BOLD signal during recognition memory. *Brain*, 132, 383–91.

Yushkevich, P. A., Avants, B. B., Pluta, J., *et al.* (2008). Shape-based alignment of hippocampal subfields: evaluation in postmortem MRI. *Med Image Comput Comput Assist Interv*, 11, 510–17.

Yushkevich, P. A., Avants, B. B., Pluta, J., *et al.* (2009). A high-resolution computational atlas of the human hippocampus from postmortem magnetic resonance imaging at 9.4 T. *Neuroimage*, 44, 385–98.

Yushkevich, P. A., Wang, H., Pluta, J., *et al.* (2010). Nearly automatic segmentation of hippocampal subfields in in vivo focal T2-weighted MRI. *Neuroimage*, 53, 1208–24.

Zhang, J., Yarowsky, P., Gordon, M. N., *et al.* (2004). Detection of amyloid plaques in mouse models of Alzheimer's disease by magnetic resonance imaging. *Magn Reson Med*, 51, 452–7.

CHAPTER 8

Sleep, Memory, and the Hippocampus

Ariane Foret, Laura Mascetti, Caroline Kussé, Mathieu Jaspar, Vincenzo Muto, Anahita Shaffii-Le Bourdiec, Luca Matarazzo, and Pierre Maquet

Introduction 144
Basic facts about sleep 145
Behavioural evidence for a favourable effect of sleep on declarative memories 146
The synaptic downscaling hypothesis 146
The hypothesis of a hippocampo-neocortical dialogue during sleep 147
Hippocampal activity during immediate post-training wakefulness 149

ABSTRACT Memory consolidation unfolds across time intervals that span periods of wakefulness and sleep. It is not surprising that both states of vigilance influence memory consolidation. The respective role of neural mechanisms taking place during wakefulness and during sleep is being characterized but our understanding of memory consolidation mechanisms remains fragmentary. Future research should strive to develop an integrated view of offline memory processing, explaining behaviour in terms of its underlying neural and molecular mechanisms.

KEYWORDS hippocampus, sleep, memory, consolidation, neuroimaging

Introduction

More than a century ago, Ebbinghaus noticed that syllables learned just before sleep were better remembered than when they were encoded before an equal period of wakefulness (Ebbinghaus 1885). This observation was repeatedly confirmed and suggested that sleep improves the retention of recent memories (Heine 1914; Jenkins and Dallenbach 1924). To explain this effect, two hypotheses were put forward. One theory assumed that sleep passively protects recent memories from interference, potentially because of the associated loss of consciousness, increased sensory threshold, and reduced motor output. In contrast, another hypothesis postulated that active processes taking place during sleep consolidate memory traces, such that they become both more reliable in the long term and more robust to interference. These active processes are currently conceived within two main, non-exclusive conceptual frameworks. The first one suggests that sleep locally and ubiquitously maintains synaptic homeostasis; the second assumes that sleep promotes the reorganization of memories within hippocampo-neocortical networks. This chapter summarizes the experimental evidence supporting an active participation of sleep in consolidating recent memories at different levels of description, from behaviour down to molecular processes. Further emphasis is put on hippocampus-dependent memories although similar findings were collected for procedural motor or perceptual learning.

From these findings, one should not conclude that memories are processed exclusively during sleep and some evidence for memory processing during immediate post-training wakefulness will also be examined. Collectively, these data suggest that encoding triggers a cascade of events which take place during both wakefulness and sleep, and which maintain, process, and consolidate recent memory traces, thereby optimizing behavioural adaptation to an ever-changing environment.

Basic facts about sleep

Sleep is a normal, reversible, periodically recurring behavioural state characterized by a decreased responsiveness to external stimuli, a diminished motor activity, and a characteristic position (in mammals usually, but not necessarily recumbent with eyes closed). It is often preceded by the active retreat to a safe, secluded place. It is homeostatically regulated in such a way that any extension of the waking period is followed by an increase in sleep depth and (to some extent) duration. In homeotherms, neurophysiological parameters, in addition to behaviour, allow recognition of two main sleep states: a 'regular' sleep or non-rapid eye movement (NREM) sleep, which shows gradual changes in sleep depth, and a 'paradoxical' sleep, which combines nearly complete paralysis of the musculature with elevated brain activity. Paradoxical sleep has also been termed rapid eye movement (REM) sleep, because of the large and swift movements of the eyes, which only can be observed in this sleep state. In humans, four to six NREM/REM sleep cycles with a duration of about 90 minutes each are usually recorded during one night.

During deep NREM sleep a slow oscillation (< 1 Hz) emerges from cortical or thalamocortical populations (Steriade and McCarley 2005; Crunelli and Hughes 2009). Unit recordings showed that neuronal activity during NREM sleep is characterized by a fundamental oscillation of membrane potential (Steriade et al. 1993). This slow oscillation is recorded in all major types of neocortical neurons during NREM sleep and is made up of a depolarizing phase, associated with important neuronal firing ('up state'), followed by a hyperpolarizing phase, during which cortical neurons remain silent for a few hundred milliseconds ('down state'). The slow oscillation occurs synchronously in large neuronal populations, the activity of which alternates between so-called ON and OFF states (Vyazovskiy et al. 2009), in such a way that it is reflected on electroencephalographic (EEG) recordings as large amplitude low frequency waves. Delta waves (1–4 Hz) are faster waves of smaller amplitude that arise when the synchronization in the network is less marked (Esser et al. 2007; Vyazovskiy et al. 2009). The slow oscillation entrains other NREM sleep oscillations in a coalescence of rhythms so that, for instance, spindles and gamma oscillations are more likely to occur during the up phase of the slow oscillation (Steriade and McCarley 2005). The slow oscillation also organizes the neural firing of arousing structures such as the pedunculo-pontine nucleus (Mena-Segovia et al. 2008) or the locus coeruleus (Yeshenko et al. 2006).

Slow wave activity, i.e. the EEG power between 0.75–4 Hz, is a popular parameter that estimates the abundance of slow oscillation in EEG recordings. It is used to quantify the homeostatic sleep pressure accrued during wakefulness (Achermann and Borbely 2003).

Sleep spindles are a hallmark of light NREM sleep (mostly stage 2 sleep) and, in humans, consist of waxing-and-waning 11–15-Hz oscillations, lasting 0.5–3 seconds (De Gennaro and Ferrara 2003). They are generated by the thalamus, which acts as a pacemaker, and result from reciprocal rhythmic interactions between reticular and thalamocortical cells (Steriade and McCarley 2005). Post-inhibitory-rebound-spike bursts in thalamocortical cells, induced by a recurrent inhibition by reticular cells, entrain cortical populations in spindle oscillations. In turn, a corticothalamic feedback synchronizes spindle oscillations in widespread thalamic territories (Contreras et al. 1996). Because spindles entrain synchronous firing in large thalamocortical neural populations, they are in a good position to allow for the modifications in the neural representations of recent memories. In addition, at the cellular level, it was shown that repetitive spike bursts mimicking firing patterns observed during spontaneous spindles reliably induced short- and long-term potentiation in cortical neurons of rat brain slices (Rosanova and Ulrich 2005). It has also been suggested that activity in the spindle band would trigger molecular cascades involved in brain plasticity by increasing intracellular calcium levels (Sejnowski and Destexhe 2000).

Hippocampal activity during NREM sleep is characterized by sharp waves and ripples, which are synchronous to the cortical slow oscillation (Battaglia et al. 2004; Molle et al. 2006), although it is not yet clear which oscillation is driving the other (Isomura et al. 2006; Tononi et al. 2006). Similar temporal coupling between slow oscillations, spindles and ripples has been reported in foramen ovale recordings in epileptic patients during sleep (Clemens et al. 2007).

Collectively, these data suggest that the up state of the cortical slow oscillation constitutes a remarkable period during which the activity of various brain structures, in the cortex (slow waves, spindles), the hippocampus (sharp-wave, ripples) and subcortical structures is synchronized, thereby fostering functional interactions associated with systems-level memory consolidation.

Behavioural evidence for a favourable effect of sleep on declarative memories

The main behavioural findings supporting an effect of sleep on declarative memories can be summarized as follows: total sleep deprivation during the post-encoding night significantly deteriorates subsequent retrieval (Gais et al. 2007). By contrast, retention of declarative memories is enhanced when sleep occurs immediately after encoding (Gais et al. 2006). Some evidence even shows that short daytime sleep (i.e. naps) enhance the retention of declarative memories (Tucker et al. 2006; Lahl et al. 2008; Tucker and Fishbein 2008). Declarative memory consolidation benefits primarily from NREM sleep. Word-pair associates are better retained if sleep is allowed on the first part of the night (rich in deep NREM sleep) rather than in the second portion of the night (Yaroush et al. 1971; Barrett and Ekstrand 1972; Fowler et al. 1973; Plihal and Born 1999). Olfactory cues previously presented during a spatial source memory task enhance later memory performance if delivered during post-encoding NREM sleep (Rasch et al. 2007). Enhancing slow waves and spindles by transcranial direct current stimulation during the post-encoding night improves retention of word-pair associates (Marshall et al. 2004). In particular, spindle activity during NREM sleep has been repeatedly associated with improved memory retention (Gais et al. 2002; Schabus et al. 2004; Schmidt et al. 2006).

By contrast, the impact of REM sleep on declarative memories is still unsettled and probably minor in adult humans. Despite early reports of detrimental effects of selective REM sleep deprivation on verbal memory (Empson and Clarke 1970; Tilley and Empson 1978), subsequent studies failed to observe a significant effect of REM sleep suppression (Rasch et al. 2008).

In any case, the behavioural effects of sleep on memory are usually moderate: the difference in retrieval rates between sleep and waking conditions, although significant, usually amounts to a marginal fraction of the learned material only (e.g. Plihal and Born 1999; Marshall et al. 2004; Gais et al. 2006; Rasch et al. 2007). This suggests that most of the learned material can still be retained and consolidated during wakefulness, without the necessary intervention of sleep (see later section). However, the major behavioural effect of sleep on memory consolidation is revealed under conditions of interference. Indeed, sleep provides active protection against interferences resulting from encoding novel information on the same day as the initial learning (Ellenbogen et al. 2006).

The synaptic downscaling hypothesis

This hypothesis pertains to a ubiquitous albeit local sleep pressure which is deemed to increase monotonically in the brain with time spent awake, or more precisely, in proportion to neural activity accrued locally during wakefulness. Its central tenet claims that wakefulness is associated with a net increase in synaptic strength in the brain, which would become energetically unsustainable in the long term (Tononi and Cirelli 2003, 2006). During NREM sleep, slow oscillations would be associated with a downscaling of synaptic strength to a baseline level, a process beneficial for learning and memory because it ultimately increases the signal-to-noise ratio related to the learned material. The synaptic downscaling would thus be a general phenomenon occurring throughout the brain but which is modulated locally by the amount of neural activity and synaptic strength enhancement accumulated during the day. This hypothesis, which deliberately ignores the potential effect of REM sleep, is not specific to hippocampus-dependent memories and has seldom been tested with declarative memories.

In support of this hypothesis, a local increase in slow-wave activity (SWA, the power density between 0.75–4 Hz during NREM sleep) is observed over the relevant brain areas after vibratory stimulation of the hand (Kattler et al. 1994), training to a visuomotor adaptation task (Huber et al. 2004), transcranial magnetic stimulation of the motor cortex (Huber et al. 2007a), or after spike timing-dependent activity is elicited during waking by transcranial paired associative stimulation (Huber et al. 2008). In contrast, arm immobilization results in a decrease in SWA over contralateral sensorimotor areas during subsequent NREM sleep (Huber et al. 2006). Moreover, after learning lists of word-pair associates, during the surface-positive half-waves of NREM sleep slow waves, coherence is enhanced over widespread scalp areas in the slow-oscillatory and delta bands (Molle et al. 2004).

In rodents, similar increases in SWA were reported in rats exposed to enriched environment and seems associated with release of BDNF (brain-derived neurotrophic factor), a neurotrophin involved in synaptic potentiation (Huber et al. 2007b). At the cellular level, multiunit recordings showed that the mean firing rate in the cerebral cortex increases after periods of wakefulness and decreases after periods of sleep, consistent with a net change in synaptic strength (Vyazovskiy et al. 2009). Changes in firing patterns in NREM sleep correlate with changes in slow-wave activity (Vyazovskiy et al. 2009). The slope and amplitude of cortical evoked responses, taken as markers of local synaptic strength, increase after wakefulness and decrease after sleep in proportion to changes in SWA (Vyazovskiy et al. 2008). At the molecular level, in rat cortex and hippocampus, GluR1-containing AMPA receptor levels are high after a period of wakefulness and low after sleep, and changes in the phosphorylation states of AMPA receptors, CamKII and GSK3beta are consistent with synaptic potentiation during wakefulness and depression during sleep (Vyazovskiy et al. 2008).

The hypothesis of a hippocampo-neocortical dialogue during sleep

This hypothesis emerged from the theoretical difficulty for a modular sparsely-connected cortex to store multimodal memories requiring intermodular associations (for review, see Mc Naughton 2002). The hierarchical organization of the cortex provides a solution to this issue because unique response patterns are elicited in high-order modules by specific associations of lower-level inputs. In this case, a higher-level module can serve to index lower-level events because the activation of a specific higher-order pattern will lead to associative retrieval of the relevant activity patterns in lower-level modules. However, in the long term, it would be more efficient to generate direct intermodular connections specifically supporting the encoded associations. Intermodular connections would gradually be selected to build up a sparse intermodular connectivity which would eventually represent the internal space of raw sensory inputs. This process would require repeated activation of the memories to be stored, resulting in Hebbian synaptic associations. Ultimately, the retrieval of a specific association would require the recruitment of the relevant low-level modules, making the intervention of higher-order modules redundant.

Assuming the hippocampus serves as a higher-order indexing module, this hypothesis predicts that hippocampus-dependent memories become progressively independent of the limbic structures and are gradually stored in cortical circuits (Alvarez and Squire 1994; Frankland and Bontempi 2005) (Figure 8.1). Sleep would participate in this systems-level memory consolidation by promoting the functional interplay between hippocampal neural ensembles and neocortical areas, in the absence of incoming sensory information (Marr 1970).

This hypothesis predicts that new experiences should influence the activity of hippocampo-neocortical circuits during subsequent NREM sleep. At the systems level, human brain activity during sleep, as assessed by positron emission tomography, is indeed significantly modified by previous waking experience (Peigneux et al. 2004). After the exploration of a virtual tridimensional maze, the activity is enhanced during NREM sleep in occipital, parietal and mesiotemporal areas relative to baseline sleep (Peigneux et al. 2004). Moreover, the increase in hippocampal activity is linearly related to the individual gain in the ability to navigate in the maze the next day, suggesting that the changes in hippocampal activity during NREM sleep relates to the offline processing of spatial memory (Peigneux et al. 2004) (Figure 8.2). In contrast, sleep deprivation hinders the offline processing of memory traces during NREM sleep and alters the neural correlates of subsequent retrieval. For instance, recall of word-pair

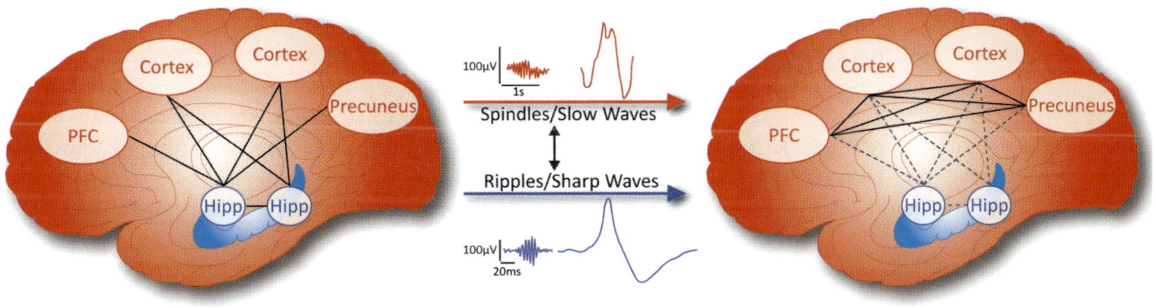

Figure 8.1 Declarative memory consolidation during sleep: the hippocampo-neocortical dialogue model. Red and blue respectively pertains to cortical and hippocampal areas and processes. Hipp, hippocampus; PFC, prefrontal cortex. Left) Recent memories are encoded in distributed in hippocampo-neocortical networks. Middle) During period of inactivity, memories will be processed offline. During sleep, the hippocampal-neocortical dialogue is thought to be supported by synchronous firing during simultaneous cortical spindles-slow waves on the one hand and hippocampal ripples-sharp waves on the other hand. Right) Memory consolidation increases the burden of memory storing in cortical circuits comprising strongly connected brain areas.

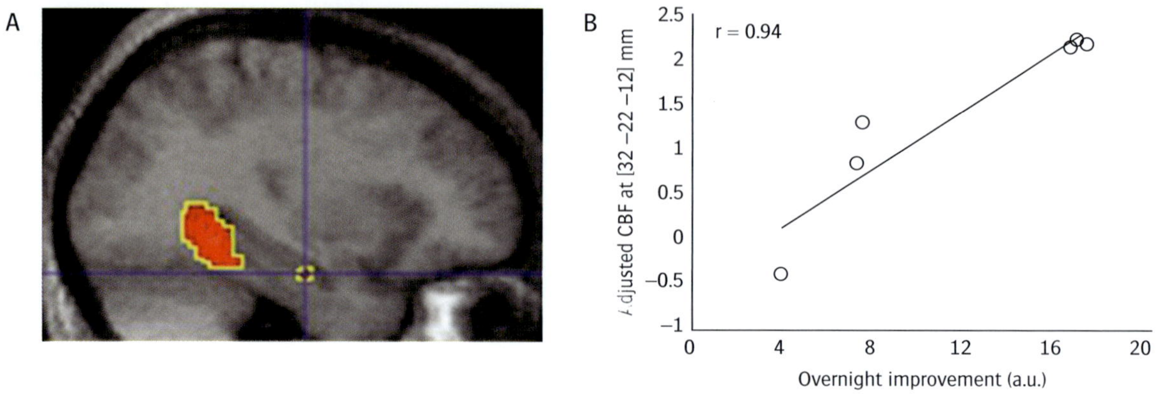

Figure 8.2 Hippocampal blood flow during NREM sleep and memory consolidation. Volunteers were trained in a virtual town inspired from Maguire et al. (1998). They had a colour three-dimensional first-person view from inside the environment in which they navigated using a joystick to control their moves. Volunteers were trained on the spatial memory task in the afternoon (pre-sleep session). They were allowed eight 15-min free exploration periods in the virtual town. Each exploration was followed by a set of three 90-s tests of route retrieval. During these tests, volunteers were designated a starting object and were instructed to reach a remote object located in another district, in no more than 90 s. After time elapsed, the distance remaining between the subject's actual location and his final destination was computed using the shortest possible path and used as a quantitative estimate of topographical knowledge (i.e. the less the remaining distance to destination the better the performance). During the subsequent night, volunteers were scanned during resting wakefulness, light and deep NREM sleep and during REM sleep using $H_2^{15}O$ positron emission tomography. On the next day (post-sleep session), subjects were administered the last fifteen 90-s tests of route retrieval performed in the pre-sleep session, at the same time of day as the initial session to avoid potential circadian differences in arousal or vigilance. A) A significant regression with overnight gain in performance was observed in the right hippocampus and parahippocampal gyrus. B) Scatter plot of the right hippocampal cerebral blood flow during NREM sleep (relative to wakefulness) with the overnight improvement in performance (difference in distance to target from pre- to post-sleep session). Reprinted from *Neuron*, 44(3), Peigneux, P., Laureys, S., Fuchs, S., et al. (2004). Are spatial memories strengthened in the human hippocampus during slow wave sleep? 535–45. © 2004, with permission from Elsevier.

associates 48 h after encoding is associated with larger hippocampal responses for words encoded before sleep than for words encoded before a night of sleep deprivation (Gais et al. 2007). Similar results were observed in experiments testing the influence of sleep deprivation on topographical memory (Orban et al. 2006) or emotional memory (Sterpenich et al. 2007). The effects of sleep deprivation during the first post-training night are long lasting and can still be recorded several months after encoding (Gais et al. 2007; Sterpenich et al. 2009), suggesting that sleep immediately following encoding irreversibly conditions the systems-level structure of long-term memory.

At the microscopic systems level, a key finding supporting this hypothesis was found in animals, namely that correlations in neural activity observed during encoding are preserved during post-training sleep (Wilson and McNaughton 1994; Kudrimoti et al. 1999). Even complete firing sequences of neural discharges recorded in hippocampal cell ensembles during wakefulness are spontaneously repeated during NREM sleep (Nadasdy et al. 1999; Lee and Wilson 2002), as well as during REM sleep (Louie and Wilson 2001). These cellular events are temporally organized by sleep oscillations, namely during NREM sleep, the cortical slow rhythm, thalamocortical spindles and hippocampal sharp waves and ripples. Simultaneous recordings in the hippocampus and prefrontal cortex in rat showed that hippocampal ripples and cortical spindles are temporally correlated (Siapas and Wilson 1998). Coordinated activity between the hippocampus and the cortex during spindle/sharp wave/ripple episodes might contribute to hippocampo-neocortical communication and memory consolidation during sleep (Siapas and Wilson 1998), and echoes the relationship observed between spindle density and behavioural gains in humans (discussed earlier). Repetition of neural firing patterns in the neocortex is synchronized to hippocampal sharp waves (Ji and Wilson 2007; Peyrache et al. 2009). Selective elimination of sharp waves/ripples during post-training sleep results in performance impairment in rats trained on a spatial memory task (Girardeau et al. 2009). These reactivations seem to depend on learning as the replayed firing patterns appear only after the acquisition of new rules (Peyrache et al. 2009). Moreover, cells in prefrontal cortex fire consistently within 100 ms after hippocampal cells during NREM sleep (Wierzynski et al. 2009). This result provides the first evidence at the single cell-pair level for highly consistent directional interactions between these areas within the window of plasticity (Wierzynski et al. 2009). Collectively, these data support the hypothesis that during sleep hippocampo-neocortical interactions may progressively transfer the burden of memory from hippocampal ensembles to long-term neocortical stores.

Hippocampal activity during immediate post-training wakefulness

From the preceding paragraph, it appears that memory consolidation is increasingly framed in terms of processes taking place during selective brain rhythms and not specifically during a particular vigilance state. Importantly, some brain oscillations critical for memory consolidation are not univocally associated with sleep. For instance, hippocampal sharp waves and ripples are not exclusively observed during NREM sleep but also in awake immobility or during consummatory behaviours (eating, drinking) (Buzsaki 1996). In addition, human behavioural data show that most of the learned material can be retained even after sleep deprivation (see earlier section), suggesting that memory consolidation also takes place during post-encoding wakefulness, although it does result in the same neural fate as during sleep (Sterpenich et al. 2007). These facts open the possibility that memories are not only maintained

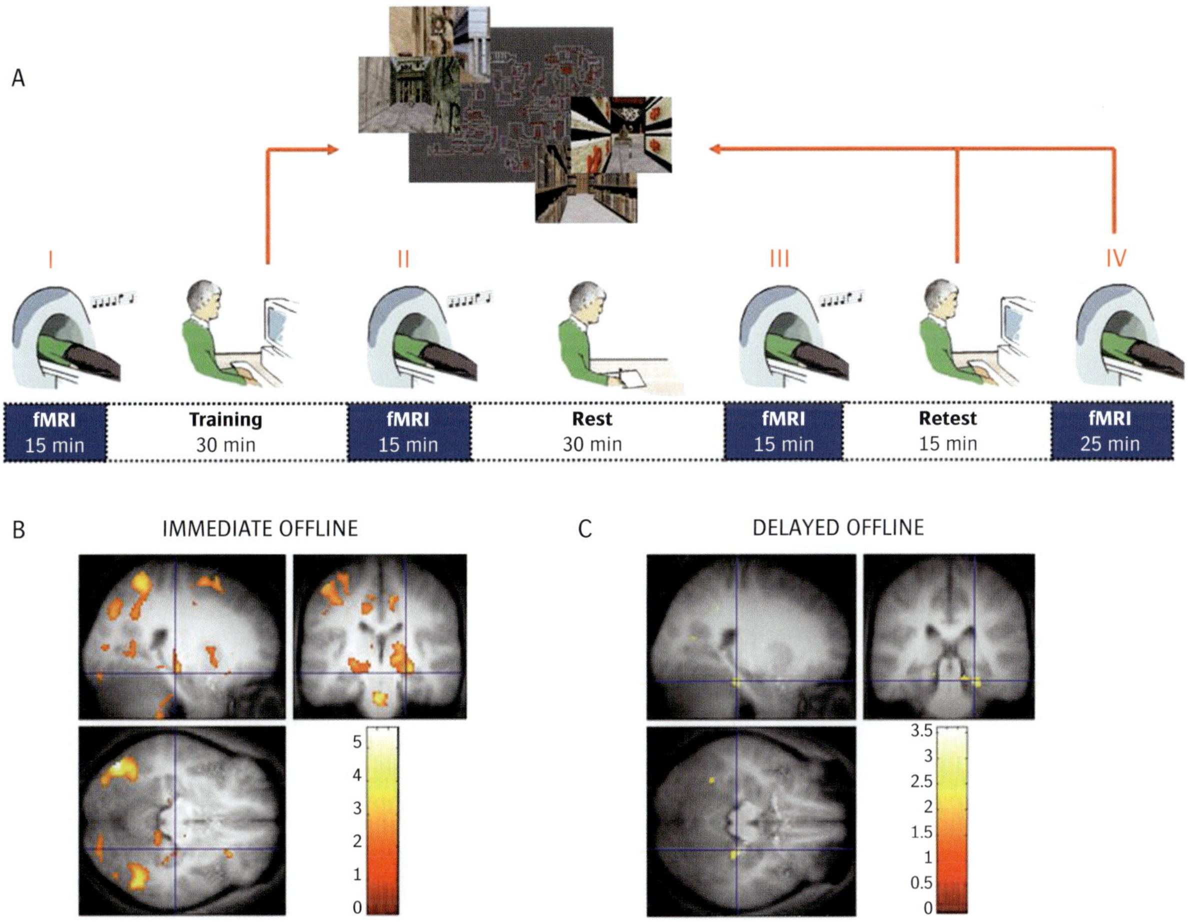

Figure 8.3 Task-specific modulation of regional brain responses by prior learning during wakefulness. Experimental design. All volunteers underwent four consecutive fMRI scanning sessions (I–IV) within a half-day. During baseline (scanning session I), they performed an auditory oddball task during which they mentally counted the number of high-pitched tones interspersed in a stream of low-pitched tones. This auditory task served as a probe to assess the changes in responses induced by subsequent training. Subjects were then trained during 30 min outside of the scanner a spatial memory navigation task (training). Immediately after the end of the training session, they were scanned again while performing the auditory oddball task (session II, first post-training session). They were then allowed a further 30-min break outside of the scanner without any further practice (rest). Thereafter, they were scanned once again while performing the auditory oddball task (session III, second post-training session). Afterwards, memory performance was tested outside of the scanner (retest). Finally, subjects underwent a fourth fMRI session (IV), during which they explored virtual environments, to determine the set of brain areas associated with task practice. Increased hippocampal responses to the unrelated probe task (auditory oddball) after spatial learning (Session II > Session I). This result shows that responses to an unrelated cognitive task (auditory oddball) are modulated by recent spatial learning. Increased brain responses in the parahippocampal gyrus after a further 30-min break (Session III > Session II). This result shows that the modulation of responses to an unrelated cognitive task (auditory oddball) by recent spatial learning changes over time, suggesting that the memory trace is being processed during post-training wakefulness. Reproduced from Peigneux, P., Orban, P., Balteau, E. et al. (2006). Offline persistence of memory-related cerebral activity during active wakefulness. *PLoS Biol*, 4, E100. © 2006.

during initial post-training wakefulness, but are also likely to be extensively processed during this period of time. Accordingly, in rodents, pairwise correlations in firing rates detected during training accounted for a significant proportion of the variance during subsequent sleep (Wilson and McNaughton 1994; Skaggs and McNaughton 1996) but also during quiet wakefulness (Kudrimoti et al. 1999). Likewise, in monkeys exposed to reaching tasks, the correlation structure of neural firing patterns induced by the task was preserved within and between motor, somatosensory, and parietal cortices during subsequent resting wakefulness (Hoffman and McNaughton 2002). In humans, during the resting wakefulness following an associative memory task (object-face or scene-face association), functional interactions between the hippocampus and occipital areas were enhanced in proportion to later associative memory performance (Tambini et al. 2010). Even more compelling are the results showing that brain activity elicited during a new learning episode modulates responses to an unrelated cognitive task, during the waking period following the end of training, in proportion to later behavioural performance (Peigneux et al. 2006). Within the first post-training hours, this modulation evolves in learning-related cerebral structures, which establish and reinforce functional connections with other brain regions. These processes indicate that the human brain has already extensively processed recent memories during the first hours of post-training wakefulness, even when simultaneously coping with unrelated cognitive demands (Peigneux et al. 2006) (Figure 8.3).

Acknowledgements

Personal research reported in this review was supported by the Belgian Fonds National de la Recherche Scientifique (FNRS), Fondation Médicale Reine Elisabeth (FMRE), Research Fund of the University of Liège and 'Interuniversity Attraction Poles Programme – Belgian State – Belgian Science Policy'. AF, ASLB, CK, LaMa, LuMa, and PM are supported by the FNRS.

References

Achermann, P., and Borbely, A. A. (2003). Mathematical models of sleep regulation. *Front Biosci,* 8, s683–93.

Alvarez, P., and Squire, L. R. (1994). Memory consolidation and the medial temporal lobe: a simple network model. *Proc Natl Acad Sci U S A,* 91, 7041–5.

Barrett, T. R., and Ekstrand, B. R. (1972). Effect of sleep on memory. 3. Controlling for time-of-day effects. *J Exp Psychol,* 96, 321–7.

Battaglia, F. P., Sutherland, G. R., and McNaughton, B. L. (2004). Hippocampal sharp wave bursts coincide with neocortical 'up-state' transitions. *Learn Mem,* 11, 697–704.

Buzsaki, G. (1996). The hippocampo-neocortical dialogue. *Cereb Cortex,* 6, 81–92.

Clemens, Z., Molle, M., Eross, L., *et al.* (2007). Temporal coupling of parahippocampal ripples, sleep spindles and slow oscillations in humans. *Brain,* 130, 2868–78.

Contreras, D., Destexhe, A., Sejnowski, T. J., *et al.* (1996). Control of spatiotemporal coherence of a thalamic oscillation by corticothalamic feedback. *Science,* 274, 771–4.

Crunelli, V., and Hughes, S. W. (2009). The slow (<1 Hz) rhythm of non-REM sleep: a dialogue between three cardinal oscillators. *Nat Neurosci,* 13, 9–17.

De Gennaro, L., and Ferrara, M. (2003). Sleep spindles: an overview. *Sleep Med Rev,* 7, 423–40.

Ebbinghaus, H. (1885). *Über das Gedächtnis, Untersuchungen zur Experimentellen Psychologie.* Leipzig, Duncker und Humblot.

Ellenbogen, J. M., Hulbert, J. C., Stickgold, R., *et al.* (2006). Interfering with theories of sleep and memory: sleep, declarative memory, and associative interference. *Curr Biol,* 16, 1290–4.

Empson, J. A., and Clarke, P. R. (1970). Rapid eye movements and remembering. *Nature,* 227, 287–8.

Esser, S. K., Hill, S. L., and Tononi, G. (2007). Sleep homeostasis and cortical synchronization: I. Modeling the effects of synaptic strength on sleep slow waves. *Sleep,* 30, 1617–30.

Fowler, M. J., Sullivan, M. J., and Ekstrand, B. R. (1973). Sleep and memory. *Science,* 179, 302–4.

Frankland, P. W., and Bontempi, B. (2005). The organization of recent and remote memories. *Nat Rev Neurosci,* 6, 119–30.

Gais, S., Albouy, G., Boly, M., *et al.* (2007). Sleep transforms the cerebral trace of declarative memories. *Proc Natl Acad Sci U S A,* 104, 18778–83.

Gais, S., Lucas, B., and Born, J. (2006). Sleep after learning aids memory recall. *Learn Mem,* 13, 259–62.

Gais, S., Molle, M., Helms, K., el. (2002). Learning-dependent increases in sleep spindle density. *J Neurosci,* 22, 6830–4.

Girardeau, G., Benchenane, K., Wiener, S. I., *et al.* (2009). Selective suppression of hippocampal ripples impairs spatial memory. *Nat Neurosci,* 12, 1222–3.

Heine, R. (1914). Über des Wiedererkennen und Rückwirkende Hemmung. *Zeitschrift für Psychologie,* 68, 161–236.

Hoffman, K. L., and McNaughton, B. L. (2002). Coordinated reactivation of distributed memory traces in primate neocortex. *Science,* 297, 2070–3.

Huber, R., Esser, S. K., Ferrarelli, F., *et al.* (2007a). TMS-induced cortical potentiation during wakefulness locally increases slow wave activity during sleep. *PLoS ONE,* 2, e276.

Huber, R., Ghilardi, M. F., Massimini, M., *et al.* (2006). Arm immobilization causes cortical plastic changes and locally decreases sleep slow wave activity. *Nat Neurosci,* 9, 1169–76.

Huber, R., Ghilardi, M. F., Massimini, M., *et al.* (2004). Local sleep and learning. *Nature,* 430, 78–81.

Huber, R., Maatta, S., Esser, S. K., *et al.* (2008). Measures of cortical plasticity after transcranial paired associative stimulation predict changes in electroencephalogram slow-wave activity during subsequent sleep. *J Neurosci,* 28, 7911–8.

Huber, R., Tononi, G., and Cirelli, C. (2007b). Exploratory behavior, cortical BDNF expression, and sleep homeostasis. *Sleep,* 30, 129–39.

Isomura, Y., Sirota, A., Ozen, S., *et al.* (2006). Integration and segregation of activity in entorhinal-hippocampal subregions by neocortical slow oscillations. *Neuron,* 52, 871–882.

Jenkins, J. G., and Dallenbach, K. M. (1924). Obliviscence during sleep and waking. *Am J Psychol,* 35, 605–612.

Ji, D., and Wilson, M. A. (2007). Coordinated memory replay in the visual cortex and hippocampus during sleep. *Nat Neurosci,* 10, 100–7.

Kattler, H., Dijk, D. J., and Borbely, A. A. (1994). Effect of unilateral somatosensory stimulation prior to sleep on the sleep EEG in humans. *J Sleep Res,* 3, 159–164.

Kudrimoti, H. S., Barnes, C. A., and McNaughton, B. L. (1999). Reactivation of hippocampal cell assemblies: effects of behavioral state, experience, and EEG dynamics. *J Neurosci,* 19, 4090–101.

Lahl, O., Wispel, C., Willigens, B., *et al.* (2008). An ultra short episode of sleep is sufficient to promote declarative memory performance. *J Sleep Res,* 17, 3–10.

Lee, A. K., and Wilson, M. A. (2002). Memory of sequential experience in the hippocampus during slow wave sleep. *Neuron,* 36, 1183–94.

Louie, K., and Wilson, M. A. (2001). Temporally structured replay of awake hippocampal ensemble activity during rapid eye movement sleep. *Neuron,* 29, 145–56.

Maguire, E. A., Burgess, N., Donnett, J. G., *et al.* (1998). Knowing where and getting there: A human navigation network. *Science,* 280, 921–4.

Marr, D. (1970). A theory for cerebral neocortex. *Proc R Soc Lond B Biol Sci,* 176, 161–234.

Marshall, L., Molle, M., Hallschmid, M., *et al.* (2004). Transcranial direct current stimulation during sleep improves declarative memory. *J Neurosci,* 24, 9985–92.

Mc Naughton, B. L. (2002). Off-line reprocessing of recent memory and its role in memory consolidation: a progress report. In Maquet, P., Smith, C., and Stickgold, R. (Eds.) *Sleep and brain plasticity,* pp. 225–46. Oxford: Oxford University Press.

Mena-Segovia, J., Sims, H. M., Magill, P. J., *et al.* (2008). Cholinergic brainstem neurons modulate cortical gamma activity during slow oscillations. *J Physiol,* 586, 2947–60.

Molle, M., Marshall, L., Gais, S., *et al.* (2004). Learning increases human electroencephalographic coherence during subsequent slow sleep oscillations. *Proc Natl Acad Sci U S A,* 101, 13963–8.

Molle, M., Yeshenko, O., Marshall, L., *et al.* (2006). Hippocampal sharp wave-ripples linked to slow oscillations in rat slow-wave sleep. *J Neurophysiol,* 96, 62–70.

Nadasdy, Z., Hirase, H., Czurko, A., *et al.* (1999). Replay and time compression of recurring spike sequences in the hippocampus. *J Neurosci,* 19, 9497–507.

Orban, P., Rauchs, G., Balteau, E., *et al.* (2006). Sleep after spatial learning promotes covert reorganization of brain activity. *Proc Natl Acad Sci U S A,* 103, 7124–9.

Peigneux, P., Laureys, S., Fuchs, S., *et al.* (2004). Are spatial memories strengthened in the human hippocampus during slow wave sleep? *Neuron,* 44, 535–45.

Peigneux, P., Orban, P., Balteau, E., *et al.* (2006). Offline persistence of memory-related cerebral activity during active wakefulness. *PLoS Biol,* 4, e100.

Peyrache, A., Khamassi, M., Benchenane, K., *et al.* (2009). Replay of rule-learning related neural patterns in the prefrontal cortex during sleep. *Nat Neurosci,* 12, 919–26.

Plihal, W., and Born, J. (1999). Effects of early and late nocturnal sleep on priming and spatial memory. *Psychophysiology,* 36, 571–82.

Rasch, B., Buchel, C., Gais, S., *et al.* (2007). Odor cues during slow-wave sleep prompt declarative memory consolidation. *Science,* 315, 1426–9.

Rasch, B., Pommer, J., Diekelmann, S., *et al.* (2008). Pharmacological REM sleep suppression paradoxically improves rather than impairs skill memory. *Nat Neurosci,* 12, 396–7.

Rosanova, M., and Ulrich, D. (2005). Pattern-specific associative long-term potentiation induced by a sleep spindle-related spike train. *J Neurosci,* 25, 9398–405.

Schabus, M., Gruber, G., Parapatics, S., *et al.* (2004). Sleep spindles and their significance for declarative memory consolidation. *Sleep,* 27, 1479–85.

Schmidt, C., Peigneux, P., Muto, V., *et al.* (2006). Encoding difficulty promotes postlearning changes in sleep spindle activity during napping. *J Neurosci,* 26, 8976–82.

Sejnowski, T. J., and Destexhe, A. (2000). Why do we sleep? *Brain Res,* 886, 208–23.

Siapas, A. G., and Wilson, M. A. (1998). Coordinated interactions between hippocampal ripples and cortical spindles during slow-wave sleep. *Neuron,* 21, 1123–8.

Skaggs, W. E., and McNaughton, B. L. (1996). Replay of neuronal firing sequences in rat hippocampus during sleep following spatial experience. *Science,* 271, 1870–3.

Steriade, M., and McCarley, R. W. (2005). *Brain control of wakefulness and sleep,* New York: Kluwer Academic.

Steriade, M., Nunez, A., and Amzica, F. (1993). A novel slow (< 1 Hz) oscillation of neocortical neurons in vivo: depolarizing and hyperpolarizing components. *J Neurosci,* 13, 3252–65.

Sterpenich, V., Albouy, G., Boly, M., *et al.* (2007). Sleep-related hippocampo-cortical interplay during emotional memory recollection. *PLoS Biol,* 5, e282.

Sterpenich, V., Albouy, G., Darsaud, A., *et al.* (2009). Sleep promotes the neural reorganization of remote emotional memory. *J Neurosci,* 29, 5143–52.

Tambini, A., Ketz, N., and Davachi, L. (2010). Enhanced brain correlations during rest are related to memory for recent experiences. *Neuron,* 65, 280–90.

Tilley, A. J., and Empson, J. A. (1978). REM sleep and memory consolidation. *Biol Psychol,* 6, 293–300.

Tononi, G., and Cirelli, C. (2003). Sleep and synaptic homeostasis: a hypothesis. *Brain Res Bull,* 62, 143–50.

Tononi, G., and Cirelli, C. (2006). Sleep function and synaptic homeostasis. *Sleep Med Rev,* 10, 49–62.

Tononi, G., Massimini, M., and Riedner, B. A. (2006). Sleepy dialogues between cortex and hippocampus: who talks to whom? *Neuron,* 52, 748–9.

Tucker, M. A., and Fishbein, W. (2008). Enhancement of declarative memory performance following a daytime nap is contingent on strength of initial task acquisition. *Sleep,* 31, 197–203.

Tucker, M. A., Hirota, Y., Wamsley, E. J., *et al.* (2006). A daytime nap containing solely non-REM sleep enhances declarative but not procedural memory. *Neurobiol Learn Mem,* 86, 241–7.

Vyazovskiy, V. V., Cirelli, C., Pfister-Genskow, M., *et al.* (2008). Molecular and electrophysiological evidence for net synaptic potentiation in wake and depression in sleep. *Nat Neurosci,* 11, 200–8.

Vyazovskiy, V. V., Olcese, U., Lazimy, Y. M., *et al.* (2009). Cortical firing and sleep homeostasis. *Neuron,* 63, 865–78.

Wierzynski, C. M., Lubenov, E. V., Gu, M., *et al.* (2009). State-dependent spike-timing relationships between hippocampal and prefrontal circuits during sleep. *Neuron,* 61, 587–96.

Wilson, M. A., and McNaughton, B. L. (1994). Reactivation of hippocampal ensemble memories during sleep. *Science,* 265, 676–9.

Yaroush, R., Sullivan, M. J., and Ekstrand, B. R. (1971). Effect of sleep on memory. II. Differential effect of the first and second half of the night. *J Exp Psychol,* 88, 361–6.

Yeshenko, O., Moelle, M., Marshall, L., *et al.* (2006). Locus coeruleus firing during SWS is time locked to slow oscillations: possible contribution of the noradrenergic system to off-line information processing in rats. *J Sleep Res,* 15(s1), 11.

CHAPTER 9

The Ageing Hippocampus

Lan T. Hoang, James P. Lister, and Carol A. Barnes

Introduction 152
Cross-species cognitive ageing 153
Spatial memory 154
Neuroanatomical changes in the hippocampus 154
Biophysical properties of aged neurons 158
Electrophysiological measures of connectivity 159
Neural plasticity 160
Age-related changes in gene expression 162
Dynamic alterations of Arc transcription—treatment approaches 163
Dynamics of aged neural ensembles 164
Summary 168

ABSTRACT This chapter reviews alterations observed in the hippocampus during normative ageing in the absence of pathology. Topics include discussions of age-related spatial memory deficits across mammalian species, as well as neuroanatomical, biophysical, and electrophysiological changes observed in the aged hippocampus. Further, the impact that these changes might have on plasticity mechanisms and dynamic processes such as gene expression and epigenetic modifications over the lifespan is discussed. Special emphasis is given to the fact that normative ageing is distinct from neuropathological states such as Alzheimer's disease. Moreover, this overview highlights how the study of normal ageing in the hippocampus has furthered the understanding of the specificity of pathological processes as well as presented possible avenues for the development of interventions, treatments, or therapeutic strategies for optimizing cognition during the normal life course.

KEYWORDS spatial memory, neuroanatomical changes, electrophysiology, plasticity, immediate-early genes, DNA methylation, multiple single-unit recording, place cells

Introduction

Understanding the biology of normative lifespan development, particularly the trajectories expected for brain ageing and cognition, is an increasingly exciting area of scientific discovery. Twentieth-century ideas that global neural and mental deterioration are an inevitable part of the ageing process have been replaced with a deeper empirically-driven appreciation for the subtle, selective and adaptive process of brain and cognitive ageing. This is a fortunate coincidence, as 2011 was the year that marked the entry of the first of the baby boom generation into an age cohort considered to be 'old' (i.e. 65 years). Over the past few decades it has become clear that the study of ageing in the absence of disease is fundamental to understanding pathological conditions in which age is a major risk factor. The acknowledgment that 'ageing is not a disease' is a fundamental paradigm shift, and provides important benchmarks for discriminating pathological from developmentally normative states. Another crucial conceptual advance in this timeframe has been the appreciation that a singular defect, such as significant neuron loss, does not describe the ageing process well, in any brain region. In fact, brain systems appear to age independently and at different rates—even subareas within a single brain region can show differential age-related neurobiological change (for example, some hippocampal neurons show synapse loss, some do not). Thus, when developing treatments to facilitate or optimize performance of the ageing brain, it will be necessary both to target specifically the systems and cellular operations that are actually dysfunctional, and to avoid perturbing those processes that are not altered.

Appreciating this complexity will allow the development of targeted strategies necessary not only for optimization of normal ageing trajectories, but also for understanding how to approach disease processes coincident with brain ageing. This overview serves to emphasize why it is important to catalogue the details of normative brain and cognitive ageing in some depth—the next questions that can be asked include: What is 'successful' ageing? What biological processes distinguish between a mammal that is ageing well, and one that is ageing suboptimally? How can this knowledge be used to generate experiments aimed at slowing negative brain/behaviour outcomes in ageing? For focus, this chapter, and this handbook emphasize the hippocampus. This structure's importance for encoding, storage, consolidation, and retrieval of episodic experience is virtually undisputed. Equally clear, however, is the fact that the hippocampus does not work in isolation. The interface between the hippocampus and the rest of the brain must be understood if the circuit dynamics responsible for the varied phenotypes of cognitive ageing are to be discovered. This goal may be achievable—but for the present, a detailed multilevel understanding of even one brain structure is a strong first step in this endeavour. The present chapter provides a review of some of the data available on hippocampal ageing, primarily (but not exclusively) from non-primate species, where the most detailed information has been obtained.

Cross-species cognitive ageing

Behavioural assays that effectively measure hippocampus-dependent memory across the lifespan have provided the tools necessary to understand whether the cognitive consequences of ageing are similar across animals. Measurement methods that do not include language are required to conduct such cross-species experimental comparisons. Fortuitously, one well-known consequence of damage to the hippocampus in mammals is a spatial memory impairment, essentially as outlined in the now classic work by O'Keefe and Nadel (1978), *The Hippocampus as a Cognitive Map*. Here we discuss several examples of human and non-human animal experiments that suggest similarities in the kinds of hippocampus-dependent memory deficits that can be observed over the lifespan. These behavioural data can then be used to ground questions concerning the neurobiological underpinnings of this age-related cognitive phenotype.

Given that there are behavioural changes during ageing and across species that implicate altered hippocampal function (discussed later in more detail), what other variables should be considered when attempting to draw comparisons across mammals concerning normative ageing? One huge consideration is that humans appear to be alone in spontaneously showing the neurological pattern of the markers that define Alzheimer's disease (AD; amyloid plaques and neurofibrillary tangles). There are examples of sparse, but similar pathological signposts in other animals—for instance Cork et al. (1988) examined a population of five bears from the genus *Ursus maritimus*. One 28-year-old polar bear showed senile plaques in the hippocampus and neocortex (no tangles), and one 30-year-old Asian black bear exhibited well-formed intraneuronal neurofibrillary tangles within the CA1 and entorhinal cortical regions (no plaques). Härtig et al. (2000) examined the brains of a number of aged animals for the presence of hyperphosphorylated tau protein. Sparsely scattered abnormally phosphorylated tau was found in Campbell's guenon, rhesus macaques, baboons, spectacled bear, guanaco, and reindeer, while aged bisons showed considerable immunoreactivity for abnormal tau, including neuropil threads. Additionally, Braak et al. (1994) examined groups of old dogs, cats, horses, goats, and sheep. None of the animals examined in this study showed amyloid deposition indicative of plaques, but the old sheep and goats both showed a few neurons in the hippocampus and entorhinal cortex that had intraneuronal neurofibrillary tangles. In other groups of dogs, on the other hand, diffuse amyloid plaques and vascular angiopathies are detected, some of these markers being age-related (Colle et al. 2000), and correlated with maintenance behaviours (e.g. eating, drinking). Neurofibrillary tangles were not observed in this group of dogs either.

While ageing rats and mice do not show spontaneous plaque or tangle accumulation, the rodent that comes closest to showing human-like AD pathology is the Chilean *Octodon degus* (degu), whose Aβ shows an approximately 98% amino acid homology with human Aβ (e.g. Inestrosa et al. 2005; van Groen et al. 2011). Even though there is extensive white matter pathology and presumably significant hippocampal dysfunction in the degu, these pathological symptoms develop at earlier 'equivalent ages' in the degu than in sporadic cases of human AD, and degus do not develop significant neuronal loss as do humans (van Groen et al. 2011). Thus, while examples of the neuropathological hallmarks of AD pathology can be found in other species, these animals do not spontaneously present with a similar disease process. Regardless of whether such pathology can or cannot be detected—it appears that there are animals within each of these different species, including humans, that are free from AD—emphasis in the present chapter will be on humans and animals that are not likely to have neural markers of this disease.

Spatial memory

As previously suggested, to understand the cognitive consequences of ageing in neural circuits, it is extremely helpful to have animal models that allow mechanistic investigation. The first step is to identify a behaviour that is sensitive to experimental or accidental damage of the structure, and that can also be shown to change across age. For the hippocampus, spatial memory and navigation ability can be examined across different animals and used to compare to neurobiological functions of the hippocampus.

Several illustrative examples of spatial memory deficits in older primates are provided here. Uttl and Graf (1993), for example, examined the ability of people between the ages of 15–74 years to remember the locations of exhibits in a science museum through which they had just navigated (Figure 9.1A). In this 'real life-type' situation, declines could be detected in the ability to recall where the exhibits were as early as 55 years of age. In a laboratory situation analogous to a number of rodent experiments (discussed later), Newman and Kazsniak (2000) tested young and older adults for their ability to learn the location of a landmark in a large tent-like environment that had a number of distinguishing cues around its perimeter (Figure 9.1B). When the landmark was removed, and the participants were asked to place it back into the correct location, the elderly adults were not as accurate as were the younger ones. Virtual reality spatial environments have also been used to evaluate navigational abilities in 'virtual mazes' across age. Older adults, compared with younger adults, learned such mazes more slowly, making many repeat entries into interconnected hallways leading to dead ends (Figure 9.1C; Moffat et al. 2001), and also use spatial strategies less often and less effectively in performance of virtual navigation tasks (Iaria et al. 2003). A particularly good example of a deficit in this sort of memory in old nonhuman primates comes from an experiment conducted by Rapp et al. (1997). In this study, monkeys were tethered (using something rather like a retractable dog leash) to the centre of a large platform, and allowed to make one choice at a time between eight different food wells located around the circumference of the platform (Figure 9.1D) (Rapp et al. 1997). The monkey's task was to collect the highly favoured food from each well without revisiting a well from which the food had already been taken. Delays could be imposed between choices, and at the longer delays, older monkeys were particularly disadvantaged at remembering 'where to go next'.

Non-primate species also show age-related performance changes in hippocampus-sensitive spatial memory tasks. For example, old dogs learn a spatial delayed non-matching-to-sample task (Figure 9.2A) more slowly, and once criterion is met, perform more poorly at long delay intervals than do younger dogs (Head et al. 1995). The literature on age-related changes in spatial memory in the rodent is quite extensive, and include early studies showing age differences in rats on the performance of complex, multiple arm T-mazes (e.g. Stone 1929; and see Verzar-McDougall 1957; Goodrick 1968; Penner and Barnes 2007) and radial 8-arm mazes (Figure 9.2B; e.g. Barnes et al. 1980; Wallace et al. 1980; deToledo-Morrell et al. 1984; Beatty et al. 1985), as well as less accurate spatial navigation of old rats and mice in the Barnes maze (Figure 9.2C; e.g. Barnes 1979; Bach et al. 1999) and in the spatial version of the Morris water maze (Figure 9.2D; e.g. Gage et al. 1984; Rapp et al. 1987; Gallagher and Pelleymounter 1988). The overall conclusion that can be drawn from these experiments is that some of the cognitive changes that occur during the normal ageing process implicate the hippocampus.

Neuroanatomical changes in the hippocampus

One obvious place to begin the search for underlying causes of cognitive decline with ageing is to examine anatomical features of the aged brain. As previously mentioned, significant cell loss is not a prominent feature of normative ageing. Early investigations of hippocampal cell density and dendritic branching patterns suggested extensive deterioration in the human hippocampus and entorhinal cortex (Scheibel 1976, 1979). We now know that these studies were confounded by at least two methodological problems. The first was that both demented and non-demented individuals were included in the 'aged' group, and this mixing of diseased and normal participants cannot appropriately reflect normative ageing. The second problem was that the study did not use cell quantification methods in which unbiased counting of the number of objects in a three-dimensional structure could be made. These 'stereological' methods can ensure that objects are sampled with a probability that is proportional to their number and not their size (West 1993a). Subsequent studies employing stereological techniques concluded that significant cell loss in the hippocampus was not characteristic of normal ageing in humans (West 1993b), non-human primates (Keuker et al. 2003) or rodents (Rapp and Gallagher 1996; Rasmussen et al. 1996; Figure 9.3).

In terms of dendritic branching, investigations that employed stereological principles and accounted for the

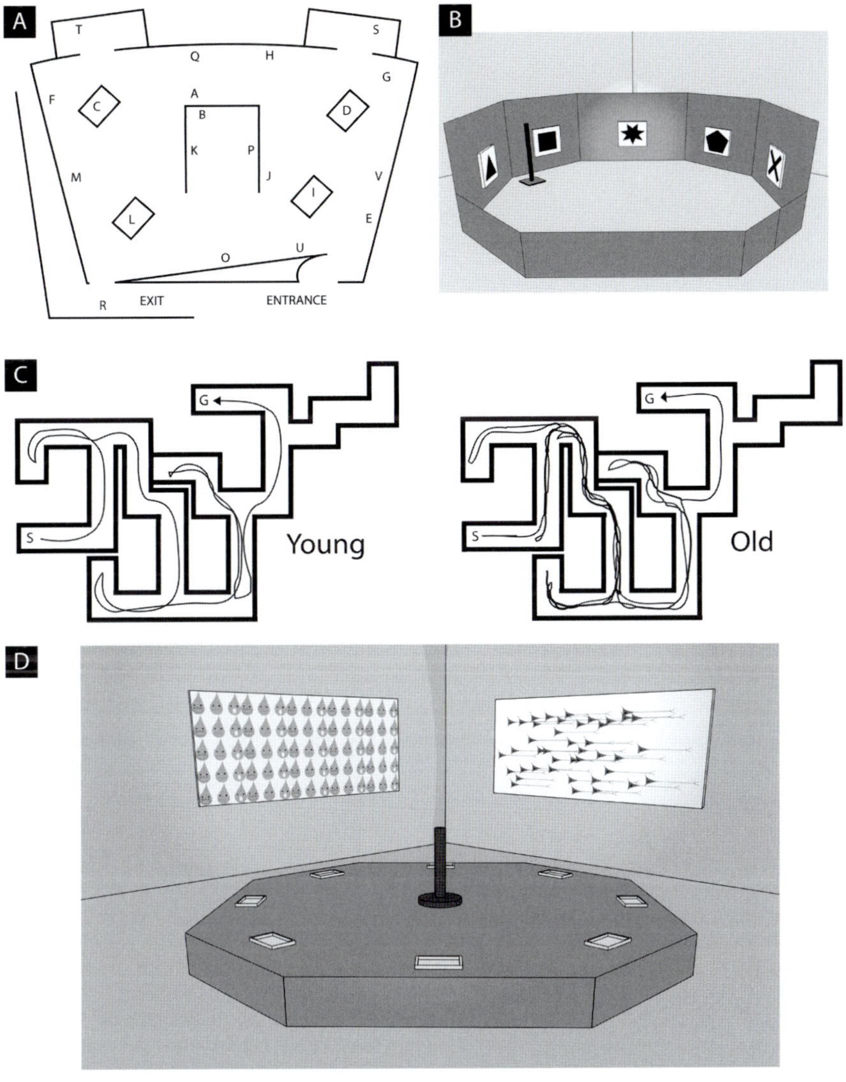

Figure 9.1 Tests of hippocampus-dependent spatial memory in human and non-human primates. A) Floor plan of exhibit locations (denoted by letters) in a science museum. Humans were asked to remember the locations of these exhibits after actually visiting the museum. Older subjects remembered fewer locations compared to young subjects. Reproduced from Uttl, B. and Graf, P. (1993). Episodic spatial memory in adulthood. *Psychology and Aging*, 8, 257–73, © 1993.B) Illustration of a spatial memory apparatus used in humans where participants were asked to navigate a large octagonal tent-like enclosure with distinct cues placed periodically around the periphery of the tent (depicted here with nearest walls partially removed to display interior). Participants were asked to locate and learn the position of a landmark, in this case a pole. When the landmark was removed after learning, older and younger adults were asked to return it to the correct location. Older adults showed less accuracy in this task. Reproduced from *Aging, Neuropsychology, and Cognition*, Newman, M. C. and Kaszniak, A. W. Spatial memory and ageing: Performance on a human analog of the Morris water maze, 2000. Taylor & Francis, reprinted by permission of the publisher (Taylor & Francis Group, http://www.informaworld.com). C) Diagram of a virtual environment in which human subjects are asked to learn to navigate a path through a maze given start and goal locations. Young adults (*left*) traverse the virtual maze with fewer errors whereas older adults (*right*) made repeated errors and learned to navigate the maze more slowly. Reprinted from *Neurobiology of Aging*, 22 (5), Moffat, S. D., Zonderman, A. B., and Resnick, S. M. Age differences in spatial memory in a virtual environment navigation task, 787–96, Copyright (2001), with permission from Elsevier. D) Illustration of an apparatus used to investigate spatial memory in non-human primates. Monkeys were tethered to a pole in the center of a platform and required to sequentially retrieve food rewards from wells at the perimeter of the maze through the use of distal visual cues in the testing room. When long delays were imposed between choices, the older monkeys made more repetitive errors (returning to locations where reward was already collected) Reproduced from Rapp, P.R., Kansky, M.T., and Roberts, J.A. (1997). Impaired spatial information processing in aged monkeys with preserved recognition memory. *Neuroreport*, 8, 1923–8, published 1997 by LLW.

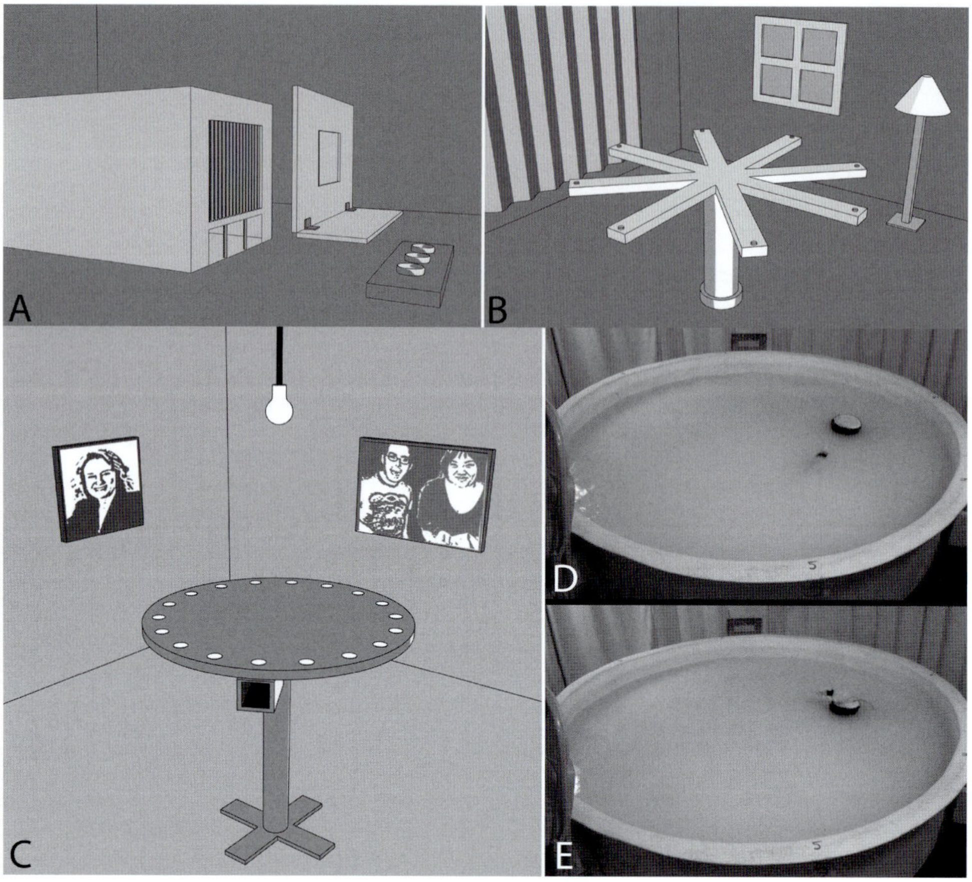

Figure 9.2 Tests of hippocampus-dependent spatial memory in non-primate species. A) Schematic of a Wisconsin General Testing Apparatus (WGTA) modified for use in canines. The WGTA consists of three basic parts: a chamber housing the animal, with one wall containing a grating and gaps for the animal to reach through (or in the case of canines, poke their heads through); a hinged door to allow the researcher to present the test to the animal; and a test tray containing food wells that may be covered by objects. Dogs are required to sit inside the main box of the WGTA and learn a spatial delayed non-matching-to-sample task. In this task, dogs had to displace an object over the correct food location to obtain reward. Older dogs learned the task more slowly and performed more poorly at long delay intervals than did young dogs. Reproduced from Head, E., Mehta, R., Hartley, J., et al. (1995). Spatial learning and memory as a function of age in the dog. *Behavioral Neuroscience*, 109, 851–8, published 1995 by the American Psychological Association. B) Illustration of a radial 8-arm maze used to investigate spatial learning and memory in rodents. Rodents begin trials on the centre platform of the maze and are required to learn and navigate the various maze arms using the distal visual cues in the testing room. In general, aged rodents learn this task more slowly and if delays are imposed between traversals of maze arms, the performance of older rodents declines. C) Illustration of the Barnes maze (Barnes 1979) used to investigate spatial memory in rodents, used with permission. Rats or mice begin trials in a start box (not depicted) in the centre of the platform in a well lit testing room and are required to learn the location of the dark escape box located under one of many holes on the outer perimeter of the maze using distal visual cues in the testing room. Reproduced from Barnes, C.A. (1979). Memory deficits associated with senescence: a neurophysiological and behavioral study in the rat. *Journal of Comparative and Physiological Psychology*, 93, 74–104, published 1979 by the American Psychological Association. (D, E) Images of the spatial version of the Morris water maze used to assess spatial learning and memory in rodents. Mice or rats are required to navigate a pool of opaque water using distal visual cues in the testing room to locate a submerged escape platform. Note that the platform was not submerged in photographs to enhance visibility. The raised platform is used in visual discrimination trials to assess acuity. (D) Image of a rat swimming towards the escape platform and (E) image of a rat that has successfully located the escape platform. Older rodents display longer latencies and path lengths to learn the location of the escape platform.

participants' mental disease status found significantly enhanced dendritic branching in layer II of the parahippocampal gyrus in normal older adults (Buell and Coleman 1979, 1981). Additionally, increased dendritic extent in the dentate gyrus itself has been reported in aged humans (Flood et al. 1985, 1987a). In areas CA1, CA3, and the subiculum of the hippocampus, increases in dendritic branching have been reported to occur in both young and aged individuals suggesting that growth may occur in these brain areas throughout the lifespan (Flood et al. 1987b). A lack of dendritic deterioration in the hippocampus with age has also been observed in rodents. No significant

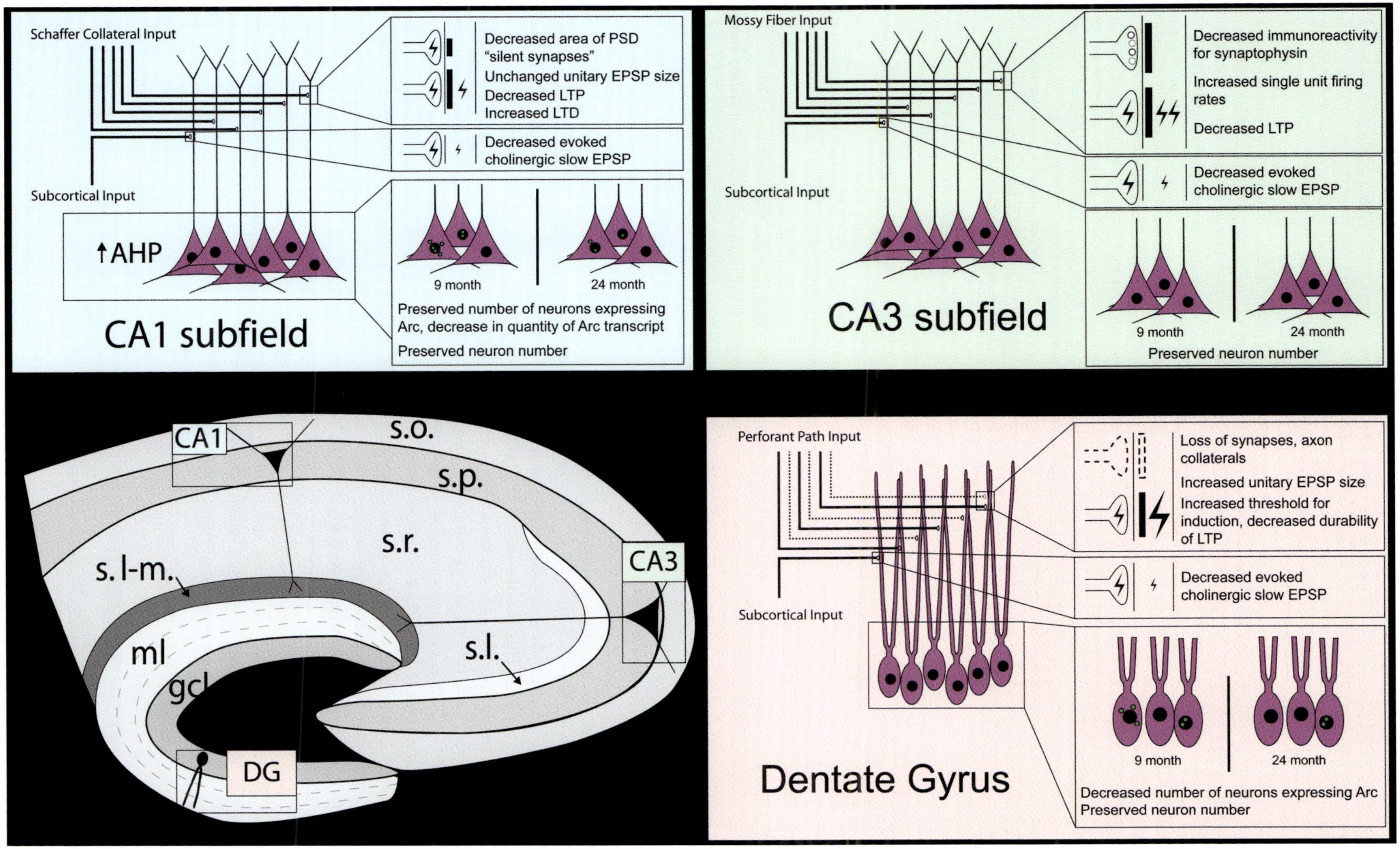

Figure 9.3 Summary of hippocampal changes with age. Summary of the anatomical, biophysical, and electrophysiological properties observed in three subregions of the hippocampus during ageing. *Lower left*) Anatomical schematic of a coronal section of the dorsal hippocampus. The relative locations of areas CA1, CA3, and the dentate gyrus (DG) are shown. Additionally, the location of the cell bodies of principal neurons for each hippocampal subregion discussed in the chapter is depicted. Granule cells in the dentate gyrus (DG), whose cell bodies are located in the granule cell layer (GCL), have dendrites that terminate in the molecular layer of the DG (ML). Mossy fibres originating from granule cells extend through the stratum lucidium (s.l.) of CA3. The cell bodies of pyramidal neurons in ▶

Figure 9.3 *continued*
areas CA3 and CA1 are located in the stratum pyramidale (s.p.) of the hippocampus and have apical dendrites that extend into the stratum lacunosum-moleculare (s-l-m) and the stratum radiatum (s.r.). The basal dendrites of areas CA3 and CA1 arborize in the stratum oriens (s.o.). In the aged hippocampus, all three subregions exhibit preserved neuron number and a decreased evoked cholinergic slow EPSP. Otherwise, the pattern of alterations that occur during normative ageing in areas CA1, CA3 and the dentate gyrus are not consistent between subregions. *Lower right*) The dentate gyrus, which receives input from the entorhinal cortex via the perforant pathway, exhibits a pronounced loss of synapses and axon collaterals from the medial entorhinal cortical projection. Perhaps compensatory, the remaining synapses have an increased unitary EPSP. Also, these synapses show an increased threshold for LTP induction as well as decreased durability of LTP with age suggesting that it is harder for these neurons to encode new information. Indeed, gene expression imaging studies have shown a decrease in the number of aged dentate gyrus neurons (24 months) that express Arc (an immediate early gene involved in memory and plasticity) following behaviour compared to young (9 months) rats. *Upper right*) The pyramidal cells in area CA3 receive input from dentate gyrus granule cells via mossy fibre axons as well as subcortical input. A decrease in the immunoreactivity for synaptophysin has been found during ageing which suggests a loss of synapses. Despite possible synaptic loss, single unit firing rates in CA3 have been observed to increase in the aged animal. Although decreased LTP has been observed in this region in aged animals, little is known about age-associated alterations in Arc expression. *Upper left*) The pyramidal cells in area CA1 receive input from CA3 pyramidal cells via Schaffer collateral axons as well as subcortical input. The postsynaptic density (PSD) sizes of a subset of these synapses are reduced, resulting in electrophysiological 'silence' of some synaptic contacts. The remaining functional synapses show a preservation in the size of the unitary excitatory postsynaptic potential (EPSP), but decreases in long-term potentiation (LTP) and increases in long-term depression (LTD). Furthermore, aged CA1 neurons have an increased after-hyperpolarizing potential (AHP) that might compensate for increased excitatory drive from CA3. Gene expression imaging studies have revealed that the same number of aged (24 months) CA1 neurons express Arc after behaviour compared to young animals (9 months) but the amount of Arc transcript per cell is decreased with age.

age-related changes were found in dendritic length of granule cells (Flood 1993) or dendritic extent in area CA1 (Turner and Dupree 1991; Pyapali and Turner 1996).

Investigations of potential alterations in dendritic spine density across the lifespan suggest that the changes that do occur are region specific. In the hippocampus, there is no significant reduction in spine density in the dentate gyrus of humans or rats (Curcio and Hinds 1983; de Ruiter and Uylings 1987) nor is there a change in CA1 of rats (Markham et al. 2005). Reductions in spine density have been reported, however, in the subiculum of aged non-human primates (Uemura 1985).

Although principal cell numbers, dendritic branching, and dendritic spine density are not substantially altered in the aged hippocampus, aged animals do exhibit alterations in synapse number (Figure 9.3). This change in cell connectivity has important implications for the function of aged hippocampal networks. Early investigations using electron microscopy reported a 27% decrease in axodendritic synapse number of the layer II entorhinal cortical projection to the dentate gyrus (the so-called 'perforant path'; Bondareff and Geinisman 1976; Geinisman et al. 1977). These results have been replicated using stereological methods and indicate a reduction in numbers of spinous synapses in the inner and middle molecular layer of the dentate gyrus in aged compared to young rats (Geinisman et al. 1992). Moreover, both perforated and non-perforated axospinous synapses were reduced during ageing, not axodendritic synapses. Furthermore, the reductions in synaptic contacts was specific to aged, spatial memory-impaired animals as aged rats with no memory impairment were not different from the young rats. These data support the idea that axospinous synapse loss is a feature frequently found in the aged dentate gyrus.

Because layer II entorhinal cortical projections reach both the granule cells of the dentate gyrus and CA3 pyramidal cells, it might be expected that there would also be reduced synapse numbers in this dendritic region of CA3 during ageing. While serial-section electron microscopic studies of synapse numbers have not been conducted in the CA3 region, confocal microscopy has been used to investigate the immunoreactivity of synaptophysin, which is a presynaptic vesicle marker (Figure 9.3; Smith et al. 2000a). Consistent with likely synapse loss of entorhinal inputs, aged rats with spatial memory deficits exhibited significant reductions in synaptophysin in the CA3 lacunosum moleculare compared with young rats.

Unlike the granule cells and CA3 pyramidal cells, CA1 pyramidal cells do not lose synaptic contacts with age in the Schaffer collateral projection region (Geinisman et al. 2004). Additional analysis of the morphology of the Schaffer collateral synapses of old rats in the CA1 region, however, has revealed a difference between young and aged animals—a subset of the aged axospinous perforated synapses show a reduced postsynaptic density size (Nicholson et al. 2004).

Biophysical properties of aged neurons

Although a number of morphological changes occur in the hippocampus of aged, memory-impaired rats, sharp electrode recordings taken *in vitro* from hippocampal pyramidal and granule cells suggest that old neurons are remarkably healthy. For example, properties such as the resting membrane potential, membrane time constant, input resistance, threshold to action potential, and the amplitude and duration of the Na^+ action potential are not altered over the lifespan (reviewed in Barnes 1994; Rosenzweig and Barnes 2003). Naturally, there is at least one exception to this general observation, which includes an increased amplitude of the after-hyperpolarizing potential (AHP) in CA1 pyramidal cells (Landfield and Pitler 1984; Disterhoft et al. 1996; Thibault and Landfield 1996). The AHP is the phase of a neuron's action potential where the cell's membrane potential falls below the normal resting potential and contributes, in part, to the refractory period during which a neuron cannot fire another action potential. The implication of this biophysical change would be that old CA1 pyramidal cells should remain refractory longer than younger cells (this is discussed in more detail in the 'Dynamics of neural ensembles' section). Age-related increases in the AHP are thought to be due to alterations in Ca^{2+} conductance in aged neurons as the outward K^+ current that underlies the AHP is Ca^{2+}-dependent (Figure 9.3; Thibault and Landfield 1996). Additionally, it has been shown that CA1 pyramidal neurons have an increased density of L-type Ca^{2+} channels (Thibault and Landfield 1996). Because of the long-lasting activation kinetics of L-type channels, their increased density is hypothesized to lead to disruptions in Ca^{2+} homeostasis in old pyramidal cells (Thibault and Landfield 1996; reviewed in Kumar et al. 2009). Hippocampal granule cells, on the other hand, do not exhibit the same repolarizing currents as do pyramidal cells, nor is there a difference observed between young and old rats in the magnitude of the hyperpolarizing afterpotential that terminates the action potential in this cell type (Barnes and McNaughton 1980).

Electrophysiological measures of connectivity

In addition to measuring the biophysical properties of individual hippocampal cells, other physiological indicators of the input-output characteristics of the circuit have been evaluated in young and old rats. These include an extracellular indicator of input size—the presynaptic 'fibre potential'. For example, the compound action potential (fibre potential) of axon collaterals of layer II entorhinal cortical cells can be measured in the dentate gyrus following stimulation of the perforant pathway (a region where entorhinal cortical axons converge on their way to the hippocampus). The amplitude of this fibre potential response is proportional to the numbers of these collaterals that discharge action potentials following stimulus delivery. The fibre potential can therefore be considered to reflect a measure of the 'input' to a given region. The extracellularly-recorded field 'excitatory postsynaptic potential' (EPSP), on the other hand, reflects the summed currents activated in the postsynaptic cells following input. This membrane depolarization is a complex function of existing pre- and postsynaptic factors, and can be thought of as an indicator of the 'strength' of the activated synapses for a given input size (i.e. the numbers of axons carrying input to the region). Finally, if enough synapses are activated to reach EPSP amplitudes above threshold for action potential discharge, then 'population spikes' are superimposed on the field EPSP, and can be quantified. The population spike is proportional to the numbers of cells (e.g. granule cells or CA1 pyramidal cells) discharging action potentials following a given stimulus. Thus, the population spike can be considered a measure of the 'output' of the circuit in question.

In terms of ageing, the amplitude of the presynaptic fibre potential at the perforant path—granule cell synapse is decreased in old rats compared to young (Figure 9.3; Barnes and McNaughton 1980; Foster et al. 1991; Barnes et al. 2000). The stimulus current levels necessary to elicit a fibre potential response in old rats does not change (i.e. there is no change in threshold to elicit an action potential), the change is in the ability to recruit more axons at higher stimulus intensities (Barnes and McNaughton 1980). Given that there is no age-related change in the numbers of cells in the entorhinal cortex (Merrill et al. 2001), the electrophysiological demonstration of a reduced fibre potential amplitude can most easily be explained as entorhinal axon collateral pruning (fewer perforant path axon collaterals). These data are consistent with a loss of synapses onto granule cells from the medial entorhinal cortex—and suggests that axon branches are lost along with their synaptic contacts. For CA1, on the other hand, the amplitude of the Schaffer collateral fibre potential does not different between age groups (Kerr et al. 1991; Barnes et al. 1997a, 2000; Rosenzweig et al. 1997; Potier et al. 2000). These electrophysiological data are consistent with the anatomical observation that Schaffer collateral CA1 synapse numbers do not differ across age (Geinisman et al. 2004), and thus presumably there is no axon pruning in this region.

Consistent with the anatomical observation of fewer synapses and the electrophysiological observation of reduced amplitude of the presynaptic fibre potential, the amplitude of the field EPSP in the dentate gyrus is reduced in old compared to young rats (Barnes 1979; Barnes and McNaughton 1980). The relative synaptic strength for a given input can be determined, by plotting the amplitude of the extracellular field EPSP (a measure of synaptic strength) as a function of presynaptic fibre potential amplitude (a measure of input). When this is done for the perforant path—granule cell synapse the granule cell postsynaptic response is *greater* in the *old* rats than in the young, for a given compound action potential size (Figure 9.3; Barnes and McNaughton

1980; Foster et al. 1991). It is also possible to stimulate one or a few axons to evoke 'unitary EPSPs' which can be measured intracellularly with sharp electrodes. This method gives a more direct measure of synaptic strength. The hypothesis tested with respect to granule cells in ageing rats was that unitary EPSP sizes of old rats should be larger than in young rats if the conclusions derived from the field potential recording data were correct. In fact, this is what was observed—single perforant path axon stimulation resulted in larger postsynaptic depolarizations (unitary EPSPs) in old rats than in young (Foster et al. 1991). Among a number of possibilities, these data are consistent with the idea that a compensatory process (synaptic strengthening) is engaged that keeps the overall magnitude of depolarizing input onto the granule cells relatively constant across age.

In area CA1, on the other hand, there is no synapse loss with ageing and no reduction of presynaptic fibre potential amplitude, as discussed above. Nevertheless, the amplitude of the field EPSP following Schaffer collateral stimulation is reduced in old rats (Figure 9.3; Landfield et al. 1986; Barnes et al. 1992; Bauman et al. 1992; Deupree et al. 1993; Potier et al. 2000). One hypothesis that may explain reduced synaptic strength at this synapse in the old rats is the observation that old rats showed reduced postsynaptic density sizes in subsets of their axospinous perforated synaptic contacts (Nicholson et al. 2004). Because the size of intracellularly recorded unitary EPSP sizes do not change in old CA1 pyramidal cells (Figure 9.3; Barnes et al. 1992), it is possible that the synapses with smaller postsynaptic densities are silent. Thus, for CA1, it appears that there is a loss of 'functional synapses' rather than a loss of incoming axons or synaptic contacts.

Taken together, these findings support the idea that age-associated alterations in synapses of the hippocampus is subregion specific with direct synaptic loss occurring in the dentate gyrus and synaptic loss of function or silencing occurring in CA1. The relationship between EPSP amplitude (synaptic strength) and population spike amplitude (output) is also different between subregions with respect to ageing. While this relationship has not been reported to change across age in CA1, a clear excitability shift occurs in old rat granule cells. In field potential recordings, old rats exhibit larger population spikes for a given EPSP amplitude, and intracellular data suggest a reduced granule cell discharge threshold following synaptic activation (Barnes and McNaughton 1980).

While subregion selectivity of age-related change is more the 'rule', there are at least two other electrophysiological changes that occur with ageing across all three hippocampal subfields. The first is the cholinergic slow EPSP that can be induced by high-frequency (50 Hz) stimulation of the stratum oriens in CA1 or CA3, or in the granule cell layer for the dentate gyrus where cholinergic fibres are particularly rich. There is an attenuation of the amplitude of this atropine-sensitive response by about 50% in each of the hippocampal subregions in aged rats (Figure 9.3; Shen and Barnes 1996). Another age-related change that occurs in CA1 and CA3 pyramidal cells and in hippocampal granule cells is an increase in gap junction-mediated electrotonic coupling between the cells in old rats (Barnes et al. 1987). The former effect (reduced cholinergic response) would be expected to reduce excitability; the latter effect (increased gap junctional connections) would be expected to increase excitability. Clearly there is substantial evidence of biological adaptation and rearrangement that occurs with ageing—and much more to be learned about which of these changes may be compensatory to keep information processing in the systems stable.

Neural plasticity

What is the impact on plasticity mechanisms of the altered morphology, biophysical properties, and circuit connectivity of aged hippocampal neurons just reviewed? Among the main contenders for the biological processes that the brain likely uses to encode and store information, two well-described forms of synaptic plasticity, long-term potentiation (LTP) and long-term depression (LTD), have been examined in some detail in the ageing rodent. LTP is the long-lasting strengthening of synapses induced by patterned electrical stimulation of inputs. LTP can be divided into induction and maintenance phases, both of which have been investigated to test the hypothesis that defects in this process may underlie memory deficits associated with ageing. Although there have been apparently contradictory results reported, this has largely occurred because of differences between studies in stimulation parameters used as well as the hippocampal region under investigation. In general, aged animals exhibit deficits in both LTP induction and maintenance.

Even though there is clear synapse loss (Geinisman et al. 1992) and reduced field EPSP sizes in the dentate gyrus of old rats (Barnes and McNaughton 1980; Foster et al. 1991), robust LTP induction can occur at the perforant-path granule cell synapse (Barnes 1979; Diana et al. 1994a, 1994b). For CA3, there is no LTP induction deficit in the layer II entorhinal cortical projection to these pyramidal cells (Dieguez and Barea-Rodriguez 2004). Finally, in spite of the reduced EPSP size and loss of functional synapses in area CA1, old rats can also show strong LTP induction at the Schaffer collateral CA3–CA1 synapse (Landfield and Lynch 1977; Landfield et al. 1978). Successful LTP induction at all major synaptic sites in the hippocampus of aged animals is achieved when the stimulation protocols used are well above threshold intensities required to induce LTP. When stimulus parameters are closer to the LTP induction threshold, however, aged animals can show deficits at the perforant path-granule cell synapse (Barnes et al. 2000) as well as the Schaffer collateral CA3–CA1 synapse (Deupree et al. 1993; Moore et al. 1993). For the dentate gyrus, this defect is due to an actual increase in the threshold intensity necessary to induce LTP. This was shown by using a 'paired-depolarization' stimulation protocol, in which weak presynaptic stimulation is combined with direct intracellularly-applied depolarizing current in granule cells. When this procedure is conducted, more current must be injected before LTP is produced in old granule cells (Barnes et al. 2000). These data are consistent with the interpretation that granule cells in the aged dentate gyrus have an increased LTP induction threshold.

In contrast, aged pyramidal cells in CA1 do not have an altered threshold for LTP induction elicited by the paired-depolarization procedure (Barnes et al. 1996a). Nevertheless, old rats still exhibit less LTP when weak stimulation parameters are used (Moore et al. 1993; Rosenzweig et al. 1997; Tombaugh et al. 2002). The mechanism for the age-related deficit using peri-threshold stimulation parameters in CA1 is distinct from that observed in the dentate gyrus. Although Schaffer collateral axons in young and aged rats respond and can similarly follow trains of high-frequency stimulation used in induction protocols, CA1 neurons in aged rats exhibit weaker temporal summation of the depolarizing currents. This defective EPSP summation leads to less overall depolarization during the high-frequency stimulation episode. It is this reduced depolarization that is responsible for old pyramidal cells exhibiting reduced LTP with weak stimulation.

LTP induction typically requires sufficient depolarization of AMPA receptors to eject the Mg^{2+} block from the pore of the N-methyl-D-aspartate (NMDA) receptor. Once activated, the NMDA receptor (a ligand-gated cation channel selective for glutamate) allows considerable influx of Ca^{2+} ions into the postsynaptic cell. Once in the cell, Ca^{2+} acts as a major signalling molecule affecting a wide range of processes including membrane excitability and gene regulation. The major sources of intracellular Ca^{2+} include influx through this NMDA receptor as well as through voltage-gated Ca^{2+} channels (such as the L-type Ca^{2+} channel) and intracellular stores. A number of factors suggest that the plasticity deficits observed during ageing could be caused, in part, by changes in Ca^{2+} regulation. These include the suggestion that NMDA receptor function declines (Gonzalez et al. 1991; Pittaluga et al. 1993), the density of L-type Ca^{2+} channels increase (Thibault and Landfield 1996), and the sequestration and buffering of intracellular Ca^{2+} decreases during ageing (reviewed in Verkhratsky and Toescu 1998; Kumar et al. 2009).

Because Ca^{2+} concentrations are hypothesized to determine the threshold for synaptic modification, disrupted calcium regulation in old hippocampal cells could affect whether a synapse will be potentiated or depressed (Bear et al. 1987; Bear and Malenka 1994), altering the probability that synaptic activity will induce LTP or LTD. Indeed, Ca^{2+} imaging studies in aged CA1 neurons show no differences between young and old rats in Ca^{2+} levels at rest, but greater Ca^{2+} concentrations following stimulation (Thibault et al. 2001). Moreover, it has been shown that aged rats are more susceptible to LTD than are young rats (Norris et al. 1996) and that inhibition of Ca^{2+} release from intracellular stores via the application of cyclopiazonic acid can 'recalibrate' LTD induction thresholds within CA1 of aged animals more towards normally-occurring levels in younger rats (Kumar and Foster 2005).

Once LTP has been induced, the continued expression of the increased synaptic strengthening or maintenance of LTP can be measured. In the short term (i.e. over an hour), decay of LTP is not different between young and aged animals (Landfield and Lynch 1977; Landfield et al. 1978). Conversely, aged animals show deficits in LTP maintenance in the dentate (Barnes and McNaughton 1980) and CA3 over a longer time course (Diana et al. 1994a, 1994b). Furthermore, the impairment of LTP maintenance in aged animals often correlates with deficits in hippocampus-dependent memory (reviewed by Barnes 1979; deToledo-Morrell et al. 1988; Barnes and McNaughton 1980).

The general deficits in LTP maintenance observed in aged animals have been shown to be accompanied by a greater vulnerability or susceptibility to induction of LTD (Norris et al. 1996; Foster and Norris 1997). The balance between LTP and LTD in aged animals is likely due to a variety of interactions between synaptic activity and age-associated alterations in signaling pathways. As discussed previously, age-related changes in the handling of Ca^{2+} are likely to affect plasticity and the balance of LTP/LTD in aged neurons. Additionally, functional decreases in the NMDA receptor in aged animals (Pittaluga et al. 1993) as well as a reduced NMDA receptor mediated response (Rao et al. 1993) have been reported. Indeed, Wenk and Barnes (2000) reported fewer NMDA receptor binding sites in CA1, CA3, and the subiculum of aged rats, although no age differences were found in the dentate gyrus. Recent work suggests that shifts in the intracellular redox state also contributes to the decline in NMDA receptor responses (Bodhinathan et al. 2010). In this study, it was shown that the oxidizing agent xanthine/xanthine oxidase decreased the NMDA receptor-mediated response at the CA3-CA1 synapse in young hippocampal slices but not aged. When a reducing agent (dithiothreitol) was applied, the NMDA receptor response was selectively enhanced to a greater extent in aged slices and facilitated the induction of LTP.

Additional age-related differences that involve the NMDA receptor include alterations in downstream signalling pathways, particularly the cAMP/CREB pathway (discussed further in the 'Gene expression' section). Briefly, the cAMP/CREB pathway is involved in initiating gene transcription and thus protein synthesis, both required for LTP maintenance. Although relatively few studies have aimed to directly investigate the molecular mechanisms of reduced LTP maintenance in ageing, it has been shown that upregulation of the cAMP pathway attenuates age-related plasticity impairments (Bach et al. 1999). Specifically, the cAMP pathway was upregulated by dopamine D1/D5 receptor agonists.

The dopamine D1/D5 receptor was again implicated in plasticity processes in work by Huang and Kandel (2006). They reported a form of protein synthesis-dependent LTP that could be induced by low frequency stimulation of CA1 in aged but not young hippocampal slices. Further, Huang and Kandel (2006) reported that this potentiation was mediated by the D1/D5 dopamine receptor and was blocked by emetine (an irreversible protein synthesis blocker). Interestingly, this form of LTP involves the NMDA receptor as well as voltage-dependent calcium channels (such as the L-type calcium channel). Given that Ca^{2+} conductance is known to change with age, it could be that differences in Ca^{2+} levels contributed to the difference in LTP induction compared to the young group. Moreover, memory has been shown to be enhanced in old animals when voltage-dependent calcium channels are blocked (Deyo et al. 1989; Scriabine et al. 1989; Straube et al. 1990; McMonagle-Strucko and Fanelli 1993), suggesting that the inhibition of low frequency-induced plasticity could actually be beneficial during ageing.

To summarize, LTP induction and maintenance generally declines with age. Although there are no 'fixes' for age-related anatomical changes in the hippocampus such as synapse loss, hippocampal LTP involves a variety of different receptors and intracellular signalling cascades that can be manipulated to alleviate age-associated deficits in plasticity. These manipulations include pharmacological applications such as cyclopiazonic acid to prevent the release of Ca^{2+} from internal stores (Kumar and Foster 2005), reducing agents to balance the internal redox state and restore NMDA receptor function (Bodhinathan et al. 2010), dopamine D1/D5 agonists to upregulate the cAMP pathway (Bach et al. 1999), and inhibition of low frequency stimulation by emetine (Huang and Kandel 2006). In short, there are a number of viable 'treatments' that could be explored further with the aim of ameliorating plasticity deficits that may improve hippocampal-dependent memory in the aged.

Age-related changes in gene expression

The maintenance of LTP requires gene expression and *de novo* protein synthesis. Studies investigating protein synthesis alterations during ageing have generally concluded that long-term memory impairments in old animals can be attributed to decreases in protein translation (Davis et al. 1981; Mizumori et al. 1985; reviewed in Schimanski and Barnes 2010). Further examination of specific proteins that have roles in synaptic plasticity focused on the mRNAs that encode these proteins as well as the transcription factors that regulate mRNA expression. Given that aged animals exhibit alterations in LTP induction, it is not surprising that age-related changes in gene expression have also been observed.

Immediate-early genes (IEGs) are genes that do not require *de novo* protein synthesis or the activation of other responsive genes. A critical event for the expression of IEGs is the phosphorylation of cAMP-responsive-element-binding protein (CREB). When phosphorylated, CREB promotes the transcription of IEG mRNA that is then translated into protein. Based on the functional role of the translated protein, IEGs can be grouped into those that encode transcription factors which regulate gene expression and those that encode effector proteins which directly affect the cell (Guzowski 2002; reviewed by Sheng and Greenberg 1990). Early investigations of IEGs that function as transcription factors found that the IEG c-fos is rapidly induced in the neurons of animals following seizure (Morgan et al. 1987). Subsequent studies found that c-fos expression can be up-regulated by stimuli that elicit LTP (Dragunow et al. 1989; Worley et al. 1993; Herdegen and Leah 1998) and has been hypothesized to have a role in learning and memory (Grimm and Tischmeyer 1997; Guzowski and McGaugh 1997). Changes in c-fos expression over the lifespan likely occur, however, the data are variable. Some studies report no age-related changes in c-fos (Desjardins et al. 1997), while others report significant declines in the proportion of cells that express c-fos after learning (Touzani et al. 2003). Furthermore, using a reverse Northern strategy it was found that c-fos was *elevated* in old compared to young rats following an LTP induction paradigm in an awake preparation (Lanahan et al. 1997).

Another IEG that encodes a transcription factor is zif268. Like c-fos, zif268 has been shown to be upregulated after stimuli that induce LTP (Cole et al. 1989; Richardson et al. 1992). Similar to the experiments that have examined c-fos, the data on age-related alterations in zif268 expression are mixed. For example, it has been suggested that resting levels of zif268 in aged animals is reduced (Yau et al. 1996). The proportion of cells that express zif268 at rest, however, does not correlate with performance on a spatial learning task in the Morris water maze (Desjardins et al. 1997). Conversely, more recent studies indicate that reduced levels of zif268 in aged CA2 pyramidal cells at rest correlate with spatial learning (Yau et al. 1996). In general, it is thought that the expression of zif268 is necessary but not sufficient for the maintenance of LTP and long-term memory (Jones et al. 2001).

Among the effector IEGs are neuronal activity-regulated pentraxin (Narp; Tsui et al. 1996), Homer 1a (Brakeman et al. 1997), and activity-regulated cytoskeletal gene (Arc; Link et al. 1995; Lyford et al. 1995). Whereas c-fos and zif268 encode transcription factors and are located near the nucleus, the effector IEGs translocate closer to the synapse. Homer 1a interacts with several proteins within the postsynaptic density and may play an important role in modifying glutamatergic signalling pathways (Xiao et al. 2000). Narp mRNA translocates to the synapse after transcription where it is released and is thought to cluster glutamate receptors critical in postsynaptic membrane depolarization (α-amino-3-hydroxyl-5-methyl-4-isoxazole-propionate or AMPA receptors) (Reti et al. 2002) and Arc mRNA also localizes to the active region of the dendrite (Steward et al. 1998). There is some indication that Narp expression is downregulated in area CA1 of spatial memory-impaired rats (Blalock et al. 2003), but the data are sparse and little is known about alterations in Narp across the lifespan. Likewise, changes in Homer 1a expression in CA1 of aged rats does not appear to occur (Penner and Barnes, unpublished data) but more in-depth investigations are needed. The specialized localization pattern of Arc mRNA, however, has made it the most studied IEG in terms of learning, memory, and plasticity (reviewed in Plath et al. 2006; Bramham et al. 2008; Tzingounis and Nicoll 2006).

The observation that *Arc* is targeted to dendrites suggests that it may contribute to structural plasticity (Lyford et al. 1995). In fact, it has been shown that *Arc* is only targeted to that portion of the dendrite that has undergone synaptic modification (Steward et al. 1998). Furthermore, Arc blockade selectively impairs LTP maintenance and long-term memory without affecting LTP induction (Guzowski et al. 2000). *Arc* has also been shown to be induced in a selective way by behaviour such as spatial exploration (Guzowski et al. 1999). In particular, *Arc* is thought to be selectively expressed in hippocampal neurons that show patterned neural activity highly correlated with a rat's position in an environment (namely, the 'place field' of a cell; Guzowski et al. 1999). About 40% of CA1 pyramidal cells and 2% of granule cells show place-specific firing in a given environment (discussed in more detail in 'Dynamics of neural ensembles' section; Jung and McNaughton 1993; Wilson and McNaughton 1993) which is reflected in the proportion of neurons that express Arc for the same exploratory behaviour (Guzowski et al. 1999). In addition to activity-dependence, the transcriptional kinetics of Arc allow activated neurons to be mapped across two distinct epochs of behaviour, as well as across multiple brain regions. Of note is that Homer 1a also seems to be activity-dependent and the transcriptional kinetics of H1a are also well known, although Homer 1a transcription occurs at a longer time delay than Arc. A powerful variant of anatomical methods that monitor the expression of neural-activity-dependent IEGs, such as Arc or Homer 1a, has been called cellular compartment analysis of temporal activity by fluorescence *in situ* hybridization (catFISH; Guzowski et al. 2000; Plath et al. 2006).

There is a complex set of changes in Arc expression in the hippocampus during normative ageing. Utilizing fluorescence *in situ* hybridization (specifically catFISH), the number of cells that express Arc can be determined after one or two distinct behavioural treatments. For example, aged rats show a reduced number of granule cells that express Arc after two bouts of spatial exploration (Small et al. 2004; Penner et al. 2011). In areas CA1 and CA3, however, the same numbers of cells express Arc following equivalent behaviours across age (Small et al. 2004). Further investigation of Arc expression in the aged hippocampus used quantitative methods (real-time quantitative reverse transcriptase polymerase chain reaction or RT-PCR) in which gene expression levels within single cells can be determined (Penner et al. 2011). Penner et al. (2011) used the catFISH method on one hemisphere to count the numbers of cells that express Arc following behaviour, and RT-PCR on the other hemisphere to obtain an estimate of the numbers of transcripts of Arc expressed in dissected hippocampal subregions. The data revealed that while there is no change in the number of cells that express *Arc* in aged CA1 pyramidal cells (consistent with Small et al. 2004), there is a reduction in the amount of Arc transcribed per CA1 cell. For granule cells, RT-PCR revealed that aged granule cells transcribe less Arc per cell (Penner et al. 2011). Therefore, even though Arc is transcribed during learning in the hippocampus of aged rats, the total effective output is diminished. This reduction in Arc transcript could affect the ability of dendrites to be engaged in plasticity and ultimately LTP maintenance and long-term memory.

Dynamic alterations of Arc transcription—treatment approaches

Transcription can be powerfully affected by epigenetic processes such as DNA methylation and histone modifications. To understand the mechanisms that may be responsible for attenuated Arc transcription, Penner et al. (2011) measured DNA methylation of the Arc gene. DNA methylation involves the donation of methyl groups across CG-rich gene regulatory regions generally found near or within promoter regions of mammalian genes.

Methylation of these CpG 'islands' usually confers the transcriptional silencing of the affected genes. While it was previously thought that DNA methylation was a static process (Kangaspeska et al. 2008; Métivier et al. 2008; Ma et al. 2009), recent work suggests that it can be an active process that can change dynamically by such events as learning and memory (Graff and Mansuy 2008; Roth et al. 2009). Moreover, disruption of the methylation process can result in long-term memory consolidation deficits (Miller and Sweatt 2007; Lubin et al. 2008).

In area CA1, it was found that spatial exploration resulted in a decrease in DNA methylation of the promoter region of the Arc gene in both adult and aged animals (Penner et al. 2011). In the intragenic region of the Arc gene in CA1, however, aged animals exhibited a significant decrease in methylation whereas adult animals showed an *increase* in methylation following behaviour. Given that Arc mRNA levels in CA1 are attenuated in aged rats, and that DNA methylation is typically thought to result in transcriptional silencing, the results of an age-related decrease in methylation is unexpected. Recent work, however, indicates that, under some conditions, DNA methylation may be indicative of active transcription (Chahrour et al. 2008). Therefore, it seems that optimal transcription of Arc may require two events in CA1 pyramidal cells, demethylation at the promoter region and methylation at the intragenic region of the gene.

Like in CA1, the dentate gyrus of adult and aged animals showed increases in methylation of the promoter region of the Arc gene after behaviour. In the intragenic region, however, the pattern of methylation in the dentate gyrus is completely opposite of what was observed in CA1. Namely, aged animals exhibit an *increase* in methylation of the intragenic region of the Arc gene in the dentate gyrus whereas adult animals show a reduction. Given that granule cells in the dentate gyrus are functionally distinct from CA1 pyramidal cells, it could be that the implications of increased methylation during ageing in these cells are also different.

Though sparse data from area CA3 exist with respect to dynamic methylation, it has been shown that antiepileptic agents such as sodium valproate produce dose-dependent spatial memory improvements in aged rats (Koh et al. 2010). Though the interpretation of Koh et al. (2010) was that sodium valproate attenuated a posited over-excitation or excess of neural activity in CA3 pyramidal cells, the compound is also a potent histone deacetylase (HDAC) inhibitor. The acetylation of histones 'loosens' DNA binding to histones and facilitates transcription. Therefore, HDACs are thought to silence or inhibit transcription essentially like the downstream process of DNA methylation (described earlier). When HDACs are inhibited, transcription should be restored. If aged CA3 pyramidal cells are similar to aged CA1 pyramidal cells in that they transcribe less Arc per cell, a treatment with an HDAC inhibitor to increase transcription of genes important for learning would indeed result in an increase in performance in aged animals. This type of treatment may provide a novel avenue to 'intervene' in the process of memory loss that occurs during ageing.

Although the interpretations of data involving DNA methylation and histone acetylation in the hippocampus during normative ageing is still somewhat speculative, it is clear that the transcription and expression of genes are necessary for optimal learning and memory function. Further, as the transcription of genes is known to generally be affected by the ageing process, a better understanding of the dynamic mechanisms that control gene activation will provide critical insights that may lead to the development of strategies or therapeutic interventions designed for more successful ageing—such as the one alluded to previously.

Dynamics of aged neural ensembles

Principal cells in the hippocampus show elevated firing rates that correlate with an animal's position in a given environment. This is why O'Keefe and Dostrovsky (1971) named them 'place cells'. When a rat is not within this region of space, hippocampal cells are nearly silent compared to rates of 10–20 Hz recorded as the rat moves through the cell's receptive field (known as its 'place field'). These hippocampal cells include both the pyramidal cells of the CA1 and CA3 subfields (O'Keefe and Dostrovsky 1971) as well as the granule cells of the dentate gyrus (Jung and McNaughton 1993; Leutgeb et al. 2007). With the development of multiple single-unit recording technologies, the firing of many hippocampal neurons can be recorded simultaneously in freely behaving animals (Wilson and McNaughton 1993). If the population of recorded hippocampal cells is large enough, the trajectory of an animal through an environment can be reconstructed from cell firing patterns alone (Wilson and McNaughton 1993).

When place fields from many individual place cells are composited over a schematic of the environment, the resulting activity-based 'map' (Figure 9.4) is an example of a distributed representation in neuronal encoding—the representation of the environment is encoded by a collection of cells. This has several advantages over local representations in which one neuron encodes the entire representation. One advantage of distributed encoding schemes particularly relevant to the current discussion is that individual place cells can exhibit place fields in multiple,

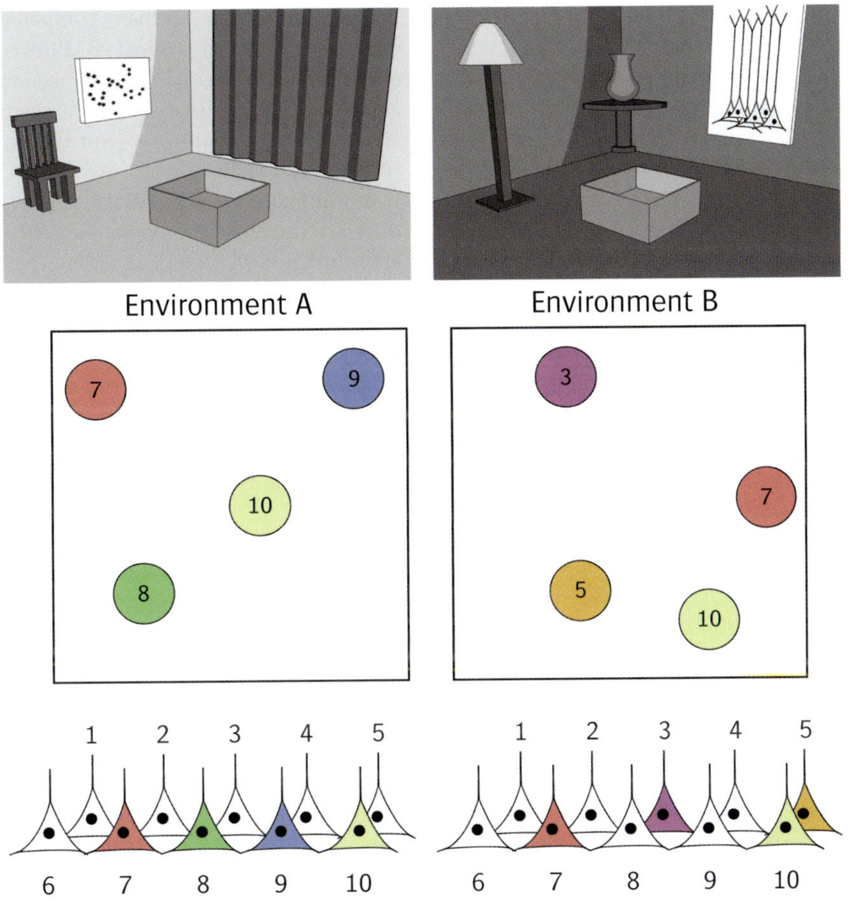

Figure 9.4 Neuronal representation of environments. Illustration of how a subpopulation of hippocampal place cells might collectively 'encode' or represent two different environments, A and B, illustrated at the top of the figure. Each environment consists of a box that a rat can explore, placed in a physically and visually separate room. The hippocampus will encode separate maps for each environment and recall the appropriate map upon revisiting an environment. *Left*) An example of how pyramidal place cells might encode a box in Environment A. Each numbered place cell and the region of Environment A in which the place cell selectively fires (its place field) are denoted. For example, place cell 7 has a place field in the upper left corner of Environment A whereas place cell 8 has a place field in the lower left corner of the environment. In a box that measures 60 cm on each side, approximately 40% of place cells in the dorsal CA1 subregion of the rat hippocampus will be activated during thorough exploration. *Right*) An illustration depicting 'remapping' of hippocampal place cells. In Environment B, a statistically independent population of CA1 neurons is recruited to encode this environment. In this example, two neurons active in Environment A are now silent, and two previously silent neurons are now active. Note that some neurons are active in both environments but did not fire in the same locations of the box between environments (Neurons 7 and 10). When place cells are recycled to encode multiple representations, there is no correlation in the location of firing between environments that share similar features—for instance, a place cell that fires in the upper left corner of Environment A (Neuron 7) does not also fire in the upper left corner of Environment B. Additionally, individual place cells do not express place fields that are fixed in relation to each other across all environments—two cells that express adjacent place fields in one environment, and are both active in a second environment, do not necessarily fire in adjacent locations in that second environment (note the relative positions of place fields for Neurons 7 and 10 in Environment A vs. Environment B.) This underscores the fact that the hippocampus distributes representations across populations of neurons and that unique representations are defined by collections of place cells as opposed to any single cell. By recycling place cells in combination with distributing the representation of an environment across a subpopulation of neurons, many different environments can thus be encoded.

physically distinct environments, and the location of one neuron's place field is not fixed relative to the location of any other neuron's place field across distinct environments. Put another way, if two neurons exhibit neighbouring place fields in one environment, they do not necessarily exhibit neighbouring place fields in any other environment. Thus the spatial distribution of place fields from any arbitrary group of place cells mapped in a composite will change (or 'remap') from one environment to another (Figure 9.4). This suggests that, for any given environment, a random subpopulation of hippocampal place cells collectively can encode the animal's location, consistent with the *'cognitive map'* hypothesis of hippocampal function (O'Keefe and Nadel 1978). By distributing the representation of an environment across a subpopulation of neurons, and recycling individual place cells in the representations of different environments, a much larger number of environments may be encoded in memory than if a local representation scheme of memory encoding was implemented.

Place fields in area CA1 are established the first time an animal enters an environment (Hill 1978) and this feature does not dramatically change with age (e.g. Shen et al. 1997). Other basic properties of CA1 pyramidal cells are also not affected by the ageing process. These include mean and maximum firing rates as well as spike characteristics such as amplitude and width (Barnes et al. 1983; Shen et al. 1997). As previously mentioned (in the Biophysical properties of aged neurons' section), aged CA1 neurons are observed to have larger after-hyperpolarizing potentials (AHP) as measured *in vitro*. This larger AHP suggests that aged CA1 cells should be less excitable as this greater hyperpolarization brings them further from action potential threshold. When CA1 pyramidal neurons are recorded *in vivo*, however, there is no difference in the firing rates of these cells with age (e.g. Barnes et al. 1983; Markus et al. 1994; Barnes et al. 1997b; Smith et al. 2000b). While firing characteristics in CA1 do not change in freely behaving aged rats, aged CA3 pyramidal neurons are reported to have slightly higher firing rates (Wilson et al. 2005). Given this pattern of data, an interesting hypothesis might be suggested—namely, that the increased CA1 AHP in the older rat is a protective or compensatory change that could serve to normalize the greater drive from CA3 and prevent hyperactivity in CA1. There are few *in vivo* recordings of single granule cells, and unfortunately, the effect of age on the properties of granule cells is unknown.

Despite basic place cell properties remaining largely intact over the lifespan, changes in ensemble dynamics do occur with age. These changes have important implications for the behaviour of aged animals. For example, it has been shown that the maintenance of place field maps differs between young and old animals. In normal young rats, hippocampal maps generated in a given environment can remain stable for months (Thompson and Best 1990). For young rats, the same 'hippocampal map' is retrieved on multiple visits to the same environment. For the most part aged rats also show stable map retrieval within and between episodes in an environment. Occasionally, however, if an old animal is removed from an environment and returned sometime later, the original map in CA1 is not retrieved. Instead, an independent population of CA1 place cells is activated, even in a very familiar room (Figure 9.5A; Barnes et al. 1997b). This so called 'remapping' phenomenon is thought to be due to deficient LTP in aged animals. Specifically, it appears that the maintenance of stable maps relies on an NMDA-dependent LTP-like mechanism (Kentros et al. 1998) and protein synthesis (Agnihotri et al. 2004), as blockade of either results in map retrieval errors upon returning to a previously visited environment.

Whereas spatial representations in CA1 of old rats appear less stable (Barnes et al. 1997b), the maps in CA3 seem more rigid in old animals compared to their young counterparts. If an animal is allowed to explore one environment, is then returned to its home cage for a time, and subsequently placed into a distinct and different environment, old animals tend to 'reuse' the same map (Figure 9.5B; Wilson et al. 2005). This is not the case for young animals, as CA3 spatial representations in these animals are statistically independent between two different environments (Wilson et al. 2005). Overall, the end result is that CA1 pyramidal cells in old rats tend to remap when they should not, and old CA3 pyramidal cells do not remap when they should. These data suggest that brain mechanisms of pattern separation and completion are altered in old hippocampal circuits (reviewed in Marr 1971; Burke and Barnes 2010) for which there is increasing corroborative evidence from human behaviour and MRI ageing experiments (Toner et al. 2009; Stark et al. 2010; Yassa et al. 2011).

In addition to deficits in place field maintenance and stability, aged animals also exhibit impairments in place field plasticity. In young rats, CA1 place fields expand asymmetrically as animals repeatedly traverse the same path (such as on a circular track). During place field expansion, the centre of the place field shifts 'backward' opposite to the animal's movement trajectory (Mehta et al. 1997). As the animal moves along the path, place cells become active in a particular sequence, such that the firing of one principal cell will always precede the firing of the cell that has the next place field along the track. The connections between the cells with adjacent place fields appear to become strengthened, undergoing an LTP-like change. Because of the stronger connections between cells with overlapping place fields, in subsequent traversals, cells along the route begin to fire earlier than in initial traversals, resulting in an asymmetrical backward expansion of the place field (reviewed in Rosenzweig and Barnes 2003). This expansion plasticity is blocked in young animals by the NMDA receptor antagonist CPP (3-(2-carboxypiperazin-4-yl)propyl-1-phosphonic acid) (Ekstrom et al. 2001), a treatment that also blocks LTP.

Shen et al. (1997) showed that this behaviourally-induced plasticity process is defective in old rats, consistent with the idea that aged animals have impaired NMDA receptor activity. Putting these observations together, Burke et al. (2008) hypothesized that it should be possible to restore place field expansion in aged animals, if the deficit was due to an NMDA-dependent, LTP-like mechanism. Memantine, a low affinity non-competitive NMDA receptor antagonist, has been shown to improve LTP durability and spatial memory (Barnes et al. 1996b) and is, in fact, an FDA-approved agent for alleviating cognitive impairments in AD. When Burke et al. (2008) examined place field expansion plasticity in old rats treated with memantine, they found that it restored the backward shift of the place fields, although treatment with an AMPA receptor agonist (the ampakine CX516) did not.

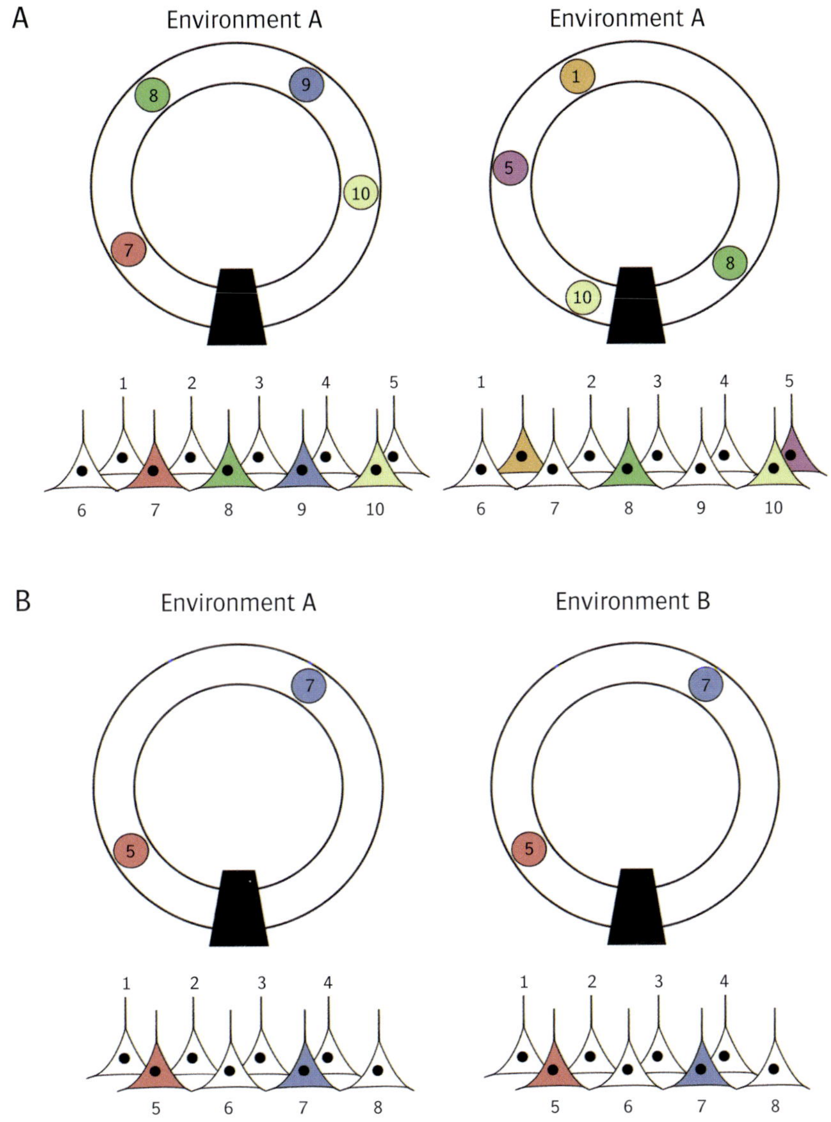

Figure 9.5 Age-related alterations in hippocampal place field maps. Illustration of age-associated deficits in hippocampal place field maps in areas CA1 and CA3. A) Illustration of remapping observed in aged CA1 place cell ensembles. In normal young animals, CA1 place maps are stable across multiple visits to the same environment. In aged animals, however, place cells in area CA1 occasionally 'remap' or exhibit a different pattern of firing between two visits to the same familiar environment. *Left*) Hippocampal neurons establish place fields upon visiting a new environment; for instance, place cell 7 in area CA1 of an aged animal selectively fires in the lower left region of the circular track in Environment A whereas place cell 8 has a place field in the upper left region of the track. *Right*) When the aged animal visits this same environment sometime later, some place cells that were previously active are now silent, and other previously silent neurons express place fields. Additionally, neurons active in both visits to the same environment now fire in different locations (Neuron 8). This remapping suggests that the aged CA1 appears to treat the second visit to the same track as an entry to a novel environment in contrast to normal function in the young hippocampus when presented with two distinct environments, as depicted in Figure 9.4. B) Illustration of rigidity observed in aged CA3 place cell ensembles. In normal young animals, CA3 place maps change or remap between different environments. In aged animals, however, place cells in area CA3 do not remap when they should and thus display 'rigidity'. *Left*) Place cells in the CA3 subfield of an aged animal establish location-specific firing at select points along the circular track. *Right*) When the aged animal is placed in a different environment, however, the first CA3 place map is retained. It is important to note that the CA3 subfield displays a slightly lower percentage of active place cells in a given environment compared to the percentage active in the CA1 subfield (25% vs. 30–40%, respectively, in the dorsal hippocampus of the rat, for an environment measuring 3600 cm^2) (Jung and McNaughton 1993; Wilson and McNaughton 1993).

Given that the NMDA receptor antagonist CPP blocks place field expansion in young animals (Ekstrom et al. 2001), it may seem paradoxical that antagonizing NMDA receptors with a different compound would facilitate plasticity. Memantine, however, differs from other NMDA receptor antagonists because of its fast kinetics and strong voltage dependency. This property allows memantine to quickly leave the NMDA channel after glutamate activation (reviewed in Parsons et al. 2007) and is hypothesized to increase the signal-to-noise ratio of significant inputs. In aged animals, this may be especially important as Ca^{2+} dysregulation, particularly in CA1 pyramidal neurons, is thought to underlie LTP deficits at synapses in that region (Kumar and Foster 2005) and could also play a role in altered place field expansion plasticity. If this is correct, at least part of memantine's effect on hippocampal circuit plasticity and dynamics could be due to modification of Ca^{2+} influx though the NMDA receptor, effectively ameliorating disrupted Ca^{2+} homeostasis that is known to occur during ageing.

Gathering what we know about the aged hippocampus (i.e. synapse loss or silence, dysregulated Ca^{2+} homeostasis, impairments in plasticity, etc.), commercially available pharmacological treatments such as memantine, that seem to ameliorate a systems level age-related plasticity deficit, promise encouraging avenues for treating cognitive impairment. Although the mechanisms of action of memantine as a treatment for cognitive decline have been hypothesized to be through its action on glutamaterigic systems, this agent also has an affinity for the $\alpha 7$ nicotinic acetylcholine receptor (nAChR) which also conducts Ca^{2+} in the hippocampus (Aracava et al. 2005). This raises the possibility that, along with Memantine's effect of reducing the amount of Ca^{2+} that enters neurons through the NMDA receptor, it could also act to restore normal plasticity mechanisms in aged neurons by reducing Ca^{2+} entry through the $\alpha 7$ nACh receptor. These observations underscore the importance of gaining a thorough understanding of both the basic mechanisms the brain uses to maintain optimal plasticity, as well as the avenues though which any given agent *actually* works to modify cognition.

Summary

Over the past several decades there has been enormous progress in our understanding of how the hippocampus, a structure critical for memory, is altered across the lifespan in a number of mammalian species. While the hippocampus is certainly 'front and centre' with respect to vulnerability to a variety of diseases that also tend to occur in older individuals, we have begun to appreciate that this structure has an ageing-related, but disease-independent biological trajectory that can be considered to be normative, or characteristic of advanced stages of development. Small and his colleagues (Small et al. 1999, 2002; Wu et al. 2008) have articulated this point with the example of AD, in which it is clear that the earliest pathological changes are detected in entorhinal cortex and the CA1 region of the hippocampus (Gómez-Isla et al. 1996; West et al. 2000). The hippocampal granule cells, on the other hand, are relatively preserved early on in the disease (e.g. Thal et al. 2002) when affected brains are compared to age-matched controls without AD. Using various molecular and metabolic imaging indices as a measure of functional status, older humans, non-human primates and rodents all show pronounced and selective reductions in activity markers in the dentate gyrus compared to their younger counterparts before changes in any other subfield are detectable (Small et al. 2000, 2004). This is not to say that there are no age-related changes in the CA1 region in aged mammals—this chapter has documented a number of the changes that do occur—rather, it appears that by some measures, it is the dentate gyrus and hilar region that is particularly affected first during advancing age. This discrepancy in the locus of alterations between the beginning stages of AD and normal ageing makes it tempting to suggest that AD has a distinct time course of pathological changes that simply are superimposed on normative ageing processes.

By examining cognition and brain function of animals that do not show the classic neuropathological pattern that defines AD, the phenotypic characteristics of advanced age can begin to be understood in some detail. Although the goal of eradicating disease is paramount, appreciating how the ageing process confers vulnerability to pathological states is likely to inform and direct such important discoveries. An exciting promise for the future is that we will not only be able to slow the progression of or eliminate some diseases, but we will also begin to understand phenomena like 'cognitive reserve' and successful ageing. The concept of cognitive reserve has been suggested to confer resilience to neuropathological damage. Certain patients with extensive AD pathology postmortem show very little clinical or behavioural manifestations of the disease (Katzman et al. 1988), and this observation has raised the hypothesis that certain 'proxies' including education and lifestyle factors may contribute to optimal cognitive performance during the lifespan even in the presence of pathology (Stern 2002). Although more work is needed to understand the features of cognitive reserve that might have implications for disease intervention, harnessing this knowledge could have the remarkable consequence of increasing the quality of life of older individuals across world populations—a remarkably different vision from that of only a half century ago, when normative ageing was equated with dementia.

Acknowledgements

This work is supported by McKnight Brain Research Foundation; NIA grant AG003376; and NIA grant AG033460.

References

Agnihotri, N.T., Hawkins, R.D., Kandel, E.R., and Kentros, C. (2004). The long-term stability of new hippocampal place fields requires new protein synthesis. *Proc Natl Acad Sci U S A*, 101, 3656–61.

Aracava, Y., Pereira, E.F., and Albuquerque, E.X. (2005). Mematine blocks alpha7* nicotinic acetylcholine receptors more potently than n-methyl-D-aspartate receptors in rat hippocampal neurons. *J Pharmacol Exp Ther*, 312, 1195–205.

Bach, M.E., Barad, M., Son, H., *et al.* (1999). Age-related defects in spatial memory are correlated with defects in the late phase of hippocampal long-term potentiation in vitro and are attenuated by drugs that enhance the cAMP signaling pathway. *Proc Natl Acad Sci U S A*, 96, 5280–5.

Barnes, C.A. (1979). Memory deficits associated with senescence: a neurophysiological and behavioral study in the rat. *J Comp Physiol Psychol*, 93, 74–104.

Barnes, C.A. (1994). Normal aging: Regionally specific changes in hippocampal synaptic transmission. *Trends Neurosci*, 17, 13–18.

Barnes, C.A., Danysz, W., and Parsons, C.G. (1996b). Effects of the uncompetitive NMDA receptor antagonist memantine on hippocampal long-term potentiation, short-term exploratory modulation and spatial memory in awake, freely moving rats. *Eur J Neurosci*, 8, 565–71.

Barnes, C.A. and McNaughton, B.L. (1980). Physiological compensation for loss of afferent synapses in rat hippocampal granule cells during senescence. *J Physiol (Lond)*, 309, 473–85.

Barnes, C.A., McNaughton, B.L., and O'Keefe, J. (1983). Loss of place specificity in hippocampal complex spike cells of senescent rat. *Neurobiol Aging*, 4, 113–19.

Barnes, C.A., Nadel, L., and Honig, W.K. (1980). Spatial memory deficit in senescent rats. *Can J Psychol*, 34, 29–39.

Barnes, C.A., Rao, G., Foster, T.C., and McNaughton, B.L. (1992). Region-specific age effects on AMPA sensitivity: electrophysiological evidence for loss of synaptic contacts in hippocampal field CA1. *Hippocampus*, 2, 457–68.

Barnes, C.A., Rao, G., and Houston, F.P. (2000). LTP induction threshold change in old rats at the perforant path – granule cell synapse. *Neurobiol Aging*, 21, 613–20.

Barnes, C.A., Rao, G., and McNaughton, B.L. (1987). Increased electrotonic coupling in aged rat hippocampus: A possible mechanism for cellular excitability changes. *J Comp Neurol*, 259, 547–58.

Barnes, C.A., Rao, G., and McNaughton, B.L. (1996a). Functional integrity of NMDA-dependent LTP induction mechanisms across the lifespan of F-344 rats. *Learn Memory*, 3, 124–37.

Barnes, C.A., Rao, G., and Shen, J. (1997a). Age-related decrease in the N-methyl-D-aspartateR-mediated excitatory postsynaptic potential in hippocampal region CA1. *Neurobiol Aging*, 18, 445–52.

Barnes, C.A., Suster, M.S., Shen, J., and McNaughton, B.L. (1997b). Multistability of cognitive maps in the hippocampus of old rats. *Nature*, 388, 272–5.

Bauman, L.A., Mahle, C.D., Boissard, C.G., and Gribkoff, V.K. (1992). Age-dependence of effects of A1 adenosine receptor antagonism in rat hippocampal slices. *J Neurophysiol*, 68, 629–38.

Bear, M.F., Cooper, L.N., and Ebner, F.F. (1987). A physiological basis for a theory of synapse modification. *Science*, 237, 42–8.

Bear, M.F. and Malenka, R.C. (1994). Synaptic plasticity: LTP and LTD. *Curr Opin Neurobiol*, 4, 389–99.

Beatty, W.W., Bierley, R.A., and Boyd, J.G. (1985). Preservation of accurate spatial memory in aged rats. *Neurobiol Aging*, 6, 219–25.

Blalock, E.M., Chen, K.C., Sharrow, K., *et al.* (2003). Gene microarrays in hippocampal aging: statistical profiling identifies novel processes correlated with cognitive impairment. *J Neurosci*, 23, 3807–19.

Bodhinathan, K., Kumar, A., and Foster, T.C. (2010). Intracellular redox state alters NMDA receptor response during aging through Ca2+/calmodulin-dependent protein kinase II. *J Neurosci*, 30, 1914–24.

Bondareff, W., and Geinisman, Y. (1976). Loss of synapses in the dentate gyrus of the senescent rat. *Am JAnat*, 145, 129–36.

Braak, H., Braak, E., and Strothjohann, M. (1994). Abnormally phosphorylated tau protein related to the formation of neurofibrillary tangles and neuropil threads in the cerebral cortex of sheep and goat. *Neurosci Lett*, 171, 1–4.

Brakeman, P.R., Lanahan, A.A., O'Brien, R., *et al.* (1997). Homer: a protein that selectively binds metabotropic glutamate receptors. *Nature*, 386, 284–8.

Bramham, C.R., Worley, P.F., Moore, M.J., and Guzowski, J.F. (2008). The immediate early gene arc/arg3.1: regulation, mechanisms, and function. *J Neurosci*, 28, 11760–7.

Buell, S.J. and Coleman, P.D. (1979). Dendritic growth in the aged human brain and failure of growth in senile dementia. *Science*, 206, 854–6.

Buell, S.J. and Coleman, P.D. (1981). Quantitative evidence for selective dendritic growth in normal human aging but not in senile dementia. *Brain Res*, 214, 23–41.

Burke, S.N. and Barnes, C. A. (2010). Senescent synapses and hippocampal circuit dynamics. *Trends Neurosci*, 33, 153–61.

Burke, S.N., Maurer, A.P., Yang, Z., Navratilova, Z., and Barnes, C.A. (2008). Glutamate receptor-mediated restoration of experience-dependent place field expansion plasticity in aged rats. *Behav Neurosci*, 122, 535–48.

Chahrour, M., Jung, S.Y., Shaw, C., *et al.* (2008). MeCP2, a key contributor to neurological disease, activates and represses transcription. *Science*, 320, 1224–9.

Cole, A.J., Saffen, D.W., Baraban, J.M., and Worley, P.F. (1989). Rapid increase of an immediate early gene messenger RNA in hippocampal neurons by synaptic NMDA receptor activation. *Nature*, 340, 474–6.

Colle, M.A., Hauw, J.J., Crespeau, F., *et al.* (2000). Vascular and parenchymal Abeta deposition in the aging dog: correlation with behavior. *Neurobiol Aging*, 21, 695–704.

Cork, L.C., Powers, R.E., Selkoe, D.J., Davies, P., Geyer, J.J., and Price, D.L. (1988). Neurofibrillary tangles and senile plaques in aged bears. *J Neuropathol Exp Neurol*, 47, 629–41.

Curcio, C.A. and Hinds, J. W. (1983). Stability of synaptic density and spine volume in dentate gyrus of aged rats. *Neurobiol Aging*, 4, 77–87.

Davis, H.P., Rosenzweig, M.R., Kinkade, P.T., and Bennett, E.L. (1981). Effects of anisomycin on retention of the passive-avoidance habit as a function of age. *Exp Aging Res*, 7, 33–44.

de Ruiter, J.P. and Uylings, H.B. (1987). Morphometric and dendritic analysis of fascia dentate granule cells in human aging and senile dementia. *Brain Res*, 402, 217–29.

Desjardins, S., Mayo, W., Vallee, M., *et al.* (1997). Effect of aging on the basal expression of c-Fos, c-Jun, and Egr-1 proteins in the hippocampus. *Neurobiol Aging*, 18, 37–44.

deToledo-Morrell, L., Geinisman, Y., and Morrell, F. (1988). Age-dependent alterations in hippocampal synaptic plasticity: relation to memory disorders. *Neurobiol Aging*, 9, 581–90.

deToledo-Morrell, L., Morrell, F., and Fleming, S. (1984). Age-dependent deficits in spatial memory are related to impaired hippocampal kindling. *Behav Neurosci*, 98, 902–7.

Deupree, D.L., Bradley, J., and Turner, D.A. (1993). Age-related alterations in potentiation in the CA1 region in F344 rats. *Neurobiol Aging*, 14, 249–58.

Deyo, R.A., Straube, K.T., and Disterhoft, J.F. (1989). Nimodipine facilitates associative learning in aging rabbits. *Science*, 243, 809–11.

Diana, G., Domenici, M.R., Loizzo, A., Scotti de Carolis, A., and Sagratella, S. (1994a). Age and strain differences in rat place learning and hippocampal dentate gyrus frequency-potentiation. *Neurosci Lett*, 171, 113–6.

Diana, G., Scotti de Carolis, A., Frank, C., Domenici, M.R., and Sagratella, S. (1994b). Selective reduction of hippocampal dentate frequency potentiation in aged rats with impaired place learning. *Brain Res Bull*, 35, 107–11.

Dieguez, D., Jr. and Barea-Rodriguez, E.J. (2004). Aging impairs the late phase of long-term potentiation at the medial perforant path-CA3 synapse in awake rats. *Synapse*, 52, 53–61.

Disterhoft, J.F., Thompson, L.T., Moyer, J.R., Jr., and Mogul, D.J. (1996). Calcium-dependent afterhyperpolarization and learning in young and aging hippocampus. *Life Sci*, 59, 413–20.

Dragunow, M., Abraham, W.C., Goulding, M., Mason, S.E., Robertson, H.A., and Faull, R.L. (1989). Long-term potentiation and the induction of c-fos mRNA and proteins in the dentate gyrus of unanesthetized rats. *Neurosci Lett*, 101, 274–80.

Ekstrom, A.D., Meltzer, J., McNaughton, B.L., and Barnes, C.A. (2001). NMDA receptor antagonism blocks experience-dependent expansion of hippocampal 'place fields'. *Neuron*, 31, 631–8.

Flood, D.G. (1993). Critical issues in the analysis of dendritic extent in aging humans, primates and rodents. *Neurobiol Aging*, 14, 649–54.

Flood, D.G., Buell, S.J., Defiore, C.H., Horwitz, G.J., and Coleman, P.D. (1985). Age-related dendritic growth in dentate gyrus of human brain is followed by regression in the 'oldest old'. *Brain Res*, 345, 366–8.

Flood, D.G., Buell, S.J., Horwitz, G.J., and Coleman, P.D. (1987a). Dendritic extent in human dentate gyrus granule cells in normal aging and senile dementia. *Brain Res*, 402, 205–16.

Flood, D.G., Guarnaccia, M., and Coleman, P.D. (1987b). Dendritic extent in human CA2-3 hippocampal pyramidal neurons in normal aging and senile dementia. *Brain Res*, 409, 88–96.

Foster, T.C., Barnes, C.A., Rao, G., and McNaughton, B.L. (1991). Increase in perforant path quantal size in aged F-344 rats. *Neurobiol Aging*, 12, 441–8.

Foster, T.C. and Norris, C.M. (1997). Age-associated changes in Ca(2+)-dependent processes: relation to hippocampal synaptic plasticity. *Hippocampus*, 7, 602–12.

Gage, F.H., Dunnett, S.B., and Bjorklund, A. (1984). Spatial learning and motor deficits in aged rats. *Neurobiol Aging*, 5, 43–8.

Gallagher, M. and Pelleymounter, M.A. (1988). Spatial learning deficits in old rats: A model for memory decline in the aged. *Neurobiol Aging*, 9, 549–56.

Geinisman, Y., Bondareff, W., and Dodge, J.T. (1977). Partial deafferentation of neurons in the dentate gyrus of the senescent rat. *Brain Res*, 134, 541–5.

Geinisman, Y., deToledo-Morrell, L., Morrell, F., Persina, I.S., and Rossi, M. (1992). Age-related loss of axospinous synapses formed by two afferent systems in the rat dentate gyrus as revealed by the unbiased stereological dissector technique. *Hippocampus*, 2, 437–44.

Geinisman, Y., Ganeshina, O., Yoshida, R., Berry, R.W., Disterhoft, J.F., and Gallagher, M. (2004). Aging, spatial learning, and total synapse number in the rat CA1 stratum radiatum. *Neurobiol Aging*, 25, 407–16.

Gómez-Isla, T., Price, J.L., McKeel, D.W., Jr, Morris, J.C., Growdon, J.H., and Hyman, B.T. (1996). Profound loss of layer II entorhinal cortex neurons occurs in very mild Alzheimer's disease. *J Neurosci*, 16, 4491–500.

Gonzales, R.A., Brown, L.M., Jones, T.W., Trent, R.D., Westbrook, S.L., and Leslie, S.W. (1991). N-methyl-D-aspartate mediated responses decrease with age in Fischer 344 rat brain. *Neurobiol Aging*, 12, 219–25.

Goodrick, C.L. (1968). Learning, retention, and extinction of a complex maze habit for mature-young and senescent Wistar albino rats. *J Gerontol*, 23, 298–304.

Graff, J. and Mansuy, I.M. (2008). Epigenetic codes in cognition and behaviour. *Behav Brain Res*, 192, 70–87.

Grimm, R. and Tischmeyer, W. (1997). Complex patterns of immediate early gene induction in rat brain following brightness discrimination training and pseudotraining. *Behav Brain Res*, 84, 109–16.

Guzowski, J.F. (2002). Insights into immediate-early gene function in hippocampal memory consolidation using antisense oligonucleotide and fluorescent imaging approaches. *Hippocampus*, 12, 86–104.

Guzowski, J.F., Lyford, G. L., Stevenson, G. D., *et al.* (2000). Inhibition of activity-dependent arc protein expression in the rat hippocampus impairs the maintenance of long-term potentiation and the consolidation of long-term memory. *J Neurosci*, 20, 3993–4001.

Guzowski, J.F. and McGaugh, J. L. (1997). Antisense oligodeoxynucleotide-mediated disruption of hippocampal cAMP response element binding protein levels impairs consolidation of memory for water maze training. *Proc Natl Acad Sci U S A*, 94, 2693–8.

Guzowski, J.F., McNaughton, B.L., Barnes, C.A., and Worley, P.F. (1999). Environment-specific expression of the immediate-early gene Arc in hippocampal neuronal ensembles. *Nat Neurosci*, 2, 1120–4.

Härtig, W., Klein, C., Brauer, K., *et al.* (2000). Abnormally phosphorylated protein tau in the cortex of aged individuals of various mammalian orders. *Acta Neuropathol*, 100, 305–12.

Head, E., Mehta, R., Hartley, J., *et al.* (1995). Spatial learning and memory as a function of age in the dog. *Behav Neurosci*, 109, 851–8.

Herdegen, T. and Leah, J.D. (1998). Inducible and constitutive transcription factors in the mammalian nervous system: control of gene expression by Jun, Fos and Krox, and CREB/ATF proteins. *Brain Res. Brain Res Reviews*, 28, 370–490.

Hill, A.J. (1978). First occurrence of hippocampal spatial firing in a new environment. *Exp Neurol*, 62, 282–97.

Huang, Y.Y. and Kandel, E.R. (2006). Age-related enhancement of a protein synthesis-dependent late phase of LTP induced by low frequency paired-pulse stimulation in hippocampus. *Learn Memory*, 13, 298–306.

Iaria, G., Petrides, M., Dagher, A., Pike, B., and Bohbot, V.D. (2003). Cognitive strategies dependent on the hippocampus and caudate nucleus in human navigation: variability and change with practice. *J Neurosci*, 23, 5945–52.

Inestrosa, N.C., Reyes, A.E., Chacón, M.A., *et al.* (2005). Human-like rodent amyloid-beta-peptide determines Alzheimer pathology in aged wild-type Octodon degu. *Neurobiol Aging*, 26, 1023–8.

Jones, M.W., Errington, M.L., French, P.J., *et al.* (2001). A requirement for the immediate early gene Zif268 in the expression of late LTP and long-term memories. *Nat Neurosci*, 4, 289–96.

Jung, M.W. and McNaughton, B.L. (1993). Spatial selectivity of unit activity in the hippocampal granular layer. *Hippocampus*, 3, 165–82.

Kangaspeska, S., Stride, B., Metivier, R., *et al.* (2008). Transient cyclical methylation of promoter DNA. *Nature*, 452, 112–5.

Katzman, R. Terry, R., DeTeresa, R., *et al.* (1988). Clinical, pathological, and neurochemical changes in dementia: a subgroup with preserved mental status and numerous neocortical plaques. *Ann Neurol*, 23, 138–44.

Kentros, C., Hargreaves, E., Hawkins, R.D., Kandel, E.R., Shapiro, M., and Muller, R.V. (1998). Abolition of long-term stability of new hippocampal place cell maps by NMDA receptor blockade. *Science*, 280, 2121–6.

Kerr, D.S., Campbell, L.W., Applegate, M.D., Brodish, A., and Landfield, P.W. (1991). Chronic stress-induced acceleration of electrophysiologic and morphometric biomarkers of hippocampal aging. *J Neurosci*, 11, 1316–24.

Keuker, J.I., Luiten, P.G., and Fuchs, E. (2003). Preservation of hippocampal neuron numbers in aged rhesus monkeys. *Neurobiol Aging*, 24, 157–65.

Koh, M.T., Haberman, R.P., Foti, S., McCown, T.J., and Gallagher, M. (2010). Treatment strategies targeting excess hippocampal activity benefit aged rats with cognitive impairment. *Neuropsychopharmacology*, 35, 1016–25.

Kumar, A., Bodhinathan, K., and Foster, T. C. (2009). Susceptibility to calcium dysregulation during brain aging. *Front Aging Neurosci*, 1, 2.

Kumar, A. and Foster, T.C. (2005). Intracellular calcium stores contribute to increased susceptibility to LTD induction during aging. *Brain Res*, 1031, 125–8.

Lanahan, A., Lyford, G., Stevenson, G.S., Worley, P.F., and Barnes, C.A. (1997). Selective alteration of long-term potentiation-induced transcriptional response in hippocampus of aged, memory-impaired rats. *J Neurosci*, 17, 2876–85.

Landfield, P.W. and Lynch, G. (1977). Impaired monosynaptic potentiation in in vitro hippocampal slices from aged, memory-deficient rats. *J Gerontol*, 32, 523–33.

Landfield, P.W., McGaugh, J.L., and Lynch, G. (1978). Impaired synaptic potentiation processes in the hippocampus of aged, memory-deficient rats. *Brain Res*, 150, 85–101.

Landfield, P.W. and Pitler, T.A. (1984). Prolonged Ca^{2+}-dependent afterhyperpolarizations in hippocampal neurons of aged rats. *Science*, 226, 1089–92.

Landfield, P.W., Pitler, T.A., and Applegate, M.D. (1986). The effects of high Mg^{2+}-to-Ca^{2+} ratios on frequency potentiation in hippocampal slices of young and aged rats. *J Neurophysiol*, 56, 797–811.

Leutgeb, J.K., Leutgeb, S., Moser, M.B., and Moser, E.I. (2007). Pattern separation in the dentate gyrus and CA3 of the hippocampus. *Science*, 315, 961–6.

Link, W., Konietzko, U., Kauselmann, G., *et al.* (1995). Somatodendritic expression of an immediate early gene is regulated by synaptic activity. *Proc Natl Acad Sci U S A*, 92, 5734–8.

Lubin, F.D., Roth, T.L., and Sweatt, J.D. (2008). Epigenetic regulation of BDNF gene transcription in the consolidation of fear memory. *J Neurosci*, 28, 10576–86.

Lyford, G.L., Yamagata, K., Kaufmann, W.E., *et al.* (1995). Arc, a growth factor and activity-regulated gene, encodes a novel cytoskeleton-associated protein that is enriched in neuronal dendrites. *Neuron*, 14, 433–45.

Ma, D.K., Jang, M.H., Guo, J.U., *et al.* (2009). Neuronal activity-induced Gadd45b promotes epigenetic DNA demethylation and adult neurogenesis. *Science*, 323, 1074–7.

Markham, J.A., McKian, K.P., Stroup, T.S., and Juraska, J.M. (2005). Sexually dimorphic aging of dendritic morphology in CA1 of hippocampus. *Hippocampus*, 15, 97–103.

Markus, E.J., Barnes, C.A., McNaughton, B.L., Gladden, V.L., and Skaggs, W.E. (1994). Spatial information content and reliability of hippocampal CA1 neurons: effects of visual input. *Hippocampus*, 4, 410–21.

Marr, D. (1971). Simple memory: a theory for archicortex. *Philos Trans R Soc Lond B Biol Sci*, 262, 23–81.

McMonagle-Strucko, K. and Fanelli, R.J. (1993). Enhanced acquisition of reversal training in a spatial learning task in rats treated with chronic nimodipine. *Pharmacol Biochem Behav*, 44, 827–35.

Mehta, M.R., Barnes, C.A., and McNaughton, B.L. (1997). Experience-dependent, asymmetric expansion of hippocampal place fields. *Proc Natl Acad Sci U S A*, 94, 8918–21.

Merrill, D.A., Chiba, A.A., and Tuszynski, M.H. (2001). Conservation of neuronal number and size in the entorhinal cortex of behaviorally characterized aged rats. *J Comp Neurol*, 438, 445–56.

Métivier, R., Gallais, R., Tiffoche, C., *et al.* (2008). Cyclical DNA methylation of a transcriptionally active promoter. *Nature*, 452, 45–50.

Miller, C.A. and Sweatt, J.D. (2007). Covalent modification of DNA regulates memory formation. *Neuron*, 53, 857–69.

Mizumori, S.J., Rosenzweig, M.R., and Bennett, E.L. (1985). Long-term working memory in the rat: effects of hippocampally applied anisomycin. *Behav Neurosci*, 99, 220–32.

Moffat, S.D., Zonderman, A.B., and Resnick, S.M. (2001). Age differences in spatial memory in a virtual environment navigation task. *Neurobiol Aging*, 22, 787–96.

Moore, C.I., Browning, M.D., and Rose, G.M. (1993). Hippocampal plasticity induced by primed burst, but not long-term potentiation, stimulation is impaired in area CA1 of aged Fischer 344 rats. *Hippocampus*, 3, 57–66.

Morgan, J.I., Cohen, D.R., Hempstead, J.L., and Curran, T. (1987). Mapping patterns of c-fos expression in the central nervous system after seizure. *Science*, 237, 192–7.

Newman, M.C. and Kaszniak, A.W. (2000). Spatial memory and aging: Performance on a human analog of the Morris Water Maze Task. *Aging Neuropsychol Cogn*, 7, 86–93.

Nicholson, D.A., Yoshida, R., Berry, R.W., Gallagher, M., and Geinisman, Y. (2004). Reduction in size of perforated postsynaptic densities in hippocampal axospinous synapses and age-related spatial learning impairments. *J Neurosci*, 24, 7648–53.

Norris, C.M., Korol, D.L., and Foster, T.C. (1996). Increased susceptibility to induction of long-term depression and long-term potentiation reversal during aging. *J Neurosci*, 16, 5382–92.

O'Keefe, J. and Dostrovsky, J. (1971). The hippocampus as a spatial map. Preliminary evidence from unit activity in the freely-moving rat. *Brain Res*, 34, 171–5.

O'Keefe, J. and Nadel, L. (1978). *The Hippocampus as a Cognitive Map*. Oxford: Clarendon Press.

Parsons, C.G., Stoffler, A., and Danysz, W. (2007). Memantine: a NMDA receptor antagonist that improves memory by restoration of homeostasis in the glutamatergic system – too little activation is bad, too much is even worse. *Neuropharmacology*, 53, 699–723.

Penner, M.R. and Barnes, C.A. (2007). Memory changes with age: Neurobiological correlates. In Kesner, R.P. and Martinez, J.L., Jr. (Eds.) *Neurobiology of Learning and Memory* (2nd edn.), pp. 483–518. Amsterdam: Elsevier.

Penner, M.R., Roth, T.L., Chawla, M.K., *et al.* (2011). Age-related changes in Arc transcription and DNA methylation within the hippocampus. *Neurobiol Aging*, 32, 2198–210.

Pittaluga, A., Fedele, E., Risiglione, C., and Raiteri, M. (1993). Age-related decrease of the NMDA receptor-mediated noradrenaline release in rat hippocampus and partial restoration by D-cycloserine. *Eur J Pharmacol*, 231, 129–34.

Plath, N., Ohana, O., Dammermann, B., *et al.* (2006). Arc/Arg3.1 is essential for the consolidation of synaptic plasticity and memories. *Neuron*, 52, 437–44.

Potier, B., Poindessous-Jazat, F., Dutar, P., and Billard, J.M. (2000). NMDA receptor activation in the aged rat hippocampus. *Exp Gerontol*, 35, 1185–99.

Pyapali, G.K. and Turner, D.A. (1996). Increased dendritic extent in hippocampal CA1 neurons from aged F344 rats. *Neurobiol Aging*, 17, 601–11.

Rao, G., Barnes, C.A., and McNaughton, B. L. (1993). Effects of age on L-glutamate-induced depolarization in three hippocampal subfields. *Neurobiol Aging*, 14, 27–33.

Rapp, P.R. and Gallagher, M. (1996). Preserved neuron number in the hippocampus of aged rats with spatial learning deficits. *Proc Natl Acad Sci U S A*, 93, 9926–30.

Rapp, P.R., Kansky, M.T., and Roberts, J.A. (1997). Impaired spatial information processing in aged monkeys with preserved recognition memory. *Neuroreport*, 8, 1923–8.

Rapp, P.R., Rosenberg, R.A., and Gallagher, M. (1987). An evaluation of spatial information processing in aged rats. *Behav Neurosci*, 101, 3–12.

Rasmussen, T., Schliemann, T., Sorensen, J.C., Zimmer, J., and West, M.J. (1996). Memory impaired aged rats: no loss of principal hippocampal and subicular neurons. *Neurobiol Aging*, 17, 143–7.

Reti, I.M., Reddy, R., Worley, P.F., and Baraban, J.M. (2002). Prominent Narp expression in projection pathways and terminal fields. *J Neurochem*, 82, 935–44.

Richardson, C.L., Tate, W.P., Mason, S.E., Lawlor, P. A., Dragunow, M., and Abraham, W.C. (1992). Correlation between the induction of an immediate early gene, zif/268, and long-term potentiation in the dentate gyrus. *Brain Res*, 580, 147–54.

Rosenzweig, E.S. and Barnes, C.A. (2003). Impact of aging on hippocampal function: plasticity, network dynamics, and cognition. *Prog Neurobiol*, 69, 143–79.

Rosenzweig, E.S., Rao, G., McNaughton, B.L., and Barnes, C.A. (1997). Role of temporal summation in age-related long-term potentiation-induction deficits. *Hippocampus*, 7, 549–58.

Roth, T.L., Lubin, F.D., Funk, A.J., and Sweatt, J.D. (2009). Lasting epigenetic influence of early-life adversity on the BDNF gene. *Biol Psychiatry*, 65, 760–9.

Scheibel, A.B. (1979). The hippocampus: organizational patterns in health and senescence. *Mech Ageing Devel*, 9, 89–102.

Scheibel, M.E., Lindsay, R.D., Tomiyasu, U., and Scheibel, A.B. (1976). Progressive dendritic changes in the aging human limbic system. *Exp Neurol*, 53, 420–30.

Schimanski, L.A. and Barnes, C.A. (2010). Neural protein synthesis during aging: effects on plasticity and memory. *Front Aging Neurosci*, 2, 26.

Scriabine, A., Schuurman, T., and Traber, J. (1989). Pharmacological basis for the use of nimodipine in central nervous system disorders. *FASEB J*, 3, 1799–1806.

Shen, J. and Barnes, C.A. (1996). Age-related decrease in cholinergic synaptic transmission in three hippocampal subfields. *Neurobiol Aging*, 17, 439–51.

Shen, J., Barnes, C.A., McNaughton, B.L., Skaggs, W.E., and Weaver, K.L. (1997). The effect of aging on experience-dependent plasticity of hippocampal place cells. *J Neurosci*, 17, 6769–82.

Sheng, M. and Greenberg, M.E. (1990). The regulation and function of c-fos and other immediate early genes in the nervous system. *Neuron*, 4, 477–85.

Small, S.A., Chawla, M.K., Buonocore, M., Rapp, P.R., and Barnes, C.A. (2004). Imaging correlates of brain function in monkeys and rats isolates a hippocampal subregion differentially vulnerable to aging. *Proc Natl Acad Sci U S A*, 101, 7181–6.

Small, S.A., Perera, G.M., DeLaPaz, R., Mayeux, R., and Stern, Y. (1999). Differential regional dysfunction of the hippocampal formation among elderly with memory decline and Alzheimer's disease. *Ann Neurol*, 45, 466–72.

Small, S.A., Tsai, W.Y., DeLaPaz, R., Mayeux, R., and Stern, Y. (2002). Imaging hippocampal function across the human life span: Is memory decline normal or not? *Ann Neurol*, 51, 290–5.

Small, S.A., Wu, E.X., Bartsch, D., *et al.* (2000). Imaging physiologic dysfunction of individual hippocampal subregions in humans and genetically modified mice. *Neuron*, 28, 653–64.

Smith, T.D., Adams, M.M., Gallagher, M., Morrison, J.H. and Rapp, P.R. (2000a). Circuit-specific alterations in hippocampal synaptophysin immunoreactivity predict spatial learning impairment in aged rats. *J Neurosci*, 20, 6587–93.

Smith, A.C., Gerrard, J.L., Barnes, C.A., and McNaughton, B.L. (2000b). Effect of age on burst firing characteristics of rat hippocampal pyramidal cells. *NeuroReport*, 11, 3865–71.

Stark, S.M., Yassa, M.A., and Stark, C.E. (2010). Individual differences in spatial pattern separation performance associated with healthy aging in humans. *Learn Memory*, 17, 284–8.

Stern, Y. (2002). What is cognitive reserve? Theory and research application of the reserve concept. *J Int Neuropsychol Soc*, 8, 448–60.

Steward, O., Wallace, C.S., Lyford, G.L., and Worley, P.F. (1998). Synaptic activation causes the mRNA for the IEG Arc to localize selectively near activated postsynaptic sites on dendrites. *Neuron*, 21, 741–51.

Stone, C.P. (1929). The age factor in animal learning: II. Rats in a multiple light discrimination box and a difficult maze. *Genet Psychol Monog*, 6, 125–201.

Straube, K.T., Deyo, R.A., Moyer, J.R., Jr., and Disterhoft, J.F. (1990). Dietary nimodipine improves associative learning in aging rabbits. *Neurobiol Aging*, 11, 659–61.

Thal, D.R., Rub, U., Orantes, M. and Braak, H. (2002). Phases of A beta-deposition in the human brain and its relevance for the development of AD. *Neurology*, 58, 1791–1800.

Thibault, O., Hadley, R., and Landfield, P.W. (2001). Elevated postsynaptic [Ca2+]i and L-type calcium channel activity in aged hippocampal neurons: relationship to impaired synaptic plasticity. *J Neurosci*, 21, 9744–56.

Thibault, O. and Landfield, P. W. (1996). Increase in single L-type calcium channels in hippocampal neurons during aging. *Science*, 272, 1017–20.

Thompson, L.T. and Best, P.J. (1990). Long-term stability of the place-field activity of single units recorded from the dorsal hippocampus of freely behaving rats. *Brain Res*, 509, 299–308.

Tombaugh, G.C., Rowe, W.B., Chow, A.R., Michael, T.H., and Rose, G.M. (2002). Theta-frequency synaptic potentiation in CA1 in vitro distinguishes cognitively impaired from unimpaired aged Fischer 344 rats. *J Neurosci*, 22, 9932–40.

Toner, C.K., Pirogovsky, E., Kirwan, C.B., and Gilbert, P.E. (2009). Visual object pattern separation deficits in nondemented older adults. *Learn Memory*, 16, 338–42.

Touzani, K., Marighetto, A., and Jaffard, R. (2003). Fos imaging reveals ageing-related changes in hippocampal response to radial maze discrimination testing in mice. *Eur J Neurosci*, 17, 628–40.

Tsui, C.C., Copeland, N.G., Gilbert, D.J., Jenkins, N.A., Barnes, C., and Worley, P.F. (1996). Narp, a novel member of the pentraxin family, promotes neurite outgrowth and is dynamically regulated by neuronal activity. *J Neurosci*, 16, 2463–78.

Turner, D.A. and Deupree, D.L. (1991). Functional elongation of CA1 hippocampal neurons with aging in Fischer 344 rats. *Neurobiol Aging*, 12, 201–10.

Tzingounis, A.V. and Nicoll, R.A. (2006). Arc/Arg3.1: linking gene expression to synaptic plasticity and memory. *Neuron*, 52, 403–7.

Uemura, E. (1985). Age-related changes in the subiculum of Macaca mulatta: Synpatic density. *Exp Neurol*, 87, 403–11.

Uttl, B. and Graf, P. (1993). Episodic spatial memory in adulthood. *Psychol Aging*, 8, 257–73.

van Groen, T., Kadish, I., Popovic, N., *et al.* (2011). Age-related brain pathology in Octodon degu: Blood vessel, white matter and Alzheimer-like pathology. *Neurobiol Aging*, 32, 1651–61.

Verkhratsky, A. and Toescu, E.C. (1998). Calcium and neuronal ageing. *Trends Neurosci*, 21, 2–7.

Verzar-Mcdougall, E.J. (1957). Studies in learning and memory in ageing rats. *Gerontologia*, 1, 65–85.

Wallace, J.E., Krauter, E.E., and Campbell, B.A. (1980). Animal models of declining memory in the aged: short-term and spatial memory in the aged rat. *J Gerontol*, 35, 355–63.

Wenk, G.L. and Barnes, C.A. (2000). Regional changes in the hippocampal density of AMPA and NMDA receptors across the lifespan of the rat. *Brain Res*, 885, 1–5.

West, M.J. (1993a). New stereological methods for counting neurons. *Neurobiol Aging*, 14, 275–85.

West, M.J. (1993b). Regionally specific loss of neurons in the aging human hippocampus. *Neurobiol Aging*, 14, 287–93.

West, M.J., Kawas, C.H., Martin, L.J., and Troncoso, J.C. (2000). The CA1 region of the human hippocampus is a hot spot in Alzheimer's disease. *Ann N Y Acad Sci*, 908, 255–9.

Wilson, I.A., Ikonen, S., Gallagher, M., Eichenbaum, H., and Tanila, H. (2005). Age-associated alterations of hippocampal place cells are subregion specific. *J Neurosci*, 25, 6877–86.

Wilson, M.A. and McNaughton, B.L. (1993). Dynamics of the hippocampal ensemble code for space. *Science*, 261, 1055–8.

Worley, P.F., Bhat, R.V., Baraban, J.M., Erickson, C.A., McNaughton, B.L., and Barnes, C.A. (1993). Thresholds for synaptic activation of transcription factors in hippocampus: correlation with long-term enhancement. *J Neurosci*, 13, 4776–86.

Wu, W., Brickman, A.M., Luchsinger, J., *et al.* (2008). The brain in the age of old: The hippocampal formation is targeted differentially by diseases of late-life. *Ann Neurol*, 64, 698–706.

Xiao, B., Tu, J.C., and Worley, P.F. (2000). Homer: a link between neural activity and glutamate receptor function. *Curr Opin Neurobiol*, 10, 370–4.

Yassa, M.A. Lacy, J.W., Stark, S.M., Albert, M.S., Gallagher, M., and Stark, C.E. (2011). Pattern separation deficits associated with increased hippocampal CA3 and dentate gyrus activity in nondemented older adults. *Hippocampus*, 21, 968–79.

Yau, J.L., Olsson, T., Morris, R.G., Noble, J., and Seckl, J.R. (1996). Decreased NGFI-A gene expression in the hippocampus of cognitively impaired aged rats. *Brain Res Mol Brain Res*, 42, 354–7.

CHAPTER 10
Neurodevelopmental Disorders

Alexander Prehn-Kristensen, Richard Cooke,
Lioba Baving, and Cynthia Schumann

Hippocampal size and intelligence 175
Neurodevelopmental outcome in children born preterm and the hippocampus 176
Febrile seizures and hippocampal sclerosis 179
Embryo fetal alcohol syndrome 180
Effects of cocaine on the developing hippocampus 181
Autism spectrum disorder 181
Angelman syndrome 186
Down syndrome 187
Fragile X syndrome 188
Attention-deficit hyperactivity disorder 189
Gilles de la Tourette syndrome 190
Conclusions 191

ABSTRACT Neurodevelopmental disorders are a diverse group of behaviourally-relevant multifactorial disorders related to an impairment of growth, maturation, or development of the central nervous system. The role of the hippocampus, however, has only recently been acknowledged. These disorders affect memory, emotion, cognitive abilities, and behaviour. Individuals with neurodevelopmental disorders might have comorbid disorders including intellectual impairment, seizures, and anxiety. The hippocampus is known to be vulnerable to hypoxia, stress, and malnutrition, and thus plays a clinically relevant role in children born preterm, after having experienced febrile seizures, fetal alcohol, or cocaine. Hippocampal dysfunctions are also known to be inherent as seen in Angelman syndrome, fragile-X syndrome, or Down syndrome. Complex interactions between environmental conditions and genetically predispositions are discussed in relation to neurodevelopmental disorders such as autism spectrum disorder, attention-deficit hyperactivity disorder, or Gilles de la Tourette syndrome in which hippocampal alterations are reported as well. However, even in normal brain development, there is a relationship between hippocampal structure and cognitive function as pointed out at the beginning of this chapter. Here, the structure–function relationship of the hippocampus within these complex syndromes and findings of developmental impairments are reviewed. The introduction of magnetic resonance imaging (MRI), particularly with volumetric analysis, but also of functional imaging methods such as functional MRI (fMRI), positron emission tomography (PET), or single-photon emission computed tomography (SPECT) facilitated the study of hippocampal changes associated with deficits in children in recent years.

KEYWORDS hippocampus, neurodevelopment, intelligence, born preterm, febrile seizures, hippocampal sclerosis, alcohol, cocaine, autism spectrum disorders, Angelman syndrome, Down syndrome, fragile X syndrome, attention-deficit hyperactivity disorder, Gilles de la Tourette syndrome

Hippocampal size and intelligence

There have been numerous concepts that the size of the brain correlates with information processing capacity. MRI studies have found an average correlation of brain size and intelligence of 0.40 (Willerman et al. 1991; Andreasen et al. 1993; Raz et al. 1993; Egan et al. 1994; Wickett et al. 1994; Reiss et al. 1996). An enriched environment increases the size of certain brain regions, in part, because of neurogenesis and increased complexity of connections (Diamond 2001; Grossman et al. 2002). Although twin studies reveal strong genetic influences on overall brain and grey matter volume (Andreasen et al. 1993; Wilke et al. 2003), the size of some structures, such as the hippocampus, are less heritable (Bartley et al. 1997; Sullivan et al. 2001; Thompson et al. 2001). It is well documented that the hippocampus can undergo environmentally-induced increases in size (Maguire et al. 2000, 2003) with numerous reports of postnatal neurogenesis and synaptogenesis (Kempermann et al. 1997; Nilsson et al. 1999; Brown et al. 2003; Jabes et al. 2010). As discussed in Chapter 2 by Pierre Lavenex in this book, the hippocampal formation is far from mature at birth and undergoes a protracted period of postnatal neuron addition and maturation (Jabes et al. 2011). Similar studies are also being carried out to explore postnatal neurogenesis and volume increases in humans (Schumann and Amaral, personal communication).

One might hypothesize that the size of structures such as the hippocampus, which are intimately involved in learning and memory, would be more likely to correlate with intelligence measures than, for example, structures such as the amygdala, which are involved in the regulation of emotions. This is, in fact, the case. In typically developing children, Schumann and colleagues (2007) found a strong positive correlation of hippocampal volume with measures of full scale and verbal intelligence quotient (IQ) even after correcting for the height, age, and grey matter volume of the participant (Figure 10.1). The volume of the hippocampus ranged from ~4.5–6.5 cm^3 and predicted 36% of the variance in full scale IQ and 47% of the variance in verbal IQ in typically developing males 8–18 years of age. Study participants with a verbal IQ above 120 had a 25% larger hippocampus than those with a verbal IQ below 100 (Schumann et al. 2007). In adults, Andreasen et al. (1993) also found a positive correlation of hippocampal size with IQ measures. The relationship of size and IQ is not common to all structures in the temporal lobe; Schumann and colleagues (2007) found no relation between the volume of the amygdala and intelligence. This finding may presumably be accounted for by the nature of the IQ exam, which measures multiple domains of information processing that are performed by the hippocampus, such as memory formation (Squire & Zola 1996; Zola et al. 2000; Manns et al. 2003a). A larger hippocampus could reasonably be associated with a greater ability to store and retrieve information that would be valuable in completing the exam. Functions performed by the amygdala, such as danger detection and the production of a fear response, are less likely to contribute to a higher score on an IQ exam.

These data are of interest in relation to the nature versus nurture question of intelligence. Do some children have a higher IQ because they have a bigger hippocampus from birth and are therefore better able to learn and remember information? Or, does the size of the hippocampus reflect

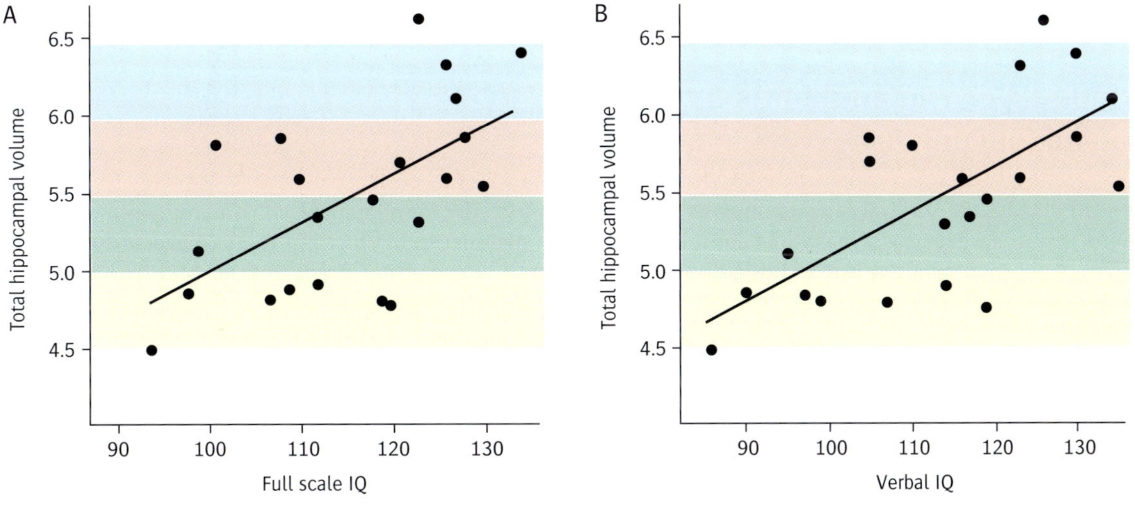

Figure 10.1 Linear regression scatterplot showing that hippocampal volume measured in cubic centimetres (cm^3) in typically developing male children positively correlates with (A) full scale ($r^2 = 0.3$) and (B) verbal IQ ($r^2 = 0.47$). Reproduced from Schumann, C. M., Hamstra, J., Goodlin-Jones, B. L., Kwon, H., Reiss, A. L., and Amaral, D. G. (2007). Hippocampal size positively correlates with verbal IQ in male children. *Hippocampus*, 17(6), 486–93. © 2007, with permission from John Wiley and Sons.

the quality and quantity of a child's educational experience? Data in support of either of these perspectives can be found in the literature. There is substantial evidence, for example, that the hippocampus can undergo environmentally-induced increases in neuron number when rodents are placed in an enriched environment (Kempermann et al. 1997; Nilsson et al. 1999; Brown et al. 2003). Experience-induced neurogenesis in the hippocampus positively correlates with improvements on learning and memory tasks. Kempermann and Gage (2002) found a significant correlation between the level of new neurons in the dentate gyrus of the hippocampus and the level of performance on hippocampal-dependent tasks that involve the acquisition of new information. Species that develop complex spatial maps, such as food-caching birds, have larger hippocampi than members of the same species who do not engage in food caching (Clayton and Krebs 1994), even though they have comparable sized hippocampi at birth (Clayton 1998). Studies of experienced London taxi cab drivers who have memorized the complex roadway system of the city also have larger hippocampal volume than age-matched controls or newly licensed drivers (Maguire et al. 2000, 2003). It is not clear whether the process of learning the roadways induces an enlargement of the hippocampus or that only those individuals with a larger hippocampus, who are capable of memorizing the complexities of the roadways, ever succeed at the job.

One interesting aspect of the Schumann et al. (2007) findings is that hippocampal volume strongly correlated with verbal, but not performance, IQ scores. The performance, or non-verbal, IQ score is commonly determined by having an individual perform two activities involving spatial reasoning, pattern recognition, and visuomotor/ coordination skills (Wechsler 1999). The verbal IQ score is commonly produced from two verbal tasks involving verbal knowledge, expressive vocabulary, and abstract verbal reasoning ability. As discussed elsewhere in this book, the hippocampus is responsible for the encoding of declarative memory: both episodic information, i.e. memories of personal past experiences, and semantic information, i.e. general facts. Patients with bilateral hippocampal lesions, such as the wel-known patient H.M., have intact visuomotor and perceptual abilities as well as procedural (or non-declarative) learning (Milner 1962; Corkin 1968; Gabrieli et al. 1986; Shadmehr et al. 1998). However, H.M. and other hippocampal-damaged patients are severely impaired in declarative memory (Scoville & Milner 1957; Milner 1962; Gabrieli et al. 1998; Manns et al. 2003b). Thus, the hippocampus is responsible for learning factual information that is tested as part of the verbal IQ score. The hippocampus is generally not required, however, for skill learning, which contributes to the performance IQ score.

Given that the volume of the hippocampus accounts for a substantial proportion of the variability in verbal IQ, it would be of interest to determine if differences in hippocampal volume precede formal education or are a result of it. This could only be accomplished through a longitudinal MRI analysis that begins as early as possible after birth. Data of this type are currently being collected and will afford an opportunity to test the hypothesis that hippocampal differences are observed prior to formal education. If hippocampal volume differences are observed prior to schooling, children with extremely small hippocampi may merit from educational practices that are less dependent on declarative memory and capitalize on procedural learning. To the extent that individual IQ scores may predict future success, so might early individual estimates of hippocampal volume.

Neurodevelopmental outcome in children born preterm and the hippocampus

In recent decades the survival rates amongst the smallest preterm infants have increased many-fold due to the development of neonatal intensive care techniques. As a result, paediatricians are aware that a considerable proportion of these children later show major or minor neurodevelopmental sequelae. Major difficulties such as cerebral palsy and neurosensory deficits are easily recognized, but lesser problems such as dyspraxia, visuospatial problems, and cognitive and behavioural disorders such as attention deficit usually require screening for (Foulder-Hughes and Cooke 2003). The hippocampus is known to be vulnerable to hypoxia, stress, and malnutrition, all likely to be present in fetal growth restriction and following preterm birth. Hippocampal injuries during early lifetime can seriously affect the development of the hippocampus and its functions (Bachevalier and Vargha-Khadem 2005; De Haan et al. 2006). The introduction of MRI, particularly with volumetric analysis, has enabled the study of hippocampal changes associated with later deficits in preterm children in recent years (for an example see Figure 10.2).

Structural studies

Volumetric measurements of the hippocampus are not easy because of its highly irregular shape, and differences in technique (manually versus automatically) and age grouping probably account for the variation in normative values in some studies (Honeycutt and Smith 1995).

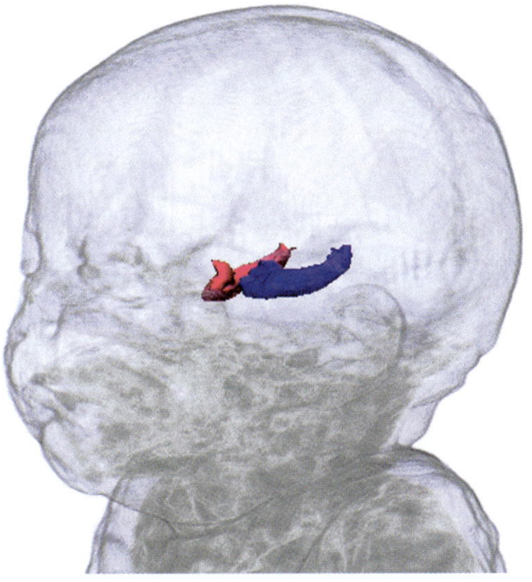

Figure 10.2 Three-dimensional image of neonate hippocampi. Image kindly provided by Dr. Deanne Thompson, Murdoch Children's Research Institute, Royal Children's Hospital, Victoria, Australia. © Deanne K. Thompson, 2011.

Volumetry in preterm is even more challenging due to the lack of myelination since the temporal lobe myelinates later than other lobes. The rate of development in both preterm and term infants of the hippocampus has been assessed from 23 weeks through to adolescence and adulthood. In a study of cranial ultrasound scans made in 158 children born before 35 completed weeks of gestation, and free of intracranial pathology, the degree of completeness of hippocampal inversion was examined (Bajic et al. 2010). Hippocampal inversion is the folding of the hippocampal structures around the hippocampal sulcus into the temporal lobe. A rounded form on the scan (the ratio between the horizontal and vertical diameters of the hippocampal body being less than or equal to 1) was taken to indicate incomplete hippocampal inversion. Fifty per cent of infants at 23–24 weeks, 24% at 25–28 weeks, and 14% at 29–36 weeks showed incomplete inversion. The last frequency is similar to that in the adult population (19%). Bilateral incomplete inversion was seen mainly before 25 weeks gestation. Measuring hippocampal volumes from 1 month postnatal to 15 years in 50 children, Pfluger et al. (1999) produced normative growth curves for right and left sides in males and females. The right hippocampus was larger than the left and the development of the hippocampus through childhood faster in girls. Giedd et al. (1997) also showed a more rapid growth of the hippocampus in girls between 4–18 years. A similar study in Japan evaluated 42 normal children aged 3 weeks to 14 years with volumetric MRI (Utsunomiya et al. 1999). They showed that hippocampal volume increased rapidly up to 2 years of age and more slowly after that. In 91% of cases the hippocampus was larger on the right side. Right–left asymmetry was seen from an early age. Another group examined volumetric magnetic resonance images from 32 term and 184 preterm infants at term (Thompson et al. 2009). Both groups showed right asymmetry at term. No gender differences were noted at this stage, and the preterm group showed significantly less asymmetry than the term infants (Figure 10.3). The timing of myelination in the hippocampus has been studied in 48 term and preterm infants using T1 and T2-weighted images (Sie et al. 1997). Myelin was seen to be present by 30–34 weeks postmenstrual age, and little further progress noted before postconceptual age of 46 weeks.

Isaacs et al. (2000) have shown that preterm infants in adolescence have a smaller hippocampus bilaterally than their term peers on volumetric MRI, despite equivalent head size. They also showed specific deficits in several aspects of everyday memory both on objective testing and by parental questionnaires. They also had a specific deficit in numeracy. A similar later study involving 87 very low birth weight adolescents and eight term controls showed a significantly lower hippocampal ratio in the preterm group (Abernethy et al. 2002). In those preterm infants with a lower intelligence quotient (more than 1 SD below the mean), a smaller left hippocampus and a lower hippocampal ratio were seen. Preterm infants with attention deficit hyperactivity disorder (Connors' Teacher Rating Scale) had a significantly smaller hippocampus bilaterally. The same authors studied a later cohort at 7 years of 105 preterm infants born in Liverpool at less than 33 weeks of gestation (Abernethy et al. 2004). Similar techniques were used but no significant relationships between intelligence or dyspraxia and hippocampal volumes were found. The mean volumes measured were considerably larger than in the previous cohort, and this difference may simply reflect the greater proportion of intrauterine growth restriction present in that cohort. Small but statistically significant differences in T2 relaxation times for the right but not the left hippocampus were found in preterm 7-year-olds with lower intelligence (Abernethy et al. 2003). This difference implied a deficit in hippocampal myelination in those infants with a lower intelligence.

Isaacs et al. (2003) investigated memory function and hippocampal volumes in two groups of adolescents, one born preterm and the other consisting of adolescents with developmental amnesia, and an age-matched term control group. Preterm adolescents showed a mean bilateral reduction in hippocampal volume of 8–9% compared with the controls, but the group with developmental amnesia had a mean bilateral reduction of 40%. The two test groups had similar overall intelligence and short-term memory scores as the controls. However, the developmental amnesia group had poorer scores on many verbal and delayed memory tasks, but the preterm group performed less well on only a few memory measures such as route following and prospective memory. The authors concluded that disabling memory defects are mainly associated with severe and early reduction in volume of the hippocampus by at

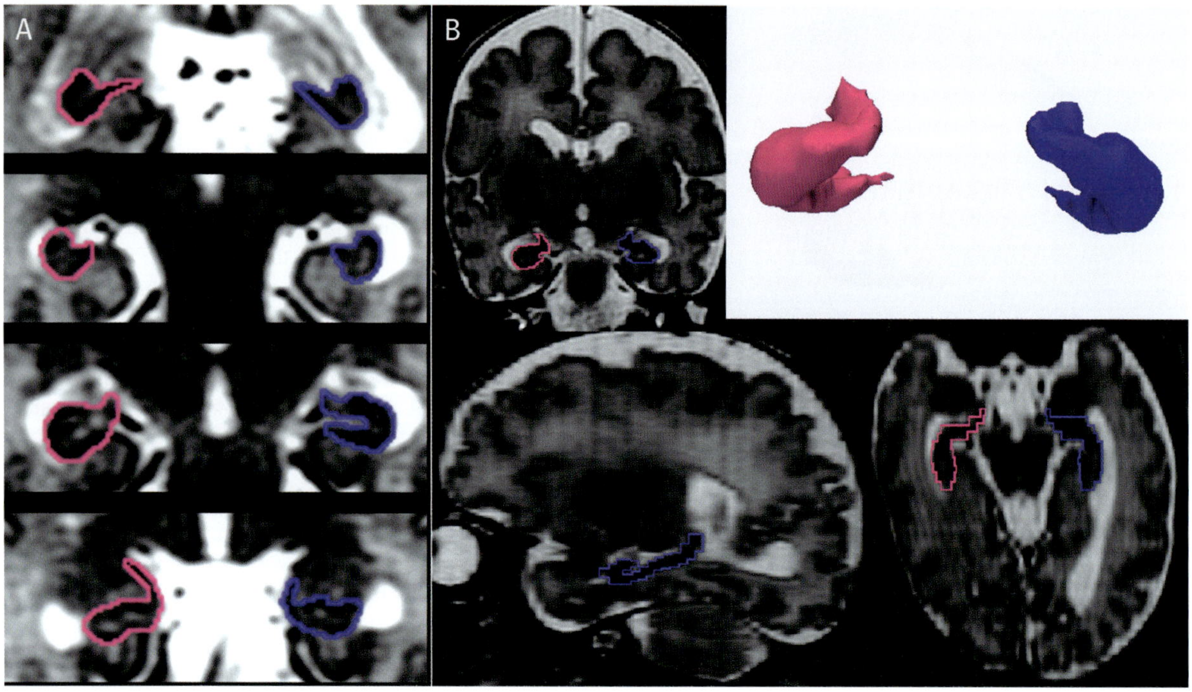

Figure 10.3 MR-determined hippocampal asymmetry in full-term and preterm neonates. A) Segmentation boundaries of four coronal sections through the hippocampal tail, body, and head. B) The hippocampal segmentation boundaries shown in all three orthogonal views: axial, sagittal, and coronal, as well as a three-dimensional representation of the left (blue) and right (pink) hippocampal volumes. Reproduced from Thompson, D. K., Wood, S. J., Doyle, L. W., Warfield, S. K., Egan, G. F., and Inder, T. E. (2009). MR-determined hippocampal asymmetry in full-term and preterm neonates. *Hippocampus*, 19 (2), 118–23. © 2009, with permission from John Wiley and Sons.

least 20–30% each side. They were unable to determine whether this was sufficient alone to cause the deficits seen, or whether a reduction in volume of other regions of the brain was also required.

In a more recent study, volumetric magnetic resonance studies at term equivalent were obtained on 156 infants born at 29 weeks of gestation or fewer or with a birth weight < 1250 grams (Beauchamp et al. 2008). The relationship between hippocampal volumes and performance on a working memory task (delayed alternation) at 2 years of age was examined. Those children who performed poorly had significantly smaller hippocampal volumes than those preterm children who had shown an intact working memory. The hippocampal changes in preterm infants with poor memory function in later years are present by term equivalent.

Functional studies

Learning in adults born very preterm has been studied using fMRI (Lawrence et al. 2010a). On a verbal paired-associate learning task, the preterm adults scored similarly to the term controls. On imaging, however, the preterms showed increased brain activation in the left parahippocampal and precentral gyri during encoding and in the precentral gyrus during recall. They also had decreased grey matter volume in the hippocampus bilaterally, and increased grey matter volume in the left parahippocampal gyrus. The preterms also showed (compared with controls) activation in the left parahippocampal gyrus which correlated to the grey matter volume on that side. These changes were most marked in the least mature at birth. It was thought that these functional changes associated with structural changes reflected a process of neural reorganization after early brain injury.

Using the Bayley Scales of Infant Development 184 preterm infants and 32 term control infants were evaluated at 2 years of age (Thompson et al. 2008). They had previously had volumetric MRI at term equivalent. There was a mean reduction in hippocampal volume in the preterms of 3.4%, but hippocampal volumes did not differ from term controls if head size was allowed for. A number of perinatal variables were significantly associated with lower hippocampal volumes in the preterms. These were moderate to severe white matter injury, exposure to postnatal steroids, and indomethacin treatment for persistent ductus arteriosus. The Mental and Psychomotor scores from the Bayley Scales were significantly related to hippocampal volumes when corrected for head size and sex, but less so after further correction for white matter injury and exposure to postnatal steroids. Other studies have shown that the use of hydrocortisone rather than dexamethasone in the postnatal period for managing chronic lung disease does not appear to have adverse effects on hippocampal development and later cognition (Lodygensky et al. 2005; Rademaker et al. 2006).

Febrile seizures and hippocampal sclerosis

Febrile seizures are a common form of generalized seizure seen in children usually under the age of 5 years. They occur in about 4% of the population, are often familial, and are usually of short duration and uncomplicated. The seizures appear to be associated with a sudden rise in body temperature, most commonly at the onset of a viral upper respiratory illness. If they do not recur within a year they usually will not do so again, and the long-term prognosis is excellent. A proportion of febrile seizures are classified as complex, and are of more than 20 min in duration and may result in neurological signs such as temporary hemiplegia. These complex febrile seizures are often associated with earlier cerebral injury, often from the perinatal period, and also later neurological sequelae. Febrile status epilepticus is usually focal.

Fifty years ago Meyer suggested that sclerosis of Ammon's horn in the hippocampus (mesial temporal sclerosis, MTS), associated with temporal lobe epilepsy (TLE), was in fact the result of earlier ischaemic injury to the hippocampus as the result of periods of hypoxia during febrile seizures in early childhood (Cavanagh and Meyer 1956). In those cases of TLE with sclerosis, 64% had experienced prior status epilepticus, most often recorded as 'teething convulsions'. Debate has continued ever since as to whether seizures are the cause or the result of the sclerosis.

The advent of widespread availability of MRI allowed the relationship between early febrile seizures and the age of unprovoked seizure onset to be correlated with hippocampal volumes *in vivo* (Cascino 1995). In an MRI study of 53 children aged between 2–17 years with TLE, 30 were shown to have MTS. Fourteen of these children were under 10 years of age and some as young as 2, suggesting that MTS was often underdiagnosed previously and that it is likely to be the cause of rather than to be caused by TLE (Harvey et al. 1995). A major US study in which 49 patients had temporal lobe resection for TLE, compared hippocampal volumetrics by MRI preoperatively with neuropathology and outcomes (Luby et al. 1995). Although hippocampal atrophy was common, it was not necessarily indicative of significant neuronal cell loss, nor a factor determining a successful outcome. In a further volumetric MRI study of 35 cases of complex partial seizures confirmed by video-electroencephalography, those patients with a past history of prolonged febrile seizures had a smaller ipsilateral hippocampal volumes and ipsilateral/contralateral ratio than those without a history of prolonged febrile seizures (Theodore et al. 1999). The duration of the patient's epilepsy was also independently related to the extent of hippocampal atrophy suggesting progressive hippocampal damage after the initial insult. Later, but similar studies came to different conclusions. A series of 77 consecutive surgical patients with TLE and proven MTS had their hippocampal volumes measured by MRI and were compared with normal controls (Bower et al. 2000). Volumes were not different between those patients with a history of febrile seizures and those without. The authors believed that the hippocampal sclerosis was a pre-existing abnormality.

An extensive pathological study of hippocampal sclerosis in 572 specimens from patients with TLE and 73 from patients with extratemporal seizures, showed that hippocampal sclerosis was strongly linked with an initial precipitating injury (IPI) in both TLE and extratemporal seizures (Mathern et al. 2002). In TLE cases, IPI occurred early and often involved seizures, but pathological appearances in cases with early or late IPI were indistinguishable. In TLE patients hippocampal neuron loss was related to the duration of the epilepsy, and continued for 30 years or more. Severe seizures in the youngest patients were not associated with neuronal damage, but there were signs of abnormal hippocampal development. The authors concluded that most neuronal loss occurred with the initial IPI, but that as a consequence of repeated seizures, hippocampal neuronal loss continues over the next 30 years. To some extent this view is supported by animal studies in which experimental prolonged febrile seizures were induced (Bender et al. 2004). Neuronal loss was not seen in the hippocampus and neurogenesis remained unchanged. Synaptic reorganization and other longer term changes were seen, however. An interesting case report describes the MRI appearances of the hippocampi of a 23-month-old boy 3 days after a prolonged focal febrile seizure (Sokol et al. 2003). An increased T2-weighted signal and increased volume was seen in the right hippocampus. Two months later a repeat scan showed hippocampal sclerosis. There had been four previous described cases in the literature. A similar study, published more recently, described the appearances of a markedly hyperintense hippocampus on T2-weighted images made within 72 h of febrile status epilepticus in young children (Provenzale et al. 2008). Repeat MRI at a mean of 9 months later showed MTS.

The varied conclusions from the studies described may be due to retrospective nature and variation in definitions for febrile seizures. MTS may arise from a variety of causes, and there is certainly evidence that there are syndrome-specific genes for febrile seizures and MTS (Cendes 2004). The association of febrile seizures with TLE is complex, and probably due to interactions between several genetic and environmental factors.

Embryo fetal alcohol syndrome

Prenatal alcohol exposure can result in severe neurological, behavioural, and growth abnormalities (Jones and Smith 1973; Jones et al. 1973; Niccols 2007). In animals, prenatal exposure to even moderate levels of alcohol (in rats, daily exposure to 110 mL of 5% ethanol during gestation) can affect hippocampal neuroanatomy and long-term hippocampal plasticity (Sutherland et al. 1997; Berman and Hannigan 2000). In the same way, literature on human studies indicates that prenatal exposure to alcohol can result in abnormal hippocampal development and function. Fetal alcohol spectrum disorders (FASD) is used to capture the spectrum of human individuals affected by prenatal alcohol exposure, ranging from non-dysmorphic individuals, who only display neuropsychological disorders (PAE, prenatal alcohol exposure; ARND, alcohol-related neurodevelopmental disorder), to fetal alcohol syndrome (FAS), which is characterized by central nervous system (CNS) damage (e.g. encephalopathy) and facial dysmorphism. Thereby, chronic as well as even acute alcohol exposure in specific periods of gestation (particularly during the first trimester) can harm nervous tissue of the fetus sustainable (Aversi-Ferreira and Nascimento 2008).

Structural studies

Evidence of altered hippocampal structure in humans exposed to alcohol *in utero* is not as clear as that reported in animal literature (Miki et al. 2008; Samudio-Ruiz et al. 2009). Magnetic resonance studies suggest that at least some children prenatally exposed to alcohol display reduced hippocampal volume (Autti-Rämö et al. 2002). While a sparing hippocampal volume in proportion to limbic brain areas was reported in FAS children/adolescents (Archibald et al. 2001), in another study, an absolute but not a relative decrease in hippocampal volume was found (Astley et al. 2009b). Interestingly, in the latter study, the hippocampal volume decreases with the magnitude of severity. With respect to the hemispheres, there are signs of pronounced asymmetry (right > left) in the hippocampal volumes in children with FAS/PAE (Riikonen et al. 1999; Autti-Rämö et al. 2002; Willoughby et al. 2008). However, here too the findings are not very solid (Riikonen et al. 2005). In fact, some reviews concluded that data of reduced hippocampal volume in FASD are equivocal (McGee and Riley 2006; Norman et al. 2009). It is noteworthy, however, that the differences in hippocampal volume between FASD and healthy controls become more prominent with maturation (Willoughby et al. 2008).

Functional studies

While SPECT in FAS did not reveal any differences in the hippocampal serotonergic/dopaminergic system nor in blood flow (Riikonen et al. 2005), a tendency towards a lower hippocampal choline concentration was reported in a recent study (Astley et al. 2009a). Particularly, the choline concentration in the left hippocampus decreases with decreasing size of several brain regions. Choline is a precursor of the neurotransmitter acetylcholine which is critically involved in hippocampus-related memory formation. In line with this, increased recruitment of frontal memory circuits was observed in an fMRI study involving verbal associative, hippocampus-dependent memory in PAE. This result was considered to be the result of compensation of dysfunctional medial temporal memory systems (Sowell et al. 2007).

Several studies described memory deficits in prenatal alcohol-exposed children and adults (Manji et al. 2009), which can be associated with hippocampal dysfunction. For example, it has been reported that children with FAS displayed a distortion in the recall of the locations of objects (Uecker and Nadel 1996, 1998) and show deficits in a virtual Morris water maze task (Hamilton et al. 2003). In older FASD children impairment of spatial memory performance correlates with a decrease in hippocampal volume (Willoughby et al. 2008). Adolescents who had been exposed to alcohol prenatally display reduced performance in visuospatial as well as in declarative memory tasks (Olson et al. 1998). Problems in spatial memory tasks persist into adulthood: adults with FAS revealed impaired verbal and non-verbal memory (Kerns et al. 1997; Coles et al. 2010). However, since other brain regions besides the hippocampus are more involved in verbal memory performance, the latter results should be interpreted merely as an indication rather than as evidence for hippocampal memory dysfunction in adults with FASD.

Limitations

Although the mentioned findings emphasize deficits in hippocampus-dependent memory in FASD, it has to be taken into consideration that PAE affects cognitive functions more globally than just the hippocampal system. In particular, deficits in encoding and verbal memory also contribute to spatial memory deficits as these have been reported in FASD (Kaemingk and Halverson 2000; Pei et al. 2008). Among other neuropsychological abnormities, one of the core neuropsychological symptoms of FASD is attention deficit (Nanson and Hiscock 1990; Lee et al. 2004). Therefore, it is not surprising that most patients with FASD also seem to fulfil the criteria of ADHD (Bhatara et al. 2006; Rasmussen et al. 2010). This leads to the concern that comparisons between FASD and healthy controls can not be made independently of ADHD symptoms, as was done in the majority of the mentioned studies. Since it can be assumed that ADHD may have specific hippocampus alterations (see later section), more comprehensive conclusions were possible by introducing drug-naïve ADHD patients without FASD as controls into the study designs.

Effects of cocaine on the developing hippocampus

Chronic intake of drugs, such as cocaine, amphetamine, morphine, heroin, and alcohol, decreases hippocampal neurogenesis. A large number of human studies have shown that drug abuse affects several cognitive functions including memory performance such as spatial and episodic memory which requires the active engagement of the hippocampus. In addition, substance-induced dysfunction of episodic and contextual memory may become a discrete problem when cocaine users are exposed to drug-associated social and environmental cues after drug withdrawal (Grant et al. 1996). Here, we will focus on cocaine which is one of the most addictive and commonly abused psychoactive substances.

Psychostimulants such as cocaine enhance monoaminergic transmission in the CNS. Cocaine acts primarily at the presynaptic level to block reuptake of dopamine, norepinephrine, and serotonin by specific presynaptic plasma membrane transporters (Torres et al. 2003; Elliott and Beveridge 2005). This leads to elevated levels of circulating catecholamine and an excessive synaptic response resulting hypertension, tachycardia, vasoconstriction, agitation, euphoria, and excitation.

In the 1980s, cocaine became one of the most frequently abused illicit drugs during pregnancy. Prenatal exposure to cocaine disturbs the fetal blood supply, changes placental gene expression, and affects the development of the monoaminergic system (Salisbury et al. 2009). Children who were prenatally exposed to cocaine display mild deficits in sustaining attention and tend to have behavioural problems (Ackerman et al. 2010). In some individuals, prenatal cocaine exposure affects short-term memory, memory for words and faces, and working memory (Mayes et al. 2007; Bethancourt et al. 2009; Hurt et al. 2009; Li et al. 2009; Richardson et al. 2009; Betancourt et al. 2011).

Chronic, but not short-term cocaine exposure reduces proliferation of progenitor cells in the human and animal hippocampus. These molecular changes in hippocampal cells, however, are not long-lasting: cocaine intake does not significantly affect progenitor cell survival or differentiation, and effects of cocaine-induced changes in progenitor proliferation are reversible after periods of abstinence (Venkatesan et al. 2007; Canales 2010; Sudai et al. 2010). Impairment of hippocampus-dependent memory performance has been frequently reported in animals; however, factors of age, dosage, and time of abstinence should be taken into account, since these factors can mediate cocaine's effect on memory (Canales 2010).

Chronic exposure to cocaine mainly results in deficits in executive functions such as attention, decision-making, or problems solving, and the most prominent cocaine-induced changes in human brain function are associated with the prefrontal cortex (PFC) (Beveridge et al. 2008). However, disturbed memory functions are also a part of the cognitive deficits observed in chronic cocaine users (Jovanovski et al. 2005). Working memory, which depends on the integration of the PFC and the hippocampus, is impaired in some cocaine-dependent individuals but not in all (Jovanovski et al. 2005; Pace-Schott et al. 2008; Moeller et al. 2010). Somewhat pronounced memory performance deficits derive from memory performance, in which the hippocampus plays a larger role: non-addicted and addicted cocaine users displayed impairment in spatial, short-term, and verbal memory. An association between the lifetime amount of cocaine abuse and memory deficits has been reported (Ardila et al. 1991; O'Malley et al. 1992; Hoff et al. 1996; Reske et al. 2010). Individuals addicted to cocaine display a reduced learning curve in memorizing word lists (Fox et al. 2009), and, in the same vein, cocaine users show poorer performance in attention and word recognition than compared to controls. While attention significantly improved after 2 weeks of abstinence, memory deficits remained (Pace-Schott et al. 2008). Functional imaging studies revealed that cocaine reduces the metabolism in the primate hippocampus and adjacent regions (Beveridge et al. 2006). In humans who consume cocaine, reduced glucose metabolism in the PFC was observed in which the reduction predicted performance in verbal and visual memory tasks (Goldstein et al. 2004). As shown by fMRI resting-state measurements, chronic cocaine user display reduced connectivity between functional networks related to the mesocorticolimbic system (Gu et al. 2010). The authors discuss whether such reduced connectivity between the PFC and both the hippocampus and the amygdala may reflect a difficulty in learning of negative consequences as seen in cocaine users. Taken together, these results emphasize dysfunctions of the hippocampus, the PFC and their interaction in memory processes.

Autism spectrum disorder

Autism spectrum disorder is a common, heterogeneous, *behaviorally defined*, neurodevelopmental disorder affecting approximately one in 100 children. Symptoms typically become evident before 3 years of age and are generally associated with varying degrees of developmental delay. Clinical features include deficits in social interaction and verbal and non-verbal communication as well as restricted or stereotyped patterns of behaviour (American Psychiatric Association, DSM IV). Individuals with autism might also have comorbid disorders including intellectual impairment (Fombonne 2006), seizures (Tuchman and Rapin 2002), and anxiety (White et al. 2009). Increasingly, researchers refer to 'the autisms' rather than a single autism phenotype (Geschwind and Levitt 2007; Amaral et al. 2008). There is no consistent genetic aetiology of autism even though estimates of heritability are as high as 90% (Levitt and Campbell 2009). The autism spectrum disorders are likely to involve multiple genes and complex interactions between genetic risk and environmental factors.

There is no clear and consistent neuropathology that has emerged for autism, however, postmortem and structural MRI studies highlight the frontal and temporal cortices, amygdala, and cerebellum as predominantly pathological in autism (see Amaral et al. 2008 for review). As reviewed in this section, there is also evidence, although somewhat inconsistent, that the anatomical development of the hippocampus in individuals with autism deviates from that of typical development. It is unclear though whether these abnormalities are related to impairments, or even possibly strengths, in hippocampal-associated functions such as spatial or declarative memory in people with autism (Ameli et al. 1988; Rumsey and Hamburger 1988; DeLong 1992; Bennetto et al. 1996; Dawson et al. 2001; Minshew and Goldstein 2001; Bowler et al. 2004; Caron et al. 2004).

Postmortem brain tissue studies

Bauman and Kemper carried out the initial groundbreaking studies during the mid 1980s to early 1990s to systematically describe pathology in a sample of postmortem brains from people with autism (Bauman and Kemper 1985; Bauman et al. 1985). Over this series of studies, they collected and described six cases of idiopathic autism, five of which had mental retardation and four of which had epilepsy. In a side-by-side comparison of each autism case with an age- and sex-matched control, they observed that neurons in the amygdala and hippocampus of most autism cases appeared unusually small and more densely-packed (i.e. increased numbers of small neurons per unit volume) than in age-matched controls. They noted that this was particularly evident in the CA3 region in the hilum of the fascia dentate (also known as CA4) and the deep layer of the entorhinal cortex. In a 29-year-old male with autism, they observed that the clear zone of the entorhinal cortex referred to as the lamina dessicans, which normally disappears in early childhood, was still present (Bauman and Kemper 1985). In addition, further analyses using rapid Golgi technique revealed stunted dendritic arbours and a limited amount of secondary and tertiary branching in CA1 and CA4 pyramidal neurons (Raymond et al. 1989). Bauman and Kemper (1994) concluded from these studies that the pattern of small, closely packed neurons and decreased complexity of dendritic arbours resembles that typically seen during an earlier stage of brain maturation and might, therefore, represent a curtailment of normal development.

Bailey and colleagues (1998) carried out a similar comprehensive qualitative and semiquantitative study of six cases of autism with mental retardation, four of which had seizures, compared to seven controls. In a quantitative assessment of neuronal density of the hippocampus in CA1, CA3, and CA4 at the level of the lateral geniculate body, they found an increased density of neurons in two of the six autism cases, although this was not found to be significant overall. Interestingly, in a 4-year-old with autism, Bailey and colleagues noted the temporal lobes to be hyperconvoluted with an upturned hippocampus, which may be due to the need to accommodate an abnormally enlarged temporal cortex as suggested by MRI studies (Schumann et al. 2010).

Recent studies from Blatt and colleagues propose widespread abnormalities in the gamma-aminobutyric acid (GABA)ergic system of the brain in individuals with autism (Blatt et al. 2001; Oblak et al. 2010). In an earlier autoradiographic study of the hippocampus, they found a reduction in [^{3}H]muscimol binding to $GABA_A$ receptors in the CA1 pyramidal cell layer and in [^{3}H]flunitrazepam binding to benzodiazepine sites in CA2 and subiculum in adult autistic cases versus age-matched controls (Blatt et al. 2001). In a follow up study, Guptill et al. (2007) predicted that the decrease in [^{3}H]flunitrazepam-labelled benzodiazepine sites was due to a decrease in the number of binding sites rather than an altered affinity (Figure 10.4), possibly resulting in altered inhibitory functioning of $GABA_A$ receptors in the hippocampal circuitry of the autistic brain. A recent immunohistochemistry study evaluated subpopulations of GABAergic neurons using calcium binding proteins in the hippocampus in a small sample of five individuals with autism (Lawrence et al. 2010b). They found an increased density of immunoreactive interneurons to calbindin in the dentate gyrus, calretinin in CA1, and parvalbumin in CA1 and CA3 in the autism cases relative to five age-matched male controls, further suggesting alterations in the GABAergic system in autism (Guptill et al. 2007).

Magnetic resonance imaging studies

There have been relatively few studies of autism to date that have published volumetric analyses of the hippocampus and a consistent story has yet to emerge. In general, studies have found either no difference (Piven et al. 1998; Haznedar et al. 2000; Nicolson et al. 2006; Dager et al. 2007) or a moderate enlargement (Sparks et al. 2002; Bigler et al. 2003; Rojas et al. 2004; Schumann et al. 2004; Palmen et al. 2006; Groen et al. 2010) in hippocampal volume between autism and typically developing control subjects.

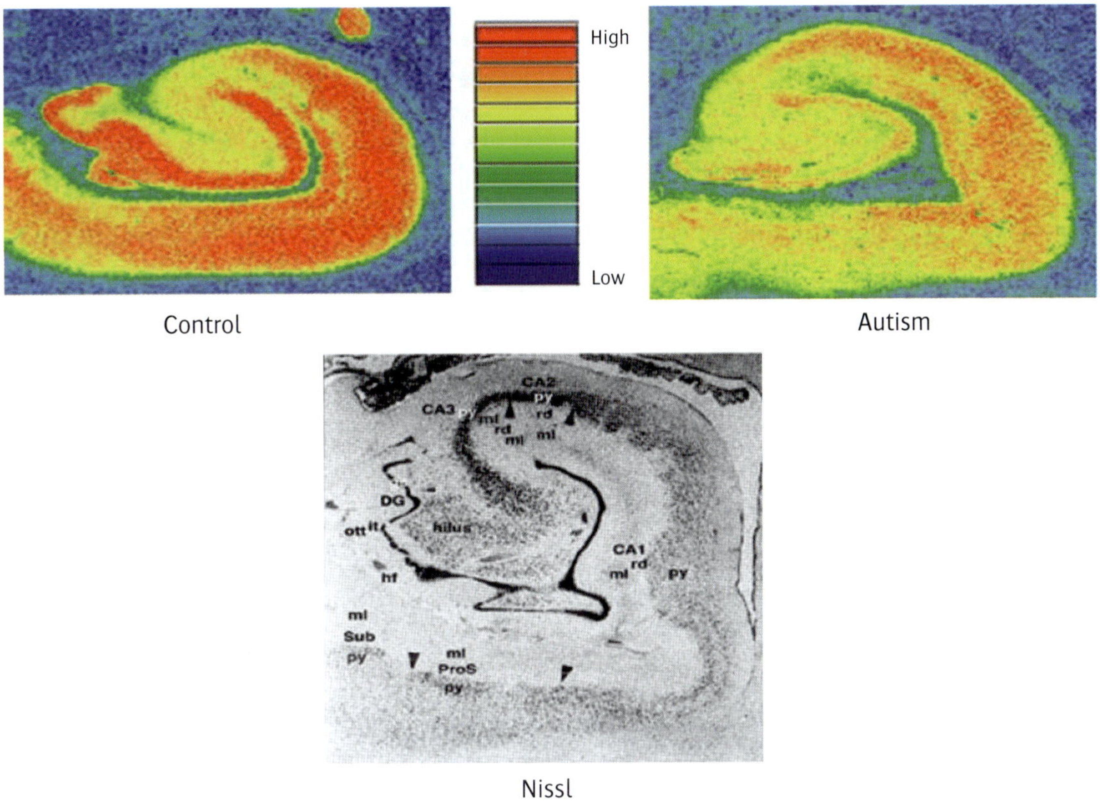

Figure 10.4 Tissue sections from a control and autistic brain through the mid-level of the human hippocampal formation. The colour index chart indicates a range of high to low binding of [3H]flunitrazepam-labelled benzodiazepine binding sites. Note the increased binding in the control section (left) compared to the age-matched autistic section (right) especially within the dentate gyrus, CA, and subicular subfields indicated. A comparable Nissl stained section indicating hippocampal subfields and lamina is centred below the sections for reference. DG, dentate gyrus; gc, granule cell layer of the dentate gyrus; hf, hippocampal fissure; it, inner third of the molecular layer of the dentate gyrus; ml, molecular layer; ott, outer two-thirds of the molecular layer of the dentate gyrus; ProS, prosubiculum; py, pyramidal layer (stratum pyramidale); rd, stratum radiatum; Sub, subiculum. With kind permission from Springer Science+Business Media: *Journal of Autism and Developmental Disorders*, [3H]-Flunitrazepam-labeled benzodiazepine binding sites in the hippocampal formation in autism: A multiple concentration autoradiographic study, 37(5), 2006, 911–20, Guptill, J. T.

One exception is Aylward et al. (1999a), who reported a decrease in hippocampal volume in adolescents and adults with autism after controlling for total cerebral volume. Saitoh et al. (2001) found only the cross-sectional area of the dentate gyrus and CA4, measured in three contiguous 5-mm sections, to be smaller than normal in children and adults with autism.

There are several possible explanations for the inconsistent reports of hippocampal volume in autism. The first is that most MRI studies to date have been conducted on a wide age range of subjects which might contribute to the disparate findings across studies as is the case with other brain structures. We now know that the brain in many individuals with autism undergoes an abnormal developmental time course that appears to include a period of early overgrowth followed by a deceleration in age-related growth, which is particularly noted in the frontal and temporal cortices and the amygdala (see Amaral et al. 2008 for review). It is also important to note that different brain regions and tissue types may undergo different age trajectories. For example, frontal lobe enlargement appears to be limited to very early childhood (~4 years of age) whereas amygdala enlargement appears to extend a few years later (~10 years of age) in cases of autism relative to typical controls. Various brain structures also undergo different growth trajectories in typical children as well. For example, the amygdala in typically developing boys continues to grow into adolescence; this age related growth does not occur in children with autism (Giedd et al. 1996, 1997; Schumann et al. 2004). Therefore it is possible that the results of a volumetric study on the hippocampus may also heavily depend on the age of the participants. There is currently little evidence to determine if the hippocampus follows a similar aberrant developmental trajectory that appears to occur in other brain regions, such as the amygdala, in children with autism. The sex of the participants is another factor that may alter study findings, in that the same brain structure may develop at different rates in males and females. Giedd and colleagues (1996) have demonstrated that prolonged hippocampal age-related growth through adolescence occurs in typically developing girls, but does not occur in typically developing boys or those with autism (Giedd et al. 1996, 1997; Schumann et al. 2004).

An analogous second explanation is that many MRI studies statistically 'correct' for total brain volume measures in order to determine if the size of the hippocampus is abnormal in size, not just in autism, but also relative to the rest of the brain. For example, Sparks et al. (2002) reported that the hippocampus was ~9% larger in young children between 3–5 years of age with autism relative to typically developing controls. This finding was considered significant before correction of total brain volume. However, if the overall brain volume is also larger in the children with autism, which is often reported at that age, then 'corrected' hippocampal differences are likely rendered insignificant. In contrast, the majority of MRI studies on adolescents and adults with autism report that overall brain volume is no longer significantly different in size relative to typically developing individuals. Therefore, if a similar hippocampal volumetric study was carried out in adolescents with autism with similar results, the 9% increase in volume after correction with total cerebral volume would more likely not be altered by correction and therefore be reported as significant. For example, Schumann and colleagues (2004) measured the volume of the hippocampus in male children 7–19 years of age compared to age-matched typically developing male controls. They found that the male children with autism had a ~10% larger hippocampal volume relative to typically developing children (Figure 10.5), similar to the findings of the Sparks study. Total brain volume in this study was not found to be significantly different between autism and control subjects, and therefore, even after correction for total cerebral volume, hippocampal volume was reported as significantly enlarged in autism subjects relative to controls. Similarly, Palmen and collegues (2006) found the hippocampus to be significantly larger in subjects with autism age 7–25 years of age relative to controls; however in this study they found total brain volume to also be increased. Therefore after 'correction', hippocampal volume was no longer considered significantly different. In these studies the hippocampus was found to be ~8–10% larger, though the interpretation of findings was different based on the size of the rest of the brain.

Recent studies suggest that macroscopic hippocampal pathology in autism may be better described by the shape of the structure rather than the overall volume in order to identify subtle regional differences in the anatomy between groups (Nicolson et al. 2006; Dager et al. 2007). Nicolson and colleagues (2006) carried out surface modelling in male children with autism (mean age ~10 years) in which each subject's hippocampus was traced and converted to a three-dimensional surface mesh representation, averaged across subjects, and compared to the control mean at each surface point (Figure 10.6). Although Nicolson et al. (2006) did not detect a difference in overall hippocampal volume between those with autism relative to controls; they did find subtle increases in the anterior portion of the hippocampus. After controlling for total brain size, they found that patients with autism had reductions in the volume of the right medial posterior hippocampus and a regional increase in thickness of the right medial anterior hippocampus. These findings are difficult to interpret though given that a number of patients in the study were taking psychotropic medications, which may alone lead to progressive reductions in hippocampal volume (Giedd et al. 1999). Dager and colleagues (2007) also performed a similar

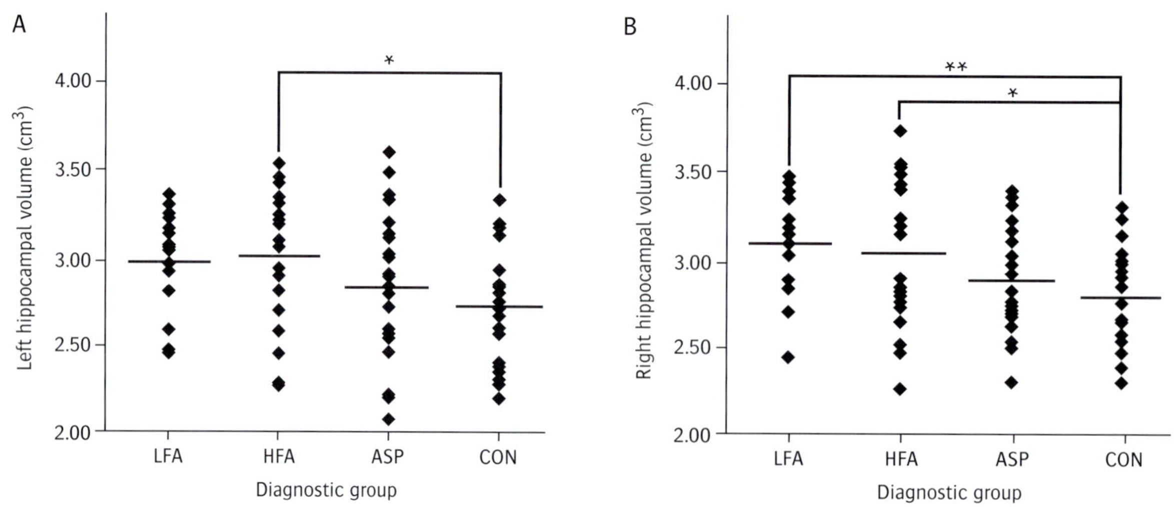

Figure 10.5 Absolute left (*a*) and right (*b*) hippocampal volume (in cubic centimetres) by diagnostic group for subjects 7.5–18.5 years of age (*p_0.05; **p_0.01; LFA, low function autism; HFA, high function autism; ASP, Asperger syndrome) Reproduced from Schumann, C. M., Hamstra, J., Goodlin-Jones, B. L. *et al.* (2004). The amygdala is enlarged in children but not adolescents with autism; the hippocampus is enlarged at all ages. *Journal of Neuroscience*, 24, 6392–401. © 2004 with permission from the Society for Neuroscience.

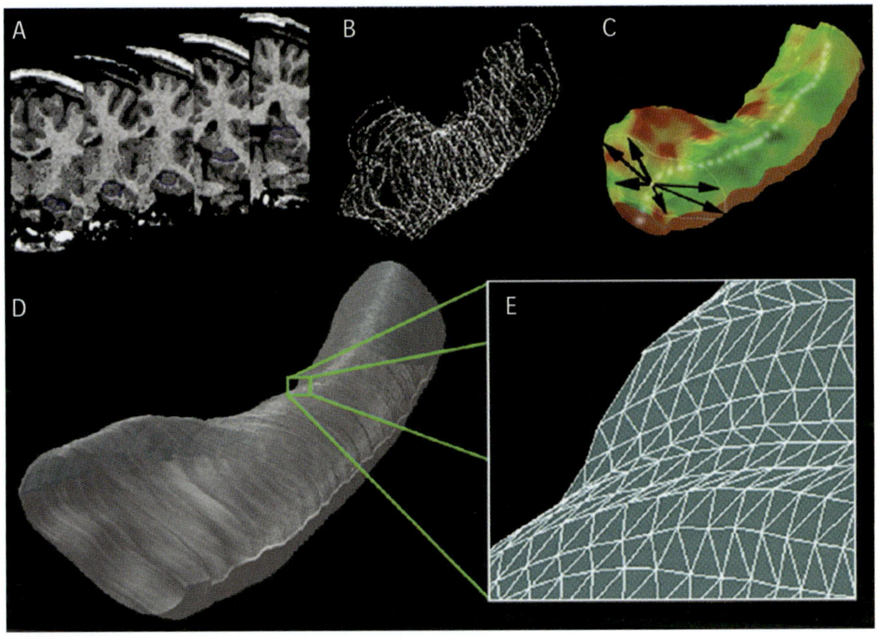

Figure 10.6 Hippocampal surface modelling. Each individual's hippocampus is traced in coronal MRI sections (A) and converted to a mesh surface representation (B and C) in which the radial size of the hippocampus is measured from a centreline and plotted in colour on the surface to index radial atrophy. Arrows in (C) represent vectors from the centreline to various points on the hippocampal surface. These meshes are averaged across subjects (D), and atrophy or expansion relative to the control mean is computed at each surface point (E). Reprinted from *Psychiatry Research: Neuroimaging*, 148(1), Nicolson, R., DeVito, T. J., Vidal, C. N., et al. (2006). Detection and mapping of hippocampal abnormalities in autism, 11–21. © 2006, with permission from Elsevier.

analysis of hippocampal shape in children 3–4 years of age with an autism spectrum disorder relative to typically developing controls. In this study, the authors did not find a difference in overall hippocampal volume, but did suggest an upward bending of the head and the tail of the hippocampus coupled with inward deformation of the medial aspect (i.e. in the vicinity of the subiculum). Interestingly, Dager et al. (2007) also suggested that shape changes may reflect physical constraints from structural alterations of adjacent structures, such as the amygdala, rather than specific hippocampal pathology. This hypothesis is analogous to findings from Bailey and colleagues (1998) in which the hippocampus was noted as severely upturned in a 4-year-old boy with autism that may be due to the need to accommodate an abnormally large temporal cortex (Schumann et al. 2010).

A number of recent studies have used proton magnetic resonance spectroscopy (^{1}H-MRS) to characterize concentrations of brain metabolites as an indication of pathology in cellular chemistry (see Dager et al. 2008 for review). These include glutamine/glutamate (Glu+Gln); *N*-acetylaspartate (NAA), a marker of neuronal density and/or mitochondrial function; choline-containing compounds (Cho), a measure of membrane synthesis/turnover; creatine and phosphocreatine (Cr+PCr), a measure of cellular density energy metabolism; and *myo*-inositol, a major osmolite and precursor to several brain metabolites. Page et al. (2006) examined high-functioning adults with autism and reported a significant increase in the concentration of Glu+Gln and Cr+PCr in the amygdala-hippocampal region but not in the parietal region. Interestingly, elevated hippocampal glutamine levels have been reported in temporal lobe epilepsy patients which might contribute to an ongoing state of increased cerebral excitability/excitotoxicity susceptible to seizure generation (Petroff et al. 2002). Therefore, findings from Page et al. (2006) of elevated glutamine levels may provide a mechanism for the high incidence of seizures in autism. Endo et al. (2007) found significantly reduced levels of NAA and Cr in the right amygdala-hippocampal region in a small sample of children with autism relative to controls, which also correlated with the severity of social and emotional impairments in the autism subjects. The finding of *reduced* NAA levels is surprising given that *increased* levels of NAA generally reflect an increase in the density of neurons, as was described by Bauman and Kemper (1994) in the hippocampus of autism cases. However, reduced NAA levels are a consistent finding in the brain of children with autism (Friedman et al. 2003) and do not support the hypothesis of densely packed neurons arising from disrupted neuronal apoptosis or synaptic pruning (Dager et al. 2008). A limitation of interpreting both the Page et al. and Endo et al. studies is that they combined the amygdala and hippocampus as one region of analysis. A recent study that focused on the left hippocampus in autistic adults with severe aggression found elevated concentrations of Cho and Cr+PCr in the autistic subjects compared to controls which also positively correlated with their level of aggression (Suzuki et al. 2010).

Studies of hippocampal connectivity are limited. Conturo and colleagues (2008) used diffusion tensor imaging (DTI) and tracking (DTT) to investigate the integrity of white matter pathways involved in face processing, including the hippocampo-fusiform pathway, in high-functioning individuals with autism and pairwise matched controls. DTI provides a tool for quantifying the microscopic mobility or diffusion of water in tissue; DTT adds the ability to evaluate the fibre coherence and myelination in the pathway where water preferentially diffuses along the direction of the fibres. Surprisingly, they found a decrease in the diffusivity of water in the right hippocampo-fusiform pathway, which typically reflects more myelination and is generally associated with higher function. However, in this study, the finding of decreased diffusivity was correlated with lower face recognition and performance IQ scores. Conturo et al. (2008) suggest that their findings might be explained by densely-packed smaller neurons (as noted by Bauman and Kemper 1994) leading to decreased axon diameters, causing slower neural conduction speed, and therefore resulting in decreased function.

Functional implications

There is minimal evidence that the hippocampus is functionally pathological in autism. Several studies have shown that declarative memory function is fairly normal in higher-functioning autistic subjects (Ameli et al. 1988; Rumsey and Hamburger 1988; Bennetto et al. 1996; Minshew and Goldstein 2001). Dawson et al. (2001) studied visual object recognition in children with autism using 'delayed non-matching to sample' (DNMS) and initially found that the children with autism were impaired on this task. However, this task required the child to form a stimulus-reward association (Diamond 1995), which may alone be impaired in autism without the memory component. Therefore, Dawson et al. (2001) performed a 'paired comparison task' study to selectively test object recognition without requiring the child to form a stimulus–reward association and found that children with autism did not display a memory impairment.

In contrast to the notion that autism may be associated with impaired hippocampal memory function, Kanner (1943) originally described children with autism as having an extraordinary ability to learn geography and recite long lists of items or facts. Caron et al. (2004) found that individuals with autism are not only equally capable of learning a route on a map but are superior to typically developing controls on the speed and accuracy at which they memorize and reproduce the map. An intriguing, although highly speculative, possibility is that hippocampal 'pathology' in autism may be a precursor to, or a consequence of, autistic symptomatology. For example, it is possible that the increased size of the hippocampus could indicate a use-dependent expansion of hippocampal connections rather than result in a functional impairment. This hypothesis might explain why significant hippocampal enlargement is not detected in very young children 2–3 years of age (Dager et al. 2007) but is found in some studies of adolescents and adults with autism (Schumann et al. 2004). There is substantial evidence in the animal literature that the volume of the hippocampus is correlated with spatial memory function. Species that develop complex spatial maps, such as food-caching birds, have larger hippocampi than members of the same species who do not engage in food caching (Clayton and Krebs 1994). Similar data have been demonstrated for London taxi-cab drivers who must memorize the complex roadway system of the city (Maguire et al. 2000, 2003). Since there is some evidence of enhanced spatial or episodic memory function in individuals with autism, it is possible that an abnormally large hippocampus might result from, or yield, enhanced function. However, this hypothesis is highly speculative and in need of comprehensive assessment. To date, fMRI studies have not reported any significant alterations in the function of the hippocampus in individuals with autism (Dalton et al. 2008).

Hippocampal pathology in autism may not directly result in obvious behavioural impairments, but may be associated with the occurrence of seizure disorder. Approximately 30% of individuals with autism also have some form of epilepsy with a strong correlation between intellectual impairment and seizure prevalence (see Tuchman et al. 2010 for review). Abnormalities in the GABAergic system in the hippocampus, such as an imbalance in the availability of $GABA_A$ receptor subunits as found by the series of studies from Blatt and Colleagues (Blatt et al. 2001; Guptill et al. 2007), could alter receptor activity and therefore reduce the threshold for the development of seizures in children with autism.

Angelman syndrome

In autism spectrum disorder, chromosome 15 has been of particular interest, as numerous reports have located abnormalities within the 15q11–13 region (Cook et al. 1997; Peters et al. 2004; Trillingsgaard and Østergaard 2004; Bucan et al. 2009). This region is also assumed to cause Angelman syndrome (AS), and there are reports of noticeable comorbidity between autism spectrum disorder and AS (Veltman et al. 2005). AS is a hereditary disease that occurs in one of every 15,000–20,000 births. It is usually characterized by mental retardation, epilepsy, a puppet-like gait, a happy disposition, hyperactivity, sleep disorders, and lack of speech (Angelman 1965; Williams et al. 2006). The most common genetic defect leading to AS is a ~4-Mb maternal deletion in chromosomal region 15q11–13 causing an absence of Ube3a (Ube3a: animal, UBE3A: human) expression in the maternally imprinted brain regions. Disruption of maternal expression of the UBE3A gene results in the deletion of the protein product of UBE3A, namely E6-associated protein (E6-AP). E6-AP is a cellular ubiquitin ligase enzyme which promotes the degradation of four target enzymes, including the protein p53. Maternal Ube3a-deficient mice display a phenotype similar to that seen in human AS patients, although milder in form, and these mutants show cytoplasmic accumulation of the p53, one of the target proteins of E6-AP, in the Purkinje cells and in neurons of the hippocampus (Albrecht et al. 1997; Jiang et al. 1998; Yashiro et al. 2009). Furthermore, it has been described that maternal Ube3a-deficient mice display affected dendritic spine development on cerebellar Purkinje cells and on pyramidal neurons of the hippocampus and cortex (Dindot et al. 2008). In adult mice, maternal Ube3a-gene deletion alters hippocampal neurogenesis and impairs functional synaptic plasticity within the DG (Mardirossian et al. 2009). Ube3a mutants displayed deficits in hippocampus-dependent contextual fear conditioning, as well as hippocampal long-term potentiation (LTP) impairments despite normal baseline synaptic transmission and neuroanatomy (Jiang et al. 1998). In Ube3a null mutants, Weeber and colleagues observed increased phosphorylation in parts of the hippocampal phosphor-calcium/calmodulin-dependent protein kinase II (CaMKII) which promotes LTP and may cause the observed learning deficits (Elgersma et al. 2004, Weeber et al. 2003). Interestingly, AS symptoms including memory deficits can be reversed in Ube3a mice when increased CaMKII phosphorylation is inhibited (van Woerden et al. 2007). The CaMKII dysfunction is possibly caused by malregulation of p53; however, the underlying mechanism is still unclear (Shalin et al. 2006). Moreover, it is unclear whether AS is limited to variations in the Ube3a gene, since Jiang and colleagues recently described that a maternal deletion from Ube3a to Gabrb3 results in pronounced AS symptoms which resemble AS pathology more than just a deletion of Ube3a. This also included impaired contextual fear conditioning and spatial learning (Jiang et al. 2010). Nevertheless, the Ube3a model for AS provides suitable insight for molecular basis of hippocampal brain plasticity in mammals.

Down syndrome

Down syndrome (DS) is caused by a trisomy of the human chromosome 21 (HS21), occurring in one of every 650–1000 live births (Bittles et al. 2007). The additional copy of HS21 is proposed to result in multiple defects which characterize phenotypes of DS (Wiseman et al. 2009). DS, the leading genetic cause of mental retardation, is characterized by a reduced number of cortical neurons and brain size. Cognitive disabilities in DS appear to result mainly from two pathological processes: impairment of neurogenesis and Alzheimer-like degeneration (Contestabile et al. 2010).

Long-term and spatial memory deficits characterize patients with DS, and animal studies provide evidence of inherent hippocampal dysfunction caused by genes located on the HS21 chromosome (Rachidi and Lopes 2007). Early onset of Alzheimer's-like neurohistopathology is systematically observed by the fourth decade in DS (Dalton and Crapper-McLachlan 1986) which may be caused by a number of genes, predominantly the APP and S100B genes, also located on HS21 (Roizen and Patterson 2003; Rachidi and Lopes 2007). However, alterations in the hippocampal structure and function are observed much earlier during childhood.

Structural studies

Early disruption of hippocampal neurogenesis and apoptosis has been described prenatally in fetuses with DS. Histopathological findings in fetal hippocampal tissue displayed not only reduced proliferation in the dentate gyrus but also a higher incidence of cell apoptosis in the dentate gyrus, as well as in germinal zones of the hippocampus (Contestabile et al. 2007; Guidi et al. 2008). Postnatal investigations around month 18 show a delayed myelination that manifests primarily in nerve tracts linking the frontal and temporal lobes (Wisniewski 1990; Koo et al. 1992). Smaller hippocampal volumes in DS children and adolescents also indicate early developmental alteration as examined by MR-volumetry in humans, as well as by histopathological findings in animal models (Pinter et al. 2001; Lorenzi and Reeves 2006). The differences in hippocampal volume even increase with age. Several structural studies in non-demented human adults with DS reported age-related and disproportional (i.e. corrected for global brain size) decrease in hippocampal volume (Aylward et al. 1999b; Lawlor et al. 2001; Krasuski et al. 2002; Teipel et al. 2003, 2004). Moreover, while almost all individuals with DS display characteristics of AD such as plaques and tangles by the age of 40 years, amyloid accumulation already begins in childhood (Leverenz and Raskind 1998).

Functional studies

Infants with DS show relatively normal ability in hippocampus-dependent learning and memory until the hippocampus becomes functional at the age of 16–18 months (Nadel 2003). At that age, infants with DS display deficits in hippocampus-dependent spatial memory performance (Mangan 1992 cited after Nadel 2003). Children with DS demonstrated poorer performance in hippocampus-dependent (such as spatial or paired association learning) but not in prefrontally mediated memory tasks (e.g. working memory) (Pennington et al. 2003). These findings implicate distinct alterations in hippocampal memory function in DS. Likewise, adolescents with DS showed normal learning ability for tasks requiring implicit memory processing, in which the hippocampus plays a smaller role, but exhibited poor performance in spatial memory tasks (Carlesimo et al. 1997). Deficits in hippocampal memory performance persist into adulthood (Vicari et al. 2000), and senile plaques and tangles occur with increasing age, which cause Alzheimer-like dementia including forgetfulness, impaired short-term memory, learning problems, and deficits in visuospatial organization (Devenny et al. 2000; Ballas 2008; Krinsky-McHale et al. 2008).

These reports seem to reflect distinct deficits in hippocampus-dependent learning and memory in DS, and it is suggested that there are specific problems in the hippocampus spatial cognitive system in DS. However, it must also be taken into account that alterations in other brain regions also contribute to memory deficits in DS (Nadel 2003). Furthermore, the role of cerebral network plasticity in terms of the functional reorganization and compensatory mechanisms is unclear.

DS is often used as a model for AD. In fact, by the age of 40 years almost 100% of individuals with DS show neuropathology similar to that associated with AD, however, less than 50% show dementia as seen in AD (Nadel 2003) (Figure 10.7). Why some individuals with DS develop dementia while others do not, still remains unclear (Lott and Dierssen 2010).

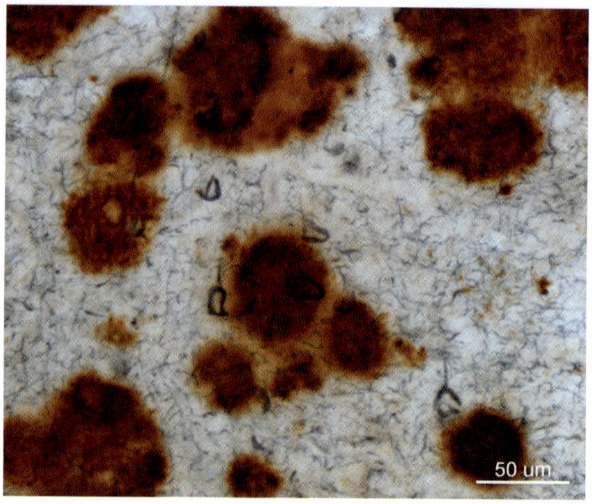

Figure 10.7 Alzheimer's disease neuropathology in individuals with Down syndrome. Immunolabelling of the midfrontal cortex of a 46-year-old person with Down syndrome with anti-amyloid-β (using 6E10 against abeta1-16 [brown]) reveals extensive plaque accumulation. Tau pathology, using an anti-phosphorylated tau antibody (AT8 [blue]), shows neurofibrillary tangle staining. Image kindly provided by Dr Elizabeth Head, Sanders-Brown Center on Aging, University of Kentucky, USA.

While associations between abnormal hippocampal structure and functions have been obtained in animal DS models, proof of an association has only been found indirectly in humans. In older but not non-demented DS patients, the loss of left temporal lobe volume correlated positively with decline in memory performance (Krasuski et al. 2002). One PET study revealed reduced glucose metabolism in parietal and temporal neocortices of older demented DS persons, similar to that observed in AD patients (Schapiro et al. 1992). In contrast, increased glucose metabolism has been observed in older patients with DS which correlated negatively with decreased temporal lobe volume (Haier et al. 2003, 2008). The authors interpreted these results as a compensatory brain function at an early stage of dementia development. In younger DS patients, no significant abnormalities of glucose metabolism were observed (Schapiro et al. 1990). SPECT studies revealed hypoperfusion in regions including the temporal brain in some children with DS (Gökçora et al. 1999; Aydin et al. 2007). Whether or not abnormal perfusion is associated with reduced memory performance in demented adults with DS, however, is unclear (Deb et al. 1992; Jones et al. 1997).

Fragile X syndrome

Fragile X is the most commonly inherited form of intellectual disability, affecting approximately 1 in 4000 males and 1 in 8000 females (Crawford et al. 2001). This disorder is caused by the disrupted expression of the fragile X mental retardation (FMR1) gene located on the long arm of the X chromosome (Verkerk et al. 1991). This region includes a trinucleotide CGG repeat expansion that results in 30 copies of the trinucleotide repeat in typically developing individuals, 50–200 repeats in premutation carriers, and over 200 repeats in individuals with the full mutation. The full mutation results in hypermethylation of the FMR1 gene promoter and the subsequent reduction of FMRP protein, which in turn has a profound impact on brain development. FMRP mRNA has been noted to have particularly high levels during human fetal development in the hippocampus (Abitbol et al. 1993). Individuals with the full fragile X mutation generally have intellectual disability, with other behavioural features including gaze aversion, social difficulties, impulsivity, anxiety, and hyperarousal (Reiss and Freund 1992). Fragile X is also the most commonly known genetic disorder associated with autism with many overlapping symptoms between the two diagnoses (Feinstein and Reiss 1998; Bailey 2004; Kaufmann 2004; Dalton et al. 2008).

Structural MRI studies

Consistent changes in the brains of individuals with fragile X include enlargement of the caudate nucleus, a decrease of the cerebellar vermis and superior temporal gyrus volume, with less consistent changes reported in amygdala and hippocampal volumes (Lightbody and Reiss 2009). Reiss and colleagues (1994) initially suggested age-related increases in bilateral hippocampal volume in individuals with fragile X. Another study of 10 children 3–12 years of age with fragile X reported an approximate 10% increase in hippocampal volume relative to typically developing age matched control children (Kates et al. 1997). However, studies of adults with fragile X have not always found an increased hippocampal size relative to control adults (Jakala et al. 1997; Greco et al. 2011). Hazlett et al. (2009) found a 26% enlargement of hippocampal volume in children with fragile X compared to another developmentally delayed control group, but no differences compared to typically developing children or those with autism. Clearly more studies are needed to determine if the hippocampus is abnormal in size in individuals with fragile X and if those changes are age-related. Postmortem brain tissue studies are also needed to clarify the underlying neurobiology of possible hippocampal pathology in fragile X.

Functional MRI studies

Studies of hippocampal function typically find reduced hippocampal activation during memory tasks in individuals with fragile X. In girls with the full mutation of fragile X, Greicius et al. (2004) found reduced activation in the hippocampus during a visual memory encoding task compared to age matched, typically developing girls. In men with the fragile X premutation, Koldewyn et al. (2008) did not find a difference in hippocampal volume, but did report a reduction in left hippocampal activation during a recall task relative to controls. Interestingly, in this study, activation in the left hippocampus was negatively correlated with both FMR1 mRNA level and psychiatric symptomatology in the premutation group. These studies support the theory that increased levels of FMR1 mRNA affect hippocampal function, since the hippocampus is one of the major sites of FMR-1 expression in the human fetal brain (Abitbol et al. 1993), and contributes to memory impairments, even in individuals with the fragile X premutation.

Although previous fMRI studies have shown a decrease in hippocampal activation during memory-specific tasks in individuals with fragile X, there is some indication that the hippocampus might display hyperactivation in non-memory related tasks such as emotional recognition. Dalton and colleagues (2008) measured hippocampal activation and tracked eye movements in nine adolescents with fragile X and compared them to 15 typically developing and 14 autistic spectrum adolescents during an emotion discrimination task. Participants were asked to decide whether a picture of a human face displayed an emotion or appeared neutral. They found that the fragile X group displayed significantly greater activation in the left hippocampus during the emotion recognition task compared to both the autism and typically developing groups. Dalton et al. (2008) suggests that, although individuals with fragile X and autism share some behavioural characteristics, the underlying neural circuitry used in face processing may be different between the two groups. In addition, the increased hippocampal activation might suggest increased task difficulty for the fragile X group (Dalton et al. 2008).

Attention-deficit hyperactivity disorder

With a prevalence rate of 3–6%, attention-deficit hyperactivity disorder (ADHD) is the most frequently diagnosed psychiatric disorder in children and adolescents. It is characterized by patterns of inattentiveness, hyperactivity, and impulsivity persisting into adultness (American Psychiatric Association 2000). Although alterations in prefrontal brain areas are most prominent (Brennan and Arnsten 2008) in ADHD, reports of abnormal hippocampal structure and function are increasing.

Structural studies

Several structural imaging studies report decreased volume of the temporal cortex (see Cherkasova and Hechtman 2009), and it is described that temporal lobe volume reductions as a part of the general brain volume reduction correlate with ADHD symptom severity (Castellanos et al. 2002). While no differences between ADHD patients and controls in hippocampus volume were revealed in early structural studies, in more recent MR studies conflicting results were demonstrated: Brieber and colleagues observed a decrease in bilateral hippocampus volume (Brieber et al. 2007) in ADHD. Likewise, a reduced left hippocampus was observed in a sample of young male patients with conduct disorder, of which almost 75% suffered additionally from ADHD (Hübner et al. 2008). However, the authors also noted that the severity of conduct disorder severity correlates with hippocampal volume reduction. Plessen and colleagues, however, reported on an enlarged anterior dorsal and lateral part of the hippocampal surface (not volume) which correlated slightly negatively with symptom severity (Plessen et al. 2006). The authors interpreted these findings as a compensatory response. In adults with ADHD, in turn, no differences between ADHD and controls in hippocampus volume were observed (Perlov et al. 2008).

Functional studies

ADHD is characterized by a dopaminergic dysfunction, which is not only observed in fronto-striatal brain circuits (Bush et al. 2005), but also in temporal brain regions. Dopamine D4-receptors show a differential distribution with the highest density in the PFC, followed by the hippocampus (Lahti and Lahti 1997; Oak et al. 2000). The D4-receptor gene is postulated to be a candidate gene for ADHD (Bellgrove and Mattingley 2008; Gizer et al. 2009). Moreover, whether a higher expression of D4-receptors (and its variant D4.7) may be a substrate for novel-seeking behaviour in ADHD (Ebstein et al. 1996; LaHoste et al. 1996; Becker et al. 2005) has been suggested. In contrast, a decreased binding action of hippocampal D2/D3 receptor has been shown in ADHD (Volkow et al. 2007).

Further functional imaging studies revealed hyper- as well as hypoactivation in the hippocampus: a higher normalized glucose metabolism in the hippocampus during an auditory task requiring low cognitive effort was reported in adolescent girls with ADHD (Ernst et al. 1997). Also, an increase in hippocampal activation in boys with ADHD was observed during sustained attention, which was interpreted as a compensation of dysfunctional frontostriatal brain circuits (Rubia et al. 2009). Hippocampal deactivations, in contrast, were found in adolescents with ADHD during response inhibition (Schulz et al. 2004), as well as during decision-making in adults with ADHD (Ernst et al. 2003).

Young ADHD patients show deficits in visuospatial memory (Rommelse et al. 2008; Reck et al. 2010) and display poorer performance in declarative memory tasks as recognition of low emotional stimulus material is affected in young patients with ADHD (Krauel et al. 2007, 2009a, 2009b). It is noteworthy that the reduced declarative memory performance in ADHD was not associated with hippocampal activity. Abnormal enhanced hippocampal activation, however, was observed during encoding of visual stimuli in the context of a working memory task in children and adolescents with ADHD (Prehn-Kristensen et al. 2011b). At the same time, the consolidation of declarative memory during sleep, which requires hippocampal reactivation, is reduced in children and adolescents with ADHD (Prehn-Kristensen et al. 2011a). Although these results suggest deficits in hippocampus-associated memory function in ADHD, direct evidence of abnormal hippocampal activity during clear hippocampus-related memory performance is still lacking.

Limitations

Methylphenidate (MPH) is the first-choice treatment of ADHD (Arnsten 2006). MPH in extreme high doses not only disturbs hippocampus-sensitive memory performance, as was observed in juvenile and adult rats which received either a single dose of 10 mg/kg or long-term administration of at least 3 mg/kg twice daily (Chuhan and Taukulis 2006; LeBlanc-Duchin and Taukulis 2007), but might also impair hippocampal neurogenesis, as was observed in juvenile rats which received chronic administration of 2 mg/kg twice daily (Lagace et al. 2006). However, it has to be taken in consideration that MPH treatment in humans with ADHD never reaches comparable high doses (in general less than 1 mg/kg daily); it is also noteworthy that long-term MPH administration in ADHD at therapeutic doses (less than 1 mg/kg daily) does not seem to affect regional cerebral blood flow in the temporal cortex in human (Lee et al. 2005). Due to the high rate of ADHD patients receiving MPH, ADHD patients with a history of chronic MPH-intake are frequently included in study protocols. In order to make more assured conclusions regarding hippocampal structure and function in ADHD without eventual pharmacological effects of MPH, the lifetime intake of MPH in study designs should be taken into account.

Gilles de la Tourette syndrome

The core symptoms of Gilles de la Tourette syndrome (TS) are waxing and waning of vocal and motor tics which usually begin between the ages of 3–8 years. TS is considered to be a dysfunction of basal ganglia circuits projecting into frontal and limbic brain areas and is associated with behavioural symptoms such as obsessive–compulsive symptoms or attention deficit (Leckman 2002). Apart from some deficits in executive functions, which can be ascribed to frontostriatal dysfunction (Eddy et al. 2009), there are signs of reduced hippocampus-associated memory function in TS.

Structural studies

While no differences in hippocampus volume were found in adult TS patients (Garraux et al. 2006; Müller-Vahl et al. 2009), a reduced volume in the left hippocampus in children with TS was reported. Moreover, an association between reduced volume and tic severity was described (Ludolph et al. 2006). It is discussed whether these different results reflect effects of age (Kassubek et al. 2006). In contrast to the former studies, Peterson and colleagues (2007) reported an enlarged hippocampus volume in TS which was ascribed to an enlarged surface of the head and medial hippocampus. The volume of these regions correlated negatively with tic severity and decreased with age. The authors concluded that hippocampal hypertrophy represents a compensatory, neuromodulatory effect which could reflect associative learning-based strategies that attenuate severity (Peterson et al. 2007).

Functional studies

While a tic severity-related reduction in metabolic activation and hypoperfusion in temporal brain regions (including hippocampus) in adult TS is frequently found (Moriarty et al. 1995; Eidelberg et al. 1997; Stern et al. 2000), studies in children with TS showing abnormal hippocampal function are rare. Nevertheless, it is reported that children with TS display an asymmetric perfusion (right > left) in the medial temporal lobe as indicated by SPECT (Chiu et al. 2001). Moreover, pimozide, a neuroleptic drug which not only reduces D2-receptor activity but also decreases serotonin turnover in the hippocampus (Tecott et al. 1986; Messiha 1991), reduces tic symptoms in children and adolescents with TS (Sallee et al. 1997). In line with this, PET studies revealed reduced D2-receptor availability in the hippocampus (Gilbert et al. 2006) as well as hippocampal activity during tic suppression in adults with TS (Peterson et al. 1998). Additionally, a high density of endogenous cannabinoid receptors CB1 was found in the hippocampus (Herkenham et al. 1990). Given that cannabis consumption as well as the application of isolated delta-9-THC can reduce tic severity in TS (Hemming and Yellowlees 1993; Müller-Vahl et al. 1998; Müller-Vahl et al. 2002; Müller-Vahl et al. 2003), it is suggested that the hippocampal cannabinoid receptor system is involved in TS symptoms (Müller-Vahl et al. 1998).

Most pronounced memory deficits in patients with TS are found by using tasks which involve frontostriatal brain circuits, such as those requiring strategic, working, and procedural memory performance (Stebbins et al. 1995). Deficits in hippocampus-associated memory performance are more subtle: two studies showed that declarative memory seemed to be unaffected in children and adults with TS (Stebbins et al. 1995; Marsh et al. 2004) and in adults with TS verbal memory performance did not differ from healthy controls (Lavoie et al. 2007). However, Channon and colleagues demonstrated reduced deficits in word list memory in TS (Channon et al. 2003) and, in the same way, there are reports of reduced memory performance in story recall and spatial memory in children and adults with TS (Bornstein 1991; Channon et al. 2003; Watkins et al. 2005; Lavoie et al. 2007; Rasmussen et al. 2009).

Limitations

Findings of hippocampus-associated memory deficits in TS should be interpreted with care since it can not be ruled out that low memory performance may also be caused by lack of effort, attentional, or strategic cognitive functions during encoding and retrieval processes (Eddy et al. 2009). Furthermore, the high comorbidity of TS with ADHD has to be taken into account (Roessner et al. 2007). For example, Marsh and colleagues observed reduced declarative memory function in children and adults with TS; after controlling for ADHD, however, these deficits vanished (Marsh et al. 2004).

Conclusions

In summary, we emphasize that hippocampal alterations in neurodevelopmental disorders are more frequent than commonly described. Considering the spectrum of disorders, this account of disorders is of course not complete (for review, see Geuze et al. 2005). Notably, hippocampal alterations cannot explain the whole of pathology in these disorders, since alterations might be more prominent in other brain regions or being globally expressed. However, hippocampal involvement needs to be acknowledged to comprehend the spectrum of the behavioural alterations in patients with neurodevelopmental disorders.

Acknowledgements

A. Prehn-Kristensen and L. Baving were supported by the Deutsche Forschungsgemeinschaft (KFO 163 'Cognitive Control of Memory Functions', SFB 654, 'Plasticity and Sleep'). R. Cooke was supported by The Medical Research Council and Children Nationwide.

References

Abernethy, L.J., Palaniappan, M., and Cooke, R.W. (2002). Quantitative magnetic resonance imaging of the brain in survivors of very low birth weight. *Arch Dis Child*, 87, 279–83.

Abernethy, L.J., Klafkowski, G., Foulder-Hughes, L., and Cooke, R.W. (2003). Magnetic resonance imaging and T2 relaxometry of cerebral white matter and hippocampus in children born preterm. *Pediatr Res*, 54, 868–74.

Abernethy, L.J., Cooke, R.W., and Foulder-Hughes, L. (2004). Caudate and hippocampal volumes, intelligence, and motor impairment in 7-year-old children who were born preterm. *Pediatr Res*, 55, 884–93.

Abitbol, M., Menini, C., Delezoide, A.L., Rhyner, T., Vekemans, M., and Mallet, J. (1993). Nucleus basalis magnocellularis and hippocampus are the major sites of FMR-1 expression in the human fetal brain. *Nat Genet*, 4, 147–53.

Ackerman, J.P., Riggins, T., and Black, M.M. (2010). A review of the effects of prenatal cocaine exposure among school-aged children. *Pediatrics*, 125, 554–65.

Albrecht, U., Sutcliffe, J.S., Cattanach, B.M., *et al*. (1997). Imprinted expression of the murine Angelman syndrome gene, Ube3a, in hippocampal and Purkinje neurons. *Nat Genet*, 17, 75–8.

Amaral, D.G., Schumann, C.M., and Nordahl, C.W. (2008). Neuroanatomy of autism. *Trends Neurosci*, 31, 137–45.

Ameli, R., Courchesne, E., Lincoln, A., Kaufman, A.S., and Grillon, C. (1988). Visual memory processes in high-functioning individuals with autism. *J Autism Dev Disord*, 18, 601–15.

American Psychiatric Association (2000). *Diagnostic and Statistical Manual of Mental disorders, 4th ed. (Text revision): DSM IV-TR*. Washington, DC: American Psychiatric Association.

Andreasen, N.C., Flaum, M., Swayze, V., 2nd., *et al*. (1993). Intelligence and brain structure in normal individuals. *Am J Psychiatry*, 150, 130–4.

Angelman, H. (1965). 'Puppet' Children - A Report on Three Cases. *Dev Med Child Neurol*, 7, 681–88.

Archibald, S.L., Fennema-Notestine, C., Gamst, A., Riley, E.P., Mattson, S.N., and Jernigan, T.L. (2001). Brain dysmorphology in individuals with severe prenatal alcohol exposure. *Dev Med Child Neurol*, 43, 148–54.

Ardila, A., Rosselli, M., and Strumwasser, S. (1991). Neuropsychological deficits in chronic cocaine abusers. *Int J Neurosci*, 57, 73–9.

Arnsten, A.F. (2006). Stimulants: Therapeutic actions in ADHD. *Neuropsychopharmacology*, 31, 2376–83.

Astley, S.J., Richards, T., Aylward, E.H., *et al*. (2009a). Magnetic resonance spectroscopy outcomes from a comprehensive magnetic resonance study of children with fetal alcohol spectrum disorders. *Magn Reson Imaging*, 27, 760–78.

Astley, S.J., Aylward, E.H., Olson, H.C., *et al*. (2009b). Magnetic resonance imaging outcomes from a comprehensive magnetic resonance study of children with fetal alcohol spectrum disorders. *Alcohol Clin Exp Res*, 33, 1671–89.

Autti-Rämö, I., Autti, T., Korkman, M., Kettunen, S., Salonen, O., and Valanne, L. (2002). MRI findings in children with school problems who had been exposed prenatally to alcohol. *Dev Med Child Neurol*, 44, 98–106.

Aversi-Ferreira, T.A., and Nascimento, G.N.L. (2008). The effect of acute and chronic exposure to ethanol on the developing encephalon: a review. *Revista Brasileira de Saúde Materno Infantil*, 8, 241–9.

Aydin, M., Kabakus, N., Balci, T.A., and Ayar, A. (2007). Correlative study of the cognitive impairment, regional cerebral blood flow, and electroencephalogram abnormalities in children with Down's syndrome. *Int J Neurosci*, 117, 327–36.

Aylward, E., Minshew, N., Goldstein, G., *et al*. (1999a). MRI volumes of amygdala and hippocampus in non-mentally retarded autistic adolescents and adults. *Neurology*, 53, 2145–50.

Aylward, E.H., Li, Q., Honeycutt, N.A., *et al*. (1999b). MRI volumes of the hippocampus and amygdala in adults with Down's syndrome with and without dementia. *Am J Psychiatry*, 156, 564–8.

Bachevalier, J., and Vargha-Khadem, F. (2005). The primate hippocampus: ontogeny, early insult and memory. *Curr Opin Neurobiol*, 15, 168–74.

Bailey, A., Luthert, P., Dean, A., *et al*. (1998). A clinicopathological study of autism. *Brain*, 121, 889–905.

Bailey, D.B., Jr. (2004). Newborn screening for fragile X syndrome. *Ment Retard Dev Disabil Res Rev*, 10, 3–10.

Bajic, D., Ewald, U., and Raininko, R. (2010). Hippocampal development at gestation weeks 23 to 36. An ultrasound study on preterm neonates. *Neuroradiology*, 52, 489–94.

Ballas, P. (2008). ADHD and sleep deprivation in school-aged children. *Curr Psychiatry Rep*, 10, 1–2.

Bartley, A.J., Jones, D.W., and Weinberger, D.R. (1997). Genetic variability of human brain size and cortical gyral patterns. *Brain*, 120(Pt 2), 257–69.

Bauman, M.L., and Kemper, T.L. (1985). Histoanatomic observations of the brain in early infantile autism. *Neurology*, 35, 866–74.

Bauman, M.L., LeMay, M., Bauman, R.A., and Rosenberger, P. (1985). Computerized tomographic (CT) observations of the posterior fossa in early infantile autism. *Neurology*, 35(Suppl 1), 247.

Bauman, M.L., and Kemper, T.L. (1994). *The Neurobiology of Autism*. Baltimore, MD: Johns Hopkins University Press.

Beauchamp, M.H., Thompson, D.K., Howard, K., *et al*. (2008). Preterm infant hippocampal volumes correlate with later working memory deficits. *Brain*, 131, 2986–94.

Becker, K., Laucht, M., El-Faddagh, M., and Schmidt, M.H. (2005). The dopamine D4 receptor gene exon III polymorphism is associated with novelty seeking in 15-year-old males from a high-risk community sample. *J Neural Transm*, 112, 847–58.

Bellgrove, M.A., and Mattingley, J.B. (2008). Molecular genetics of attention. *Ann N Y Acad Sci*, 1129, 200–12.

Bender, R.A., Dube, C., and Baram, T.Z. (2004). Febrile seizures and mechanisms of epileptogenesis: insights from an animal model. *Adv Exp Med Biol*, 548, 213–25.

Bennetto, L., Pennington, B.F., and Rogers, S.J. (1996). Intact and impaired memory functions in autism. *Child Dev*, 67, 1816–35.

Berman, R.F., and Hannigan, J.H. (2000). Effects of prenatal alcohol exposure on the hippocampus: spatial behavior, electrophysiology, and neuroanatomy. *Hippocampus*, 10, 94–110.

Betancourt, L.M., Yang, W., Brodsky, N.L., *et al*. (2011). Adolescents with and without gestational cocaine exposure: Longitudinal analysis of inhibitory control, memory and receptive language. *Neurotoxicol Teratol*, 33, 36–46.

Bethancourt, J.A., Camarena, Z.Z., and Britton, G.B. (2009). Exposure to oral methylphenidate from adolescence through young adulthood produces transient effects on hippocampal-sensitive memory in rats. *Behav Brain Res*, 202, 50–7.

Beveridge, T.J., Smith, H.R., Daunais, J.B., Nader, M.A., and Porrino, L.J. (2006). Chronic cocaine self-administration is associated with altered functional activity in the temporal lobes of non human primates. *Eur J Neurosci*, 23, 3109–18.

Beveridge, T.J., Gill, K.E., Hanlon, C.A., and Porrino, L.J. (2008). Review. Parallel studies of cocaine-related neural and cognitive impairment in humans and monkeys. *Philos Trans R Soc Lond B Biol Sci*, 363, 3257–66.

Bhatara, V., Loudenberg, R., and Ellis, R. (2006). Association of attention deficit hyperactivity disorder and gestational alcohol exposure: an exploratory study. *J Atten Disord*, 9, 515–22.

Bigler, E.D., Tate, D.F., Neeley, E.S., *et al.* (2003). Temporal lobe, autism, and macrocephaly. *Am J Neuroradiol*, 24, 2066–76.

Bittles, A.H., Bower, C., Hussain, R., and Glasson, E.J. (2007). The four ages of Down syndrome. *Eur J Public Health*, 17, 221–25.

Blatt, G.J., Fitzgerald, C.M., Guptill, J.T., Booker, A.B., Kemper, T.L., and Bauman, M.L. (2001). Density and distribution of hippocampal neurotransmitter receptors in autism: an autoradiographic study. *J Autism Dev Disord*, 31, 537–43.

Bornstein, R.A. (1991). Neuropsychological performance in adults with Tourette's syndrome. *Psychiatry Res*, 37, 229–36.

Bower, S.P., Kilpatrick, C.J., Vogrin, S.J., Morris, K., and Cook, M.J. (2000). Degree of hippocampal atrophy is not related to a history of febrile seizures in patients with proved hippocampal sclerosis. *J Neurol Neurosurg Psychiatry*, 69, 733–8.

Bowler, D.M., Gardiner, J.M., and Berthollier, N. (2004). Source memory in adolescents and adults with Asperger's syndrome. *J Autism Dev Disord*, 34, 533–42.

Brennan, A.R., and Arnsten, A.F. (2008). Neuronal mechanisms underlying attention deficit hyperactivity disorder: the influence of arousal on prefrontal cortical function. *Ann N Y Acad Sci*, 1129, 236–45.

Brieber, S., Neufang, S., Bruning, N., *et al.* (2007). Structural brain abnormalities in adolescents with autism spectrum disorder and patients with attention deficit/hyperactivity disorder. *J Child Psychol Psychiatry*, 48, 1251–58.

Brown, J., Cooper-Kuhn, C.M., Kempermann, G., *et al.* (2003). Enriched environment and physical activity stimulate hippocampal but not olfactory bulb neurogenesis. *Eur J Neurosci*, 17, 2042–6.

Bucan, M., Abrahams, B.S., Wang, K., *et al.* (2009). Genome-wide analyses of exonic copy number variants in a family-based study point to novel autism susceptibility genes. *PLoS Genet*, 5, e1000536.

Bush, G., Valera, E.M., and Seidman, L.J. (2005). Functional neuroimaging of attention-deficit/hyperactivity disorder: a review and suggested future directions. *Biol Psychiatry*, 57, 1273–84.

Canales, J.J. (2010). Comparative neuroscience of stimulant-induced memory dysfunction: role for neurogenesis in the adult hippocampus. *Behav Pharmacol*, 21, 379–93.

Carlesimo, G.A., Marotta, L., and Vicari, S. (1997). Long-term memory in mental retardation: evidence for a specific impairment in subjects with Down's syndrome. *Neuropsychologia*, 35, 71–79.

Caron, M.J., Mottron, L., Rainville, C., and Chouinard, S. (2004). Do high functioning persons with autism present superior spatial abilities? *Neuropsychologia*, 42, 467–81.

Cascino, G.D. (1995). Clinical correlations with hippocampal atrophy. *Magn Reson Imaging*, 13, 1133–6.

Castellanos, F.X., Lee, P.P., Sharp, W., *et al.* (2002). Developmental trajectories of brain volume abnormalities in children and adolescents with attention-deficit/hyperactivity disorder. *J Am Med Assoc*, 288, 1740–8.

Cavanagh, J.B., and Meyer, A. (1956). Aetiological aspects of Ammon's horn sclerosis associated with temporal lobe epilepsy. *Br Med J*, 2, 1403–7.

Cendes, F. (2004). Febrile seizures and mesial temporal sclerosis. *Curr Opin Neurol*, 17, 161–4.

Channon, S., Crawford, S., Vakili, K., and Robertson, M.M. (2003). Real-life-type problem solving in Tourette syndrome. *Cogn Behav Neurol*, 16, 3–15.

Cherkasova, M.V., and Hechtman, L. (2009). Neuroimaging in attention-deficit hyperactivity disorder: beyond the frontostriatal circuitry. *Can J Psychiatry*, 54, 651–64.

Chiu, N.T., Chang, Y.C., Lee, B.F., Huang, C.C., and Wang, S.T. (2001). Differences in 99mTc-HMPAO brain SPET perfusion imaging between Tourette's syndrome and chronic tic disorder in children. *Eur J Nucl Med*, 28, 183–90.

Chuhan, Y.S., and Taukulis, H.K. (2006). Impairment of single-trial memory formation by oral methylphenidate in the rat. *Neurobiology of Learning and Memory*, 85, 125–31.

Clayton, N.S., and Krebs, J.R. (1994). Hippocampal growth and attrition in birds affected by experience. *Proc Natl Acad Sci USA*, 91, 7410–4.

Clayton, N.S. (1998). Memory and the hippocampus in food-storing birds: a comparative approach. *Neuropharmacology*, 37, 441–52.

Coles, C.D., Lynch, M.E., Kable, J.A., Johnson, K.C., and Goldstein, F.C. (2010). Verbal and nonverbal memory in adults prenatally exposed to alcohol. *Alcoholism, Clinical and Experimental Research*, 34, 897–906.

Contestabile, A., Fila, T., Ceccarelli, C., *et al.* (2007). Cell cycle alteration and decreased cell proliferation in the hippocampal dentate gyrus and in the neocortical germinal matrix of fetuses with Down syndrome and in Ts65Dn mice. *Hippocampus*, 17, 665–78.

Contestabile, A., Benfenati, F., and Gasparini, L. (2010). Communication breaks-Down: from neurodevelopment defects to cognitive disabilities in Down syndrome. *Prog Neurobiol*, 91, 1–22.

Conturo, T.E., Williams, D.L., Smith, C.D., Gultepe, E., Akbudak, E., and Minshew, N.J. (2008). Neuronal fiber pathway abnormalities in autism: an initial MRI diffusion tensor tracking study of hippocampo-fusiform and amygdalo-fusiform pathways. *J Int Neuropsychol Soc*, 14, 933–46.

Cook, E.H., Jr., Lindgren, V., Leventhal, B.L., *et al.* (1997). Autism or atypical autism in maternally but not paternally derived proximal 15q duplication. *Am J Hum Genet*, 60, 928–34.

Corkin, S. (1968). Acquisition of motor skill after bilateral medial temporal-lobe excision. *Neuropsychologia*, 6, 225–64.

Crawford, D.C., Acuna, J.M., and Sherman, S.L. (2001). FMR1 and the fragile X syndrome: human genome epidemiology review. *Genet Med*, 3, 359–71.

Dager, S.R., Wang, L., Friedman, S.D., *et al.* (2007). Shape mapping of the hippocampus in young children with autism spectrum disorder. *Am J Neuroradiol*, 28, 672–7.

Dager, S.R., Corrigan, N.M., Richards, T.L., and Posse, S. (2008). Research applications of magnetic resonance spectroscopy to investigate psychiatric disorders. *Top Magn Reson Imaging*, 19, 81–96.

Dalton, A.J., and Crapper-McLachlan, D.R. (1986). Clinical expression of Alzheimer's disease in Down's syndrome. *Psychiatr Clin N Am*, 9, 659–70.

Dalton, K.M., Holsen, L., Abbeduto, L., and Davidson, R.J. (2008). Brain function and gaze fixation during facial-emotion processing in fragile X and autism. *Autism Res*, 1, 231–9.

Dawson, G., Osterling, J., Rinaldi, J., Carver, L., and McPartland, J. (2001). Brief report: Recognition memory and stimulus-reward associations: indirect support for the role of ventromedial prefrontal dysfunction in autism. *J Autism Dev Disord*, 31, 337–41.

De Haan, M., Mishkin, M., Baldeweg, T., and Vargha-Khadem, F. (2006). Human memory development and its dysfunction after early hippocampal injury. *Trends Neurosci*, 29, 374–81.

Deb, S., de Silva, P.N., Gemmell, H.G., Besson, J.A., Smith, F.W., and Ebmeier, K.P. (1992). Alzheimer's disease in adults with Down's syndrome: the relationship between regional cerebral blood flow equivalents and dementia. *Acta Psychiatr Scand*, 86, 340–45.

DeLong, G.R. (1992). Autism, amnesia, hippocampus, and learning. *Neurosci Biobehav Rev*, 16, 63–70.

Devenny, D.A., Krinsky-McHale, S.J., Sersen, G., and Silverman, W.P. (2000). Sequence of cognitive decline in dementia in adults with Down's syndrome. *J Intellect Disabil Res*, 44(Pt 6), 654–65.

Diamond, A. (1995). Evidence of robust recognition memory early in life even when assessed by reaching behavior. *J Exp Child Psychol*, 59, 419–56.

Diamond, M.C. (2001). Response of the brain to enrichment. *Anais da Academia Brasileira de Ciencias*, 73, 211–20.

Dindot, S.V., Antalffy, B.A., Bhattacharjee, M.B., and Beaudet, A.L. (2008). The Angelman syndrome ubiquitin ligase localizes to the synapse and nucleus, and maternal deficiency results in abnormal dendritic spine morphology. *Hum Mol Genet*, 17, 111–18.

Ebstein, R.P., Novick, O., Umansky, R., *et al.* (1996). Dopamine D4 receptor (D4DR) exon III polymorphism associated with the human personality trait of Novelty Seeking. *Nat Genet*, 12, 78–80.

Eddy, C.M., Rizzo, R., and Cavanna, A.E. (2009). Neuropsychological aspects of Tourette syndrome: a review. *J Psychosom Res*, 67, 503–13.

Egan, V., Chiswick, A., Santosh, C., Naidu, K., Rimmington, J.E., and Best, J.J.K. (1994). Size isn't everything – a study of brain volume, intelligence and auditory-evoked potentials. *Pers Indiv Differ*, 17, 357–67.

Eidelberg, D., Moeller, J.R., Antonini, A., *et al.* (1997). The metabolic anatomy of Tourette's syndrome. *Neurology*, 48, 927–34.

Elgersma, Y., Sweatt, J.D., and Giese, K.P. (2004). Mouse genetic approaches to investigating calcium/calmodulin-dependent protein kinase II function in plasticity and cognition. *J Neurosci*, 24, 8410–5.

Elliott, J.M., and Beveridge, T.J. (2005). Psychostimulants and monoamine transporters: upsetting the balance. *Curr Opin Pharmacol*, 5, 94–100.

Endo, T., Shioiri, T., Kitamura, H., *et al.* (2007). Altered chemical metabolites in the amygdala-hippocampus region contribute to autistic symptoms of autism spectrum disorders. *Biol Psychiatry*, 62, 1030–7.

Ernst, M., Cohen, R.M., Liebenauer, L.L., Jons, P.H., and Zametkin, A.J. (1997). Cerebral glucose metabolism in adolescent girls with attention-deficit/hyperactivity disorder. *J Am Acad Child Adolesc Psychiatry*, 36, 1399–406.

Ernst, M., Kimes, A.S., London, E.D., *et al.* (2003). Neural substrates of decision making in adults with attention deficit hyperactivity disorder. *Am J Psychiatry*, 160, 1061–70.

Feinstein, C., and Reiss, A.L. (1998). Autism: the point of view from fragile X studies. *J Autism Dev Disord*, 28, 393–405.

Fombonne, E. (2006). Past and Future Perspectives on Autism Epidemiology. In J.L.R. Rubenstein and S.O. Moldin (ed.) *Understanding Autism: From Basic Neuroscience to Treatment*, pp. 25–48. Boca Ratan, FL: Taylor & Francis.

Foulder-Hughes, L.A., and Cooke, R.W. (2003). Motor, cognitive, and behavioural disorders in children born very preterm. *Dev Med Child Neurol*, 45, 97–103.

Fox, H.C., Jackson, E.D., and Sinha, R. (2009). Elevated cortisol and learning and memory deficits in cocaine dependent individuals: relationship to relapse outcomes. *Psychoneuroendocrinology*, 34, 1198–207.

Friedman, S.D., Shaw, D.W., Artru, A.A., *et al.* (2003). Regional brain chemical alterations in young children with autism spectrum disorder. *Neurology*, 60, 100–7.

Gabrieli, J., Merrill, F., and Corkin, S. (1986). Acquisition of perceptual skills in amnesia due to bilateral medial temporal-lobe (H.M.) or diencephalic lesions. *Soc Neurosci Abstr*, 12, 20.

Gabrieli, J.D., Poldrack, R.A., and Desmond, J.E. (1998). The role of left prefrontal cortex in language and memory. *Proc Natl Acad Sci USA*, 95, 906–13.

Garraux, G., Goldfine, A., Bohlhalter, S., Lerner, A., Hanakawa, T., and Hallett, M. (2006). Increased midbrain gray matter in Tourette's syndrome. *Ann Neurol*, 59, 381–85.

Geschwind, D.H., and Levitt, P. (2007). Autism spectrum disorders: developmental disconnection syndromes. *Curr Opin Neurobiol*, 17, 103–11.

Geuze, E., Vermetten, E., and Bremner, J.D. (2005). MR-based in vivo hippocampal volumetrics: 2. Findings in neuropsychiatric disorders. *Mol Psychiatry*, 10, 160–84.

Giedd, J.N. (1997). Normal development. *Child and Adolescent Psychiatric Clinics of North America*, 6, 265–82.

Giedd, J.N., Snell, J.W., Lange, N., *et al.* (1996). Quantitative magnetic resonance imaging of human brain development: ages 4–18. *Cerebral Cortex*, 6, 551–60.

Giedd, J.N., Castellanos, F.X., Rajapakse, J.C., Vaituzis, A.C., and Rapoport, J.L. (1997). Sexual dimorphism of the developing human brain. *Prog Neuro-Psychopharmacol Biol Psychiatry*, 21, 1185–201.

Giedd, J., Jeffries, N., Blumenthal, J., *et al.* (1999). Childhood-onset schizophrenia: progressive brain changes during adolescence. *Biol Psychiatry*, 46, 892–98.

Gilbert, D.L., Christian, B.T., Gelfand, M.J., Shi, B., Mantil, J., and Sallee, F.R. (2006). Altered mesolimbocortical and thalamic dopamine in Tourette syndrome. *Neurology*, 67, 1695–97.

Gizer, I.R., Ficks, C., and Waldman, I.D. (2009). Candidate gene studies of ADHD: a meta-analytic review. *Human Genet*, 126, 51–90.

Gökçora, N., Atasever, T., Karabacak, N.I., Vural, G., and Gucuyener, K. (1999). Tc-99m HMPAO brain perfusion imaging in young Down's syndrome patients. *Brain Develop*, 21, 107–12.

Goldstein, R.Z., Leskovjan, A.C., Hoff, A.L., *et al.* (2004). Severity of neuropsychological impairment in cocaine and alcohol addiction: association with metabolism in the prefrontal cortex. *Neuropsychologia*, 42, 1447–58.

Grant, S., London, E.D., Newlin, D.B., *et al.* (1996). Activation of memory circuits during cue-elicited cocaine craving. *Proc Natl Acad Sci USA*, 93, 12040–5.

Greco, C.M., Navarro, C.S., Hunsaker, M.R., *et al.* (2011). Neuropathologic features in the hippocampus and cerebellum of three older men with fragile X syndrome. *Mol Autism*, 2, 2.

Greicius, M.D., Boyett-Anderson, J.M., Menon, V., and Reiss, A.L. (2004). Reduced basal forebrain and hippocampal activation during memory encoding in girls with fragile X syndrome. *Neuroreport*, 15, 1579–83.

Groen, W., Teluij, M., Buitelaar, J., and Tendolkar, I. (2010). Amygdala and hippocampus enlargement during adolescence in autism. *J Am Acad Child Adolesc Psychiatry*, 49, 552–60.

Grossman, A.W., Churchill, J.D., Bates, K.E., Kleim, J.A., and Greenough, W.T. (2002). A brain adaptation view of plasticity: is synaptic plasticity an overly limited concept? *Prog Brain Res*, 138, 91–108.

Gu, H., Salmeron, B.J., Ross, T.J., *et al.* (2010). Mesocorticolimbic circuits are impaired in chronic cocaine users as demonstrated by resting-state functional connectivity. *Neuroimage*, 53, 593–601.

Guidi, S., Bonasoni, P., Ceccarelli, C., *et al.* (2008). Neurogenesis impairment and increased cell death reduce total neuron number in the hippocampal region of fetuses with Down syndrome. *Brain Pathol*, 18, 180–97.

Guptill, J.T., Booker, A.B., Gibbs, T.T., Kemper, T.L., Bauman, M.L., and Blatt, G.J. (2007). [3H]-flunitrazepam-labeled benzodiazepine binding sites in the hippocampal formation in autism: a multiple concentration autoradiographic study. *J Autism Dev Disord*, 37, 911–20.

Haier, R.J., Alkire, M.T., White, N.S., *et al.* (2003). Temporal cortex hypermetabolism in Down syndrome prior to the onset of dementia. *Neurology*, 61, 1673–79.

Haier, R.J., Head, K., Head, E., and Lott, I.T. (2008). Neuroimaging of individuals with Down's syndrome at-risk for dementia: evidence for possible compensatory events. *Neuroimage*, 39, 1324–32.

Hamilton, D.A., Kodituwakku, P., Sutherland, R.J., and Savage, D.D. (2003). Children with Fetal Alcohol Syndrome are impaired at place learning but not cued-navigation in a virtual Morris water task. *Behav Brain Res*, 143, 85–94.

Harvey, A.S., Grattan-Smith, J.D., Desmond, P.M., Chow, C.W., and Berkovic, S.F. (1995). Febrile seizures and hippocampal sclerosis: frequent and related findings in intractable temporal lobe epilepsy of childhood. *Pediatric Neurol*, 12, 201–6.

Hazlett, H.C., Poe, M.D., Lightbody, A.A., *et al.* (2009). Teasing apart the heterogeneity of autism: Same behavior, different brains in toddlers with fragile X syndrome and autism. *J Neurodev Disord*, 1, 81–90.

Haznedar, M.M., Buchsbaum, M.S., Wei, T.-C., *et al.* (2000). Limbic circuitry in patients with autism spectrum disorders studied with positron emission tomography and magnetic resonace imaging. *Am J Psychiatry*, 157, 1994–2001.

Hemming, M., and Yellowlees, P.M. (1993). Effective treatment of Tourette's syndrome with marijuana. *J Psychopharmacol*, 7, 389–91.

Herkenham, M., Lynn, A.B., Little, M.D., *et al.* (1990). Cannabinoid receptor localization in brain. *Proc Natl Acad Sci USA*, 87, 1932–36.

Hoff, A.L., Riordan, H., Morris, L., *et al.* (1996). Effects of crack cocaine on neurocognitive function. *Psychiatry Res*, 60, 167–76.

Honeycutt, N.A., and Smith, C.D. (1995). Hippocampal volume measurements using magnetic resonance imaging in normal young adults. *J Neuroimaging*, 5, 95–100.

Hübner, T., Vloet, T.D., Marx, I., *et al.* (2008). Morphometric brain abnormalities in boys with conduct disorder. *J Am Acad Child Adolesc Psychiatry*, 47, 540–47.

Hurt, H., Betancourt, L.M., Malmud, E.K., *et al.* (2009). Children with and without gestational cocaine exposure: a neurocognitive systems analysis. *Neurotoxicol Teratol*, 31, 334–41.

Isaacs, E.B., Lucas, A., Chong, W.K., *et al.* (2000). Hippocampal volume and everyday memory in children of very low birth weight. *Pediatr Res*, 47, 713–20.

Isaacs, E.B., Vargha-Khadem, F., Watkins, K.E., Lucas, A., Mishkin, M., and Gadian, D.G. (2003). Developmental amnesia and its relationship to degree of hippocampal atrophy. *Proc Natl Acad Sci USA*, 100, 13060–3.

Jabes, A., Lavenex, P.B., Amaral, D.G., and Lavenex, P. (2010). Quantitative analysis of postnatal neurogenesis and neuron number in the macaque monkey dentate gyrus. *Eur J Neurosci*, 31, 273–85.

Jabes, A., Lavenex, P.B., Amaral, D.G., and Lavenex, P. (2011). Postnatal development of the hippocampal formation: A stereological study in macaque monkeys. *J Comp Neurol*, 519, 1051–70.

Jakala, P., Hanninen, T., Ryynanen, M., *et al.* (1997). Fragile-X: neuropsychological test performance, CGG triplet repeat lengths, and hippocampal volumes. *J Clin Invest*, 100, 331–8.

Jiang, Y.H., Armstrong, D., Albrecht, U., *et al.* (1998). Mutation of the Angelman ubiquitin ligase in mice causes increased cytoplasmic p53 and deficits of contextual learning and long-term potentiation. *Neuron*, 21, 799–811.

Jiang, Y.H., Pan, Y., Zhu, L., *et al.* (2010). Altered ultrasonic vocalization and impaired learning and memory in Angelman syndrome mouse model with a large maternal deletion from Ube3a to Gabrb3. *PLoS ONE*, 5, e12278.

Jones, A.M., Kennedy, N., Hanson, J., and Fenton, G.W. (1997). A study of dementia in adults with Down's syndrome using 99Tc(m)-HMPAO SPET. *Nucl Med Commun*, 18, 662–7.

Jones, K.L., Smith, D.W., Ulleland, C.N., and Streissguth, P. (1973). Pattern of malformation in offspring of chronic alcoholic mothers. *Lancet*, 1, 1267–71.

Jones, K.L., and Smith, D.W. (1973). Recognition of the fetal alcohol syndrome in early infancy. *Lancet*, 302, 999–1001.

Jovanovski, D., Erb, S., and Zakzanis, K.K. (2005). Neurocognitive deficits in cocaine users: a quantitative review of the evidence. *J Clin Expl Neuropsychol*, 27, 189–204.

Kaemingk, K.L., and Halverson, P.T. (2000). Spatial memory following prenatal alcohol exposure: more than a material specific memory deficit. *Child Neuropsychol*, 6, 115–28.

Kanner, L. (1943). Autistic disturbances of affective contact. *Nervous Child*, 2, 217–50.

Kassubek, J., Juengling, F.D., and Ludolph, A.G. (2006). Heterogeneity of voxel-based morphometry findings in Tourette's syndrome: an effect of age? *Ann Neurol*, 59, 872–3.

Kates, W.R., Abrams, M.T., Kaufmann, W.E., Breiter, S.N., and Reiss, A.L. (1997). Reliability and validity of MRI measurement of the amygdala and hippocampus in children with fragile X syndrome. *Psychiatry Res*, 75, 31–48.

Kaufmann, L.J. (2004). State suicide prevention plans. *NCSL Legisbrief*, 12, 1–2.

Kempermann, G., Kuhn, H.G., and Gage, F.H. (1997). More hippocampal neurons in adult mice living in an enriched environment. *Nature*, 386, 493–95.

Kempermann, G., and Gage, F.H. (2002). Genetic determinants of adult hippocampal neurogenesis correlate with acquisition, but not probe trial performance, in the water maze task. *Eur J Neurosci*, 16, 129–36.

Kerns, K.A., Don, A., Mateer, C.A., and Streissguth, A.P. (1997). Cognitive deficits in nonretarded adults with fetal alcohol syndrome. *J Learning Disabil*, 30, 685–93.

Koldewyn, K., Hessl, D., Adams, J., *et al.* (2008). Reduced hippocampal activation during recall is associated with elevated FMR1 mRNA and psychiatric symptoms in men with the fragile X premutation. *Brain Imaging Behav*, 2, 105–16.

Koo, B.K., Blaser, S., Harwood-Nash, D., Becker, L.E., and Murphy, E.G. (1992). Magnetic resonance imaging evaluation of delayed myelination in Down syndrome: a case report and review of the literature. *J Child Neurol*, 7, 417–21.

Krasuski, J.S., Alexander, G.E., Horwitz, B., Rapoport, S.I., and Schapiro, M.B. (2002). Relation of medial temporal lobe volumes to age and memory function in nondemented adults with Down's syndrome: implications for the prodromal phase of Alzheimer's disease. *Am J Psychiatry*, 159, 74–81.

Krauel, K., Duzel, E., Hinrichs, H., Santel, S., Rellum, T., and Baving, L. (2007). Impact of emotional salience on episodic memory in attention-deficit/hyperactivity disorder: A functional magnetic resonance imaging study. *Biol Psychiatry*, 61, 1370–79.

Krauel, K., Duzel, E., Hinrichs, H., Rellum, T., Santel, S., and Baving, L. (2009a). Emotional memory in ADHD patients with and without comorbid ODD/CD. *J Neural Transm*, 116, 117–20.

Krauel, K., Duzel, E., Hinrichs, H., *et al.* (2009b). Electrophysiological correlates of semantic processing during encoding of neutral and emotional pictures in patients with ADHD. *Neuropsychologia*, 47, 1873–82.

Krinsky-McHale, S.J., Devenny, D.A., Kittler, P., and Silverman, W. (2008). Selective attention deficits associated with mild cognitive impairment and early stage Alzheimer's disease in adults with Down syndrome. *Am J Ment Retard*, 113, 369–86.

Lagace, D.C., Yee, J.K., Bolanos, C.A., and Eisch, A.J. (2006). Juvenile administration of methylphenidate attenuates adult hippocampal neurogenesis. *Biol Psychiatry*, 60, 1121–30.

LaHoste, G.J., Swanson, J.M., Wigal, S.B., *et al.* (1996). Dopamine D4 receptor gene polymorphism is associated with attention deficit hyperactivity disorder. *Mol Psychiatry*, 1, 121–4.

Lahti, R.A., and Lahti, A.C. (1997). Images in neuroscience. Clin Genet. V. Association of genetic and personality characteristics. *Am J Psychiatry*, 154, 1496.

Lavoie, M.E., Thibault, G., Stip, E., and O'Connor, K.P. (2007). Memory and executive functions in adults with Gilles de la Tourette syndrome and chronic tic disorder. *Cogn Neuropsychiatry*, 12, 165–81.

Lawlor, B.A., McCarron, M., Wilson, G., and McLoughlin, M. (2001). Temporal lobe-oriented CT scanning and dementia in Down's syndrome. *Int J Geriatr Psychiatry*, 16, 427–29.

Lawrence, E.J., McGuire, P.K., Allin, M., *et al.* (2010a). The very preterm brain in young adulthood: the neural correlates of verbal paired associate learning. *J Pediatr*, 156, 889–95.

Lawrence, Y.A., Kemper, T.L., Bauman, M.L., and Blatt, G.J. (2010b). Parvalbumin-, calbindin-, and calretinin-immunoreactive hippocampal interneuron density in autism. *Acta Neurol Scand*, 121, 99–108.

LeBlanc-Duchin, D., and Taukulis, H.K. (2007). Chronic oral methylphenidate administration to periadolescent rats yields prolonged impairment of memory for objects. *Neurobiology of Learning and Memory*, 88, 312–20.

Leckman, J.F. (2002). Tourette's syndrome. *Lancet*, 360, 1577–86.

Lee, J.S., Kim, B.N., Kang, E., *et al.* (2005). Regional cerebral blood flow in children with attention deficit hyperactivity disorder: comparison before and after methylphenidate treatment. *Human Brain Mapp*, 24, 157–64.

Lee, K.T., Mattson, S.N., and Riley, E.P. (2004). Classifying children with heavy prenatal alcohol exposure using measures of attention. *J Int Neuropsychol Soc*, 10, 271–77.

Leverenz, J.B., and Raskind, M.A. (1998). Early amyloid deposition in the medial temporal lobe of young Down syndrome patients: a regional quantitative analysis. *Exp Neurol*, 150, 296–304.

Levitt, P., and Campbell, D.B. (2009). The genetic and neurobiologic compass points toward common signaling dysfunctions in autism spectrum disorders. *J Clin Invest*, 119, 747–54.

Li, Z., Coles, C.D., Lynch, M.E., *et al.* (2009). Prenatal cocaine exposure alters emotional arousal regulation and its effects on working memory. *Neurotoxicol Teratol*, 31, 342–8.

Lightbody, A.A., and Reiss, A.L. (2009). Gene, brain, and behavior relationships in fragile X syndrome: evidence from neuroimaging studies. *Dev Disabil Res Rev*, 15, 343–52.

Lodygensky, G.A., Rademaker, K., Zimine, S., *et al.* (2005). Structural and functional brain development after hydrocortisone treatment for neonatal chronic lung disease. *Pediatrics*, 116, 1–7.

Lorenzi, H.A., and Reeves, R.H. (2006). Hippocampal hypocellularity in the Ts65Dn mouse originates early in development. *Brain Res*, 1104, 153–59.

Lott, I.T., and Dierssen, M. (2010). Cognitive deficits and associated neurological complications in individuals with Down's syndrome. *Lancet Neurol*, 9, 623–33.

Luby, M., Spencer, D.D., Kim, J.H., deLanerolle, N., and McCarthy, G. (1995). Hippocampal MRI volumetrics and temporal lobe substrates in medial temporal lobe epilepsy. *Magn Reson Imaging*, 13, 1065–71.

Ludolph, A.G., Juengling, F.D., Libal, G., Ludolph, A.C., Fegert, J.M., and Kassubek, J. (2006). Grey-matter abnormalities in boys with Tourette syndrome: magnetic resonance imaging study using optimised voxel-based morphometry. *Br J Psychiatry*, 188, 484–85.

Maguire, E.A., Gadian, D.G., Johnsrude, I.S., *et al.* (2000). Navigation-related structural change in the hippocampi of taxi drivers. *Proc Natl Acad Sci USA*, 97, 4398–403.

Maguire, E.A., Spiers, H.J., Good, C.D., Hartley, T., Frackowiak, R.S., and Burgess, N. (2003). Navigation expertise and the human hippocampus: a structural brain imaging analysis. *Hippocampus*, 13, 250–9.

Mangan, P.A. (1992). Spatial memory abilities and abnormal development of the hippocampal formation in Down syndrome. Unpublished Doctoral Dissertation. Tucson, AZ: University of Arizona.

Manji, S., Pei, J., Loomes, C., and Rasmussen, C. (2009). A review of the verbal and visual memory impairments in children with foetal alcohol spectrum disorders. *Dev Neurorehabil*, 12, 239–47.

Manns, J.R., Hopkins, R.O., and Squire, L.R. (2003a). Semantic memory and the human hippocampus. *Neuron*, 38, 127–33.

Manns, J.R., Hopkins, R.O., Reed, J.M., Kitchener, E.G., and Squire, L.R. (2003b). Recognition memory and the human hippocampus. *Neuron*, 37, 171–80.

Mardirossian, S., Rampon, C., Salvert, D., Fort, P., and Sarda, N. (2009). Impaired hippocampal plasticity and altered neurogenesis in adult Ube3a maternal deficient mouse model for Angelman syndrome. *Exp Neurol*, 220, 341–8.

Marsh, R., Alexander, G.M., Packard, M.G., *et al.* (2004). Habit learning in Tourette syndrome: a translational neuroscience approach to a developmental psychopathology. *Arch Gen Psychiatry*, 61, 1259–68.

Mathern, G.W., Adelson, P.D., Cahan, L.D., and Leite, J.P. (2002). Hippocampal neuron damage in human epilepsy: Meyer's hypothesis revisited. *Prog Brain Res*, 135, 237–51.

Mayes, L., Snyder, P.J., Langlois, E., and Hunter, N. (2007). Visuospatial working memory in school-aged children exposed in utero to cocaine. *Child Neuropsychol*, 13, 205–18.

McGee, C.L., and Riley, E.P. (2006). Brain imaging and fetal alcohol spectrum disorders. *Annali dell'Istituto Superiore di Sanita*, 42, 46–52.

Messiha, F.S. (1991). Genetics, haloperidol and pimozide: a comparative study in two mouse strains. *Neurotoxicology*, 12, 47–54.

Miki, T., Yokoyama, T., Sumitani, K., *et al.* (2008). Ethanol neurotoxicity and dentate gyrus development. *Congenit Anom*, 48, 110–17.

Milner, B. (1962). *Les troubles de la memoire accompagnant des lesions Hippocampiques bilaterales*. Paris: CNRS.

Minshew, N.J., and Goldstein, G. (2001). The pattern of intact and impaired memory functions in autism. *J Child Psychol Psychiatry*, 42, 1095–101.

Moeller, F.G., Steinberg, J.L., Schmitz, J.M., *et al.* (2010). Working memory fMRI activation in cocaine-dependent subjects: association with treatment response. *Psychiatry Res*, 181, 174–82.

Moriarty, J., Costa, D.C., Schmitz, B., Trimble, M.R., Ell, P.J., and Robertson, M.M. (1995). Brain perfusion abnormalities in Gilles de la Tourette's syndrome. *Br J Psychiatry*, 167, 249–54.

Müller-Vahl, K.R., Kolbe, H., Schneider, U., and Emrich, H.M. (1998). Cannabinoids: possible role in patho-physiology and therapy of Gilles de la Tourette syndrome. *Acta Psychiatr Scand*, 98, 502–06.

Müller-Vahl, K.R., Schneider, U., Koblenz, A., *et al.* (2002). Treatment of Tourette's syndrome with Delta 9-tetrahydrocannabinol (THC): a randomized crossover trial. *Pharmacopsychiatry*, 35, 57–61.

Müller-Vahl, K.R., Schneider, U., Prevedel, H., *et al.* (2003). Delta 9-tetrahydrocannabinol (THC) is effective in the treatment of tics in Tourette syndrome: a 6-week randomized trial. *J Clin Psychiatry*, 64, 459–65.

Müller-Vahl, K.R., Kaufmann, J., Grosskreutz, J., Dengler, R., Emrich, H.M., and Peschel, T. (2009). Prefrontal and anterior cingulate cortex abnormalities in Tourette Syndrome: evidence from voxel-based morphometry and magnetization transfer imaging. *BMC Neuroscience*, 10, 47.

Nadel, L. (2003). Down's syndrome: a genetic disorder in biobehavioral perspective. *Genes, Brain, Behav*, 2, 156–66.

Nanson, J.L., and Hiscock, M. (1990). Attention deficits in children exposed to alcohol prenatally. *Alcohol, Clin Exp Res*, 14, 656–61.

Niccols, A. (2007). Fetal alcohol syndrome and the developing socio-emotional brain. *Brain Cogn*, 65, 135–42.

Nicolson, R., DeVito, T.J., Vidal, C.N., *et al.* (2006). Detection and mapping of hippocampal abnormalities in autism. *Psychiatry Res*, 148, 11–21.

Nilsson, M., Perfilieva, E., Johansson, U., Orwar, O., and Eriksson, P.S. (1999). Enriched environment increases neurogenesis in the adult rat dentate gyrus and improves spatial memory. *J Neurobiol*, 39, 569–78.

Norman, A.L., Crocker, N., Mattson, S.N., and Riley, E.P. (2009). Neuroimaging and fetal alcohol spectrum disorders. *Dev Disabil Res Rev*, 15, 209–17.

O'Malley, S., Adamse, M., Heaton, R.K., and Gawin, F.H. (1992). Neuropsychological impairment in chronic cocaine abusers. *Am J Drug Alcohol Abuse*, 18, 131–44.

Oak, J.N., Oldenhof, J., and Van Tol, H.H. (2000). The dopamine D(4) receptor: one decade of research. *Eur J Pharmacol*, 405, 303–27.

Oblak, A.L., Gibbs, T.T., and Blatt, G.J. (2010). Decreased GABA(B) receptors in the cingulate cortex and fusiform gyrus in autism. *J Neurochem*, 114, 1414–23.

Olson, H.C., Feldman, J.J., Streissguth, A.P., Sampson, P.D., and Bookstein, F.L. (1998). Neuropsychological deficits in adolescents with fetal alcohol syndrome: clinical findings. *Alcohol Clin Exp Res*, 22, 1998–2012.

Pace-Schott, E.F., Morgan, P.T., Malison, R.T., *et al.* (2008). Cocaine users differ from normals on cognitive tasks which show poorer performance during drug abstinence. *Am J Drug Alcohol Abuse*, 34, 109–21.

Page, L.A., Daly, E., Schmitz, N., *et al.* (2006). In vivo 1H-magnetic resonance spectroscopy study of amygdala-hippocampal and parietal regions in autism. *Am J Psychiatry*, 163, 2189–92.

Palmen, S.J., Durston, S., Nederveen, H., and Van Engeland, H. (2006). No evidence for preferential involvement of medial temporal lobe structures in high-functioning autism. *Psychol Med*, 36, 827–34.

Pei, J.R., Rinaldi, C.M., Rasmussen, C., Massey, V., and Massey, D. (2008). Memory patterns of acquisition and retention of verbal and nonverbal information in children with fetal alcohol spectrum disorders. *Can J Clin Pharmacol*, 15, e44–56.

Pennington, B.F., Moon, J., Edgin, J., Stedron, J., and Nadel, L. (2003). The neuropsychology of Down syndrome: evidence for hippocampal dysfunction. *Child Dev*, 74, 75–93.

Perlov, E., Philipsen, A., Tebartz van Elst, L., *et al.* (2008). Hippocampus and amygdala morphology in adults with attention-deficit hyperactivity disorder. *J Psychiatry Neurosci*, 33, 509–15.

Peters, S.U., Beaudet, A.L., Madduri, N., and Bacino, C.A. (2004). Autism in Angelman syndrome: implications for autism research. *Clin Genet*, 66, 530–6.

Peterson, B.S., Skudlarski, P., Anderson, A.W., *et al.* (1998). A functional magnetic resonance imaging study of tic suppression in Tourette syndrome. *Arch Gen Psychiatry*, 55, 326–33.

Peterson, B.S., Choi, H.A., Hao, X., *et al.* (2007). Morphologic features of the amygdala and hippocampus in children and adults with Tourette syndrome. *Arch Gen Psychiatry*, 64, 1281–91.

Petroff, O.A., Errante, L.D., Rothman, D.L., Kim, J.H., and Spencer, D.D. (2002). Glutamate-glutamine cycling in the epileptic human hippocampus. *Epilepsia*, 43, 703–10.

Pfluger, T., Weil, S., Weis, S., *et al.* (1999). Normative volumetric data of the developing hippocampus in children based on magnetic resonance imaging. *Epilepsia*, 40, 414–23.

Pinter, J.D., Brown, W.E., Eliez, S., Schmitt, J.E., Capone, G.T., and Reiss, A.L. (2001). Amygdala and hippocampal volumes in children with Down syndrome: a high-resolution MRI study. *Neurology*, 56, 972–74.

Piven, J., Bailey, J., Ranson, B.J., and Arndt, S. (1998). No difference in hippocampus volume detected on magnetic resonance imaging in autistic individuals. *J Autism Dev Disord*, 28, 105–10.

Plessen, K.J., Bansal, R., Zhu, H., *et al.* (2006). Hippocampus and amygdala morphology in attention-deficit/hyperactivity disorder. *Arch Gen Psychiatry*, 63, 795–807.

Prehn-Kristensen, A., Göder, R., Fischer, J., *et al.* (2011a). Reduced sleep-associated consolidation of declarative memory in attention-deficit/hyperactivity disorder. *Sleep Medicine*, 12(7), 672–9.

Prehn-Kristensen, A., Krauel, K., Hinrichs, H., *et al.* (2011b). Methylphenidate does not improve interference control during a working memory task in young patients with attention-deficit hyperactivity disorder. *Brain Res*, 1388, 56–68.

Provenzale, J.M., Barboriak, D.P., VanLandingham, K., MacFall, J., Delong, D., and Lewis, D.V. (2008). Hippocampal MRI signal hyperintensity after febrile status epilepticus is predictive of subsequent mesial temporal sclerosis. *AJR. Am J Roentgenol*, 190, 976–83.

Rachidi, M., and Lopes, C. (2007). Mental retardation in Down syndrome: from gene dosage imbalance to molecular and cellular mechanisms. *Neurosci Res*, 59, 349–69.

Rademaker, K.J., Rijpert, M., Uiterwaal, C.S., *et al.* (2006). Neonatal hydrocortisone treatment related to 1H-MRS of the hippocampus and short-term memory at school age in preterm born children. *Pediatr Res*, 59, 309–13.

Rasmussen, C., Soleimani, M., Carroll, A., and Hodlevskyy, O. (2009). Neuropsychological functioning in children with Tourette syndrome (TS). *J Can Acad Child Adolesc Psychiatry*, 18, 307–15.

Rasmussen, C., Benz, J., Pei, J., *et al.* (2010). The impact of an ADHD co-morbidity on the diagnosis of FASD. *J Can Acad Child Adolesc Psychiatry*, 17, e165–76.

Raymond, G., Bauman, M.L., and Kemper, T. (1989). The hippocampus in autism: Golgi analysis. *Ann Neurol*, 26, 483–84.

Raz, N., Torres, I., Spencer, W., Millman, D., Baertschi, J., and Sarpel, G. (1993). Neuroanatomical correlates of age-sensitive and age-invariant cognitive abilities. *Intelligence*, 17, 407–22.

Reck, S.G., Hund, A.M., and Landau, S. (2010). Memory for object locations in boys with and without ADHD. *J Atten Disord*, 13, 505–15.

Reiss, A.L., and Freund, L. (1992). Behavioral phenotype of fragile X syndrome: DSM-III-R autistic behavior in male children. *Am J Med Genet*, 43, 35–46.

Reiss, A.L., Lee, J., and Freund, L. (1994). Neuroanatomy of fragile X syndrome: the temporal lobe. *Neurology*, 44, 1317–24.

Reiss, A.L., Abrams, M.T., Singer, H.S., Ross, J.L., and Denckla, M.B. (1996). Brain development, gender and IQ in children: A volumetric imaging study. *Brain*, 119, 1763–74.

Reske, M., Eidt, C.A., Delis, D.C., and Paulus, M.P. (2010). Nondependent stimulant users of cocaine and prescription amphetamines show verbal learning and memory deficits. *Biol Psychiatry*, 68, 762–9.

Richardson, G.A., Goldschmidt, L., and Willford, J. (2009). Continued effects of prenatal cocaine use: preschool development. *Neurotoxicol Teratol*, 31, 325–33.

Riikonen, R., Salonen, I., Partanen, K., and Verho, S. (1999). Brain perfusion SPECT and MRI in foetal alcohol syndrome. *Dev Med Child Neurol*, 41, 652–59.

Riikonen, R.S., Nokelainen, P., Valkonen, K., *et al.* (2005). Deep serotonergic and dopaminergic structures in fetal alcoholic syndrome: a study with nor-beta-CIT-single-photon emission computed tomography and magnetic resonance imaging volumetry. *Biol Psychiatry*, 57, 1565–72.

Roessner, V., Becker, A., Banaschewski, T., Freeman, R.D., and Rothenberger, A. (2007). Developmental psychopathology of children and adolescents with Tourette syndrome—impact of ADHD. *Eur Child Adolesc Psychiatry*, 16 Suppl 1, 24–35.

Roizen, N.J., and Patterson, D. (2003). Down's syndrome. *Lancet*, 361, 1281–89.

Rojas, D.C., Smith, J.A., Benkers, T.L., Camou, S.L., Reite, M.L., and Rogers, S.J. (2004). Hippocampus and amygdala volumes in parents of children with autistic disorder. *Am J Psychiatry*, 161, 2038–44.

Rommelse, N.N., Van der Stigchel, S., Witlox, J., *et al.* (2008). Deficits in visuo-spatial working memory, inhibition and oculomotor control in boys with ADHD and their non-affected brothers. *J Neural Transm*, 115, 249–60.

Rubia, K., Smith, A.B., Halari, R., *et al.* (2009). Disorder-specific dissociation of orbitofrontal dysfunction in boys with pure conduct disorder during reward and ventrolateral prefrontal dysfunction in boys with pure ADHD during sustained attention. *Am J Psychiatry*, 166, 83–94.

Rumsey, J.M., and Hamburger, S.D. (1988). Neuropsychological findings in high-functioning men with infantile autism, residual state. *J Clin Expl Neuropsychol*, 10, 201–21.

Saitoh, O., Karns, C., and Courchesne, E. (2001). Development of the hippocampal formation from 2 to 42 years: MRI evidence of smaller area dentata in autism. *Brain*, 124, 1317–24.

Salisbury, A.L., Ponder, K.L., Padbury, J.F., and Lester, B.M. (2009). Fetal effects of psychoactive drugs. *Clinics Perinatol*, 36, 595–619.

Sallee, F.R., Nesbitt, L., Jackson, C., Sine, L., and Sethuraman, G. (1997). Relative efficacy of haloperidol and pimozide in children and adolescents with Tourette's disorder. *Am J Psychiatry*, 154, 1057–62.

Samudio-Ruiz, S.L., Allan, A.M., Valenzuela, C.F., Perrone-Bizzozero, N.I., and Caldwell, K.K. (2009). Prenatal ethanol exposure persistently impairs NMDA receptor-dependent activation of extracellular signal-regulated kinase in the mouse dentate gyrus. *J Neurochem*, 109, 1311–23.

Schapiro, M.B., Grady, C.L., Kumar, A., *et al.* (1990). Regional cerebral glucose metabolism is normal in young adults with Down syndrome. *J Cereb Blood Flow Metab*, 10, 199–206.

Schapiro, M.B., Haxby, J.V., and Grady, C.L. (1992). Nature of mental retardation and dementia in Down syndrome: study with PET, CT, and neuropsychology. *Neurobiol Aging*, 13, 723–34.

Schulz, K.P., Fan, J., Tang, C.Y., *et al.* (2004). Response inhibition in adolescents diagnosed with attention deficit hyperactivity disorder during childhood: an event-related FMRI study. *Am J Psychiatry*, 161, 1650–57.

Schumann, C.M., Hamstra, J., Goodlin-Jones, B.L., *et al.* (2004). The amygdala is enlarged in children but not adolescents with autism; the hippocampus is enlarged at all ages. *J Neurosci*, 24, 6392–401.

Schumann, C.M., Hamstra, J., Goodlin-Jones, B.L., Kwon, H., Reiss, A.L., and Amaral, D.G. (2007). Hippocampal size positively correlates with verbal IQ in male children. *Hippocampus*, 17, 486–93.

Schumann, C.M., Bloss, C.S., Barnes, C.C., *et al.* (2010). Longitudinal magnetic resonance imaging study of cortical development through early childhood in autism. *J Neurosci*, 30, 4419–27.

Scoville, W.B., and Milner, B. (1957). Loss of recent memory after bilateral hippocampal lesions. *J Neurol Neurosurg Psychiatry*, 20, 11–21.

Shadmehr, R., Brandt, J., and Corkin, S. (1998). Time-dependent motor memory processes in amnesic subjects. *J Neurophysiol*, 80, 1590–7.

Shalin, S.C., Egli, R., Birnbaum, S.G., Roth, T.L., Levenson, J.M., and Sweatt, J.D. (2006). Signal transduction mechanisms in memory disorders. *Prog Brain Res*, 157, 25–41.

Sie, L.T., van der Knaap, M.S., van Wezel-Meijler, G., and Valk, J. (1997). MRI assessment of myelination of motor and sensory pathways in the brain of preterm and term-born infants. *Neuropediatrics*, 28, 97–105.

Sokol, D.K., Demyer, W.E., Edwards-Brown, M., Sanders, S., and Garg, B. (2003). From swelling to sclerosis: acute change in mesial hippocampus after prolonged febrile seizure. *Seizure*, 12, 237–40.

Sowell, E.R., Lu, L.H., O'Hare, E.D., *et al.* (2007). Functional magnetic resonance imaging of verbal learning in children with heavy prenatal alcohol exposure. *Neuroreport*, 18, 635–39.

Sparks, B.F., Friedman, S.D., Shaw, D.W., *et al.* (2002). Brain sructural abnormalities in young children with autism spectrum disorder. *Neurology*, 59, 184–92.

Squire, L.R., and Zola, S.M. (1996). Memory, memory impairment, and the medial temporal lobe. *Cold Spring Harb Symp Quant Biol*, 61, 185–95.

Stebbins, G.T., Singh, J., Weiner, J., Wilson, R.S., Goetz, C.G., and Gabrieli, J.D.E. (1995). Selective impairments of memory functioning in unmedicated adults with Gilles de la Tourette's syndrome. *Neuropsychology*, 9, 329–37.

Stern, E., Silbersweig, D.A., Chee, K.Y., *et al.* (2000). A functional neuroanatomy of tics in Tourette syndrome. *Arch Gen Psychiatry*, 57, 741–48.

Sudai, E., Croitoru, O., Shaldubina, A., *et al.* (2010). High cocaine dosage decreases neurogenesis in the hippocampus and impairs working memory. *Addict Biol*, 16, 251–60.

Sullivan, E.V., Pfefferbaum, A., Swan, G.E., and Carmelli, D. (2001). Heritability of hippocampal size in elderly twin men: equivalent influence from genes and environment. *Hippocampus*, 11, 754–62.

Sutherland, R.J., McDonald, R.J., and Savage, D.D. (1997). Prenatal exposure to moderate levels of ethanol can have long-lasting effects on hippocampal synaptic plasticity in adult offspring. *Hippocampus*, 7, 232–38.

Suzuki, K., Nishimura, K., Sugihara, G., *et al.* (2010). Metabolite alterations in the hippocampus of high-functioning adult subjects with autism. *Int J Neuropsychopharmacology*, 13, 529–34.

Tecott, L.H., Kwong, L.L., Uhr, S., and Peroutka, S.J. (1986). Differential modulation of dopamine D2 receptors by chronic haloperidol, nitrendipine, and pimozide. *Biol Psychiatry*, 21, 1114–22.

Teipel, S.J., Schapiro, M.B., Alexander, G.E., *et al.* (2003). Relation of corpus callosum and hippocampal size to age in nondemented adults with Down's syndrome. *Am J Psychiatry*, 160, 1870–8.

Teipel, S.J., Alexander, G.E., Schapiro, M.B., Moller, H.J., Rapoport, S.I., and Hampel, H. (2004). Age-related cortical grey matter reductions in non-demented Down's syndrome adults determined by MRI with voxel-based morphometry. *Brain*, 127, 811–24.

Theodore, W.H., Bhatia, S., Hatta, J., *et al.* (1999). Hippocampal atrophy, epilepsy duration, and febrile seizures in patients with partial seizures. *Neurology*, 52, 132–6.

Thompson, D.K., Wood, S.J., Doyle, L.W., *et al.* (2008). Neonate hippocampal volumes: prematurity, perinatal predictors, and 2-year outcome. *Ann Neurol*, 63, 642–51.

Thompson, D.K., Wood, S.J., Doyle, L.W., Warfield, S.K., Egan, G.F., and Inder, T.E. (2009). MR-determined hippocampal asymmetry in full-term and preterm neonates. *Hippocampus*, 19, 118–23.

Thompson, P.M., Cannon, T.D., Narr, K.L., *et al.* (2001). Genetic influences on brain structure. *Nat Neurosci*, 4, 1253–8.

Torres, G.E., Gainetdinov, R.R., and Caron, M.G. (2003). Plasma membrane monoamine transporters: structure, regulation and function. *Nat Rev Neurosci*, 4, 13–25.

Trillingsgaard, A., and Østergaard, J. (2004). Autism in Angelman syndrome: an exploration of comorbidity. *Autism*, 8, 163–74.

Tuchman, R., and Rapin, I. (2002). Epilepsy in autism. *Lancet Neurology*, 1, 352–8.

Tuchman, R., Cuccaro, M., and Alessandri, M. (2010). Autism and epilepsy: historical perspective. *Brain Devel*, 32, 709–18.

Uecker, A., and Nadel, L. (1996). Spatial locations gone awry: object and spatial memory deficits in children with fetal alcohol syndrome. *Neuropsychologia*, 34, 209–23.

Uecker, A., and Nadel, L. (1998). Spatial but not object memory impairments in children with fetal alcohol syndrome. *Am J Ment Retard*, 103, 12–18.

Utsunomiya, H., Takano, K., Okazaki, M., and Mitsudome, A. (1999). Development of the temporal lobe in infants and children: analysis by MR-based volumetry. *AJNR*, 20, 717–23.

van Woerden, G.M., Harris, K.D., Hojjati, M.R., *et al.* (2007). Rescue of neurological deficits in a mouse model for Angelman syndrome by reduction of alphaCaMKII inhibitory phosphorylation. *Nat Neurosci*, 10, 280–2.

Veltman, M.W., Craig, E.E., and Bolton, P.F. (2005). Autism spectrum disorders in Prader-Willi and Angelman syndromes: a systematic review. *Psychiatr Genet*, 15, 243–54.

Venkatesan, A., Nath, A., Ming, G.L., and Song, H. (2007). Adult hippocampal neurogenesis: regulation by HIV and drugs of abuse. *CMLS*, 64, 2120–32.

Verkerk, A.J., Pieretti, M., Sutcliffe, J.S., *et al.* (1991). Identification of a gene (FMR-1) containing a CGG repeat coincident with a breakpoint cluster region exhibiting length variation in fragile X syndrome. *Cell*, 65, 905–14.

Vicari, S., Bellucci, S., and Carlesimo, G.A. (2000). Implicit and explicit memory: a functional dissociation in persons with Down syndrome. *Neuropsychologia*, 38, 240–51.

Volkow, N.D., Wang, G.J., Newcorn, J., *et al.* (2007). Depressed dopamine activity in caudate and preliminary evidence of limbic involvement in adults with attention-deficit/hyperactivity disorder. *Arch Gen Psychiatry*, 64, 932–40.

Watkins, L.H., Sahakian, B.J., Robertson, M.M., *et al.* (2005). Executive function in Tourette's syndrome and obsessive-compulsive disorder. *Psychol Med*, 35, 571–82.

Wechsler, D. (1999). *Wechsler Abbreviated Scale of Intelligence*. San Antonio: Psychological Corporation.

Weeber, E.J., Jiang, Y.H., Elgersma, Y., *et al.* (2003). Derangements of hippocampal calcium/calmodulin-dependent protein kinase II in a mouse model for Angelman mental retardation syndrome. *J Neurosci*, 23, 2634–44.

White, S.W., Oswald, D., Ollendick, T., and Scahill, L. (2009). Anxiety in children and adolescents with autism spectrum disorders. *Clin Psychol Rev*, 29, 216–29.

Wickett, J.C., Vernon, P.A., and Lee, D.H. (1994). In-Vivo Brain Size, Head Perimeter, and Intelligence in a Sample of Healthy Adult Females. *Person Indiv Differ*, 16, 831–38.

Wilke, M., Sohn, J.H., Byars, A.W., and Holland, S.K. (2003). Bright spots: correlations of gray matter volume with IQ in a normal pediatric population. *Neuroimage*, 20, 202–15.

Willerman, L., Schultz, R., Rutledge, J.N., and Bigler, E.D. (1991). Invivo brain size and intelligence. *Intelligence*, 15, 223–28.

Williams, C.A., Beaudet, A.L., Clayton-Smith, J., *et al.* (2006). Angelman syndrome 2005: updated consensus for diagnostic criteria. *Am J Med Genet. Part A*, 140, 413–8.

Willoughby, K.A., Sheard, E.D., Nash, K., and Rovet, J. (2008). Effects of prenatal alcohol exposure on hippocampal volume, verbal learning, and verbal and spatial recall in late childhood. *J Int Neuropsychol Soc*, 14, 1022–33.

Wiseman, F.K., Alford, K.A., Tybulewicz, V.L., and Fisher, E.M. (2009). Down syndrome – recent progress and future prospects. *Hum Mol Genet*, 18, R75–83.

Wisniewski, K.E. (1990). Down syndrome children often have brain with maturation delay, retardation of growth, and cortical dysgenesis. *Am J Med Genet Suppl*, 7, 274–81.

Yashiro, K., Riday, T.T., Condon, K.H., *et al.* (2009). Ube3a is required for experience-dependent maturation of the neocortex. *Nat Neurosci*, 12, 777–83.

Zola, S.M., Squire, L.R., Teng, E., Stefanacci, L., Buffalo, E.A., and Clark, R.E. (2000). Impaired recognition memory in monkeys after damage limited to the hippocampal region. *J Neurosci*, 20, 451–63.

CHAPTER 11

The Hippocampus in Neurological Disease

Thorsten Bartsch

The vulnerability of the hippocampus in neurological disease 200
MRI correlates of hippocampal damage 203
Hippocampal ischaemia 203
Anoxic encephalopathy 204
Limbic encephalitis 205
Epilepsy, status epilepticus, and postictal changes 208
Traumatic brain injury 209
Hypoglycaemic encephalopathy 210
Transient global amnesia 210
Multiple sclerosis 215

ABSTRACT The hippocampus, and in particular the CA1 region, is vulnerable to metabolic stress. This affection can be observed in acute neurological disorders such as ischaemia, limbic encephalitis, hypoglycaemic encephalopathy, epilepsy, multiple sclerosis, and transient global amnesia. The basis of this regional susceptibility is, however, only partly understood but may include genetically-determined, glutamate-dependent, and calcium-mediated mechanisms of neuronal excitotoxicity and oxidative stress. Additionally, inflammatory responses and immunologically-mediated mechanisms may contribute to the response to acute metabolic stress but may also be involved in ageing and neurodegenerative disorders. The structure–function relationship of the hippocampus in pathological states can be monitored *in vivo* using magnetic resonance imaging (MRI). Here, the cellular and imaging correlates and pattern of hippocampal damage is reviewed in terms of studying the evolution and sequelae of hippocampal CA1 diffusion and ADC changes in acute neurological disorders. The evolution of diffusion changes over the course of the disease show that CA1 diffusion lesions can be most prominently detected within the first 3 days after the injury, frequently leading to a 'delayed neuronal death' of CA1 neurons. The cellular correlates of hippocampal damage, as reflected by MRI are incorporated into a pathophysiological framework highlighting the clinical neurobiology of the hippocampus.

KEYWORDS transient global amnesia, hippocampal stroke, hippocampus, limbic encephalitis, selective vulnerability, CA1, memory

The vulnerability of the hippocampus in neurological disease

It has been known for over 100 years that pathological conditions, such as ischaemia, hypoglycaemia, oxidative stress, and anoxia, can cause selective lesions in the hippocampus. This vulnerability is most clearly evident in CA1 hippocampal neurons and can be observed in various neurological diseases. The exact mechanisms of this region-specific selective vulnerability are not well understood but may include glutamate-dependent and calcium-mediated mechanisms of neuronal cytotoxicity. In addition to hippocampal CA1 neurons, Purkinje cells in the cerebellum and neurons in the caudate nucleus, outer putamen, and cortical layers III and IV show a similar vulnerability to noxious influences, albeit to a lesser degree than pyramidal cells of the hippocampus (Nikonenko et al. 2009). In monkeys, graded neuronal damage occurs after variable durations of global ischaemia; ischaemia for up to 5 min results in almost no damage, whereas 10–15 min of ischaemia leads to the selective degeneration of CA1 pyramidal cells. A longer duration of ischaemia results in damage to structures outside of the hippocampus, such as neocortical layers III, V, and VI and the striatum (Tabuchi et al. 1992, 1995). Notably, a dorsal–ventral gradient in ischaemia-induced degeneration can be observed, with decreasing cell death from the dorsal-septal to the ventral-temporal regions (Ashton et al. 1989).

A brief episode of cerebral ischaemia results in the selective cell death of hippocampal CA1 neurons. The structural changes that follow the insult are most prominent after a delay of 3–4 days and thus have been termed 'delayed neuronal death' (Pulsinelli et al. 1982; Schmidt-Kastner and Freund 1991; Kirino 2000). As outlined by Michaelis in Chapter 4, this phenomenon of 'delayed neuronal death' reflects susceptibility to harmful conditions and is most commonly associated with deleterious cellular mechanisms in the final pathway that lead to extracellular glutamate and excessive intracellular Ca^{2+} accumulation by the increased activation of N-methyl-D-aspartate (NMDA) receptors. Although this model—which has been most thoroughly outlined in the pathophysiology of ischaemic stroke—may account for the earlier stages of this cascade, downstream molecular processes (probably genetically determined) are involved in the later, delayed phase leading to the completion of delayed neuronal death (Schreiber and Baudry 1995; Newrzella et al. 2007; Lavenex et al. 2011). Transient ischaemia results in the prolonged up-regulation of caspase-3 mRNA in rat hippocampal CA1 neurons (Ni et al. 1998), as death of the CA1 pyramidal neurons is due to apoptosis and not necrosis (Kirino et al. 1986; Nitatori et al. 1995; Kirino 2000). With regard to cell death, cytochrome c seems to be involved in the delayed death of hippocampal CA1 neurons in rats following transient global ischaemia (Sugawara et al. 1999). Hence, recent models point to a pivotal role for genomically determined differences in the tolerance to glutamate and the distribution of glutamate receptors and to an increase in excitatory synaptic transmission in the pathophysiological cascade (Ikegaya and Matsuki 2002; Quintana 2006).

Acute ischaemia leads to energy depletion due to a loss of oxygen and glucose with subsequent impaired adenosine triphosphate (ATP) generation, increased membrane permeability and the facilitated production of reactive oxygen species (ROS) (Kreisman et al. 2000). As a consequence of altered ATP supplies, changes in ion gradients lead to the accumulation of extracellular glutamate, and the resulting intracellular accumulation of Ca^{2+} results in cytotoxicity. With regard to sustained Ca^{2+} influx as a key mechanism in neuronal death, it has been shown that transient ischaemia leads to reduced GluR2expression in CA1 hippocampal neurons, which in turn leads to increased Ca^{2+} permeability through kainate/AMPAR (alpha-amino-3-hydroxyl-5-methyl-4-isoxazole-propionate) receptor subunits in CA1 neurons (Pellegrini-Giampietro et al. 1992). The paucity of calcium-binding proteins, such as chromogranin, could also play a role in the accumulation of Ca^{2+}, as chromogranin is extremely weak in CA1 and CA3 neurons within the human hippocampus (Munoz 1990). Glutamate excitotoxicity is a key mechanism in CA1 neuronal death and can be blocked using NMDA receptor (NMDAR) and AMPA receptor (AMPAR) antagonists, thereby conferring neuroprotection (Bonde et al. 2005). Furthermore, the destruction of CA3 projections prevents the delayed neuronal death of CA1 neurons following ischaemia, which highlights the critical role played by glutamatergic Schaffer collaterals in hippocampal ischaemia (Onodera et al. 1986; Buchan and Pulsinelli 1990). CA1 neurons express a considerably higher density of NMDA NR2 subunits than do CA3 or DG neurons (Butler et al. 2010). An NMDAR-dependent cytosolic Ca^{2+}accumulation results in mitochondrial calcium overload with subsequent damage to mitochondrial function and increased production of reactive oxygen species (ROS), which lead to CA1-selective neurodegeneration (Stanika et al. 2010). Oxidative stress is the result of an imbalance between the generation of deleterious reactive oxygen and nitrogen species and the cell's compensatory mechanisms (Bao et al. 2009). The differential regional vulnerability of cerebral structures is at least partly a reflection of regional differences in antioxidant enzymes (Wang et al. 2005, 2007, 2009; Bao et al. 2009; Wang and Michaelis 2010). For example, hypoglycaemia induces changes in the activity of several antioxidant enzymes, while largely sparing the activity of superoxide dismutase in the cerebellum (Singh et al. 2004). This factor may explain the resilience of the cerebellum to oxidative stress (Sathyasaikumar et al. 2007). Another prominent role in the pathophysiological cascade leading to cell death in hippocampal pyramidal neurons is the putative cellular toxicity of Zn^{2+} (Stanika et al. 2010).

Although the final pathways in noxious conditions, such as ischaemia, hypoglycaemia, and status epilepticus, may all lead to neuronal hippocampal necrosis, it has been suggested by Siesjo et al. that these clinical conditions may have different initial pathophysiological cascades. During ischaemia, for example, the cellular metabolic rate rapidly and dramatically declines, which profoundly impairs ion homeostasis and cellular redox systems and causes acidosis to rapidly develop. In the hypoglycaemic state, cellular oxygen consumption continues, with a loss of high-energy phosphates and ion homeostasis, whereas redox systems become oxidized and acidosis is severely reduced. During a seizure, the cellular metabolic rate increases, although the cellular energy state and ion homeostasis are largely preserved. This preservation suggests that necrosis of hippocampal cells occurs despite sufficient oxygenation and might not be strictly related to energy failure (Siesjo and Wieloch 1986).

The pathophysiological cascade of structural changes in hippocampal neurons following ischaemia has been reviewed by Nikonenko et al. (2009). With regard to hippocampal neurons, ischaemic damage results in cellular swelling that includes mitochondrial swelling and subsequent disintegration. Hippocampal ischaemia leads to further decreases in protein synthesis, protease-dependent degeneration, and apoptosis. It has been suggested that distinct populations of hippocampal cells are differentially vulnerable to ischaemia, as hippocampal interneurons are believed to be more resistant to the effects of ischaemia than larger projection neurons (Harry and Lefebvre d'Hellencourt 2003; Avignone et al. 2005). Notably, CA1 neurons are most susceptible to ischaemia, whereas CA3 and dentate gyrus neurons remain largely unaffected (Kovalenko et al. 2006), a finding that may be attributed—at least in part—to the protective effect of acidosis in CA3 neurons (Cronberg et al. 2005). Similarly, endoplasmic reticulum stress-induced neurodegeneration shows a similar distribution in the subregions of the cornu ammonis (Kosuge et al. 2008; Osada et al. 2010).

Ischaemia alters the level of synaptic plasticity in CA1 neurons (Calabresi et al. 2003). Long-term potentiation (LTP) is physiologically induced by NMDA receptors, whereas the expression of LTP is dependent on AMPARs. However, the type of LTP that emerges during ischaemia—the so-called anoxic LTP—is exclusively mediated by NMDARs and is dependent on an increase in intracellular calcium (Crepel et al. 1993). The emergence of anoxic LTP can be blocked by inhibitors of protein kinase C (PKC) and can be attenuated by inhibitors of nitric oxide synthase (NOS). Even during a weak ischaemic insult, rapid structural changes can be observed in the hippocampal CA1 stratum radiatum excitatory synapses. These changes include a depletion of synaptic vesicles and alterations in their spatial arrangement, with a subsequent degeneration of synaptic contacts (Quintana 2006). Downstream postischaemic changes include dendritic rearrangements in the CA1 neurons. Some of these morphological changes are at least partly mediated by and dependent on calcium ions and NMDARs, which suggests that the anoxic LTP may contribute to these postischaemic changes. This possibility implies that similar changes may be involved in the mechanisms of cell death or the adaptive plasticity associated with ischaemia (Kovalenko et al. 2006). The regions that are not directly affected by the injury often exhibit enhanced neuronal plasticity, the functional effects of which are poorly understood (Xu and Pulsinelli 1994, 1996; Gao et al. 1998). Wang et al. showed that transient ischaemia in the rat causes delayed cell death in the CA1 field of the hippocampus, whereas the dentate gyrus is largely spared. However, synaptic plasticity in the dentate gyrus—which is reflected by synaptic field responses and LTP—is also only transiently affected by ischaemia, whereas synaptic plasticity in the CA1 region is permanently abolished. Disrupting neurogenesis in the dentate gyrus with radiation, however, prevented these neurons from recovering (S. Wang et al. 2005).

It appears that injury stimulates cell proliferation in the dentate gyrus, thereby giving rise to newly generated neurons and supporting the activity-dependent role of adult neurogenesis in the mechanisms of plasticity and compensation (Schwamm et al. 1998; Bueters et al. 2008; Miles and Kernie 2008). With regard to the mechanisms of plasticity following hippocampal damage, it has been shown—at least in the rat— that global ischaemia results in an almost complete loss of CA1 neurons, which is reflected by robust deficits in spatial memory. The reappearance of nearly 40% of the CA1 neurons within 90 days of the insult was associated with a restoration of hippocampal spatial memory function, which indicates that the newly formed neurons that originate in the lateral periventricular region have the potential to restore hippocampal function (Bendel et al. 2005). Notably, exposure to a mild transient ischaemic event some days before sustained periods of ischaemia induces an ischaemic tolerance in CA1 neurons that prevents the emergence of anoxic LTP and exerts a protective effect against neuronal death (Kawai et al. 1998). This finding suggests that the postischaemic pathological effects on synaptic plasticity that are mediated by NMDARs may also have the capacity to induce changes that ultimately restore neuronal activity (Calabresi et al. 2003; Crepel et al. 2003).

Ischaemia also leads to the transient activation of inhibitory synaptic transmission in the CA1 region, which may protect CA1 neurons and delay neuronal damage (Liang et al. 2009). Although many studies have described structural changes in response to ischaemia, the electrophysiological correlates are mainly unknown. It has recently been shown that inducing unilateral hippocampal ischaemia by blocking the initial ascending part of the longitudinal hippocampal artery in rats induces a high-frequency discharge (HFD, 300 Hz over 7–24 s) in the CA1 neurons that correlates with profound synchronous neuronal activity. This high-frequency discharge is followed by a DC voltage shift and a decrease in activity in the low and high gamma bands. Notably, the contralateral (non-affected) hippocampus displays similar high-frequency changes but not the DC voltage shifts or the long-lasting changes in gamma-band activity (Barth and Mody 2011).

Although neurons are the cells most strongly affected by hippocampal ischaemia, glial cells are also affected to a lesser degree. Glial cells have fewer glutamate receptors, higher ATP levels, and higher antioxidant activity than neurons. The effect of ischaemia on astrocytes could enhance neuronal degeneration, as CA1 astrocytes are more vulnerable to ischaemia than astrocytes in the dentate gyrus (Ouyang et al. 2007). The authors have suggested that increased oxidative stress and functional impairments in the GLT-1 glutamate transporter in astrocytes are critical to the delayed neuronal death of CA1 neurons (Ouyang et al. 2007).

MRI correlates of hippocampal damage

Acute neurological disorders that affect the hippocampus, including ischaemic stroke, encephalitis, seizures, and hypoglycaemia, result in structural changes that can be detected using MRI. On a cellular level, these processes result in the breakdown of energy-dependent Na/K-ATPases with subsequent cellular swelling and cytotoxic oedema. The intracellular accumulation of lactate and other osmotically active substances during anaerobic glycolysis further enhances this cytotoxic oedema. The changes that occur during the course of various pathological processes result in a reduced net diffusion of water molecules by Brownian motion, which can be measured as a reduction in the apparent diffusion coefficient (ADC) using MRI (Sotak 2004). Thus, the ADC reflects the degree of abnormal water movement in injured tissue that can be caused by a net shift of water from the extracellular to intracellular compartments and by a reduced transmembrane permeability caused by the collapse of energy-dependant transmembrane ion gradients (Liu et al. 2001; Mascalchi et al. 2005). Hence, diffusion-weighted MRI is a sensitive tool for detecting changes in the ADC during neuropathological states (Sotak 2004). The selective vulnerability of hippocampal neurons (with CA1 neurons being more vulnerable than CA3 neurons) was demonstrated in a diffusion MRI model that used a hippocampal slice to showed that water diffusion changes and pathological swelling correlate with changes in the ADC (Shepherd et al. 2003, 2007). The characteristics and kinetics of extracellular diffusion in hippocampal CA1 tissue in the normal and ischaemic state have thus far been studied only in animal models. These models have revealed differences in the tortuosity of extracellular diffusion between neocortical and CA1 tissues (Hrabetova 2005) that may contribute to the selective vulnerability of CA1 neurons to metabolic stress (Kawasaki et al. 1990; Schmidt-Kastner and Freund 1991). In ischaemic hippocampal CA1 slices, the compartmental diffusion showed an increase in tortuosity and a decrease in the diffusion coefficient (Perez-Pinzon et al. 1995; Patlak et al. 1998).

Diffusion tension MR imaging (DTI) displays water diffusion in nervous tissue that is correlated with the three-dimensional coherent orientation of the tissue's microstructure. Recently, attempts to increase the resolution of diffusion-weighted imaging for detecting pathological states on a cytoarchitectural level in the hippocampus have combined diffusion-tensor methods with microscopy (Shepherd et al. 2007).

Hippocampal ischaemia

The hippocampal region is one of the central nervous system (CNS) structures that is most sensitive to ischaemic damage and patients with infarctions often present with memory disturbances. In the 1970s, Benson et al. coined the phrase *amnestic stroke*; since then, several case series have confirmed that posterior cerebral artery (PCA) insults can lead to transient or persistent amnesia (Victor et al. 1961; Benson et al. 1974; Ott and Saver 1993; Szabo et al. 2009). A common finding in these cases is the unilateral or bilateral involvement of the hippocampus. The amnesia is often accompanied by (and complicated by) hemianopia depending on the extension of the infarct lesion. MRI provides a good localization of the extent of the insult. An MRI study that evaluated hippocampal involvement in 14 PCA-stroke patients found that half of the patients with left or bilateral hippocampal infarctions exhibited a memory deficit (Takahashi et al. 1997). An anatomical classification of hippocampal lesions in ischaemia was performed by Szabo et al. Four distinct patterns of hippocampal lesions could be identified (Figure 11.1). They included the (1) complete, (2) lateral (correlating to a selective CA1 affection), (3) partial (either posterior or anterior regions of the hippocampus), and (4) small focal lesions on the hippocampus (Szabo et al. 2009). Amnestic deficits were observed in 11 out of 57 patients, although a careful neuropsychological screening of 20 patients revealed deficits in verbal and non-verbal episodic long-term memory in case of left and right hippocampal stroke, respectively (Szabo et al. 2009). Notably, no isolated infarcts of the hippocampus were detected. The body of the hippocampus is supplied primarily by the PCA, which branches into the proximal posterior cerebral artery, the anterior and posterior hippocampal arteries, and finally into the longitudinal terminal segments of the hippocampal arteries. The head of the hippocampus is supplied by the anterior choroidal artery (AChA), which branches from the internal carotidal artery (ICA) (Szabo et al. 2009). Strategic insults within the territory supplied by the thalamo-perforate arteries in the medial thalamus can result in diencephalic amnesia (Graff-Radford et al. 1990).

Using high-resolution positron emission tomography (PET), postischaemic neuronal death was visualized in the primate hippocampus using ligands for the benzodiazepine and muscarinic cholinergic receptors. Decreased binding of both these ligands was observed 7 days after ischaemia, which corresponds to postmortem cellular degenerative changes (Zhao et al. 2002). The development of postischaemic hippocampal lesions was studied by measuring the

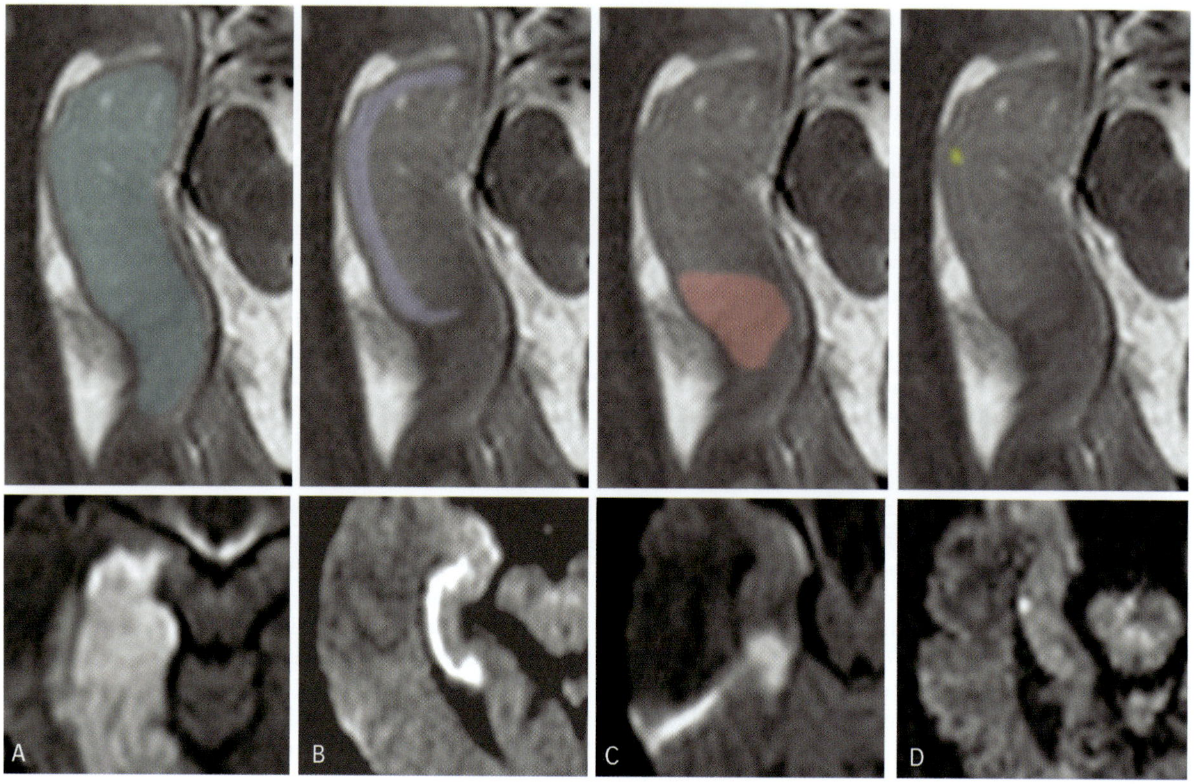

Figure 11.1 Anatomical classification of ischaemia-induced hippocampal lesions. Four distinct patterns of hippocampal lesions can be identified (depending on the extent of the lesion), including the complete hippocampus (A, green), the lateral (B, blue) and posterior (C, red) sections of the hippocampus and small, focal dot-like lesions in the lateral hippocampus (D, yellow) that correspond to the CA1 area (Szabo et al. 2009). The distribution of each lesion is presented as a schematic drawing (top row) and as a DWI hyperintense acute ischaemic lesion (bottom row). © Images courtesy of Dr K. Szabo, Department of Neurology, University Hospital Mannheim, University of Heidelberg, Germany.

time course of the 7T MRI ADC in gerbils. The ADC decreased from day 1 to day 4, and normalization followed within the next few weeks. Histopathologically, these changes correlated with apoptotic changes in the hippocampal pyramidal neurons (Zhu et al. 2000; Liu et al. 2001; Yamada et al. 2003). This time course resembles that which is observed in neocortical ischaemia, which also shows an initial decrease within the first 2–3 days followed by pseudo-normalization (or elevation) over the subsequent weeks (Schwamm et al. 1998; Yang et al. 1999; Fiebach et al. 2002; Munoz Maniega et al. 2004; Schulz et al. 2007). A similar time course has also been observed for hippocampal CA1 lesions in transient global amnesia (Bartsch 2007; Weon et al. 2008).

Anoxic encephalopathy

The pathophysiological cascade in ischaemic stroke leads to further exacerbated structural changes during extensive hypoxia (as occurs during cardiorespiratory arrest, for example) (Horn and Schlote 1992). Previous neuropathological studies of patients who survived the initial incident showed severe neuronal necrosis in the CA1 region of the hippocampus together with lesions observed in the cortex and putamen (Petito et al. 1987). In particular, the hippocampal damage was correlated with the duration of the postarrest period, which is likely a reflection of the particular time course of the delayed neuronal death in the CA1 area (Petito et al. 1987). Notably, some studies revealed a similar pattern of delayed neuronal death in cerebellar Purkinje cells (Ng et al. 1989; Horn and Schlote 1992). Similarly detailed neuropathological studies showed selective impairment of CA1 neurons was followed by significant cognitive and memory deficits (Zola-Morgan et al. 1986, 1992; Kartsounis et al. 1995). Some studies in rodents have shown a uniform pattern of neuronal damage in the hippocampus, whereas others have found a differential pattern (Auer et al. 1984; Ng et al. 1989). In cases of anoxic encephalopathy, serial MRIs have revealed the early development of post-ischaemic damage in the cerebellum, neocortex, and caudate putamen, whereas the temporal course of the DWI changes in the hippocampus indicated the delayed development of diffusion-restricted lesions (Smith et al. 1984; Konaka et al. 2007). Fujioka et al. addressed the question of specific volume loss following reperfusion over the course of cardiorespiratory arrest. Patients who were in a vegetative state after cardiorespiratory arrest had up to a 30% decrease in hippocampal volume within 21 days of the incident; this finding was in contrast to the volume changes in the temporal lobe (excluding the hippocampus), which decreased by only 10% or less (Fujioka et al. 2000). From a potential therapeutic-regenerative perspective, inhibited expression of the transient receptor potential melastatin 7 (TRPM7) channel, a receptor potential channel that is critically involved in cell survival, led to resistance to delayed neuronal death in CA1 neurons *in vivo* and preserved neuronal structure and function in the wake of brain ischaemia (Sun et al. 2009).

Limbic encephalitis

Herpes simplex limbic encephalitis

Herpes simplex encephalitis (HSE) is the most common fatal sporadic encephalitis of the brain. The majority of cases are caused by the herpes simplex virus-1 (HSV-1). Herpes simplex encephalitis is thought to be triggered by the retrograde transmission of latent, dormant herpes viruses from peripheral neurons to the CNS, although the exact immunological mechanisms are poorly understood. The clinical symptoms include memory loss, seizures and a confused state. The gold standard for diagnosis is the detection of HSV DNA in the cerebrospinal fluid (CSF) using a polymerase chain reaction (PCR) test. Given the predilection of the hippocampus and the temporal lobes to this infection, a role in the transmission has been attributed to the olfactory nerve. To date, the only effective therapy is the nucleoside analogue acyclovir. Neuropathologically, herpes viruses cause fulminant haemorrhagic and necrotizing meningoencephalitis and severe oedema. Necrotizing inflammation and haemorrhagic tissue necrosis are the predominant findings in the temporal lobe structures. MRI typically shows a pattern of T2-hyperintense cortical lesions in the mediobasal temporal lobes, including the hippocampus, inferior frontal lobes, insular cortex, and cingulate gyrus (Demaerel et al. 1992; Herweh et al. 2007). Commonly, the hippocampus is also involved in the progression of the disease (Schaefer et al. 2000). The picture with regard to diffusion-weighted imaging (DWI) studies of HSE is less clear, as hyperintense DWI abnormalities have been reported in the initial stages of the disease (Ohta et al. 1999; Kuker et al. 2004) and are occasionally accompanied by a reduction in the ADC that is likely the result of cytotoxic oedema (Sener 2001; Heiner and Demaerel 2003; McCabe et al. 2003; Duckworth et al. 2005). Serial MRIs in patients with HSV have revealed a reduction in the ADC in neocortical tissue within the first 5–6 days after the onset of the disease (Hatipoglu et al. 2008; Sawlani 2009). However, increased ADC has also been reported, which likely reflects combined vasogenic and extracellular oedema (Sener 2001; Samann et al. 2003).

The severity of the disease seems to be correlated with the development of the imaging parameters, as patients in severe stages of the disease predominantly show cytotoxic oedema, whereas patients with favourable outcomes display vasogenic oedema (Sener 2001). In terms of cortical hyperintensity in HSE, the changes in the diffusion-weighted images emerge within the first few days after the onset of disease and become stable within 14 days (Kuker et al. 2004). The analysis of diffusion changes in the hippocampus in patients with HSE is often complicated by the postictal diffusion abnormalities that frequently occur during the course of the disease (Szabo et al. 2005).

A reduction in both the mean diffusivity and the ADC is seen in the temporal lobe within the first few days following the onset of HSE, which is similar to the evolution of lesions in cerebral ischaemia. Later stages (beyond 14 days), however, were characterized by an elevated mean diffusivity (Noguchi et al. 2010). Both of these stages seem to correlate with a progression from cytotoxic oedema to inflammatory vasogenic oedema, with the latter reflecting increased permeability of the blood–brain barrier over the course of the immune response to the infection (Herweh et al. 2007). MRI showing DWI hyperintensities and ADC reductions in the initial stages likely reflects cytotoxic oedema in the affected tissue, which is similar to cerebral ischaemia (Fiebach et al. 2002). In cerebral ischaemia, cytotoxic oedema reflects the swelling of neural and glial tissue, which decreases the extracellular space but increases its tortuosity and anisotropy in the CA1 region (Shepherd et al. 2003; Hrabetova 2005; Shepherd et al. 2007).

In cases of HSE, experimental evidence suggests that cytotoxic oedema and cellular swelling may also be the result of the direct cellular intrusion and replication of viral components that compromises cellular integrity (Herweh et al. 2007). Furthermore, oxidative stress may further aggravate these cellular changes during the course of the disease.

DTI allows for the *in vivo* study of neural tissue; its histological organization and directionality (anisotropy) are measured from the full-diffusion tensor and fractional anisotropy.

DTI is more sensitive to tissue pathology that is not detectable using conventional MRI. Additional studies of recovered HSE patients have examined the microstructural integrity of their white matter using fractional anisotropy (FA), mean diffusivity (MD), and axial radial diffusivity (RD) derived from the diffusion-tensor (DT). In HSE patients, reduced fractional anisotropy has also been found in the white matter tracts contralateral to the affected side, which may serve as an aggravating factor in the memory disturbances seen in these cases (Grydeland et al. 2010).

Autoimmune limbic encephalitis

Patients with autoimmune limbic encephalitis present with subacute memory disturbances that are accompanied by various neuropsychiatric symptoms, such as disorientation, confusion, temporal lobe seizures, and behavioural abnormalities, which are indicative of impaired limbic structures. The memory impairment shows particular deficits in episodic memory. In 1960, Brierley et al. described three cases of subacute encephalitis that presented with memory disturbances and neuropsychiatric symptoms. At the postmortem examination, inflammation and cellular necrosis was noted; this finding was most striking in the hippocampus, but was also present in other temporal lobe structures (Brierley 1960). In two of the patients, lung cancer was also found. In 1968, Corsellis et al. reported three additional similar cases and coined the term limbic encephalitis, emphasizing the association between an underlying malignancy and the limbic encephalitis (Corsellis et al. 1968). Since the close of the last century, additional insight into this specific form of limbic encephalitis has been obtained by the observation of antibodies against neuronal structures in patients who presented with neoplastic-associated limbic encephalitis (Dalmau 2009). Recent postmortem studies have shown neuronal loss, T-cell infiltration, and astrogliosis in the CA1—and, to a lesser degree, the CA3 and CA4—regions; the CA2 was relatively spared (Bien et al. 2007; Malter et al. 2010). MRI usually reveals a predilection for limbic structures (e.g. the hippocampus, amygdala, hypothalamus and insular and cingulate cortices), but the hippocampus is often predominantly affected (Bien et al. 2007) (Figure 11.2). Currently, two types of autoimmune limbic encephalitis can be differentiated: paraneoplastic limbic encephalitis (PLE) and non-paraneoplastic limbic encephalitis (NPLE) (Tuzun and Dalmau 2007). PLE occurs in some patients with cancers of the lung, thymus, breast, and testes. In some cases, the disease progression can be stabilized by the resection of the tumour if it is detected. PLE is associated with the presence of onconeural (paraneoplastic) antibodies, such as anti-Hu, anti-Ma2 (anti-Ta), CRMP5/CV2, and ANNA-3. These antibodies are detectable in small-cell lung carcinomas, testicular tumours and lymphoma. In NPLE, auto-antibodies can be found without evidence of an underlying malignancy. This condition is

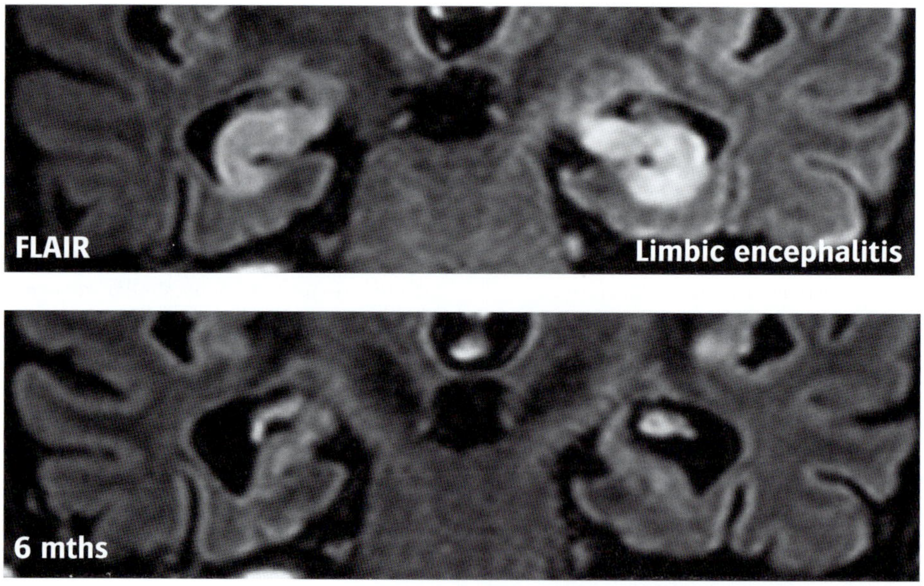

Figure 11.2 Magnetic resonance images (coronal FLAIR) depicting the highly-selective involvement of both hippocampi in subacute non-paraneoplastic autoimmune limbic encephalitis with bilateral swelling and oedema. The patient presented with memory disturbances, confusion and seizures in the initial stages of the disease. Follow-up imaging 6 months after the encephalitis showed a prominent degeneration and atrophy of both hippocampi, which corresponded to profound memory deficits in this patient as a functional sequela of the limbic encephalitis. © Images courtesy of Prof. Olav Jansen, Dept. of Neuroradiology, University Hospital Kiel, Germany.

also potentially treatable using immunosuppressive drugs (Chan et al. 2007). In LE patients, the antibodies against the neuronal surface antigens attack voltage-gated potassium channels (VGKC (LGI1- and CASPR2)), glutamic acid decarboxylase (GAD) and NMDA, AMPA, and GABA(B) receptors. In cases of seronegative limbic encephalitis, however, it is likely that the limbic encephalitis is caused by as-yet unknown anti-neuronal antibodies. For patients in whom neither antibodies nor an underlying malignancy is found, the diagnosis is based on clinical symptomatology, CSF findings and imaging.

The current model implies that antibodies produced in the course of immunological reactions to abnormal tissue (such as tumours) cross-react with healthy neuronal or non-neuronal surface tissue within the CNS. For example, antibodies against NMDARs result in the degeneration of this glutamate receptor subtype in the hippocampus, which impairs hippocampal synaptic function and leads to memory disturbances, psychiatric symptoms, and seizures (Hughes et al. 2010; Moscato et al. 2010).

Potassium channel complex antibody-associated encephalitis is the most common form of NPLE. In the majority of patients, the T2-weighted and fluid-attenuated inversion recovery (FLAIR) MRI shows temporal lobe abnormalities. The time course of MR changes in limbic encephalitis has been studied in detail by Urbach et al. (Urbach et al. 2006). MRI showed a predilection for hippocampal involvement that usually extended to the parahippocampal or fusiform gyrus and, in some patients, remote areas such as the thalamus. In some patients, the hippocampus was selectively affected. The hippocampi showed prominent swelling in the T2-weighted imaging typically seen within the first 3 months of disease onset. The DWI hyperintensities of the hippocampus and ADC changes observed were indicative of a T2-shine-through effect (Irani et al. 2010). Typically, atrophic temporal-mesial or hippocampal structures were observed in the later stages (1 year after disease onset) (Sener 2002; Urbach et al. 2006; Bien et al. 2007). Notably, the hippocampi can become successively affected (Chatzikonstantinou et al. 2009).

A more detailed analysis of memory impairment was performed by Chan et al. and Kapur et al., who showed that limbic encephalitis patients with antibodies against voltage-gated potassium channels (VGKC-Ab) had anterograde amnesia and extensive retrograde amnesia without a temporal gradient. The MRI showed MR abnormalities that were restricted to the hippocampus. These findings were considered an indication of a permanent role for the hippocampus in the consolidation process (Chan et al. 2007). By contrast, findings of only mild, temporally limited retrograde amnesia in hippocampal limbic encephalitis were taken as evidence for a time-limited role for the hippocampus in memory consolidation (Kapur and Brooks 1999).

Human herpesvirus-6 limbic encephalitis

Human herpesvirus-6 (HHV6) belongs to the family of herpes viruses and is a ubiquitous neurotropic virus that is latent in most adults. The virus is commonly inoculated within the first years of life and usually causes a non-specific febrile state. Rarely, CNS infections with HHV6 have been reported, but HHV6-associated encephalitis has recently been considered an emerging syndrome in patients with compromised immunological competency, such as transplant recipients. Similar to herpes simplex encephalitis, HHV6 encephalitis involves the structures of the temporal lobes and is sometimes difficult to identify in its initial stages, either clinically or through imaging techniques (Gorniak et al. 2006; Chamberlain and Chowdhary 2008). HHV6 encephalitis can be a complication in patients who have HIV, are undergoing allogeneic stem cell transplantation (SCT) or bone marrow transplantation (BMT), or who have received organ transplants. In MRI, patients with HHV6 encephalitis or post-transplant acute limbic encephalitis (PALE) show hyperintensities of the uncus, amygdala, entorhinal area, and hippocampus on T2, FLAIR, and DWI sequences. These hyperintensities feature involvement of the CA1 region of the hippocampus (Wainwright et al. 2001; Noguchi et al. 2006; Seeley et al. 2007; Noguchi et al. 2010). Imaging studies have shown a reduced ADC in the hippocampus in the initial stages that evolves into elevated ADC values and high signal intensity on T2-weighted images in the later stages (Wainwright et al. 2001). Atrophy of the hippocampus is typically observed as a structural sequela of the infection (Noguchi et al. 2006). Again, the progression of the reduced ADC towards elevated values in the early stages of the disease suggests a dynamic course of the diffusivity from cytotoxic oedema to a vasogenic oedema. A postmortem study of brain tissue from HHV6 encephalitis patients has shown necrotic changes in both hippocampi that were characterized by severe neuronal loss and reactive astrocytosis, predominantly in the CA1 and CA4 regions (Seeley et al. 2007). Further study has revealed bilateral subacute hippocampal sclerosis. Notably, immunoreactivity to viral HHV6 proteins in neurons and glial cells was restricted to the hippocampus, including various layers of Ammon's horn, as immunoreactivity was absent in the neocortical tissue or tissue that was shown to be unaffected by MRI (Wainwright et al. 2001; Noguchi et al. 2006). In other postmortem studies, the hippocampus has had the highest HHV6 load—in terms of an overexpression of the HHV6 viral antigen—in hippocampal neurons and astrocytes, whereas other extralimbic areas such as the cortex, thalamus, and cerebellar cortex, have been negative for the viral antigen. Notably, it has been suggested that the hippocampus is particularly vulnerable to the reactivation of HHV6, although the neuroimmunological mechanisms that underlie this tropism remain elusive (Fotheringham et al. 2007).

Epilepsy, status epilepticus, and postictal changes

Epilepsy is a neurological disorder that is characterized by recurrent seizures. Seizures are correlates of abnormal, excessive, or synchronous neuronal activity within the CNS. Diffusion-weighted MRI visualizes the degree of water movement abnormality in injured tissue. These abnormalities can result from a net shift of water from the extracellular to intracellular compartment due to reduced transmembrane permeability caused by the collapse of energy-dependant transmembrane ion gradients (Liu et al. 2001; Mascalchi et al. 2005). Accordingly, this method has been employed to gain insight into the pathological processes that occur both during and after epilepsy, including status epilepticus (Figure 11.3).

The time course of peri-ictal restricted diffusion can be divided into four stages (Wieshmann et al. 1997; Yu and Tan 2008). During both acute and active seizures, the metabolic demands of the affected cells increase dramatically and lead to an increase in cerebral blood flow without concomitant changes in the ADC. The subacute yet active stage is characterized by vasogenic extracellular oedema with a concomitant increase in the ADC (Wall et al. 2000), although various studies have not been able to detect this stage. Frequently, however, complex changes in the ADC over time—such as decreases, generalized changes and increases in the ADC—can be observed in the hippocampus from 2–210 min following a seizure, depending on the type of seizure and the latency period between the seizure and scanning (Hufnagel et al. 2003). There are limited data describing the time course of dynamic postictal ADC changes (Yu and Tan 2008; Di Bonaventura et al. 2009). Ictal or peri-ictal increases in the ADC are likely a reflection of vasogenic extracellular oedema in the hippocampus (Yu and Tan 2008). This peri-ictal stage is typically followed by a postictal stage with a decrease in the ADC that parallels that of cytotoxic intracellular oedema and lasts 1–3 days (Helpern and Huang 1995; Kim et al. 2001; Gong et al. 2008). It is thought that this cytotoxic component may reflect the consequences of intrinsic glutamate-mediated seizure activity rather than the effects of hypoxia or cerebral ischaemia. In rats with kainate-induced

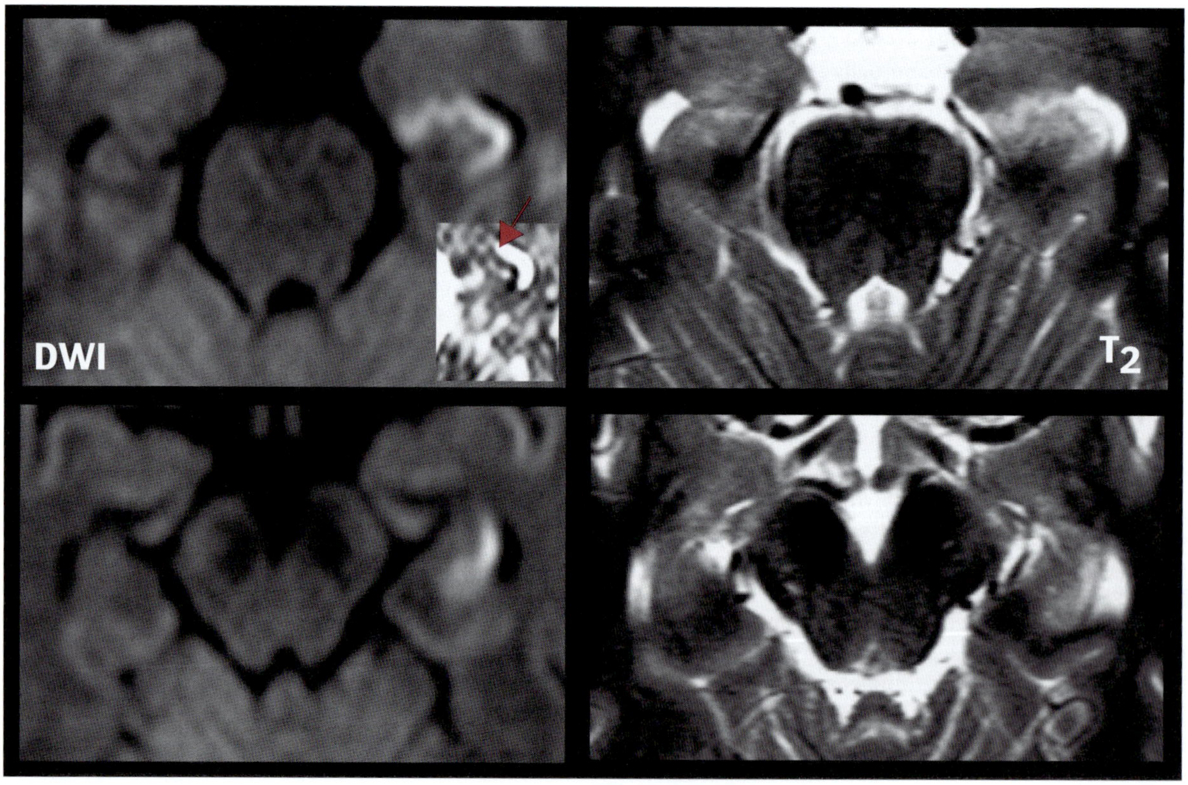

Figure 11.3 Postictal diffusion changes in a patient experiencing status epilepticus. The patient presented with a status of complex partial seizure with impaired consciousness and unresponsiveness. The axial MRI performed 72 h after the onset of the seizures shows left-sided diffusion abnormalities with a decreased ADC value (arrow in inset) and swelling of the hippocampal cortex in the CA1 region. © Images courtesy of Prof. Olav Jansen, Dept. of Neuroradiology, University Hospital Kiel, Germany.

seizures, MRI shows a decrease in the ADC of the limbic and hippocampal structures within the first 72 h (Nakasu et al. 1995; Engelhorn et al. 2007). Reduced ADC has also been observed in children experiencing new-onset prolonged seizures (Farina et al. 2004). Animal data indicate an association between the postictal ADC decrease and subsequent neuronal loss. The interictal DW changes seem to reflect neuroglial damage in the course of the seizures but do not directly reflect epileptogenicity (Engelhorn et al. 2007). In patients with complex partial status epilepticus, a decrease in the ADC within the epileptogenic hippocampal zones and subcortical areas of propagation has been observed within 3 days of seizure onset. These changes were associated with a close spatial correlation between focal hyperperfusion in areas of ADC/DWI change and increased T2 signal intensity that indicates swelling (Diehl et al. 2001; Kim et al. 2001; Szabo et al. 2005). A transient reduction in the ADC and a corresponding lactate peak in the MR spectroscopy was described in 12 patients within 48 h of acute symptomatic seizures or status epilepticus, whereas the ADC in the hippocampus was increased interictally (Parmar et al. 2006). Additional chronic changes in the ADC can be characterized as a combination of vasogenic extracellular and cytotoxic intracellular oedema (Yu and Tan 2008).

The cytotoxic cellular changes that result from cellular swelling and glutamate-induced cytotoxicity may lead to neuronal death and cellular gliosis in the hippocampus. This process may be paralleled by increased ADC values as a reflection of the increased interstitial water content (Olney et al. 1983; Sloviter 1983). Microglial activation has also been observed following status epilepticus (Avignone et al. 2008). Prolonged status epilepticus in rodents causes moderate neuronal necrosis in the cerebral cortex (layers III–IV), the hippocampus (CA1 and CA4 pyramidal cells), and the thalamus, whereas CA3 pyramidal cells and dentate neurons are spared (Nevander et al. 1984, 1985; Siesjo and Wieloch 1986; Ingvar et al. 1988). The magnitude of neuronal death is dependent on the duration of the seizure activity, as the CA3 and the dentate granule cell layers—and the caudate-putamen, which is considered to be resistant to seizure-induced cell injury—are damaged by seizure activity lasting 40 min or longer (Fujikawa 1996). A reduced ADC in the hippocampus and cortical laminar necrosis of the hippocampus was observed following complex partial status epilepticus, which suggests combined damage to neurons and glia (Heinrich et al. 2007). Severe status epilepticus may lead to neuronal death in Sommer's sector (the prosubiculum and CA1 region) and in the CA3 region (Sommer 1880; DeGiorgio et al. 1992).

Traumatic brain injury

Traumatic brain injury (TBI) is the most common form of acquired brain injury; based on the magnitude of the impact, TBI can result in neurological and/or cognitive deficits. Due to the intrinsic vulnerability of the hippocampus, structural sequelae in hippocampal circuits and deficits in hippocampus-dependant cognitive domains are often observed. Traditionally, hippocampal degeneration and neuronal loss following traumatic brain injury have been attributed to intracranial hypertension and impaired cerebral perfusion. Consistent with this hypothesis, hippocampal volume is reduced following traumatic brain injury in humans, and impaired memory is correlated with this volume reduction (Tate and Bigler 2000). Although even mild experimental traumatic injury most often involves the CA1 region, the changes that are observed do not involve changes in intracranial pressure or perfusion. Given that TBI-induced lesions can be selective for the hippocampus, it has been suggested that trauma-induced selective neuronal degeneration in the CA1 region may involve mechanisms other than those mentioned above and include glutamate-associated excitotoxic components (Kotapka et al. 1991). Indeed, an impact to the surface of the brain may result in the selective neuronal loss of hippocampal hilar neurons on both the ipsilateral and contralateral sides. However, the dentate gyrus and pyramidal neurons can be unaffected. The affected hilar neurons show abnormal hyperexcitability in response to electrical stimulation that is correlated with the extent of the neuronal loss. It has been suggested that this phenomenon of selective hippocampal vulnerability may provide a rationale for the emergence of impaired memory following TBI (Lowenstein et al. 1992). On a structural level, TBI leads to a massive deafferentation of CA1 neurons and a subsequent impaired synaptic plasticity due to a reduction in afferent inputs from the CA3 neurons. This deafferentation has the potential for regeneration, but it may also account for the acute and chronic perturbations of the hippocampal circuits that lead to clinically relevant memory impairments in these patients (Norris and Scheff 2009). Both traumatic brain injury and haemorrhagic hypotension lead to decreased expression of neuroprotective genes in the CA1 region compared to adjacent uninjured neurons (Hellmich et al. 2005). Notably, TBI induces the remodelling of hippocampal circuits by stimulating neurogenesis within the hippocampus, as brain injury leads to the proliferation and migration of quiescent progenitor cells outside of the subgranular layer of the dentate gyrus (Yu et al. 2008).

Hypoglycaemic encephalopathy

Hypoglycaemia is an abrupt decrease in serum glucose levels to less than 50 mg/dL and presents with various symptoms ranging from impaired or lost consciousness (including coma) to transient focal neurological deficits. These CNS deficits are intrinsically correlated with the time course and magnitude of the hypoglycaemic state, the presence of seizures and the age of the patients. Recent experimental neuropathology studies have demonstrated necrosis of the neurons in brain regions that are known to be vulnerable to hypoxia/ischaemia, including the cerebral cortex, striatum and hippocampus (Auer et al. 1984, 1985). Within the hippocampus, dense neuronal necrosis and calcium staining has been observed in the dentate gyrus, and selective neuronal necrosis in the CA1 region is also evident; however, the CA3 region is relatively resistant to experimental hypoglycaemia (Paschen et al. 1991). Irreversible neuronal damage from hypoglycaemia emerges rapidly within 4–12 h following the onset of hypoglycaemia (Auer et al. 1984, 1985). Notably, experimentally induced hypoglycaemia does not necessarily lead to an infarction. Clinically, comatose patients exposed to prolonged hypoglycaemia show similar patterns of neuronal necrosis without infarcts, which is consistent with the lack of cerebral acidosis in hypoglycaemia. Furthermore, in contrast to the ischaemia findings, the cerebellum is usually spared; a certain involvement of the dentate gyrus can be observed, however, which is consistent with the latter structure considered relatively resistant to hypoxia (Auer et al. 1989). Using MRI, DWI abnormalities with selectively reduced ADC in the CA1 region have been observed (Boeve et al. 1995; Holemans et al. 2001; Ko et al. 2008). Following a prolonged state of hypoglycaemia, lesions have been found in the internal capsule, basal ganglia, cerebral cortex, corona radiate, and hippocampus, whereas the cerebellum and the brain stem are relatively unaffected. The majority of lesions corresponded to the regions of reversible cytotoxic oedema as discerned by the DWI (Cho et al. 2006; Kang et al. 2010). Clearly, the development of neurological sequelae and the subsequent clinical outcome are dependent on the severity and duration of the neuroglycopenic state. Notably, young patients with type 1 diabetes, who have greater exposure to repeated episodes of severe hypoglycaemia, exhibit an increase in hippocampal volume (Hershey et al. 2010). The pathophysiological mechanisms—including selective vulnerability—are not entirely understood but are thought to involve mechanisms such as energy failure, which lead to the arrest of protein synthesis and the activation of several oxidative-stress pathways with the subsequent generation of ROS and reactive nitrogen species (RNS) (Haces et al. 2010). Moreover, a lack of glucose leads to cellular calcium influx and intracellular alkalosis, which causes selective neuronal death in the hippocampus (Auer 2004). Increased glutamate levels may exert cytotoxic effects on neuronal elements that are mirrored by cytotoxic oedema in the MRI (Kang et al. 2010).

Transient global amnesia

The syndrome of transient global amnesia (TGA) was described independently in 1956 by M. Bender and M. Guyotat and J. Courjon; however, as early as the end of the 19th century, Paul Sollier, Theodule Ribot and R. Benon described transient amnestic states as being suggestive of TGA (Ribot 1882; Sollier 1892; Benon 1909; Bender 1956; Guyotat and Courjon 1956; Pearce and Bogousslavsky 2009; Gil et al. 2010). In 1964, a decade after Bender's report, Charles Miller Fisher and Raymond D. Adams published a larger case series and introduced the term TGA (Fisher and Adams 1964). Since then, the clinical features of TGA have been well described, although its exact aetiology and pathophysiology are incompletely understood (Bartsch et al. 2006).

TGA is an amnestic syndrome that is defined by the sudden onset of antero- and retrograde amnesia that lasts up to 24 h. The clinical and neuropsychological aspects of this syndrome have been characterized in recent studies, and several aetiological factors, including migraine-related mechanisms, focal ischaemia, venous flow abnormalities and epileptic phenomena, are believed to be involved in its pathophysiology (Lewis 1998; Sander and Sander 2005; Bartsch 2007). Although the core amnestic syndrome is usually shorter than 24 h, there is general agreement that mild subclinical neuropsychological deficits with concomitant vegetative symptoms can persist for days following the episode (Quinette et al. 2003; Bartsch et al. 2006; Jager et al. 2009). Notably, some studies have suggested that a subclinical impairment of memory function can persist for several months following an acute episode (Hodges and Oxbury 1990; Gallassi et al. 1993; Borroni et al. 2004; Guillery-Girard et al. 2006). However, a recent meta-analysis that included 25 studies found no significant long-term performance differences between patients and healthy controls (Jager et al. 2009).

The diagnosis of TGA is primarily based on clinical findings and can be made only when the following diagnostic criteria established by Caplan and Hodges are fulfilled: (1) the presence of anterograde amnesia that is witnessed by an observer; (2) no clouding of consciousness or loss of personal identity; (3) cognitive impairment that is limited to amnesia; (4) no focal neurological or epileptic signs; (5) no recent history of head trauma or seizures; and (6) the resolution of symptoms within 24 h (Caplan 1985; Hodges 1990). Temporally graded retrograde amnesia is consistently present during the acute attack, although this feature is not included in the current diagnostic criteria. Mild vegetative symptoms, such as headache, nausea and/or dizziness, can also be present during the acute phase.

Pathophysiological concepts

Since its initial description, several pathophysiological mechanisms, including migraine-related mechanisms, hypoxic-ischaemic events, venous flow abnormalities, psychological mechanisms and epilepsy-related activity, have been suggested as being associated with the pathophysiology of TGA (Sander and Sander 2005). As a substantial number of patients report Valsalva-associated manoeuvres prior to the onset of TGA, recent studies have investigated the possibility of increased venous pressure leading to hippocampal venous congestion with subsequent ischaemia (Lewis 1998; Akkawi et al. 2001). Several studies have reported a higher rate of retrograde flow patterns in the jugular vein during the Valsalva manoeuvre, in combination with a higher rate of insufficient jugular vein valves, in TGA patients (Lewis 1998; Nedelmann et al. 2005; Schreiber et al. 2005; Chung et al. 2006; Akkawi et al. 2007). Using a combination of ultrasonography and magnetic resonance venography in TGA patients with and without a preceding Valsalva manoeuvre, no correlation between intracranial venous drainage patterns and jugular valve insufficiency was found, which argues against a direct relationship between jugular flow and TGA (Schreiber et al. 2005; Bettermann 2006).

TGA, the hippocampus, and stroke

Stroke-like mechanisms (that is, a haemodynamic or thromboembolic aetiology) have always been thought to be involved in the aetiology of TGA. However, several systematic case control studies have compared patients with TGA to those with transient ischaemic attacks (TIAs) and failed to find an association between the frequency of stroke risk factors and TGA, although an increased risk for TGA has been associated with a history of migraine (Hodges and Warlow 1990; Quinette et al. 2006; Enzinger et al. 2008). Compared to patients who have experienced TIA, patients with TGA have a better prognosis for future cerebrovascular events (Miller et al. 1987; Hodges and Warlow 1990; Berli et al. 2009).

TGA and migraine-related mechanisms

Pathophysiological mechanisms that link migraine (both with and without aura) with TGA have been widely discussed in the literature, and several studies have reported a higher incidence of migraine in patients with TGA (Olesen and Jorgensen 1986; Schmidtke and Ehmsen 1998; Quinette et al. 2006). In rare cases, patients have had an active migraine attack within the months prior to the TGA or exhibited migrainous features during an acute TGA episode. Interestingly, the prevalence of headache during the acute TGA episode is not associated with a history of migraine (Quinette et al. 2006). A recent study of 142 patients found a prior history of migraine to be a risk factor for TGA patients under the age of 56 years (Quinette et al. 2006). One of the key mechanisms underlying migraine pathophysiology is the phenomenon called 'cortical spreading depression' (CSD), which was first described by Leão in 1944 and reflects glutamate-mediated transient neuronal and glial depolarization that is followed by a long-lasting suppression of neuronal activity. A large body of experimental and clinical findings strongly suggests that CSD is the neurophysiological correlate of the migraine aura. CSD can also be elicited in the hippocampus, where it propagates across the cortical surface and modulates the excitability of CA1 neurons and alters the distribution of excitatory transmitter receptors, including glutamate receptors (Wernsmann et al. 2006; Haghir et al. 2009). Furthermore, cortical spreading depression is accompanied by a reduction in the ADC, which indicates decreased cellular diffusion (Bradley et al. 2002). More importantly, it has been shown in the rodent hippocampus that cortical spreading depression can lead to a hypoxic state in the CA1 field with subsequent cellular damage to the CA1 neurons (Pomper et al. 2006). Cortical spreading depression-related mechanisms in the hippocampus may thus affect CA1 neuronal function and structural integrity and—in addition to the genuine vulnerability of CA1 neurons—may represent a neurophysiological trigger for eliciting cellular metabolic changes that result in TGA (Gorji 2001).

Neuroimaging in TGA: PET and SPECT

Quantitative imaging of changes in regional cerebral glucose, oxygen metabolism and cerebrovascular blood flow in TGA has been performed using PET and SPECT. In some studies, changes in the mesiotemporal flow have been described (reviewed in Bartsch 2010). However, most studies have also described concomitantly reduced or increased CBF changes in various other anatomical structures, such as the unilateral or bilateral thalamic, prefrontal, frontal, amygdala, striatal, cerebellar, occipital, and pre- and postcentral areas. Some studies have found either no mesiotemporal changes (or no changes at all) or global hypoperfusion (Fujii et al. 1989; Eustache et al. 1997; Yamane et al. 2008; Chung et al. 2009). In summary, the imaging data that have been derived from the SPECT/PET studies are difficult to compare and interpret. Most of the changes have been found to be normalized upon follow-up examination. This variability in the PET and SPECT results is probably related to methodological differences, such as the imaging protocol and resolution and the latency of scanning. The imaging was also frequently performed days or weeks after the acute TGA and thus did not capture the initial pathophysiological event (Eustache et al. 2000).

A correlation between SPECT and MRI in the period 24–73 h after TGA onset has been examined in only three studies of seven patients. In five patients, a predominant hypoperfusion in the cerebellar vermis was observed in combination with punctate DWI lesions in the hippocampus, whereas the two other studies found bilateral hypoperfusion in the temporal lobes using SPECT (Di Filippo and Calabresi 2007; Yang et al. 2009).

Functional MRI was performed in two patients during the acute TGA episode to assess memory function and cerebral activation patterns. In both patients, a greatly reduced activation of the temporal lobe structures was associated with the encoding of visual scenes or the recognition of old scenes (LaBar et al. 2002; Westmacott et al. 2008).

MRI in TGA

Structural imaging using MRI has been used to detect abnormalities in memory-relevant structures of the mesiotemporal region. Recent data using high-resolution MRI has shown that focal hyperintense lesions that correlate with restricted diffusion in the lateral hippocampus can now be reliably detected (Figure 11.4) (Sedlaczek et al. 2004; Bartsch et al. 2006; Bartsch 2007; Lee et al. 2007; Bartsch et al. 2008; Weon et al. 2008; Yang et al. 2008). A detailed analysis of the locations of the hippocampal lesions has shown that almost all of the lesions were selectively located in the area corresponding to the CA1 sector (the Sommer sector) of the hippocampal cornu ammonis (Figure 11.5) (Sedlaczek et al. 2004; Bartsch et al. 2006). The lesions in diffusion-weighted imaging are accompanied by corresponding lesions in the T2-weighted images (as Bartsch et al., 2011). Focal hyperintense lesions range from 1–5 mm in size. Recent neuroimaging data, including our own data, have shown that the detectability of hippocampal DWI lesions in TGA patients depends on the timing of the imaging. The highest detectability is within 48–72 h of the onset of symptoms, and earlier imaging may not detect these lesions (Huber et al. 2002; Sedlaczek et al. 2004). The detectability of lesions is also dependent on the sensitivity of the MR scanner (Sedlaczek et al. 2004; Weon et al. 2008). The temporal evolution of hippocampal lesions has been studied by measuring the time course of the ADC of hippocampal DWI lesions in TGA patients using serial 3T MRI (Bartsch 2007; Weon et al. 2008). In contrast to the acute amnestic syndrome (which typically lasts no longer than 12 h), the DWI lesions evolved after 24–48 h and were detectable up to 7–10 d after the TGA episode. Correspondingly, the ADC values were at a minimum between 24–72 h and returned to normal around day 10 (Bartsch 2007). In addition, weak DWI lesions have been retrospectively detected in patients who were scanned early in the episode (within 6 h of the onset of symptoms) (Sedlaczek et al. 2004; Bartsch 2007). Although the course of the diffusion changes in TGA mimics the temporal progression of ischaemic lesions, structural sequelae in the CA1 region that are detectable with high-resolution MRI have not yet been observed (Cianfoni et al. 2005; Bartsch et al. 2006; Bartsch 2010).

In addition to DWI, which provides information regarding the structural integrity of brain tissue, magnetic resonance spectroscopy (MRS) allows assessing *in vivo* metabolic changes in cerebral structures. Complementary imaging studies combining MRI and focal MRS of CA1 DWI/T2 lesions in TGA have revealed a distinct double lactate peak (Bartsch et al. 2008). Lactate is a marker of anaerobic glycolysis and thus indicates that there is acute metabolic stress in the CA1 neurons during a TGA episode. The lactate peak was confined to the DWI lesion and could not be detected in the peri-focal tissues, suggesting that the changes in CA1 neurons indicating metabolic stress are indeed highly focal and not a reflection of a globally altered metabolic status in the hippocampus. With regard to the hippocampus, elevated lactate and reduced N-acetylaspartate (NAA) have also been observed in patients with prolonged seizures, chronic temporal lobe epilepsy and status epilepticus (Connelly et al. 1998; Lee et al. 2005; Parmar et al. 2006). Thus, these results do not preclude the possibility that the observed metabolic changes in TGA reflect a common pathway of CA1 neurons in response to acute metabolic cellular stress.

The development of hippocampal CA1 lesions in the course of a TGA episode strongly suggests a transient perturbation of the hippocampal or mesiotemporal memory circuits. Clinical and experimental data show that hippocampal CA1 neurons are critically involved in the process of memory consolidation, thereby constituting a relay function in direct and polysynaptic intrahippocampal circuits. On a cellular network level, the CA1 neurons of the cornu ammonis receive information input from the parahippocampal post-rhinal and medial entorhinal cortex (layer III) via the perforant pathways and Schaffer collaterals from the CA3 region. The postsynaptic output from the CA1 region projects via the subiculum to the deep layers of the entorhinal cortex, and lesions in the CA1 region affect the output relay function of these pathways. In humans, clinical data suggest that lesions in the hippocampus result in clinically significant memory impairment (Squire et al. 2004).

Pathophysiological mechanisms of hippocampal dysfunction in TGA

The CA1 sector (the Sommer sector) of the hippocampal cornu ammonis is selectively vulnerable to the metabolic

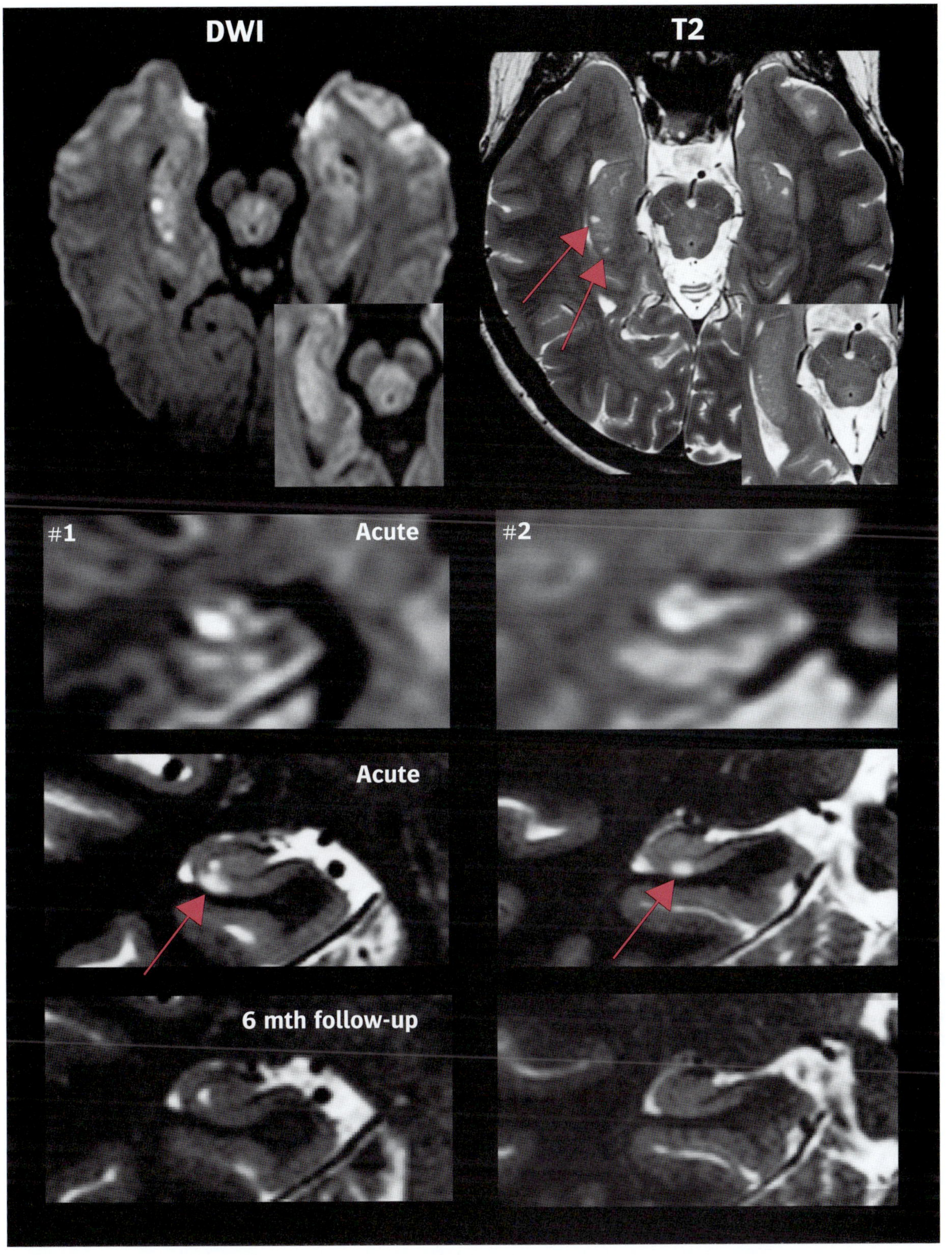

Figure 11.4 A representative 3T MRI in a patient with TGA, showing the typical lesions in the diffusion-weighted imaging and the corresponding lesions in the T2-weighted images (red arrows). Note the two unilateral lesions in the CA1 sector of the cornu ammonis in this case. The images from a 6-month follow-up showed complete resolution of the lesions (insets). To identify identical slices, the thickness of the follow-up images was 1 mm. From The American Association for the Advancement of Science, 328 (5984), Bartsch, T., et al., *Focal Lesions of Human Hippocampal CA1 Neurons in Transient Global Amnesia Impair Place Memory*, pp. 1412–5 © 2010 American Association of Science. Reprinted with permission from AAAS

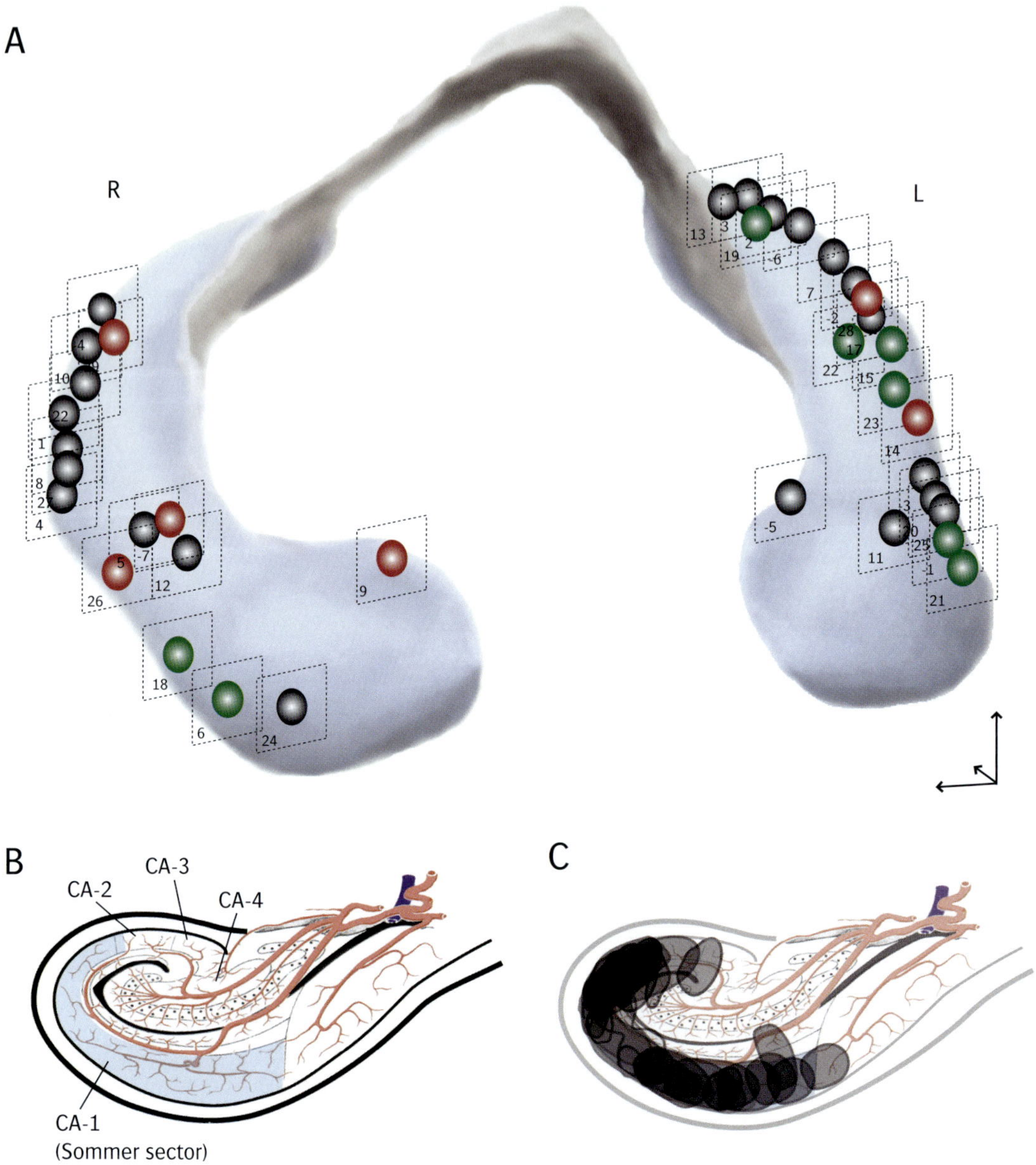

Figure 11.5 A) A model of the hippocampus showing the distribution of the MRI lesions that are found in TGA. The green spots indicate lesions with impaired verbal memory, and the red spots indicate non-verbal memory deficits. The black spots indicate normal memory testing (> 24 h). B) An anatomical template showing the various subfields of the hippocampal cornu ammonis according to Lorente de No. C) The distribution of the DWI/T2 lesions within the cornu ammonis show that the lesions are confined to the CA1 subfield (Bartsch et al. 2006). Reproduced from Bartsch, T., Alfke, K., Stingele, R., et al. (2006). Selective affection of hippocampal CA-1 neurons in patients with transient global amnesia without long-term sequelae, *Brain*, 129, 2874–84, © 2006 with permission from Oxford University Press.

and oxidative stress caused by the hypoxaemia, β-amyloid-induced neurotoxicity and ischaemia that is mediated by glutamate overload and calcium influx (Calabresi et al. 2003; Kosuge et al. 2008). A number of studies have described stressful emotional, physical and behavioural situations that precede the onset of a TGA episode (Hodges 1990; Hodges and Warlow 1990; Merriam et al. 1992; Quinette et al. 2006). In this context, recent animal data obtained from studying stress responses in the hippocampus have shown that acute emotional and behavioural stress impairs LTP and enhances long-term depression (LTD) in CA1 neurons, which leads to the disruption of hippocampus-dependent memory (Howland and Wang 2008). Acute stress may thus modulate the CA1 synaptic mechanisms that are involved in learning and memory (Kim et al. 1996; Yang et al. 2005; Howland and Wang 2008).

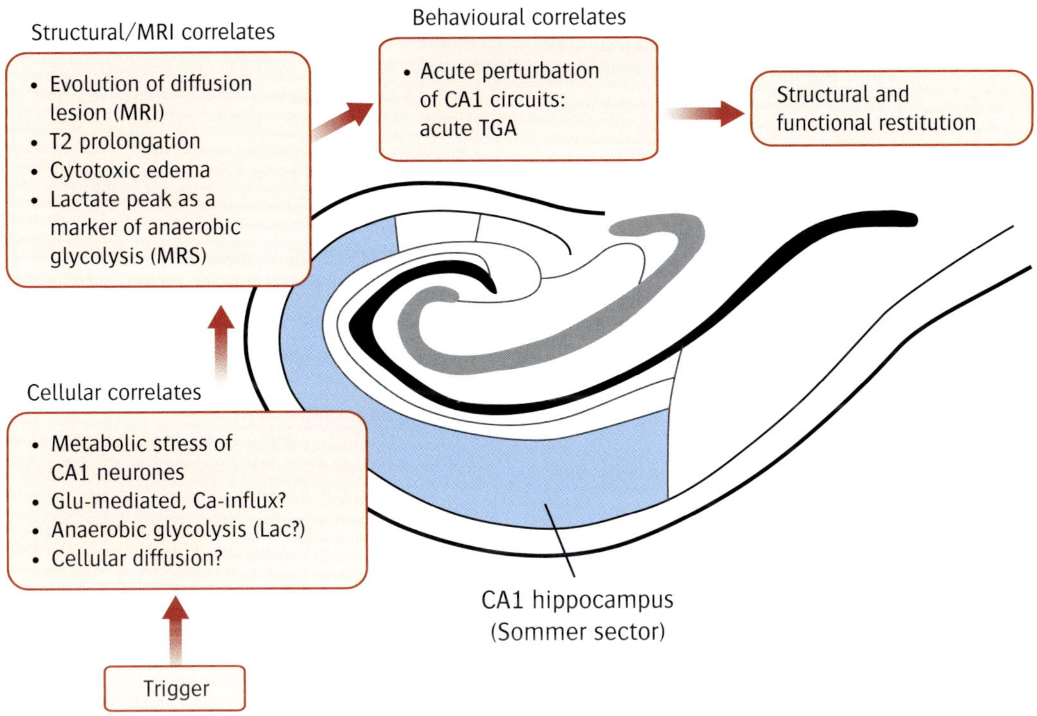

Figure 11.6 A schematic model highlighting the cellular mechanisms and pathophysiological cascade in the CA1 field of the cornu ammonis that are involved in the progression of hippocampal lesions in TGA. They may also account for other neurological diseases in which the hippocampus is affected. Reprinted from *The Lancet Neurology*, 9 (2), Bartsch, T. and Deuschl, G. (2010). Transient global amnesia: functional anatomy and clinical implications, 205–14, © (2010), with permission from Elsevier.

Enhanced glutamatergic transmission and increased Ca influx in the CA1 neurons in response to stress are mediated by increased levels of corticotropin-releasing hormone (CRH), neurosteroids, β-adrenoceptor agonists and corticosterine acting via CA1 mineralocorticoid (MR) and glucocorticoid (GR) receptors (Joëls 2009). As suggested, this enhanced calcium exposure may represent a risk factor for increased metabolic vulnerability and thus may potentially impair the structural integrity of the CA1 neurons (Joëls 2009). It is tempting to assume that these mechanisms may also be involved in a pathophysiological cascade that leads to the selective impairment of hippocampal CA1 neurons and the subsequent perturbation of memory pathways, resulting in an acute TGA (Figure 11.6). It has been suggested that the specific susceptibility of the hippocampus to behavioural stress may play also a role in the memory deficits that occur in conditions such as post-traumatic stress syndrome and depression. Given the diversity of the preceding events, however, and in light of the homogenous imaging findings, it is likely that a spectrum of triggers is capable of inducing the pathophysiological changes that occur in TGA.

Multiple sclerosis

Multiple sclerosis (MS) is an inflammatory demyelinating disease that affects the central nervous system and is commonly associated with white-matter damage. Recently, grey-matter demyelination has been recognized as an important target of the neuroinflammatory cascade that leads to neuronal damage, cell death, synaptic damage, and subsequent neurological disability (Wegner and Stadelmann 2009). A majority of MS patients exhibit cognitive dysfunction during the course of the disease, which suggests involvement of the hippocampal memory system. MRI studies have shown that bilateral demyelinating lesions in the hippocampal region account for the memory deficits in severely-affected patients (Brainin et al. 1988). A correlative analysis has shown an association between a volumetric decrease in the hippocampi of patients with MS and hippocampus-dependent memory performance (Anderson et al. 2010). Using high-resolution hippocampal subfield volumetry, decreased volume was found in the

CA1 area and the subiculum subregions. Notably, patients with MS and depressive symptoms also have smaller CA2, CA3, and DG volumes and increased cortisol levels (Gold et al. 2010). Recently, novel *in vivo* imaging techniques, such as the three-dimensional double inversion-recovery (3D-DIR) technique, have been used to visualize hippocampal lesions in MS patients (Roosendaal et al. 2008). In addition to decreased hippocampal volume, deficits in the resting-state functional connectivity between the hippocampus and anterior cingulate gyrus, thalamus and prefrontal cortex have been found in MS patients as a reflection of deficits in large-scale cognitive circuits (Roosendaal et al. 2010).

Neuropathologically, hippocampal demyelination has been thought to play a critical role in hippocampal dysfunction, as postmortem specimens frequently show demyelination in the cornu ammonis (Geurts et al. 2007). Autopsy studies have shown neuronal populations that are decreased by approximately one-third in the CA1 and CA3–2 regions. The size of the neurons was also decreased in the CA1 region and was correlated with hippocampal atrophy (Papadopoulos et al. 2009).

Experimental autoimmune encephalomyelitis (EAE) is an established animal model of multiple sclerosis that presents as an acute or chronically relapsing inflammatory and demyelinating autoimmune disease. In EAE, neurodegenerative changes (decreased CA1 pyramidal layer volume, loss of inhibitory interneurons and increased death of neurons and glia) have been observed. Diffuse demyelination in the hippocampus and behavioural deficits in a spatial learning task have also been observed (Ziehn et al. 2010). Histopathological and molecular changes in demyelinated hippocampi were recently studied by Dutta et al., who showed that demyelinated hippocampi rarely exhibit neuronal loss but rather have decreased synaptic density. Notably, proteins in the hippocampus that are indicative of axonal transport, synaptic plasticity, neuronal survival and glutamate neurotransmission were affected in contrast to demyelinated motor cortex neurons. The reduced density of the synaptophysin-positive pre-synaptic terminals in the CA1 regions and dentate gyrus is indicative of demyelination-associated synaptic pruning. These findings reveal possible dysfunction in the synaptic mechanisms that are associated with learning (such as LTP) and suggest a critical role of myelin in the normal function of neuronal subpopulations within the hippocampus (Dutta et al. 2011).

References

Akkawi, N.M., Agosti, C., Rozzini, L., Anzola, G.P., and Padovani, A. (2001). Transient global amnesia and disturbance of venous flow patterns. *Lancet*, 357, 957.

Akkawi, N.M., Agosti, C., Borroni, B., and Padovani, A. (2007). Detection of intracranial venous reflux in patients of transient global amnesia. *Neurology*, 68, 163; author reply 163.

Anderson, V.M., Fisniku, L.K., Khaleeli, Z., *et al.* (2010). Hippocampal atrophy in relapsing-remitting and primary progressive MS: a comparative study. *Mult Scler*, 16, 1083–90.

Ashton, D., Van Reempts, J., Haseldonckx, M., and Willems, R. (1989). Dorsal-ventral gradient in vulnerability of CA1 hippocampus to ischemia: a combined histological and electrophysiological study. *Brain Res*, 487, 368–72.

Auer, R.N. (2004). Hypoglycemic brain damage. *Metab Brain Dis*, 19, 169–75.

Auer, R.N., Wieloch, T., Olsson, Y., and Siesjo, B.K. (1984). The distribution of hypoglycemic brain damage. *Acta Neuropathol*, 64, 177–191.

Auer, R.N., Kalimo, H., Olsson, Y., and Siesjo, B.K. (1985). The temporal evolution of hypoglycemic brain damage. II. Light- and electron-microscopic findings in the hippocampal gyrus and subiculum of the rat. *Acta Neuropathol*, 67, 25–36.

Auer, R.N., Hugh, J., Cosgrove, E., and Curry, B. (1989). Neuropathologic findings in three cases of profound hypoglycemia. *Clin Neuropathol*, 8, 63–8.

Avignone, E., Frenguelli, B.G., and Irving, A.J. (2005). Differential responses to NMDA receptor activation in rat hippocampal interneurons and pyramidal cells may underlie enhanced pyramidal cell vulnerability. *Eur J Neurosci*, 22, 3077–90.

Avignone, E., Ulmann, L., Levavasseur, F., Rassendren, F., and Audinat, E. (2008). Status epilepticus induces a particular microglial activation state characterized by enhanced purinergic signaling. *J Neurosci*, 28, 9133–44.

Bao, X., Pal, R., Hascup, K.N., *et al.* (2009). Transgenic expression of Glud1 (glutamate dehydrogenase 1) in neurons: in vivo model of enhanced glutamate release, altered synaptic plasticity, and selective neuronal vulnerability. *J Neurosci*, 29, 13929–44.

Barth, A.M., and Mody, I. (2011). Changes in hippocampal neuronal activity during and after unilateral selective hippocampal ischemia in vivo. *J Neurosci*, 31, 851–60.

Bartsch, T., and Deuschl, G. (2010). Transient global amnesia: Functional anatomy and clinical implications. *Lancet Neurol*, 9, 205–14.

Bartsch, T., Alfke K., Stingele R., *et al.* (2006). Selective affection of hippocampal CA-1 neurons in patients with transient global amnesia without long-term sequelae. *Brain*, 129, 2874–84.

Bartsch, T., Alfke, K., Deuschl, G., and Jansen, O. (2007). Evolution of hippocampal CA-1 diffusion lesions in transient global amnesia. *Ann Neurol*, 62, 475–80.

Bartsch, T., Alfke K., Wolff S., *et al.* (2008). Focal MR spectroscopy of hippocampal CA-1 lesions in transient global amnesia. *Neurology*, 70, 1030–5.

Bartsch, T., Schönfeld, R., Müller, F. J., *et al.* (2010). Focal lesions of human hippocampal CA1 neurons in transient global amnesia impair place memory. *Science*, 328(5984), 1412–15.

Bartsch, T., Döhring, J., Rohr, A., Jansen, O., and Deuschl, G. (2011). CA1 neurons in the human hippocampus are critical for autobiographical memory, mental time travel, and autonoetic consciousness. *Proc Natl Acad Sci USA*, 108(42), 17562–7.

Bendel, O., Bueters, T., von Euler, M., *et al.* (2005). Reappearance of hippocampal CA1 neurons after ischemia is associated with recovery of learning and memory. *J Cereb Blood Flow Metab*, 25, 1586–1595.

Bender, M.B. (1956). Syndrome of isolated episode of confusion with amnesia. *Journal of the Hillside Hospital*, 5, 212–15.

Benon, R. (1909). Les ictus amnésiques dans les démences 'organiques'. *Ann Méd Psychol*, 67, 207–19.

Benson, D.F., Marsden, C.D., and Meadows, J.C. (1974). The amnesic syndrome of posterior cerebral artery occlusion. *Acta neurologica Scandinavica*, 50, 133–45.

Berli, R., Hutter, A., Waespe, W., and Bachli, E.B. (2009). Transient global amnesia – not so rare after all. *Swiss Med Wkly*, 139, 288–92.

Bettermann, K. (2006). Transient global amnesia: the continuing quest for a source. *Arch Neurol*, 63, 1336–8.

Bien, C.G., Urbach, H., Schramm, J., *et al.* (2007). Limbic encephalitis as a precipitating event in adult-onset temporal lobe epilepsy. *Neurology*, 69, 1236–44.

Boeve, B.F., Bell, D.G., and Noseworthy, J.H. (1995). Bilateral temporal lobe MRI changes in uncomplicated hypoglycemic coma. *Can J Neurol Sci*, 22, 56–8.

Bonde, C., Noraberg, J., Noer, H., and Zimmer, J. (2005). Ionotropic glutamate receptors and glutamate transporters are involved in necrotic neuronal cell death induced by oxygen-glucose deprivation of hippocampal slice cultures. *Neuroscience*, 136, 779–4.

Borroni, B., Agosti, C., Brambilla, C., *et al.* (2004). Is transient global amnesia a risk factor for amnestic mild cognitive impairment? *J Neurol*, 251, 1125–7.

Bradley, D.P., Smith, J.M., Smith, M.I., *et al.* (2002). Cortical spreading depression in the feline brain following sustained and transient stimuli studied using diffusion-weighted imaging. *J Physiol*, 544, 39–56.

Brainin, M., Goldenberg, G., Ahlers, C., *et al.* (1988). Structural brain correlates of anterograde memory deficits in multiple sclerosis. *J Neurol*, 235, 362–5.

Brierley, J., Corsellis, JA., and Hierons, RN. (1960). Subacute encephalitis of later adult life mainly affecting the limbic area. *Brain*, 83, 358–68.

Buchan, A.M., and Pulsinelli, W.A. (1990). Septo-hippocampal deafferentation protects CA1 neurons against ischemic injury. *Brain Res*, 512, 7–14.

Bueters, T., von Euler, M., Bendel, O., von Euler, G. (2008). Degeneration of newly formed CA1 neurons following global ischemia in the rat. *Exp Neurol*, 209, 114–24.

Butler, T.R., Self, R.L., and Smith, K.J., *et al.* (2010). Selective vulnerability of hippocampal cornu ammonis 1 pyramidal cells to excitotoxic insult is associated with the expression of polyamine-sensitive N-methyl-D-asparate-type glutamate receptors. *Neuroscience*, 165, 525–34.

Calabresi, P., Centonze, D., Pisani, A., Cupini, L., and Bernardi, G. (2003). Synaptic plasticity in the ischaemic brain. *Lancet Neurol*, 2, 622–9.

Caplan, L. (1985). Transient global amnesia. In Vinken, P.J., Bruyn, G.W., Klawans, H.L. (Eds.) *Handbook of Clinical Neurology*, pp. 205–18. Amsterdam: Elsevier.

Chamberlain, M.C., and Chowdhary S. (2008). Post-transplant acute limbic encephalitis: clinical features and relationship to HHV6. *Neurology*, 70, 491–2; author reply 492–3.

Chan, D., Henley, S.M., Rossor, M.N., and Warrington, E.K. (2007). Extensive and temporally ungraded retrograde amnesia in encephalitis associated with antibodies to voltage-gated potassium channels. *Arch Neurol*, 64, 404–10.

Chatzikonstantinou, A., Szabo, K., Ottomeyer, C., Kern, R., and Hennerici, M.G. (2009). Successive affection of bilateral temporomesial structures in a case of non-paraneoplastic limbic encephalitis demonstrated by serial MRI and FDG-PET. *J Neurol*, 256, 1753–5.

Cho, S.J., Minn, Y.K., Kwon, K.H. (2006). Severe hypoglycemia and vulnerability of the brain. *Arch Neurol-Chicago*, 63, 138.

Chung, C.P., Hsu, H.Y., Chao, A.C., *et al.* (2006). Detection of intracranial venous reflux in patients of transient global amnesia. *Neurology*, 66, 1873–7.

Chung, Y.A., Jeong, J., Yang, D.W., *et al.* (2009). A Tc-99m SPECT study of regional cerebral blood flow in patients with transient global amnesia. *Neuroimage*, 47, 50–5.

Cianfoni, A., Tartaglione, T., Gaudino, S., *et al.* (2005). Hippocampal magnetic resonance imaging abnormalities in transient global amnesia. *Arch Neurol*, 62, 1468–9.

Connelly, A., Van Paesschen, W., Porter, D.A., *et al.* (1998). Proton magnetic resonance spectroscopy in MRI-negative temporal lobe epilepsy. *Neurology*, 51, 61–6.

Corsellis, J.A., Goldberg, G.J., and Norton, A.R. (1968). 'Limbic encephalitis' and its association with carcinoma. *Brain*, 91, 481–96.

Crepel, V., Hammond, C., Chinestra, P., Diabira, D., and Ben-Ari, Y. (1993). A selective LTP of NMDA receptor-mediated currents induced by anoxia in CA1 hippocampal neurons. *J Neurophysiol*, 70, 2045–55.

Crepel, V., Epsztein, J., and Ben-Ari, Y. (2003). Ischemia induces short- and long-term remodeling of synaptic activity in the hippocampus. *J Cell Mol Med*, 7, 401–7.

Cronberg, T., Jensen, K., Rytter, A., and Wieloch, T. (2005). Selective sparing of hippocampal CA3 cells following in vitro ischemia is due to selective inhibition by acidosis. *Eur J Neurosci*, 22, 310–16.

Dalmau, J. (2009). Recognizing paraneoplastic limbic encephalitis. *J Clin Oncol*, 27, e230–1; author reply e232.

DeGiorgio, C.M., Tomiyasu, U., Gott, P.S., and Treiman, D.M. (1992). Hippocampal pyramidal cell loss in human status epilepticus. *Epilepsia*, 33, 23–7.

Demaerel, P., Wilms, G., Robberecht, W., *et al.* (1992). MRI of herpes simplex encephalitis. *Neuroradiology*, 34, 490–3.

Di Bonaventura, C., Bonini, F., Fattouch, J., *et al.* (2009). Diffusion-weighted magnetic resonance imaging in patients with partial status epilepticus. *Epilepsia*, 50(Suppl 1), 45–52.

Di Filippo, M., and Calabresi, P. (2007). Ischemic bilateral hippocampal dysfunction during transient global amnesia. *Neurology*, 69, 493.

Diehl, B., Najm, I., Ruggieri, P., *et al.* (2001). Postictal diffusion-weighted imaging for the localization of focal epileptic areas in temporal lobe epilepsy. *Epilepsia*, 42, 21–8.

Duckworth, J.L., Hawley, J.S., Riedy, G., and Landau, M.E. (2005). Magnetic resonance restricted diffusion resolution correlates with clinical improvement and response to treatment in herpes simplex encephalitis. *Neurocrit Care*, 3, 251–3.

Dutta, R., Chang, A., Doud, M.K., *et al.* (2011). Demyelination causes synaptic alterations in hippocampi from multiple sclerosis patients. *Ann Neurol*, 69(3), 445–54.

Engelhorn, T., Weise, J., Hammen, T., *et al.* (2007). Early diffusion-weighted MRI predicts regional neuronal damage in generalized status epilepticus in rats treated with diazepam. *Neuroscience Lett*, 417, 275–80.

Enzinger, C., Thimary, F., Kapeller, P., *et al.* (2008). Transient global amnesia: diffusion-weighted imaging lesions and cerebrovascular disease. *Stroke*, 39, 2219–25.

Eustache, F., Desgranges, B., Petit-Taboue, M.C., *et al.* (1997). Transient global amnesia: implicit/explicit memory dissociation and PET assessment of brain perfusion and oxygen metabolism in the acute stage. *J Neurol Neurosurg Psychiatry*, 63, 357–67.

Eustache, F., Desgranges, B., Aupee, A.M., Guillery, B., and Baron, J.C. (2000). Functional neuroanatomy of amnesia: positron emission tomography studies. *Microsc Res Tech*, 51, 94–100.

Farina, L., Bergqvist, C., Zimmerman, R.A., *et al.* (2004). Acute diffusion abnormalities in the hippocampus of children with new-onset seizures: the development of mesial temporal sclerosis. *Neuroradiology*, 46, 251–7.

Fiebach, J.B., Jansen, O., Schellinger, P.D., *et al.* (2002). Serial analysis of the apparent diffusion coefficient time course in human stroke. *Neuroradiology*, 44, 294–8.

Fisher, C.M., and Adams, R.D. (1964). Transient global amnesia. *Acta Neurol Scand Suppl*, 40,(Suppl 9), 1–83.

Fotheringham, J., Akhyani, N., Vortmeyer, A., *et al.* (2007). Detection of active human herpesvirus-6 infection in the brain: correlation with polymerase chain reaction detection in cerebrospinal fluid. *J Infect Dis*, 195, 450–4.

Fujii, K., Sadoshima, S., Ishitsuka, T., *et al.* (1989). Regional cerebral blood flow and metabolism in patients with transient global amnesia: a positron emission tomography study. *J Neurol Neurosurg Psychiatry*, 52, 622–30.

Fujikawa, D.G. (1996). The temporal evolution of neuronal damage from pilocarpine-induced status epilepticus. *Brain Res*, 725, 11–22.

Fujioka, M., Nishio, K., Miyamoto, S., *et al.* (2000). Hippocampal damage in the human brain after cardiac arrest. *Cerebrovasc Dis*, 10, 2–7.

Gallassi, R., Stracciari, A., Morreale, A., *et al.* (1993). Transient global amnesia: neuropsychological findings after single and multiple attacks. *Eur Neurol*, 33, 294–8.

Gao, T.M., Pulsinelli, W.A., and Xu, Z.C. (1998). Prolonged enhancement and depression of synaptic transmission in CA1 pyramidal neurons induced by transient forebrain ischemia in vivo. *Neuroscience*, 87, 371–83.

Geurts, J.J., Bo, L., Roosendaal, S.D., *et al.* (2007). Extensive hippocampal demyelination in multiple sclerosis. *J Neuropathol Exp Neurol*, 66, 819–27.

Gil, R., Abdul-Samad, F., Mathis, S., Neau, J.P. (2010). [Was there a confusion before 1950 between global transient global amnesia and psychogenic amnesia?]. *Rev Neurol (Paris)*, 166, 699–703.

Gold, S.M., Kern, K.C., O'Connor, M.F., *et al.* (2010). Smaller cornu ammonis 2-3/dentate gyrus volumes and elevated cortisol in multiple sclerosis patients with depressive symptoms. *Biol Psychiatry*, 68, 553–9.

Gong, G., Shi, F., Concha, L., Beaulieu, C., and Gross, D.W. (2008). Insights into the sequence of structural consequences of convulsive status epilepticus: a longitudinal MRI study. *Epilepsia*, 49, 1941–5.

Gorji, A. (2001). Spreading depression: a review of the clinical relevance. *Brain Res Brain Res Rev*, 38, 33–60.

Gorniak, R.J., Young, G.S., Wiese, D.E., Marty, F.M., and Schwartz, R.B. (2006). MR imaging of human herpesvirus-6-associated encephalitis in 4 patients with anterograde amnesia after allogeneic hematopoietic stem-cell transplantation. *AJNR Am J Neuroradiol*, 27, 887–91.

Graff-Radford, N.R., Tranel, D., Van Hoesen, G.W., and Brandt, J.P. (1990). Diencephalic amnesia. *Brain*, 113(Pt 1), 1–25.

Grydeland, H., Walhovd, K.B., Westlye, L.T., *et al.* (2010). Amnesia following herpes simplex encephalitis: diffusion-tensor imaging uncovers reduced integrity of normal-appearing white matter. *Radiology*, 257, 774–81.

Guillery-Girard, B., Quinette, P., Desgranges, B., *et al.* (2006). Long-term memory following transient global amnesia: an investigation of episodic and semantic memory. *Acta Neurol Scand*, 114, 329–33.

Guyotat, M., and Courjon, J. (1956). Les ictus amnésiques. *J Med Lyon*, 37, 697–701.

Haces, M.L., Montiel, T., and Massieu, L. (2010). Selective vulnerability of brain regions to oxidative stress in a non-coma model of insulin-induced hypoglycemia. *Neuroscience*, 165, 28–38.

Haghir, H., Kovac, S., Speckmann, E.J., Zilles, K., and Gorji, A. (2009). Patterns of neurotransmitter receptor distributions following cortical spreading depression. *Neuroscience*, 163(4), 1340–52.

Harry, G.J., and Lefebvre d'Hellencourt, C. (2003). Dentate gyrus: alterations that occur with hippocampal injury. *Neurotoxicology*, 24, 343–56.

Hatipoglu, H.G., Sakman, B., and Yuksel, E. (2008). Magnetic resonance and diffusion-weighted imaging findings of herpes simplex encephalitis. *Herpes*, 15, 13–17.

Heiner, L., and Demaerel, P. (2003). Diffusion-weighted MR imaging findings in a patient with herpes simplex encephalitis. *Eur J Radiol*, 45, 195–8.

Heinrich, A., Runge, U., Kirsch, M., and Khaw, A.V. (2007). A case of hippocampal laminar necrosis following complex partial status epilepticus. *Acta Neurol Scand*, 115, 425–8.

Hellmich, H.L., Garcia, J.M., Shimamura, M., *et al.* (2005). Traumatic brain injury and hemorrhagic hypotension suppress neuroprotective gene expression in injured hippocampal neurons. *Anesthesiology*, 102, 806–14.

Helpern, J.A., and Huang, N. (1995). Diffusion-weighted imaging in epilepsy. *Magn Reson Imaging*, 13, 1227–31.

Hershey, T., Perantie, D.C., Wu, J., *et al.* (2010). Hippocampal volumes in youth with type 1 diabetes. *Diabetes*, 59, 236–41.

Herweh, C., Jayachandra, M.R., Hartmann, M., *et al.* (2007). Quantitative diffusion tensor imaging in herpes simplex virus encephalitis. *J Neurovirol*, 13, 426–32.

Hodges, J.R., and Oxbury, S.M. (1990). Persistent memory impairment following transient global amnesia. *J Clin Exp Neuropsychol*, 12, 904–20.

Hodges, J.R., and Warlow C.P. (1990). The aetiology of transient global amnesia. A case-control study of 114 cases with prospective follow-up. *Brain*, 113(Pt 3), 639–57.

Hodges, J.R., and Warlow, C.P. (1990). Syndromes of transient amnesia: Towards a classification. A study of 153 cases. *J Neurol Neurosurg Psychiatry*, 53, 834–43.

Holemans, X., Dupuis, M., Misson, N., and Vanderijst, J.F. (2001). Reversible amnesia in a Type 1 diabetic patient and bilateral hippocampal lesions on magnetic resonance imaging (MRI). *Diabetic Med*, 18, 761–3.

Horn, M., and Schlote, W. (1992). Delayed neuronal death and delayed neuronal recovery in the human brain following global ischemia. *Acta Neuropathol*, 85, 79–87.

Howland, J.G., and Wang, Y.T. (2008). Synaptic plasticity in learning and memory: stress effects in the hippocampus. *Prog Brain Res*, 169, 145–58.

Hrabetova, S. (2005). Extracellular diffusion is fast and isotropic in the stratum radiatum of hippocampal CA1 region in rat brain slices. *Hippocampus*, 15, 441–50.

Huber, R., Aschoff, A.J., Ludolph, A.C., and Riepe, M.W. (2002). Transient global amnesia. Evidence against vascular ischemic etiology from diffusion weighted imaging. *J Neurol*, 249, 1520–4.

Hufnagel, A., Weber, J., Marks, S., *et al.* (2003). Brain diffusion after single seizures. *Epilepsia*, 44, 54–63.

Hughes, E.G., Peng, X., Gleichman, A.J., *et al.* (2010). Cellular and synaptic mechanisms of anti-NMDA receptor encephalitis. *J Neurosci*, 30, 5866–75.

Ikegaya, Y., and Matsuki, N. (2002). Regionally selective neurotoxicity of NMDA and colchicine is independent of hippocampal neural circuitry. *Neuroscience*, 113, 253–6.

Ingvar, M., Morgan, P.F., and Auer, R.N. (1988). The nature and timing of excitotoxic neuronal necrosis in the cerebral cortex, hippocampus and thalamus due to flurothyl-induced status epilepticus. *Acta Neuropathol*, 75, 362–9.

Irani, S.R., Bera, K., Waters, P., *et al.* (2010). N-methyl-D-aspartate antibody encephalitis: temporal progression of clinical and paraclinical observations in a predominantly non-paraneoplastic disorder of both sexes. *Brain*, 133, 1655–67.

Jager, T., Bazner H., Kliegel M., Szabo K., Hennerici, M.G. (2009). The transience and nature of cognitive impairments in transient global amnesia: a meta-analysis. *J Clin Exp Neuropsychol*, 31, 8–19.

Joëls, M. (2009). Stress, the hippocampus, and epilepsy. *Epilepsia*, 50, 586–97.

Kang, E.G., Jeon, S.J., Choi, S.S., Song, C.J., and Yu, I.K. (2010). Diffusion MR imaging of hypoglycemic encephalopathy. *AJNR Am J Neuroradiol*, 31, 559–64.

Kapur, N., and Brooks, D.J. (1999). Temporally-specific retrograde amnesia in two cases of discrete bilateral hippocampal pathology. *Hippocampus*, 9, 247–54.

Kartsounis, L.D., Rudge P., and Stevens, J.M. (1995). Bilateral lesions of CA1 and CA2 fields of the hippocampus are sufficient to cause a severe amnesic syndrome in humans. *J Neurol Neurosurg Psychiatry*, 59, 95–98.

Kawai, K., Nakagomi T., Kirino T., Tamura, A., and Kawai, N. (1998). Preconditioning in vivo ischemia inhibits anoxic long-term potentiation and functionally protects CA1 neurons in the gerbil. *J Cereb Blood Flow Metab*, 18, 288–96.

Kawasaki, K., Traynelis, S.F., and Dingledine, R. (1990). Different responses of CA1 and CA3 regions to hypoxia in rat hippocampal slice. *J Neurophysiol*, 63, 385–94.

Kim, J.A., Chung, J.I., Yoon, P.H., *et al.* (2001). Transient MR signal changes in patients with generalized tonicoclonic seizure or status epilepticus: periictal diffusion-weighted imaging. *AJNR Am J Neuroradiol*, 22, 1149–60.

Kim, J.J., Foy, M.R., and Thompson, R.F. (1996). Behavioral stress modifies hippocampal plasticity through N-methyl-D-aspartate receptor activation. *Proc Natl Acad Sci USA*, 93, 4750–53.

Kirino, T. (2000). Delayed neuronal death. *Neuropathology*, 20(Suppl), S95–7.

Kirino, T., Tamura, A., and Sano, K. (1986). A reversible type of neuronal injury following ischemia in the gerbil hippocampus. *Stroke*, 17, 455–9.

Ko, S.B., Bae, H.J., Lee, S.H., and Yoon, B.W. (2008). Teaching NeuroImage: hippocampal involvement in a patient with hypoglycemic coma. *Neurology*, 71, e63.

Konaka, K., Miyashita, K., and Naritomi, H. (2007). Changes in diffusion-weighted magnetic resonance imaging findings in the acute and subacute phases of anoxic encephalopathy. *J Stroke Cerebrovasc Dis*, 16, 82–3.

Kosuge, Y., Imai, T., Kawaguchi, M., *et al.* (2008). Subregion-specific vulnerability to endoplasmic reticulum stress-induced neurotoxicity in rat hippocampal neurons. *Neurochem Int*, 52, 1204–11.

Kotapka, M.J., Gennarelli, T.A., Graham, D.I., *et al.* (1991). Selective vulnerability of hippocampal neurons in acceleration-induced experimental head injury. *J Neurotrauma*, 8, 247–58.

Kovalenko, T., Osadchenko, I., Nikonenko, A., *et al.* (2006). Ischemia-induced modifications in hippocampal CA1 stratum radiatum excitatory synapses. *Hippocampus*, 16, 814–25.

Kreisman, N.R., Soliman, S., Gozal, D. (2000). Regional differences in hypoxic depolarization and swelling in hippocampal slices. *J Neurophysiol*, 83, 1031–8.

Kuker, W., Nagele, T., Schmidt, F., Heckl, S., and Herrlinger, U. (2004). Diffusion-weighted MRI in herpes simplex encephalitis: a report of three cases. *Neuroradiology*, 46, 122–5.

LaBar, K.S., Gitelman, D.R., Parrish, T.B., and Mesulam, M.M. (2002). Functional changes in temporal lobe activity during transient global amnesia. *Neurology*, 58, 638–41.

Lavenex, P., Sugden, S.G., Davis, R.R., Gregg, J.P., and Lavenex, P.B. (2011). Developmental regulation of gene expression and astrocytic processes may explain selective hippocampal vulnerability. *Hippocampus*, 21, 142–9.

Lee, H.Y., Kim, J.H., Weon, Y.C., *et al.* (2007). Diffusion-weighted imaging in transient global amnesia exposes the CA1 region of the hippocampus. *Neuroradiology*, 49, 481–7.

Lee, S.K., Kim, D.W., Kim, K.K., *et al.* (2005). Effect of seizure on hippocampus in mesial temporal lobe epilepsy and neocortical epilepsy: an MRS study. *Neuroradiology*, 47, 916–23.

Lewis, S.L. (1998). Aetiology of transient global amnesia. *Lancet*, 352, 397–9.

Liang, R., Pang, Z.P., Deng, P., and Xu, Z.C. (2009). Transient enhancement of inhibitory synaptic transmission in hippocampal CA1 pyramidal neurons after cerebral ischemia. *Neuroscience*, 160, 412–18.

Liu, K.F., Li, F., Tatlisumak, T., *et al.* (2001). Regional variations in the apparent diffusion coefficient and the intracellular distribution of water in rat brain during acute focal ischemia. *Stroke*, 32, 1897–905.

Lowenstein, D.H., Thomas, M.J., Smith, D.H., and McIntosh, T.K. (1992). Selective vulnerability of dentate hilar neurons following traumatic brain injury: a potential mechanistic link between head trauma and disorders of the hippocampus. *J Neurosci*, 12, 4846–53.

Malter, M.P., Helmstaedter, C., Urbach, H., Vincent, A., and Bien, C.G. (2010). Antibodies to glutamic acid decarboxylase define a form of limbic encephalitis. *Ann Neurol*, 67, 470–8.

Mascalchi, M., Filippi, M., Floris, R., *et al.* (2005). Diffusion-weighted MR of the brain: methodology and clinical application. *Radiol Med (Torino)*, 109, 155–97.

McCabe, K., Tyler, K., and Tanabe, J. (2003). Diffusion-weighted MRI abnormalities as a clue to the diagnosis of herpes simplex encephalitis. *Neurology*, 61, 1015–16.

Merriam, A.E., Wyszynski, B., and Betzler, T. (1992). Emotional arousal-induced transient global amnesia. A clue to the neural transcription of emotion? *Psychosomatics*, 33, 109–13.

Miles, D.K., and Kernie, S.G. (2008). Hypoxic-ischemic brain injury activates early hippocampal stem/progenitor cells to replace vulnerable neuroblasts. *Hippocampus*, 18, 793–806.

Miller, J.W., Petersen, R.C., Metter, E.J., Millikan, C.H., and Yanagihara, T. (1987). Transient global amnesia: clinical characteristics and prognosis. *Neurology*, 37, 733–7.

Moscato, E.H., Jain, A., Peng, X., *et al.* (2010). Mechanisms underlying autoimmune synaptic encephalitis leading to disorders of memory, behavior and cognition: insights from molecular, cellular and synaptic studies. *Eur J Neurosci*, 32, 298–309.

Munoz, D.G. (1990). The distribution of chromogranin A-like immunoreactivity in the human hippocampus coincides with the pattern of resistance to epilepsy-induced neuronal damage. *Ann Neurol*, 27, 266–75.

Munoz Maniega, S., Bastin, M.E., Armitage, P.A., *et al.* (2004). Temporal evolution of water diffusion parameters is different in grey and white matter in human ischaemic stroke. *J Neurol Neurosurg Psychiatry*, 75, 1714–18.

Nakasu, Y., Nakasu, S., Morikawa, S., *et al.* (1995). Diffusion-weighted MR in experimental sustained seizures elicited with kainic acid. *AJNR Am J Neuroradiol*, 16, 1185–92.

Nedelmann, M., Eicke, B.M., and Dieterich, M. (2005). Increased incidence of jugular valve insufficiency in patients with transient global amnesia. *J Neurol*, 252, 1482–6.

Nevander, G., Ingvar, M., Auer, R., and Siesjo, B.K. (1984). Irreversible neuronal damage after short periods of status epilepticus. *Acta Physiol Scand*, 120, 155–7.

Nevander, G., Ingvar, M., Auer, R., and Siesjo, B.K. (1985). Status epilepticus in well-oxygenated rats causes neuronal necrosis. *Annals of neurology*, 18, 281–90.

Newrzella, D., Pahlavan, P.S., Kruger, C., *et al.* (2007). The functional genome of CA1 and CA3 neurons under native conditions and in response to ischemia. *BMC Genomics*, 8, 370.

Ng, T., Graham, D.I., Adams, J.H., and Ford, I. (1989). Changes in the hippocampus and the cerebellum resulting from hypoxic insults: frequency and distribution. *Acta Neuropathol*, 78, 438–43.

Ni, B., Wu, X., Su, Y., *et al.* (1998). Transient global forebrain ischemia induces a prolonged expression of the caspase-3 mRNA in rat hippocampal CA1 pyramidal neurons. *J Cereb Blood Flow Metab*, 18, 248–56.

Nikonenko, A.G., Radenovic, L., Andjus, P.R., and Skibo, G.G. (2009). Structural Features of Ischemic Damage in the Hippocampus. *Anat Rec*, 292, 1914–21.

Nitatori, T., Sato N., Waguri S., *et al.* (1995). Delayed neuronal death in the CA1 pyramidal cell layer of the gerbil hippocampus following transient ischemia is apoptosis. *Journal of Neuroscience*, 15, 1001–11.

Noguchi, T., Mihara, F., Yoshiura, T., *et al.* (2006). MR imaging of human herpesvirus-6 encephalopathy after hematopoietic stem cell transplantation in adults. *AJNR Am J Neuroradiol*, 27, 2191–5.

Noguchi, T., Yoshiura, T., Hiwatashi, A., *et al.* (2010). CT and MRI findings of human herpesvirus 6-associated encephalopathy: comparison with findings of herpes simplex virus encephalitis. *AJR Am J Roentgenol*, 194, 754–60.

Norris, C.M., and Scheff, S.W. (2009). Recovery of afferent function and synaptic strength in hippocampal CA1 following traumatic brain injury. *Journal of neurotrauma*, 26, 2269–78.

Ohta, K., Funaki, M., Tanaka, M., and Suzuki, N. (1999). Early cerebellar involvement on diffusion-weighted magnetic resonance images in herpes simplex encephalitis. *J Neurol*, 246, 736–8.

Olesen, J., and Jorgensen M.B. (1986). Leao's spreading depression in the hippocampus explains transient global amnesia. A hypothesis. *Acta Neurol Scand*, 73, 219–20.

Olney, J.W., deGubareff, T., and Sloviter, R.S. (1983). 'Epileptic' brain damage in rats induced by sustained electrical stimulation of the perforant path. II. Ultrastructural analysis of acute hippocampal pathology. *Brain research bulletin*, 10, 699–712.

Onodera, H., Sato, G., and Kogure, K. (1986). Lesions to Schaffer collaterals prevent ischemic death of CA1 pyramidal cells. *Neuroscience letters*, 68, 169–74.

Osada, N., Kosuge, Y., Ishige, K., and Ito, Y. (2010). Characterization of neuronal and astroglial responses to ER stress in the hippocampal CA1 area in mice following transient forebrain ischemia. *Neurochem Int*, 57, 1–7.

Ott, B.R., and Saver, J.L. (1993). Unilateral amnesic stroke. Six new cases and a review of the literature. *Stroke; a journal of cerebral circulation*, 24, 1033–42.

Ouyang, Y.B., Voloboueva, L.A., Xu, L.J., and Giffard, R.G. (2007). Selective dysfunction of hippocampal CA1 astrocytes contributes to delayed neuronal damage after transient forebrain ischemia. *J Neurosci*, 27, 4253–60.

Papadopoulos, D., Dukes, S., Patel, R., *et al.* (2009). Substantial archaeocortical atrophy and neuronal loss in multiple sclerosis. *Brain Pathol*, 19, 238–53.

Parmar, H., Lim, S.H., Tan, N.C., and Lim, C.C. (2006). Acute symptomatic seizures and hippocampus damage: DWI and MRS findings. *Neurology*, 66, 1732–5.

Paschen, W., Bengtsson, F., Rohn, G., *et al.* (1991). Cerebral polyamine metabolism in reversible hypoglycemia of rat: relationship to energy metabolites and calcium. *J Neurochem*, 57, 204–15.

Patlak, C.S., Hospod, F.E., Trowbridge, S.D., and Newman, G.C. (1998). Diffusion of radiotracers in normal and ischemic brain slices. *J Cereb Blood Flow Metab*, 18, 776–802.

Pearce, J.M., and Bogousslavsky, J. (2009). 'Les ictus amnesiques' and transient global amnesia. *Eur Neurol*, 62, 188–92.

Pellegrini-Giampietro, D.E., Zukin, R.S., Bennett, M.V., Cho, S., and Pulsinelli, W.A. (1992). Switch in glutamate receptor subunit gene expression in CA1 subfield of hippocampus following global ischemia in rats. *Proc Natl Acad Sci USA*, 89, 10499–503.

Perez-Pinzon, M.A., Tao, L., and Nicholson, C. (1995). Extracellular potassium, volume fraction, and tortuosity in rat hippocampal CA1, CA3, and cortical slices during ischemia. *J Neurophysiol*, 74, 565–73.

Petito, C.K., Feldmann, E., Pulsinelli, W.A., and Plum, F. (1987). Delayed hippocampal damage in humans following cardiorespiratory arrest. *Neurology*, 37, 1281–6.

Pomper, J.K., Haack, S., Petzold, G.C., *et al.* (2006). Repetitive spreading depression-like events result in cell damage in juvenile hippocampal slice cultures maintained in normoxia. *J Neurophysiol*, 95, 355–68.

Pulsinelli, W.A., Brierley, J.B., and Plum, F. (1982). Temporal profile of neuronal damage in a model of transient forebrain ischemia. *Annals of neurology*, 11, 491–8.

Quinette, P., Guillery, B., Desgranges, B., *et al.* (2003). Working memory and executive functions in transient global amnesia. *Brain*, 126, 1917–34.

Quinette, P., Guillery-Girard, B., Dayan, J., *et al.* (2006). What does transient global amnesia really mean? Review of the literature and thorough study of 142 cases. *Brain*, 129, 1640–58.

Quintana, P., Stefano, A., Hakkoum, D., and Muller, D. (2006). Glutamate receptor changes associated with transient anoxia/hypoglycaemia in hippocampal slice cultures. *Eur J of Neurosci*, 23, 975–83.

Ribot, T. (1882). *Diseases of memory*. New York: Appleton.

Roosendaal, S.D., Hulst, H.E., and Vrenken, H., *et al.* (2010). Structural and functional hippocampal changes in multiple sclerosis patients with intact memory function. *Radiology*, 255, 595–604.

Roosendaal, S.D., Moraal, B., Vrenken, H., *et al.* (2008). In vivo MR imaging of hippocampal lesions in multiple sclerosis. *J Magn Reson Imaging*, 27, 726–31.

Samann, P.G., Schlegel, J., Muller, G., *et al.* (2003). Serial proton MR spectroscopy and diffusion imaging findings in HIV-related herpes simplex encephalitis. *AJNR Am J Neuroradiol*, 24, 2015–19.

Sander, K., and Sander, D. (2005). New insights into transient global amnesia: recent imaging and clinical findings. *Lancet Neurol*, 4, 437–44.

Sathyasaikumar, K.V., Swapna, I., Reddy, P.V., *et al.* (2007). Fulminant hepatic failure in rats induces oxidative stress differentially in cerebral cortex, cerebellum and pons medulla. *Neurochem Res*, 32, 517–24.

Sawlani, V. (2009). Diffusion-weighted imaging and apparent diffusion coefficient evaluation of herpes simplex encephalitis and Japanese encephalitis. *J Neurol Sci*, 287, 221–6.

Schaefer, P.W., Grant, P.E., and Gonzalez, R.G. (2000). Diffusion-weighted MR imaging of the brain. *Radiology*, 217, 331–45.

Schmidt-Kastner, R., and Freund, T.F. (1991). Selective vulnerability of the hippocampus in brain ischemia. *Neuroscience*, 40, 599–636.

Schmidtke, K., and Ehmsen, L. (1998). Transient global amnesia and migraine. A case control study. *Eur Neurol*, 40, 9–14.

Schreiber, S.J., Doepp, F., Klingebiel, R., and Valdueza, J.M. (2005). Internal jugular vein valve incompetence and intracranial venous anatomy in transient global amnesia. *J Neurol Neurosurg Psychiatry*, 76, 509–13.

Schreiber, S.S., and Baudry M. (1995). Selective neuronal vulnerability in the hippocampus–a role for gene expression? *Trends in neurosciences*, 18, 446–51.

Schulz, U.G., Flossmann, E., Francis, J.M., Redgrave, J.N., and Rothwell, P.M. (2007). Evolution of the diffusion-weighted signal and the apparent diffusion coefficient in the late phase after minor stroke: a follow-up study. *J Neurol*, 254, 375–83.

Schwamm, L.H., Koroshetz, W.J., Sorensen, A.G., *et al.* (1998). Time course of lesion development in patients with acute stroke: serial diffusion- and hemodynamic-weighted magnetic resonance imaging. *Stroke*, 29, 2268–76.

Sedlaczek, O., Hirsch, J.G., Grips, E., *et al.* (2004). Detection of delayed focal MR changes in the lateral hippocampus in transient global amnesia. *Neurology*, 62, 2165–70.

Seeley, W.W., Marty, F.M., Holmes, T.M., *et al.* (2007). Post-transplant acute limbic encephalitis: clinical features and relationship to HHV6. *Neurology*, 69, 156–65.

Sener, R.N. (2001). Herpes simplex encephalitis: diffusion MR imaging findings. *Comput Med Imaging Graph*, 25, 391–7.

Sener, R.N. (2002). MRI and diffusion MRI in nonparaneoplastic limbic encephalitis. *Comput Med Imaging Graph*, 26, 339–42.

Shepherd, T.M., Thelwall, P.E., Blackband, S.J., *et al.* (2003). Diffusion magnetic resonance imaging study of a rat hippocampal slice model for acute brain injury. *J Cereb Blood Flow Metab*, 23, 1461–70.

Shepherd, T.M., Ozarslan, E., Yachnis, A.T., King, M.A., and Blackband, S.J. (2007). Diffusion tensor microscopy indicates the cytoarchitectural basis for diffusion anisotropy in the human hippocampus. *AJNR Am J Neuroradiol*, 28, 958–64.

Siesjo, B.K., and Wieloch T. (1986). Epileptic brain damage: pathophysiology and neurochemical pathology. *Adv Neurol*, 44, 813–47.

Singh, P., Jain, A., and Kaur, G. (2004). Impact of hypoglycemia and diabetes on CNS: correlation of mitochondrial oxidative stress with DNA damage. *Mol Cell Biochem*, 260, 153–9.

Sloviter, R.S. (1983). 'Epileptic' brain damage in rats induced by sustained electrical stimulation of the perforant path. I. Acute electrophysiological and light microscopic studies. *Brain research bulletin*, 10, 675–97.

Smith, M.L., Auer, R.N., and Siesjo, B.K. (1984). The density and distribution of ischemic brain injury in the rat following 2–10 min of forebrain ischemia. *Acta Neuropathol*, 64, 319–32.

Sollier, P. (1892). *Les troubles de la mémoire*. Paris: Rueff.

Sommer, W. (1880). Erkrankungen des Ammonshornes als aetiologisches Prinzip der Epilepsie. *Arch Psychiatr*, 10, 631–75.

Sotak, C.H. (2004). Nuclear magnetic resonance (NMR) measurement of the apparent diffusion coefficient (ADC) of tissue water and its relationship to cell volume changes in pathological states. *Neurochem Int*, 45, 569–82.

Squire, L.R., Stark, C.E., and Clark, R.E. (2004). The medial temporal lobe. *Annu Rev Neurosci*, 27, 279–306.

Stanika, R.I., Winters, C.A., Pivovarova, N.B., and Andrews, S.B. (2010). Differential NMDA receptor-dependent calcium loading and mitochondrial dysfunction in CA1 vs. CA3 hippocampal neurons. *Neurobiol Dis*, 37, 403–11.

Sugawara, T., Fujimura, M., Morita-Fujimura, Y., Kawase, M., and Chan, P.H. (1999). Mitochondrial release of cytochrome c corresponds to the selective vulnerability of hippocampal CA1 neurons in rats after transient global cerebral ischemia. *J Neurosci*, 19, RC39.

Sun, H.S., Jackson, M.F., Martin, L.J., *et al.* (2009). Suppression of hippocampal TRPM7 protein prevents delayed neuronal death in brain ischemia. *Nat Neurosci*, 12, 1300–7.

Szabo, K., Forster, A., Jager, T., *et al.* (2009). Hippocampal lesion patterns in acute posterior cerebral artery stroke: clinical and MRI findings. *Stroke*, 40, 2042–5.

Szabo, K., Poepel, A., Pohlmann-Eden, B., *et al.* (2005). Diffusion-weighted and perfusion MRI demonstrates parenchymal changes in complex partial status epilepticus. *Brain*, 128, 1369–76.

Tabuchi, E., Endo, S., Ono, T., *et al.* (1992). Hippocampal neuronal damage after transient forebrain ischemia in monkeys. *Brain Res Bull*, 29, 685–90.

Tabuchi, E., Ono, T., Nishijo, H., Endo, S., and Kuze, S. (1995). Ischemic neuronal damage specific to monkey hippocampus: histological investigation. *Brain Res Bull*, 37, 73–87.

Takahashi, S., Higano, S., Kurihara, N., *et al.* (1997). Correlation of lesions in the hippocampal region noted on MR images with clinical features. *Eur Radiol*, 7, 281–6.

Tate, D.F., and Bigler, E.D. (2000). Fornix and hippocampal atrophy in traumatic brain injury. *Learn Mem*, 7, 442–6.

Tuzun, E., and Dalmau, J. (2007). Limbic encephalitis and variants: classification, diagnosis and treatment. *Neurologist*, 13, 261–71.

Urbach, H., Soeder, B.M., Jeub, M., *et al.* (2006). Serial MRI of limbic encephalitis. *Neuroradiology*, 48, 380–6.

Victor, M., Angevine, J.B., Jr., Mancall, E.L., Fisher, C.M. (1961). [Memory loss with lesions of hippocampal formation. Report of a case with some remarks on the anatomical basis of memory]. *Arch Neurol-Chicago*, 5, 244–63.

Wainwright, M.S., Martin, P.L., Morse, R.P., *et al.* (2001). Human herpesvirus 6 limbic encephalitis after stem cell transplantation. *Annals of neurology*, 50, 612–19.

Wall, C.J., Kendall, E.J., and Obenaus, A. (2000). Rapid alterations in diffusion-weighted images with anatomic correlates in a rodent model of status epilepticus. *AJNR Am J Neuroradiol*, 21, 1841–52.

Wang, S., Kee, N., Preston, E., and Wojtowicz, J.M. (2005). Electrophysiological correlates of neural plasticity compensating for ischemia-induced damage in the hippocampus. *Exp Brain Res*, 165, 250–60.

Wang, X., and Michaelis, E.K. (2010). Selective neuronal vulnerability to oxidative stress in the brain. *Front Aging Neurosci*, 2, 12.

Wang, X., Pal, R., Chen, X.W., *et al.* (2007). Genome-wide transcriptome profiling of region-specific vulnerability to oxidative stress in the hippocampus. *Genomics*, 90, 201–12.

Wang, X., Zaidi, A., Pal, R., *et al.* (2009). Genomic and biochemical approaches in the discovery of mechanisms for selective neuronal vulnerability to oxidative stress. *BMC Neurosci*, 10, 12.

Wang, X., Pal, R., Chen, X.W., *et al.* (2005). High intrinsic oxidative stress may underlie selective vulnerability of the hippocampal CA1 region. *Brain Res Mol Brain Res*, 140, 120–6.

Wegner, C., and Stadelmann, C. (2009). Gray matter pathology and multiple sclerosis. *Curr Neurol Neurosci Rep*, 9, 399–404.

Weon, Y.C., Kim, J.H., Lee, J.S., Kim, S.Y. (2008). Optimal diffusion-weighted imaging protocol for lesion detection in transient global amnesia. *AJNR Am J Neuroradiol*, 29, 1324–8.

Wernsmann, B., Pape, H.C., Speckmann, E.J., and Gorji, A. (2006). Effect of cortical spreading depression on synaptic transmission of rat hippocampal tissues. *Eur J Neurosci*, 23, 1103–10.

Westmacott, R., Silver, F.L., and McAndrews, M.P. (2008). Understanding medial temporal activation in memory tasks: evidence from fMRI of encoding and recognition in a case of transient global amnesia. *Hippocampus*, 18, 317–25.

Wieshmann, U.C., Symms, M.R., and Shorvon, S.D. (1997). Diffusion changes in status epilepticus. *Lancet*, 350, 493–4.

Xu, Z.C., and Pulsinelli W.A. (1994). Responses of CA1 pyramidal neurons in rat hippocampus to transient forebrain ischemia: an in vivo intracellular recording study. *Neuroscience letters*, 171, 187–191.

Xu, Z.C., and Pulsinelli, W.A. (1996). Electrophysiological changes of CA1 pyramidal neurons following transient forebrain ischemia: an in vivo intracellular recording and staining study. *J Neurophysiol*, 76, 1689–97.

Yamada, I., Kuroiwa T., Endo, S., Miyasaka, N. (2003). Temporal evolution of apparent diffusion coefficient and T2 value following transient focal cerebral ischemia in gerbils. *Acta Neurochir Suppl*, 86, 147–51.

Yamane, Y., Ishii, K., Shimizu, K., *et al.* (2008). Global cerebral hypoperfusion in a patient with transient global amnesia. *J Comput Assist Tomogr*, 32, 415–17.

Yang, C.H., Huang, C.C., and Hsu, K.S. (2005). Behavioral stress enhances hippocampal CA1 long-term depression through the blockade of the glutamate uptake. *J Neurosci*, 25, 4288–93.

Yang, Q., Tress, B.M., Barber, P.A., *et al.* (1999). Serial study of apparent diffusion coefficient and anisotropy in patients with acute stroke. *Stroke*, 30, 2382–90.

Yang, Y., Kim, S., Kim, J.H. (2008). Ischemic evidence of transient global amnesia: location of the lesion in the hippocampus. *J Clin Neurol*, 4, 59–66.

Yang, Y., Kim, J.S., Kim, S., *et al.* (2009). Cerebellar hypoperfusion during transient global amnesia: An MRI and oculographic study. *J Clin Neurol*, 5, 74–80.

Yu, J.T., and Tan, L. (2008). Diffusion-weighted magnetic resonance imaging demonstrates parenchymal pathophysiological changes in epilepsy. *Brain Res Rev*, 59, 34–41.

Yu, T.S., Zhang, G., Liebl, D.J., and Kernie, S.G. (2008). Traumatic brain injury-induced hippocampal neurogenesis requires activation of early nestin-expressing progenitors. *J Neurosci*, 28, 12901–12.

Zhao, L., Yamashima, T., Wang, X.D., *et al.* (2002). PET imaging of ischemic neuronal death in the hippocampus of living monkeys. *Hippocampus*, 12, 109–18.

Zhu, L., Saito, N., Abe, O., *et al.* (2000). Changes in the apparent diffusion coefficient of water and T2 relaxation time in gerbil hippocampus after mild ischemia. *Neuroreport*, 11, 3333–6.

Ziehn, M.O., Avedisian, A.A., Tiwari-Woodruff, S., and Voskuhl, R.R. (2010). Hippocampal CA1 atrophy and synaptic loss during experimental autoimmune encephalomyelitis, EAE. *Lab Invest*, 90, 774–86.

Zola-Morgan, S., Squire, L.R., and Amaral, D.G. (1986). Human amnesia and the medial temporal region: enduring memory impairment following a bilateral lesion limited to field CA1 of the hippocampus. *J Neurosci*, 6, 2950–67.

Zola-Morgan, S., Squire, L.R., Rempel, N.L., Clower, R.P., and Amaral, D.G. (1992). Enduring memory impairment in monkeys after ischemic damage to the hippocampus. *J Neurosci*, 12, 2582–96.

CHAPTER 12

Alzheimer's Disease

Maija Pihlajamäki and Hilkka Soininen

Introduction 223
Alzheimer's disease 224
Structural MRI 226
Functional MRI 231
Hippocampal structure and function in AD and MCI 236
Concluding remarks 237

ABSTRACT The clinical spectrum of hippocampal dysfunction encompasses a wide range of neuropsychiatric symptoms. The present chapter aims at addressing the hippocampal involvement in the most common form of memory disorders, that is Alzheimer's disease (AD), focusing on neurobiological changes revealed by structural and functional imaging. AD results in progressive brain atrophy and inevitable neurological deterioration. Routine clinical evaluation of cognitively impaired elderly subjects includes structural computed tomography or magnetic resonance imaging (MRI), and, for example, the presence of atrophy in the hippocampus and other medial temporal lobe memory structures strongly supports the diagnosis of AD. Clinical functional MRI has proved to be an interesting tool in investigating the neural correlates of cognitive impairment characteristic of AD *in vivo* and has provided novel insights into the pathognomonic alterations in the hippocampal formation and related whole-brain memory networks. In addition to clinical AD, this chapter reviews recent advances in our understanding of the neuroimaging correlates of subjects at increased risk to develop AD, such as subjects with amnestic mild cognitive impairment and cognitively intact elderly subjects carrying the apolipoprotein E ε4 allele, focusing again on the most intensively studied structure, the hippocampus. Large-scale, worldwide, multimodal imaging studies on predictors of AD are ongoing and there is great hope that imaging of the hippocampus and related memory structures would facilitate early diagnosis of AD and other dementias as well as improve treatment options of these devastating diseases in the near future.

KEYWORDS ageing, Alzheimer's disease, amyloid, atrophy, dementia, functional magnetic resonance imaging, hippocampus, magnetic resonance imaging, medial temporal lobe, memory, mild cognitive impairment, neurofibrillary tangles, volumetry, voxel-based morphometry

Introduction

Alzheimer's disease (AD) is the leading cause of dementia in the elderly worldwide, and the number of patients will increase in the coming decades. The neuropathological cascade of AD (Braak and Braak 1991) begins years, even decades, before obvious manifestation of clinical symptoms, of which, progressive loss of episodic memory is the most characteristic. Mild cognitive impairment (MCI), and in particular the amnestic subtype of MCI, is currently the most commonly used concept to describe the intermediary state between normal ageing and clinical AD (Petersen et al. 1995, 2009). It is assumed that the novel disease-modifying therapies would be most efficacious in preventing progression of AD if administered early in the course of this devastating disease (Cummings et al. 2007). Various magnetic resonance imaging (MRI) techniques combined with sophisticated data analysis methods provide promising tools in the detection of individuals at high risk for developing clinical AD. Recent revisions of the criteria for

AD and MCI strongly emphasize the use of imaging biomarkers in clinical diagnostics of these disorders (Dubois et al. 2007, 2010; Frisoni et al. 2010).

Structural MRI provides tools to investigate alterations in brain structure *in vivo*. Structural MRI studies in AD patients have reported significant atrophy of the medial temporal lobe (MTL) structures, particularly of the hippocampus, while in MCI the degree of atrophy in these structures is intermediate between that detected in healthy ageing and AD (Convit et al. 1997; De Santi et al. 2001; Dickerson et al. 2001; Du et al. 2001; Jack et al. 1997; Juottonen et al. 1999; Killiany et al. 2002; Pennanen et al. 2004; Xu et al. 2000). Several studies have shown that structural MRI estimates of tissue damage in vulnerable brain regions along the perforant pathway, such as the hippocampus, reflect underlying neuropathological changes characteristic to AD (Burton et al. 2009; Csernansky et al. 2004; Gosche et al. 2002; Jack et al. 2002; Zarow et al. 2005). There is plenty of data to support the crucial role of the hippocampus, as well as of the neighbouring entorhinal cortex, in declarative memory (Eichenbaum 2000; Hasselmo and McClelland 1999; Lavenex and Amaral 2000; Squire and Zola 1996; Suzuki 2007; Tulving and Markowitsch, 1998) and the significant role of the MTL pathology underlying the characteristic episodic memory impairment encountered in AD (Braak and Braak 1991; deToledo-Morrell et al. 2004; Gómez-Isla et al. 1996; Hyman et al. 1984; Kordower et al. 2001; Mortimer et al. 2004; Scheff et al. 2006). Because of the reasons mentioned previously, the volume of the hippocampus measured with structural MRI is increasingly used as a biomarker for AD. In addition to having an established position as a diagnostic contributor for AD, evaluation of hippocampal atrophy is a promising tool in the prediction of conversion from MCI to AD as well as one of the surrogate markers used in drug trials.

Optimally, individuals with imminent AD would be identified at a minimally symptomatic state, that is before irreversible alterations in brain structure emerge in order to leave scope for the use of potential novel treatment options. It appears likely that brain structures vulnerable to AD would show evidence of dysfunction prior to significant cell death and macroscopic structural changes. Therefore, there has been considerable interest in recent years to evaluate the usefulness of functional MRI (fMRI) and other functional imaging methods in the investigation and early detection of AD. Typically, fMRI experiments compare the MRI signal during one cognitive condition (e.g. encoding novel stimuli) to a control task (e.g. viewing familiar stimuli) or to a passive baseline condition (e.g. visual fixation). Previous fMRI studies in clinical AD patients compared to elderly control subjects have consistently reported decreased hippocampal activity during novel encoding (Dickerson et al. 2005; Golby et al. 2005; Grön and Riepe 2004; Hämäläinen et al. 2007a; Kato et al. 2001; Machulda et al. 2003; Pariente et al. 2005; Remy et al. 2005; Rombouts et al. 2000; Small et al. 1999; Sperling et al. 2003a), whereas in MCI both decreased and increased hippocampal activations have been detected (Dickerson et al. 2005; Hämäläinen et al. 2007a; Johnson et al. 2006a; Kircher et al. 2007; Machulda et al. 2003; Mandzia et al. 2009; Petrella et al. 2006; Woodard et al. 2009). To date, there are few studies evaluating the value of hippocampal fMRI in the prediction of conversion from MCI to AD, the findings of which will be discussed later.

In this chapter, we review recent advances in our understanding of the brain structural and functional MRI correlates of clinical AD, amnestic MCI, and the genetic risk states for AD in relation to normal ageing, focusing on the most intensively studied structure, the hippocampus. Structural MRI analytic methods included in this review are primarily volumetric and voxel-based morphometric (VBM) measures as well as hippocampal shape analysis. Regarding fMRI, we focus on clinically relevant studies reporting changes in task-related hippocampal activity during various memory paradigms. Other functional or molecular imaging modalities such as positron emission tomography (PET) will be commented whenever relevant in this context. Finally, a few studies relating structural and functional imaging findings of the hippocampus to each other, and the potential of these imaging techniques to predict subsequent cognitive decline from MCI to clinical AD are discussed.

Alzheimer's disease

Clinical features of AD

AD, which was originally described in 1907 by the German physician Alois Alzheimer (Alzheimer 1907; Maurer et al. 1997), is the most common form of dementia. In 2006, it was estimated that the worldwide prevalence of AD was 27 million, and researchers have predicted that the prevalence of AD will quadruple by 2050 (Brookmeyer et al. 1998). AD is a progressive neurodegenerative syndrome which typically begins roughly between the ages of 40–90 years with insidious impairment of episodic memory (i.e. memory for past personal experiences in a particular spatial and temporal context) followed by abnormalities in executive, visuospatial and language functions as well as neurobehavioral problems.

Presentation and clinical course of AD can be variable. Accordingly, AD can be heterogeneous in terms of genetic background (Bertram et al. 2007), neuropathological and neuroradiological patterns (Henry-Feugeas 2007; Jagust et al. 2008; Jellinger 2002) as well as response to treatment (Kaduszkiewicz et al. 2005). The most common form of AD seen in the elderly is often termed sporadic or late-onset AD as opposed to the relatively rare early-onset forms of the disease, which are caused by autosomal dominant mutations in genes such as the presenilin 1 and 2 and amyloid precursor protein genes (Tanzi and Bertram 2001). For late-onset AD, the main known genetic risk factor is still the apolipoprotein E (APOE) ε4 allele in chromosome 19 (Bertram et al. 2007; Saunders et al. 1993), while many other candidate genes remain under study. Neuropathologically, the disease is characterized by the accumulation of extracellular deposits of β-amyloid (Aβ) plaques, intracellular neurofibrillary tangles (NFTs) consisting of hyperphosphorylated tau protein, and brain atrophy with regional synaptic, neuronal, and axonal loss (Braak and Braak 1991).

Clinical diagnosis of AD

The diagnosis of AD relies on clinical judgement. Sets of criteria have been developed to define the disease accurately. Perhaps the most widely used criteria for defining AD were developed by the National Institute of Neurological and Communicative Disorders and Stroke/Alzheimer's Disease and Related Disorders Association (NINCDS-ADRDA; McKhann et al. 1984). As knowledge on the biological basis of AD has increased, a recent proposal for new research criteria for AD suggests that biomarkers could be used as supportive features in the diagnostics to improve the specific identification of AD (Dubois et al. 2007, 2010). According to this suggestion, the diagnosis of AD requires meeting the core criterion of significant episodic memory impairment, and at least one or more of the supportive biomarker criteria.

Neuropsychologically, episodic memory is often assessed by testing for delayed recall—that is, by asking patients to retain new visual/verbal material for a few minutes while simultaneously inhibiting active rehearsal of the studied items by a distracting task. Tests that control for an effective registration of the items to be remembered and then probe response to cueing are also sensitive indicators of impaired storage abilities and associative function of the hippocampus. As opposed to episodic memory, semantic memory is often less impaired and remote personal memories and procedural knowledge tend to be well spared until later in the course of AD. On the other hand, if impairment of episodic memory is not a prominent early complaint, potential causes of cognitive disorders other than AD should be considered. Long-term declarative memory is dependent on a number of neocortical structures, but MTL structures and the hippocampus in particular play a pivotal role in supporting episodic memory. Accordingly, hippocampal atrophy is one of the most widely documented supportive biomarkers of AD, in addition to abnormal cerebrospinal fluid (CSF) findings (low Aβ and/or increased tau or phosphotau), a specific pattern of glucose hypometabolism as indicated by [^{18}F]fluorodeoxyglucose (FDG) PET, or proven AD autosomal dominant mutation within the immediate family.

Broadly speaking, the concept of 'mild cognitive impairment' (MCI) refers to subjects with cognitive impairment beyond that expected for their age and education but who are not demented (Petersen et al. 1995, 2001, 2009; Winblad et al. 2004). During recent years, MCI has had a number of definitions. Currently widely used diagnostic criteria for the amnestic subtype of MCI are as follows: (1) memory complaint, preferably corroborated by an informant; (2) objective memory impairment; (3) normal general cognitive function; (4) intact activities of daily living; and (5) not demented (Petersen et al. 2001). Persons who present with amnestic MCI have an increased risk of developing clinical AD with an annual conversion rate of 12–15%, in contrast to 1–2% risk of conversion in age-matched healthy elderly individuals (Petersen et al. 1999). Features supporting an increased risk of rapid conversion from MCI to AD include significant hippocampal atrophy on MRI, poor performance on tests of episodic memory, and the presence of an APOE ε4 allele. Importantly, however, not all subjects with MCI progress to dementia/AD, but a high number of MCI subjects remain stable or may even revert back to normal during follow-up (Ganguli et al. 2004; Gauthier et al. 2006; Larrieu et al. 2002; Petersen 2004).

Finally, the most recent 'new lexicon' for concepts related to AD suggests that AD itself should be considered as a symptomatic clinicobiological entity that encompasses both predementia (i.e. prodromal AD) and dementia phases. In this context, in contrast to the preceding paragraph, the term 'MCI' should in fact only be used to refer to individuals who do not meet the proposed criteria for prodromal AD, in that they either have (non-amnestic) memory symptoms that are not characteristic of AD or because they are negative for biological markers of AD (Dubois et al. 2010). In this chapter, the term MCI refers, however, to the concept defined by Petersen et al. (2001) as most of the imaging and other studies discussed later in the chapter are based on this conceptual framework.

Neuropathological diagnosis of AD

A diagnosis of *definite* AD can only be done by *postmortem* neuropathological evaluation. The currently used criteria are based on the initial work of Khachaturian (1985). The neuropathological criteria were redefined in 1991 by the Consortium to Establish a Registry for Alzheimer's Disease (CERAD), providing definitions as normal and possible, probable or definite AD (Mirra et al. 1991). These criteria were based on the presence of the extracellular Aβ plaques. Another neuropathological hallmark of AD are the intracellular NFTs. Also during 1991, Braak and Braak suggested that the progression of AD could be defined by the propagation of NFTs in the brain (Braak and Braak 1991). Consequently, AD pathology was divided into six stages based on the location of NFTs in the brain: the transentorhinal stages I–II, the limbic stages III–IV and the isocortical stages V–VI. In other words, deposition of the NFTs appears to begin in the MTL, in the transentorhinal and entorhinal cortices with predilection for layers 2 and 5, spreading next to the hippocampal regions CA1 and subiculum, and further to the temporal and other neocortical regions (Delacourte et al. 1999). In contrast, deposition of Aβ is initially mainly seen in the cortex, then in the hippocampus and later in other brain regions (Thal et al. 2002). Within the hippocampal formation, Aβ accumulation is often modest and located in the CA1 and subicular regions as well as in the molecular layer of dentate gyrus.

Clinically, AD is staged as mild, moderate, or severe according to the degree of difficulty of the symptoms. The clinical status of AD has been linked to the Braak stages (see Figure 12.1). The transentorhinal stage is considered to represent clinically silent cases, the limbic stage corresponds to incipient AD, and by the isocortical stage a fully developed clinical AD is present (Braak and Braak 1995). In addition to disease severity, the number of NFTs and synapse or neuron loss are known to correlate with the degree of global cognitive impairment. More recently, the National Institute on Aging and Reagan Institute (NIA-R) working group developed consensus criteria for the neuropathological assessment of AD (The National Institute on Aging, and Reagan Institute working group on diagnostic criteria for the neuropathological assessment of Alzheimer's disease, 1997). These criteria recommend that both the CERAD assessment for Aβ pathology, as well as the neuropathological staging according to the Braak and Braak criteria should be performed during the diagnostic process.

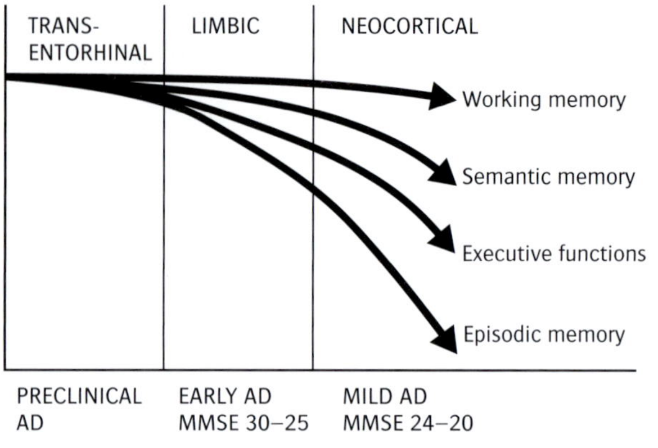

Figure 12.1 Clinical and neuropathological stages of Alzheimer's disease (AD) and roughly corresponding Mini-Mental State Examination (MMSE) scores.

Recent concepts on the pathological cascade of AD suggest that the earliest AD-related neuropathological changes may appear in the hippocampus and other MTL substructures years, or even decades prior to the manifestation of full-blown AD (Ohm et al. 1995; Selkoe 2001; Selkoe 2002). On the other hand, similar neuropathological changes can sometimes be detected even in cognitively normal elderly subjects, which reflects the problem of studying 'healthy' ageing. In subjects with amnestic MCI, a number of neuropathological and imaging studies have demonstrated that 60% or more of MCI cases already present considerable neuropathological alterations, in fact often enough to qualify for diagnosis of AD (Bennett et al. 2005; Davis et al. 1999; DeKosky et al. 2002; Markesbery et al. 2006; Morris and Price 2001; Tiraboschi et al. 2000).

Structural MRI

Structural MRI in AD and MCI

A structural imaging study, either an MRI or a computed tomography (CT) scan of the brain should be performed at least once during the diagnostic process of AD. MRI, with its excellent tissue contrast and spatial resolution ≤ 1 mm^3, allows better discrimination of mildly affected AD patients from normal subjects or patients at early stages of other neurodegenerative diseases than is possible with CT. In general, brain/hippocampal atrophy can be estimated through procedures with varying levels of human input, such as visual rating, manual labelling or volumetry (see Figure 12.2) of the structures of interest, or by automated algorithms on a voxel-by-voxel basis and by using digital anatomic atlases. In an AD patient, typical findings on a high-resolution T1-weighted MRI scan include hippocampal atrophy with variable amounts of cortical atrophy, which is often prominent in the temporal and parietal cortices.

The hippocampus and neighbouring MTL structures have been the focus of much research due to their important role in memory and in the development of AD. The MTL is known to be affected by the neuropathological hallmarks of AD such as NFTs, selective neuronal loss and synaptic alterations very early in the course of the disease (Braak and Braak 1991; Gómez-Isla et al. 1996, 1997; Hyman et al. 1984; Kordower et al. 2001; Scheff et al. 2006, 2007). In agreement with the hierarchical progression of neurofibrillary pathology, the first AD-related changes on structural MRI are usually detected in the MTL memory structures. Several studies have used MRI to measure the volume of MTL structures, especially the hippocampus, in AD patients (Convit et al. 1997; De Santi et al. 2001;

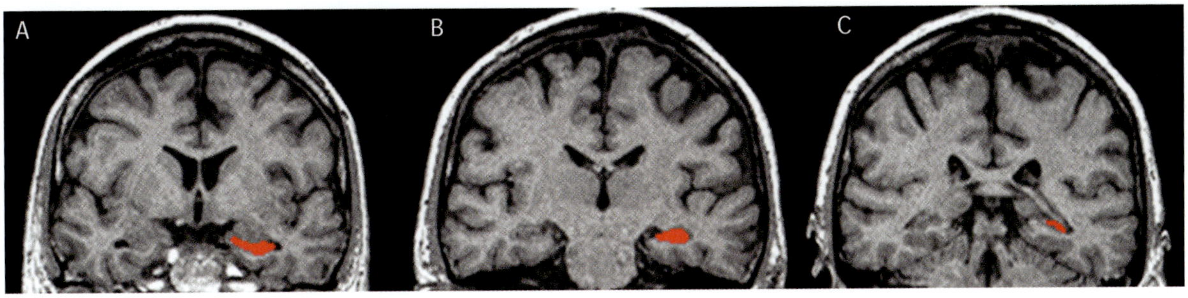

Figure 12.2 Example high-resolution T1-weighted coronal MRI slices demonstrating one of the commonly used manual hippocampal volumetric procedures (Laakso et al. 1996; Juottonen et al. 1999; Pennanen et al. 2004) at the level of the anterior (A), middle (B), and posterior (C) regions of the hippocampus, where the slices A and C demonstrate, correspondingly, the most anterior and final slices to be included in the calculation of the total three-dimensional volume.

Deweer et al. 1995; Dickerson et al. 2001; Du et al. 2001; Jack et al. 1997, 2002; Jagust et al. 2008; Juottonen et al. 1999; Killiany et al. 2002; Laakso et al. 1996; Pennanen et al. 2004; van de Pol et al. 2006; Wolf et al. 2001; Xu et al. 2000). Hippocampal atrophy as extensive as 40% of the volume has been reported in clinical AD patients, whereas in amnestic MCI the hippocampal volume is typically reduced by 10-15% compared to elderly controls (Bobinski et al. 1998; Convit et al. 1993; de Leon et al. 1989; de Leon et al. 1993, 1997; Dickerson et al. 2001; Du et al. 2001; Frisoni et al. 1999; Jack et al. 1992; Seab et al. 1988; Shi et al. 2009; van de Pol et al. 2006; van der Flier 2005). Alternatively, quick and easy visual rating scales applied to high-resolution coronal T1-weighted MR images can be used for routine clinical purposes to evaluate hippocampal atrophy in AD patients (Scheltens et al. 1992).

Notably, previous volumetric MRI studies have reported atrophy of MTL structures, such as the hippocampus and entorhinal and parahippocampal cortices preclinically, in subjects with ageing-associated cognitive decline, MCI or 'incipient AD' compared to cognitively intact elderly controls (De Santi, 2001; Dickerson, 2001; Du, 2001; Jack, 1999; Killiany, 2000, 2002; Pantel, 2003). A characteristic structural finding in MCI is atrophy of the entorhinal cortex, often accompanied by hippocampal atrophy (Becker et al. 2006; De Santi et al. 2001; Dickerson et al. 2001; Du et al. 2001; Jack et al. 1999; Killiany et al. 2002; Pennanen et al. 2004; Wolf et al. 2004). A cross-sectional volumetric MRI study in 65 MCI subjects relative to 59 elderly controls and 48 AD patients provided evidence that during the course of AD, hippocampal atrophy is preceded by entorhinal atrophy (Pennanen et al. 2004). In that study, the most efficient overall clinical classification between elderly controls and MCI subjects was achieved using entorhinal volumes, whereas hippocampal volume was better than entorhinal volume in discriminating individuals with MCI from AD patients. In other words, entorhinal volume loss was the dominant structural feature at the stage of MCI, followed by pronounced macroscopic hippocampal atrophy in patients with clinical AD (see Figures 12.3 and 12.4).

Serial structural MRI scans over time within individual subjects have demonstrated that shrinkage of the hippocampus as well as temporoparietal cortices occurs at a much faster rate in individual MCI and AD patients than in normal ageing (DeCarli et al. 2007; Jack et al. 1999; Scahill et al. 2002). In mild AD, for example, rates of hippocampal atrophy are estimated to be 3–6% per year, compared to approximately 0.5–2% per year in normal ageing. Interestingly, across the course of AD, hippocampal atrophy is not a linear phenomenon but is reported to accelerate significantly both at the transitional phase from normality to cognitive impairment and from MCI to clinical AD dementia, with a possible deceleration between these transitions (Fotenos et al. 2005; Jack et al. 2008). Rate of hippocampal atrophy in also known to correlate with change in cognitive performance, suggesting validity of structural MRI measures as markers of progression over a wide range of disease severity (Jack et al. 2004; Morra et al. 2009; Ridha et al. 2008; Thompson et al. 2004).

Recently, several different new methods have been developed to evaluate not only the quantitative volume but also the 3-dimensional shape and subfields of the hippocampus in the presence of AD and MCI relative to controls, and also in prediction of future conversion from MCI to AD (Apostolova et al. 2006, 2010; Csernansky et al. 2005; Gerardin et al. 2009; Li et al. 2007; Qiu et al. 2009; Scher et al. 2007; Wang et al. 2003).

Methods enabling reliable automatic discrimination between AD, MCI, and normal ageing—or, between AD and other brain pathologies—would be of great clinical value. Findings of these studies have suggested that, among several brain structures, surface inward-deformation (i.e. atrophy) in MCI and AD is most prominent in the anterior hippocampal regions, the other significant region being basolateral amygdala (Qiu et al. 2009). Within the hippocampal formation, notable inward deformation has been detected on the lateral side of the hippocampus possibly corresponding to volume loss in the CA1 subdivision, consistent with earlier neuropathological findings (Bobinski et al. 1995, 1998; Scheff et al. 2006, 2007; West et al. 1994, 2000, 2004; Zarow et al. 2005).

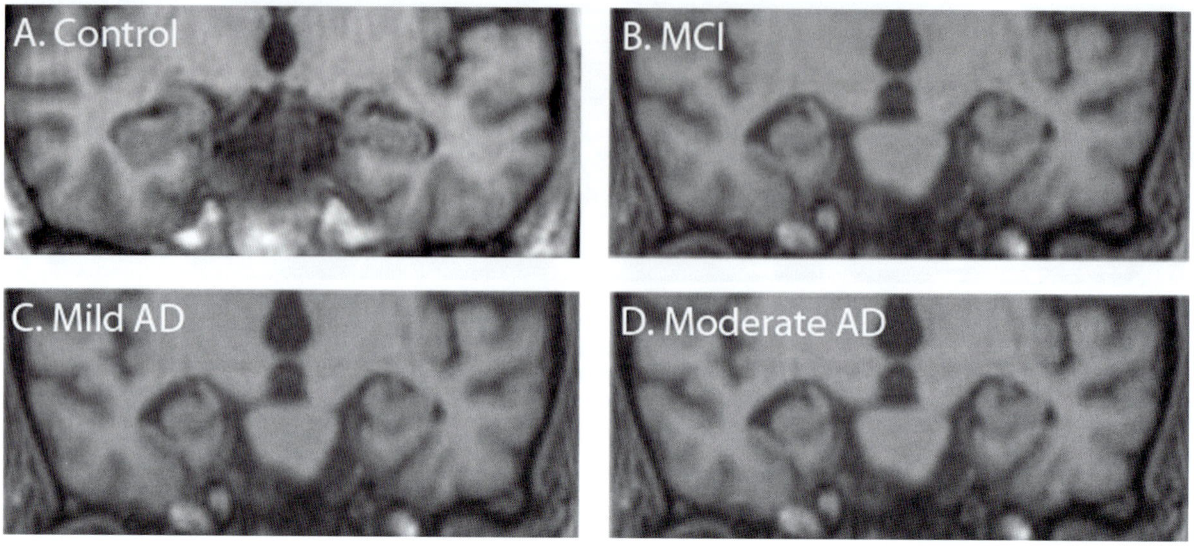

Figure 12.3 A) A coronal T1-weighted MRI scan of a cognitively normal older individual. B) MR imaging of the brain of a subject with mild cognitive impairment (MCI), and patients with mild (C) and moderate (D) Alzheimer's disease (AD) reveals symmetrical, progressive hippocampal atrophy, and also more generalized cortical atrophy, along the advancing course of the disease. In MCI, atrophy of the hippocampus is on average greater than in normal control subjects, however, overlap with cognitively normal elderly subjects on a single initial MRI scan is notable.

Discrimination accuracy between AD patients, MCI subjects and elderly controls based on multidimensional classification of hippocampal shape features has been reported to be superior to that of hippocampal volumetry (Gerardin et al. 2009). Accordingly, alterations in the shape and volume of the hippocampal CA1 region have been suggested to be potential novel tools to assist in the early diagnosis of AD (Mueller et al. 2010; Scher et al. 2007).

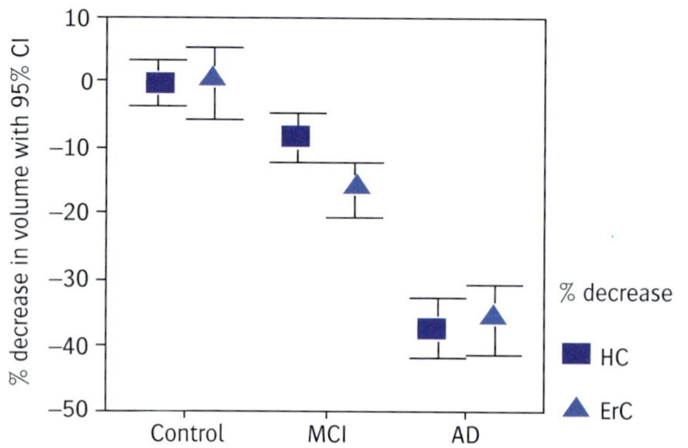

Figure 12.4 Mean decrease of hippocampal (HC) and entorhinal cortical (ErC) volumes in patients with mild cognitive impairment (MCI) and Alzheimer's disease (AD) in percentages (%) (Data from: Pennanen et al. 2004). Cognitive changes typically seen in amnestic MCI and patients with mild AD often present the classical 'hippocampal pattern' of recent memory loss.

Structural MRI in ageing and APOE ε4 carriers

Although the volume of the hippocampus may decline to some extent during normal ageing, the difference between ageing and AD is that the neuronal loss observed in the ageing MTL is not as severe as in clinical AD (Drachman, 2006). Findings of hippocampal atrophy in relation to ageing are somewhat controversial. Some studies have reported that the MTL is well preserved during ageing (Good et al. 2001), whereas others have shown age-related shrinkage in the MTL structures (Convit et al. 1995; Du et al. 2006; Jack et al. 1997; Mueller and Weiner 2009; Pruessner et al. 2001; Scahill et al. 2003; Schuff et al. 1999; Tisserand et al. 2004). Longitudinal MRI studies have, however, provided evidence that hippocampal atrophy increases along with ageing, particularly in the head of the hippocampus and subiculum (Mueller et al. 1998; Scahill et al. 2003; Wang et al. 2003). Inconsistent findings in previous studies on hippocampal atrophy in ageing may partly be explained by the fact that there is large interindividual variability in the hippocampal volume across the adult life span (Lupien et al. 2007), and partly by the possibility that some of the 'healthy' elderly subjects may in fact already have on-going early stage pathological processes.

Regarding the impact of the APOE ε4 allele—the most important genetic risk factor for late-onset AD—there is evidence of increased MTL atrophy in healthy APOE ε4 carriers in comparison to non-carriers both in cross-sectional and longitudinal studies (Cohen et al. 2001; den Heijer et al. 2002; Geroldi et al. 1999; Moffat et al. 2000;

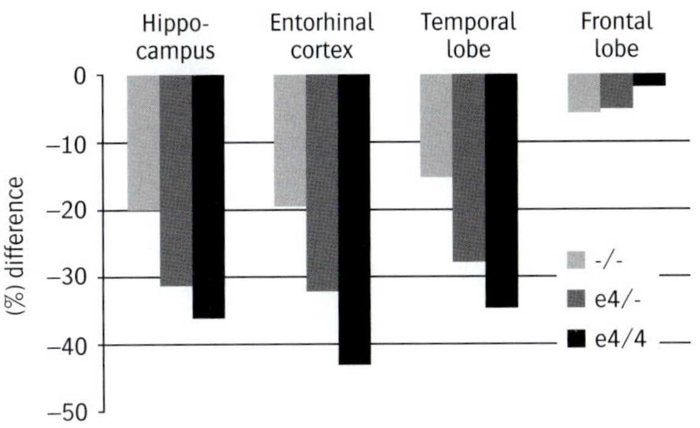

Figure 12.5 The APOE ε4 allele is associated, in a dose-dependent manner, with less frontal and more medial temporal lobe atrophy even in patients with clinical Alzheimer's disease (AD). AD patients carrying one or two APOE epsilon 4 alleles (e4 and e4/4, correspondingly) demonstrate pronounced atrophy of the hippocampus and entorhinal cortex, which is in agreement with the more severe memory impairment observed in the ε4 carriers. Difference in volume in percentages (%) is presented in relation to the corresponding regional volumes in cognitively normal control subjects (Data from: Geroldi et al. 1999).

Mueller and Weiner 2009) as well as decreased asymmetry of hippocampal volumes in nondemented elderly carriers (Soininen et al. 1995). In MCI, the the APOE ε4 allele is also associated with increased atrophy of the hippocampus and/or other MTL structures (Farlow et al. 2004; Fleisher et al. 2005; Pennanen et al. 2006). Even in clinical AD patients, the APOE ε4 genotype may increase the amount of hippocampal atrophy in a dose-dependent manner (Lehtovirta et al. 1995; Geroldi et al. 1999; see Figure 12.5). In addition to volumetric measurements, voxel-based studies have observed APOE ε4- related MTL atrophy in AD patients (Boccardi et al. 2004).

In summary, the APOE ε4 genotype seems to predispose patients to vulnerability of the hippocampus and related MTL structures as well as memory loss. Interestingly enough, patients negative for the APOE ε4 allele are more likely to present atrophy of brain areas outside the MTL such as frontal and temporoparietal cortices. APOE ε4 seems to modulate regional vulnerability of the hippocampus and other brain regions to AD pathology and macroscopic atrophy (van der Flier et al. 2011).

Structural MRI in prediction of AD

As reviewed earlier, studies in patients with early AD and in subjects at either cognitive or genetic risk for AD have demonstrated that hippocampal, as well as entorhinal volumes typically are smaller than those measured in control subjects. Hippocampal atrophy is also correlated with the degree of memory impairment (Grundman et al. 2003; Mortimer et al. 2004). Since some of the MCI subjects will remain stable and some will progress to dementia with time, great interest has been focused on attempts to identify the features predicting future conversion from MCI to clinical AD. Several studies have demonstrated the potential of baseline MRI hippocampal volume to predict conversion during follow-up (Apostolova et al. 2006, 2010; Csernansky et al. 2005; DeCarli et al. 2007; den Heijer et al. 2006; deToledo-Morrell et al. 2004; Devanand et al. 2007; Fleisher et al. 2008; Geroldi et al. 2006; Grundman et al. 2002; Jack et al. 1999, 2005; Korf et al. 2004; Tapiola et al. 2008; Visser et al. 2002).

In addition to volumetric measures, recent MRI studies utilizing whole-brain VBM or cortical thickness (CTH) analysis have confirmed the importance of hippocampal/MTL atrophy in predicting subsequent cognitive decline. Whole-brain structural MRI studies have also provided evidence of the significance of brain areas outside the MTL in evaluating the likelihood of future conversion to clinical dementia. In the recent VBM and CTH studies (see Figure 12.6), the most significant differences in structural imaging between the stable and progressive MCI subjects were observed in the posteromedial parietal and prefrontal cortices, patients later progressing to dementia showing prominent atrophy in these brain regions already at baseline (Bakkour et al. 2009; Bozzali et al. 2006; Chetelat et al. 2002, 2005; Hämäläinen et al. 2007b; Julkunen et al. 2009, 2010; Tisserand et al. 2004).

A recent MRI study (Liu et al. 2010) analysed volumes of 24 cortical regions and thicknesses of 34 cortical regions aiming to define predictors of conversion in 100 amnestic MCI subjects, 118 AD patients, and 94 age-matched healthy controls using automated image processing softwares applied to baseline images. Based on cognitive testing at baseline and a year later, 21 subjects converted from MCI to AD. In this study, in addition to the hippocampus, amygdala and caudate volumes were significantly smaller in progressive MCI subjects than in controls and stable MCI subjects. The authors concluded that hippocampal volume measures, together with a few other regional volumes, were a powerful tool in identifying AD patients at the very earliest stage of the disease (Liu et al. 2010)

When investigating features of hippocampal atrophy more specifically, there is some evidence that the shrinkage of the CA1 subfield in particular may correspond to conversion (Apostolova et al. 2006, 2010; Csernansky et al. 2005; Scher et al. 2007). Additionally, it has been demonstrated using serial MRI—instead of one-time baseline imaging—that the hippocampal atrophy rate increases in relation to the cognitive decline and further progression to AD (Fotenos et al. 2005; Jack et al. 2004, 2005, 2008; Rusinek et al. 2003, 2004; van de Pol et al. 2007).

From these studies, it can be concluded that analysis of the hippocampal total volume, shape and subfields looks promising in helping to identify a group among the MCI subjects, which is at high risk for developing AD. Structural alterations in the hippocampus seem to be an excellent topographical marker linking the molecular pathology of AD to the clinical and cognitive decline.

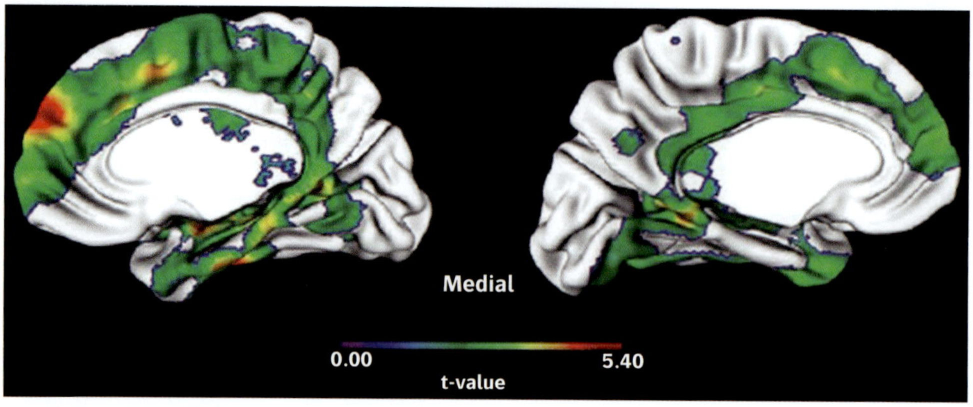

Figure 12.6 Hippocampal, medial temporal, and other cortical brain regions demonstrating significantly ($p < 0.05$, false discovery rate (FDR)-corrected) reduced cortical thickness at baseline MRI scanning in the progressive compared to stable MCI subjects. Reproduced from Julkunen, V., Niskanen, E., Muehlboeck, S., *et al.* (2009). Cortical thickness analysis to detect progressive mild cognitive impairment: a reference to Alzheimer's disease. *Dementia and Geriatric Cognitive Disorders*, 28, 404–12 with permission from S. Karger AG. Differences are illustrated with colour-labelled t-values scaling from 2.2 to the maximum $t = 5.4$.

Hippocampal atrophy and cortical alterations in AD

In addition to hippocampal atrophy, another characteristic imaging finding in patients with early AD is hypometabolism of the posteromedial association cortices as consistently demonstrated in FDG-PET studies (Herholz et al. 2002; Minoshima et al. 1997; Mosconi et al. 2008b; Rapoport 1991), FDG-PET being a measure of energy utilization in the brain that predominantly marks synaptic activity. Given the strong anatomic connectivity between the hippocampus, other MTL structures and the posteromedial cortices (Kobayashi and Amaral 2003, 2007; Parvizi et al. 2006), posteromedial metabolic abnormalities have been proposed to reflect remote effects of hippocampal pathology such as neurofibrillary tangles and loss of neurons and synapses (Blaizot et al. 2002; Bradley et al. 2002; Meguro et al. 1999, 2001).

Another potential explanation for the disrupted posteromedial cortical function in AD has been provided by recent PET studies using a tracer called [11C]Pittsburgh Combound B, or PIB (Klunk et al. 2004). These studies have suggested that the posteromedial cortices are among the earliest brain areas containing Aβ plaques (Forsberg et al. 2007; Mintun et al. 2006). Functional imaging studies such as single-photon emission computed tomography (SPECT) or FDG-PET are not used routinely in AD diagnostics but may be of use in selected cases and particular differential diagnostic questions, especially in cases without significant atrophy in structural imaging. Similar, although less pronounced metabolic deficits in FDG-PET have been demonstrated in asymptomatic APOE ε4 carriers (Reiman et al. 1996) with further studies having suggested that assessment of hippocampal atrophy, FDG-PET findings and APOE ε4 status might be a useful combination in predicting future conversion to AD from MCI (Drzezga et al. 2005; Mosconi et al. 2008a).

Furthermore, studies using VBM to assess structural alteration at the whole-brain level have suggested that several cortical areas outside the hippocampus present atrophy in MCI and AD compared to healthy elderly subjects (Chetelat et al. 2002; Hämäläinen et al. 2007b; Karas et al. 2004; Pennanen et al. 2005). A recent study compared regional grey matter atrophy, as measured by VBM, directly to regional metabolic alterations in FDG-PET (Chetelat et al. 2008). This study showed that particularly in the posteromedial cortices, the resting-state metabolic abnormalities significantly exceed the level of structural atrophy. The authors speculated that excessive hypometabolism in the posterior association cortices may reflect disconnection with the MTL due to atrophy, which again supports the importance of the early involvement of the hippocampus and related MTL structures in AD.

Hippocampal atrophy as a possible biomarker for AD

Biomarkers are tools that aid in understanding the prediction, diagnosis, progression, or outcome of the treatment of a disease (Mayeux 2004). When considering AD, an ideal biomarker for clinical purposes should detect a fundamental feature of AD neuropathology, be diagnostically sensitive and specific, and be precise with good reproducibility for monitoring treatment effects (Kantarci and Jack 2004). Currently, possible imaging biomarkers for AD include structural MRI of the MTL and assessment of glucose metabolism with FDG-PET, PIB-PET or other molecular imaging methods being future biomarker candidates.

Regarding MRI of the MTL structures, hippocampal atrophy is the most investigated and best validated marker of AD starting from the stage of emerging cognitive impairment onwards to moderate dementia. As reviewed above, hippocampal atrophy is an inevitable concomitant of the underlying degenerative processes and also associated with the 'hippocampal pattern' of recent memory loss characteristic to early AD. In other words, hippocampal atrophy is considered to map 'upstream' to the Braak staging of NFT pathology and 'downstream' to cognitive alterations (Frisoni et al. 2010). Control and AD subjects can be distinguished well with rating of MTL/hippocampal atrophy (de Leon et al. 1989; Scheltens et al. 1992; Seab et al. 1988; Wahlund et al. 1999, 2000); discrimination accuracy of greater than 85% between AD patients and controls has been reported using hippocampal volumetric measures (Laakso et al. 2000; Pennanen et al. 2004). According to several studies, the controls and MCI subjects as well as subjects converting from MCI to AD can, however, be better distinguished by entorhinal than by hippocampal volume (Dickerson et al. 2001; Killiany et al. 2002; Pennanen et al. 2004). Based on the evidence that the rate of change in hippocampal volume is related to disease progression, although in a non-linear way, hippocampal atrophy is also considered a valid parameter to be used as an outcome measure in clinical trials of disease-modifying therapies.

Despite the considerable advances reviewed previously, it still remains a challenge to assess, at a single-subject level, whether an elderly individual with minor memory problems in fact already has prodromal AD or when she/he might convert to AD. Regarding hippocampal volume, overlap with healthy elderly control subjects on an initial MRI scan is great. It is also worth noting that as a biomarker hippocampal atrophy is not a pathognomonic feature for AD but lacks specificity to this disease. Therefore, for clinical diagnostic purposes at the level of individual patients, a combination of biomarkers providing complementary clinical, neuropsychological, topographical (hippocampal atrophy) and pathophysiological (CSF amyloid and tau concentrations) information is recommendable (Dubois et al. 2010).

In summary, hippocampal atrophy is a key diagnostic marker for early AD. It is included in the core diagnostic features in the most recently suggested new criteria for AD (Dubois et al. 2007, 2010). Manual tracing on MR images that is the present gold standard, is laborious and may result in heterogeneous volumetric estimates across different studies/laboratories depending on the adopted segmentation protocol. Therefore, several approaches to develop semi-automated or template-based fully automated measures have been introduced and applied in recent years (Colliot et al. 2008; Khan et al. 2008; Liu et al. 2010; Lötjönen et al. 2010; Morra et al. 2008, 2009; Schuff et al. 2009; van de Pol et al. 2007). Application and validity of these novel automated measures remains to be tested in future large trials.

Functional MRI

Introduction to functional MRI in AD

During the past 15 years, there has been a huge interest worldwide in fMR imaging of human cognition. FMRI has proved to be a very useful tool in investigating the neural correlates of intact human memory and other higher cognitive functions *in vivo*. FMRI activation studies during various cognitive tasks such as memory have often aimed to answer questions relevant to neuropsychiatric disorders even if performed in young healthy control subjects (see Figure 12.7). Clinical fMRI research into the pathophysiology of age-associated neurodegenerative diseases has become established more recently.

The fMRI technique most widely used to identify cerebral activation is based on imaging of the endogenous blood-oxygen-level-dependent (BOLD) contrast (Kwong et al. 1992; Ogawa et al. 1992). In a nutshell, the relative decrease in the amount of deoxygenated haemoglobin enhances the MRI signal locally in brain areas activated during a particular cognitive task. The BOLD contrast thus represents haemodynamic changes indirectly reflecting underlying cellular activity. The BOLD fMRI signal, and neurovascular coupling linking cellular activity to haemodynamic changes, is likely to undergo changes during healthy ageing and during pathological processes related to neurodegenerative dementias (Bangen et al. 2009; D'Esposito et al. 2003; Iadecola et al. 2004). Some of the changes that may occur even in healthy elderly subjects include altered cerebrovascular ultrastructure, reduced elasticity of vessels and increased atherosclerosis. In AD, the presence of Aβ in the brain and cerebral vasculature, together with altered cholinergic activity, impairs synaptic, neuronal and glial function, and may thus lead to attenuated BOLD response (D'Esposito et al. 2003; Iadecola et al. 2004; Selkoe 2002). Both increased and decreased BOLD fMRI as well as PET activation responses have, however, been reported in AD and MCI compared to elderly controls, which does not support the view of attenuation of the BOLD signal solely due to vascular reasons (Bäckman et al. 1999; Golby et al. 2005; Grady et al. 2003; Sperling et al. 2003).

Perhaps the greatest advantage of fMRI is that the key memory structures in MCI and AD patients can be imaged while they are attempting to do the type of cognitive processing that is causing them difficulty clinically, such as memory encoding. On the other hand, if the patients are not able to cooperate and adequately perform the cognitive task anymore, this major advantage of fMRI activation

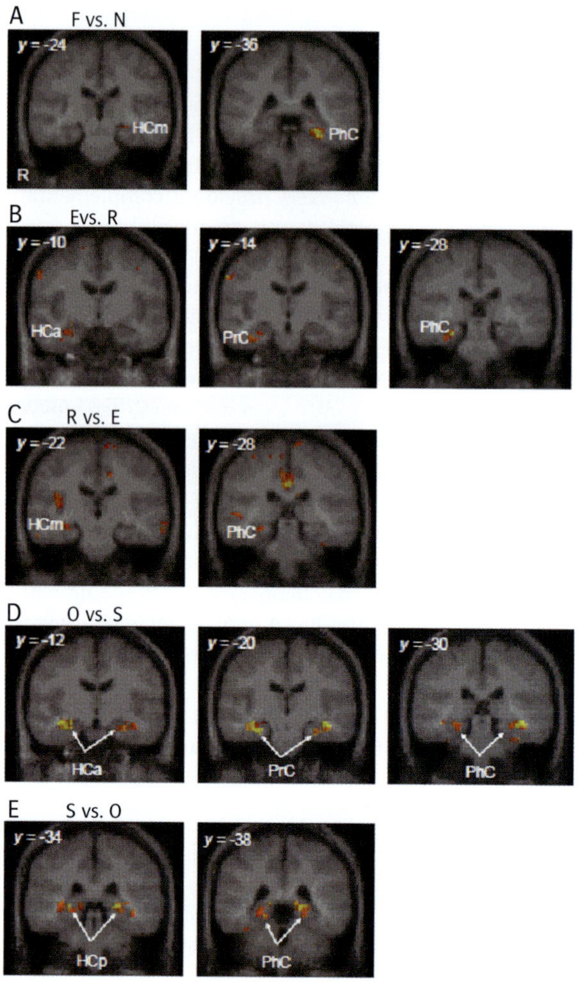

Figure 12.7 Examples of medial temporal lobe (MTL) fMRI responses in 14 young healthy control subjects during cognitive tasks relevant to early Alzheimer's disease (Pihlajamäki et al. 2000, 2003, 2004). The MTL activation areas illustrated in the figure depict the following cognitive contrasts: (A) category fluency vs. listing of numbers; (B) encoding vs. retrieval of picture-pairs; (C) retrieval vs. encoding of picture-pairs; (D) encoding novel objects vs. spatial changes; and finally, (E) encoding spatial changes vs. novel objects. The anterior hippocampus (HCa) was activated during encoding of picture-pairs (B) and processing of novel objects (D), the middle hippocampus (HCm) during retrieval of semantically related words (A) and picture-pairs (C), and the posterior hippocampus (HCp) during processing of spatial changes (E). Activation of the perirhinal cortex (PrC) was observed during encoding of picture-pairs (B) and processing of novel objects (D). The parahippocampal cortex (PhC) was activated in all of these statistical comparisons, but during processing of novel objects (D) the PhC activation area was more anterior than during processing of spatial changes (E). R, right; F, category fluency; N, listing of numbers; E, encoding; R, retrieval; O, novel object; S, spatial change. Y-coordinates refer to the Talairach standard coordinates corresponding to each coronal MRI slice.

studies is lost. Reproducibility of fMRI activation in older and cognitively impaired subjects has also not yet been widely established (Putcha et al. 2011). Despite these potential challenges, BOLD fMRI has produced invaluable information already by now and continuous technical development will likely enable even deeper understanding of the human brain function in the future.

Functional MRI in AD and MCI

'Memory' refers to the mind's ability to retain and retrieve past experiences. Successful memory formation is a complex cognitive process that is thought to require interplay of several large-scale neural networks, in particular, the hippocampus, related MTL structures, and heteromodal association cortices (Eichenbaum 2000; Hasselmo and McClelland 1999; Lavenex and Amaral 2000; Mesulam 1998; Squire and Zola-Morgan 1991; Squire and Zola 1996; Suzuki 2007; Tulving and Markowitsch, 1998). Activation of the hippocampal and parahippocampal regions during memory encoding has been demonstrated in several fMRI studies in young healthy subjects (Brewer et al. 1998; Chua et al. 2007; Pihlajamäki et al. 2003, 2004; Sperling et al. 2001, 2003; Stern et al. 1996; Wagner et al. 1998).

The hallmark of early AD is the inability to form new enduring episodic memories. At the same time, mild AD patients typically present with neuropathological changes such as synaptic alterations, selective neuronal loss and neurofibrillary tangles in the MTL structures (Braak and Braak 1991; Gomez-Isla et al. 1997; Hyman et al. 1984; Kordower et al. 2001; Scheff et al. 2006, 2007). In contrast, the frontal lobes, for example, are considered to be relatively spared in early AD in terms of distribution of histopathological changes, volume loss, or perfusion deficits (Braak and Braak 1991; Johnson and Albert 2000; Lehtovirta et al. 1996;).

Given the prominence of MTL pathology in early AD, the pioneering fMRI studies on AD focused on utilizing episodic memory tasks and investigating alterations in hippocampal activation (Kato et al. 2001; Machulda et al. 2003; Rombouts et al. 2000; Small et al. 1999; Sperling et al. 2003a). To date, there are several fMRI studies, which have consistently reported diminished or absent hippocampal and/or parahippocampal activation in AD compared to healthy older controls, during encoding novel information (Dickerson et al. 2005; Golby et al. 2005; Grön and Riepe, 2004; Hämäläinen et al. 2007a; Kato et al. 2001; Machulda et al. 2003; Pariente et al. 2005; Remy et al. 2005; Rombouts et al. 2000; Small et al. 1999; Sperling et al. 2003a). The studies with clinical AD patients have used a number of encoding stimuli, including faces (Rombouts et al. 2005; Small et al. 1999), face-name pairs (Pariente et al. 2005; Sperling et al. 2003a), line-drawings (Hämäläinen et al. 2007a; Rombouts et al. 2000), scenes (Golby et al. 2005), and geometric shapes (Kato et al. 2001).

The term 'MCI' refers to the transitional state between the cognitive changes of healthy ageing and the fully developed clinical features of neurodegenerative dementia (Petersen et al. 2001). Subjects with amnestic MCI are an important group to investigate, as they are at increased risk for developing dementia, AD in particular. MCI is, however, known to be a very heterogeneous population both clinically and pathologically (Jicha et al. 2006). Consonant with the notion of heterogeneity, results of fMRI studies in MCI subjects relative to controls or AD patients have been variable regarding findings of hippocampal activation, ranging from hyperactivation during encoding tasks (Dickerson et al. 2004, 2005; Hämäläinen et al. 2007a; Kircher et al. 2007; Woodard et al. 2009; Yassa et al. 2010) to hypoactivation both during encoding (Johnson et al. 2006a; Machulda et al. 2003; Mandzia et al. 2009) and retrieval (Mandzia et al. 2009; Petrella et al. 2006). It has also been shown that greater baseline hippocampal activity in MCI subjects predicts a greater degree and rate of cognitive decline during follow-up, even when accounting for the baseline cognitive status, age, education, gender, hippocampal volume and APOE status (Miller et al. 2008). In the whole-brain analysis, the hippocampal formation was the only brain region predicting cognitive decline.

In addition to the heterogeneity of the MCI population, some of the diversity of fMRI results in MCI subjects may be explained by differences in subject selection criteria and the level of clinical severity, fMRI paradigms and their difficulty as well as functional imaging and data analysis methods. However, typically MCI subjects with significantly impaired memory have, similar to AD patients, shown decreased hippocampal activity compared with controls (Machulda et al. 2003; Petrella et al. 2006), consonant with the notion that these subjects in fact likely already have prodromal AD.

Functional MRI in ageing and APOE ε4 carriers

Several experiments in cognitively intact older control subjects have demonstrated that elderly individuals are able to activate their hippocampus during (successful) associative encoding largely to the same degree as young subjects (Miller et al. 2008; Rand-Giovannetti et al. 2006; Sperling et al. 2003a, 2003b). It has been suggested that age-related changes in memory performance may primarily be due to alterations in cortical regions or in the connectivity between the MTL and neocortical regions. With regards to the hippocampal function, age-related differences have also been reported. It has, for example, been shown that the elderly subjects had less hippocampal activity during working memory, visual attention and episodic retrieval tasks, but instead displayed increased parahippocampal activity during episodic retrieval (Cabeza et al. 2004). This finding was later confirmed with a study

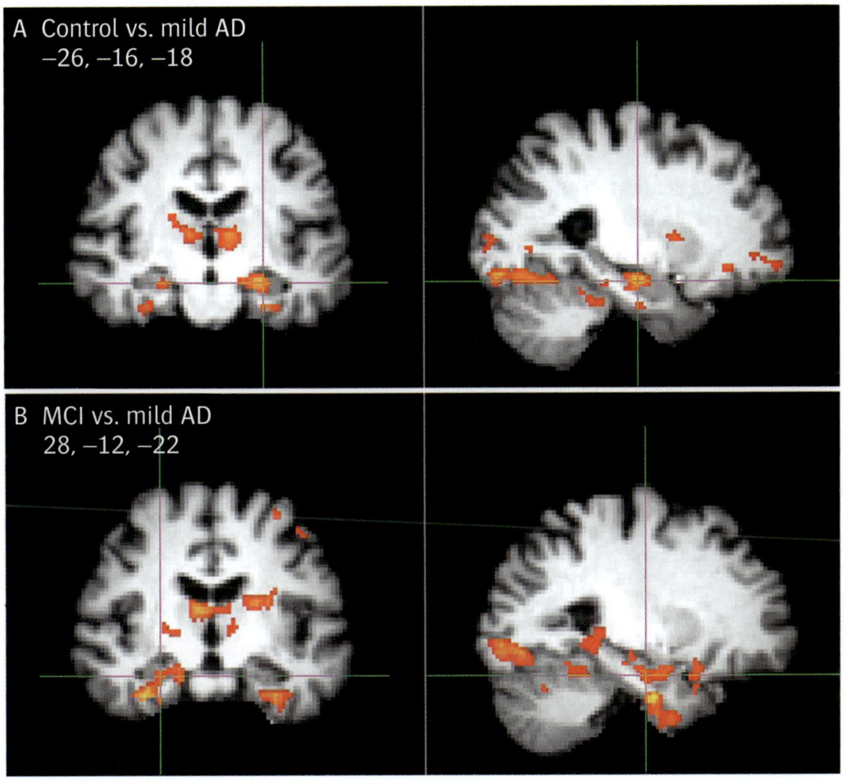

Figure 12.8 Typically, in fMRI experiments during encoding novel stimuli compared to repeated stimuli, both healthy elderly control subjects (A) and subjects with mild cognitive impairment (MCI) (B) demonstrate greater hippocampal activity ($p < 0.05$, corrected for multiple comparisons) than patients with mild Alzheimer's disease (AD). The x, y, z coordinates refer to the corresponding MNI standard coordinates. Coronal and sagittal slices are presented both in the panel (A) and (B). Left in the figure is right in the brain.

revealing decreased recollection-related activity in the hippocampus together with increased familiarity-related activation in the neighbouring rhinal cortex (Daselaar et al. 2006a). Age-related reductions in hippocampal activity have also been demonstrated during semantic classification tasks (Dennis et al. 2007). It has been speculated that elderly subjects compensate for the emerging hippocampal deficits by relying more on the MTL cortical regions, possibly through frontal top-down mechanisms (Daselaar et al. 2006b).

As in MCI subjects, results of fMRI studies comparing activation in APOE ε4 carriers versus their non-carrier counterparts have been diverse. Increased hippocampal and cortical activation has been reported by several groups and has usually been interpreted to reflect compensatory neural mechanisms (Bondi et al. 2005; Bookheimer et al. 2000; Burggren et al. 2002; Dickerson et al. 2005; Fleisher et al. 2005; Han et al. 2007; Pihlajamäki and Sperling 2009; Smith et al. 2002; Wishart et al. 2006). Cognitive tasks used in these studies have covered classical memory encoding tasks using novel and repeated pictures or face-name pairs (Bondi et al. 2005; Dickerson et al. 2005), encoding and retrieval of word-pairs (Bookheimer et al. 2000; Fleisher et al. 2005), a letter fluency task (Smith et al. 2002), and an auditory verbal *n*-back working memory tasks (Wishart et al. 2006). At the same time, several studies have demonstrated reduced functional brain activity in the hippocampus and other brain areas in cognitively normal ε4 carriers (Borghesani et al. 2007; Lind et al. 2006; Smith et al. 1999; Trivedi et al. 2006) during tasks such as visuospatial encoding (Borghesani et al. 2007), visual naming and letter fluency (Smith et al. 1999), encoding novel and familiar items (Trivedi et al. 2006) and semantic categorization (Lind et al. 2006). Several of the above mentioned studies have carefully matched the at-risk subjects with their APOE ε4 non-carrier counterparts regarding age, gender, and cognitive performance.

It is currently difficult to draw firm conclusions of the APOE ε4 effects on BOLD fMRI activation pattern. In the most recent large-scale fMRI studies, more complex patterns of alterations in brain activation differentially affected by APOE ε4 and family history of AD have been suggested (Bassett et al. 2006; Johnson et al. 2006b). Imaging elderly asymptomatic or very mildly impaired subjects at-risk for AD by virtue of genetics and/or family history is an intriguing and important field but more research and follow-up studies are needed to understand the interactions between the various genotypes, emerging neuropathological changes, age and several other factors and their effects on the fMRI activity patterns in the hippocampus and elsewhere in the brain.

The Hippocampus and repetition suppression in AD

Interestingly, there is also evidence of increased hippocampal fMRI activity in AD during specific contrasts, primarily involving the brain response to repetitive stimuli. Reduction in the anterior MTL neuronal firing rate (i.e. individual neurons or populations of neurons) to repeated stimuli, sometimes termed repetition suppression (Desimone 1996), is well documented in rodents and non-human primates but is also thought to be one of the key mechanisms underlying certain memory functions in humans. In a recent human fMRI study, Golby et al. (2005) reported impaired repetition suppression paralleled by more MTL activation in AD patients than in older controls during processing of repeated scenes. Another recent study provided evidence that the normal suppression of MTL activity to repeated face-name pairs as compared to visual fixation is impaired in AD (Pihlajamäki et al. 2008; see Figure 12.9). Similar findings have been reported in individuals with amnestic MCI as well (Johnson et al. 2004). Furthermore, failure of the hippocampus and surrounding MTL cortices

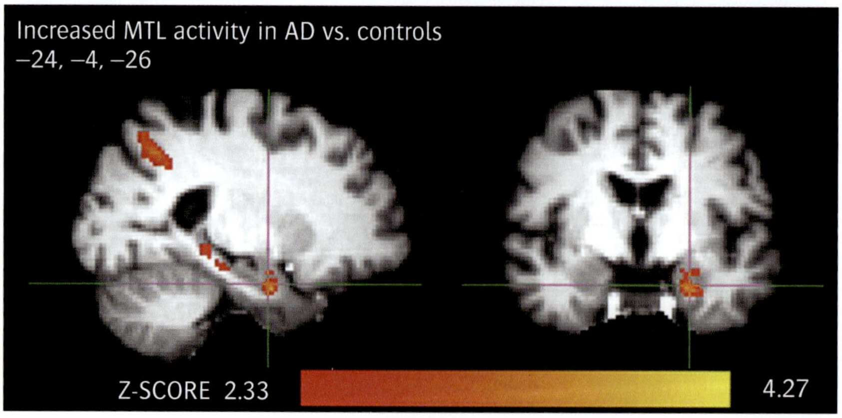

Figure 12.9 Increased fMRI activity in the hippocampus and surrounding medial temporal lobe (MTL) cortical areas in patients with Alzheimer's disease (AD) relative to elderly control subjects during processing of repeated face-name pairs ($p < 0.05$, corrected for multiple comparison) (Data from: Pihlajamäki et al. 2008). The x, y, z coordinates refer to the corresponding MNI standard coordinates. Left in the figure is right in the brain.

to discriminate familiar from novel information at encoding has been related to both poor associative recognition memory and as poor performance in neuropsychological tests of episodic memory across a range of age and cognitive impairment (Pihlajamäki et al. 2011; see Figure 12.10). In that particular study, cognitively normal young and elderly subjects showed preserved suppression of hippocampal activity to stimulus repetition, whereas cognitively impaired elderly and AD patients showed reduced suppression of hippocampal activity.

The strongest correlation between the impaired repetition suppression and poor memory performance was found in

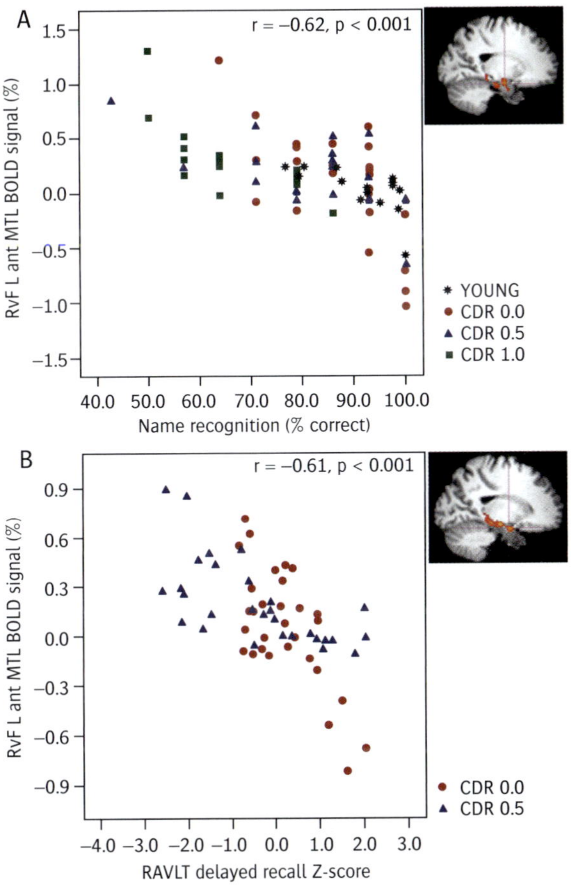

Figure 12.10 A) The left (L) anterior medial temporal lobe (MTL) activity during processing Repeated face-name pairs versus Fixation (RvF) is negatively correlated with the post-scan name recognition memory test performance ($p < 0.05$, cluster-corrected) across 90 subjects ranging from young healthy subjects to patients with mild Alzheimer's disease ($r = -0.62$, $p < 0.001$, Pearson correlation) (Pihlajamäki et al. 2011). B) Similarly, there is a significant inverse relation between the MTL repetition suppression and free recall performance: fMRI activity of the left anterior and middle hippocampus and neighbouring rhinal and parahippocampal cortices demonstrate significant negative correlation with the Rey Auditory Verbal Learning Test (RAVLT) delayed recall performance across the 60 cognitively normal (Clinical Dementia Rating, CDR, 0) and mildly impaired (CDR 0.5) elderly individuals ($r = -0.61$, $p < 0.001$, Pearson correlation).

the anterior MTL regions, which are known to be sensitive to encoding novel associative stimuli (Chua et al. 2007; Daselaar et al. 2006a; Henson et al. 2003; Pihlajamäki et al. 2003; Sperling et al. 2001, 2003a) and to suppress neural and BOLD fMRI activity to repeated stimuli (Brozinsky et al. 2005; Johnson et al. 2004, 2008; Rand-Giovannetti et al. 2006) and present AD pathological changes in the early stages of the disease (Braak and Braak 1991; Gomez Isla et al. 1997; Hyman et al. 1984; Kordower et al. 2001; Scheff et al. 2006, 2007). It can be speculated that the anterior MTL may have lost its inherent sensitivity to recognize familiar information, or—vice versa—sensitivity to novel encoding in the cognitively impaired elderly subjects and in AD patients (Golby et al. 2005; Johnson et al. 2004, 2008; Pihlajamäki et al. 2008, 2011) corresponding to the characteristic cognitive deficit in incipient AD that is the inability to form and retain new memories.

In summary, previous fMRI studies in AD patients, compared to healthy elderly individuals, have reported decreased MTL activity during processing of novel versus repeated information (Dickerson et al. 2005; Golby et al. 2005; Rombouts et al. 2000; Sperling et al. 2003b). These findings of decreased hippocampal activity comparing novel to repeated stimuli in AD or MCI patients in the fMRI contrast are, in fact, likely to be explained at least to some degree by failure of hippocampal repetition suppression to repeatedly presented stimuli as reviewed above (Golby et al. 2005; Johnson et al. 2004; Pihlajamäki et al. 2008, 2011), in addition to alterations in novel encoding. Episodic memory impairment typically seen in mild AD may not only manifest as decreased but also as increased fMRI activity of the hippocampus and other MTL memory structures.

Functional MRI in prediction of cognitive decline

Previous cross-sectional fMRI studies in subjects with MCI/prodromal AD have reported variable results, ranging from hypoactivation to hyperactivation compared to cognitively normal older individuals. It has been hypothesized that subjects in early phases of prodromal AD may present a short period of paradoxical hippocampal hyperactivation, which is then followed by loss of hippocampal activation along with cognitive decline. To our knowledge, there is one longitudinal fMRI study (O'Brien et al. 2010) performed in 51 non-demented older individuals with either baseline Clinical Dementia Rating (CDR) scores of 0 or 0.5. In this study, subjects classified as CDR 0 at baseline showed no difference in hippocampal fMRI activity during a face-name associative memory paradigm over 2 years. In contrast, those subjects who were CDR 0.5 at baseline demonstrated a significant decrease in hippocampal fMRI activity, and furthermore, subjects with more rapid decline demonstrated both the highest hippocampal activation at baseline and the greatest loss of hippocampal activation during follow-up. The findings remained significant after accounting for age, hippocampal volume, and APOE status. The authors concluded that cognitive decline is associated with loss of hippocampal activation. The results support previous findings of paradoxically increased activation before the impending clinical decline, and suggest that fMRI may prove valuable in tracking very early progression of brain dysfunction on the trajectory towards clinical AD but prior to the point of irreversible neuronal loss and macroscopic atrophy.

Other studies with clinical follow-up data on MCI subjects have reported that increased hippocampal and/or entorhinal fMRI activity at baseline in MCI compared to healthy elderly controls may indicate higher likelihood of subsequent cognitive decline (Dickerson et al. 2004; Miller et al. 2008). Similar findings of a temporary period with abnormal enhanced MTL activity during a preclinical stage of AD have been reported in subjects at genetic risk for AD by virtue of the APOE ε4 (Bookheimer et al. 2000) and in subjects carrying the autosomal dominant presenilin 1 mutation (Mondadori et al. 2006). It is hoped that these intriguing findings become confirmed and extended by large multimodal imaging studies including structural, functional and molecular imaging.

Hippocampal structure and function in AD and MCI

There are not many studies combining structural and functional MRI during memory tasks to investigate hippocampal function in AD and MCI to date. A recent study in 21 elderly controls, 14 MCI subjects, and 15 mild AD patients reported fMRI results during associative encoding of word-picture pairs and results of brain structural analysis using both hippocampal volumetric and VBM methods (Hämäläinen et al. 2007a). The main aim of the study was to investigate the structural changes underlying alterations in memory-related fMRI activity and also to examine what the additional value provided by functional over structural imaging is (Jauhiainen et al. 2009). In this study, the subjects with amnestic multiple-domain MCI, with mean Mini-Mental State Examination scores of 25.6, showed significantly more atrophy in the left anterior hippocampus, MTL, and also in superior frontal regions as compared to controls. Interestingly, these MCI subjects also showed increased fMRI activity in the posterior hippocampus and adjacent parahippocampal and fusiform cortices (see Figure 12.11). In other words, the MCI subjects did not present fMRI activation in the anterior hippocampus typically activated in healthy young and elderly subjects during novel associative encoding (Chua et al. 2007; Pihlajamäki et al. 2003; Sperling et al. 2003a) but in the more posterior MTL regions. These structural and functional MRI results suggest that the posteriorly dislocated, possibly compensatory MTL activation during a memory encoding task is due to structural atrophy of the anterior MTL regions (Hämäläinen et al. 2007a). The results are in line with the results by Dickerson et al. (2005), who demonstrated increased anterior-middle MTL activity during a face-name encoding paradigm in very mild MCI subjects with no significant MTL atrophy in terms of hippocampal and entorhinal volumetric measurements.

The hypothesis of the increased hippocampal activity in MCI being compensatory by nature was further supported by the finding that there was a negative correlation between the left hippocampal volume and left parahippocampal activation (Hämäläinen et al. 2007a) such that the MCI subjects with hippocampal atrophy elicited greater parahippocampal fMRI activity. There was no correlation between the hippocampal structure and parahippocampal function in the elderly controls or AD patients, but the inverse relationship between emerging MTL atrophy and evidence of increased compensatory fMRI activity was restricted to the MCI subjects. This is in agreement with the findings by Dickerson et al. (2004) who showed that, in MCI subjects, greater clinical impairment in terms of CDR Sum-of-Boxes scores was related to smaller hippocampal volumes but greater parahippocampal fMRI activation. These results altogether show that structural and functional MRI methods provide complementary information and together advance our understanding of the MTL pathophysiology underlying the earliest cognitive symptoms of MCI and early AD.

As discussed in the previous paragraphs, it seems that there is a temporary phase of abnormal MTL hyperactivity along the course of MCI, which may be an indicator of compensatory neural mechanisms recruited in MCI subjects in order to keep behavioural performance during a memory task close to the level of cognitively intact elderly subjects. It is also possible that the observed MTL hyperactivation reflects cholinergic or other neurotransmitter upregulation in MCI patients (DeKosky et al. 2002). Alternatively, increased hippocampal activity may be a marker of other AD-related pathobiological processes, such as aberrant sprouting (Hashimoto and Masliah 2003) or impaired synaptic transmission (Stern et al. 2004). Altogether, alterations in hippocampal activity in MCI in the setting of minimal MTL atrophy are in line with previous laboratory and animal data suggesting that pathophysiologic alterations revealed by fMRI may precede structural abnormalities early in the course of AD (Selkoe 2002; Walsh and Selkoe 2004). Around the conversion from MCI to clinical AD, the ability to compensate for the emerging MTL pathology and related memory problems is lost, which is

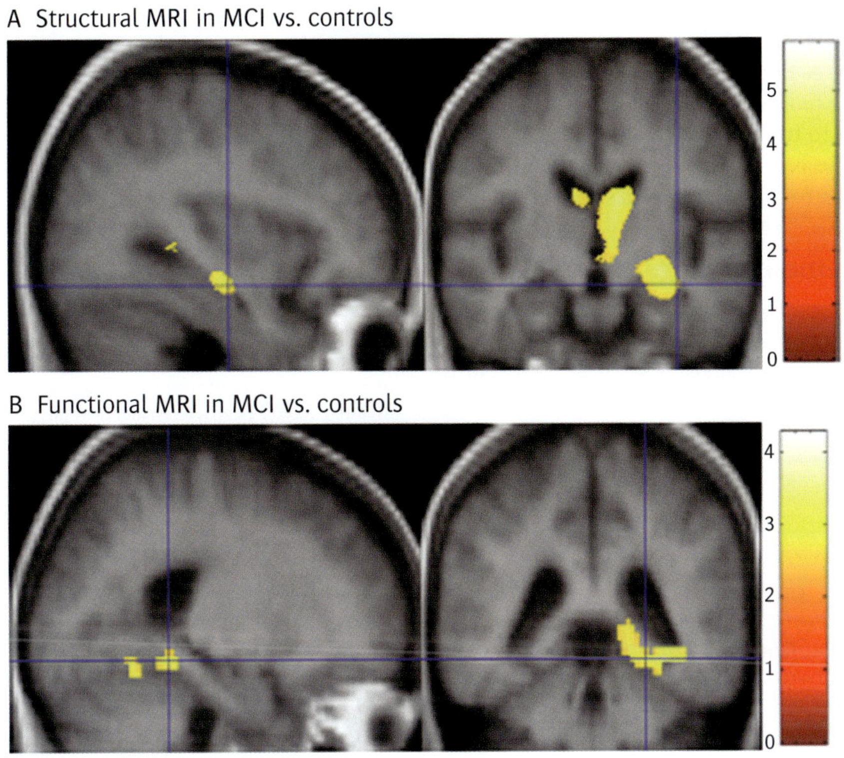

Figure 12.11 A) Structural MRI combined with voxel-based morphometric (VBM) analysis demonstrates decreased grey matter density in subjects with Mild Cognitive Impairment (MCI), compared to elderly control subjects, in the left anterior medial temporal lobe (MTL). B) When comparing the fMRI activation findings during associative encoding of novel word-picture pairs, significantly increased left MTL activity is observed in MCI relative to elderly controls. Colour bars present the corresponding T-values within the significantly atrophied (A) or activated (B) brain areas in the SPM between-group comparisons. Threshold for statistical significance in both VBM and fMRI analyses was $p < 0.05$, cluster-corrected. Left in the figure is right in the brain.

then seen as poor task performance and disrupted hippocampal fMRI activity during the classical novel encoding paradigms. It is possible that in subjects with 'early' MCI, challenging fMRI paradigms are needed to induce the MTL hyperactivity, whereas is clinically more impaired 'late' MCI subjects, such as in the study by Hämäläinen et al. (2007a), relatively easy tasks suffice to challenge the MTL memory structures.

Concluding remarks

In summary, there have been a number of promising clinically relevant imaging and other studies targeting alterations in the hippocampal structure and function in AD, MCI, and subjects at-risk for AD relative to normal ageing. Hippocampal atrophy, for example, is now considered an established diagnostic marker at the mild cognitive impairment stages during the course of AD. Functional imaging during memory paradigms has shown evidence of specific alterations in the hippocampus and related whole-brain memory networks that may be able to differentiate the process of very early AD from normal ageing. The greatest potential of multimodal imaging of the hippocampal formation most likely lies in the study of very early stages of dementias, at the point of emerging neuronal dysfunction without significant macroscopic brain atrophy, and perhaps in the evaluation of pharmacological efficacy for novel therapies. In the context of early diagnostics of AD, the most interesting and challenging target group to be investigated continues to be the elderly subjects with subtle memory impairment as these subjects would still have preserved brain function left and thus scope for therapeutic interventions.

The recent advances in molecular imaging techniques such as the *in vivo* detection of Aβ pathology with PIB-PET provide novel opportunities to evaluate alterations in hippocampal function and episodic memory performance in non-demented elderly individuals perhaps with varying levels of amyloid burden. Carefully designed future studies using multimodal imaging are hoped to yield us new tools that aid in the early identification of subjects likely to develop dementia, at the point when disease modifying therapies would be most efficacious. Further longitudinal studies are needed to track the evolution of hippocampal alterations over the course of the cognitive continuum from healthy ageing to clinical dementias such as AD.

Acknowledgements

The original work from our laboratory described in this chapter was supported by Health Research Council of the Academy of Finland, grant #121038 (HS) and grants #108188 and #214050 (MP), and Kuopio University Hospital EVO grants #477311 and #5772720. We express special gratitude to the subjects who participated in this study.

References

Alzheimer, A. (1907). Über eine eigenartige Erkrankung der Hirnrinde. *Allgem Z Psychiatr Psych-Gerisch Med*, 64, 146–8.

Apostolova, L.G., Dutton, R.A., Dinov, I.D., *et al.* (2006). Conversion of mild cognitive impairment to Alzheimer disease predicted by hippocampal atrophy maps. *Arch Neurol*, 63, 693–99.

Apostolova, L.G., Mosconi, L., Thompson, P.M., *et al.* (2010). Subregional hippocampal atrophy predicts Alzheimer's dementia in the cognitively normal. *Neurobiol Aging*, 31, 1077–88.

Bäckman, L., Andersson, J.L., Nyberg, L., Winblad, B., Nordberg, A., and Almkvist, O. (1999). Brain regions associated with episodic retrieval in normal aging and Alzheimer's disease. *Neurology*, 52, 1861–70.

Bakkour, A., Morris, J.C., and Dickerson, B.C. (2009). The cortical signature of prodromal AD: regional thinning predicts mild AD dementia. *Neurology*, 72, 1048–55.

Bangen, K.J., Restom, K., Liu, T.T., *et al.* (2009). Differential age effects on cerebral blood flow and BOLD response to encoding: Associations with cognition and stroke risk. *Neurobiol Aging*, 30, 1276–87.

Bassett, S.S., Yousem, D.M., Cristinzio, C., *et al.* (2006). Familial risk for Alzheimer's disease alters fMRI activation patterns. *Brain*, 129, 1229–39.

Becker, J.T., Davis, S.W., Hayashi, K.M., *et al.* (2006). Imaging Methods and Analysis in Geriatrics Research Group. Three-dimensional patterns of hippocampal atrophy in mild cognitive impairment. *Arch Neurol*, 63, 97–101.

Bennett, D.A., Schneider, J.A., Bienias, J.L., Evans, D.A., and Wilson, R.S. (2005). Mild cognitive impairment is related to Alzheimer disease pathology and cerebral infarctions. *Neurology*, 64, 834–41.

Bertram, L., McQueen, M.B., Mullin, K., Blacker, D., and Tanzi, R.E. (2007). Systematic meta-analyses of Alzheimer disease genetic association studies: the AlzGene database. *Nat Genet*, 39, 17–23.

Blaizot, X., Meguro, K., Millien, I., Baron, J.C., and Chavoix, C. (2002). Correlations between visual recognition memory and neocortical and hippocampal glucose metabolism after bilateral rhinal cortex lesions in the baboon: implications for Alzheimer's disease. *J Neurosci*, 22, 9166–70.

Bobinski, M., Wegiel, J., Wisniewski, H.M., *et al.* (1995). Atrophy of hippocampal formation subdivisions correlates with stage and duration of Alzheimer disease. *Dementia*, 6, 205–10.

Bobinski, M., de Leon M.J., Tarnawski, M., *et al.* (1998). Neuronal and volume loss in CA1 of the hippocampal formation uniquely predicts duration and severity of Alzheimer disease. *Brain Res*, 805, 267–9.

Boccardi, M., Sabattoli, F., Testa, C., Beltramello, A., Soininen, H., and Frisoni, G.B. (2004). APOE and modulation of Alzheimer's and frontotemporal dementia. *Neurosci Lett*, 356, 167–70.

Bondi, M.W., Houston, W.S., Eyler, L.T., and Brown, G.G. (2005). fMRI evidence of compensatory mechanisms in older adults at genetic risk for Alzheimer disease. *Neurology*, 64, 501–8.

Bookheimer, S.Y., Strojwas, M.H., Cohen, M.S., *et al.* (2000). Patterns of brain activation in people at risk for Alzheimer's disease. *N Engl J Med*, 343, 450–6.

Borghesani, P.R., Johnson, L.C., Shelton, A.L., *et al.* (2008). Altered medial temporal lobe responses during visuospatial encoding in healthy APOE*4 carriers. *Neurobiol Aging*, 29, 981–91.

Bozzali, M., Filippi, M., Magnani, G., *et al.* (2006). The contribution of voxel-based morphometry in staging patients with mild cognitive impairment. *Neurology*, 67, 453–60.

Braak, H. and Braak, E. (1991). Neuropathological stageing of Alzheimer-related changes. *Acta Neuropathol (Berl)*, 82, 239–59.

Braak, H. and Braak, E. (1995). Staging of Alzheimer's disease-related neurofibrillary changes. *Neurobiol Aging*, 16, 271–8.

Bradley, K. M., O'Sullivan, V. T., Soper, N. D., *et al.* (2002). Cerebral perfusion SPET correlated with Braak pathological stage in Alzheimer's disease. *Brain*, 125, 1772–81.

Brewer, J.B., Zhao, Z., Desmond, J.E., Glover, G.H., and Gabrieli, J.D. (1998). Making memories: brain activity that predicts how well visual experience will be remembered. *Science*, 281, 1185–7.

Brookmeyer R., Gray S. and Kawas C. (1998). Projections of Alzheimer's disease in the United States and the public health impact of delaying disease onset. *Am J Public Health*, 88, 1337–42.

Brozinsky, C.J., Yonelinas, A.P., Kroll, N.E., and Ranganath, C. (2005). Lag-sensitive repetition suppression effects in the anterior parahippocampal gyrus. *Hippocampus*, 15, 557–61.

Burggren, A.C., Small, G.W., Sabb, F.W., and Bookheimer, S.Y. (2002). Specificity of brain activation patterns in people at genetic risk for Alzheimer disease. *Am J Geriatr Psychiatry*, 10, 44–51.

Burton, E.J., Barber, R., Mukaetova-Ladinska, E.B., *et al.* (2009). Medial temporal lobe atrophy on MRI differentiates Alzheimer's disease from dementia with Lewy bodies and vascular cognitive impairment: a prospective study with pathological verification of diagnosis. *Brain*, 132, 195–203.

Cabeza, R., Daselaar, S.M., Dolcos, F., Prince, S.E., Budde, M., and Nyberg, L. (2004). Task-independent and task-specific age effects on brain activity during working memory, visual attention and episodic retrieval. *Cereb Cortex*, 14, 364–75.

Chetelat, G., Desgranges, B., De La Sayette, V., Viader, F., Eustache, F., and Baron, J.C. (2002). Mapping gray matter loss with voxel-based morphometry in mild cognitive impairment. *Neuroreport*, 13, 1939–43.

Chetelat, G., Landeau, B., Eustache, F., *et al.* (2005). Using voxel-based morphometry to map the structural changes associated with rapid conversion in MCI: a longitudinal MRI study. *Neuroimage*, 27, 934–46.

Chetelat, G., Desgranges, B., Landeau, B., *et al.* (2008). Direct voxel-based comparison between grey matter hypometabolism and atrophy in Alzheimer's disease. *Brain*, 131, 60–71.

Chua, E.F., Schacter, D.L., Rand-Giovannetti, E., and Sperling, R.A. (2007). Evidence for a specific role of the anterior hippocampal region in successful associative encoding. *Hippocampus*, 17, 1071–80.

Cohen, R.M., Small, C., Lalonde, F., Friz, J., and Sunderland, T. (2001). Effect of apolipoprotein E genotype on hippocampal volume loss in aging healthy women. *Neurology*, 57, 2223–8.

Colliot, O., Chételat, G., Chupin, M., *et al.* (2008). Discrimination between Alzheimer disease, mild cognitive impairment, and normal aging by using automated segmentation of the hippocampus. *Radiology*, 248, 194–201.

Convit, A., de Leon, M.J., Golomb, J., *et al.* (1993). Hippocampal atrophy in early Alzheimer's disease: anatomic specificity and validation. *Psychiatr Q*, 64, 371–87.

Convit, A., de Leon, M.J., Hoptman, M.J., Tarshish, C., De Santi, S., and Rusinek H. (1995). Age-related changes in brain: I. Magnetic resonance imaging measures of temporal lobe volumes in normal subjects. *Psychiatr Q*, 66, 343–55.

Convit, A., De Leon, M.J., Tarshish, C., *et al.* (1997). Specific hippocampal volume reductions in individuals at risk for Alzheimer's disease. *Neurobiol Aging*, 18, 131–8.

Csernansky, J. G., Hamstra, J., Wang, L., *et al.* (2004). Correlations between antemortem hippocampal volume and postmortem neuropathology in AD subjects. *Alzheimer Dis Assoc Disord*, 18, 190–5.

Csernansky, J.G., Wang, L., Swank, J., *et al.* (2005). Preclinical detection of Alzheimer's disease: hippocampal shape and volume predict dementia onset in the elderly. *Neuroimage*, 25, 783–92.

Cummings, J.L., Doody, R., and Clark, C. (2007). Disease-modifying therapies for Alzheimer disease: challenges to early intervention. *Neurology*, 69, 1622–34.

Daselaar, S.M., Fleck M.S., and Cabeza R. (2006a). Triple dissociation in the medial temporal lobes: Recollection, familiarity, and novelty. *J Neurophysiol*, 96, 1902–11.

Daselaar, S.M., Fleck, M.S., Dobbins, I.G., Madden, D.J., and Cabeza, R. (2006b). Effects of healthy aging on hippocampal and rhinal memory functions: an event-related fMRI study. *Cereb Cortex*, 16, 1771–82.

Davis K.L., Mohs R.C., Marin, D., *et al.* (1999). Cholinergic markers in elderly patients with early signs of Alzheimer's disease. *JAMA*, 281, 1401–6.

DeCarli, C., Frisoni, G.B., Clark, C.M., *et al.* (2007). Qualitative estimates of medial temporal atrophy as a predictor of progression from mild cognitive impairment to dementia. *Arch Neurol*, 64, 108–15.

DeKosky, S.T., Ikonomovic, M.D., Styren, S.D., *et al.* (2002). Upregulation of choline acetyltransferase activity in hippocampus and frontal cortex of elderly subjects with mild cognitive impairment. *Ann Neurol*, 51, 145–55.

Delacourte, A., David, J.P., Sergeant, N., *et al.* (1999). The biochemical pathway of neurofibrillary degeneration in aging and Alzheimer's disease. *Neurology*, 52, 1158–65.

de Leon, M.J., George, A.E., Stylopoulos, L.A., Smith, G., and Miller, D.C. (1989). Early marker for Alzheimer's disease: the atrophic hippocampus. *Lancet*, 2, 672–3.

de Leon, M.J., Golomb, J.,George, A.E., *et al.* (1993). The radiologic prediction of Alzheimer disease: the atrophic hippocampal formation. *Am J Neuroradiol*, 14, 897–906.

de Leon, M.J., George, A.E., Golomb, J., *et al.* (1997). Frequency of hippocampal formation atrophy in normal aging and Alzheimer's disease. *Neurobiol Aging*, 18, 1–11.

de Leon, M.J., DeSanti, S., Zinkowski, R., *et al.* (2006). Longitudinal CSF and MRI biomarkers improve the diagnosis of mild cognitive impairment. *Neurobiol Aging*, 27, 394–401.

den Heijer, T., Oudkerk, M., Launer, L.J., van Duijn, C.M., Hofman, A., and Breteler, M.M. (2002). Hippocampal, amygdalar, and global brain atrophy in different apolipoprotein E genotypes. *Neurology*, 59, 746–8.

den Heijer, T., Geerlings, M.I., Hoebeek, F.E., Hofman, A., Koudstaal, P.J., and Breteler, M.M. (2006). Use of hippocampal and amygdalar volumes on magnetic resonance imaging to predict dementia in cognitively intact elderly people. *Arch Gen Psychiatry*, 63, 57–62.

Dennis, N.A., Daselaar, S., and Cabeza, R. (2007). Effects of aging on transient and sustained successful memory encoding activity. *Neurobiol Aging*, 28, 1749–58.

De Santi, S., de Leon, M.J., Rusinek, H., *et al.* (2001). Hippocampal formation glucose metabolism and volume losses in MCI and AD. *Neurobiol Aging*, 22, 529–39.

Desimone, R. (1996). Neural mechanisms for visual memory and their role in attention. *Proc Natl Acad Sci U S A*, 93, 13494–9.

D'Esposito, M., Deouell, L.Y., and Gazzaley, A. (2003). Alterations in the BOLD fMRI signal with ageing and disease: a challenge for neuroimaging. *Nat Rev Neurosci*, 4, 863–72.

DeToledo-Morrell, L., Stoub, T.R., Bulgakova, M., *et al.* (2004). MRI-derived entorhinal volume is a good predictor of conversion from MCI to AD. *Neurobiol Aging*, 25, 1197–203.

Devanand, D.P., Pradhaban, G., Liu, X., *et al.* (2004). Hippocampal and entorhinal atrophy in mild cognitive impairment: prediction of Alzheimer disease. *Neurology*, 68, 828–36.

Deweer, B., Lehéricy, S., Pillon, B., *et al.* (1995). Memory disorders in probable Alzheimer's disease: the role of hippocampal atrophy as shown with MRI. *J Neurol Neurosurg Psychiatry*, 58, 590–7.

Dickerson, B.C., Goncharova, I., Sullivan, M.P., *et al.* (2001). MRI-derived entorhinal and hippocampal atrophy in incipient and very mild Alzheimer's disease. *Neurobiol Aging*, 22, 747–54.

Dickerson, B.C., Salat, D.H., Greve, D.N., *et al.* (2005). Increased hippocampal activation in mild cognitive impairment compared to normal aging and AD. *Neurology*, 65, 404–11.

Dickerson, B.C., Salat, D.H., Bates, J.F., *et al.* (2004). Medial temporal lobe function and structure in mild cognitive impairment. *Ann Neurol*, 56, 27–35.

Drachman, D.A. (2006). Robert Katzman and Alzheimer's disease: an appreciation. *Alzheimer Dis Assoc Disord*, 20, S29–30.

Drzezga, A., Grimmer, T., Riemenschneider, M., *et al.* (2005). Prediction of individual clinical outcome in MCI by means of genetic assessment and (18)F-FDG PET. *J Nucl Med*, 46, 1625–32.

Du, A.T., Schuff, N., Amend, D., *et al.* (2001). Magnetic resonance imaging of the entorhinal cortex and hippocampus in mild cognitive impairment and Alzheimer's disease. *J Neurol Neurosurg Psychiatry*, 71, 441–7.

Du, A.T., Schuff, N., Chao, L.L., *et al.* (2006). Age effects on atrophy rates of entorhinal cortex and hippocampus. *Neurobiol Aging*, 27, 733–40.

Dubois, B., Feldman, H.H., Jacova, C., *et al.* (2007). Research criteria for the diagnosis of Alzheimer's disease: revising the NINCDS-ADRDA criteria. *Lancet Neurol*, 6, 734–46.

Dubois, B., Feldman, H.H., Jacova, C., *et al.* (2010). Revising the definition of Alzheimer's disease: a new lexicon. *Lancet Neurol*, 9, 1118–27.

Eichenbaum, H. (2000). A cortical-hippocampal system for declarative memory. *Nat Rev Neurosci*, 1, 41–50.

Farlow, M.R., He, Y., Tekin, S., Xu, J., Lane, R., and Charles, H.C. (2004). Impact of APOE in mild cognitive impairment. *Neurology*, 63, 1898–901.

Fleisher, A., Grundman, M., Jack, C.R. Jr., *et al.* (2005). Sex, apolipoprotein E epsilon 4 status, and hippocampal volume in mild cognitive impairment. *Arch Neurol*, 62, 953–7.

Fleischer, A.S., Sun, S., Taylor, C., *et al.* (2008). Volumetric MRI vs clinical predictors of Alzheimer disease in mild cognitive impairment. *Neurology*, 70, 191–9.

Forsberg, A., Engler, H., Almkvist, O., *et al.* (2007). PET imaging of amyloid deposition in patients with mild cognitive impairment. *Neurobiol Aging*, 29, 1456–65.

Fotenos, A.F., Snyder, A.Z., Girton, L.E., Morris, J.C., and Buckner, R.L. (2005). Normative estimates of cross-sectional and longitudinal brain volume decline in aging and AD. *Neurology*, 64, 1032–9.

Frisoni, G.B., Laakso, M.P., Beltramello, A., *et al.* (1999). Hippocampal and entorhinal cortex atrophy in frontotemporal dementia and Alzheimer's disease. *Neurology*, 52, 91–100.

Frisoni, G.B., Fox, N.C., Jack, C.R. Jr., Scheltens, P., and Thompson, P.M. (2010). The clinical use of structural MRI in Alzheimer disease. *Nat Rev Neurol*, 6, 67–77.

Ganguli, M., Dodge, H.H., Shen, C., and DeKosky, S.T. (2004). Mild cognitive impairment, amnestic type: an epidemiologic study. *Neurology*, 63, 115–21.

Gauthier, S., Reisberg, B., Zaudig, M., *et al.* (2006). Mild cognitive impairment. *Lancet*, 367, 1262–70.

Gerardin, E., Chételat, G., Chupin, M., *et al.* (2009). Multidimensional classification of hippocampal shape features discriminates Alzheimer's disease and mild cognitive impairment from normal aging. *Neuroimage*, 47, 1476–86.

Geroldi, C., Pihlajamäki, M., Laakso, M.P., *et al.* (1999). APOE-epsilon4 is associated with less frontal and more medial temporal lobe atrophy in AD. *Neurology*, 53, 1825–32.

Geroldi, C., Rossi, R., Calvagna, C., *et al.* (2006). Medial temporal atrophy but not memory deficit predicts progression to dementia in patients with mild cognitive impairment. *J Neurol Neurosurg Psychiatry*, 77, 1219–22.

Golby, A., Silverberg, G., Race, E., *et al.* (2005). Memory encoding in Alzheimer's disease: an fMRI study of explicit and implicit memory. *Brain*, 128, 773–87.

Gómez-Isla, T., Price, J.L., McKeel, D.W. Jr, Morris, J.C., Growdon, J.H., and Hyman, B.T. (1996). Profound loss of layer II entorhinal cortex neurons occurs in very mild Alzheimer's disease. *J Neurosci*, 16, 4491- 500.

Gómez-Isla, T., Hollister, R., West, H., *et al.* (1997). Neuronal loss correlates with but exceeds neurofibrillary tangles in Alzheimer's disease. *Ann Neurol*, 41, 17–24.

Good, C.D., Johnsrude, I., Ashburner, J., Henson, R.N., Friston, K.J., and Frackowiak, R.S. (2001). Cerebral asymmetry and the effects of sex and handedness on brain structure: a voxel-based morphometric analysis of 465 normal adult human brains. *Neuroimage*, 14, 685–700.

Gosche, K.M., Mortimer, J.A., Smith, C.D., Markesbery, W.R., and Snowdon, D.A. (2002). Hippocampal volume as an index of Alzheimer neuropathology: findings from the Nun Study. *Neurology*, 58, 1476–82.

Grady, C.L., McIntosh, A.R., Beig, S., Keightley, M.L., Burian, H., and Black, S.E. (2003). Evidence from functional neuroimaging of a compensatory prefrontal network in Alzheimer's disease. *J Neurosci*, 23, 986–93.

Grön, G. and Riepe, M.W. (2004). Neural basis for the cognitive continuum in episodic memory from health to Alzheimer disease. *Am J Geriatr Psychiatry*, 12, 648–52.

Grundman, M., Sencakova, D., Jack, C.R. Jr., *et al.* (2002). Brain MRI hippocampal volume and prediction of clinical status in a mild cognitive impairment trial. *J Mol Neurosci*, 19, 23–7.

Grundman, M., Jack, C.R. Jr., Petersen, R.C., *et al.* (2003). Hippocampal volume is associated with memory but not nonmemory cognitive performance in patients with mild cognitive impairment. *J Mol Neurosci*, 20, 241–8.

Han, S.D., Houston, W.S., Jak AJ., *et al.* (2007). Verbal paired-associate learning by APOE genotype in non-demented older adults: fMRI evidence of a right hemispheric compensatory response. *Neurobiol Aging*, 28, 238–47.

Hashimoto, M. and Masliah, E. (2003). Cycles of aberrant synaptic sprouting and neurodegeneration in Alzheimer's and dementia with Lewy bodies. *Neurochem Res*, 28, 1743–56.

Hasselmo, M.E. and McClelland, J.L. (1999). Neural models of memory. *Curr Opin Neurobiol*, 9, 184–8.

Hämäläinen, A., Pihlajamäki, M., Tanila, H., *et al.* (2007a). Increased fMRI responses during encoding in mild cognitive impairment. *Neurobiol Aging*, 28, 1889–903.

Hämäläinen, A., Tervo, S., Grau-Olivares, M., *et al.* (2007b). Voxel-based morphometry to detect brain atrophy in progressive mild cognitive impairment. *Neuroimage*, 37, 1122–31.

Henry-Feugeas, M.C. (2007). MRI of the 'Alzheimer syndrome'. *J Neuroradiol*, 34, 220–7.

Henson, R.N., Cansino, S., Herron, J.E., Robb, W.G., and Rugg, M.D. (2003). A familiarity signal in human anterior medial temporal cortex? *Hippocampus*, 13, 301–4.

Herholz, K., Salmon, E., Perani, D., *et al.* (2002). Discrimination between Alzheimer dementia and controls by automated analysis of multicenter FDG PET. *Neuroimage*, 17, 302–16.

Hyman, B.T., Van Hoesen, G.W., Damasio, A.R., and Barnes, C.L. (1984). Alzheimer's disease: cell-specific pathology isolates the hippocampal formation. *Science*, 225, 1168–70.

Iadecola C. (2004). Neurovascular regulation in the normal brain and in Alzheimer's disease. *Nat Rev Neurosci*, 5, 347–60.

Jack, C.R.Jr., Petersen, R.C., O'Brien, P.C., and Tangalos, E.G. (1992). MR-based hippocampal volumetry in the diagnosis of Alzheimer's disease. *Neurology*, 42, 183–8.

Jack, C.R. Jr., Petersen, R.C., Xu, Y.C., *et al.* (1997). Medial temporal atrophy on MRI in normal aging and very mild Alzheimer's disease. *Neurology*, 49, 786–94.

Jack, C.R.Jr., Petersen, R.C., Xu, Y.C., *et al.* (1999). Prediction of AD with MRI-based hippocampal volume in mild cognitive impairment. *Neurology*, 52, 1397–403.

Jack, C.R. Jr., Dickson, D.W., Parisi, J.E., *et al.* (2002). Antemortem MRI findings correlate with hippocampal neuropathology in typical aging and dementia. *Neurology*, 58, 750–7.

Jack, C.R.Jr., Shiung, M.M., Gunter, J.L., *et al.* (2004). Comparison of different MRI brain atrophy rate measures 'with clinical disease progression in AD. *Neurology*, 62, 591–600.

Jack, C.R.Jr., Shiung, M.M., Weigand, S.D., *et al.* (2005). Brain atrophy rates predict subsequent clinical conversion in normal elderly and amnestic MCI. *Neurology*, 65, 1227–31.

Jack, C.R. Jr., Weigand, S.D., Shiung, M.M., *et al.* (2008). Atrophy rates accelerate in amnestic mild cognitive impairment. *Neurology*, 70, 1740–52.

Jagust, W.J., Zheng, L., Harvey, D.J., *et al.* (2008). Neuropathological basis of magnetic resonance images in aging and dementia. *Ann Neurol*, 63, 72–80.

Jauhiainen, A.M., Pihlajamäki, M., Tervo, S., *et al.* (2009). Discriminating accuracy of medial temporal lobe volumetry and fMRI in mild cognitive impairment. *Hippocampus*, 19, 166–75.

Jellinger, K.A. (2002). Alzheimer disease and cerebrovascular pathology: an update. *J Neural Transm*, 109, 813–36.

Jicha, G.A., Parisi, J.E., Dickson, D.W., *et al.* (2006). Neuropathologic outcome of mild cognitive impairment following progression to clinical dementia. *Arch Neurol*, 63, 674–81.

Johnson, K.A. and Albert, M.S. (2000). Perfusion abnormalities in prodromal AD. *Neurobiol Aging*, 21, 289–92.

Johnson, S.C., Baxter, L.C., Susskind-Wilder L., *et al.* (2004). Hippocampal adaptation to face repetition in healthy elderly and mild cognitive impairment. *Neuropsychologia*, 42, 980–9.

Johnson, S.C., Schmitz, T.W., Moritz, C.H., *et al.* (2006a). Activation of brain regions vulnerable to Alzheimer's disease: the effect of mild cognitive impairment. *Neurobiol Aging*, 27, 1604–12.

Johnson, S.C., Schmitz, T.W., Trivedi, M.A., *et al.* (2006b). The influence of Alzheimer disease family history and apolipoprotein E epsilon4 on mesial temporal lobe activation. *J Neurosci*, 26, 6069–76.

Johnson, S.C., Schmitz, T.W., Asthana, S., Gluck, M.A., and Myers, C. (2008). Associative learning over trials activates the hippocampus in healthy elderly but not mild cognitive impairment. *Neuropsychol Dev Cogn B Aging Neuropsychol Cogn*, 15, 129–45.

Julkunen, V., Niskanen, E., Muehlboeck, S., *et al.* (2009). Cortical thickness analysis to detect progressive mild cognitive impairment: a reference to Alzheimer's disease. *Dement Geriatr Cogn Disord*, 28, 404–12.

Julkunen, V., Niskanen, E., Koikkalainen, J., *et al.* (2010). Differences in cortical thickness in healthy controls, subjects with mild cognitive impairment, and Alzheimer's disease patients: A longitudinal study. *J Alzheimers Dis*, 21(4), 1141–51.

Juottonen, K., Laakso, M.P., Partanen, K., and Soininen, H. (1999). Comparative MR analysis of the entorhinal cortex and hippocampus in diagnosing Alzheimer disease. *Am J Neuroradiol*, 20, 139–44.

Kaduszkiewicz, H., Zimmermann, T., Beck-Bornholdt, H.P., and van den Bussche, H. (2005). Cholinesterase inhibitors for patients with Alzheimer's disease, systematic review of randomised clinical trials. *BMJ*, 331, 321–7.

Kantarci, K. and Jack, C.R.Jr. (2004). Quantitative magnetic resonance techniques as surrogate markers of Alzheimer's disease. *NeuroRx*, 1, 196–205.

Karas, G.B., Scheltens, P., Rombouts, S.A., *et al.* (2004). Global and local gray matter loss in mild cognitive impairment and Alzheimer's disease. *Neuroimage*, 23, 708–16.

Kato, T., Knopman, D., and Liu, H. (2001). Dissociation of regional activation in mild AD during visual encoding: a functional MRI study. *Neurology*, 57, 812–6.

Khachaturian, Z.S. (1985). Diagnosis of Alzheimer's disease. *Arch Neurol*, 42, 1097–105.

Khan, A.R., Wang, L., and Beg, M.F. (2008). FreeSurfer-initiated fully-automated subcortical brain segmentation in MRI using large deformation diffeomorphic metric mapping. *Neuroimage*, 41, 735–46.

Killiany, R.J., Gomez-Isla, T., Moss, M., *et al.* (2000). Use of structural magnetic resonance imaging to predict who will get Alzheimer's disease. *Ann Neurol*, 47, 430–9.

Killiany, R.J., Hyman, B.T., Gomez-Isla, T., *et al.* (2002). MRI measures of entorhinal cortex vs hippocampus in preclinical AD. *Neurology*, 58, 1188–96.

Kircher, T.T., Weis, S., Freymann, K., *et al.* (2007). Hippocampal activation in patients with mild cognitive impairment is necessary for successful memory encoding. *J Neurol Neurosurg Psychiatry*, 78, 812–18.

Klunk, W.E., Engler, H., Nordberg, A., *et al.* (2004). Imaging brain amyloid in Alzheimer's disease with Pittsburgh Compound-B. *Ann Neurol*, 55, 306–19.

Kobayashi, Y. and Amaral, D.G. (2003). Macaque monkey retrosplenial cortex: II. Cortical afferents. *J Comp Neurol*, 466, 48–79.

Kobayashi, Y. and Amaral, D.G. (2007). Macaque monkey retrosplenial cortex: III. Cortical efferents. *J Comp Neurol*, 502, 810–33.

Kordower, J.H., Chu, Y., Stebbins, G.T., *et al.* (2001). Loss and atrophy of layer II entorhinal cortex neurons in elderly people with mild cognitive impairment. *Ann Neurol*, 49, 202–13.

Korf, E.S., Wahlund, L.O., Visser, P.J., and Scheltens, P. (2004). Medial temporal lobe atrophy on MRI predicts dementia in patients with mild cognitive impairment. *Neurology*, 63, 94–100.

Kwong, K.K., Belliveau, J.W., Chesler, D.A., *et al.* (1992). Dynamic magnetic resonance imaging of human brain activity during primary sensory stimulation. *Proc Natl Acad Sci U S A*, 89, 5675–9.

Laakso, M.P., Partanen, K., Riekkinen, P., *et al.* (1996). Hippocampal volumes in Alzheimer's disease, Parkinson's disease with and without dementia, and in vascular dementia: An MRI study. *Neurology*, 46, 678–81.

Laakso, M.P., Hallikainen, M., Hänninen, T., Partanen, K., and Soininen, H. (2000). Diagnosis of Alzheimer's disease: MRI of the hippocampus vs delayed recall. *Neuropsychologia*, 38, 579–84.

Larrieu, S., Letenneur, L., Orgogozo, J.M., *et al.* (2002). Incidence and outcome of mild cognitive impairment in a population-based prospective cohort. *Neurology*, 59, 1594–9.

Lavenex, P. and Amaral, D.G. (2000). Hippocampal-neocortical interaction: a hierarchy of associativity. *Hippocampus*, 10, 420–30.

Lehtovirta, M., Laakso, M.P., Soininen, H., *et al.* (1995). Volumes of hippocampus, amygdala and frontal lobe in Alzheimer patients with different apolipoprotein E genotypes. *Neuroscience*, 67, 65–72.

Lehtovirta, M., Soininen, H., Laakso, M.P., *et al.* (1996). SPECT and MRI analysis in Alzheimer's disease: relation to apolipoprotein E epsilon 4 allele. *J Neurol Neurosurg Psychiatry*, 60, 644–9.

Li, S., Shi, F., Pu, F., *et al.* (2007). Hippocampal shape analysis of Alzheimer disease base'd on machine learning methods. *Am J Neuroradiol*, 28, 1339–45.

Lind, J., Persson, J., Ingvar, M., *et al.* (2006). Reduced functional brain activity response in cognitively intact apolipoprotein E epsilon4 carriers. *Brain*, 129, 1240–8.

Liu, Y., Paajanen, T., Zhang, Y., *et al.* (2010). Analysis of regional MRI volumes and thicknesses as predictors of conversion from mild cognitive impairment to Alzheimer's disease. *Neurobiol Aging*, 31, 1375–85.

Lupien, S.J., Evans, A., Lord, C., *et al.* (2007). Hippocampal volume is as variable in young as in older adults: implications for the notion of hippocampal atrophy in humans. *Neuroimage*, 34, 479–85.

Lötjönen, J.M., Wolz, R., Koikkalainen, J.R., *et al.* (2010). Fast and robust multi-atlas segmentation of brain magnetic resonance images. *Neuroimage*, 49, 2352–65.

Machulda, M.M., Ward, H.A., Borowski, B., *et al.* (2003). Comparison of memory fMRI response among normal, MCI, and Alzheimer's patients. *Neurology*, 61, 500–6.

Mandzia, J.L., McAndrews, M.P., Grady, C.L., Graham, S.J., and Black, S.E. (2009). Neural correlates of incidental memory in mild cognitive impairment: an fMRI study. *Neurobiol Aging*, 30, 717–30.

Markesbery, W.R., Schmitt, F.A., Kryscio, R.J., Davis, D.G., Smith, C.D., and Wekstein, D.R. (2006). Neuropathologic substrate of mild cognitive impairment. *Arch Neurol*, 63, 38–46.

Maurer, K., Volk, S., and Gerbaldo, H. (1997). Auguste D and Alzheimer's disease. *Lancet*, 349, 1546–9.

Mayeux, R. (2004). Biomarkers: potential uses and limitations. *NeuroRx*, 1, 182–8.

McKhann, G., Drachman, D., Folstein, M., Katzman, R., Price, D., and Stadlan, E.M. (1984). Clinical diagnosis of Alzheimer's disease: report of the NINCDS-ADRDA Work Group under the auspices of Department of Health and Human Services Task Force on Alzheimer's Disease.' *Neurology*, 34, 939–44.

Meguro, K., Blaizot, X., Kondoh, Y., Le Mestric, C., Baron, J.C., and Chavoix, C. (1999). Neocortical and hippocampal glucose hypometabolism following neurotoxic lesions of the entorhinal and perirhinal cortices in the non-human primate as shown by PET. Implications for Alzheimer's disease. *Brain*, 122, 1519–31.

Meguro, K., LeMestric, C., Landeau, B., Desgranges, B., Eustache, F., and Baron, J.C. (2001). Relations between hypometabolism in the posterior association neocortex and hippocampal atrophy in Alzheimer's disease: a PET/MRI correlative study. *J Neurol Neurosurg Psychiatry*, 71, 315–21.

Mesulam, M.M. (1998). From sensation to cognition. *Brain*, 121, 1013–52.

Miller, S.L., Fenstermacher, E., Bates, J., Blacker, D., Sperling, R.A., and Dickerson, B.C. (2008). Hippocampal activation in adults with mild cognitive impairment predicts subsequent cognitive decline. *J Neurol Neurosurg Psychiatry*, 79, 630–5.

Minoshima, S., Giordani, B., Berent, S., Frey, K.A., Foster, N.L., and Kuhl, D.E. (1997). Metabolic reduction in the posterior cingulate cortex in very early Alzheimer's disease. *Ann Neurol*, 42, 85–94.

Mintun, M.A., Larossa, G.N., Sheline, Y.I., *et al*. (2006). [11C]PIB in a nondemented population, potential antecedent marker of Alzheimer disease. *Neurology*, 67, 446–52.

Mirra, S.S., Heyman, A., McKeel, D., *et al*. (1991). The Consortium to Establish a Registry for Alzheimer's Disease (CERAD). Part II. Standardization of the neuropathologic assessment of Alzheimer's disease. *Neurology*, 41, 479–86.

Moffat, S.D., Szekely, C.A., Zonderman, A.B., Kabani, N.J., and Resnick, S.M. (2000). Longitudinal change in hippocampal volume as a function of apolipoprotein E genotype. *Neurology*, 55, 134–6.

Mondadori, C.R., Buchmann, A., Mustovic, H., *et al*. (2006). Enhanced brain activity may precede the diagnosis of Alzheimer's disease by 30 years. *Brain*, 129, 2908–22.

Morra, J.H., Tu, Z., Apostolova, L.G., *et al*. (2008). Validation of a fully automated 3D hippocampal segmentation method using subjects with Alzheimer's disease mild cognitive impairment, and elderly controls. *Neuroimage*, 43, 59–68.

Morra, J.H., Tu, Z., Apostolova, L.G., *et al*. (2009). Automated 3D mapping of hippocampal atrophy and its clinical correlates in 400 subjects with Alzheimer's disease, mild cognitive impairment, and elderly controls. *Hum Brain Mapp*, 30, 2766–88.

Morris, J.C. and Price, A.L. (2001). Pathologic correlates of nondemented aging, mild cognitive impairment, and early-stage Alzheimer's disease. *J Mol Neurosci*, 17, 101–18.

Mortimer, J.A., Gosche, K.M., Riley, K.P., *et al*. (2004). Delayed recall, hippocampal volume and Alzheimer neuropathology: findings from the Nun Study. *Neurology*, 62, 428–32.

Mosconi., L., De Santi, S., Li, J., *et al*. (2008a). Hippocampal hypometabolism predicts cognitive decline from normal aging. *Neurobiol Aging*, 29, 676–92.

Mosconi, L., Tsui, W.H., Herholz, K., *et al*. (2008). Multicenter standardized 18F-FDG PET diagnosis of mild cognitive impairment, Alzheimer's disease, and other dementias. *J Nucl Med*, 49, 390–8.

Mueller, E.A., Moore, M.M., Kerr, D.C., *et al*. (1998). Brain volume preserved in healthy elderly through the eleventh decade. *Neurology*, 51, 1555–62.

Mueller, S.G. and Weiner, M.W. (2009). Selective effect of age, Apo e4, and Alzheimer's disease on hippocampal subfields. *Hippocampus*, 19, 558–64.

Mueller, S.G., Schuff, N., Yaffe, K., Madison, C., Miller, B., and Weiner, M.W. (2010). Hippocampal atrophy patterns in mild cognitive impairment and Alzheimer's disease. *Hum Brain Mapp*, 31, 1339–47.

O'Brien, J.L., O'Keefe, K.M., LaViolette, P.S., *et al*. (2010). Longitudinal fMRI in elderly reveals loss of hippocampal activation with clinical decline. *Neurology*, 74, 1969–76.

Ogawa, S., Tank, D.W., Menon, R., *et al*. (1992). Intrinsic signal changes accompanying sensory stimulation: functional brain mapping with magnetic resonance imaging. *Proc Natl Acad Sci U S A*, 89, 5951–5.

Ohm, T.G., Kirca, M., Bohl, J., Scharnagl, H., Gross, W., and März W. (1995). Apolipoprotein E polymorphism influences not only cerebral senile plaque load but also Alzheimer-type neurofibrillary tangle formation. *Neuroscience*, 66, 583–7.

Pantel, J., Kratz, B., Essig, M., and Schroder, J. (2003). Parahippocampal volume deficits in subjects with aging-associated cognitive decline. *Am J Psychiatry*, 160, 379–82.

Pariente, J., Cole, S., Henson, R., *et al*. (2005). Alzheimer's patients engage an alternative network during a memory task. *Ann Neurol*, 58, 870–9.

Parvizi, J., Van Hoesen, G.W., Buckwalter, J., and Damasio, A. (2006). Neural connections of the posteromedial cortex in the macaque. *Proc Natl Acad Sci U S A*, 103, 1563–8.

Pennanen, C., Kivipelto, M.,Tuomainen, S., *et al*. (2004). Hippocampus and entorhinal cortex in mild cognitive impairment and early AD. *Neurobiol Aging*, 25, 303–10.

Pennanen, C., Testa, C., Laakso, M.P., *et al*. (2005). A voxel based morphometry study on mild cognitive impairment. *J Neurol Neurosurg Psychiatry*, 76, 11–4.

Pennanen, C., Testa, C. and Boccardi, M., *et al*. (2006). The effect of apolipoprotein polymorphism on brain in mild cognitive impairment, a voxel-based morphometric study. *Dement Geriatr Cogn Disord*, 22, 60–6.

Petersen, R.C., Smith, G.E., Ivnik, R.J., *et al*. (1995). Apolipoprotein E status as a predictor of the development of Alzheimer's disease in memory-impaired individuals. *JAMA*, 273, 1274–8.

Petersen, R.C., Smith, G.E., Waring, S.C., Ivnik, R.J., Tangalos, E.G., and Kokmen, E. (1999). Mild cognitive impairment: clinical characterization and outcome. *Arch Neurol*, 56, 303–8.

Petersen, R.C., Doody, R., Kurz, A., *et al*. (2001). Current concepts in mild cognitive impairment. *Arch Neurol*, 58, 1985–92.

Petersen, R.C. (2004). Mild cognitive impairment as a diagnostic entity. *J Intern Med*, 256, 183–94.

Petersen, R.C., Roberts, R.O., Knopman, D.S., *et al*. (2009). Mild cognitive impairment: ten years later. *Arch Neurol*, 66, 1447–55.

Petrella, J.R., Krishnan, S., Slavin, M.J., Tran, T.T., Murty, L., and Doraiswamy, P.M. (2006). Mild cognitive impairment: evaluation with 4-T functional MR imaging. *Radiology*, 240, 177–86.

Pihlajamäki, M., Tanila, H., Hänninen, T., *et al*. (2000). Verbal fluency activates the left medial temporal lobe: a functional magnetic resonance imaging study. *Ann Neurol*, 47, 470–6.

Pihlajamäki, M., Tanila, H., Hänninen, T., *et al*. (2003). Encoding of novel picture pairs activates the perirhinal cortex: an fMRI study. *Hippocampus*, 13, 67–80.

Pihlajamäki, M., Tanila, H., Könönen, M., *et al*. (2004). Visual presentation of novel objects and new spatial arrangements of objects differentially activates the medial temporal lobe subareas in humans. *Eur J Neurosci*, 19, 1939–49.

Pihlajamäki, M., DePeau, K.M., Blacker, D., and Sperling, R.A. (2008). Impaired medial temporal repetition suppression is related to failure of parietal deactivation in Alzheimer disease. *Am J Geriatr Psychiatry*, 16, 283–92.

Pihlajamäki, M. and Sperling, R.A. (2009). Functional MRI assessment of task-induced deactivation of the default mode network in Alzheimer's disease and at-risk older individuals. *Behav Neurol*, 21, 77–91.

Pihlajamäki, M., O'Keefe, K., O'Brien, J., Blacker, D., and Sperling, R.A. (2011). Failure of repetition suppression and memory encoding in aging and Alzheimer's disease. *Brain Imaging Behav*, 5, 36–44.

Pruessner, J.C., Collins, D.L., Pruessner, M., and Evans, A.C. (2001). Age and gender predict volume decline in the anterior and posterior hippocampus in early adulthood. *J Neurosci*, 21, 194–200.

Putcha, D., O'Keefe, K., LaViolette, P., *et al.* (2011). Reliability of functional magnetic resonance imaging associative encoding memory paradigms in non-demented elderly adults. *Hum Brain Mapp*, 32(12), 2027–44.

Qiu, A., Fennema-Notestine, C., Dale, A.M., and Miller, M.I.; Alzheimer's Disease Neuroimaging Initiative. (2009). Regional shape abnormalities in mild cognitive impairment and Alzheimer's disease. *Neuroimage*, 45, 656–61.

Rand-Giovannetti, E., Chua, E.F., Driscoll, A.E., *et al.* (2006). Hippocampal and neocortical activation during repetitive encoding in older persons. *Neurobiol Aging*, 27, 173–82.

Rapoport, S. I. (1991). Positron emission tomography in Alzheimer's disease in relation to disease pathogenesis: a critical review. *Cerebrovasc Brain Metab Rev*, 3, 297–335.

Reiman, E.M., Caselli, R.J., Yun, L.S., *et al.* (1996). Preclinical evidence of Alzheimer's disease in persons homozygous for the e4 allele for apolipoprotein E. *N Engl J Med*, 334, 752–8.

Remy, F., Mirrashed, F., Campbell, B., and Richter, W. (2005). Verbal episodic memory impairment in Alzheimer's disease: a combined structural and functional MRI study. *Neuroimage*, 25, 253–66.

Ridha, B.H., Anderson, V.M., Barnes, J., *et al.* (2008). Volumetric MRI and cognitive measures in Alzheimer disease: comparison of markers of progression. *J Neurol*, 255, 567–74.

Rombouts, S.A., Barkhof, F., Veltman, D.J., *et al.* (2000). Functional MR imaging in Alzheimer's disease during memory encoding. *Am J Neuroradiol*, 21, 1869–75.

Rombouts, S.A., Barkhof, F., Goekoop, R., Stam, C.J., and Scheltens, P. (2005). Altered resting state networks in mild cognitive impairment and mild Alzheimer's 'disease: an fMRI study. *Hum Brain Mapp*, 26, 231–9.

Rusinek, H., De Santi, S., Frid, D., *et al.* (2003). Regional brain atrophy rate predicts future cognitive decline: 6-year longitudinal MR imaging study of normal aging. *Radiology*, 229, 691–6.

Rusinek, H., Endo, Y., De Santi, S., *et al.* (2004). Atrophy rate in medial temporal lobe during progression of Alzheimer disease. *Neurology*, 63, 2354–9.

Saunders, A.M., Strittmatter, W.J., Schmechel, D., *et al.* (1993). Association of apolipoprotein E allele epsilon 4 with late-onset familial and sporadic Alzheimer's disease. *Neurology*, 43, 1467–72.

Scahill, R.I., Schott, J.M., Stevens, J.M., Rossor, M.N., and Fox, N.C. (2002). Mapping the evolution of regional atrophy in Alzheimer's disease: unbiased analysis of fluid-registered serial MRI. *Proc Natl Acad Sci U S A*, 99, 4703–7.

Scahill, R.I., Frost, C., Jenkins, R., Whitwell, J.L., Rossor, M.N., and Fox, N.C. (2003). A longitudinal study of brain volume changes in normal aging using serial registered magnetic resonance imaging. *Arch Neurol*, 60, 989–94.

Scheff, S.W., Price, D.A., Schmitt, F.A., and Mufson, E.J. (2006). Hippocampal synaptic loss in early Alzheimer's disease and mild cognitive impairment. *Neurobiol Aging*, 27, 1372–84.

Scheff, S.W., Price, D.A., Schmitt, F.A., DeKosky, S.T., and Mufson, E.J. (2007). Synaptic alterations in CA1 in mild Alzheimer disease and mild cognitive impairment. *Neurology*, 68, 1501–8.

Scheltens, P., Leys, D.,Barkhof, F., *et al.* (1992). Atrophy of medial temporal lobes on MRI in 'probable' Alzheimer's disease and normal ageing: diagnostic value and neuropsychological correlates. *J Neurol Neurosurg Psychiatry*, 55, 967–72.

Scher, A.I., Xu, Y., Korf, E.S., *et al.* (2007). Hippocampal shape analysis in Alzheimer's disease: a population-based study. *Neuroimage*, 36, 8–18.

Seab, J.P., Jagust, W.J., Wong, S.T., Roos, M.S., Reed, B.R., and Budinger, T.F. (1988). Quantitative NMR measurements of hippocampal atrophy in Alzheimer's disease. *Magn Reson Med*, 8, 200–8.

Selkoe, D.J. (2001). Alzheimer's disease: genes, proteins, and therapy. *Physiol Rev*, 81, 741–66.

Selkoe, D.J. (2002). Alzheimer's disease is a synaptic failure. *Science*, 298, 789–91.

Shi, F., Liu, B., Zhou, Y., Yu, C., and Jiang, T. (2009). Hippocampal volume and asymmetry in mild cognitive impairment and Alzheimer's disease: Meta-analyses of MRI studies. *Hippocampus*, 19, 1055–64.

Schuff, N., Amend, D.L., Knowlton, R., Norman, D., Fein, G., and Weiner, M.W. (1999). Age-related metabolite changes and volume loss in the hippocampus by magnetic resonance spectroscopy and imaging. *Neurobiol Aging*, 20, 279–85.

Schuff, N., Woerner, N., Boreta, L., *et al.* (2009). MRI of hippocampal volume loss in early Alzheimer's disease in relation to ApoE genotype and biomarkers. *Brain*, 132, 1067–77.

Small, S.A., Perera, G.M., DeLaPaz, R., Mayeux, R., and Stern, Y. (1999). Differential regional dysfunction of the hippocampal formation among elderly with memory decline and Alzheimer's disease. *Ann Neurol*, 45, 466–72.

Smith, C.D., Andersen, A.H., Kryscio, R.J., *et al.* (1999). Altered brain activation in cognitively intact individuals at high risk for Alzheimer's disease. *Neurology*, 53, 1391–6.

Smith, C.D., Andersen, A.H., Kryscio, R.J., *et al.* (2002). Women at risk for AD show increased parietal activation during a fluency task. *Neurology*, 58, 1197–202.

Soininen, H., Partanen, K., Pitkänen, A., *et al.* (1995). Decreased hippocampal volume asymmetry on MRIs in nondemented elderly subjects carrying the apolipoprotein E epsilon 4 allele. *Neurology*, 45, 391–2.

Sperling, R.A., Bates, J.F., Cocchiarella, A.J., Schacter, D.L., Rosen, B.R., and Albert, M.S. (2001). Encoding novel face-name associations: a functional MRI study. *Hum Brain Mapp*, 14, 129–39.

Sperling, R.A., Bates, J.F., Chua, E.F., *et al.* (2003a). fMRI studies of associative encoding in young and elderly controls and mild Alzheimer's disease. *J Neurol Neurosurg Psychiatry*, 74, 44–50.

Sperling, R., Chua, E.,Cocchiarella, A., *et al.* (2003b). Putting names to faces: successful encoding of associative memories activates the anterior hippocampal formation. *Neuroimage*, 20, 1400–10.

Stern, C.E., Corkin, S., González, R.G., *et al.* (1996). The hippocampal formation participates in novel picture encoding: evidence from functional magnetic resonance imaging. *Proc Natl Acad Sci U S A*, 93, 8660–5.

Stern, E.A., Bacskai, B.J., Hickey, G.A., Attenello, F.J., Lombardo, J.A., and Hyman, B.T. (2004). Cortical synaptic integration in vivo is disrupted by amyloid-beta plaques. *J Neurosci*, 24, 4535–40.

Squire, L.R. and Zola-Morgan, S. (1991). The medial temporal lobe memory system. *Science*, 253, 1380–6.

Squire, L.R. and Zola, S.M. (1996). Structure and function of declarative and nondeclarative memory systems. *Proc Natl Acad Sci U S A*, 93, 13515–22.

Suzuki, W.A. (2007). Making new memories: the role of the hippocampus in new associative learning. *Ann N Y Acad Sci*, 1097, 1–11.

Tanzi, R.E. and Bertram, L. (2001). New frontiers in Alzheimer's disease genetics. *Neuron*, 32, 181–4.

Tapiola, T., Pennanen, C., Tapiola, M., *et al.* (2008). MRI of hippocampus and entorhinal cortex in mild cognitive impairment: a follow-up study. *Neurobiol Aging*, 29, 31–8.

Thal, D.R., Rub, U., Orantes, M., Braak H. (2002). Phases of A beta-deposition in the human brain and its relevance for the development of AD. *Neurology*, 58, 1791–800.

Thompson, P. M., Hayashi, K.M., De Zubicaray, G.I., *et al.* (2004). Mapping hippocampal and ventricular change in Alzheimer disease. *Neuroimage*, 22, 1754–66.

Tiraboschi, P., Hansen, L.A., Alford, M., *et al.* (2000). The decline in synapses and cholinergic activity is asynchronous in Alzheimer's disease. *Neurology*, 55, 1278–83.

Tisserand, D.J., van Boxtel, M.P., Pruessner, J.C., Hofman, P., Evans, A.C., and Jolles, J. (2004). A voxel-based morphometric study to determine individual differences in gray matter density associated with age and cognitive change over time. *Cereb Cortex*, 14, 966–73.

Trivedi, M.A., Schmitz, T.W., Ries, M.L., *et al.* (2006). Reduced hippocampal activation during episodic encoding in middle-aged individuals at genetic risk of Alzheimer's disease, a cross-sectional study. *BMC Med*, 4, 1.

Tulving, E. and Markowitsch, H.J. (1998). Episodic and declarative memory: role of the hippocampus. *Hippocampus*, 8, 198–204.

van de Pol, L.A., Hensel, A., Barkhof, F., Gertz, H.J., Scheltens, P., and van der Flier, W.M. (2006). Hippocampal atrophy in Alzheimer disease: age matters. *Neurology*, 66, 236–8.

van de Pol, L.A., van der Flier, W.M., Korf, E.S., Fox, N.C., Barkhof, F., and Scheltens, P. (2007). Baseline predictors of rates of hippocampal atrophy in mild cognitive impairment. *Neurology*, 69, 1491–7.

van der Flier, W.M., van Straaten, E.C., Barkhof, F., *et al.* (2005). Medial temporal lobe atrophy and white matter hyperintensities are associated with mild cognitive deficits in non-disabled elderly people: the LADIS study. *J Neurol Neurosurg Psychiatry*, 76, 1497–500.

van der Flier, W.M., Pijnenburg, Y.A., Fox, N.C., and Scheltens, P. (2011). Early-onset versus late-onset Alzheimer's disease: the case of the missing APOE 4 allele. *Lancet Neurol*, 10, 280–8.

Vannini, P., Almkvist, O., Dierks, T., Lehmann, C., and Wahlund, L.O. (2007). Reduced neuronal efficacy in progressive mild cognitive impairment: a prospective fMRI study on visuospatial processing. *Psychiatry Res*, 156, 43–57.

Visser, P.J., Verhey, F.R., Hofman, P.A., Scheltens, P., and Jolles, J. (2002). Medial temporal lobe atrophy predicts Alzheimer's disease in patients with minor cognitive impairment. *J Neurol Neurosurg Psychiatry*, 72, 491–7.

Wagner, A.D., Schacter, D.L., Rotte, M., *et al.* (1998). Building memories: remembering and forgetting of verbal experiences as predicted by brain activity. *Science*, 281, 1188–91.

Wahlund, L.O., Julin, P., Lindqvist, J., Scheltens, P. (1999). Visual assessment of medical temporal lobe atrophy in demented and healthy control subjects: correlation with volumetry. *Psychiatry Res*, 90, 193–9.

Wahlund, L.O., Julin, P., Johansson, S.E., Scheltens, P. (2000). Visual rating and volumetry of the medial temporal lobe on magnetic resonance imaging in dementia: a comparative study. *J Neurol Neurosurg Psychiatry*, 69, 630–5.

Walsh, D.M. and Selkoe, D.J. (2004). Deciphering the molecular basis of memory failure in Alzheimer's disease. *Neuron*, 44, 181–93.

Wang, L., Swank, J.S., Glick, I.E., *et al.* (2003). Changes in hippocampal volume and shape across time distinguish dementia of the Alzheimer type from healthy aging. *Neuroimage*, 20, 667–82.

West, M.J., Coleman, P.D., Flood, D.G., Troncoso, J.C. (1994). Differences in the pattern of hippocampal neuronal loss in normal ageing and Alzheimer's disease. *Lancet*, 344, 769–72.

West, M.J., Kawas, C.H., Martin, L.J., Troncoso, J.C. (2000). The CA1 region of the human hippocampus is a hot spot in Alzheimer's disease. *Ann N Y Acad Sci*, 908, 255–9.

West, M.J., Kawas, C.H., Stewart, W.F., Rudow, G.L., Troncoso, J.C. (2004). Hippocampal neurons in pre–clinical Alzheimer's disease. *Neurobiol Aging*, 25, 1205–12.

Winblad, B., Palmer, K., Kivipelto, M., *et al.* (2004). Mild cognitive impairment – beyond controversies, towards a consensus: report of the International Working Group on Mild Cognitive Impairment. *J Intern Med*, 256, 240–6.

Wishart, H.A., Saykin, A.J., Rabin, L.A., *et al.* (2006). Increased brain activation during working memory in cognitively intact adults with the APOE epsilon4 allele. *Am J Psychiatry*, 163, 1603–10.

Wolf, H., Grunwald, M., Kruggel, F., *et al.* (2001). Hippocampal volume discriminates between normal cognition; questionable and mild dementia in the elderly. *Neurobiol Aging*, 22, 177–86.

Wolf, H., Hensel, A., Kruggel, F., *et al.* (2004). Structural correlates of mild cognitive impairment. *Neurobiol Aging*, 25, 913–24.

Woodard, J.L., Seidenberg, M., Nielson, K.A., *et al.* (2009). Semantic memory activation in amnestic mild cognitive impairment. *Brain*, 132, 2068–78.

Xu, Y., Jack, C.R. Jr., O'Brien, P.C., *et al.* (2000). Usefulness of MRI measures of entorhinal cortex versus hippocampus in AD. *Neurology*, 54, 1760–7.

Yassa, M.A., Stark, S.M., Bakker, A., Albert, M.S., Gallagher, M., and Stark, C.E. (2010). High-resolution structural and functional MRI of hippocampal CA3 and dentate gyrus in patients with amnestic Mild Cognitive Impairment. *Neuroimage*, 51, 1242–52.

Zarow, C., Vinters, H.V., Ellis, W.G., *et al.* (2005). Correlates of hippocampal neuron number in Alzheimer's disease and ischemic vascular dementia. *Ann Neurol*, 57, 896–903.

CHAPTER 13

Temporal Lobe Epilepsy and the Hippocampus

Radwa A.B. Badawy and Graeme D. Jackson

Introduction 245
Mesial temporal sclerosis and epileptogenicity 246
Clinical picture 248
Electroencephalographic findings 250
Neuroimaging in temporal lobe epilepsy 252
Nuclear neuroimaging with positron emission tomography (PET) and single-photon emission computed tomography (SPECT) 254
Treatment 256

ABSTRACT Hippocampal sclerosis (HS) is the commonest cause of temporal lobe epilepsy which accounts for 3–50% of all epilepsy syndromes. In this chapter we review the pathological substrate of HS and mesial temporal sclerosis and how this relates to epileptogenesis and seizures. We then discuss the clinical picture of different forms of temporal lobe epilepsy and common findings on electroencephalography and imaging. Invasive and non-invasive diagnostic tools assessing epileptogenicity of the hippocampus are reviewed. Finally, we present an overview of management options available for patients with temporal lobe epilepsy.

KEYWORDS medial and lateral temporal lobe epilepsy, mesial temporal sclerosis, hippocampal sclerosis focal seizures, convulsive seizures, loss of awareness, electroencephalography, magnetic resonance imaging, functional imaging treatment of temporal lobe epilepsy

Introduction

The term 'epilepsy' encompasses a complex group of conditions characterized by episodic brain dysfunction with multifactorial causes reflecting acquired and genetic factors (Berkovic et al. 2006). The epilepsy syndromes can be subdivided into focal and generalized types depending on whether seizures arise from a localized brain area or show widespread involvement of both hemispheres from the outset. Temporal lobe epilepsy (TLE) is a focal epilepsy characterized by seizures originating in or primarily involving temporal lobe structures (Commission on Classification and Terminology of the ILAE 1981, 1989). It accounts for approximately 30–50% of all epilepsies and is divided into two major types: mesial temporal lobe epilepsy (MTLE) originating in the hippocampus, parahippocampal gyrus, and amygdala which are located in the inner aspect of the temporal lobe and lateral (neocortical) temporal lobe epilepsy (LTLE; less than 10% of cases), which is thought to originate in the neocortex on the outer surface of the temporal lobe of the brain. It should be noted, however, that because of strong interconnections, seizures beginning in either the mesial or lateral temporal areas often spread to involve both areas and also to neighbouring areas on the same side of the brain as well as the temporal lobe on the opposite side of the brain.

The aetiology of TLE is heterogeneous with many cases being due to structural lesions, although familial forms are increasingly recognized especially in LTLE (Raymond et al. 1994; Berkovic et al. 1996). Hippocampal sclerosis (HS) is the most common cause of MTLE (Gates and Cruz-Rodriguez 1990). This is mainly due to the extensive reciprocal connections between the entorhinal cortex, hippocampus, and other brain regions that make it a potential candidate for generation and propagation of seizures (Rutecki et al. 1989; Stringer and Lothman 1992). Other structural causes for TLE include tumours, vascular, malformations of cortical development, traumatic, viral and other infectious and parasitic disorders, and cerebrovascular disease.

Mesial temporal sclerosis and epileptogenicity

The terms mesial temporal sclerosis (MTS) hippocampal sclerosis (HS), and Ammon's horn sclerosis, are used almost interchangeably. However, these terms are, strictly, not synonymous since the distribution of pathological changes is different between these entities (Margerison and Corsellis 1966; Babb 1986). MTS is the histological substrate found in approximately 65% of temporal lobectomy specimens obtained in connection with TLE (Commission on Classification and Terminology of the ILAE 1981, 1989; Conference 1990; Xu et al. 2007). MTS encompasses changes in the amygdala, hippocampus, and the adjacent entorhinal cortex (Margerison and Corsellis 1966). HS consists of gliosis and neuronal loss primarily in the hilar region and CA1 pyramidal region of the hippocampus, mild cell loss (if at all) in the CA2 pyramidal region, and an intermediate lesion in the CA3 and CA4 pyramidal regions, dentate granule cells, and subiculum. Neuronal loss is accompanied by fibrillary gliosis leading to hippocampal atrophy. In Ammon's horn sclerosis, the abnormalities are restricted to areas CA1–CA4 exclusively (Falconer 1974). The CA3 region is most prone to epileptiform activity, partly because of excitatory connections with neighbouring cells (Miles and Wong 1986). Conversely, it is difficult to induce seizures in granule cells, in part due to the lack of excitatory connections with neighbouring granule cells and the presence of strong polysynaptic inhibitory synapses on granule cells (McNamara 1999). However, structural reorganization with alteration of the synaptic circuitry may transform the granule cells into an epileptogenic population (Babb and Brown 1986; Babb et al. 1991) and promote seizure initiation and/or propagation (Ribak and Dashtipour 2002).

Cellular mechanisms

The electrical properties of epileptic neurons have been studied with *in vitro* slice preparations of animal or human neocortex or hippocampus, *in vivo* animal models of seizures induced by proconvulsants such as penicillin, and *in vivo* human studies using depth electrical recordings in patients undergoing epilepsy surgery (for a detailed review refer to McCormick and Contreras (2001)). These studies have shown that there are excessive discharges in epileptic neurons. The main intracellular component of epileptic discharges is an overt depolarization, called a paroxysmal depolarizing shift (PDS) that occurs synchronously in a group of neurons (McCormick and Contreras 2001). PDS has a slow component of sustained depolarization lasting tens of milliseconds with superadded rapid sharp depolarizations called sustained repetitive or burst firing. Under normal physiologic conditions, in the CA3 region of the hippocampus (Schwartzkroin and Haglund 1986), there is a subset of pyramidal neurons that can intrinsically generate a burst firing pattern in response to a brief depolarization. In the epileptic brain it is in these regions that the PDS tends to be seen first (Connors 1984; Jefferys 1994). Sustained burst (repetitive) firing is generated through the activation in the dendrites of slow action potentials mediated by Na^+ and Ca^{2+} currents, providing a sustained depolarization of the neuron (Traub and Jefferys 1994). The termination of the PDS and neuronal burst firing is predominantly mediated by activation of outward K^+ currents and possibly through inactivation of inward currents (McCormick and Contreras 2001). In addition, termination of the PDS and burst firing may be mediated by contributions from both $GABA_A$ and $GABA_B$ inhibitory conductances (McCormick 1989; Scanziani et al. 1994) or may involve decoupling of gap-junction mediated currents, which are sensitive to changes in extracellular pH (Perez-Velazquez et al. 1994).

Whether TLE is the cause or result of MTS is controversial. It seems likely that seizures are the initial cause of hippocampal injury, which then develops into HS. After HS is present, it then is independently epileptogenic and can be the cause of most of the subsequent intractable seizures (Jackson et al. 1998, 1999; Briellmann et al. 2005, 2007). Whatever caused the initial seizure may or may not persist and be independently intractable (Scott et al. 2003), or cause occasional seizures after hippocampal surgery.

Experimentally, animal models of status epilepticus and more than 100 brief convulsions induced by kindling seizures are able to reproduce this pattern of injury (Cavazos et al. 1994). The available evidence suggests that status epilepticus and chronic epilepsy with recurrent seizures are associated with neuronal injury and reactive changes in human and animal hippocampi such that HS is generally considered an acquired lesion (Holmes 2002). In addition, many reports emphasize the association between initial preceding events and development of MTS (Sagar and Oxbury 1987). Patients with chronic refractory TLE associated with HS often (60–80% of cases) report a significant event that happened several months or years prior to the onset of habitual seizures. The most common initial precipitating injury is prolonged or complicated febrile seizures during infancy or early childhood (Falconer 1974; Sagar and Oxbury 1987; Luna et al., 1989; French et al. 1993).

In humans, it has been proposed that there is a critical period (up to 5 years) when the hippocampus is susceptible to damage from prolonged seizures (Mathern et al. 1994, 1996; Shinnar 1998). A pre-existing hippocampal abnormality could be a risk factor for the development of the febrile convulsions in the first place (Fernandez et al. 1998). A pre-existing inherited unilateral hippocampal abnormality, which appears to be casually related to the febrile convulsions, could explain why HS is often unilateral. Genetic predisposition may also play a role in the pathogenesis of HS, which traditionally has not been regarded to have a major genetic contribution (Berkovic and Jackson 2000). Either this initial injury causes the HS and associated brain injury in a single step, or it predisposes the hippocampus to subsequent injury where HS develops (Jackson et al. 1999).

Other risk factors include severe perinatal events (documented intraventricular haemorrhage, skull fracture, severe birth asphyxia, Apgar score of 0–3 at 5 min or umbilical artery pH < 7.0), severe postnatal head injury (documented coma for over 2 h, documented contusion, or cerebral haemorrhage), significant cerebral infection (documented viral encephalitis, bacterial meningitis or cerebral abscess), stroke (intracerebral haemorrhage or cerebral infarction), and cerebral tumour (French et al. 1993; Mathern et al. 1995). However, these risk factors have been associated with a variety of types of epilepsy and may not all be a risk factor for the development of HS.

There is usually an interval between the initial precipitating event and the onset of habitual seizures (Mathern et al. 1995). This interval is called the silent or latent period. During this interval, the epileptogenic lesion may develop, suggesting that HS is at least in the beginning, a progressive and dynamic disease (French et al. 1993; Jeong et al. 1999). The age at initial injury has been reported to be inversely correlated with the latent interval (Mathern et al. 1995). The postulated changes that occur during this interval are the formation of aberrant mossy fibre axon connections without evidence of seizure-induced cell death (Mathern et al. 2002). Mossy fibres are the axons of the dentate granule cells and typically project into the hilus and the CA3 pyramidal neurons. As neurons in the hilus are lost, their feedback projection into the dentate granule cells also degenerates. Denervation due to loss of the hilar projection induces sprouting of the neighbouring mossy fibre axons. The net consequence of this phenomenon is formation of recurrent excitatory collaterals, increasing the net excitatory drive of dentate granule neurons (Sutula et al. 1988). Moreover, prolonged seizure discharges stimulate dentate granule cell neurogenesis as a potential repair mechanism, leading to aberrant connections (Parent and Lowenstein 1997). Such alterations may affect normal brain development and further promote epileptogenesis, whereas seizure activity during later adolescence may strengthen recurrent excitation and induce excitotoxic cell damage, i.e. MTS (Sutula et al. 1992; Nadler 2003).

Anecdotally, however, there are examples of individuals who experienced frequent focal and generalized seizures but in whom MRI findings and postmortem neuropathological examination were remarkable for the normality of both hippocampus and neocortex, at least using qualitative assessment (Holtkamp et al. 2004; Cohen-Gadol et al. 2005). Furthermore, HS was not a common finding in a large-scale population study conducted on young patients with childhood onset non-symptomatic focal epilepsy (Berg et al. 2011). The authors suggested that this could be interpreted as either that the hippocampal pathology has not yet progressed enough and cases will develop HS and pharmacoresistance in the future or that the phenomenon of TLE with HS is actually very rare in the general population of people with epilepsy. This is further corroborated by the reports of typical features of HS on MRI developing in some patients in adult life (Jackson et al. 1999).

Neuropathology

The most established scoring system used in the neuropathological work-up of MTS is the Wyler score (Wyler et al. 1992). This score is based on the percentage of neuronal loss within hippocampal subfields CA1–CA4. Threshold values are defined either by 10% (Wyler score 1 = mild MTS) or 50% of neuronal loss. Classification includes also five grades (W0 = normal, W1 = mild, W2 = moderate, W3 = classical HS, and W4 = severe HS). End folium sclerosis was subsumed into W2. An extension and revision of the Wyler score was published by Proper (Proper et al. 2001) including mossy fibre sprouting. Recently Blümcke and his colleagues proposed a classification system for MTS based on a cluster analysis of microscopically examined hippocampal specimens obtained from surgically treated MTLE patients (Blümcke et al. 2007) (Figure 13.1). They identified five distinct neuropathological subgroups and found that they were associated with specific clinical histories and/or postsurgical outcome. The first group included the surgical specimens that failed to demonstrate specific neuronal cell loss despite electrophysiological evidence for seizure generation within the mesial temporal lobe. A small group presented with segmental cell loss mainly affecting the hilar region (CA4) and most likely correspond to end folium sclerosis. Then another small cohort of patients presenting with segmental cell loss restricted to sector CA1 (CA1-sclerosis). In their classification system, they designated these two less abundant subgroups as atypical variants of HS, i.e. MTS type 3 and MTS type 2, respectively. The vast majority of cases, however, revealed severe cell loss of CA1 pyramidal neurons, whereas all other segments of the hippocampus proper showed neuronal cell loss to different degrees, i.e. MTS type 1.

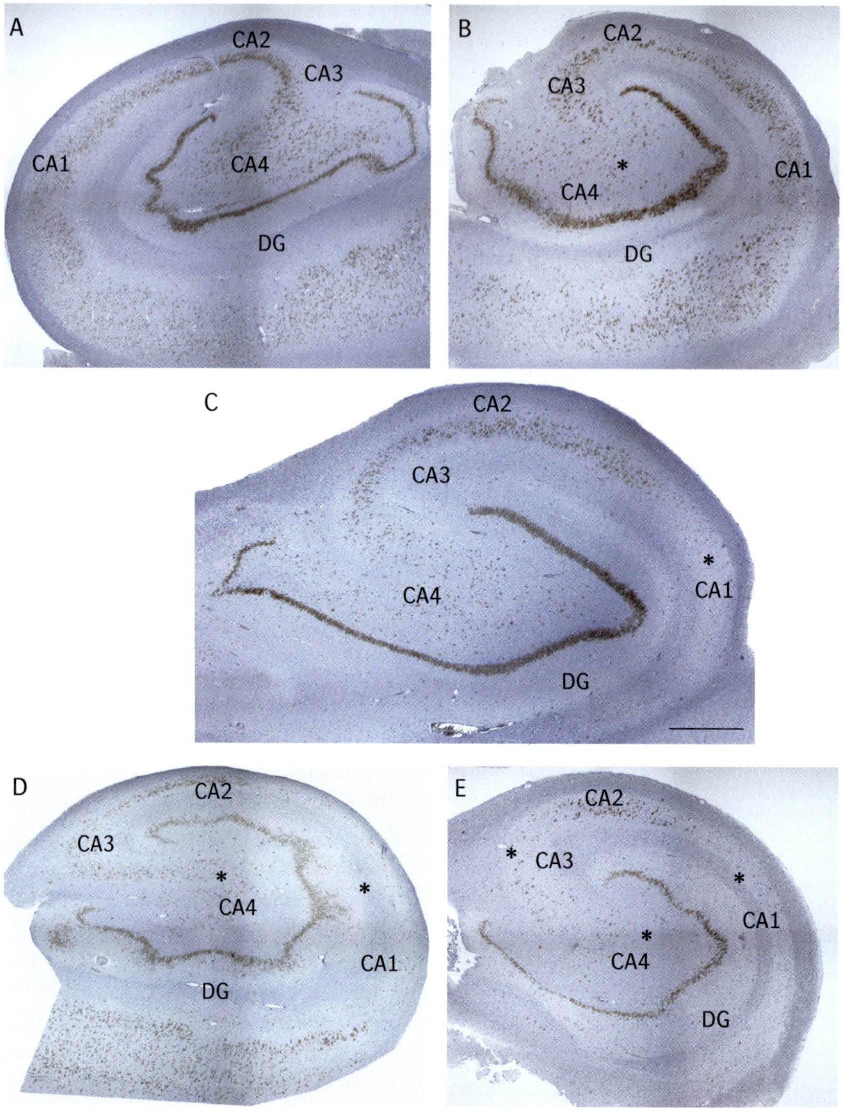

Figure 13.1 Histopathological findings and classification in patients with mesial temporal sclerosis (MTS). A) No MTS, B) MTS type 3 (end folium sclerosis), C) MTS type 2 (CA1-sclerosis), D) MTS type 1a (classic HS), and E) MTS type 1b (severe HS). Scale bars: 1 mm. Asterisks index regions with predictive cell loss patterns. Reproduced from Blümcke, I. (2007). A new clinico-pathological classification system for mesial temporal sclerosis, *Acta Neuropathologica*, 113(3), 235–44. © 2007, with permission from Springer Science + Business Media.

Only the degree of CA2 pyramidal cell loss determined a difference in this cohort and they differentiated them, therefore, into MTS type 1a (with moderate CA2 cell loss) and MTS type 1b (with severe CA2 cell loss).

Clinical picture

Seizure semiology

Seizures originating in temporal lobe structures are either focal seizures with retained awareness (previously called simple partial seizures or 'aura'), focal seizures with impairment of awareness or focal dyscognitive seizures (previously called complex partial seizures), and focal seizures with generalized tonic–clonic convulsions (secondarily generalized tonic–clonic seizures) (Berg et al. 2010).

Focal seizures with retained awareness involve small areas of the temporal lobe such as the amygdala or the hippocampus. These have been reported to occur as part of the seizure semiology in 20–90% of patients with MTLE, particularly when the underlying cause is HS (Gupta et al. 1983; Engel 1996; Luders et al. 1998). The reported variations stem from study methodology and from patient selection. Some patients may report focal seizures with retained awareness during a specific time in their life. Not uncommonly, patients may report variations in symptoms depending on the type of drug treatment and some patients may lose their auras with specific drug regimens. The symptoms felt by the person, and the signs observable by others, during seizures, which begin in the temporal lobe depend upon the specific regions of the temporal lobe and neighbouring brain areas affected by the seizure. In patients with HS, the most common symptom is the 'abdominal aura' or rising epigastric sensation, or 'butterflies in the stomach'. This sensation is described as 'rising' into the chest and neck (Holmes et al. 1988; Palmini and Gloor 1992). An unprovoked feeling of fear is common and may be present simultaneously with the epigastric aura (Cendes 2003). Olfactory illusions and hallucinations may also occur. Acharya et al found that that these are more commonly associated with temporal lobe tumours than with other causes of temporal lobe epilepsy (Acharya et al. 1998). Auditory hallucinations consist of a buzzing sound, a voice or voices, or muffling of ambient sounds. This type of aura is more common with LTLE. Patients may report distortions of shape, size, and distance of objects. These visual illusions are unlike the visual hallucinations associated with occipital lobe seizure in that no formed elementary visual image is noted, such as the visual image of a face that may be seen with seizures arising from the fusiform or the inferior temporal gyrus. Things may appear shrunken (micropsia) or larger (macropsia) than usual. Tilting of structures has also been reported. Vertigo has been described with seizures in the posterior superior temporal gyrus (Engel 1996; Luders et al. 1998). Other symptoms include psychic phenomena such as a feeling of déjà vu or jamais vu, a sense of familiarity or unfamiliarity, respectively. Patients may also experience depersonalization (i.e. feeling of detachment from oneself) or derealization (i.e. surroundings appear unreal). Fear or anxiety usually is associated with seizures arising from the amygdala. Sometimes, the fear is strong, described as an 'impending sense of doom.' In addition, patients may describe a sense of dissociation or autoscopy, in which they report seeing their own body from outside. Autonomic phenomena characterized by changes in heart rate, piloerection, and sweating are also commonly reported (Engel 1996; Luders et al. 1998).

The second type of seizures is the focal seizures with impairment of awareness or focal dyscognitive seizure. The phenomenology of this type of seizures includes a disruption of normal awareness, intrusion in the ability to respond to the environment, automatic behaviour, and, commonly, amnesia. Seizures may typically begin with arrest of motor activity and a blank stare with impaired awareness and responsiveness. Motionless stare has been reported in many patients during temporal lobe attacks (Delgado-Escueta and Walsh 1985). Then involuntary, automatic motor behaviours (automatisms) develop. Patients may continue their ongoing motor activity or react to their surroundings in a semi-purposeful manner (i.e. reactive automatisms). They can also have repetitive stereotyped manual automatisms. Ictal automatisms characteristically consist of prominent oroalimentary signs that include lip smacking, chewing, salivation, and swallowing. Speech disturbances (speech arrest, vocalization, humming, among others), are also common (Jackson et al. 2005). Gestural changes and mimetic behaviour such as grimacing, smiling, or laughing may be observed. Many patients also develop hand (fumbling, pulling, picking type behaviour) or ambulatory automatisms (French et al. 1993; Luders et al. 1993). Nose-wipe at the end of a complex partial seizure is usually with the hand ipsilateral to the side of the seizure focus and occurs at the ictal-to-postictal transition (Meletti et al. 2003).

Motor phenomena—in particular, contralateral dystonic posturing of the upper limb—are reliable lateralizing findings of TLE onset (Kotagal et al. 1989). Often, the patient may manifest automatisms of the contralateral hand during the ictal phase. Late and complex head turning is usually not a reliable sign of lateralization in temporal lobe because it probably reflects secondary spread (Quesney 1986).

The third type is seizures which begin in the temporal lobe but then spread to the whole brain manifesting as generalized tonic–clonic seizures. These begin with a focal seizure with or without retained awareness phase initially, but then the arms, trunk, and legs stiffen in either a flexed or extended position. After this, coarse jerking of the limbs and trunk occur (Abou-Khalil et al. 1993). This type is not frequently seen in patients with TLE.

Following each of these seizures, there is some period of recovery in which the patient is confused and neurological function is altered. This is called the postictal state. The degree and length of the impairment directly correlates with the severity of the three seizure types listed previously. Focal seizures with retained awareness often last less than 60 s, focal seizures with impairment of awareness often last less than 2 min, and focal seizures with generalized tonic–clonic convulsions usually last less than 3 min. The postictal state in the latter two types often lasts much longer than the seizure ictus itself. Postictal aphasia reliably lateralizes seizure onset to the dominant temporal lobe. Postictal anterograde or retrograde amnesia also occurs but is not reliably lateralizing (Jackson et al. 2005).

An attempt has been made to subclassify TLE according to the region of ictal onset (Weiser 1988). This classification is based on depth EEG studies and is limited by the inherent characteristics of the sampling problems of depth electrodes, since only a few electrodes are used for recording a seizure originating at a distance but rapid spread may be misinterpreted as originating at the nearest depth electrode.

This subclassification of seizures has not been widely accepted but the clinical features that were described for seizure onset in different parts of the temporal lobe are instructive. *Hippocampal-amygdalar* seizures are the most common. Auras occur in 80–90% of patients and include déjà vu, epigastric sensations, fear, and other affective sensations. Consciousness is preserved initially but is lost with contralateral seizure spread. The *temporal polar* seizures are very similar to hippocampal-amygdalar seizures. However, prominent early autonomic changes and oroalimentary automatisms are more common in this latter type. Auditory hallucinations are rare and may be more frequent in seizures arising from the lateral posterior neocortical region. Posterior *lateral temporal neocortical* seizures (LTLE) are also characterized by vestibular and complex visual hallucinations. *Opercular-insular* onset seizures (also LTLE) often manifest with auditory and visceral auras and prominent visceromotor phenomena. Finally, seizures with *frontobasal-cingulate* origin (again sometimes LTLE, depending of the site of seizure origin) typically have prominent motor automatisms and loss of awareness. Seizures with these characteristics often arise from the frontal lobe and spread posteriorly to involve mesial structures, and should not be classified under the temporal group (Jackson et al. 2005).

Electroencephalographic findings

Interictal electroencephalographic (EEG) studies in patients with MTLE usually demonstrate anterior temporal or unifocal temporal lobe sharp activity (Quesney 1986). This activity is usually located in the anterior temporal region (F7/F8 and T3/T4 electrodes) or basal temporal electrodes (most commonly T1/T2 and in research settings T9/T10 and F9/F10) (Figure 13.2). It is not uncommon for a patient with MTLE to have a normal EEG. The yield of the EEG can be increased on a repeat study with prolonged recordings, and, in certain patients, activation with sleep deprivation can be useful. Using anterior temporal and sphenoidal electrodes may also provide a higher yield of positive EEGs (Gloor et al. 1982; Weiser et al. 1993). Interictal features of LTLE are less consistent and specific than those of MTLE. In LTLE, interictal epileptiform discharges localize to the temporal region but tend to be

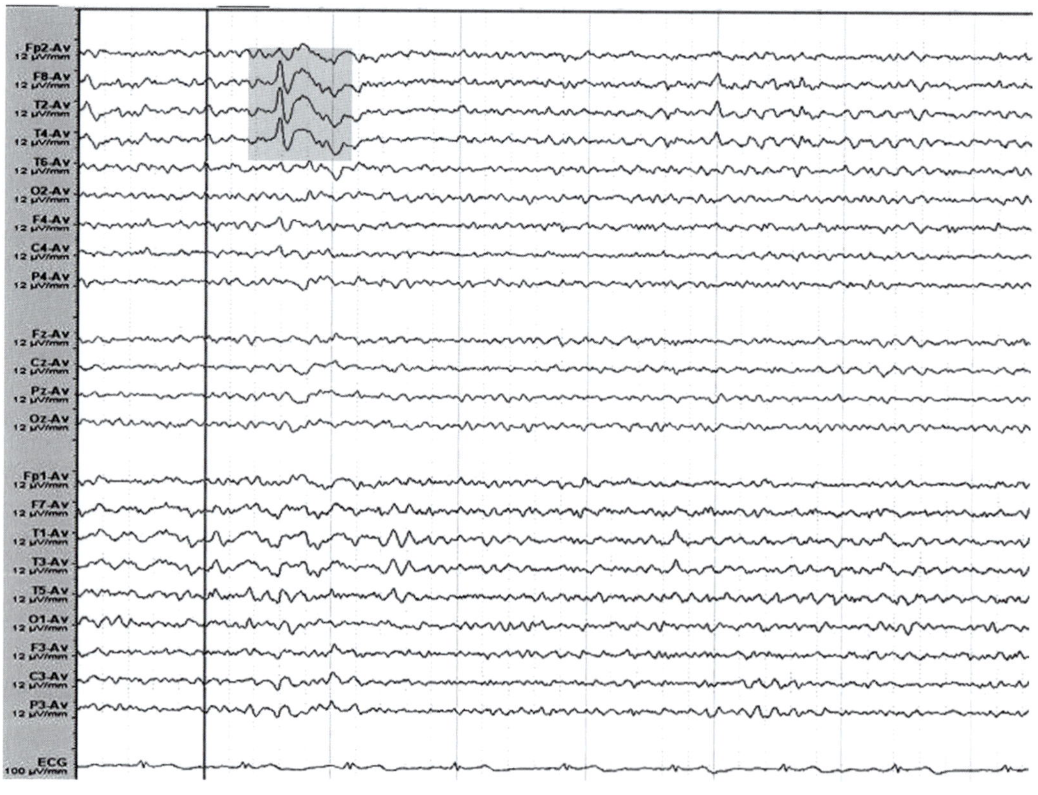

Figure 13.2 Scalp interictal epileptiform discharge recorded over the right temporal region (F8, T2, T4 electrodes) in a patient with MTLE.

more broadly distributed than in MTLE (Quesney 1986). Topographic voltage analysis indicates that most epileptiform discharges seen in patients with LTLE have a relatively broad negative region over the ipsilateral temporal scalp and lack the contralateral central-parietal scalp positivity seen in MTLE (Ebersole 1996). Focal slowing and bilaterally independent temporal epileptiform discharges appear to be equally prevalent in the two syndromes (Burgerman et al. 1995).

Ictal recordings from patients with typical temporal lobe epilepsy usually exhibit 5–7 Hz, rhythmic, sharp theta activity, maximal in the anterior or basal temporal electrodes on the side of seizure origin within 30 s or so after onset of clinical symptoms and signs (Figure 13.3) (Gil-Nagel and Risinger 1997; Foldvary et al. 2001). This finding occurs in 82–94% of patients with MTLE and lateralizes seizure onset correctly in more than 95% (Risinger et al. 1989; Walczak et al. 1992). Ictal EEG changes are only rarely seen in scalp electrodes when clinical symptoms first appear (Foldvary et al. 1997). The ictal temporal-sphenoidal discharge has less lateralizing value in patients with bilateral temporal epileptiform discharges (Walczak et al. 1992). Although quite sensitive, temporal-sphenoidal ictal discharges are probably less specific for MTLE than are the interictal discharges, which are also seen in LTLE (Ebersole and Pacia 1996; Gil-Nagel and Risinger 1997) and, occasionally, in extratemporal epilepsy. Propagation of the ictal discharge from extratemporal to temporal areas presumably accounts for the latter observation. Focal slowing or attenuation, usually on the side of seizure onset, occurs postictally in about 70% of patients with MTLE (Walczak et al. 1992). When present, this postictal slowing is a reliable lateralizing finding.

The temporal-sphenoidal pattern so common in MTLE appears less often in LTLE (Foldvary et al. 1997; Gil-Nagel and Risinger 1997) where it is more common to see widely distributed, often hemispheric ictal discharges. It is also much more common to see parasagittal spread, either ipsilateral or contralateral to the epileptogenic temporal lobe in LTLE (Foldvary et al. 1997). In addition, absence of a scalp ictal discharge is more often in LTLE than in MTLE (Ebersole and Pacia 1996). Other differences are (1) generally slower frequency, (2) less stability of frequency and voltage, and (3) appearance later in the evolution of the seizure (Ebersole and Pacia 1996; Foldvary et al. 1997) in LTLE. It should be noted however that there is a considerable overlap in EEG findings between MTLE and LTLE which precludes confident distinction of the two syndromes by ictal EEG alone.

Invasive EEG recording may be used to localize the origin of seizures in cases with insufficient information from scalp EEG. In TLE, bilateral depth electrodes are preferred over bilateral subdural strips for lateralizing a hippocampal focus. Subdural electrodes are used to define the extent and distribution of a neocortical epileptogenic zone within the temporal lobe (Luders et al. 1989). Depth electrodes are mostly used for definition of the side of seizure onset in patients in whom scalp EEG raised the question of bilateral mesial temporal epilepsy. This can be clearly determined with depth electrodes orthogonally or

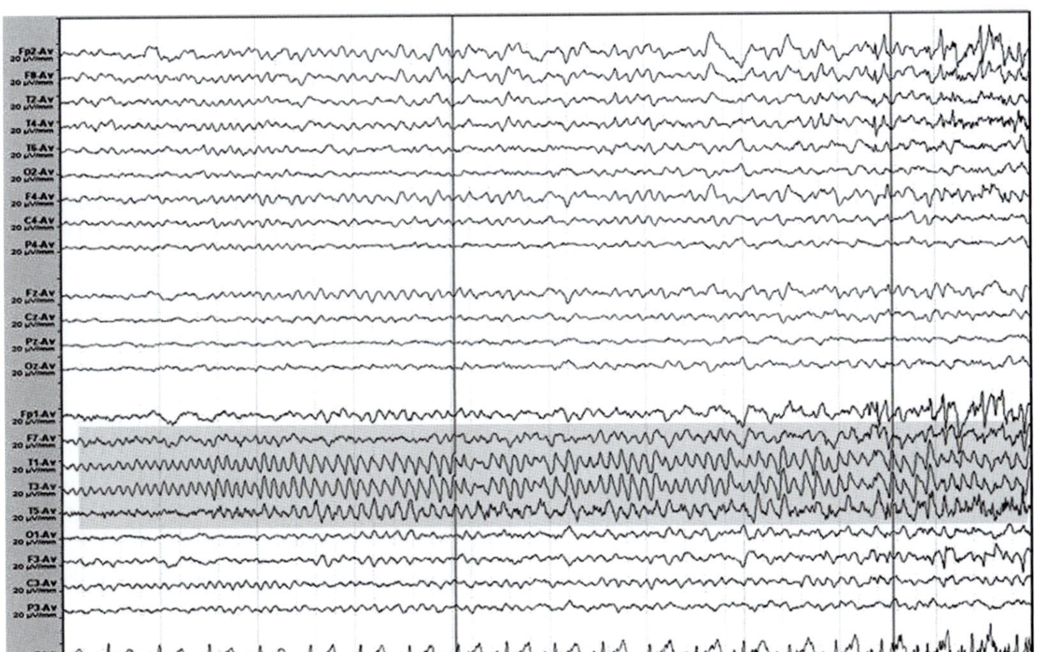

Figure 13.3 Scalp ictal epileptiform discharge recorded over the left temporal region showing rhythmic activity starting at the T1 and T3 electrodes then seen at the F7–T5 electrodes in a patient with MTLE.

parasagittally implanted into the anterior and posterior hippocampus on both sides (Wyllie 1997). Depth electrodes are multicontact electrode arrays typically with 4–12 contacts of an MRI-compatible nichrome or platinum alloy (So et al. 1989). They are typically inserted into the brain through twist-drill skull holes under stereotactic guidance, the usual targets being the medial temporal region (Pillay et al. 1992).

Neuroimaging in temporal lobe epilepsy

Magnetic resonance imaging

Neuroimaging has an important role in the investigation and treatment of patients with epilepsy and all patients with electroclinical evidence of symptomatic or focal seizures should have a structural neuroimaging study. MRI is the imaging procedure of choice in the investigation of patients with epilepsy. The advantages of MRI include the use of non-ionizing radiation, higher sensitivity and higher specificity than a computed tomography (CT) scan, multiplanar imaging capability, improved contrast of soft tissue, and high anatomical resolution. The sensitivity of MRI in detecting abnormalities in patients with epilepsy depends on the pathological substrate of the epilepsy (e.g. developmental malformations, tumours, and HS), the MRI techniques, and the experience of the interpreting physician. In addition, MRI is indicated if seizures persist in the presence of a previously normal CT scan or when there are progressive neurological changes. A repeated MRI is also indicated at 2–5 years in the context of a previously normal MRI if seizures persist.

As discussed earlier, the most common aetiology of intractable TLE is HS. MRI is capable of detecting this abnormality with high sensitivity and specificity (Jackson et al. 2005). The typical MRI findings of HS include atrophy of the hippocampus on T1-weighted imaging that is typically seen in 90–95% of cases. The atrophy is most prominent in the hippocampal body. On fluid-attenuated inversion recovery (FLAIR) imaging, increased signal is observed within the hippocampus. FLAIR sequences are ideally suited to identify signal changes within the hippocampus since gliotic changes have increased water content appearing as increased signal on T2-weighted MRI (Figure 13.4). One must be aware though that the baseline signal of the hippocampus on FLAIR MRI is greater than of the cortex, and can be mistakenly interpreted as bilateral HS. In these cases, thin-slice coronal T2-weighted images should be reviewed for confirmation. Hippocampal atrophy and increased signal are not always seen together in the same patient. For example, some patients have increased T2-weighted signal without accompanying atrophy. The increased T2 signal is felt to reflect gliosis rather than neuronal cell loss. Another frequent finding is decreased signal on inversion recovery sequences (Jackson et al. 1993). The detection of these abnormalities should be carried out with optimized imaging techniques, which include angulated coronal sections obtained perpendicular to the long axis of the hippocampal structures (Jackson et al. 1993). A volumetric sequence yielding 1- to 1.5-mm images without gaps is extremely useful for studying the hippocampal structures and detecting small areas of focal cortical dysplasia. These images can be reconstructed in any plane with multiple angulations in a relatively short time.

The MRI findings in patients with HS can be subdivided into three main groups (Jackson et al. 2005):

1. Classical HS: the sole lesion is HS with volume loss and signal change.
2. HS *plus* other damage: there is an identified abnormality in addition to HS, but the abnormality is not a definitely different lesion (such as temporal pole white matter signal abnormality, temporal lobe or hemisphere atrophy).
3. Distinct abnormality *plus* HS (dual pathology): HS is present in addition to another clearly identified lesion (such as a focal tumour or definite dysplasia).

High-resolution MRI is 80–90% sensitive for identifying HS by qualitative interpretation. Bilateral hippocampal

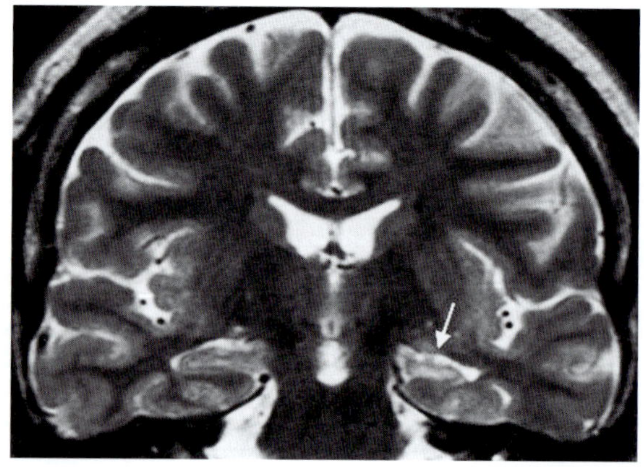

Figure 13.4 Classical hippocampal sclerosis. Coronal T2 showing left hippocampal sclerosis (arrow). The abnormal signal appears in the centre of the structure, in the atrophic hippocampal tissue surrounded by the alveus.

atrophy is seen in 10–20% of patients, but it may be difficult to visualize unless quantitative measures are done. Care must be taken, however, not to attribute all cases of increased signal within the hippocampus to HS. While increased T2-weighted signal is indicative of HS it is not always accompanied by hippocampal atrophy, the hippocampus should never be abnormally enlarged since this indicates a low-grade neoplasm, hamartoma, or hippocampal dysplasia and not HS.

When evaluating the medial temporal structures (hippocampus, amygdala, entorhinal cortex, and parahippocampal gyrus), one should evaluate the size, signal, shape, and dual pathology as approximately 15–20% of patients with HS have dual pathology (another pathologic substrate within or outside the temporal lobe in addition to HS) (Li et al. 1995). Dual pathology is more commonly observed with malformations of cortical development and developmental lesions such as porencephaly. It is less commonly seen with vascular malformations (7%) and neoplasms (2%). With vascular malformations, HS is more likely to be present if the lesion is in close proximity to the hippocampus (Li et al. 1995; Kwan and Brodie 2000).

Diffusion-weighted imaging (DWI) is a MRI method that produces *in vivo* images of biological tissues weighted with the local microstructural characteristics of water diffusion. It can provide information about damage to parts of the nervous system and thus can be used to assess the effect of seizures on the hippocampus. Animal models of seizures using DWI have shown focal decrease in the apparent diffusion coefficient (ADC) during the acute phase of seizures followed by elevation of the ADC in chronic HS (Zhong et al. 1993). A decreased ADC has also been observed in patients who developed HS following status epilepticus and acute symptomatic seizures and single seizures (Kim et al. 2001; Farina et al. 2004; Parmar et al. 2006).

Functional magnetic resonance imaging

Functional MRI (fMRI) is a technique that maps the physiological or metabolic consequences of altered electrical activity in the brain. Compared to EEG, it has the advantage of a spatial resolution of a few millimetres although it has a lower temporal resolution (few seconds) (Matthews and Jezzard 2004). FMRI allows the exploration of the local vascular response, and indirectly the coupled neural response to particular stimuli. It utilizes differences in the paramagnetic properties of oxy- and deoxy-haemoglobin, known as blood oxygen level dependent (BOLD) contrast, to generate images of cerebral activity (Ogawa and Lee 1990). During cortical activation, cerebral blood flow to the eloquent cortex increases focally as a response to the stimulus, but oxygen extraction changes little. This causes a relative increased concentration of oxyhaemoglobin and a relatively decreased concentration of deoxyhaemoglobin draining the activated cortex. Deoxyhaemoglobin is paramagnetic; it exerts magnetic susceptibility effects on local tissue, which are detected by T2-weighted imaging as decreased signal intensity. Oxyhaemoglobin, on the other hand, is diamagnetic and has little effect on T2-weighted images. Thus, cortical activation results in a relative decrease of the lowered signal intensity produced by the decreased concentration of deoxyhaemoglobin, which leads to a relative increase in signal in the activated cortex relative to contiguous cortex (Logothetis et al. 2001, 2003). One of the earliest and best-validated clinical applications of fMRI was, and remains, pre-surgical assessment of brain function in patients with epilepsies. There is a substantial body of evidence that shows that fMRI is a good technique for localizing different body representations in the primary motor (Reddy et al. 2000; McConnell et al. 2003; Zang et al. 2003; Fera et al. 2004; Newton et al. 2005; Araujo et al. 2006) and somatosensory cortex (Stippich et al. 1998; Wilke et al. 2009; Simonyan and Ludlow 2010), as well as for localizing and lateralizing language function (Wildgruber et al. 1996; Binder 1997; Binder et al. 1994, 1997; Tomczak et al. 2000; Araujo et al. 2006; Chen et al. 2008) prior to surgery. This diagnostic information permits function-preservation and safe treatment.

Proton magnetic resonance spectroscopic imaging

Magnetic resonance spectroscopy (MRS) is based on the principle that *N*-acetyl aspartate (NAA) is found primarily within neurons and precursor cells; a reduction in NAA is usually regarded as indicating loss or dysfunction of neurons (Connelly et al. 1994; Gadian et al. 1994; Cross et al. 1996). The choline (Cho) resonance reflects the contribution of several cytosolic choline-containing compounds and can be considered as a marker for cell membrane turnover. The creatine (CR) signal is usually stable over time and therefore is often used as an internal standard to which the resonance intensities of the other metabolites are normalized. Patients with TLE have reductions in the NAA/(Cho + CR) ratio (Connelly et al. 1994; Gadian et al. 1994; Cross et al. 1996). This reduction has been shown to correlate with the presence of HS and to correctly lateralize the side of seizure onset in up to 97% of patients (Kuzniecky et al. 1999; Lu et al. 2006). Lactate is one of the markers of cytotoxic oedema and anaerobic glycolysis and can also be detected using MRS. The presence of lactate on MRS was reported in patients studied during status epilepticus or shortly after acute symptomatic seizures and later on developed HS (Parmar et al. 2006). The authors concluded that the presence of lactate in those patients may represent acute neuronal damage in the hippocampus.

Nuclear neuroimaging with positron emission tomography (PET) and single-photon emission computed tomography (SPECT)

Functional neuroimaging with ictal and interictal SPECT as well as PET rank among the established functional imaging tests for the presurgical evaluation of epileptic onset zone in patients with intractable partial epilepsy. The sensitivity of these methods is excellent in TLE, in particular if a multimodal platform is used, which combines the functional imaging with the additional morphological information of MRI (la Fougere et al. 2009). Functional neuroimaging with SPECT and PET reflects seizure related changes of cerebral perfusion, glucose metabolism, and neuroreceptor status. It has an important role in the presurgical evaluation of epilepsies, as these methods are able to detect epileptogenic foci in morphologically inconspicuous areas and because they are, in contrast to EEG-based methods, not primarily dependent on the electrical activity of the brain.

SPECT

Although brain perfusion SPECT allows only quantitative estimations through tracer uptake ratios, it is widely used to measure regional cerebral blood flow (rCBF) and is an accepted adjunctive technique in the presurgical evaluation of patients with refractory focal epilepsy (Rowe et al. 1991; Newton et al. 1993, 1994; O'Brien et al. 2004). It is particularly helpful in patients with normal MRI findings, as well as in patients with abnormal MRI findings and a non-localizing EEG. Its application is based on the assumption that the increased ictal neuronal activity occurring during epileptic seizures is associated with increased metabolism and regional cerebral blood flow. For brain perfusion SPECT investigations, iodine-123- and technetium-99m-labeled radiopharmaceuticals are commercially available (Catafau 2001). These radiotracers are taken up rapidly by the brain during the first pass and, after entering neural cells, are converted rapidly to hydrophilic compounds that are trapped intracellularly and are stable for several hours. Thus, SPECT scanning can provide a semi-quantitative image of cerebral blood flow 30–60 s after intravenous injection (Neirinckx 1987).

The superiority of ictal SPECT compared with interictal SPECT for identification of the location or the lateralization of epileptic seizures has been demonstrated by several studies of patients with TLE, indicating sensitivities between 73–97% for ictal SPECT and only 50% for interictal SPECT (Newton et al. 1993; Newton et al. 1994; Devous et al. 1998; Spanaki et al. 1999; Weil et al. 2001; Zaknun et al. 2008). Cortical and subcortical rCBF changes during seizures may begin with hyperperfusion in the epileptic zone followed by rapid extension to other regions through seizure spread and generalization. However, because of its low temporal resolution, ictal SPECT hyperperfusion patterns often contain both the ictal onset zone and the propagation pathways; however, the latter region does not need to be resected to render a patient seizure free (la Fougere et al. 2009).

Ictal SPECT is performed during the seizure to help delineate the epileptogenic zone. Since seizures are associated with increased glucose metabolism (metabolism is closely coupled to cerebral blood flow), ictal SPECT scans show increased perfusion in the region of seizure onset. However, obtaining a true ictal injection is important, since with late injections, the areas of increased perfusion may represent seizure spread rather than seizure onset (Van Paesschen et al. 2007). Furthermore, the so-called 'postictal switch' phenomenon, which results in false localization or lateralization of the ictal seizure, can be observed when the delay between seizure onset and tracer application is too long (Newton et al. 1992). The sensitivity of ictal SPECT is increased significantly when ictal and interictal images are subtracted. An area of increased blood flow during the seizure that demonstrates decreased blood flow during the interictal period is more likely to be the site of seizure onset and correlates highly (approximately 90% sensitivity) with MRI abnormalities. If this subtracted image is then superimposed on high-resolution MRI, this further increases the sensitivity and specificity of the interpretation (Spanaki et al. 1999).

PET

PET with fluorine-18 fluorodeoxyglucose ([18F]FDG), as well as other receptor ligands, is an important tool to define the ictal onset zone and to better understand the neurobiology and functional alterations induced by various forms of epilepsy. The glucose analogue [18F]FDG is an indirect marker of neuronal activity and allows absolute quantification of cerebral glucose metabolism when additional arterial (ideally) or venous blood samples are taken. Interictal [18F]FDG-PET reveals interictal hypometabolism of the epileptogenic temporal lobe in more than 85% of cases and thus has an established role in the non-invasive localization of epileptogenic foci and also reflects dynamic seizure-related changes in cerebral cellular functions (la Fougere et al. 2009). The epileptogenic focus in the interictal phase usually appears as a hypometabolic area on [18F]FDG-PET. In TLE, comparison with the contralateral side has been shown to be useful for localization of the epileptic zone. However, [18F]FDG uptake frequently extends beyond the ictal-onset zone defined electrophysiologically and the epileptogenic region defined pathologically, which makes precise focus localization difficult

(Arnold et al. 1996). The degree of hypometabolism does not correlate with the degree of cell loss or the degree of hippocampal atrophy identified by MRI (Won et al. 1999).

Ictal [18F]FDG-PET cannot be routinely performed because of logistic difficulties due mainly to the short half-life of the 18F label (110 min). In addition, interpretation is difficult because of the prolonged cerebral [18F]FDG uptake, which may lead to contamination of the signal by seizure propagation (Barrington et al. 1998).

Neurotransmitters are directly responsible for modulating synaptic activity and PET also allows quantification of specific ligand–receptor relationships that seem to be important for epileptogenesis and spread of epileptic activity. Several PET receptor ligands have been used to investigate the neurochemical basis of the epilepsies and have partially entered clinical routine. In fact, PET scans with the benzodiazepine antagonist C-flumazenil (FMZ) may have greater sensitivity in identifying the epileptogenic region than FDG-PET (Savic et al. 1995, 1996) (Figure 13.5). FMZ labels central GABA receptors and reduced FMZ binding correlates with neuron loss in the hippocampus. FMZ-PET also shows enhanced sensitivity in patients with malformations of cortical development.

PET imaging with serotonin precursors or serotonin agonists has recently been used with the hope of improving the detection of the epileptogenic zone. Alpha methyl L-tryptophan PET (AMT-PET), like L-tryptophan, is a serotonin precursor that can help measure brain serotonin synthesis rates. Like tryptophan, AMT is metabolized into serotonin, but unlike tryptophan, it is not converted into protein. AMT is converted to alpha methyl serotonin, but unlike serotonin, it is not metabolized by monoamine oxidase (Chugani and Chugani 2000). AMT has also been studied in patients who failed epilepsy surgery and was able to identify increased AMT uptake in residual epileptogenic cortex as identified by intracranial EEG. However, it could only identify the epileptogenic region in 43% of patients.

Serotonin 5-HT1A receptor binding has also been studied with the PET ligands (18F)FCWAY and (11C)WAY. Toczek et al. found reduced 5-HT1A binding in the medial and lateral temporal regions ipsilateral to the epileptogenic temporal lobe (Toczek et al. 2003). Savic and colleagues reported similar findings, but they also reported reduced binding in limbic connections such as the cingulate cortex and the insula (Savic et al. 2004).

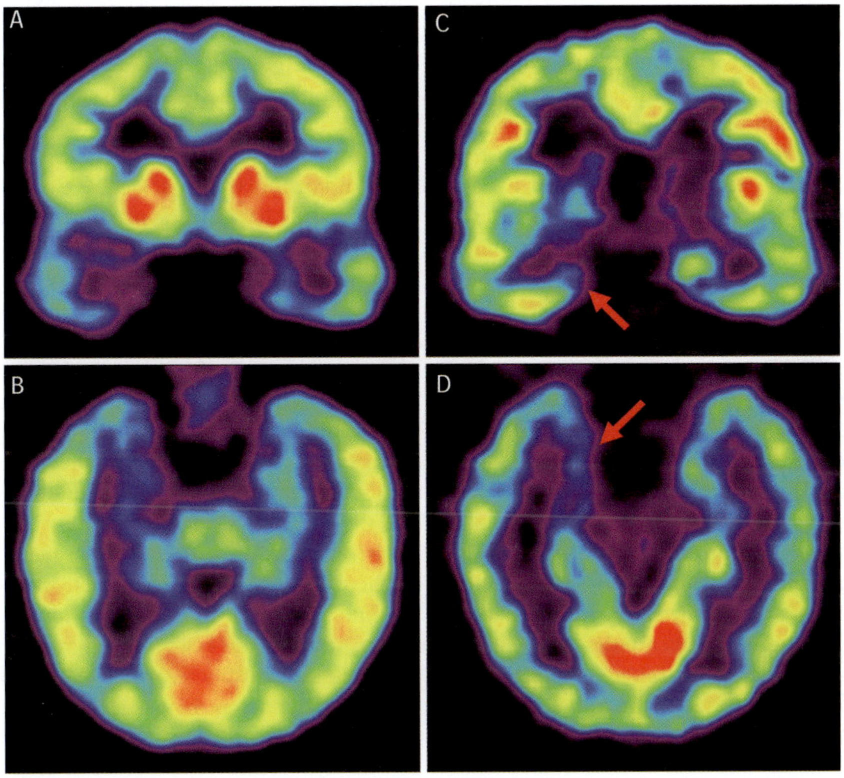

Figure 13.5 [18F]FDG (A: coronal, B: transversal) and [18F]ethylflumazenil (C: coronal, D: transversal) PET scans of a patient with right TLE showing relatively decreased glucose metabolism and reduced binding to the cBZR-GABAA receptor complex in the mesial temporal lobe (red arrows) Reprinted from *Epilepsy & Behavior,* 15(1), la Fougère, C., Rominger, A., Förster, S., Geisler, J., and Bartenstein, P., PET and SPECT in epilepsy: A critical review, 50–5, Copyright (2009), with permission from Elsevier.

Treatment

About 47–60% of patients with TLE are controlled effectively by antiepileptic drugs (AEDs) (Forsgren et al. 1996; King et al. 1998; Cascino 2008). In TLE, the most commonly used older AEDs are phenytoin, carbamazepine, primidone, valproate, and phenobarbital. Newer drugs, such as gabapentin, topiramate, levetiracetam, lamotrigine, pregabalin, tiagabine, lacosamide, and zonisamide promise similar effectiveness, possibly with fewer side effects. Felbamate and vigabatrin are newer AEDs, but can have serious adverse effects so they are not considered first-line AEDs. Nearly all AEDs function by decreasing the excitation of neurons (e.g. by blocking fast or slow sodium channels or modulating calcium channels) or by enhancing the inhibition of neurons (e.g. by potentiating the effects of inhibitory neurotransmitters like GABA) (Rogawski and Loscher 2004). Unfortunately, many patients with TLE will not experience adequate seizure control with medication (Kwan and Brodie 2000) and about 40% of those patients will continue to have seizures in spite of several trials of AEDs. Seizures are more likely to be refractory to AEDs in patients with HS (Semah et al. 1998).

For patients with refractory TLE resective surgery should be considered. In well-selected cases, 70–80% of patients with MTLE become seizure free after anterior temporal lobectomy, therefore a thorough presurgical assessment for the feasibility of surgery is essential (Jeong et al. 1999). This includes MRI, interictal and ictal EEG to localize the epileptogenic zone. The most important predictor of good outcome after surgery is the presence of unilateral HS on MRI (Radhakrishnan et al. 1998; Jeong et al. 2005). Patients with unilateral HS and scalp interictal epileptic discharges concordant with location of ictal onset have an excellent surgical outcome in 94%. On the other hand, only 60% of patients with unilateral HS and bitemporal EEG have a good outcome (Radhakrishnan et al. 1998).

The hippocampus and associated medial temporal lobe structures subserve episodic memory function and amnestic syndromes are accordingly the most significant cognitive deficits resulting from temporal lobectomy. It is therefore imperative to perform extensive neuropsychological testing preoperatively. Preoperative assessment of language and memory functions can also be performed by a Wada test. In a Wada test, amobarbital—a shortacting barbiturate—is injected via a catheter inserted transfemorally into one of the internal carotid arteries (ICAs). In this manner, most of one cerebral hemisphere is anesthetized, including the vascular territories of the anterior cerebral artery, the middle cerebral the artery, and the anterior choroidal artery. Initially used for lateralization of language, a role for Wada test was reported in the evaluation of memory function in patients with intractable TLE being considered for resective surgery. The technique varies widely across centres, but, in general, memory is assessed by presenting the patient with a number of items during the period of hemispheric inactivation and testing recall or recognition of these items after the effect of the drug has worn off (Acharya and Dinner 1997). In some patients, however, a general lowering of consciousness during the first minutes of ICA anaesthesia may seriously hamper the patient's ability to cooperate in behavioural tests. As a result, it may be difficult to estimate the degree to which the patient actually perceives stimuli presented and hence to evaluate the capacity of the hemisphere contralateral to the injection side to maintain language and memory functions. One main factor that was proposed to explain the reduced awareness was anterior cross-flow of amobarbital through the anterior communicating artery to the vascular territory of the contralateral ACA and, consequently, perfusion of the medial frontal areas bilaterally. To overcome this complication, various modifications of the Wada test were developed. For the assessment of memory functions a good procedure is superselective anesthesia via the posterior cerebral artery (PCA) (Jack et al. 1989; Wieser et al. 1997; Stabell et al. 2004). The PCA provides arterial supply to the middle and posterior parts of the hippocampus and adjacent mesial temporal structures, which are included in most cases of temporal lobectomy, and which are generally acknowledged as essential in the establishment of new declarative memories (Squire and Cave 1991). The PCA amobarbital procedure has the advantage of deactivating temporal lobe structures essential for formation of new memories, with minimal effects on language, attention, and motor functions (Jack et al. 1989; Wieser et al. 1997; Stabell et al, 2004). If the patient performs poorly on neuropsychological testing during the test, the surgical team may advise the patient against surgery or may offer a more limited operation (Spencer et al. 2000).

Once the temporal epileptic focus has been determined, surgery can be performed, which is usually a standard anterior temporal lobectomy. This generally involves removing up to 3 cm of the hippocampus medially and often the lower two-thirds of the amygdala. The uppermost part of the amygdala is usually spared to avoid injury to the optic tract on that side. The exact dimensions of cortical resection vary with each patient, because many epilepsy patients have fairly small anterior temporal lobes on the affected side. The preservation of language function often restricts the extent of removal, especially from the dominant hemisphere, which is typically the left hemisphere. The area of language function nearly always lies behind the rolandic fissure which usually lies about 5 cm from the anterior temporal tip. Therefore, in a standard resection approximately 2 inches, or 5 cm, of tissue is removed. When only a more medial excision is required, however, microsurgical techniques allow the surgeon to perform a very small anterior temporal lobectomy in which, only the hippocampus and the lower portion of the amygdala are removed without extensive lateral white matter. This procedure prevents visual field defects and can be accomplished by an angled anterior-to-posterior approach. Recently, many centres also map language function more fully with functional MRI studies prior to resection (Benke et al. 2006; Briellmann et al. 2006; Weber et al. 2006; Swanson et al. 2007).

If a patient is not an optimal candidate for epilepsy surgery, then AEDs not previously tried, the vagus nerve stimulator (Snively et al. 1998), or AEDs in clinical research trials might be alternative treatments. For children, the ketogenic diet (Chapman 2007) may also be tried. Other possible future therapies such as brain cortex responsive neural stimulators or deep brain stimulation (Loddenkemper et al. 2001; Kanner 2004) are showing promising results for treatment of TLE and other forms of epilepsy.

Recently, high-frequency deep brain stimulation of the hippocampus or the anterior nuclei of the thalamus was applied in patients with intractable temporal lobe seizures; it has been suggested that the neuromodulatory stimulation of the hippocampus and the anterior nuclei of the thalamus inhibits local and propagated epileptogenic activity.

References

Abou-Khalil, B., Andermann, E., Andermann, F., Olivier, A., and Quesney, L. F. (1993). Temporal lobe epilepsy after prolonged febrile convulsions: excellent outcome after surgical treatment. *Epilepsia*, 34(5), 878–83.

Acharya, J. N., and Dinner, D. S. (1997). Use of the intracarotid amobarbital procedure in the evaluation of memory. *J Clin Neurophysiol*, 14(4), 311–25.

Acharya, V., Acharya, J., and Luders, H. (1998). Olfactory epileptic auras. *Neurology*, 51(1), 56–61.

Araujo, D., de Araujo, D. B., Pontes-Neto, O. M., *et al.* (2006). Language and motor FMRI activation in polymicrogyric cortex. *Epilepsia*, 47(3), 589–92.

Arnold, S., Schlaug, G., Niemann, H., *et al.* (1996). Topography of interictal glucose hypometabolism in unilateral mesiotemporal epilepsy. *Neurology*, 46(5), 1422–30.

Babb, T. L. (1986). Metabolic, morphologic and electrophysiologic profiles of human temporal lobe foci: an attempt at correlation. *Adv Exp Med Biol*, 203, 115–25.

Babb, T. L., and Brown, W. J. (1986). Neuronal, dendritic, and vascular profiles of human temporal lobe epilepsy correlated with cellular physiology in vivo. *Adv Neurol*, 44, 949–66.

Babb, T. L., Kupfer, W. R., Pretorius, J. K., Crandall, P. H., and Levesque, M. F. (1991). Synaptic reorganization by mossy fibers in human epileptic fascia dentata. *Neuroscience*, 42(2), 351–63.

Barrington, S. F., Koutroumanidis, M., Agathonikou, A., *et al.* (1998). Clinical value of 'ictal' FDG-positron emission tomography and the routine use of simultaneous scalp EEG studies in patients with intractable partial epilepsies. *Epilepsia*, 39(7), 753–66.

Benke, T., Koylu, B., Visani, P., *et al.* (2006). Language lateralization in temporal lobe epilepsy: a comparison between fMRI and the Wada Test. *Epilepsia*, 47(8), 1308–19.

Berg, A. T., Berkovic, S. F., Brodie, M. J., *et al.* (2010). Revised terminology and concepts for organization of seizures and epilepsies: Report of the ILAE Commission on Classification and Terminology, 2005–2009. *Epilepsia*, 51(4), 676–85.

Berg, A. T., Pardoe, H. R., Fulbright, R. K., Schuele, S. U., and Jackson, G. D. (2011). Hippocampal size anomalies in a community-based cohort with childhood onset epilepsy. *Neurology*, 76(16), 1415–21.

Berkovic, S. F., and Jackson, G. D. (2000). The hippocampal sclerosis whodunit: enter the genes. *Ann Neurol*, 47(5), 557–8.

Berkovic, S. F., McIntosh, A., Howell, R. A., Mitchell, A., Sheffield, L. J., and Hopper, J. L. (1996). Familial temporal lobe epilepsy: a common disorder identified in twins. *Ann Neurol*, 40(2), 227–35.

Berkovic, S. F., Mulley, J. C., Scheffer, I. E., and Petrou, S. (2006). Human epilepsies: interaction of genetic and acquired factors. *Trends Neurosci*, 29(7), 391–7.

Binder, J. (1997). Functional magnetic resonance imaging. Language mapping. *Neurosurg Clin N Am*, 8(3), 383–92.

Binder, J. R., Frost, J. A., Hammeke, T. A., Cox, R. W., Rao, S. M., and Prieto, T. (1997). Human brain language areas identified by functional magnetic resonance imaging. *J Neurosci*, 17(1), 353–62.

Binder, J. R., Rao, S. M., Hammeke, T. A., *et al.* (1994). Functional magnetic resonance imaging of human auditory cortex. *Ann Neurol*, 35(6), 662–72.

Blümcke, I., Pauli, E., Clusmann, H., *et al.* (2007). A new clinico-pathological classification system for mesial temporal sclerosis. *Acta Neuropathol*, 113(3), 235–44.

Briellmann, R. S., Labate, A., Harvey, A. S., *et al.* (2006). Is language lateralization in temporal lobe epilepsy patients related to the nature of the epileptogenic lesion? *Epilepsia*, 47(5), 916–20.

Briellmann, R. S., Mark Wellard, R., Masterton, R. A., Abbott, D. F., Berkovic, S. F., and Jackson, G. D. (2007). Hippocampal sclerosis: MR prediction of seizure intractability. *Epilepsia*, 48(2), 315–23.

Briellmann, R. S., Wellard, R. M., and Jackson, G. D. (2005). Seizure-associated abnormalities in epilepsy: evidence from MR imaging. *Epilepsia*, 46(5), 760–6.

Burgerman, R. S., Sperling, M. R., French, J. A., Saykin, A. J., and O'Connor, M. J. (1995). Comparison of mesial versus neocortical onset temporal lobe seizures: neurodiagnostic findings and surgical outcome. *Epilepsia*, 36(7), 662–70.

Cascino, G. D. (2008). When drugs and surgery don't work. *Epilepsia*, 49(Suppl 9), 79–84.

Catafau, A. M. (2001). Brain SPECT in clinical practice. Part I: perfusion. *J Nucl Med*, 42(2), 259–71.

Cavazos, J. E., Das, I., and Sutula, T. P. (1994). Neuronal loss induced in limbic pathways by kindling: evidence for induction of hippocampal sclerosis by repeated brief seizures. *J Neurosci*, 14(5 Pt 2), 3106–21.

Cendes, F. (2003). Partial epilepsies: overview. *Arq Neuropsiquiatr*, 61(Suppl 1), 1–7.

Chapman, A. H. (2007). [Ketogenic diet for the treatment of refractory epilepsy]. *Arq Neuropsiquiatr*, 65(4A), 1091–2; author reply 1092.

Chen, T. L., Babiloni, C., Ferretti, A., *et al.* (2008). Human secondary somatosensory cortex is involved in the processing of somatosensory rare stimuli: an fMRI study. *Neuroimage*, 40(4), 1765–71.

Chugani, D. C., and Chugani, H. T. (2000). New directions in PET neuroimaging for neocortical epilepsy. *Adv Neurol*, 84, 447–456.

Cohen-Gadol, A. A., Bradley, C. C., Williamson, A., *et al.* (2005). Normal magnetic resonance imaging and medial temporal lobe epilepsy: the clinical syndrome of paradoxical temporal lobe epilepsy. *J Neurosurg*, 102(5), 902–9.

Conference, N. I. o. H. C. (1990). Surgery for epilepsy. *JAMA*, 264(6), 729–33.

Connelly, A., Jackson, G. D., Duncan, J. S., King, M. D., and Gadian, D. G. (1994). Magnetic resonance spectroscopy in temporal lobe epilepsy. *Neurology*, 44(8), 1411–17.

Connors, B. W. (1984). Initiation of synchronized neuronal bursting in neocortex. *Nature*, 310(5979), 685–7.

Cross, J. H., Connelly, A., Jackson, G. D., Johnson, C. L., Neville, B. G., and Gadian, D. G. (1996). Proton magnetic resonance spectroscopy in children with temporal lobe epilepsy. *Ann Neurol*, 39(1), 107–13.

Delgado-Escueta, A. V., and Walsh, G. O. (1985). Type I complex partial seizures of hippocampal origin: excellent results of anterior temporal lobectomy. *Neurology*, 35(2), 143–54.

Devous, M. D., Sr., Thisted, R. A., Morgan, G. F., Leroy, R. F., and Rowe, C. C. (1998). SPECT brain imaging in epilepsy: a meta-analysis. *J Nucl Med*, 39(2), 285–93.

Ebersole, J. S. (1996). New applications of EEG/MEG in epilepsy evaluation. *Epilepsy Res Suppl*, 11, 227–37.

Ebersole, J. S., and Pacia, S. V. (1996). Localization of temporal lobe foci by ictal EEG patterns. *Epilepsia*, 37(4), 386–99.

Engel, J., Jr. (1996). Introduction to temporal lobe epilepsy. *Epilepsy Res*, 26(1), 141–50.

Falconer, M. A. (1974). Mesial temporal (Ammon's horn) sclerosis as a common cause of epilepsy. Aetiology, treatment, and prevention. *Lancet*, 2(7883), 767–70.

Farina, L., Bergqvist, C., Zimmerman, R. A., Haselgrove, J., Hunter, J. V., and Bilaniuk, L. T. (2004). Acute diffusion abnormalities in the hippocampus of children with new-onset seizures: the development of mesial temporal sclerosis. *Neuroradiology*, 46(4), 251–7.

Fera, F., Yongbi, M. N., van Gelderen, P., Frank, J. A., Mattay, V. S., and Duyn, J. H. (2004). EPI-BOLD fMRI of human motor cortex at 1.5 T and 3.0 T: sensitivity dependence on echo time and acquisition bandwidth. *J Magn Reson Imaging*, 19(1), 19–26.

Fernandez, G., Effenberger, O., Vinz, B., *et al.* (1998). Hippocampal malformation as a cause of familial febrile convulsions and subsequent hippocampal sclerosis. *Neurology*, 50(4), 909–17.

Foldvary, N., Klem, G., Hammel, J., Bingaman, W., Najm, I., and Luders, H. (2001). The localizing value of ictal EEG in focal epilepsy. *Neurology*, 57(11), 2022–8.

Foldvary, N., Lee, N., Thwaites, G., *et al.* (1997). Clinical and electrographic manifestations of lesional neocortical temporal lobe epilepsy. *Neurology*, 49(3), 757–63.

Forsgren, L., Bucht, G., Eriksson, S., and Bergmark, L. (1996). Incidence and clinical characterization of unprovoked seizures in adults: a prospective population-based study. *Epilepsia*, 37(3), 224–9.

French, J. A., Williamson, P. D., Thadani, V. M., *et al.* (1993). Characteristics of medial temporal lobe epilepsy: I. Results of history and physical examination. *Ann Neurol*, 34(6), 774–80.

Gadian, D. G., Connelly, A., Duncan, J. S., *et al.* (1994). 1H magnetic resonance spectroscopy in the investigation of intractable epilepsy. *Acta Neurol Scand Suppl*, 152, 116–21.

Gates, J. R., and Cruz-Rodriguez, R. (1990). Mesial temporal sclerosis: pathogenesis, diagnosis, and management. *Epilepsia*, 31(Suppl 3), S55–66.

Gil-Nagel, A., and Risinger, M. W. (1997). Ictal semiology in hippocampal versus extrahippocampal temporal lobe epilepsy. *Brain*, 120(Pt 1), 183–92.

Gloor, P., Olivier, A., Quesney, L. F., Andermann, F., and Horowitz, S. (1982). The role of the limbic system in experiential phenomena of temporal lobe epilepsy. *Ann Neurol*, 12(2), 129–44.

Gupta, A. K., Jeavons, P. M., Hughes, R. C., and Covanis, A. (1983). Aura in temporal lobe epilepsy: clinical and electroencephalographic correlation. *J Neurol Neurosurg Psychiatry*, 46(12), 1079–83.

Holmes, G. L. (2002). Seizure-induced neuronal injury: animal data. *Neurology*, 59(9 Suppl 5), S3–6.

Holmes, M. D., Kelly, K., and Theodore, W. H. (1988). Complex partial seizures. Correlation of clinical and metabolic features. *Arch Neurol*, 45(11), 1191–3.

Holtkamp, M., Schuchmann, S., Gottschalk, S., and Meierkord, H. (2004). Recurrent seizures do not cause hippocampal damage. *J Neurol*, 251(4), 458–63.

Jack, C. R., Jr., Nichols, D. A., Sharbrough, F. W., *et al.* (1989). Selective posterior cerebral artery injection of amytal: new method of preoperative memory testing. *Mayo Clin Proc*, 64(8), 965–75.

Jackson, G. D., Berkovic, S. F., Duncan, J. S., and Connelly, A. (1993). Optimizing the diagnosis of hippocampal sclerosis using MR imaging. *AJNR Am J Neuroradiol*, 14(3), 753–62.

Jackson, G. D., Briellmann, R. S., and Kuzniecky, R. I. (2005). Temporal lobe epilepsy. In Kuzniecky, R. I. and Jackson, G. D. (Eds.) *Magnetic Resonance in Epilepsy* (2nd ed.), 99–176. London: Elsevier.

Jackson, G. D., Chambers, B. R., and Berkovic, S. F. (1999). Hippocampal sclerosis: development in adult life. *Dev Neurosci*, 21(3–5), 207–14.

Jackson, G. D., McIntosh, A. M., Briellmann, R. S., and Berkovic, S. F. (1998). Hippocampal sclerosis studied in identical twins. *Neurology*, 51(1), 78–84.

Jefferys, J. G. (1994). Experimental neurobiology of epilepsies. *Curr Opin Neurol*, 7(2), 113–22.

Jeong, S. W., Lee, S. K., Hong, K. S., Kim, K. K., Chung, C. K., and Kim, H. (2005). Prognostic factors for the surgery for mesial temporal lobe epilepsy: longitudinal analysis. *Epilepsia*, 46(8), 1273–9.

Jeong, S. W., Lee, S. K., Kim, K. K., Kim, H., Kim, J. Y., and Chung, C. K. (1999). Prognostic factors in anterior temporal lobe resections for mesial temporal lobe epilepsy: multivariate analysis. *Epilepsia*, 40(12), 1735–9.

Kanner, A. M. (2004). Deep brain stimulation for intractable epilepsy: which target and for which seizures? *Epilepsy Curr*, 4(6), 231–2.

Kim, J. A., Chung, J. I., Yoon, P. H., *et al.* (2001). Transient MR signal changes in patients with generalized tonicoclonic seizure or status epilepticus: periictal diffusion-weighted imaging. *AJNR Am J Neuroradiol*, 22(6), 1149–60.

King, M. A., Newton, M. R., Jackson, G. D., *et al.* (1998). Epileptology of the first-seizure presentation: a clinical, electroencephalographic, and magnetic resonance imaging study of 300 consecutive patients. *Lancet*, 352(9133), 1007–11.

Kotagal, P., Luders, H., Morris, H. H., *et al.* (1989). Dystonic posturing in complex partial seizures of temporal lobe onset: a new lateralizing sign. *Neurology*, 39(2 Pt 1), 196–201.

Kuzniecky, R., Hugg, J., Hetherington, H., *et al.* (1999). Predictive value of 1H MRSI for outcome in temporal lobectomy. *Neurology*, 53(4), 694–8.

Kwan, P., and Brodie, M. J. (2000). Early identification of refractory epilepsy. *N Engl J Med*, 342(5), 314–19.

La Fougere, C., Rominger, A., Forster, S., Geisler, J., and Bartenstein, P. (2009). PET and SPECT in epilepsy: a critical review. *Epilepsy Behav*, 15(1), 50–5.

Li, L. M., Fish, D. R., Sisodiya, S. M., Shorvon, S. D., Alsanjari, N., and Stevens, J. M. (1995). High resolution magnetic resonance imaging in adults with partial or secondary generalised epilepsy attending a tertiary referral unit. *J Neurol Neurosurg Psychiatry,* 59(4), 384–7.

Loddenkemper, T., Pan, A., Neme, S., *et al.* (2001). Deep brain stimulation in epilepsy. *J Clin Neurophysiol,* 18(6), 514–32.

Logothetis, N. K. (2003). The underpinnings of the BOLD functional magnetic resonance imaging signal. *J Neurosci,* 23(10), 3963–71.

Logothetis, N. K., Pauls, J., Augath, M., Trinath, T., and Oeltermann, A. (2001). Neurophysiological investigation of the basis of the fMRI signal. *Nature,* 412(6843), 150–7.

Lu, J. J., Ren, L. K., Feng, F., *et al.* (2006). Metabolic abnormalities in mesial temporal lobe epilepsy patients depicted by proton MR spectroscopy using a 3. 0t MR scanner. *Chin Med Sci J,* 21(4), 209–13.

Luders, H., Acharya, J., Baumgartner, C., *et al.* (1998). Semiological seizure classification. *Epilepsia,* 39(9), 1006–13.

Luders, H., Hahn, J., Lesser, R. P., *et al.* (1989). Basal temporal subdural electrodes in the evaluation of patients with intractable epilepsy. *Epilepsia,* 30(2), 131–42.

Luders, H. O., Engel, J. J., and Munari, C. (1993). General principles. In Engel (Ed.) *Surgical treatment of the epilepsies* (2nd ed.), pp. 37–153. New York: Raven Press.

Luna, D., Dulac, O., and Plouin, P. (1989). Ictal characteristics of cryptogenic partial epilepsies in infancy. *Epilepsia,* 30(6), 827–832.

Margerison, J. H., and Corsellis, J. A. (1966). Epilepsy and the temporal lobes. A clinical, electroencephalographic and neuropathological study of the brain in epilepsy, with particular reference to the temporal lobes. *Brain,* 89(3), 499–530.

Mathern, G. W., Babb, T. L., Leite, J. P., Pretorius, K., Yeoman, K. M., and Kuhlman, P. A. (1996). The pathogenic and progressive features of chronic human hippocampal epilepsy. *Epilepsy Res,* 26(1), 151–61.

Mathern, G. W., Babb, T. L., Vickrey, B. G., Melendez, M., and Pretorius, J. K. (1995). The clinical-pathogenic mechanisms of hippocampal neuron loss and surgical outcomes in temporal lobe epilepsy. *Brain,* 118(Pt 1), 105–18.

Mathern, G. W., Leiphart, J. L., De Vera, A., *et al.* (2002). Seizures decrease postnatal neurogenesis and granule cell development in the human fascia dentata. *Epilepsia,* 43(Suppl 5), 68–73.

Mathern, G. W., Leite, J. P., Pretorius, J. K., Quinn, B., Peacock, W. J., and Babb, T. L. (1994). Children with severe epilepsy: evidence of hippocampal neuron losses and aberrant mossy fiber sprouting during postnatal granule cell migration and differentiation. *Brain Res Dev Brain Res,* 78(1), 70–80.

Matthews, P. M., and Jezzard, P. (2004). Functional magnetic resonance imaging. *J Neurol Neurosurg Psychiatry,* 75, 6–12.

McConnell, K. A., Bohning, D. E., Nahas, Z., *et al.* (2003). BOLD fMRI response to direct stimulation (transcranial magnetic stimulation) of the motor cortex shows no decline with age. *J Neural Transm,* 110(5), 495–507.

McCormick, D. A. (1989). GABA as an inhibitory neurotransmitter in human cerebral cortex. *J Neurophysiol,* 62(5), 1018–27.

McCormick, D. A., and Contreras, D. (2001). On the cellular and network bases of epileptic seizures. *Annu Rev Physiol,* 63, 815–46.

McNamara, J. O. (1999). Emerging insights into the genesis of epilepsy. *Nature,* 399(6738 Suppl), A15–22.

Meletti, S., Cantalupo, G., Stanzani-Maserati, M., Rubboli, G., and Alberto Tassinari, C. (2003). The expression of interictal, preictal, and postictal facial-wiping behavior in temporal lobe epilepsy: a neuro-ethological analysis and interpretation. *Epilepsy Behav,* 4(6), 635–43.

Miles, R., and Wong, R. K. (1986). Excitatory synaptic interactions between CA3 neurones in the guinea-pig hippocampus. *J Physiol,* 373, 397–418.

Nadler, J. V. (2003). The recurrent mossy fiber pathway of the epileptic brain. *Neurochem Res,* 28(11), 1649–58.

Neirinckx, R. D. (1987). Evaluation of regional cerebral blood flow with 99mTc-d, 1 HM-PAO and SPECT. *Neurosurg Rev,* 10(3), 181–4.

Newton, J. M., Sunderland, A., and Gowland, P. A. (2005). fMRI signal decreases in ipsilateral primary motor cortex during unilateral hand movements are related to duration and side of movement. *Neuroimage,* 24(4), 1080–7.

Newton, M. R., Austin, M. C., Chan, J. G., McKay, W. J., Rowe, C. C., and Berkovic, S. F. (1993). Ictal SPECT using technetium-99m-HMPAO: methods for rapid preparation and optimal deployment of tracer during spontaneous seizures. *J Nucl Med,* 34(4), 666–70.

Newton, M. R., Berkovic, S. F., Austin, M. C., Rowe, C. C., McKay, W. J., and Bladin, P. F. (1992). Postictal switch in blood flow distribution and temporal lobe seizures. *J Neurol Neurosurg Psychiatry,* 55(10), 891–4.

Newton, M. R., Berkovic, S. F., Austin, M. C., Rowe, C. C., McKay, W. J., and Bladin, P. F. (1994). Ictal postictal and interictal single-photon emission tomography in the lateralization of temporal lobe epilepsy. *Eur J Nucl Med,* 21(10), 1067–71.

O'Brien, T. J., So, E. L., Cascino, G. D., *et al.* (2004). Subtraction SPECT coregistered to MRI in focal malformations of cortical development: localization of the epileptogenic zone in epilepsy surgery candidates. *Epilepsia,* 45(4), 367–76.

Ogawa, S., and Lee, T. M. (1990). Magnetic resonance imaging of blood vessels at high fields: in vivo and in vitro measurements and image simulation. *Magn Reson Med,* 16(1), 9–18.

Palmini, A., and Gloor, P. (1992). The localizing value of auras in partial seizures: a prospective and retrospective study. *Neurology,* 42(4), 801–8.

Parent, J. M., and Lowenstein, D. H. (1997). Mossy fiber reorganization in the epileptic hippocampus. *Curr Opin Neurol,* 10(2), 103–9.

Parmar, H., Lim, S. H., Tan, N. C., and Lim, C. C. (2006). Acute symptomatic seizures and hippocampus damage: DWI and MRS findings. *Neurology,* 66(11), 1732–5.

Perez-Velazquez, J. L., Valiante, T. A., and Carlen, P. L. (1994). Modulation of gap junctional mechanisms during calcium-free induced field burst activity: a possible role for electrotonic coupling in epileptogenesis. *J Neurosci,* 14(7), 4308–17.

Pillay, P. K., Barnett, G., and Awad, I. (1992). MRI-guided stereotactic placement of depth electrodes in temporal lobe epilepsy. *Br J Neurosurg,* 6(1), 47–53.

Proper, E. A., Jansen, G. H., van Veelen, C. W., van Rijen, P. C., Gispen, W. H., and de Graan, P. N. (2001). A grading system for hippocampal sclerosis based on the degree of hippocampal mossy Wber sprouting. *Acta Neuropathol (Berl),* 101, 405–9.

Quesney, L. F. (1986). Clinical and EEG features of complex partial seizures of temporal lobe origin. *Epilepsia,* 27 Suppl 2, S27–45.

Radhakrishnan, K., So, E. L., Silbert, P. L., *et al.* (1998). Predictors of outcome of anterior temporal lobectomy for intractable epilepsy: a multivariate study. *Neurology,* 51(2), 465–71.

Raymond, A. A., Fish, D. R., Stevens, J. M., Cook, M. J., Sisodiya, S. M., and Shorvon, S. D. (1994). Association of hippocampal sclerosis with cortical dysgenesis in patients with epilepsy. *Neurology,* 44(10), 1841–5.

Reddy, H., Lassonde, M., Bemasconi, N., *et al.* (2000). An fMRI study of the lateralization of motor cortex activation in acallosal patients. *Neuroreport,* 11(11), 2409–13.

Ribak, C. E., and Dashtipour, K. (2002). Neuroplasticity in the damaged dentate gyrus of the epileptic brain. *Prog Brain Res,* 136, 319–28.

Risinger, M. W., Engel, J., Jr., Van Ness, P. C., Henry, T. R., and Crandall, P. H. (1989). Ictal localization of temporal lobe seizures with scalp/sphenoidal recordings. *Neurology,* 39(10), 1288–93.

Rogawski, M. A., and Loscher, W. (2004). The neurobiology of antiepileptic drugs. *Nat Rev Neurosci,* 5(7), 553–64.

Rowe, C. C., Berkovic, S. F., Austin, M. C., McKay, W. J., and Bladin, P. F. (1991). Patterns of postictal cerebral blood flow in temporal lobe epilepsy: qualitative and quantitative analysis. *Neurology,* 41(7), 1096–103.

Rowe, C. C., Berkovic, S. F., Austin, M. C., *et al.* (1991). Visual and quantitative analysis of interictal SPECT with technetium-99m-HMPAO in temporal lobe epilepsy. *J Nucl Med,* 32(9), 1688–94.

Rutecki, P. A., Grossman, R. G., Armstrong, D., and Irish-Loewen, S. (1989). Electrophysiological connections between the hippocampus and entorhinal cortex in patients with complex partial seizures. *J Neurosurg,* 70(5), 667–75.

Sagar, H. J., and Oxbury, J. M. (1987). Hippocampal neuron loss in temporal lobe epilepsy: correlation with early childhood convulsions. *Ann Neurol,* 22(3), 334–40.

Savic, I., Lindstrom, P., Gulyas, B., Halldin, C., Andree, B., and Farde, L. (2004). Limbic reductions of 5-HT1A receptor binding in human temporal lobe epilepsy. *Neurology,* 62(8), 1343–51.

Savic, I., Svanborg, E., and Thorell, J. O. (1996). Cortical benzodiazepine receptor changes are related to frequency of partial seizures: a positron emission tomography study. *Epilepsia,* 37(3), 236–44.

Savic, I., Thorell, J. O., and Roland, P. (1995). [11C]flumazenil positron emission tomography visualizes frontal epileptogenic regions. *Epilepsia,* 36(12), 1225–32.

Scanziani, M., Debanne, D., Muller, M., Gahwiler, B. H., and Thompson, S. M. (1994). Role of excitatory amino acid and GABAB receptors in the generation of epileptiform activity in disinhibited hippocampal slice cultures. *Neuroscience,* 61(4), 823–32.

Schwartzkroin, P. A., and Haglund, M. M. (1986). Spontaneous rhythmic synchronous activity in epileptic human and normal monkey temporal lobe. *Epilepsia,* 27(5), 523–33.

Scott, R. C., Cross, J. H., Gadian, D. G., Jackson, G. D., Neville, B. G., and Connelly, A. (2003). Abnormalities in hippocampi remote from the seizure focus: a T2 relaxometry study. *Brain,* 126(Pt 9), 1968–74.

Semah, F., Picot, M. C., Adam, C., *et al.* (1998). Is the underlying cause of epilepsy a major prognostic factor for recurrence? *Neurology,* 51(5), 1256–62.

Shinnar, S. (1998). Prolonged febrile seizures and mesial temporal sclerosis. *Ann Neurol,* 43(4), 411–12.

Simonyan, K., and Ludlow, C. L. (2010). Abnormal activation of the primary somatosensory cortex in spasmodic dysphonia: An fMRI study. *Cereb Cortex,* 20(11), 2749–59.

Snively, C., Counsell, C., and Lilly, D. (1998). Vagus nerve stimulator as a treatment for intractable epilepsy. *J Neurosci Nurs,* 30(5), 286–9.

So, N., Gloor, P., Quesney, L. F., Jones-Gotman, M., Olivier, A., and Andermann, F. (1989). Depth electrode investigations in patients with bitemporal epileptiform abnormalities. *Ann Neurol,* 25(5), 423–31.

Spanaki, M. V., Spencer, S. S., Corsi, M., MacMullan, J., Seibyl, J., and Zubal, I. G. (1999). Sensitivity and specificity of quantitative difference SPECT analysis in seizure localization. *J Nucl Med,* 40(5), 730–6.

Spencer, D. C., Morrell, M. J., and Risinger, M. W. (2000). The role of the intracarotid amobarbital procedure in evaluation of patients for epilepsy surgery. *Epilepsia,* 41(3), 320–5.

Squire, L. R., and Cave, C. B. (1991). The hippocampus, memory, and space. *Hippocampus,* 1(3), 269–71.

Stabell, K. E., Bakke, S. J., Andresen, S., *et al.* (2004). Selective posterior cerebral artery amobarbital test: its role in presurgical memory assessment in temporal lobe epilepsy. *Epilepsia,* 45(7), 817–25.

Stippich, C., Freitag, P., Kassubek, J., *et al.* (1998). Motor, somatosensory and auditory cortex localization by fMRI and MEG. *Neuroreport,* 9(9), 1953–7.

Stringer, J. L., and Lothman, E. W. (1992). Reverberatory seizure discharges in hippocampal-parahippocampal circuits. *Exp Neurol,* 116(2), 198–203.

Sutula, T., He, X. X., Cavazos, J., and Scott, G. (1988). Synaptic reorganization in the hippocampus induced by abnormal functional activity. *Science,* 239(4844), 1147–50.

Sutula, T. P., Golarai, G., and Cavazos, J. (1992). Assessing the functional significance of mossy fiber sprouting. *Epilepsy Res Suppl,* 7, 251–9.

Swanson, S. J., Sabsevitz, D. S., Hammeke, T. A., and Binder, J. R. (2007). Functional magnetic resonance imaging of language in epilepsy. *Neuropsychol Rev,* 17(4), 491–504.

Toczek, M. T., Carson, R. E., Lang, L., *et al.* (2003). PET imaging of 5-HT1A receptor binding in patients with temporal lobe epilepsy. *Neurology,* 60(5), 749–56.

Tomczak, R. J., Wunderlich, A. P., Wang, Y., *et al.* (2000). fMRI for preoperative neurosurgical mapping of motor cortex and language in a clinical setting. *J Comput Assist Tomogr,* 24(6), 927–34.

Traub, R. D., and Jefferys, J. G. (1994). Are there unifying principles underlying the generation of epileptic afterdischarges in vitro? *Prog Brain Res,* 102, 383–94.

Van Paesschen, W., Dupont, P., Sunaert, S., Goffin, K., and Van Laere, K. (2007). The use of SPECT and PET in routine clinical practice in epilepsy. *Curr Opin Neurol,* 20(2), 194–202.

Walczak, T. S., Radtke, R. A., and Lewis, D. V. (1992). Accuracy and interobserver reliability of scalp ictal EEG. *Neurology,* 42(12), 2279–85.

Weber, B., Wellmer, J., Schur, S., *et al.* (2006). Presurgical language fMRI in patients with drug-resistant epilepsy: effects of task performance. *Epilepsia,* 47(5), 880–6.

Weil, S., Noachtar, S., Arnold, S., Yousry, T. A., Winkler, P. A., and Tatsch, K. (2001). Ictal ECD-SPECT differentiates between temporal and extratemporal epilepsy: confirmation by excellent postoperative seizure control. *Nucl Med Commun,* 22(2), 233–7.

Weiser, H. G. (1988). Human limbic seizures: EEG studies, origin, and patterns of spread. In Meldrum, B. S., Ferrendelli, J. A., and Weiser, H. G. (Eds.) *Anatomy of epileptogenesis,* pp. 127–38. London: John Libbey.

Weiser, H. G., Engel, J. J., Williamson, P. D., *et al.* (1993). Surgically remedial temporal lobe syndromes. In Engel, J. J. (Ed.) *Surgical treatment of the epilepsies* (2nd ed.), 49–64. New York: Raven Press.

Wieser, H. G., Muller, S., Schiess, R., *et al.* (1997). The anterior and posterior selective temporal lobe amobarbital tests: angiographic, clinical, electroencephalographic, PET, SPECT findings, and memory performance. *Brain Cogn,* 33(1), 71–97.

Wildgruber, D., Ackermann, H., Klose, U., Kardatzki, B., and Grodd, W. (1996). Functional lateralization of speech production at primary motor cortex: a fMRI study. *Neuroreport,* 7(15–17), 2791–5.

Wilke, M., Staudt, M., Juenger, H., Grodd, W., Braun, C., and Krageloh-Mann, I. (2009). Somatosensory system in two types of motor reorganization in congenital hemiparesis: Topography and function. *Hum Brain Mapp*, 30(3), 776–88.

Won, H. J., Chang, K. H., Cheon, J. E., *et al.* (1999). Comparison of MR imaging with PET and ictal SPECT in 118 patients with intractable epilepsy. *AJNR Am J Neuroradiol,* 20(4), 593–9.

Wyler, A. R., Dohan, C., Schweitzer, J. B., and Berry, A. B., III. (1992). A grading system for mesial temporal pathology (hippocampal sclerosis) from anterior temporal lobectomy. *J Epilepsy*, 5, 220–5.

Wyllie, E. (1997). Intracranial EEG and localization studies. In W. E. (Ed.) *The treatment of epilepsy: principles and practice* (2nd ed.), pp. 988–99. Baltimore, MD: Williams and Wilkins.

Xu, S., Pang, Q., Liu, Y., Shang, W., Zhai, G., and Ge, M. (2007). Neuronal apoptosis in the resected sclerotic hippocampus in patients with mesial temporal lobe epilepsy. *J Clin Neurosci,* 14(9), 835–40.

Zaknun, J. J., Bal, C., Maes, A., *et al.* (2008). Comparative analysis of MR imaging, Ictal SPECT and EEG in temporal lobe epilepsy: a prospective IAEA multi-center study. *Eur J Nucl Med Mol Imaging,* 35(1), 107–15.

Zang, Y., Jia, F., Weng, X., *et al.* (2003). Functional organization of the primary motor cortex characterized by event-related fMRI during movement preparation and execution. *Neurosci Lett,* 337(2), 69–72.

Zhong, J., Petroff, O. A., Prichard, J. W., and Gore, J. C. (1993). Changes in water diffusion and relaxation properties of rat cerebrum during status epilepticus. *Magn Reson Med,* 30(2), 241–6.

CHAPTER 14

The Hippocampus and Post-Traumatic Disorders

J. Douglas Bremner and Eric Vermetten

Trauma spectrum psychiatric disorders 262
Effects of stress on the hippocampus 263
Hippocampal function in PTSD 264
Glucocorticoids and the hippocampus 266
Hippocampal and other glucocorticoid receptor expression as risk factors for PTSD 267
Functional neuroimaging studies of the hippocampus in PTSD 267
Hippocampus and pharmacotherapy for PTSD 267
Summary and conclusions 268

ABSTRACT The hippocampus plays an important role in post-traumatic mental disorders. Alterations in the hippocampus have been shown in patients with a number of stress-related mental disorders, including post-traumatic stress disorder, borderline personality disorder, depression related to early abuse, and dissociative identity disorder. The hippocampus interacts with other brain regions, including the amygdala and prefrontal cortex, to mediate the stress response. Neuroimaging studies in patients with trauma-related mental disorders have replicated findings in animal studies by finding alterations in hippocampal function and structure. Abnormalities in the hippocampus are hypothesized to underlie, at least in part, symptoms of trauma-related psychiatric disorders. Psychotherapeutic interventions need to target the hippocampal circuit in order to maximize therapeutic effects. These can be found in medical interventions (e.g. selective serotonin reuptake inhibitors) as well as in psychotherapeutic interventions.

KEYWORDS depression, positron emission tomography, stress, post-traumatic stress disorder

Trauma spectrum psychiatric disorders

Traumatic stress can lead to several mental disorders that have been described as trauma-spectrum psychiatric disorders (Bremner 2002a). Central to trauma-spectrum disorders is the idea that stress has effects on common brain areas that account for the overlap in symptoms amongst the disorders. Trauma spectrum disorders include post-traumatic stress disorder (PTSD), some cases of depression (Franklin and Zimmerman 2001; Prigerson et al. 2001), and substance abuse (Kessler et al. 1995; Bremner et al. 1996), dissociative identity disorder (DID; Putnam et al. 1986), and borderline personality disorder (BPD; Yen et al. 2002; Battle et al. 2004). A number of studies have used imaging and neuropsychological testing to implicate alterations in the hippocampus as representing a common factor in the trauma spectrum disorders (Bremner 2005, 2007; Bremner et al. 2008). This chapter reviews brain imaging studies looking at the effects of traumatic stress on the brain, and integrates them with basic science findings on the neuroscience of stress.

PTSD affects 8% of people at some time during their lives (Kessler et al. 1995). Risk factors for the development of PTSD include lower education, young age, history of prior trauma, reaction at the time of the trauma, and absence of social support (Brewin et al. 2000). The diagnosis of PTSD requires exposure to an event which involves a threat to one's life or self-integrity, and is associated with intense fear, horror, or helplessness, as well as symptoms in three clusters, including intrusions, avoidance, and hyperarousal, and the presence of clinically significant distress or impairment. The diagnosis requires at least one symptom in the intrusion category, three in the avoidance category, and two in the hyperarousal category. Intrusive symptoms include recurrent intrusive memories, nightmares, feeling as if the event were recurring, and feeling a lot worse with reminders of the event, and having increased physiological reactivity with the event. Avoidant symptoms include avoidance of reminders of the event, avoidance of thoughts and feelings related to the event, trouble remembering an important aspect of the trauma, decreased interest in things, feeling detached or cut off from others, emotional numbing, and a sense of foreshortened future. Hyperarousal symptoms include a difficulty falling or staying asleep, irritability or outbursts of anger, difficulty concentrating, hypervigilance, and exaggerated startle response.

An understanding of the biological basis of PTSD requires an examination of the underlying neurobiology of fear as well as the factors that might contribute to an unsuccessful termination of the fear response in some individuals. Several factors may lead to an inadequate termination of a stress response, and the failure to contain the biological alterations initiated by stress may have long-term adverse consequences on emotion regulation. In particular, a prolonged continuation of biological responses following stress may lead to an inappropriate pairing of the traumatic memory with distress and may then initiate a cascade of secondary biological alterations. In a landmark paper in 1993, Charney et al. wrote on the psychobiology of PTSD and described long-lasting alterations in three psychobiological domains: 1) stress sensitization, 2) fear conditioning, and 3) failure in extinction. Their conceptual notions called for clinical empirical support which has now been provided by several excellent studies. Second is the model that McEwen has proposed for the understanding of PTSD (Charney et al. 1993). He introduced the 'allostasis' concept, i.e. maintaining homeostasis through change (McEwen 1998). In this model the cost of allostasis or allostatic load represented the outcome of wear and tear of the organism. From this point of view, PTSD is manifested in (as will be described later in this chapter), for example, noradrenergic hyper-reactivity, dysregulation of the hypothalamic–pituitary–adrenal (HPA) axis (basal hypocortisolaemia, reactive hypercortisolaemia), altered glucocorticoid receptor (GR), increased responsiveness or sensitivity to stressors, dysregulated responsiveness of amygdala, orbitofrontal dysfunction resulting in failure of extinction, and in reduced hippocampal volume. The brain is the key organ of allostatic regulation because its cognitive operations determine what is threatening and therefore stressful and also how the physiological and behavioural responses occur.

Effects of stress on the hippocampus

Symptoms of PTSD and other trauma spectrum disorders are felt to represent the behavioural manifestation of stress-induced changes in brain structure and function. Stress results in long-term changes in function of brain circuits involved in the stress response (Pitman 2001; Bremner 2002a; Vermetten and Bremner 2002a, 2002b). Studies have shown alterations in hippocampal dependent memory function in PTSD patients (Elzinga and Bremner 2002) as well as changes in the hippocampus, which plays an important role in memory (Bremner 2003). The hippocampus, in addition to playing a critical role in verbal declarative memory, is very sensitive to the effects of stress. Declarative (or explicit) memory involves the remembrance of facts or lists (Squire 2004). Stress in animals was associated with damage to neurons in the CA3 region of the hippocampus and inhibition of nerve growth (Sapolsky et al. 1990; McEwen et al. 1992; Nibuya et al. 1995; Magarinos et al. 1996; Sapolsky 1996; Gould et al. 1998). Mechanisms proposed for this effect include toxic effects of elevated cortisol, decreased brain-derived neurotrophic factor (BDNF), and/or elevated glutamate levels. These effects are reversible with antidepressant treatment (Nibuya et al. 1995; Malberg et al. 2000; Czéh et al. 2001; Santarelli et al. 2003; Lucassen et al. 2004).

The hippocampus interacts with other brain structures that have been implicated in a neural circuitry of stress including the amygdala and prefrontal cortex (McLaughlin and Baran 2009). The hippocampus is important for declarative, spatial, and contextual memory and is implicated in the regulation of stress. The amygdala is involved in memory for the emotional valence of events, and plays a critical role in the acquisition of fear responses (Davis 1992). The medial prefrontal cortex includes the anterior cingulate gyrus (Brodmann's area 32) and subcallosal gyrus (area 25) as well as orbitofrontal cortex. Studies demonstrated that the medial prefrontal cortex modulates emotional responsiveness through modulation of amygdala function (Morgan et al. 1993; Mlad and Quirk 2002; Milad et al. 2006).

Hippocampal function in PTSD

Memory deficits

Deficits in memory function in traumatized patients are consistent with alterations of hippocampal function in this population (Buckley et al. 2000; Elzinga and Bremner 2002). A number of studies have demonstrated verbal declarative memory deficits in PTSD consistent with hippocampal dysfunction (Golier and Yehuda 1998; Buckley et al. 2000; Brewin 2001; Elzinga and Bremner 2002). Several studies, using a variety of measures (including the Wechsler Memory Scale, the visual and verbal components of the Selective Reminding Test, the Auditory Verbal Learning Test, the California Verbal New Learning Test, and the Rivermead Behavioral Memory Test), found specific deficits in verbal declarative memory function, with a relative sparing of visual memory and IQ (Gil et al. 1990; Bremner et al. 1993; Uddo et al. 1993; Bremner et al. 1995a; Yehuda et al. 1995Barrett et al. 1996; Golier et al. 1997; Jenkins et al. 1998; Vasterling et al. 1998, 2002; Moradi et al. 1999; Sachinvala et al. 2000; Gilbertson et al. 2001; Roca and Freeman 2001), although see Stein et al. (1999). These studies have been conducted in both patients with PTSD related to Vietnam combat (Bremner et al. 1993; Uddo et al. 1993; Yehuda et al. 1995; Barrett et al. 1996; Golier et al. 1997; Vasterling et al. 1998, 2002; Sachinvala et al. 2000; Gilbertson et al. 2001; Roca and Freeman 2001), and to rape (Jenkins et al. 1998), in adults with early childhood abuse (Bremner et al. 1995a), and in traumatized children (Moradi et al. 1999). Some studies (Jenkins et al. 1998; Bremner et al. 2004a) but not others (Zalewski et al. 1994) showed differences between trauma-exposed individuals with PTSD and individuals without PTSD. These studies are consistent with deficits in explicit declarative memory that are mediated in part by the hippocampus in PTSD.

Volumetric alterations in PTSD and other trauma spectrum disorders

Studies have shown changes in hippocampal volume in patients with PTSD (Figure 14.1). Vietnam veterans with PTSD were originally shown by us to have 8% smaller right hippocampal volume based on magnetic resonance imaging (MRI) relative to controls matched for a variety of factors such as alcohol abuse and education ($p < 0.05$); smaller volume was correlated with deficits in verbal declarative memory function as measured with the Wechsler Memory Scale (Bremner et al. 1995b). A second study showed a 12% reduction in left hippocampal volume in 17 patients with childhood abuse-related PTSD compared to 17 case-matched controls that was significant after controlling for confounding factors (Bremner et al. 1997a). Other studies (Gurvits et al. 1996) showed bilateral hippocampal volume reductions in combat-related PTSD compared to combat veterans without PTSD and normal controls. Combat severity was correlated with volume reduction. Women with PTSD related to sexual abuse (Stein et al. 1997) showed a 5% reduction in left hippocampal volume. Other studies in PTSD have found smaller hippocampal volume and/or reductions in N-acetylaspartate (NAA), a marker of neuronal integrity (Freeman et al. 1998; Schuff et al. 2001; Gilbertson et al. 2002; Villarreal et al. 2002; Bremner et al. 2003b; Hedges et al. 2003; Lindauer et al. 2004a, 2005, 2006; Shin et al. 2004b; Irle et al. 2005; Mahmutyazicioglu

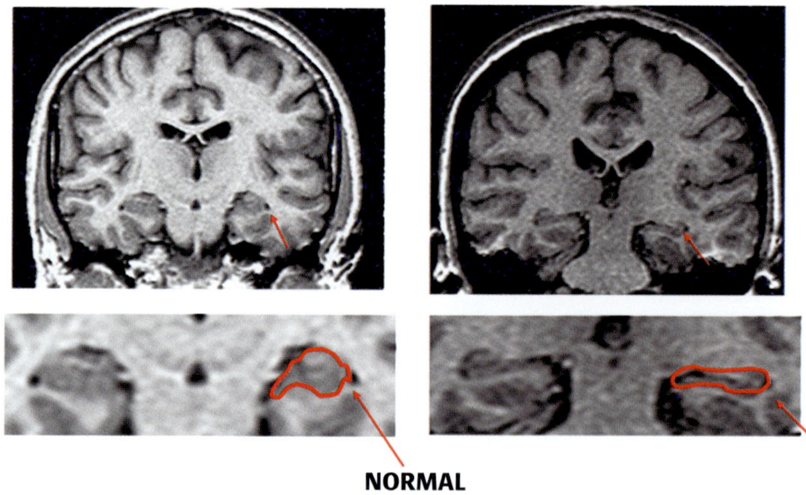

Figure 14.1 Hippocampal volume on magnetic resonance imaging (MRI) in PTSD. There is smaller hippocampal volume in this patient with PTSD compared to a control. Figure 4.2 from *Does Stress Damage the Brain? Understanding Trauma-related disorders from a mind-body perspective* by Doubglas Bremner. Copyright © 2002 by J. Douglas Bremner, M.D. Used by permission of W.W. Norton and Company, Inc.

et al. 2005; Emdad et al. 2006; Li et al. 2006; Wang et al. 2010). One study of combat veterans with PTSD found specific reductions in the CA3 subregion of the hippocampus on MRI (Wang et al. 2010).

Some studies have found smaller hippocampal volume in PTSD subjects compared to trauma exposed non-PTSD subjects (Bremner et al. 2003b) while others have not, finding reductions in both trauma-exposed non-PTSD and trauma-exposed PTSD relative to non-trauma-exposed non-PTSD subjects (Winter and Irle 2004). Studies in childhood (De Bellis et al. 1999, 2001; Carrion et al. 2001) PTSD did not find hippocampal volume reduction, although reduced NAA (indicating loss of neuronal integrity) was found in medial prefrontal cortex in childhood PTSD (De Bellis et al. 2000). Some studies of new onset or recent PTSD did not find changes in hippocampal volume (Bonne et al. 2001; Notestine et al. 2002) while others showed a reduction (Wignall et al. 2004). In two meta-analyses, data pooled from all of the published studies showed smaller hippocampal volume for both the left and the right sides, equally in adult men and women with chronic PTSD, and no change in children (Kitayama et al. 2005; Smith 2005). More recent studies of holocaust survivors with PTSD did not find a reduction in hippocampal volume (Golier et al. 2005) although PTSD patients who developed PTSD in response to an initial trauma had smaller hippocampal volume compared to those who developed PTSD after repeated trauma, suggesting a possible vulnerability of smaller hippocampal volume (Yehuda et al. 2007). Several studies have looked specifically at neurochemical alterations in the hippocampus, most using magnetic resonance spectroscopy (MRS) measurement of NAA as a marker of neuronal integrity. Studies have found reduced NAA or NAA to creatine ratio in combat veterans with PTSD (Schuff et al. 1997, 2001, 2006; Freeman et al. 1998, 2006; Kimbrell et al. 2005), and patients with civilian trauma related PTSD (Villarreal et al. 2002; Lim et al. 2003; Mohanakrishnan Menon et al. 2003; Mahmutyazicioglu et al. 2005; Li et al. 2006).

Smaller volume of the hippocampus has also been shown in patients with major depression (Bremner et al. 2000; Vythilingam et al. 2002; Sheline et al. 2003). It is not known if similar or different mechanisms are responsible for similar findings. However, smaller volumes in women with depression and early abuse compared to women with depression without early abuse suggests that stress may play a role in smaller hippocampal volume in both groups (Vythilingam et al. 2002).

As noted earlier, other 'trauma-spectrum' disorders share abnormalities in the hippocampus in common with PTSD (Figure 14.2). Studies of structural MRI have shown smaller hippocampal volume across several trauma spectrum disorders, including abuse-related PTSD (Bremner et al. 1997a; Bremner et al. 2003b), DID with early abuse (Vermetten et al. 2006), BPD with PTSD (Schmahl et al. 2009), BPD with early abuse (Driessen et al. 2000; Schmahl et al. 2003), and depression with early abuse (Vythilingam et al. 2002). In addition, in women with depression, smaller hippocampal volume was specifically associated with early abuse and not seen in depressed women without abuse, suggesting the common thread of hippocampal abnormality in these disorders. Meta-analysis of studies of hippocampal volume in adults with PTSD have not shown differences based on gender or laterality (Kitayama et al. 2005; Smith 2005).

Hippocampal activation

Several studies have shown that PTSD patients have deficits in hippocampal activation while performing a verbal

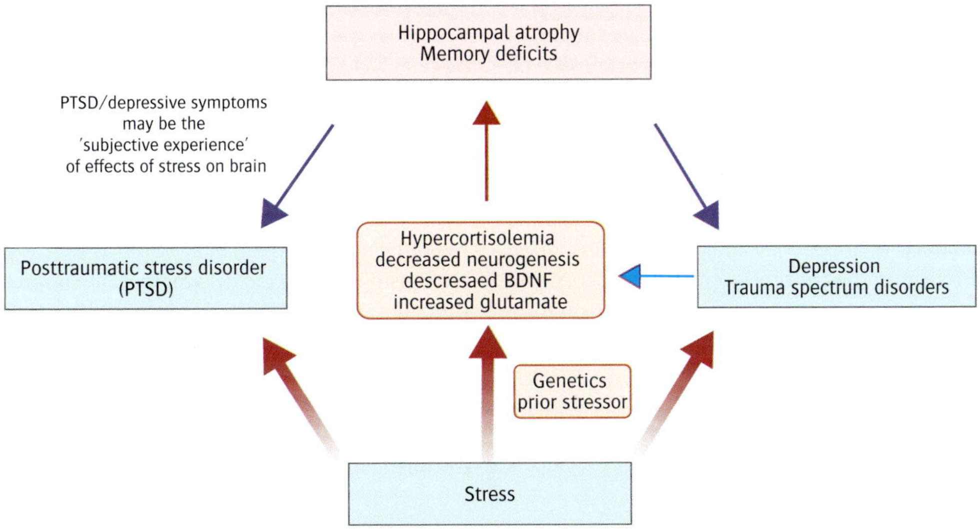

Figure 14.2 Hippocampus in PTSD and trauma spectrum disorders. Stress, acting against a backdrop of genetics and prior trauma exposure, acting through including inhibition of neurogenesis or other mechanisms, is associated with smaller hippocampal volume and impairment in cognitive function.

declarative memory task (Bremner et al. 2003b; Shin et al. 2004b) or a virtual water maze task (Astur et al. 2006). In a study with complex PTSD preferential recall of negative words was associated with increased activation in the left hippocampus and parahippocampal gyrus during both successful and false recall, suggestive of an abnormally functioning hippocampus in complex PTSD (Thomaes et al. 2009).

Conclusion

We hypothesize that stress-induced hippocampal dysfunction may mediate many of the symptoms of PTSD which are related to memory dysregulation, including both explicit memory deficits as well as fragmentation of memory in abuse survivors. A genetic contribution for risk for development of PTSD has been shown in twin studies (Goldberg et al. 1990). It is unclear at the current time whether these changes are specific to PTSD, whether certain common environmental events (e.g. stress) in different disorders lead to similar brain changes, or whether common genetic traits lead to similar outcomes (Gilbertson et al. 2002). No studies to date have been performed of postmortem brain in PTSD patients. Based on animal studies one proposed mechanism for smaller volume is stress-induced inhibition of neurogenesis (Duman et al. 2001; Santarelli et al. 2003).

Glucocorticoids and the hippocampus

Already two decades Sapolsky and others demonstrated that hippocampal damage was associated with direct exposure of glucocorticoids to the hippocampus (Sapolsky et al. 1990). Early studies in a variety of animal species (Aus der Muhlen and Ockenfels 1969) suggest that direct glucocorticoid exposure results in decreased dendritic branching (Wooley et al. 1990) and a loss of neurons (Uno et al. 1989) that are steroid and tissue specific (Packan and Sapolsky 1990). Prenatal exposure to elevated levels of glucocorticoids also results in hippocampal damage (Uno et al. 1990). Glucocorticoids appear to exert their effect through disruption of cellular metabolism (Lawrence and Sapolsky 1994) and by increasing the vulnerability of hippocampal neurons to a variety of insults, including endogenously released excitatory amino acids (Sapolsky 1986; Armanini et al. 1990). Glucocorticoids have also been shown to augment extracellular glutamate accumulation (Stein-Behrens et al. 1994). Furthermore, reduction of glucocorticoid exposure prevents the hippocampal cell loss associated with chronic stress (Landfield et al 1981; Meaney et al. 1988). Differing strains of rats may have varying glucocorticoid responses to stress, suggesting the possibility that constitutional factors may influence the glucocorticoid-mediated effects of stress on hippocampal neurons (Dhabhar et al. 1993). There are also substantial gender differences in the concentrations of glucocorticoid receptors at baseline and in response to stress, suggesting that studies related to this area which are performed exclusively in males will not be applicable to females (McCormick et al. 1994). In summary, findings to date are consistent with the idea that stress results in damage to neurons of the hippocampus, possibly through the effects of increased levels of glucocorticoids. Other factors besides glucocorticoids, such as BDNF, trkB mRNA, and nerve growth factor (NGF), which have a regulatory effect on neuronal morphology and proliferation, may also mediate stress-induced alterations in hippocampal morphology (Nibuya et al. 1995, 1996; Smith et al. 1995; Murakami et al. 2005).

Glucocorticoids also have beneficial effects besides a damaging effect on hippocampal neurons. For instance, low levels of glucocorticoids following adrenalectomy result in damage to neurons of the dentate gyrus of the hippocampus (Vaher et al. 1994). Glucocorticoids also have effects on brain function through modulation of gene expression and have a variety of effects on immunity, reproduction, bone formation, and other physiological functions. These effects may have a protective effect on the organism during certain situations of stress, but in other situations the effects of glucocorticoids may be damaging (McEwen et al. 1992). Hippocampal damage may also play a role in other aspects of the long-term dysregulation of brain function associated with stress. The hippocampus is, in general, felt to have an inhibitory effect on the HPA axis. Stress-induced damage has been shown to be associated with an increase in levels of CRF mRNA in the PVN of the hypothalamus (Herman et al. 1984) as well as a decrease in the sensitivity of rats to dexamethasone suppression of HPA function (Feldman and Conforti 1980; Magarinos et al. 1987). Consistent with this, increased levels of corticotropin-releasing factor (CRF) are found in the cerebrospinal fluid of patients with combat-related PTSD in comparison to controls (Bremner et al. 1997b), as well as in plasma (de Kloet et al., 2008).

Hippocampal and other glucocorticoid receptor expression as risk factors for PTSD

The interaction between glucocorticoids and immune function has also been investigated. The inhibition by the synthetic glucocorticoid dexamethasone of leucocyte lysozyme activity (Yehuda et al. 2004), of bacterial lipopolysaccharide-stimulated IL-6 and tumour necrosis factor-α production (Rohleder et al. 2004), and of the mononuclear leucocyte glucocorticoid receptor (GR) number per cell (Yehuda et al. 2002) is potentiated in PTSD versus healthy controls. Oddly though, the baseline GR number is reduced in all lymphocyte subpopulations (Gotovac et al. 2003), making it more likely that the potentiated dexamethasone suppression depends on increased receptor affinity or secondary mechanisms.

Altered GR expression in peripheral blood mononuclear cells (PBMCs) has been found in PTSD. We found significantly lower leucocyte GR density in veterans with and without PTSD compared to healthy controls suggesting that trauma exposure is sufficient to induce changes in GR binding characteristics. The altered glucocorticoid sensitivity in lymphocytes and other changes in measures of immunity do not appear to be secondary to changes in activity of the HPA axis, since there is no correlation between ambient plasma cortisol levels in PTSD patients and changes in lymphocyte glucocorticoid receptor number (Gotovac et al. 2003), plasma IL-1β (Spivak et al. 1997), IL-6 (Maes et al. 1999; Gill et al. 2008), or C-reactive protein (Söndergaard et al. 2004) levels.

Since it was thus far unclear whether differences in GR expression exist already prior to development of PTSD, and thus may constitute a risk factor, we undertook a prospective study to assess whether pre-deployment GR binding capacity differed between military personnel with and without PTSD symptoms after deployment. In addition it was aimed to assess and whether GR binding was a predictor of PTSD symptoms after deployment. We included soldiers who reported high levels of PTSD symptoms 6 months after deployment (n = 34) and soldiers without high levels of PTSD or depressive symptoms (n = 34). Our data showed that before deployment, GR expression in PBMCs was significantly increased in participants with high levels of PTSD symptoms after deployment as compared to matched controls. The risk for a high level of PTSD symptoms after deployment increased 4.8-fold with each increase of 1000 GR binding sites. There were no group differences in mRNA expression of GR-α, GR-P, GR-β, and FKBP5. These results demonstrated that increased pre-trauma GR binding capacity of PBMCs is a vulnerability factor for subsequent development of PTSD symptoms (Van Zuiden et al., 2012).

Functional neuroimaging studies of the hippocampus in PTSD

Imaging studies of brain function in PTSD are consistent with a role for dysfunction of the hippocampus in the disorder (Bremner 1998, 2002b; Pitman 2001; Cannistraro and Rauch 2003; Liberzon and Phan 2003; Liberzon et al. 2003; Liberzon and Martis 2006; Rauch et al. 2006). Stimulation of the noradrenergic system with yohimbine resulted in a failure of activation in the hippocampus and other prefrontal areas which was associated with an increase in anxiety and PTSD symptoms (Bremner et al. 1997c). Induction of traumatic reminders has primarily been shown to be associated with decreased medial prefrontal function, as measured with PET, SPECT, or fMRI (Shin et al. 1997, 1999, 2001, 2004a, 2005; Bremner et al. 1999a, 1999b; Liberzon et al. 1999; Semple et al. 2000; Lanius et al. 2001, 2003; Lindauer et al. 2004b; Yang et al. 2004; Britton et al. 2005; Phan et al. 2006). However, studies have shown decreased hippocampal function in women with childhood sexual abuse exposed to scripts of their traumatic events (Bremner et al. 1999a), and with retrieval of emotionally valenced words (Bremner et al. 2001) (e.g. 'rape-mutilate') in women with PTSD from early abuse (Bremner et al. 2003a).

Hippocampus and pharmacotherapy for PTSD

Studies in animals have demonstrated several agents with potentially beneficial effects on the reversibility of the glucocorticoid mediated hippocampal toxicity. It has been found that phenytoin (Dilantin®) reverses stress-induced hippocampal atrophy, probably through excitatory amino acid-induced neurotoxicity. Agents such as tianeptine and dihydroepiandrosterone (DHEA) have been described to produce similar effects (Watanabe et al. 1993). Neurons within the hippocampus were found to be unique within the brain in showing the capacity to regenerate themselves (Gould et al. 1999). Since these are *in vitro* studies that report on neurogenesis in the hippocampus through a regulation of BDNF and cAMP by selective serotonin reuptake inhibitors (SSRIs), we hypothesized that the effects of SSRIs may play a role in the reversibility of the accelerated ageing process in humans (Duman et al. 1997). There are findings indicating that a common action of antidepressant treatments is through upregulation of cAMP response element binding protein (CREB) and that this may lead to regulation of specific target genes. Such treatment effects on BDNF and trkB mRNA can have long-term effects on brain function. In these studies, however, it is evident that chronic, but not acute, administration of several different classes of antidepressants, including SSRIs and serotonin/noradrenaline reuptake inhibitors (SNRIs), increase the expression of CREB mRNA.

In a treatment study we performed with paroxetine we found that 9-month treatment not only led to a reduction in PTSD symptoms, but was accompanied by an increase in hippocampal volume of 5.6%, an increase in declarative memory performance (Vermetten et al., 2003), as well as normalization of the stress response (Vermetten et al., 2006). Other studies have shown an increase in right hippocampal and right cerebral volume but not cognitive function (Bremner et al. 2005) which are associated with an improvement in symptoms (Bremner et al. 2004b) in PTSD patients treated with phenytoin (Dilantin®), medication for the treatment of epilepsy that modulates glutamatergic function. Hippocampal atrophy may be a viable drug target for PTSD however cannot be the sole ingredient (DeCarolis and Eisch, 2010).

Summary and conclusions

Studies reviewed in this chapter demonstrate common abnormalities of the hippocampus in a range of psychiatric disorders termed trauma-spectrum disorders, which include PTSD, DID, BPD and depression related to early abuse. These studies consistently show smaller hippocampal volume and reduced markers of hippocampal neuronal integrity like NAA in PTSD. Emerging studies also consistently show smaller volume of the hippocampus in the other trauma spectrum disorders. Neuropsychological testing is also consistent deficits in declarative memory function that is consistent with hippocampal dysfunction in PTSD. Treatments that are effective for PTSD are associated with an increase in hippocampal volume and improved declarative memory function.

The hippocampus plays a critical role in memory and stress regulation, highlighting the important interplay between memory and the traumatic stress response. Promotion of nerve growth (neurogenesis) in the hippocampus may be central to the efficacy of the antidepressants. Psychotherapeutic interventions need to target the hippocampal circuit, and facilitate allostatic regulation, in order to maximize therapeutic effects.

References

Armanini, M. P., Hutchins, C., Stein, B. A., andSapolsky, R. M. (1990). Glucocorticoid endangerment of hippocampal neurons is NMDA-receptor dependent. *Brain Res*, 532, 7–11.

Astur, R. S., St Germain, S. A., Tolin, D., Ford, J., Russell, D., and Stevens, M. (2006). Hippocampus function predicts severity of post-traumatic stress disorder. *Cyberpsychol Behav*, 9 (2), 234–40.

Barrett, D. H., Green, M. L., Morris, R., Giles, W. H., and Croft, J. B. (1996). Cognitive functioning and posttraumatic stress disorder. *Am J Psychiatry*, 153(11), 1492–4.

Battle, C. L., Shea, M. T., Johnson, D. M., *et al.* (2004). Childhood maltreatment associated with adult personality disorders: findings from the Collaborative Longitudinal Personality Disorders Study. *J Personal Disord*, 18(2), 193–211.

Bonne, O., Brandes, D., Gilboa, A., *et al.* (2001). Longitudinal MRI study of hippocampal volume in trauma survivors with PTSD. *Am J Psychiatry*, 158, 1248–51.

Bremner, J. D. (1998). Neuroimaging of posttraumatic stress disorder. *Psychiatric Ann*, 28, 445–50.

Bremner, J. D. (2002a). *Does Stress Damage the Brain? Understanding Trauma-related Disorders from a Mind-Body Perspective*. New York: W.W. Norton.

Bremner, J. D. (2002b). Neuroimaging of childhood trauma. *Semin Clin Neuropsychiatry*, 7, 104–12.

Bremner, J. D. (2003). Functional neuroanatomical correlates of traumatic stress revisited 7 years later, this time with data. *Psychopharmacol Bull*, 37 (2), 6–25.

Bremner, J. D. (2005). *Brain Imaging Handbook*. New York: W.W. Norton.

Bremner, J. D. (2007). Neuroimaging in posttraumatic stress disorder and other stress-related disorders. *Neuroimaging Clinics N Am*, 17 (4), 523–38.

Bremner, J. D., Scott, T. M., Delaney, R. C., *et al.* (1993). Deficits in short-term memory in post-traumatic stress disorder. *Am J Psychiatry*, 150, 1015–19.

Bremner, J. D., Randall, P., Scott, T. M., *et al.* (1995a). Deficits in short-term memory in adult survivors of childhood abuse. *Psychiatry Res*, 59, 97–107.

Bremner, J. D., Randall, P., Scott, T. M., *et al.* (1995b). MRI-based measurement of hippocampal volume in patients with combat-related posttraumatic stress disorder. *Am J Psychiatry*, 152, 973–81.

Bremner, J. D., Southwick, S. M., Darnell, A., and Charney, D. S. (1996). Chronic PTSD in Vietnam combat veterans: Course of illness and substance abuse. *Am J Psychiatry*, 153, 369–75.

Bremner, J. D., Randall, P., Vermetten, E., *et al.* (1997a). MRI-based measurement of hippocampal volume in posttraumatic stress disorder related to childhood physical and sexual abuse: A preliminary report. *Biol Psychiatry*, 41, 23–32.

Bremner, J. D., Licinio, J., Darnell, A., *et al.* (1997b). Elevated CSF corticotropin-releasing factor concentrations in posttraumatic stress disorder. *Am J Psychiatry*, 154, 624–9.

Bremner, J. D., Innis, R. B., Ng, C. K., *et al.* (1997c). PET measurement of cerebral metabolic correlates of yohimbine administration in posttraumatic stress disorder. *Arch Gen Psychiatry*, 54, 246–56.

Bremner, J. D., Narayan, M., Staib, L. H., *et al.* (1999a). Neural correlates of memories of childhood sexual abuse in women with and without posttraumatic stress disorder. *Am J Psychiatry*, 156, 1787–95.

Bremner, J. D., Staib, L. H., Kaloupek, D., *et al.* (1999b). Neural correlates of exposure to traumatic pictures and sound in Vietnam combat veterans with and without posttraumatic stress disorder: A positron emission tomography study. *Biol Psychiatry*, 45, 806–16.

Bremner, J. D., Narayan, M., Anderson, E. R., Staib, L. H., Miller, H. L., and Charney, D. S. (2000). Hippocampal volume reduction in major depression. *Am J Psychiatry*, 157, 115–17.

Bremner, J. D., Soufer, R., McCarthy, G., *et al.* (2001). Gender differences in cognitive and neural correlates of remembrance of emotional words. *Psychopharmacol Bull*, 35, 55–87.

Bremner, J. D., Vythilingam, M., Vermetten, E., *et al.* (2003a). Neural correlates of declarative memory for emotionally valenced words in women with posttraumatic stress disorder (PTSD) related to early childhood sexual abuse. *Biol Psychiatry*, 53, 289–99.

Bremner, J. D., Vythilingam, M., Vermetten, E., *et al.* (2003b). MRI and PET study of deficits in hippocampal structure and function in women with childhood sexual abuse and posttraumatic stress disorder (PTSD). *Am J Psychiatry*, 160, 924–32.

Bremner, J. D., Vermetten, E., Afzal, N., and Vythilingam, M. (2004a). Deficits in verbal declarative memory function in women with childhood sexual abuse-related posttraumatic stress disorder (PTSD). *J Nerv Ment Dis*, 192(10), 643–9.

Bremner, J. D., Mletzko, T., Welter, S., *et al.* (2004b). Treatment of posttraumatic stress disorder with phenytoin: An open label pilot study. *J Clin Psychiatry*, 65(11), 1559–64.

Bremner, J. D., Mletzko, T., Welter, S., *et al.* (2005). Effects of phenytoin on memory, cognition and brain structure in posttraumatic stress disorder: A pilot study. *J Psychopharmacol*, 19(2), 159–65.

Bremner, J. D., Elzinga, B., Schmahl, C., and Vermetten, E. (2008). Structural and functional plasticity of the human brain in posttraumatic stress disorder. *Prog Brain Res*, 167, 171–86.

Brewin, C. R. (2001). A cognitive neuroscience account of posttraumatic stress disorder and its treatment. *Behav Res Therapy*, 39, 373–93.

Brewin, C. R., Andrews, B., and Valentine, J. D. (2000). Meta-analysis of risk factors for posttraumatic stress disorder in trauma-exposed adults. *J Consult Clin Psychol*, 68 (5), 748–66.

Britton, J. C., Phan, K. L., Taylor, S. F., Fig, L. M., and Liberzon, I. (2005). Corticolimbic blood flow in posttraumatic stress disorder during script-driven imagery. *Biol Psychiatry*, 57(8), 832–40.

Buckley, T. C., Blanchard, E. B., and Neill, W. T. (2000). Information processing and PTSD: A review of the empirical literature. *Clin Psychol Rev*, 28(8), 1041–65.

Cannistraro, P. A. and Rauch, S. L. (2003). Neural circuitry of anxiety: evidence from structural and functional neuroimaging studies. *Psychopharmacol Bull*, 37(4), 8–25.

Carrion, V.G., Weems, C. F., Eliez, S., *et al.* (2001). Attenuation of frontal asymmetry in pediatric posttraumatic stress disorder. *Biol Psychiatry*, 50, 943–51.

Charney, D. S., Deutch, A. Y., Krystal, J. H., Southwick, S. M., and Davis, M. (1993). Psychobiologic mechanisms of posttraumatic stress disorder. *Arch Gen Psychiatry*, 50, 295–305.

Czéh, B., Michaelis, T., Watanabe, T., *et al.* (2001). Stress-induced changes in cerebral metabolites, hippocampal volume, and cell proliferation are prevented by antidepressant treatment with tianeptine. *Proc Natl Acad Sci USA*, 98, 12796–801.

Davis, M. (1992). The role of the amygdala in fear and anxiety. *Ann Rev Neurosci*, 15, 353–75.

De Bellis, M.D., Keshavan, M. S., Clark, D. B., *et al.* (1999). A.E. Bennett Research Award: Developmental traumatology: Part II. Brain development. *Biol Psychiatry*, 45, 1271–84.

De Bellis, M.D., Keshavan, M. S., Spencer, S., and Hall, J. (2000). N-acetylaspartate concentration in the anterior cingulate of maltreated children and adolescents with PTSD. *Am J Psychiatry*, 157, 1175–7.

De Bellis, M.D., Hall, J., Boring, A. M., Frustaci, K., and Moritz, G. (2001). A pilot longitudinal study of hippocampal volumes in pediatric maltreatment-related posttraumatic stress disorder. *Biol Psychiatry*, 50, 305–9.

de Kloet, C. S., Vermetten, E., Geuze, E., Lentjes, E. G., Heijnen, C. J., Stalla, G. K., *et al.* (2008). Elevated plasma corticotrophin-releasing hormone levels in veterans with posttraumatic stress disorder. *Prog Brain Res*, 167, 287–91.

DeCarolis, N. A. and Eisch, A. J. (2010). Hippocampal neurogenesis as a target for the treatment of mental illness: a critical evaluation. *Neuropharmacology*, 58(6), 884–93.

Dhabhar, F. S., McEwen, B. S., and Spencer, R. L. (1993). Stress response, adrenal steroid receptor levels and corticosteroid-binding globulin levels – a comparison between Sprague-Dawley, Fischer 344 and Lewis rats. *Brain Res*, 616, 89–98.

Driessen, M., Herrmann, J., Stahl, K., *et al.* (2000). Magnetic resonance imaging volumes of the hippocampus and the amygdala in women with borderline personality disorder and early traumatization. *Arch Gen Psychiatry*, 57, 1115–22.

Duman, R. S., Heninger, G. R., and Nestler, E. J. (1997). A molecular and cellular theory of depression. *Arch Gen Psychiatry*, 54, 597–606.

Duman, R. S., Malberg, J. E., and Nakagawa, S. (2001). Regulation of adult neurogenesis by psychotropic drugs and stress. *J Pharmacol Exp Ther*, 299, 401–7.

Elzinga, B. M. and Bremner, J. D. (2002). Are the neural substrates of memory the final common pathway in PTSD? *J Affect Disord*, 70, 1–17.

Emdad, R., Bonekamp, D., Sondergaard, H. P., *et al.* (2006). Morphometric and psychometric comparisons between non-substance-abusing patients with posttraumatic stress disorder and normal controls. *Psychother Psychosom*, 75(2), 122–32.

Feldman, S., and Conforti, N. (1980). Participation of the dorsal hippocampus in the glucocorticoid feedback effect on adrenocortical activity. *Neuroendocrinology*, 30, 52–5.

Franklin, C.L. and Zimmerman, M. (2001). Posttraumatic stress disorder and major depressive disorder: Investigating the role of overlapping symptoms in diagnostic comorbidity. *J Nervous Mental Dis*, 189(8), 548–51.

Freeman, T., Kimbrell, T., Booe, L., *et al.* (2006). Evidence of resilience: neuroimaging in former prisoners of war. *Psychiatry Res*, 146(1), 59–64.

Freeman, T.W., Cardwell, D., Karson, C. N., and Komoroski, R. A. (1998). In vivo proton magnetic resonance spectroscopy of the medial temporal lobes of subjects with combat-related posttraumatic stress disorder. *Magn Reson Med*, 40, 66–71.

Gil, T., Obremskey, W., Bauer, R., *et al.* (1990). Cognitive functioning in posttraumatic stress disorder. *J Trauma Stress*, 3, 29–45.

Gilbertson, M. W., Gurvits, T. V., Lasko, N. B., Orr, S. P., and Pitman, R. K. (2001). Multivariate assessment of explicit memory function in combat veterans with posttraumatic stress disorder. *J Trauma Stress*, 14, 413–20.

Gilbertson, M. W., Shenton, M. E., Ciszewski, A., *et al.* (2002). Smaller hippocampal volume predicts pathologic vulnerability to psychological trauma. *Nature Neurosci*, 5(11), 1242–7.

Gill, J., Vythilingam, M., and Page, G. G. (2008). Low cortisol, high DHEA, and high levels of stimulated TNF-alpha, and IL-6 in women with PTSD. *J Trauma Stress*, 21(6), 530–9.

Goldberg, J., True, W. R., Eisen, S. A., and Henderson, W. G. (1990). A twin study of the effects of the Vietnam war on posttraumatic stress disorder. *JAMA*, 263, 1227–32.

Golier, J. and Yehuda, R. (1998). Neuroendocrine activity and memory-related impairments in posttraumatic stress disorder. *Dev Psychopathol*, 10(4), 857–69.

Golier, J., Yehuda, R., Cornblatt, B., Harvey, P., Gerber, D., and Levengood, R. (1997). Sustained attention in combat-related posttraumatic stress disorder. *Intgr Physiol Behav Sci*, 32(1), 52–61.

Golier, J. A., Yehuda, R., De Santi, S., Segal, S., Dolan, S., and de Leon, M. J. (2005). Absence of hippocampal volume differences in survivors of the Nazi Holocaust with and without posttraumatic stress disorder. *Psychiatry Res*, 139(1), 53–64.

Gotovac, K., Sabioncello, A., Rabatic, S., Berki, T., and Dekaris, D. (2003). Flow cytometric determination of glucocorticoid receptor (GCR) expression in lymphocyte subpopulations: lower quantity of GCR in patients with post-traumatic stress disorder (PTSD). *Clin Exp Immunol*, 131(2), 335–9.

Gould, E., Tanapat, P., McEwen, B. S., Flügge, G., and Fuchs, E. (1998). Proliferation of granule cell precursors in the dentate gyrus of adult monkeys is diminished by stress. *Proc Natl Acad Sci USA*, 95, 3168–71.

Gould, E., Beylin, A., Tanapat, P., Reeves, A., and Shors, T. J. (1999). Learning enhances adult neurogenesis in the hippocampal formation. *Nat Neurosci*, 2(3), 260–5.

Gurvits, T.G., Shenton, M. E., Hokama, H., *et al.* (1996). Magnetic resonance imaging study of hippocampal volume in chronic combat-related posttraumatic stress disorder. *Biol Psychiatry*, 40, 192–9.

Hedges, D. W., Allen, S., Tate, D. F. *et al.* (2003). Reduced hippocampal volume in alcohol and substance naive Vietnam combat veterans with posttraumatic stress disorder. *Cogn Behav Neurol*, 16(4), 219–24.

Herman, J. P., Schafer, M. K. H., Young, E. A., Thompson, R., Douglass, J., Akil, H., *et al.* (1984). Evidence for hippocampal regulation of neuroendocrine neurons of the hypothalamus-pituitary-adrenocortical axis. *Neuroscience*, 9, 3072–82.

Irle, E., Lange, C., and Sachsse, U. (2005). Reduced size and abnormal asymmetry of parietal cortex in women with borderline personality disorder. *Biol Psychiatry*, 57(2), 173–82.

Jenkins, M.A., Langlais, P. J., Delis, D., and Cohen, R. (1998). Learning and memory in rape victims with posttraumatic stress disorder. *Am J Psychiatry*, 155, 278–9.

Kessler, R.C., Sonnega, A., Bromet, E., Hughes, M., and Nelson, C. B. (1995). Posttraumatic stress disorder in the national comorbidity survey. *Arch Gen Psychiatry*, 52, 1048–60.

Kimbrell, T., Leulf, C., Cardwell, D., Komoroski, R. A., and Freeman, T. W. (2005). Relationship of in vivo medial temporal lobe magnetic resonance spectroscopy to documented combat exposure in veterans with chronic posttraumatic stress disorder. *Psychiatry Res*, 140(1), 91–4.

Kitayama, N., Vaccarino, V., Kutner, M., Weiss, P., and Bremner, J. D. (2005). Magnetic resonance imaging (MRI) measurement of hippocampal volume in posttraumatic stress disorder: A meta-analysis. *J Affect Disord*, 88(1), 79–86.

Landfield, P., Baskin, R., and Pitler, T. (1981). Brain aging correlates: retardation by hormonal-pharmacological treatments. *Science*, 214, 581–4.

Lanius, R.A., Williamson, P. C., Densmore, M., *et al.* (2001). Neural correlates of traumatic memories in posttraumatic stress disorder: A functional MRI investigation. *Am J Psychiatry*, 158, 1920–2.

Lanius, R.A., Williamson, P. C., Hopper, J., *et al.* (2003). Recall of emotional states in posttraumatic stress disorder: An fMRI investigation. *Biol Psychiatry*, 53(3), 204–10.

Lawrence, M. S., and Sapolsky, R. M. (1994). Glucocorticoids accelerate ATP loss following metabolic insults in cultured hippocampal neurons. *Brain Res*, 646, 303–6.

Li, L., Chen, S., Liu, J., Zhang, J., He, Z., and Lin, X. (2006). Magnetic resonance imaging and magnetic resonance spectroscopy study of deficits in hippocampal structure in fire victims with recent-onset posttraumatic stress disorder. *Can J Psychiatry*, 51(7), 431–7.

Liberzon, I. and Phan, K. L. (2003). Brain-imaging studies of posttraumatic stress disorder. *CNS Spectr*, 8(9), 641–50.

Liberzon, I. and Martis, B. (2006). Neuroimaging studies of emotional responses in PTSD. *Ann N Y Acad Sci*, 1071, 87–109.

Liberzon, I., Taylor, S. F., Amdur, R., *et al.* (1999). Brain activation in PTSD in response to trauma-related stimuli. *Biol Psychiatry*, 45, 817–26.

Liberzon, I., Britton, J. C., and Phan, K. L. (2003). Neural correlates of traumatic recall in posttraumatic stress disorder. *Stress*, 6(3), 151–6.

Lim, M. K., Suh, C. H., Kim, H. J., *et al.* (2003). Fire-related post-traumatic stress disorder: brain 1H-MR spectroscopic findings. *Korean J Radiol*, 4(2), 79–84.

Lindauer, R. J., Vlieger, E. J., Jalink, M., *et al.* (2004a). Smaller hippocampal volume in Dutch police officers with posttraumatic stress disorder. *Biol Psychiatry*, 56(5), 356–63.

Lindauer, R. J., Booij, J., Habraken, J. B., *et al.* (2004b). Cerebral blood flow changes during script-driven imagery in police officers with posttraumatic stress disorder. *Biol Psychiatry*, 56 (11), 853–61.

Lindauer, R. J., Vlieger, E. J., Jalink, M., *et al.* (2005). Effects of psychotherapy on hippocampal volume in out-patients with post-traumatic stress disorder: a MRI investigation. *Psychol Med*, 35 (10), 1421–31.

Lindauer, R. J., Olff, M., van Meijel, E. P., Carlier, I. V., and Gersons, B. P. (2006). Cortisol, learning, memory, and attention in relation to smaller hippocampal volume in police officers with posttraumatic stress disorder. *Biol Psychiatry*, 59(2), 171–7.

Lucassen, P. J., Fuchs, E., and Czéh, B. (2004). Antidepressant treatment with tianeptine reduces apoptosis in the hippocampal dentate gyrus and temporal cortex. *Eur J Neurosci*, 14, 161–6.

Maes, M., Lin, A. H., Delmeire, L., Van Gastel, A., Kenis, G., De Jongh, R., *et al.* (1999). Elevated serum interleukin-6 (IL-6) and IL-6 receptor concentrations in posttraumatic stress disorder following accidental man-made traumatic events. *Biol Psychiatry*, 45(7), 833–9.

Magarinos, A., Somoza, G., and DeNicola, A. (1987). Glucocorticoid negative feedback and glucocorticoid receptors after hippocampectomy in rats. *Horm Metab Res*, 19, 105–9.

Magarinos, A.M., McEwen, B. S., Flügge, G., and Fuchs, E. (1996). Chronic psychosocial stress causes apical dendritic atrophy of hippocampal CA3 pyramidal neurons in subordinate tree shrews. *J Neurosci*, 16, 3534–40.

Mahmutyazicioglu, K., Konuk, N., Ozdemir, H., Atasoy, N., Atik, L., and Gündo du, S. *et al.* (2005). Evaluation of the hippocampus and the anterior cingulate gyrus by proton MR spectroscopy in patients with post-traumatic stress disorder. *Diagn Interv Radiol*, 11(3), 125–9.

Malberg, J.E., Eisch, A. J., Nestler, E. J., and Duman, R. S. (2000). Chronic antidepressant treatment increases neurogenesis in adult rat hippocampus. *J Neurosci*, 20, 9104–10.

McCormick, C. M., Smythe, J. W., and Beers, D. (1994). Sex differences in type I corticosteroid receptor binding in selective brain areas of rats. *Ann NY Acad Sci*, 746, 431–3.

McEwen, B. S. (1998). Stress, adaptation, and disease. Allostasis and allostatic load. *Ann N Y Acad Sci*, 840, 33–44.

McEwen, B.S., Angulo, J., Cameron, H., *et al.* (1992). Paradoxical effects of adrenal steroids on the brain: Protection versus degeneration. *Biol Psychiatry*, 31, 177–99.

McLaughlin, K.J. and Baran, S.E. (2009). Chronic stress- and sex-specific neuromorphological and functional changes in limbic structures. *Mol Neurobiol*, 40, 166–82.

Meaney, M. J., Aitken, D. H., van Berkel, C., Bhatnagar, S., and Sapolsky, R. M. (1988). Effect of neonatal handling on age-related impairments associated with the hippocampus. *Science*, 239, 766–8.

Milad, M.R. and Quirk, G.J. (2002). Neurons in medial prefrontal cortex signal memory for fear extinction. *Nature*, 420, 70–3.

Milad, M. R., Rauch, S. L., Pitman, R. K., and Quirk, G. J. (2006). Fear extinction in rats: implications for human brain imaging and anxiety disorders. *Biol Psychol*, 73(1), 61–71.

Mohanakrishnan Menon, P., Nasrallah, H. A., Lyons, J. A., Scott, M. F., and Liberto, V. (2003). Single-voxel proton MR spectroscopy of right versus left hippocampi in PTSD. *Psychiatry Res*, 123(2), 101–8.

Moradi, A.R., Doost, H. T., Taghavi, M. R., Yule, W., and Dalgleish, T. (1999). Everyday memory deficits in children and adolescents with PTSD: performance on the Rivermead Behavioural Memory Test. *J Child Psychol Psychiatry*, 40, 357–61.

Morgan, C. A., Romanski, L. M., and LeDoux, J. E. (1993). Extinction of emotional learning: Contribution of medial prefrontal cortex. *Neurosci Lett*, 163, 109–13.

Muhlen, K. aus der and Ockenfels, H. (1969). [Morphological alterations in the diencephalon and telencephalon following disturbances to the feedback mechanism adenohypophysis-adrenal cortex. 3. Studies on the guinea pig after administration of cortisone and hydrocortisone]. *Z Zellforsch Mikrosk Anat*, 93(1), 126–41.

Murakami, S., Imbe, H., Morikawa, Y., Kubo, C., and Senba, E. (2005). Chronic stress, as well as acute stress, reduces BDNF mRNA expression in the rat hippocampus but less robustly. *Neurosci Res*, 53(2), 129–39.

Nibuya, M., Morinobu, S., and Duman, R. S. (1995). Regulation of BDNF and trkB mRNA in rat brain by chronic electroconvulsive seizure and antidepressant drug treatments. *J Neurosci*, 15, 7539–47.

Notestine, C. F., Stein, M. B., Kennedy, C. M., Archibald, S. L., and Jernigan, T. L. (2002). Brain morphometry in female victims of intimate partner violence with and without posttraumatic stress disorder. *Biol Psychiatry*, 51, 1089–101.

Packan, D. R. and Sapolsky, R. M. (1990). Glucocorticoid endangerment of the hippocampus: tissue, steroid and receptor specificity. *Neuroendocrinology*, 51(6), 613–18.

Phan, K. L., Britton, J. C., Taylor, S. F., Fig, L. M., and Liberzon, I. (2006). Corticolimbic blood flow during nontraumatic emotional processing in posttraumatic stress disorder. *Arch Gen Psychiatry*, 63(2), 184–92.

Pitman, R. K. (2001). Investigating the pathogenesis of posttraumatic stress disorder with neuroimaging. *J Clin Psychiatry*, 62, 47–54.

Prigerson, H. G., Maciejewski, P. K., and Rosenheck, R. A. (2001). Combat trauma: Trauma with highest risk of delayed onset and unresolved posttraumatic stress disorder symptoms, unemployment, and abuse among men. *J Nerv Ment Dis*, 189(2), 99–108.

Putnam, F.W., Guroff, J. J., Silberman, E. K., Barban, L., and Post, R. M. (1986). The clinical phenomenology of multiple personality disorder: A review of 100 recent cases. *J Clin Psychiatry*, 47, 285–93.

Rauch, S. L., Shin, L. M., and Phelps, E. A. (2006). Neurocircuitry models of posttraumatic stress disorder and extinction: human neuroimaging research–past, present, and future. *Biol Psychiatry*, 60(4), 376–82.

Roca, V. and Freeman, T.W. (2001). Complaints of impaired memory in veterans with PTSD. *Am J Psychiatry*, 158, 1738.

Rohleder, N., Joksimovic, L., Wolf, J. M., and Kirschbaum, C. (2004). Hypocortisolism and increased glucocorticoid sensitivity of pro-Inflammatory cytokine production in Bosnian war refugees with posttraumatic stress disorder. *Biol Psychiatry*, 55, 745–51.

Sachinvala, N., von Scotti, H., McGuire, M., *et al.* (2000). Memory, attention, function, and mood among patients with chronic posttraumatic stress disorder. *J Nerv Ment Dis*, 188, 818–23.

Santarelli, L., Saxe, M., Gross, C., *et al.* (2003). Requirement of hippocampal neurogenesis for the behavioral effects of antidepressants. *Science*, 301, 805–9.

Sapolsky, R. (1986). Glucocorticoid toxicity in the hippocampus: synergy with an excitotoxin. *Neuroendocrinology*, 43, 440–6.

Sapolsky, R. M. (1996). Why stress is bad for your brain. *Science*, 273, 749–50.

Sapolsky, R. M., Uno, H., Rebert, C. S., and Finch, C. E. (1990). Hippocampal damage associated with prolonged glucocorticoid exposure in primates. *J Neurosci*, 10, 2897–902.

Schmahl, C.G., Vermetten, E., Elzinga, B. M., and Douglas Bremner, J. (2003). Magnetic resonance imaging of hippocampal and amygdala volume in women with childhood abuse and borderline personality disorder. *Psychiatry Res*, 122, 193–8.

Schmahl, C., Berne, K., Krause, A., Kleindienst, N., Valerius, G., Vermetten, E., *et al.* (2009). Hippocampus and amygdala volumes in patients with borderline personality disorder with or without posttraumatic stress disorder. *J Psychiatry Neurosci*, 34(4), 289–95.

Schuff, N., Marmar, C. R., Weiss, D. S., *et al.* (1997). Reduced hippocampal volume and n-acetyl aspartate in posttraumatic stress disorder. *Ann N Y Acad Sci*, 821, 516–20.

Schuff, N., Neylan, T. C., Lenoci, M. A., *et al.* (2001). Decreased hippocampal N-acetylaspartate in the absence of atrophy in posttraumatic stress disorder. *Biol Psychiatry*, 50, 952–9.

Schuff, N., Meyerhoff, D.J., Mueller, S., *et al.* (2006). N-acetylaspartate as a marker of neuronal injury in neurodegenerative disease. *Adv Exp Med Biol*, 576, 241–62; discussion 361–3.

Semple, W.E., Goyer, P. F., McCormick, R., *et al.* (2000). Higher brain blood flow at amygdala and lower frontal cortex blood flow in PTSD patients with comorbid cocaine and alcohol abuse compared to controls. *Psychiatry*, 63, 65–74.

Sheline, Y.I., Gado, M.H., and Kraemer, H.C. (2003). Untreated depression and hippocampal volume loss. *Am J Psychiatry*, 160, 1516–18.

Shin, L.M., Kosslyn, S. M., McNally, R. J., *et al.* (1997). Visual imagery and perception in posttraumatic stress disorder: A positron emission tomographic investigation. *Arch Gen Psychiatry*, 54, 233–7.

Shin, L.M., McNally, R. J., Kosslyn, S. M., *et al.* (1999). Regional cerebral blood flow during script-driven imagery in childhood sexual abuse-related PTSD: A PET investigation. *Am J Psychiatry*, 156, 575–84.

Shin, L.M., Whalen, P. J., Pitman, R. K., *et al.* (2001). An fMRI study of anterior cingulate function in posttraumatic stress disorder. *Biol Psychiatry*, 50, 932–42.

Shin, L. M., Orr, S. P., Carson, M. A., *et al.* (2004a). Regional cerebral blood flow in the amygdala and medial prefrontal cortex during traumatic imagery in male and female Vietnam veterans with PTSD. *Arch Gen Psychiatry*, 61 (2), 168–76.

Shin, L. M., Shin, P. S., Heckers, S., *et al.* (2004b). Hippocampal function in posttraumatic stress disorder. *Hippocampus*, 14 (3), 292–300.

Shin, L. M., Wright, C. I., Cannistraro, P. A., *et al.* (2005). A functional magnetic resonance imaging study of amygdala and medial prefrontal cortex responses to overtly presented fearful faces in posttraumatic stress disorder. *Arch Gen Psychiatry*, 62(3), 273–81.

Smith, M. A., Makino, S., Kvetnansky, R., and Post, R. M. (1995). Stress and glucocorticoids affect the expression of brain-derived neurotrophic factor and neurotrophin-3 mRNA in the hippocampus. *J Neurosci*, 15, 1768–77.

Smith, M. E. (2005). Bilateral hippocampal volume reduction in adults with post-traumatic stress disorder: a meta-analysis of structural MRI studies. *Hippocampus*, 15(6), 798–807.

Squire, L. R. (2004). Memory systems of the brain: A brief history and current perspective. *Neurobiol Learning Memory*, 82, 171–7.

Söndergaard, H. P., Hansson, L. O., and Theorell, T. (2004). The inflammatory markers C-reactive protein and serum amyloid A in refugees with and without posttraumatic stress disorder. *Clin Chim Acta*, 342(1–2), 93–8.

Spivak, B., Shohat, B., Mester, R., Avraham, S., Gil-Ad, I., Bleich, A., Valevski, A., and Weizman, A. (1997). Elevated levels of serum interleukin-1 beta in combat-related posttraumatic stress disorder. *Biol Psychiatry*, 42(5), 345–8.

Stein-Behrens, B. A., Lin, W. J., and Sapolsky, R. M. (1994). Physiological elevations of glucocorticoids potentiate glutamate accumulation in the hippocampus. *J Neurochem*, 63, 596–602.

Stein, M. B., Koverola, C., Hanna, C., Torchia, M. G., and McClarty, B. (1997). Hippocampal volume in women victimized by childhood sexual abuse. *Psychol Med*, 27, 951–9.

Stein, M. B., Hanna, C., Vaerum, V., and Koverola, C. (1999). Memory functioning in adult women traumatized by childhood sexual abuse. *J Trauma Stress*, 12(3), 527–34.

Thomaes, K., Dorrepaal, E., Draijer, N. P., de Ruiter, M. B., Elzinga, B. M., van Balkom, A. J., *et al.* (2009). Increased activation of the left hippocampus region in Complex PTSD during encoding and recognition of emotional words: a pilot study. *Psychiatry Res*, 171, 44–53.

Uddo, M., Vasterling, J. J., Brailey, K., and Sutker, P. (1993). Memory and attention in posttraumatic stress disorder. *J Psychopath Behav Assess*, 15, 43–52.

Uno, H., Tarara, R., Else, J. G., Suleman, M. A., Sapolsky, R. M. (1989). Hippocampal damage associated with prolonged and fatal stress in primates. *J Neurosci*, 9, 1705–11.

Uno, H., Lohmiller, L., Thieme, C., Kemnitz, J. W., Engle, M. J., Roecker, E. B., and Farrell, P. M., (1990). Brain damage induced by prenatal exposure to dexamethasone in fetal rhesus macaques. I. Hippocampus. *Brain Res Dev Brain Res*, 53(2), 157–67.

Vaher, P. R., Luine, V. N., Gould, E., and McEwen, B. S. (1994). Effects of adrenalectomy on spatial memory performance and dentate gyrus morphology. *Brain Res*, 656, 71–8.

van Zuiden, M., Geuze, E., Willemen, H. L., Vermetten, E., Maas, M., Amarouchi, K., and Heijnen, C. J., (2012). Glucocorticoid receptor pathway components predict posttraumatic stress disorder symptom development: a prospective study. *Biol Psychiatry*, 71(4), 309–16.

Vasterling, J. J., Brailey, K., Constans, J. I., and Sutker, P. B. (1998). Attention and memory dysfunction in posttraumatic stress disorder. *Neuropsychology*, 12, 125–33.

Vasterling, J. J., Duke, L. M., Brailey, K., Constans, J. I., Allain, A. N. Jr, and Sutker, P. B. (2002). Attention, learning, and memory performance and intellectual resources in Vietnam veterans: PTSD and no disorder comparisons. *Neuropsychology*, 16, 5–14.

Vermetten, E. and Bremner, J.D. (2002a). Circuits and systems in stress. II. Applications to neurobiology and treatment of PTSD. *Depress Anxiety*, 16, 14–38.

Vermetten, E. and Bremner, J.D. (2002b). Circuits and systems in stress. I. Preclinical studies. *Depress Anxiety*, 15, 126–47.

Vermetten, E., Vythilingam, M., Southwick, S. M., Charney, D. S., and Bremner, J. D. (2003). Long-term treatment with paroxetine increases verbal declarative memory and hippocampal volume in posttraumatic stress disorder. *Biol Psychiatry*, 54 (7), 693–702.

Vermetten, E., Schmahl, C., Lindner, S., Loewenstein, R. J., and Bremner, J. D. (2006). Hippocampal and amygdalar volumes in dissociative identity disorder. *Am J Psychiatry*, 163, 1–8.

Villarreal, G., Hamilton, D. A., Petropoulos, H., *et al.* (2002). Reduced hippocampal volume and total white matter in posttraumatic stress disorder. *Biol Psychiatry*, 52, 119–25.

Vythilingam, M., Heim, C., Newport, J., *et al.* (2002). Childhood trauma associated with smaller hippocampal volume in women with major depression. *Am J Psychiatry*, 159, 2072–80.

Wang, Z., Neylan, T. C., Mueller, S. G., *et al.* (2010). Magnetic resonance imaging of hippocampal subfields in posttraumatic stress disorder. *Arch Gen Psychiatry*, 67(3), 296–303.

Watanabe, Y., Sakai, R.R., McEwen, B.S., Mendelson, S. (1993). Stress and atidepressant effects on hippocampal and cortical 5HT1A and 5HT2 receptors and transport sites for serotonin. *Brain Res*, 615, 87–94.

Wignall, E. L., Dickson, J. M., Vaughan, P., *et al.* (2004). Smaller hippocampal volume in patients with recent-onset posttraumatic stress disorder. *Biol Psychiatry*, 56(11), 832–6.

Winter, H. and Irle, E. (2004). Hippocampal volume in adult burn patients with and without posttraumatic stress disorder. *Am J Psychiatry*, 161(12), 2194–200.

Wooley, C. S., Gould, E., and McEwen, B. S. (1990). Exposure to excess glucocorticoids alters dendritic morphology of adult hippocampal pyramidal neurons. *Brain Res*, 531, 225–31.

Yang, P., Wu, M. T., Hsu, C. C., and Ker, J. H. (2004). Evidence of early neurobiological alternations in adolescents with posttraumatic stress disorder: a functional MRI study. *Neurosci Lett*, 370(1), 13–18.

Yehuda, R., Golier, J. A., Yang, R. K., and Tischler, L. (2004). Enhanced sensitivity to glucocorticoids in peripheral mononuclear leukocytes in posttraumatic stress disorder. *Biol Psychiatry*, 55(11), 1110–16.

Yehuda, R., Keefe, R. S., Harvey, P. D., *et al.* (1995). Learning and memory in combat veterans with posttraumatic stress disorder. *Am J Psychiatry*, 152, 137–9.

Yehuda, R., Halligan, S. L., Grossman, R., Golier, J. A., and Wong, C. (2002). The cortisol and glucocorticoid receptor response to low dose dexamethasone administration in aging combat veterans and holocaust survivors with and without posttraumatic stress disorder. *Biol Psychiatry*, 52(5), 393–403.

Yehuda, R., Golier, J. A., Tischler, L., *et al.* (2007). Hippocampal volume in aging combat veterans with and without post-traumatic stress disorder: Relation to risk and resilience factors. *J Psychiatr Res*, 41(5), 435–45.

Yen, S., Shea, M. T., Battle, C. L., *et al.* (2002). Traumatic exposure and posttraumatic stress disorder in borderline, schizotypal, avoidant, and obsessive-compulsive personality disorders: findings from the collaborative longitudinal personality disorders study. *J Nerv Ment Dis*, 190 (8), 510–18.

Zalewski, C., Thompson, W., and Gottesman, I. (1994). Comparison of neuropsychological test performance in PTSD, generalized anxiety disorder, and control Vietnam veterans. *Assessment*, 1, 133–42.

CHAPTER 15

The Hippocampus in Major Depression

Glenda M. MacQueen and Thomas Frodl

Introduction 273
Regional morphological variation in the hippocampus in MDD 274
Can hippocampal dysregulation be linked to MDD? 275
Clinical and demographic variables influencing hippocampal volumes in MDD 276
Biological mechanisms that may contribute to small hippocampal volume in MDD 277
The clinical relevance of small hippocampal volumes in MDD 281
Evidence that treatment for MDD may have an impact on hippocampal volumes 282
Summary 283

ABSTRACT Major depressive disorder (MDD) is characterized by one or more episodes of major depression and the absence of episodes of mania or hypomania. A major depressive episode is characterized by at least 2 weeks of pervasive low mood or loss of pleasure that occurs along with a variety of other emotional, cognitive, or neurovegetative symptoms. This chapter summarizes the clinical and neurobiological correlates and concepts of smaller hippocampal volumes in MDD. Preclinical data describe the molecular and cellular effects of chronic stress and antidepressant treatment on the hippocampus, providing plausible mechanisms through which MDD might lead to decreased hippocampal volumes. Patients with smaller hippocampal volumes are at risk for a poor clinical outcome and this may be a mechanism through which MDD appears to be an additional risk factor for developing Alzheimer's disease. The pathways through which stress may be linked to MDD, the emergence of disease chronicity or treatment resistance in MDD, and the association between MDD and memory problems may be at least partially understood by dissecting the association with depression and structural and functional changes in the hippocampus. MDD now appears to be a disorder that is associated with persistent morphological brain changes that are detectable before illness onset and which can be modified by clinical and treatment variables.

KEYWORDS hippocampus, depression, imaging, mood disorders, magnetic resonance imaging, cognition, therapy prediction, neuroplasticity

Introduction

Major depressive disorder (MDD) is characterized by one or more episodes of major depression and the absence of episodes of mania or hypomania. A major depressive episode is characterized by at least 2 weeks of pervasive low mood or loss of pleasure that occurs along with a variety of other emotional, cognitive, or neurovegetative symptoms. MDD has until recently been conceptualized as an episodic disorder associated with 'chemical imbalances' in the CNS but no permanent brain changes. Evidence has emerged in the past decade suggesting that MDD is associated with smaller hippocampal volumes that may occur in the context of changes in other discrete regions, such as the cingulate cortex, in the absence of global morphological abnormalities.

There are now several independent meta-analytic reports confirming that the volume of the hippocampus (HC) is decreased in patients with a major depressive disorder (MDD; McKinnon et al. 2009). One recent meta-analysis included 1167 patients and 1088 control subjects, across a wide range of ages from paediatric to geriatric populations: the patients had hippocampal volumes that were approximately 4–6% smaller than matched control subjects. Convergent lines of research support the notion that the HC is highly relevant for the understanding of the pathogenesis of MDD. First, MDD is clinically recognized as a highly stress-sensitive illness (Kessler 1997) and the HC is a highly stress sensitive brain region (Thomas et al. 2007). Stress, including psychological, behavioural, emotional, or psychosocial stress, is associated with structural changes to the HC (e.g. Vermetten and Bremner 2002; Malberg and Duman 2003; Pham et al. 2003). Conditions of chronic hypercortisolaemia, such as Cushing's disease, are associated with HC atrophy that is reversible following normalization of cortisol levels (Starkman 1999), supporting the hypothesis that stress-induced changes in the HC are observable in humans as well as other species. Effective antidepressant treatments may ameliorate stress-associated changes in the HC (Duman et al. 2001; Santarelli et al. 2003).

Second, when the various domains of cognitive function are assessed in patients with depression, they are most impaired on memory measures that are heavily dependent on HC function such as declarative and episodic memory (Zakzanis et al. 1998); poor memory performance is associated with small volumes of HC grey matter in non-clinical samples as well (Gatt et al. 2009). Furthermore, while there has been debate about whether mood congruent memory biases are present for implicit material, such biases can be reliably demonstrated in explicit memory tasks that are heavily dependent upon medial temporal lobe structures, including the HC (Colombel 2007).

Third, the HC is one of several regions, including the dorsomedial and dorsolateral prefrontal cortex, the anterior cingulate cortex, and the amygdala (Frodl et al. 2008d) that imaging studies have identified as components of an affective-emotional network that is dysregulated in MDD (Frodl et al. 2010c).

Despite consistency in the studies implicating the HC in MDD, questions remain regarding the pathophysiological underpinnings of these findings, the clinical factors that are associated with dysregulation in the HC, the clinical significance of HC changes in MDD, and the likelihood that medication can ameliorate HC related dysfunction in patients with MDD. We review the data relevant to each of these issues.

Regional morphological variation in the hippocampus in MDD

There may be variation in the extent to which subregions of the HC differ between patients and healthy control subjects. Zhao and colleagues used spherical harmonic shape analyses to examine for regional differences in the HC of 61 elderly subjects with depression compared to non-depressed elderly subjects (Zhao et al. 2008). They reported differences in the left HC of depressed subjects that were restricted to the midbody region of the HC. When patients were divided into remitted and non-remitted participants, the remitted patients were not different from controls, but un-remitted patients were different in the midbody and lateral regions.

Other groups have also reported that the posterior region of the HC has the greatest volume difference between patients and controls (Neumeister 2005; de Geus 2007). Posener and colleagues found alterations of the HC surface with an inward deformation in the head and tail and an outward deformation in the body of the HC (Posener 2003). One recent study reported that the right HC body and tail were smaller in female patients with familial depressive disorder compared to healthy subjects (Nifosì 2010), while another group reported that HC volumes were reduced bilaterally in the tail but only in the right HC head of depressed patients compared to controls (Malykhin et al. 2010a). This same group of investigators recently used 4.7-Tesla magnetic resonance imaging (MRI) to examine HC subfields in 11 healthy subjects (Malykhin et al. 2010b). They demonstrated anatomical differences in the distribution of various subfields of the Ammon's horn within the distinguishable parts of the HC: the largest portion of the dentate gyrus was located in the HC body, followed by the HC head and tail; the HC head had the largest part of CA1–3, followed by the HC body and tail. The HC tail had the smallest portion of the subiculum compared to the HC head. Other investigators have used shape analysis to localize subregional deformations to the subiculum and CA1 subfields extending into the CA2–3 subfields predominantly in the tail region. This study included a subset of patients who were in a first episode of depression (Cole 2010), extending the work of Ballmaier and colleagues who found small left HC volumes in patients with late onset depression that they speculated were concentrated in the subiculum and in posterior-lateral aspects of the CA1 area (Ballmaier 2008).

There are several reasons why it is difficult to ascertain the extent to which factors such as treatment exert differential effects on various HC regions. Many studies of depressed patients have reported only right and left HC volumes and do not distinguish between the head, body, or tail region; some of the discrepancy in reports of HC volume differences between depressed patients and controls may in fact have arisen because of selective measurement of portions of the HC. Many studies do not report the proportion of patients who are in treatment at the time of scanning. Until the work of Malykhin and colleagues, there have been few studies extrapolating from MRI regional findings to anatomical subfields. This work may help to link clinical fields with postmortem and preclinical studies that examine the HC subfields specifically.

Can hippocampal dysregulation be linked to MDD?

There is credible evidence that the HC is part of a network that, when dysregulated, could contribute to a variety of depressive symptoms. Hyperactivity of the hypothalamic–pituitary–adrenal (HPA) axis and the resulting increase in

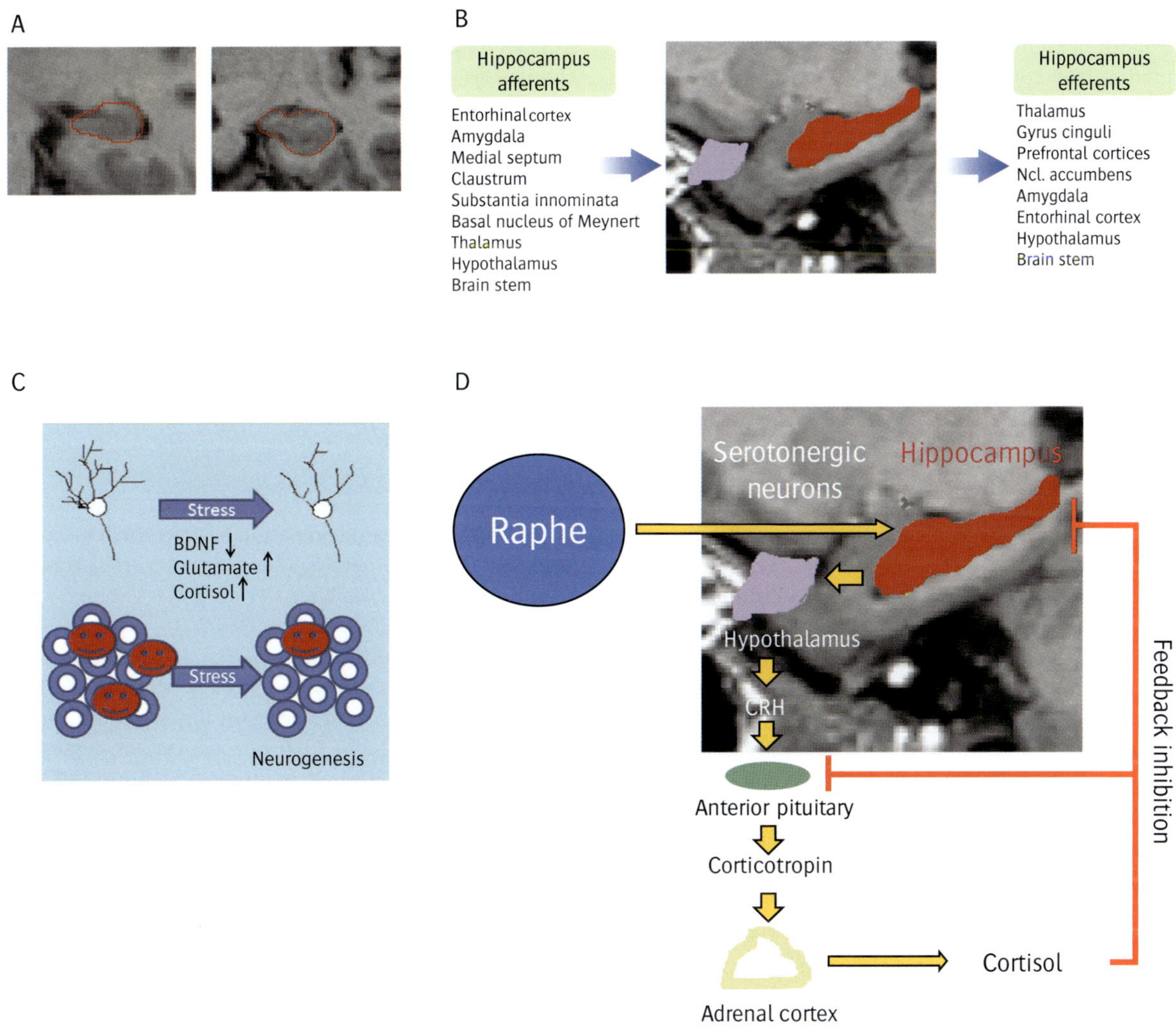

Figure 15.1 Evidence that hippocampal dysregulation is involved in MDD derives from studies that show smaller hippocampal volumes in MDD. Depicted are two MRI images that present the hippocampus within the temporal lobe region and how it is traced in a manual region of interest analysis (A). The hippocampus is involved in a neural network of affect regulation and has intensive connections from and to other brain regions (B). Experimental animal studies showed that stress affects hippocampal neurons in a way that dendrites decrease in size, length, and numbers (top row) and that neurogenesis is inhibited (bottom row) (C). Moreover, the hippocampus plays a major role in the regulation of the HPA axis, which is involved in stress response and MDD. The serotonergic system, which is involved in the neurobiology of MDD and which is mechanism of action for most of the antidepressants, interacts with the HPA stress axis in the hippocampus (D). Adapted from MacQueen, G. and Frodl, T. (2011). The hippocampus in major depression: evidence for the convergence of the bench and bedside in psychiatric research? *Molecular Psychiatry*, 16(3), 252–64, published 2011 by Nature Publishing Group.

glucocorticoid levels in the brain are associated with early stage reversible dendritic remodelling in the C1 and C3 pyramidal granule neurons, paralleled by reversible remodelling of synaptic terminal structures. With extreme or chronic stress, there are volumetric decreases in the HC formation, with an increased vulnerability to metabolic insults and even death of neurons in the CA3 region. Prolonged stress and increased levels of glucocorticoids also disrupt HC neurogenesis (Duman 2004; Dranovsky and Hen 2006). As the HC provides negative modulation to the HPA stress hormone axis through its projections to the hypothalamus, HC dysfunction may contribute to the sustained dysregulation of the stress response that is commonly observed in MDD (see Figure 15.1).

Through its prominent projections, the HC is a key regulator of prefrontal cortical function; disruption of HC function may contribute to the deficits in concentration that are identified amongst the diagnostic features of MDD as well as to the memory deficits that are commonly experiences by patients with MDD. Hippocampal afferents are critical regulators of both the nucleus accumbens and the ventral tegmental area (VTA). An indirect excitatory projection from HC to VTA may be critical for coordinating the firing of VTA cells in response to novelty (Lisman and Grace 2005). Anhedonia, another core feature of MDD, might result in part from impaired HC function resulting in reduced dopaminergic tone (Warner-Schmidt and Duman 2006).

Among the various brain areas implicated in MDD, the HC may have a particular role in the control of 5-hydroxytryptamine (serotonin, 5-HT) system–HPA axis interactions. Of particular interest is the negative regulation of 5-HT1A mRNA expression by the co-activation of mineralocorticoid and glucocorticoid receptors. Transcription of the 5-HT1A receptor gene is negatively regulated by corticosteroids in the limbic system. In rodents, adrenalectomy results in a rapid and marked increase of de novo 5-HT1A mRNA synthesis, total 5-HT1A mRNA levels, and 5-HT1A binding sites in the HC and the septum and these changes are reversed by treatment with low doses of corticosterone (Mendelson and McEwen 1992). Finally, the contributions of the HPA axis and the 5-HT system to the negative/positive control of granule cell proliferation within the dentate gyrus of the HC are probably of key importance among neurobiological mechanisms associated with MDD (for review see (Lanfumey et al. 2008)).

Clinical and demographic variables influencing hippocampal volumes in MDD

The question on the clinical and demographic variables that influence HC volumes has been addressed in detail in a recent review (McKinnon et al. 2009). In summary, small HC volumes in MDD have been linked to high levels of depressive symptoms (Vakili et al. 2000; Saylam et al. 2006), early age at illness onset (Lloyd et al. 2004; Hickie et al. 2005; Taylor et al. 2005; Janssen et al. 2007), nonresponsiveness to treatment (Vakili et al. 2000; Hsieh et al. 2002; Frodl et al. 2004a), long duration of untreated days of illness (Sheline et al. 2003), high illness burden (Sheline et al. 1996; Sheline et al. 1999; MacQueen et al. 2003; MacMaster and Kusumakar 2004), positive history of childhood abuse (Vythilingam et al. 2002), and high levels of anxiety (Rusch et al. 2001; MacMillan et al. 2003). Significant inconsistencies exist, however, particularly with respect to the relation between clinical variables (e.g. age at illness onset, duration of illness, etc.) and reductions in HC volume. For example, numerous studies report no relations between HC volumes and age at onset (Frodl et al. 2002, 2004a, 2006; Sheline et al. 2003; Hastings et al. 2004; Monkul et al. 2007). Although some studies showed a correlation between small HC volumes and more severe depressive symptomatology (Vakili et al. 2000; Saylam et al. 2006), the majority of structural studies that have investigated the relations with depression severity have not found a significant correlation between high ratings of depressive symptomatology and smaller HC volumes (Frodl et al. 2002; MacQueen et al. 2003; Caetano et al. 2004; Frodl et al. 2006).

When examined in the aggregate, McKinnon and colleagues reported that differences in HC volume were apparent only for patients with an illness duration of at least 2.5 years or more than one episode of depression (McKinnon et al. 2009). Some studies examining HC volume in patients with a long history of depression find a correlation with the amount of time spent symptomatically ill, suggesting that successful treatment and symptom reduction may halt a neurodegenerative process (Shah et al. 1998; Sheline et al. 1999; MacQueen et al. 2003). Another group reported that volume in the left HC showed a marginally significant relation with duration of illness in a group of drug-free patients with MDD (Saylam et al. 2006). Other studies, however, have not found evidence of an effect of illness duration on HC volume (Frodl et al. 2002, 2004a, 2006, 2007; Posener et al. 2003; Janssen et al. 2004; Lange and Irle 2004; Lloyd et al. 2004; O'Brien et al. 2004; Xia et al. 2004; Hickie et al. 2005; MacMaster et al. 2008). These negative findings may reflect, in part, small sample sizes (e.g. Xia et al. 2004; Monkul et al. 2007), the inclusion of patients with bipolar disorder (Hickie et al. 2005), or samples comprised of primarily young adult patients (Posener et al. 2003; Lange and Irle 2004; Monkul et al. 2007), and of patients with a low number of illness episodes (Posener et al. 2003).

Biological mechanisms that may contribute to small hippocampal volume in MDD

Genetic influences on hippocampal volumes in MDD

Hippocampal size has a moderate heritability (Lyons et al. 2001; Gilbertson et al. 2002). An association between heritability and brain size is supported by twin studies (Bartley et al. 1997) and a wide range of imaging genetic studies (Frodl et al. 2008a). A recent study examined hippocampal volumes in 112 unaffected mono and dizygotic co-twins of twins affected or not affected by depression (Baare et al. 2010). The co-twins of affected twins had smaller HC volumes than co-twins of unaffected twins; interestingly the effect was most pronounced amongst dizygotic twins. The investigators note that the differences in the pattern of findings between mono- and dizygotic twins may have resulted from the smaller sample of monozygotic twins (40 monozygotic compared to 72 dizygotic twins). They acknowledge, however, that the small HC volumes in the unaffected dizygotic twins of affected co-twins may have resulted from shared environmental factors. Supporting this notion is the fact that the high-risk twins endorsed a greater number of adverse life events in the past 12 months than did the twins at low risk (unaffected co-twin). These results are also consistent with another twin study (de Geus et al. 2007) in which twins with high trait anxiety and depression had grey matter reductions in the left posterior HC when compared to a less anxious co-twin.

Alternative phenotypic markers for genetic association studies that are more closely related to the underlying neurobiology of the disease are increasingly being used for genetic association studies. This partly derives from a need to address problems of clinical heterogeneity in psychiatric disorders, and partly from the need to delineate the functional consequences of identified risk variants at the level of brain structure and function. Imaging genetics facilitates an elucidation of the impact of genes at the level of brain that can then be extended to the pathophysiology of the disease.

Elimination of 5-HT from the synaptic cleft is mediated by a single protein, the 5-HT transporter (5-HTT), which determines the size and duration of the serotonergic responses (Lesch and Mossner 1998). The promoter region of 5-HTT has a polymorphism that results in allelic variation of functional 5-HTT expression (Lesch et al. 1996). The long (l) allele is associated with production of more 5-HTT transcription than the short (s) allele and hence more functional 5-HT uptake than the s allele (Lesch et al. 1996). The short (s) allele of the 5-HTT polymorphism is associated with anxiety, depression, and aggression-related personality traits in some reports (Lesch et al. 1996; Lesch and Mossner 1998), but a recent meta-analysis did not confirm an association between 5-HTT polymorphism and vulnerability to MDD (Risch et al. 2009). The importance of variation in 5-HTTLPR to HC volume in patients with MDD has been noted in three published reports that utilized diallelic analysis and two reports using triallelic analysis (Frodl et al. 2004b, 2008c; Taylor et al. 2005).

Brain-derived neurotrophic factor (BDNF) regulates neuronal survival, migration, phenotypic differentiation, axonal and dendritic growth, and synapse formation (Huang and Reichardt 2001). BDNF is also a key regulator of synaptic plasticity and behaviour (Lu 2003) and may be important for memory acquisition and consolidation (Tyler et al. 2002). A single-nucleotide polymorphism in the pro-domain of BDNF converts the 66th amino acid valine into methionine (Val66Met). This Val66Met polymorphism affects dendritic trafficking and synaptic localization of BDNF and impairs its secretion. The Val66Met SNP is associated with deficits in short-term episodic memory (Egan et al. 2003). Healthy met-BDNF carriers have relatively small HC volumes (Pezawas et al. 2004; Bueller et al. 2006); an effect of the Met-BDNF allele on HC volume is also apparent in patients with MDD (Frodl et al. 2007); although one recent report did not observe this association, but rather found larger differences for patients with the Val phenotype (Gonul et al. 2011). A study in healthy volunteers reported that subjects carrying the Met-BDNF allele have smaller HC volumes when they have more subthreshold symptoms of depression and a higher extend of neuroticism (Joffe et al. 2009).

Grey matter volumes in temporal lobe regions including the HC were associated with glycogen synthase kinase-3 (GSK3) polymorphisms in a large sample of depressed patients and healthy controls (Inkster et al. 2009). The underlying neurobiological background of this finding is unknown; it is interesting, however, that GSK3 activity might be associated with therapeutic effects of antidepressants and lithium (Inkster et al. 2009). Thus, these first genetic imaging studies provide evidence that genetic polymorphisms are relevant for neurobiological mechanisms in MDD and for HC volumes. These studies emphasize the potential importance of genetic influences on HC volumes.

Effects of stress on the hippocampus

The HC is vulnerable to stress, particularly during the early developmental period (Teicher et al. 2003), however, there may be species specific and time-dependent variation in how behavioural, emotional, and psychosocial stress affects the brain. While the separation of monkeys from their siblings or mothers seems not to affect HC volumes (Lyons et al. 2001; Spinelli et al. 2009), chronic stress exposure in rats or tree shrews does result in reduced HC volumes (Ohl et al. 2000; Lee et al. 2009). While transient mild stress may enhance HC function (Luine et al. 1996), chronic or severe stress disrupts HC-dependent memory in experimental animals (reviewed in Sapolsky 2003). Extended or high-dose treatment with glucocorticoids has a negative effect on HC-dependent memory in both animal models (Bodnoff et al. 1995; de Quervain et al. 1998), and humans (Newcomer et al. 1999; de Quervain et al. 2000).

Sustained elevations in stress or glucocorticoids impair the HC at the level of morphological neuroplasticity (reviewed in Sapolsky 2000). High levels of glucocorticoids (Woolley et al. 1990) or behavioural stress (Watanabe et al. 1992; Magarinos et al. 1996) result in atrophy and retraction of the apical dendrites of HC pyramidal cells. A reduction in the amount of neuropil without frank cell loss has also been observed, a finding that appears consistent with observations from postmortem studies of the HC in patients with MDD (Stockmeier et al. 2004).

Glutamatergic excess may also contribute to cell damage and even to cell death (Sapolsky 2000, 2003). Stress-induced increases in extrasynaptic glutamate may perturb the balance between synaptic and extrasynaptic NMDA tone and have a net inhibitory effect on the mechanisms of synaptic plasticity and neuronal growth and survival (reviewed in Pittenger et al. 2007). Experimental models in rats show that short- and longer-term electrophysiological correlates of memory formation in CA1 hippocampal neurons such as paired-pulse facilitation (PPF), long-term potentiation (LTP) and long-term depression (LTD) are modulated by acute and chronic stress. While PPF and LTP are downscaled probably due to a stress-induced saturation of glutamatergic transmission (Cazakoff and Howland 2010), LTD is facilitated in the ventral hippocampus, but decreased in the dorsal hippocampus (Maggio and Segal 2011). These effects can be blocked by blockade of both, NMDA and glucocorticoid receptors in CA1 (Mailliet et al. 2008).

Signalling pathways implicated in neuroplasticity target, among other downstream targets, genes for growth factors such as BDNF (Newton and Duman 2004). Acute and chronic stress reduces neurogenesis in the rodent HC (reviewed in Duman 2004; Dranovsky and Hen 2006). As neurogenesis may be required for the behavioural response to antidepressants in rodents (Santarelli et al. 2003) and impaired neurogenesis may represent a pathophysiological component of MDD (Duman 2004), suppressed neurogenesis may represent another way that the effects of stress on mechanisms of neuroplasticity contribute to the development of MDD. The specific functional role and relevance of new neurons in the HC is not firmly established, but a link between neurogenesis and the learning-related functions of the HC is an intriguing possibility (Leuner et al. 2006). Some computational theories of HC function predict a role for new neurons in HC-dependent learning (Becker et al. 2009).

There are few postmortem studies that have examined the HC specifically in patients with MDD. In one study of 19 subjects with MDD and 21 normal control subjects, Stockmeier and colleagues (Stockmeier et al. 2004) examined the density of pyramidal neurons, dentate granule cell neurons, glia, and the size of the neuronal somal area. They reported that the density of granule cells and glia in the dentate gyrus and pyramidal neurons and glia in all cornu ammonis (CA)/hippocampal subfields was significantly increased by 30–35% in MDD, while the average soma size of pyramidal neurons was significantly decreased in patients with MDD. They suggested that a reduction in neuropil in MDD may account for decreased hippocampal volume detected by neuroimaging.

The effects of early childhood stress on HC volumes

A recent meta-analysis based on 14250 participants demonstrated that early adverse life events confer significant risk of subsequent MDD (Risch et al. 2009). Youth and adults exposed to early-life adversity appear to have smaller HC volumes (Woon and Hedges 2008; for an overview on HC volume changes in post-traumatic stress disorder see also Bremner et al. (2008) and Karl et al. (2006)). Although a large population-based study from Australia did not find smaller HC volumes in subjects exposed to adverse childhood events, this study used voxel-based morphometry to conduct a whole brain analysis and the necessary corrections for multiple comparisons can decrease the probability of detecting changes in small structures like the HC (Cohen et al. 2006).

Vythilingam and colleagues reported 18% smaller left HC volumes in patients with early life trauma and MDD compared to non-abused patients with MDD who did not differ from healthy controls (Vythilingam et al. 2002). Moreover, patients with MDD and childhood emotional neglect had smaller HC volumes compared to patients with MDD without emotional neglect. The small volumes were more pronounced in the left HC and in men (Frodl et al. 2010a). In contrast, however, HC volumes did not correlate significantly with early life events in a sample of women with remitted unipolar MDD (Lenze et al. 2008) (Table 15.1). A potential explanation for the discrepancy in reports might be that standard correlational analyses of a continuous measure of early adversity might not show significant results if the underlying association between adversity and HC volumes, for example, is not described by

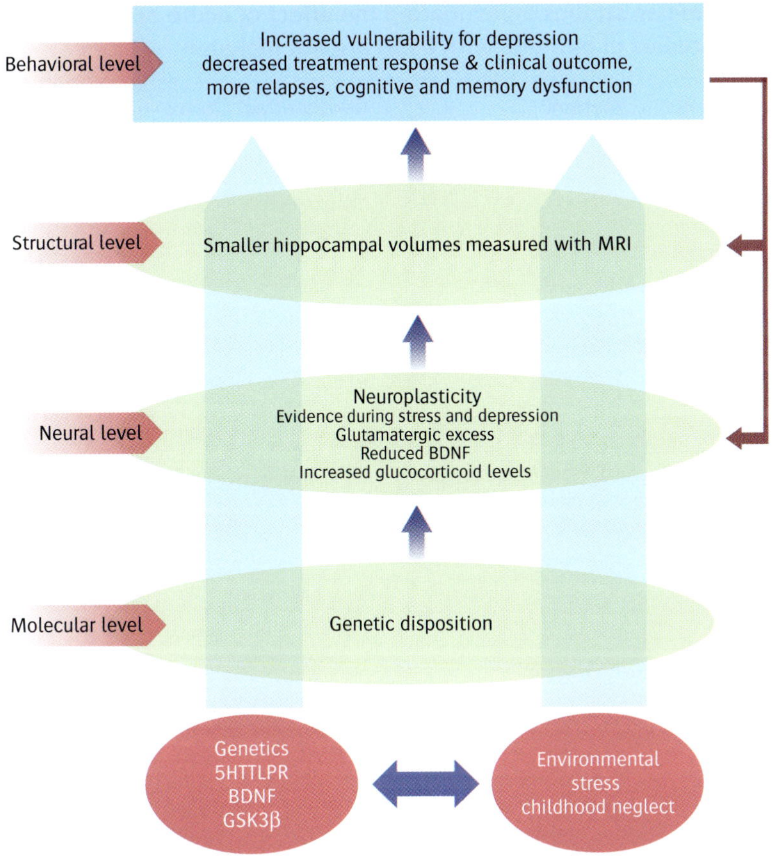

Figure 15.2 Overview of the factors influencing small HC volumes and the consequence of HC volume changes in MDD: there is evidence that genetic polymorphisms and early life stress may contribute to HC volumes prior to onset of illness. Repeated episodes of illness may further contribute to loss of HC volume via stress toxicity, reduced neurotrophic factors and excess in glutamatergic neurotransmission on hippocampal neurons. The underlying molecular basis is an ongoing matter of research. Speculatively, the changes in the HC may then contribute to treatment resistance or chronicity and may further increase the vulnerability for the disease. Consistent with this are studies reporting that small HC volumes predict poor short- and long-term responsiveness to treatment and a higher vulnerability. A more chronic course of disease and relapses seem to have further affects on neural connectivity and might result in further structural changes. Adapted from MacQueen, G. and Frodl, T. (2011). The hippocampus in major depression: evidence for the convergence of the bench and bedside in psychiatric research? *Molecular Psychiatry*, 16(3), 252–64, published 2011 by Nature Publishing Group.

a linear function. Comparing group differences between participants with and without abuse/neglect might then be an advantage for detecting relations that are best described by a non-linear function. Furthermore, gender effects have not been assessed in the majority of studies investigating the association between stress and HC volumes (although see Frodl et al. 2010a), so that further investigation is necessary to understand whether there are gender-specific effects in the impact that stress has on the HC.

In summary, there is evidence from studies of healthy subjects, patients with PTSD, and patients with MDD that early childhood stress is associated with small HC volumes. Whether variables such as the type of stress, the developmental period during which a child experiences the stress, or the chronicity and intensity of the stress can determine HC structure and function later in life needs further investigation.

Stress × gene interactions on HC volumes

Early life stress and variations in the serotonin transporter promoter polymorphism 5-HTTLPR may interact to predict development of MDD (Caspi et al. 2003). An increased risk of depression was detected in maltreated children homozygous for the S-allele (Kaufman et al. 2006). In a recent study, patients carrying the s-allele had smaller HC volumes when they had a history of emotional neglect compared to patients who had only one risk factor (environmental or genetic). Childhood stress further predicted HC alterations independent of genotype (Frodl et al. 2010b). Meta-analytic approaches suggest, however, that 5-HTTLPR may not further increase the risk for MDD in subjects who experience critical life events (Risch et al. 2009).

Table 15.1 Animal and human studies investigating the effect of acute or early life stress on hippocampal volumes. Studies of patients with PTSD are not included

Study	Subjects/patients	Method	Gender effect	Result
Frodl et al. 2010a	43 patients with MDD, 44 healthy controls	1.5 Tesla, manual tracing of whole hippocampus	More pronounced in male patients	Left hippocampal white matter was smaller in patients, who had emotional childhood neglect, compared to those without neglect. In male patients also the right hemisphere was significantly smaller when they had emotional child neglect
Lenze et al. 2008	31 women with remitted unipolar depression, 24 healthy controls	1.5 Tesla manual tracing of whole hippocampus	Only females	Childhood adversity was associated with a history of recurrent depression and with earlier age of depression onset, whereby childhood adversity and hippocampal volume were not correlated
Vythilingam et al. 2002	32 women with current unipolar depressive disorder, 21 with a history of prepubertal sexual/ physical abuse, 11 without and 14 healthy female controls	1.5 Tesla MRI, manual tracing of hippocampal volume	Only females	The depressed subjects with childhood abuse had an 18% smaller mean left hippocampal volume than the non-abused depressed subjects and a 15% smaller mean left hippocampal volume than the healthy subjects. Right hippocampal volume was similar across the three groups
Gatt et al. 2009	89 healthy subjects	1.5 Tesla, manual tracing of prefrontal cortex (lateral and medial portions), hippocampus and amygdala,	Gender used as factor in the analysis	In addition, significant BDNF–early life events interactions indicated that BDNF Met carriers exposed to greater early life events have smaller hippocampal and amygdala volumes
Ganzel et al. 2008	17 healthy adults. who were 9/11-exposed, 17 healthy controls	MRI field not indicated, voxel-based morphometry	No gender differences assessed	Lower grey matter volume in amygdala, hippocampus, insula, anterior cingulate, and medial prefrontal cortex in exposed adults compared to controls. Nonlinear (first-order quadratic) association between total number of traumas in lifetime and amygdala grey matter volume and function in the whole group
Szeszko et al. 2006	21 healthy volunteers	1.5 Tesla, manual tracing of anterior and posterior hippocampus. Association between hippocampal volume	No gender differences assessed	Greater psychological stress as measured with Derogatis Stress Profile was negatively correlated significantly and more strongly with anterior than posterior hippocampal volume.
Cohen et al. 2006	265 healthy volunteers	1.5 Tesla, voxel-based morphometry-based assessments of volumes for hippocampus, amygdala, anterior cingulate cortex (ACC), caudate	No gender differences assessed	Participants with greater than two adverse childhood events had smaller ACC and caudate nuclei than those without adverse childhood events. A significant association between total adverse childhood events and volumes for these structures was observed

In a recent study of 89 participants, significant interactions between BDNF genotype and early life stress were apparent in HC and amygdala volumes, heart rate, and working memory. The investigators used structural equation modelling (SEM) to investigate the pathways through which BDNF genotype and early life stress interact to produce effects on brain structure, body arousal, emotional stability, which in turn predict alterations in symptoms and cognition. SEM suggested that the combination of Met carrier status of the BDNF polymorphism and exposure to early life stress predicted reduced HC volumes, associated lateral prefrontal cortex volumes and, in turn, higher depression (Gatt et al. 2009). The gene × environmental interaction seems to be particularly relevant in patients with MDD. The relevance of small HC volumes as a risk factor for MDD has now been documented by both longitudinal studies and studies of subjects at high-risk for MDD as reported in the following section.

The clinical relevance of small hippocampal volumes in MDD

Alterations of hippocampal volumes in high-risk subjects

A period of high risk for MDD begins in the early teens, and risk continues to rise in a linear fashion throughout adolescence, with lifetime rates estimated to range from 15–25% by late adolescence (Hankin et al. 1998; Kessler et al. 2005).

In a recent study, small HC volumes were apparent before the manifestation of clinical symptoms of MDD in at-risk adolescents, particularly in those who experienced high levels of adversity during childhood. Both early-life adversity and smaller HC volume were associated with a higher probability of depressive episodes during prospective follow-up (Rao et al. 2010). Moreover, 23 daughters of mothers with a history of MDD had reduced HC volumes compared to 32 daughters of mothers with no history of psychopathology, indicating again that neuroanatomical anomalies associated with depression may precede the onset of a depressive episode and influence the development and course of this disorder (Chen et al. 2010). In another recent study, adults at-risk for MDD had HC volumes that were not significantly different from those already affected with MDD, suggesting that small volumes may precede the onset of MDD and might render subjects more vulnerable to MDD (Amico submitted). Early life adversity appears to contribute to both small HC volumes and increased risk of MDD; whether changes in HC volumes represent a biological mechanism through which early life adversity is transduced into risk of MDD remains to be confirmed.

Associations between clinical response and hippocampal volume

There are a number of studies reporting associations between HC volume and probability of achieving and sustaining a clinical remission (Table 15.2). In one such study of patients with geriatric depression, patients with right HC volumes in the lowest quartile of the sample were less likely to achieve remission compared to those with HC volumes in the highest quartile (Hsieh et al. 2002). Another study reported that women who responded to 8 weeks of fluoxetine had larger right HC volumes than non-responders (Vakili et al. 2000). More recently a study reported that small HC volumes predicted low rates of remission even in patients with no past treatment history, suggesting that the association between HC volumes and short-term clinical response was not simply a function of past treatment responsiveness (MacQueen et al. 2008). Frodl and colleagues reported that depressed patients who were not remitted from an episode of depression 1 year after discharge had smaller left and right HC volumes at baseline scan (Frodl et al. 2004a). The association between smaller HC volumes and poor clinical outcome was detectable after a 3-year follow-up (Frodl et al. 2008b). Kronmuller and colleagues also found associations between HC volumes

Table 15.2 The effect of hippocampal volumes on short- and long-term outcome in patients with MDD

Study	Number of patients	Outcome assessed	Gender effect	Result
Vakili et al. 2000	38	Response after 8 weeks of fluoxetine	Effect apparent only in women	Association apparent in right HC
MacQueen et al. 2008	46	Remission after 8 weeks of AD treatment	Not apparent	Association apparent bilaterally
Hsieh et al. 2002	60 elderly	Remission after 12 weeks of AD treatment	Not apparent	Association stronger in right HC
Frodl et al. 2004a	30	In remission at 1 year	Not reported	Association stronger in right HC
Kronmuller et al. 2008	57	Sustained recovery for 2 years	Effect apparent in men	Association apparent bilaterally
Frodl et al. 2008b	30	Remission for 3 years	Not reported	Association apparent bilaterally

and 2-year outcome, although the associations between large HC volume and good outcome were restricted to men (Kronmuller et al. 2008).

The relations between small HC volumes and response to treatment suggest that the association between illness duration and HC volumes may arise from the effect that small HC volumes have on outcome. That is, there may be an iterative relation between HC volume and illness burden, with small HC volumes contributing to poor clinical outcome, which in turn may put further stress on the HC leading to structural changes that increase the likelihood of a poor clinical response. Whether such relations contribute to the development of illness chronicity remains to be determined.

Associations between cognitive function and hippocampal volume

Cognitive impairment is a core feature of MDD (Hasler et al. 2004). An effect size analysis of cognitive functioning in 726 patients with MDD conducted using meta-analytic principles found that depression had the largest effect on recollection memory (Zakzanis et al. 1998). Recollection memory tasks are highly dependent on the HC in addition to some frontal regions (Squire 2004). Hippocampal volumes are correlated with verbal learning and memory performance in various neuropsychiatric conditions. Kaymak and colleagues (Kaymak et al. 2010) reported positive associations between HC volumes and recollection memory and memory of reward-associated rules in a sample of twenty medication free women experiencing a first episode of depression. In this study, the HC volumes did not correlate with basal or post-dexamethasone cortisol levels. Hippocampal volumes are also correlated with executive functioning in patients with MDD (Frodl et al. 2006; Vasic et al. 2008) and emotional memory in middle-aged healthy women (Matsuoka et al. 2007). Vermetten and colleagues reported that antidepressant treatment reversed HC volume reduction in patients with PTSD; declarative memory also improved in the treated group (Vermetten et al. 2003).

There is an epidemiological association between MDD and Alzheimer's disease (AD; Caraci et al. 2010) and this association does not seem to be restricted to late onset depression that is a prodrome to the clinical symptoms of AD (Ownby et al. 2006). Rather, meta-analyses suggest that MDD may be a risk factor for AD on its own (Ownby et al. 2006). Stress-related brain changes, decreases in adult neurogenesis, and programmed neurodegeneration may contribute to this association (Sotiropoulos et al. 2008) as may inflammatory processes that are characteristic of MDD (Leonard 2007). Moreover, both MDD and the APOE ε4 allele are risk factors for dementia, suggesting that the association between the APOE ε4 allele and small HC volumes (Kim et al. 2002) may provide a further mechanism for understanding the links between late onset depression and dementia.

Evidence that treatment for MDD may have an impact on hippocampal volumes

As data have emerged to suggest that a long history of illness is associated with small HC volumes, questions have also raised concerning the potential role of treatments in minimizing or ameliorating the apparent decline in HC volume. Given the complexity of longitudinal studies in which treatment is controlled over long periods of time, it is not surprising that there are few studies that have examined the question of whether individual treatment modalities exert an effect on HC volume over time.

Increases in HC volume following treatment with the antidepressants sertraline and paroxetine (Vermetten et al. 2003; Bremner and Vermetten 2004) have been reported in patients with PTSD. In MDD, the results are less clear, with some studies failing to find an association between antidepressant therapy and HC volume (Vythilingam et al. 2004). One study, however, found a trend towards higher HC volumes in patients who had declines in cortisol levels with pharmacotherapy (Colla et al. 2007) but see Vythilingam et al. (2004) for a contradictory pattern of findings. Investigators recently reported that HC volumes of 16 medicated patients with MDD were increased in comparison to 23 patients who were not medicated and in comparison to 34 control subjects (Malykhin et al. 2010). Right HC volumes increased in patients who took their antidepressants over the full 3-year time interval, however, the study was not designed to investigate medication effects distinct from illness effects during the naturalistic 3-year follow-up (Frodl et al. 2008b).

A recent review concluded that the inclusion of bipolar patients treated with psychotropic medications in structural imaging studies was associated with either null findings or a reduction in volume loss in key regions associated with the pathophysiological mechanisms of the disorder, including the HC (Phillips et al. 2008). There is evidence that lithium is associated with increases in HC volume with both short- and long-term treatment in patients with bipolar disorder (Yucel et al. 2007, 2008). Studies examining the impact of other medications or treatment modalities on HC volumes are lacking across diagnostic categories, but there is a recent report of increases in hippocampal volume after treatment with electroconvulsive therapy in patients with depression (Nordanskog et al. 2010). Twelve patients who were assessed before the start and completion of a course of ECT had HC volume increases in the range of approximately 4% on the right and 5% on the left.

Summary

Meta-analyses of structural imaging studies of the HC suggest that patients with MDD have smaller HC volumes compared to healthy subjects. Although the factors associated with small HC volumes are not fully understood, there is evidence that genetic polymorphisms and early life stress may contribute to HC volumes prior to onset of illness. Furthermore, subjects at risk for MDD have reduced HC volumes before the clinical onset of illness. Repeated episodes of illness may further contribute to loss of HC volume, and speculatively, the changes in the HC may then contribute to treatment resistance or chronicity. Consistent with this are studies reporting that small HC volumes predict poor short- and long-term responsiveness to treatment.

Pre-clinical and clinical studies of the HC and related structures have provided a plausible biological model for how stress may interact with genetic vulnerability to increase the risk of MDD. We now know that detecting evidence of HC disruption, as measured by small HC volumes on MRI, can allow us to identify people who are at risk of illness and patients who are at risk for treatment non-response and relapse. Clinicians should begin to consider the possibility that imaging technology, possibly also in combination to genetic data (genetic variants, family history) could be used, not for diagnosis, which may be of uncertain utility, but for matching patients with treatment intensity. If imaging can reliably identify people who are likely to require more than first-line treatment to recover, then scanning might become cost effective as it is well recognized that less than 50% of patients have an adequate response to first treatment. Identifying subgroups of patients with specific morphological markers may also facilitate the development of treatment strategies that target neuroplastic processes. With few revolutionary treatment strategies on the horizon, and the cost of MDD continuing to rise across the world, it may be time to contemplate the possibility that imaging data could contribute to clinical care by matching patients to the most appropriate level of intervention.

References

Amico, F., Koutsouleris, N., Moeller H.-J., Reiser, M., Meisenzahl, E., and Frodl, T. (submitted). Decreased hippocampal volumes in high risk subjects for major depression.

Baare, W. F., Vinberg, M., Knudsen, G. M., *et al.* (2010). Hippocampal volume changes in healthy subjects at risk of unipolar depression. *J Psychiatr Res,* 44(10), 655–62.

Ballmaier, M., Narr, K. L., Toga, A. W., *et al.* (2008). Hippocampal morphology and distinguishing late-onset from early-onset elderly depression. *Am J Psychiatry,* 165(2), 229–37.

Bartley, A. J., Jones, D. W., and Weinberger, D. R. (1997). Genetic variability of human brain size and cortical gyral patterns. *Brain,* 120(Pt 2), 257–69.

Becker, S., MacQueen, G., and Wojtowicz, J. M. (2009). Computational modeling and empirical studies of hippocampal neurogenesis-dependent memory: Effects of interference, stress and depression. *Brain Res,* 1299, 45–54.

Bodnoff, S. R., Humphreys, A. G., Lehman, J. C., *et al.* (1995). Enduring effects of chronic corticosterone treatment on spatial learning, synaptic plasticity, and hippocampal neuropathology in young and mid-aged rats. *J Neurosci,* 15(1 Pt 1), 61–9.

Bremner, J. D. and Vermetten, E. (2004). Neuroanatomical changes associated with pharmacotherapy in posttraumatic stress disorder. *Ann N Y Acad Sci,* 1032, 154–7.

Bremner, J. D., Elzinga, B., Schmahl, C., and Vermetten, E. (2008). Structural and functional plasticity of the human brain in posttraumatic stress disorder. *Prog Brain Res,* 167, 171–86.

Bueller, J. A., Aftab, M., Sen, S., *et al.* (2006). BDNF Val66Met allele is associated with reduced hippocampal volume in healthy subjects. *Biol Psychiatry,* 59(9), 812–5.

Caetano, S. C., Hatch, J. P., Brambilla, P., *et al.* (2004). Anatomical MRI study of hippocampus and amygdala in patients with current and remitted major depression. *Psychiatry Res,* 132(2), 141–7.

Caraci, F., Copani, A., Nicoletti, F., and Drago, F. (2010). Depression and Alzheimer's disease: neurobiological links and common pharmacological targets. *Eur J Pharmacol,* 626(1), 64–71.

Caspi, A., Sugden, K., Moffitt, T. E., *et al.* (2003). Influence of life stress on depression: moderation by a polymorphism in the 5-HTT gene. *Science,* 301(5631), 386–9.

Cazakoff, B. N. and Howland, J. G. (2010). Acute stress disrupts paired pulse facilitation and long-term potentiation in rat dorsal hippocampus through activation of glucocorticoid receptors. *Hippocampus,* 20(12), 1327–31.

Chen, M.C., Hamilton, J.P., and Gotlib, I.H. (2010). Decreased hippocampal volume in healthy girls at risk of depression. *Arch Gen Psychiatry,* 67, 270–6.

Cohen, R. A., Grieve, S., Hoth, K. F., *et al.* (2006). Early life stress and morphometry of the adult anterior cingulate cortex and caudate nuclei. *Biol Psychiatry,* 59(10), 975–82.

Cole, J., Toga, A. W., Hojatkashani, C., *et al.* (2010). Subregional hippocampal deformationsi n major depressive disorder. *J Affect Disord,* 126(1–2), 272–7.

Colla, M., Kronenberg, G., Deuschle, M., *et al.* (2007). Hippocampal volume reduction and HPA-system activity in major depression. *J Psychiatr Res,* 41(7), 553–60.

Colombel, F. (2007). Memory bias and depression: a critical commentary. *Encephale,* 33, 242–8.

de Geus, E. J., van't Ent, D., Wolfensberger, S. P., *et al.* (2007). Intrapair differences in hippocampal volume in monozygotic twins discordant for the risk for anxiety and depression. *Biol Psychiatry*, 61(9), 1062–71.

de Quervain, D. J., Roozendaal, B., and McGaugh, J. L. (1998). Stress and glucocorticoids impair retrieval of long-term spatial memory. *Nature*, 394(6695), 787–90.

de Quervain, D. J., Roozendaal, B., Nitsch, R. M., McGaugh, J. L., and Hock, C. (2000). Acute cortisone administration impairs retrieval of long-term declarative memory in humans. *Nat Neurosci*, 3(4), 313–4.

Dranovsky, A. and Hen, R. (2006). Hippocampal neurogenesis: regulation by stress and antidepressants. *Biol Psychiatry*, 59(12), 1136–43.

Duman, R. S. (2004). Depression: a case of neuronal life and death? *Biol Psychiatry*, 56(3), 140–5.

Duman, R. S., Nakagawa, S., and Malberg, J. (2001). Regulation of adult neurogenesis by antidepressant treatment. *Neuropsychopharmacology*, 25(6), 836–44.

Egan, M. F., Kojima, M., Callicott, J. H., *et al.* (2003). The BDNF val66met polymorphism affects activity-dependent secretion of BDNF and human memory and hippocampal function. *Cell*, 112(2), 257–69.

Frodl, T., Meisenzahl, E. M., Zetzsche, T., *et al.* (2002). Hippocampal changes in patients with a first episode of major depression. *Am J Psychiatry*, 159(7), 1112–18.

Frodl, T., Meisenzahl, E. M., Zetzsche, T., *et al.* (2004a). Hippocampal and amygdala changes in patients with major depressive disorder and healthy controls during a 1-year follow-up. *J Clin Psychiatry*, 65(4), 492–9.

Frodl, T., Meisenzahl, E. M., Zill, P., *et al.* (2004b). Reduced hippocampal volumes associated with the long variant of the serotonin transporter polymorphism in major depression. *Arch Gen Psychiatry*, 61(2), 177–83.

Frodl, T., Schaub, A., Banac, S., *et al.* (2006). Reduced hippocampal volume correlates with executive dysfunctioning in major depression. *J Psychiatry Neurosci*, 31(5), 316–23.

Frodl, T., Schüle, C., Schmitt, G., *et al.* (2007). Association of the brain-derived neurotrophic factor Val66Met polymorphism with reduced hippocampal volumes in major depression. *Arch Gen Psychiatry*, 64(4), 410–16.

Frodl, T., Moller, H. J., and Meisenzahl, E. (2008a). Neuroimaging genetics: new perspectives in research on major depression? *Acta Psychiatr Scand*, 118(5), 363–72.

Frodl, T., Jäger, M., Smajstrlova, I., *et al.* (2008b). Effect of hippocampal and amygdala volumes on clinical outcomes in major depression: a 3-year prospective magnetic resonance imaging study. *J Psychiatry Neurosci*, 33(5), 423–30.

Frodl, T., Zill, P., Baghai, T., *et al.* (2008c). Reduced hippocampal volumes associated with the long variant of the tri- and diallelic serotonin transporter polymorphism in major depression. *Am J Med Genet B Neuropsychiatr Genet*, 147B(7), 1003–7.

Frodl, T., Koutsouleris, N., Bottlender, R., *et al.* (2008d). Reduced gray matter brain volumes are associated with variants of the serotonin transporter gene in major depression. *Mol Psychiatry*, 13(12), 1093–101.

Frodl, T., Reinhold, E., Koutsouleris, N., Reiser, M., and Meisenzahl, E. M. (2010a). Interaction of childhood stress with hippocampus and prefrontal cortex volume reduction in major depression. *J Psychiatr Res*, 44(13), 799–807.

Frodl, T., Reinhold, E., Koutsouleris, N., *et al.* (2010b). Childhood stress, serotonin transporter gene and brain structures in major depression. *Neuropsychopharmacology*, 35(6), 1383–90.

Frodl, T., Bokde, A. L., Scheuerecker, J., *et al.* (2010c). Functional connectivity bias of the orbitofrontal cortex in drug-free patients with major depression. *Biol Psychiatry*, 67(2), 161–7.

Ganzel, B. L., Kim, P., Glover, G. H., and Temple, E. (2008). Resilience after 9/11: multimodal neuroimaging evidence for stress-related change in the healthy adult brain. *Neuroimage*, 40(2), 788–95.

Gatt, J. M., Nemeroff, C. B., Dobson-Stone, C., *et al.* (2009). Interactions between BDNF Val66Met polymorphism and early life stress predict brain and arousal pathways to syndromal depression and anxiety. *Mol Psychiatry*, 14(7), 681–95.

Gilbertson, M. W., Shenton, M. E., Ciszewski, A., *et al.* (2002). Smaller hippocampal volume predicts pathologic vulnerability to psychological trauma. *Nat Neurosci*, 5(11), 1242–7.

Gonul, A.S., Kitis, O., Eker, M.C., *et al.* (2011). Association of the brain-derived neurotrophic factor Val66Met polymorphism with hippocampus volumes in drug-free depressed patients. *World J Biological Psychiatry*, 12, 110–18.

Hankin, B. L., Abramson, L. Y., Moffitt, T. E., Silva, P. A., McGee, R., and Angell, K. E. (1998). Development of depression from preadolescence to young adulthood: emerging gender differences in a 10-year longitudinal study. *J Abnorm Psychol*, 107(1), 128–40.

Hasler, G., Drevets, W. C., Manji, H. K., and Charney, D. S. (2004). Discovering endophenotypes for major depression. *Neuropsychopharmacology*, 29(10), 1765–81.

Hastings, R. S., Parsey, R. V., Oquendo, M. A., Arango, V., and Mann, J. J. (2004). Volumetric analysis of the prefrontal cortex, amygdala, and hippocampus in major depression. *Neuropsychopharmacology*, 29(5), 952–9.

Hickie, I., Naismith, S., Ward, P. B., *et al.* (2005). Reduced hippocampal volumes and memory loss in patients with early- and late-onset depression. *Br J Psychiatry*, 186, 197–202.

Hsieh, M. H., McQuoid, D. R., Levy, R. M., *et al.* (2002). Hippocampal volume and antidepressant response in geriatric depression. *Int J Geriatr Psychiatry*, 17 (6), 519–25.

Huang, E. J. and Reichardt, L. F. (2001). Neurotrophins: roles in neuronal development and function. *Annu Rev Neurosci*, 24, 677–736.

Inkster, B., Nichols, T. E., Saemann, P. G., *et al.* (2009). Association of GSK3beta polymorphisms with brain structural changes in major depressive disorder. *Arch Gen Psychiatry*, 66 (7), 721–8.

Janssen, J., Hulshoff Pol, H. E., Lampe, I. K., *et al.* (2004). Hippocampal changes and white matter lesions in early-onset depression. *Biol Psychiatry*, 56 (11), 825–31.

Janssen, J., Hulshoff Pol, H. E., de Leeuw, F. E., *et al.* (2007). Hippocampal volume and subcortical white matter lesions in late life depression: comparison of early and late onset depression. *J Neurol Neurosurg Psychiatry*, 78 (6), 638–40.

Joffe, R. T., Gatt, J. M., Kemp, A. H., *et al.* (2009). Brain derived neurotrophic factor Val66Met polymorphism, the five factor model of personality and hippocampal volume: Implications for depressive illness. *Hum Brain Mapp*, 30 (4), 1246–56.

Karl, A., Schaefer, M., Malta, L. S., *et al.* (2006). A meta-analysis of structural brain abnormalities in PTSD. *Neurosci Biobehav Rev*, 30 (7), 1004–31.

Kaufman, J., Yang, B. Z., Douglas-Palumberi, H., *et al.* (2006). Brain-derived neurotrophic factor-5-HTTLPR gene interactions and environmental modifiers of depression in children. *Biol Psychiatry*, 59 (8), 673–80.

Kaymak, S. U., Demir, B., Sentürk, S., Tatar, I., Aldur, M. M., and Ulu ,B. (2010). Hippocampus, glucocorticoids and neurocognitive functions in patients with first-episode major depressive disorders. *Eur Arch Psychiatry Clin Neurosci*, 260 (3), 217–23.

Kessler, R. C. (1997). The effects of stressful life events on depression. *Annu Rev Psychol*, 48, 191–214.

Kessler, R. C., Berglund, P., Demler, O., Jin, R., Merikangas, K. R., and Walters, E. E. (2005). Lifetime prevalence and age-of-onset distributions of DSM-IV disorders in the National Comorbidity Survey Replication. *Arch Gen Psychiatry,* 62(6), 593–602.

Kim, D. H., Payne, M. E., Levy, R. M., MacFall, J. R., and Steffens, D. C. (2002). APOE genotype and hippocampal volume change in geriatric depression. *Biol Psychiatry,* 51(5), 426–9.

Kronmuller, K. T., Pantel, J., Köhler, S., *et al.* (2008). Hippocampal volume and 2-year outcome in depression. *Br J Psychiatry,* 192(6), 472–3.

Lanfumey, L., Mongeau, R., Cohen-Salmon, C., and Hamon, M. (2008). Corticosteroid-serotonin interactions in the neurobiological mechanisms of stress-related disorders. *Neurosci Biobehav Rev,* 32(6), 1174–84.

Lange, C. and Irle, E. (2004). Enlarged amygdala volume and reduced hippocampal volume in young women with major depression. *Psychol Med,* 34(6), 1059–64.

Lee, T., Jarome, T., Li, S. J., Kim, J. J., and Helmstetter, F. J. (2009). Chronic stress selectively reduces hippocampal volume in rats: a longitudinal magnetic resonance imaging study. *Neuroreport,* 20(17), 1554–8.

Lenze, S. N., Xiong, C., and Sheline, Y. I. (2008). Childhood adversity predicts earlier onset of major depression but not reduced hippocampal volume. *Psychiatry Res,* 162(1), 39–49.

Leonard, B. E. (2007). Inflammation, depression and dementia: are they connected? *Neurochem Res,* 32(10), 1749–56.

Lesch, K. P. and Mossner, R. (1998). Genetically driven variation in serotonin uptake: is there a link to affective spectrum, neurodevelopmental, and neurodegenerative disorders? *Biol Psychiatry,* 44(3), 179–92.

Lesch, K. P., Bengel, D., Heils, A., *et al.* (1996). Association of anxiety-related traits with a polymorphism in the serotonin transporter gene regulatory region. *Science,* 274(5292), 1527–31.

Leuner, B., Gould, E., and Shors, T. J. (2006). Is there a link between adult neurogenesis and learning? *Hippocampus,* 16(3), 216–24.

Lisman, J. E. and Grace, A. A. (2005). The hippocampal-VTA loop: controlling the entry of information into long-term memory. *Neuron,* 46(5), 703–13.

Lloyd, A. J., Ferrier, I. N., Barber, R., Gholkar, A., Young, A. H., and O'Brien, J. T. (2004). Hippocampal volume change in depression: late- and early-onset illness compared. *Br J Psychiatry,* 184, 488–95.

Lu, B. (2003). Pro-region of neurotrophins: role in synaptic modulation. *Neuron,* 39 (5), 735–8.

Luine, V., Martinez, C., Villegas, M., Magariños, A. M., and McEwen, B. S. (1996). Restraint stress reversibly enhances spatial memory performance. *Physiol Behav,* 59(1), 27–32.

Lyons, D. M., Yang, C., Sawyer-Glover, A. M., Moseley, M. E., and Schatzberg, A. F. (2001). Early life stress and inherited variation in monkey hippocampal volumes. *Arch Gen Psychiatry,* 58(12), 1145–51.

MacMaster, F. P. and Kusumakar, V. (2004). Hippocampal volume in early onset depression. *BMC Med,* 2, 2.

MacMaster, F. P., Mirza, Y., Szeszko, P.R., *et al.* (2008). Amygdala and hippocampal volumes in familial early onset major depressive disorder. *Biol Psychiatry,* 63(4), 385–90.

MacMillan, S., Szeszko, P. R., Moore, G. J., *et al.* (2003). Increased amygdala: hippocampal volume ratios associated with severity of anxiety in pediatric major depression. *J Child Adolesc Psychopharmacol,* 13(1), 65–73.

MacQueen, G. M. (2003). Course of illness, hippocampal function, and hippocampal volume in major depression. *Proc Natl Acad Sci USA,* 100(3), 1387–92.

MacQueen, G. and Frodl, T. (2011). The hippocampus in major depression: evidence for the convergence of the bench and bedside in psychiatric research? *Mol Psychiatry,* 16(3), 252–64.

MacQueen, G. M., Yucel, K., Taylor, V. H., Macdonald, K., and Joffe, R. (2008). Posterior hippocampal volumes are associated with remission rates in patients with major depressive disorder. *Biol Psychiatry,* 64(10), 880–3.

Magarinos, A. M., McEwen, B. S., Flügge, G., and Fuchs, E. (1996). Chronic psychosocial stress causes apical dendritic atrophy of hippocampal CA3 pyramidal neurons in subordinate tree shrews. *J Neurosci,* 16(10), 3534–40.

Maggio, N. and Segal, M. (2011). Persistent changes in ability to express long-term potentiation/depression in the rat hippocampus after juvenile/adult stress. *Biol Psychiatry,* 69(8), 748–53.

Mailliet, F., Qi, H., Rocher, C., Spedding, M., Svenningsson, P., and Jay, T. M. (2008). Protection of stress-induced impairment of hippocampal/prefrontal LTP through blockade of glucocorticoid receptors: implication of MEK signaling. *Exp Neurol,* 211(2), 593–6.

Malberg, J. E. and Duman, R. S. (2003). Cell proliferation in adult hippocampus is decreased by inescapable stress: reversal by fluoxetine treatment. *Neuropsychopharmacology,* 28(9), 1562–71.

Malykhin, N. V., Carter, R., Seres, P., and Coupland, N. J. (2010a). Structural changes in the hippocampus in major depressive disorder: contributions of disease and treatment. *J Psychiatry Neurosci,* 35(5), 337–43.

Malykhin, N. V., Lebel, R. M., Coupland, N. J., Wilman, A. H., and Carter, R. (2010b). In vivo quantification of hippocampal subfields using 4.7 T fast spin echo imaging. *Neuroimage,* 49(2), 1224–30.

Malykhin, N. V., Carter, R., Seres, P., and Coupland, N.J. (2010). Structural changes in the hippocampus in major depressive disorder: contributions of disease and treatment. *J Psychiatry Neurosci,* 35(5), 337–43.

Matsuoka, Y., Nagamine, M., Mori, E., Imoto, S., Kim, Y., and Uchitomi, Y. (2007). Left hippocampal volume inversely correlates with enhanced emotional memory in healthy middle-aged women. *J Neuropsychiatry Clin Neurosci,* 19(3), 335–8.

McKinnon, M. C., Yucel, K., Nazarov, A., and MacQueen, G. M. (2009). A meta-analysis examining clinical predictors of hippocampal volume in patients with major depressive disorder. *J Psychiatry Neurosci,* 34(1), 41–54.

Mendelson, S. D. and McEwen, B. S. (1992). Autoradiographic analyses of the effects of adrenalectomy and corticosterone on 5-HT1A and 5-HT1B receptors in the dorsal hippocampus and cortex of the rat. *Neuroendocrinology,* 55 (4), 444–50.

Monkul, E. S., Hatch, J. P., Nicoletti, M. A., *et al.* (2007). Fronto-limbic brain structures in suicidal and non-suicidal female patients with major depressive disorder. *Mol Psychiatry,* 12(4), 360–6.

Neumeister, A., Wood, S., Bonne, O., *et al.* (2005). Reduced hippocampal volume in unmedicated, remitted patients with major depression versus control subjects. *Biol Psychiatry,* 57(8), 935–7.

Newcomer, J. W., Selke, G., Melson, A. K., *et al.* (1999). Decreased memory performance in healthy humans induced by stress-level cortisol treatment. *Arch Gen Psychiatry,* 56(6), 527–33.

Newton, S. S. and Duman, R. S. (2004). Regulation of neurogenesis and angiogenesis in depression. *Curr Neurovasc Res,* 1(3), 261–7.

Nifosì, F., Toffanin, T., Follador, H., *et al.* (2010). Reduced right posterior hippocampal volume in women with recurrent familial pure depressive disorder. *Psychiatry Res,* 184(1), 23–8.

Nordanskog, P., Dahlstrand, U., Larsson, M. R., Larsson, E. M., Knutsson, L., and Johanson, A. (2010). Increase in hippocampal volume after electroconvulsive therapy in patients with depression: a volumetric magnetic resonance imaging study. *J ECT,* 26(1), 62–7.

O'Brien, J. T., Lloyd, A., McKeith, I., Gholkar, A., and Ferrier, N. (2004). A longitudinal study of hippocampal volume, cortisol levels, and cognition in older depressed subjects. *Am J Psychiatry,* 161(11), 2081–90.

Ohl, F., Michaelis, T., Vollmann-Honsdorf, G. K., Kirschbaum, C., and Fuchs, E. (2000). Effect of chronic psychosocial stress and long-term cortisol treatment on hippocampus-mediated memory and hippocampal volume: a pilot-study in tree shrews. *Psychoneuroendocrinology,* 25(4), 357–63.

Ownby, R. L., Crocco, E., Acevedo, A., John, V., and Loewenstein, D. (2006). Depression and risk for Alzheimer disease: systematic review, meta-analysis, and metaregression analysis. *Arch Gen Psychiatry,* 63(5), 530–8.

Pezawas, L., Verchinski, B. A., Mattay, V. S., *et al.* (2004). The brain-derived neurotrophic factor val66met polymorphism and variation in human cortical morphology. *J Neurosci,* 24(45), 10099–102.

Pham, K., Nacher, J., Hof, P. R., and McEwen, B. S. (2003). Repeated restraint stress suppresses neurogenesis and induces biphasic PSA-NCAM expression in the adult rat dentate gyrus. *Eur J Neurosci,* 17(4), 879–86.

Phillips, M. L., Travis, M. J., Fagiolini, A., and Kupfer, D. J. (2008). Medication effects in neuroimaging studies of bipolar disorder. *Am J Psychiatry,* 165(3), 313–20.

Pittenger, C., Sanacora, G., and Krystal, J. H. (2007). The NMDA receptor as a therapeutic target in major depressive disorder. *CNS Neurol Disord Drug Targets,* 6(2), 101–15.

Posener, J. A., Wang, L., Price, J. L., *et al.* (2003). High-dimensional mapping of the hippocampus in depression. *Am J Psychiatry,* 160 (1), 83–9.

Rao, U., Chen, L. A., Bidesi, A. S., *et al.* (2010). Hippocampal changes associated with early-life adversity and vulnerability to depression. *Biol Psychiatry,* 67(4), 357–64.

Risch, N., Herrell, R., Lehner, T., *et al.* (2009). Interaction between the serotonin transporter gene (5-HTTLPR), stressful life events, and risk of depression: a meta-analysis. *JAMA,* 301(23), 2462–71.

Rusch, B. D., Abercrombie, H. C., Oakes, T. R., Schaefer, S. M., and Davidson, R. J. (2001). Hippocampal morphometry in depressed patients and control subjects: relations to anxiety symptoms. *Biol Psychiatry,* 50(12), 960–4.

Santarelli, L., Saxe, M., Gross, C., *et al.* (2003). Requirement of hippocampal neurogenesis for the behavioral effects of antidepressants. *Science,* 301(5634), 805–9.

Sapolsky, R. M. (2000). Glucocorticoids and hippocampal atrophy in neuropsychiatric disorders. *Arch Gen Psychiatry,* 57(10), 925–35.

Sapolsky, R. M. (2003). Stress and plasticity in the limbic system. *Neurochem Res,* 28(11), 1735–42.

Saylam, C., Uçerler, H., Kiti , O., Ozand, E., and Gönül, A. S. (2006). Reduced hippocampal volume in drug-free depressed patients. *Surg Radiol Anat,* 28(1), 82–7.

Shah, P. J., Ebmeier, K. P., Glabus, M. F., and Goodwin, G. M. (1998). Cortical grey matter reductions associated with treatment-resistant chronic unipolar depression. Controlled magnetic resonance imaging study. *Br J Psychiatry,* 172, 527–32.

Sheline, Y. I., Gado, M. H., and Kraemer, H. C. (2003). Untreated depression and hippocampal volume loss. *Am J Psychiatry,* 160(8), 1516–8.

Sheline, Y. I., Sanghavi, M., Mintun, M. A., and Gado, M. H. (1999). Depression duration but not age predicts hippocampal volume loss in medically healthy women with recurrent major depression. *J Neurosci,* 19(12), 5034–43.

Sheline, Y. I., Wang, P. W., Gado, M. H., Csernansky, J. G., and Vannier, M. W. (1996). Hippocampal atrophy in recurrent major depression. *Proc Natl Acad Sci U S A,* 93(9), 3908–13.

Sotiropoulos, I., Cerqueira, J. J., Catania, C., *et al.* (2008). Stress and glucocorticoid footprints in the brain-the path from depression to Alzheimer's disease. *Neurosci Biobehav Rev,* 32(6), 1161–73.

Spinelli, S., Chefer, S., Suomi, S. J., Higley, J. D., Barr, C. S., and Stein, E. (2009). Early-life stress induces long-term morphologic changes in primate brain. *Arch Gen Psychiatry,* 66(6), 658–65.

Squire, L. R. (2004). Memory systems of the brain: a brief history and current perspective. *Neurobiol Learn Mem,* 82(3), 171–7.

Starkman, M. N., Giordani, B., Gebarski, S. S., Berent, S., Schork, M. A., and Schteingart, D. E. (1999). Decrease in cortisol reverses human hippocampal atrophy following treatment of Cushing's disease. *Biol Psychiatry,* 46(12), 1595–602.

Stockmeier, C. A., Mahajan, G. J., Konick, L. C., *et al.* (2004). Cellular changes in the postmortem hippocampus in major depression. *Biol Psychiatry,* 56(9), 640–50.

Szeszko, P. R., Betensky, J. D., Mentschel, C., *et al.* (2006). Increased stress and smaller anterior hippocampal volume. *Neuroreport,* 17(17), 1825–8.

Taylor, W. D., Steffens, D. C., Payne, M. E., *et al.* (2005). Influence of serotonin transporter promoter region polymorphisms on hippocampal volumes in late-life depression. *Arch Gen Psychiatry,* 62(5), 537–44.

Teicher, M. H., Andersen, S. L., Polcari, A., Anderson, C. M., Navalta, C. P., and Kim, D. M. (2003). The neurobiological consequences of early stress and childhood maltreatment. *Neurosci Biobehav Rev,* 27(1–2), 33–44.

Thomas, R. M., Hotsenpiller, G., and Peterson, D. A. (2007). Acute psychosocial stress reduces cell survival in adult hippocampal neurogenesis without altering proliferation. *J Neurosci,* 27(11), 2734–43.

Tyler, W. J., Alonso, M., Bramham, C. R., and Pozzo-Miller, L. D. (2002). From acquisition to consolidation: on the role of brain-derived neurotrophic factor signaling in hippocampal-dependent learning. *Learn Mem,* 9(5), 224–37.

Vakili, K., Pillay, S. S., Lafer, B., *et al.* (2000). Hippocampal volume in primary unipolar major depression: a magnetic resonance imaging study. *Biol Psychiatry,* 47(12), 1087–90.

Vasic, N., Walter, H., Höse, A., and Wolf, R. C. (2008). Gray matter reduction associated with psychopathology and cognitive dysfunction in unipolar depression: a voxel-based morphometry study. *J Affect Disord,* 109(1–2), 107–16.

Vermetten, E. and Bremner, J. D. (2002). Circuits and systems in stress. I. Preclinical studies. *Depress Anxiety,* 15(3), 126–47.

Vermetten, E., Vythilingam, M., Southwick, S. M., Charney, D. S., and Bremner, J. D. (2003). Long-term treatment with paroxetine increases verbal declarative memory and hippocampal volume in posttraumatic stress disorder. *Biol Psychiatry,* 54(7), 693–702.

Vythilingam, M., Heim, C., Newport, J., *et al.* (2002). Childhood trauma associated with smaller hippocampal volume in women with major depression. *Am J Psychiatry,* 159(12), 2072–80.

Vythilingam, M., Vermetten, E., Anderson, G. M., *et al.* (2004). Hippocampal volume, memory, and cortisol status in major depressive disorder: effects of treatment. *Biol Psychiatry,* 56(2), 101–12.

Warner-Schmidt, J. L. and Duman, R. S. (2006). Hippocampal neurogenesis: opposing effects of stress and antidepressant treatment. *Hippocampus,* 16(3), 239–49.

Watanabe, Y., Gould, E., and McEwen, B. S. (1992). Stress induces atrophy of apical dendrites of hippocampal CA3 pyramidal neurons. *Brain Res,* 588(2), 341–5.

Woolley, C. S., Gould, E., and McEwen, B. S. (1990). Exposure to excess glucocorticoids alters dendritic morphology of adult hippocampal pyramidal neurons. *Brain Res,* 531(1–2), 225–31.

Woon, F. L. and Hedges, D. W. (2008). Hippocampal and amygdala volumes in children and adults with childhood maltreatment-related posttraumatic stress disorder: a meta-analysis. *Hippocampus,* 18(8), 729–36.

Xia, J., Chen, J., Zhou, Y., *et al.* (2004). Volumetric MRI analysis of the amygdala and hippocampus in subjects with major depression. *J Huazhong Univ Sci Technolog Med Sci,* 24(5), 500–2, 6.

Yucel, K., McKinnon, M. C., Taylor, V. H., *et al.* (2007). Bilateral hippocampal volume increases after long-term lithium treatment in patients with bipolar disorder: a longitudinal MRI study. *Psychopharmacology (Berl),* 195(3), 357–67.

Yucel, K., Taylor, V. H., McKinnon, M. C., *et al.* (2008). Bilateral hippocampal volume increase in patients with bipolar disorder and short-term lithium treatment. *Neuropsychopharmacology,* 33(2), 361–7.

Zakzanis, K. K., Leach, L., and Kaplan, E. (1998). On the nature and pattern of neurocognitive function in major depressive disorder. *Neuropsychiatry Neuropsychol Behav Neurol,* 11(3), 111–19.

Zhao, Z., Taylor, W. D., Styner, M., Steffens, D. C., Krishnan, K. R. R., and MacFall, J. R. (2008). Hippocampus chape analysis and late life depression. *PLOS One,* 3(3), 1837.

CHAPTER 16
Schizophrenia

Peter Falkai, Oliver Gruber, and Andrea Schmitt

Epidemiology and aetiology of schizophrenia 288
Symptoms of schizophrenia with focus on cognitive deficits 290
Hippocampus and cognition in schizophrenia 290
The hippocampus and schizophrenia: findings from MRI studies 290
The hippocampus and schizophrenia: findings from structural postmortem studies 291
From behaviour to the molecular basis of hippocampal alterations in schizophrenia 293
Functional role of schizophrenia risk genes in the hippocampus 295
Influence of antipsychotics on the hippocampus 296

ABSTRACT Schizophrenia is a chronic neurocognitive brain disorder with unknown pathophysiology. There is evidence for an interaction between risk genes and environmental factors playing a role during early as well as late neurodevelopment leading to symptoms in early adulthood. Deficits in hippocampus-related declarative memory and decreased hippocampal volumes at the stage of first episode and during the course of the disease are key findings. Stereological postmortem studies of the hippocampus in schizophrenia revealed no alterations of neuronal cell numbers but decreased numbers of oligodendrocytes in the posterior part and a lack of astrogliosis, suggesting that schizophrenia is not a classical neurodegenerative disorder. Decreased myelination and synaptic genes support the hypothesis of impaired connectivity of the hippocampus. Dysconnectivity of the hippocampus in schizophrenia may also be related to disturbances in neurogenesis, since a reduced number of proliferating stem cells has been detected in the dentate gyrus of schizophrenia patients. Effects of several risk genes of schizophrenia like D-amino acid oxidase, neuregulin-1, dysbindin, and disrupted-in schizophrenia-1 have been demonstrated in hippocampal tissue or magnetic resonance imaging (MRI) studies. In addition, genome-wide association studies revealed new genes with widely unknown function like ZNF804 or the transcription factor TCF4. Postmortem and animal studies may elucidate the link between these risk factors and behavioural consequences such as deficits of cognition and sensorimotor gating, which are related to hippocampal function. It has, however, to be kept in mind that chronic antipsychotic treatment may have influenced results from postmortem and MRI studies.

KEYWORDS schizophrenia, hippocampus, cognition, neurobiology, MRI, gene expression, antipsychotics

Epidemiology and aetiology of schizophrenia

Schizophrenia is a debilitating, chronic neurocognitive disorder affecting mostly young adults. Its prevalence amounts to about 1% worldwide and it is considered as independent of cultural identity or ethnicity. The disorder is characterized by various abnormal cognitive, affective, and motor behavioural features. Most of the affected subjects (> 80%) are unable to maintain a self-sustaining professional life (Saha et al. 2007) and their life expectancy is shortened, in average, by 10 years (Andlin-Sobocki et al. 2005). This together with frequent hospitalizations and the significant neurocognitive impairment has a major impact on lives of those afflicted and their families and on the financial burden of European social and health systems of about €35,229,000,000 annually (Cannon et al. 2002).

Despite tremendous advances in science and medicine, schizophrenia remains an enigmatic disorder. In the absence of consistent biomarkers for schizophrenia, diagnosis still relies on the assessment of a cluster of symptoms, based on ICD-10 (International Classification of Diseases) or DSM-IV-R (Diagnostic and Statistical Manual of Mental Disorders) criteria, and the current pharmacological antipsychotic treatments have been mostly discovered on an empirical basis. They account for modern neurotransmitter hypotheses including impaired dopaminergic and serotonergic neurotransmission. Additionally, a hypofunction of the glutamatergic N-methyl-D-aspartate (NMDA) receptor has been proposed to play a role in the pathophysiology of the disease (Carlsson 1995; Olney and Farber 1995). The identification of pathophysiologically-associated biological parameters that can be utilized to unequivocally diagnose or evaluate the course and treatment of the disorder remains challenging. Three major impediments are responsible for the lack of aetiological and pathophysiological insights in schizophrenia: first, the longitudinal nature, which most likely begins *in utero* reaching into adulthood. Second, the multifactorial aetiology with interacting genetic, epigenetic, and environmental influences, and third, the inaccessibility of biomarkers for routine clinical and scientific investigations.

To date, the prevailing hypothesis of schizophrenia is that multifactorial interactions impact both early and late brain developmental changes (Figure 16.1). Several lines of evidence support this neurodevelopmental hypothesis. Among them are the association between obstetric complications and schizophrenia, brain morphological and histological abnormalities, and altered expression of genes involved in differentiation, migration, myelination, and synaptogenesis (Tkavech et al. 2003; Sawada et al. 2005; Harrison and Weinberger 2005; Schmitt et al. 2008). Among several brain regions, the hippocampus is a key structure affected in schizophrenia in a prefronto-hippocampal neuronal network (Weinberger 1999). This region is specifically sensitive for hypoxia as a common factor of different birth and obstetric complications and revealed a bilateral volume loss in schizophrenia patients with obstetric complications (e.g. Ebner et al. 2008). In addition, schizophrenia has been shown to have a strong genetic basis, with heritability estimates ranging between 50–80%. Likewise, a host of candidate gene association studies were performed, showing the association between risk genes like neuregulin-1, D-amino-acid oxidase (DAAO), dysbindin-1, or disrupted-in-schizophrenia-1 (DISC1) with schizophrenia. However the field was characterized by a partial inconsistency of findings and low replication validity (Ioannidis et al. 2001). But meanwhile the gene ZNF804A, encoding a zinc finger protein has been found to be associated with schizophrenia (O'Donovan et al. 2008). The recent studies by the International Schizophrenia Consortium (2009), Shih et al. (2004), and Stefansson et al. (2009) detected genome-wide significant single nucleotide polymorphism (SNP) associations in the extended major histocompatibility complex region on chromosome 6. The latter study also reported genome-wide significant association with markers in the neurogranin (NRGN) and the transcription factor 4 (TCF4) genes among several risk genes with only limited knowledge on their functional consequences in the brain. However, abnormal gene–environmental interaction during development can also be explained by epigenetic mechanisms, which could impact the individual risk for adverse health outcomes later in life, and such mechanisms have been demonstrated in schizophrenia (Roth et al. 2009).

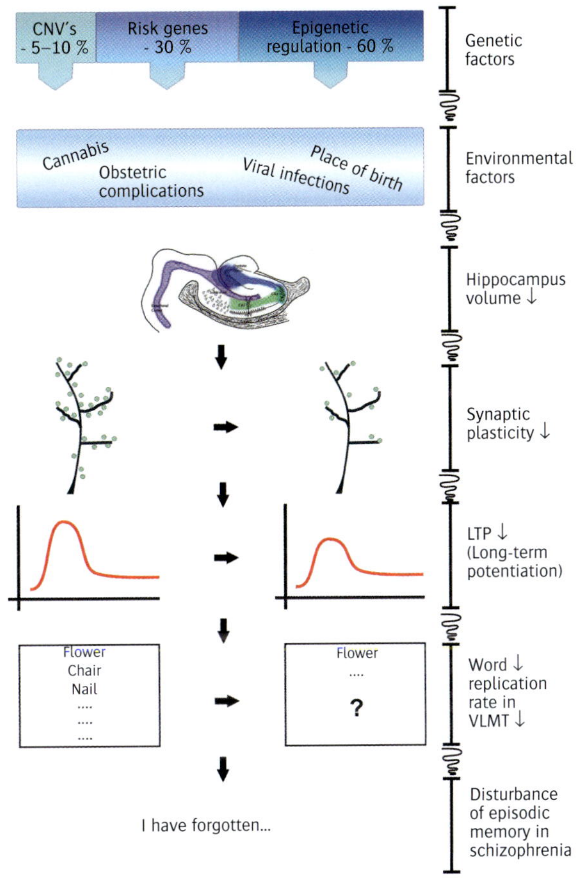

Figure 16.1 In schizophrenia, defective genes and environmental factors lead to volume deficits and disturbed connectivity of the hippocampus. Reproduced from Falkai, P., Gruber, O., Schulze, T. G., and Schmitt, A. (2011). Schizophrenie als Netzwerkstörung. *Nervenheilkunde* 30, 218–33 with permission.

Symptoms of schizophrenia with focus on cognitive deficits

Due to the cognitive decline in early adulthood, Kraepelin named schizophrenia 'dementia precox' and suggested the presence of an underlying brain pathology (Kraepelin 1919). Today, schizophrenia can be considered as an illness beginning in young adults, but related to a predisposition in brain development. In contrast to other illnesses which may display psychotic features, such as bipolar disorder, schizophrenia is often characterized by incomplete recovery of psychotic symptoms, and persistent disability. Beside psychotic features such as delusions and hallucinations, or negative and emotional symptoms such as affective flattening, disturbances of volition and social skills, schizophrenia is characterized by several cognitive dysfunctions. Especially persistent cognitive deficits are very reliable predictors for a relapse and therefore an unfavourable long-term outcome (Robinson et al. 2002; Andreasen et al. 2005). They include deficits of declarative memory along with working memory, impaired attention, and executive function, all of them being core features of the disease. Declarative memory is divided into semantic (facts, words) and episodic (personal events) components, which all are markedly debilitated in schizophrenia (Weiss and Heckers 2001).

Hippocampus and cognition in schizophrenia

The hippocampus is connected to cortical areas including the perirhinal, entorhinal, and parahippocampal cortices (Zola-Morgan and Squire 1993; Witter 1993). This hippocampal system is involved in episodic, verbal, and spatial memory as well as in connection with the amygdala in emotion memory in humans. Patients with hippocampus damage perform poorly on tests requiring them to inspect a scene with many objects and afterwards recalling the locations of individual objects (Milner et al. 1968; Jones-Gotman and Milner 1978; Smith and Milner 1981). Patients with hippocampal sclerosis and temporal epilepsy suffer from impaired verbal and non-verbal episodic memory (Tudesco Ide et al. 2010). In schizophrenia patients, and to a lesser degree in high-risk subjects and non-psychotic relatives, the hippocampus-associated verbal declarative memory is impaired (Cirillo and Seidman 2003). This underscores the genetic heritability of the cognitive core symptoms and hippocampal dysfunction in schizophrenia (Gruber et al. 2008).

The hippocampus and schizophrenia: findings from MRI studies

Smaller hippocampus volume is a well-replicated feature of schizophrenia, and appears to be related to neuronal atrophy and loss of neuropil (Harrison 2004; Steen et al., 2006). The hippocampal fissure is enlarged and hippocampal volume is already decreased at first episode (Smith et al. 2003) (Figure 16.2). However, whether hippocampal volume in schizophrenia is static, or becomes progressively smaller during the course of illness remains uncertain (Steen et al. 2006). Functional studies using MRI, positron emission tomography (PET) and single-photon emission computed tomography (SPECT) revealed hypoactivation of the hippocampus in schizophrenia patients (Weiss and Heckers 2001) and hippocampal activation was reduced during encoding and recognition of words (Jessen et al. 2003). During word and face recognition tasks, which belong to the declarative memory, schizophrenia patients showed bilateral hippocampal hypoactivation (Rametti et al. 2009). Diffusion tensor imaging (DTI) studies revealed a disrupted integrity of the hippocampal fornix and association to impaired semantic memory, pointing to disturbed

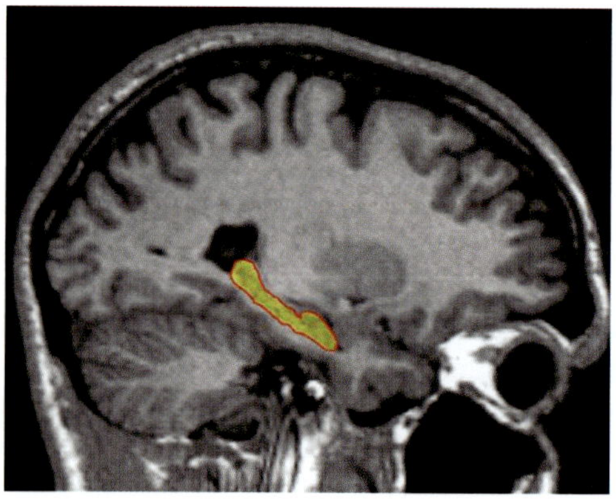

Figure 16.2 MRI scan of the hippocampus (delineated in yellow) in a schizophrenic patient. In volumetric MRI studies in schizophrenia, hippocampal volumes have been shown to be decreased.

connectivity in hippocampus-associated networks (Takei et al. 2008). In line with these findings, recent functional MRI (fMRI) investigations of schizophrenic patients during performance of working memory tasks revealed a reduced physiological deactivation of the hippocampus (Henseler et al. 2009) and a fronto-hippocampal dysconnectivity, which correlated with positive symptoms and cognitive deficits (Henseler et al. 2010).

The hippocampus and schizophrenia: findings from structural postmortem studies

Postmortem studies in schizophrenia confirmed findings from structural MRI studies and revealed volume loss in the hippocampus, which is one of the most consistent structural abnormalities (Heckers 2001). Additionally, structural postmortem studies showed a volume loss in hippocampal subregions, which may be related to positive symptoms (Bogerts et al. 1990, 1993; Bogerts 1997). The small amount of tissue reduction of about 5% shown in structural MRI studies suggests that some cellular subfractions within this structure might be lost or show reduced volumes (Wright et al. 2000; Shenton et al. 2001). However, a recent postmortem study of hippocampal grey matter applying grey-level index (GLI) revealed no cytoarchitectonic alterations and thus suggested no changes of neuronal perikarya (Hurlemann et al. 2005). This conforms to most of the histological studies which did not detect such alterations in cell density (Arnold et al. 1995; Dwork et al. 1997). Some studies investigating subtypes of neurons showed decreased density of interneurons without alterations of pyramidal neurons (Benes et al. 1998; Zhang and Reynolds 2002; Benes 2010). Likewise the density of pyramidal neurons has been shown to be decreased and first studies of glial cells revealed reduced density of glial cells in the hippocampus of male patients (Falkai and Bogerts 1986). However, earlier cell density studies using two-dimensional cell counting without considering the total volume of a region of interest may have some methodological limitations due to the influence of volume differences and tissue shrinkage by fixation procedures. Additional problems may be caused by irregular cell shape and size, non-random orientation and cutting of cells during sectioning (Williams and Rakic 1988). Confirming earlier density studies, two stereological estimates of total cell number in the hippocampus reported no differences in schizophrenia patients. But one of these only investigated total numbers of neurons but not glial cells (Heckers et al. 1991), whereas the second study did not differentiate between different cell types and included schizophrenia as well as schizoaffective patients (Walker et al. 2002).

Moreover, the hippocampus can be divided in several subregions such as cornu ammonis (CA) 1, 2, 3, subiculum, and dentate gyrus (Amaral and Lavanex 2007) (Figure 16.3). In a recent study using design-based stereology with differentiation of neurons, oligodendrocytes and astroglia in different subregions of the posterior hippocampus, we detected no alterations of neuronal numbers but decreased numbers and densities of oligodendrocytes in the left and right CA4 (Amaral and Lavanex 2007; Schmitt et al. 2009) (Figure 16.4). These results are in line with previous stereological studies of the hippocampus (Heckers et al. 1991; Walker et al. 2002). Our results indicate disturbed connectivity of the CA4 of the posterior part of the hippocampus. CA4 is the deep polymorph layer of the dentate gyrus receiving collateral mossy fibres from the granule cells (Amaral and Lavanex 2007) and additionally is among those regions specifically suggested to be involved in connectivity disturbances in schizophrenia (Harrison and Eastwood 2001). A decreased number of oligodendrocytes might be associated with deficits in the proliferation, maturation of oligodendrocytes, and myelination (Bartzokis et al. 2003). Ultrastructural investigations revealed an atrophy of myelinated axons and swelling of periaxonal oligodendrocyte processes in the hippocampus of schizophrenia patients (Uranova et al. 2007). Myelination of projections from the hippocampus and prefrontal cortex peaks in late adolescence and early adulthood, thus temporally

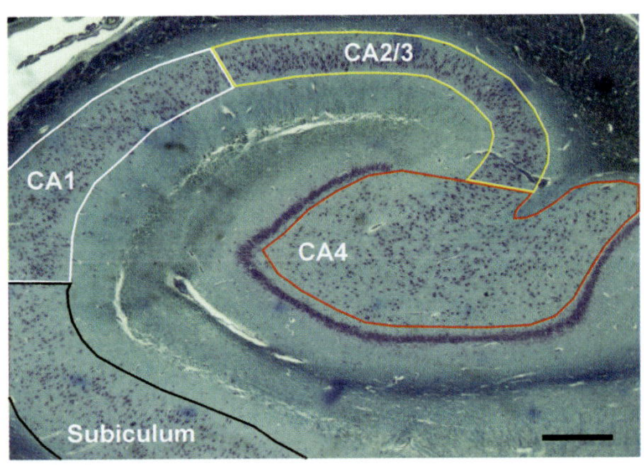

Figure 16.3 Delineated subregions of the hippocampus in a 20-µm thick frontal histologically stained section. CA4 is now named CA3 and CA2/3 is named CA2. Reproduced from Schmitt, A., Steyskal, C., Bernstein, H.-G., et al. (2009). Stereologic investigation of the porsterior part of the hippocampus in schizophrenia. *Acta Neuropathologica*, 117(4), 395–407. © 2009, with permission from Springer Science + Business Media.

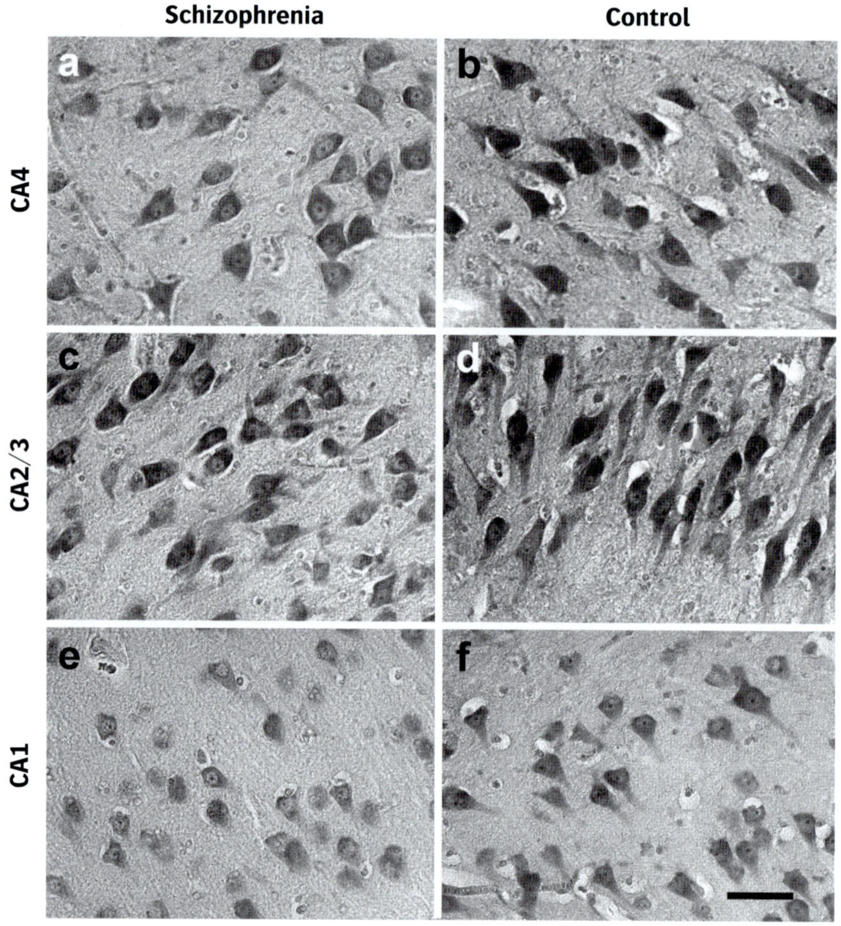

Figure 16.4 Respresentative photomicrographs from different subregions of a schizophrenic patient and a healthy control. Alterations of cell numbers cannot be assessed observing the tissue but require stereological investigations of total subregions. Reproduced from Schmitt, A., Steyskal, C., Bernstein, H.-G., et al. (2009). Stereologic investigation of the porsterior part of the hippocampus in schizophrenia. *Acta Neuropathologica*, 117(4), 395–407. © 2009, with permission from Springer Science + Business Media.

concording with the onset of schizophrenia (Yakovlev et al. 1967).

We detected effects of age on density and number of astrocytes, but no differences of astrocyte number between schizophrenia patients and controls (Schmitt et al. 2009). Our findings confirm animal studies showing increased mean astrocyte numbers in brains of aged mice (Mouton et al. 2002), as well as previous studies showing a lack of astrocytosis in schizophrenia (Casanova et al. 1990; Falkai et al. 1999), suggesting that schizophrenia is not a classical neurodegenerative disorder accompanied by astrogliosis like Alzheimer's disease. Searching for subgroups, one immunohistochemical study showed increased S100B positive subpopulations of astroglial or oligodendroglial cells in patients with paranoid schizophrenia (Steiner et al. 2008).

In recent years it has become evident that in addition to the subventricular zone (Conover and Allen 2002) the subgranular layer of the hippocampal dentate gyrus also contains progenitor cells with the capacity of proliferation until and throughout adulthood (neurogenesis). Dysconnectivity of the hippocampus in schizophrenia may also be related to disturbances in neurogenesis, since a reduced number of proliferating stem cells has been detected in the dentate gyrus of schizophrenia patients (Reif et al. 2006) as well as a reduced volume of the dentate gyrus (Falkai and Bogerts 1986) pointing into the same direction. These findings need replication, though, and the role of neurogenesis in the pathophysiology of schizophrenia remains to be investigated in animal models of risk factors of the disease.

From behaviour to the molecular basis of hippocampal alterations in schizophrenia

Patients suffering from schizophrenia exhibit deficits in sensory information processing and attention. A reliable model of sensorimotor gating is the prepulse inhibition (PPI) of acoustic startle response, which is disrupted in schizophrenic patients and restorable by antipsychotics (Braff et al. 2001; Van den Buuse et al. 2003). PPI reflects sensorimotor gating and the hippocampus is involved in startle-related circuitry (Swerdlow et al. 2001). In a neurodevelopmental animal model in rats, postnatal hippocampal lesion led to decreased PPI and schizophrenia-related behaviour such as cognitive deficits in early adulthood (Lipska et al. 1998, 2002). This model entails a maldevelopment of the prefrontal cortex until adulthood supporting the involvement of the hippocampus in complex neuronal circuits. In schizophrenia, neurodevelopmental disturbances have been hypothesized, additionally, diverse studies indicate genetic as well as environmental factors like maternal infections or perinatal hypoxia to assumedly play a role in the pathophysiology of the disease. After prenatal infection, adolescent rats exhibited reduced hippocampal volumes and alterations of gene expression (Fatemi et al. 2009). In rats exposed to neonatal hypoxia, deficits of PPI have been shown and related to the impact of this neurodevelopmental animal model for investigation of risk factors in schizophrenia (Fendt et al. 2008). Hypobaric hypoxia in adult rats is known to induce delayed neurotoxicity in the hippocampus (Shukitt-Hale et al. 1996).

The hippocampus is known to be a vulnerable region for hypoxia/ischaemia (Choi 1992). Most vulnerable are CA1 pyramidal cells, followed by CA3 and CA2 cells, whereas the dentate gyrus is quite resistant to hypoxic damage (Schmidt-Kastner et al. 1991). This regional pattern was perceptible in the results of NMDA receptor subunit NR2A downregulation in CA3 observed 3 days after repeated hypoxia (Schmitt et al. 2007). In line with our findings, NR2A and NR2B downregulation and decreased subunit immunoreactivity have been demonstrated after transient ischaemia in the hippocampus by others (Zhang et al. 1997; Hsu et al. 1998). It has been hypothesized that perinatal hypoxia induces a NMDA receptor hypofunction in key regions of schizophrenia like the hippocampus (Schmitt et al. 2007).

A dysfunction of the glutamatergic NMDA receptor has been proposed to play a role in the pathophysiology of schizophrenia (Olney and Farber 1995). NMDA receptor hypofunction gives rise to reduced gamma-aminobutyric acid (GABA)ergic inhibitory tone on glutamatergic pyramidal neurons. In the long term, these neurons could be injured by increased neurotransmission at ionotropic glutamatergic AMPA ((±)-α-amino-3-hydroxy-5-methylisoxazole-4-propionoc acid), kainite, and cholinergic receptors with subsequent cell damage and a final hypoglutamatergic state (Olney and Farber 1995; Farber et al. 1999, 2000) (Figure 16.5). The NMDA receptor is composed of different subunits responsible for different functional properties (Ishii et al. 1993; Monyer et al. 1994). The obligate NR1 subunit is combined with one or two NR2 subunits (NR2A, NR2B, NR2C and NR2D) forming the functional trimeric receptor (Figure 16.6). Studies using brain tissue obtained postmortem have examined glutamate receptors plus subunits and support the hypothesis of a glutamate dysfunction in schizophrenia. Correspondingly there exists evidence for increased density of NMDA receptor binding in the hippocampus (Dean et al. 1999) of schizophrenic patients. An endogenous NMDA receptor inhibition and feedback-upregulation of NMDA receptor binding in schizophrenic patients has been hypothesized. Gene expression of the NR1 subunit was found to be decreased in the frontal cortex (Sokolov 1998), and hippocampus (Gao et al. 2000). In contrast, the NR2B subunit was upregulated in the hippocampus (Gao et al. 2000), suggesting an imbalance in the functional properties of the receptor. The immunoreactivity and gene expression of glutamatergic AMPA receptor subunits and AMPA receptor binding was seen to be reduced in

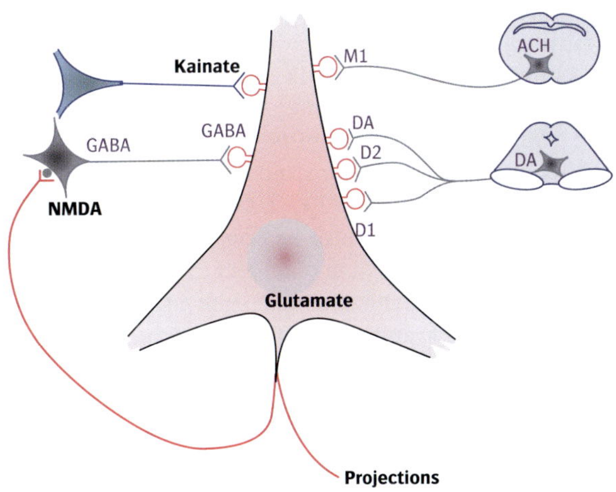

Figure 16.5 NMDA-receptor bearing GABAergic interneurons inhibit glutamatergic pyramidal neurons. If the NMDA receptor is dysfunctional, pyramidal neurons are disinhibited and degenerate by other excitatory inputs. With kind permission from Springer Science+Business Media: Schmitt, A., May, B., Müller, B., et al. (2004). Die Wirkung von Antipsychotika auf glutamaterge neurotransmission im Tiermodell. *Nervenarzt*, 75(1), 16–22, figure 3.

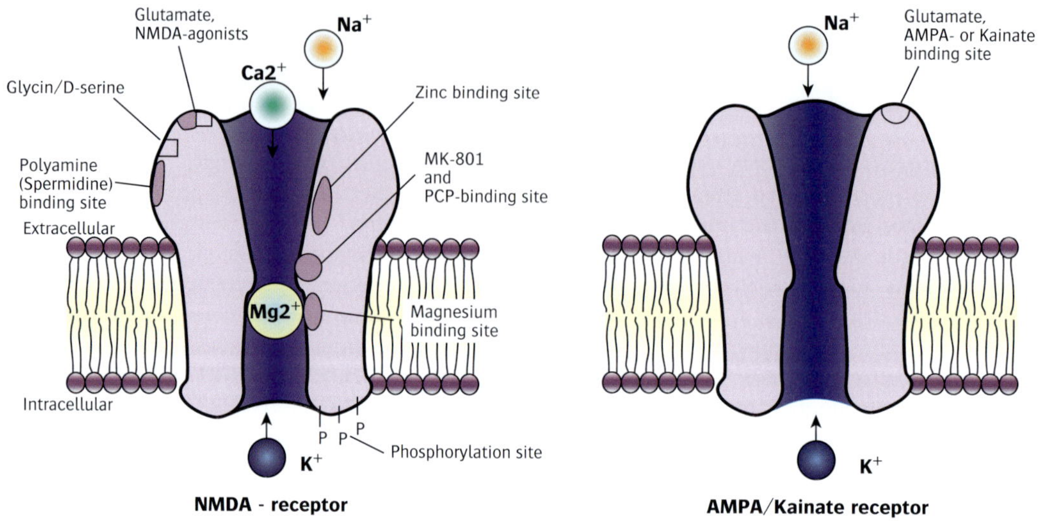

Figure 16.6 The NMDA receptor is composed of different subunits and has binding sites for agonists as well as an antagonist binding site in the ion channel. Left: With kind permission from Springer Science+Business Media: Schmitt, A., May, B., Müller, B., et al. (2004). Die Wirkung von Antipsychotika auf glutamaterge neurotransmission im Tiermodell. *Nervenarzt*, 75(1), 16–22. Right: Schmitt, A. and Braus D.F. Genetik. In Braus, D.F. (ed) *Schizophrenie*, pp. 30–1. With permission from Schattauer.

the hippocampus (Kerwin et al. 1990; Breese et al. 1995; Eastwood et al. 1997). AMPA receptors are known to modulate NMDA receptor function. Disturbed glutamate metabolism may lead to dysfunction of neurons which can be measured *in vivo* by magnetic resonance spectroscopy (MRS).

Meta-analyses of proton (H^1) MRS studies in schizophrenia revealed reduced N-acetylaspartate (NAA) levels in grey and white matter of the prefrontal cortex and hippocampus (Steen et al. 2005; Gur et al. 2007; Pajonk et al. 2010). Reductions of NAA indicate a loss of functional and structural integrity of neurons or osmolarity of cells. Altered NAA levels may also be related to decreased neuropil (dendrites and axons) in schizophrenia. Animal models reveal that NMDA receptor antagonists, along with behavioural deficits such as PPI reductions, induce reduced NAA in the cortex (Abbott and Bustillo 2006). Alterations of NAA may be related to dysfunctions of neurons with altered glutamatergic neurotransmission (Goff and Coyle 2001) resulting in a final hypoglutamatergic state of corticostriatal projection (Olney and Farber 1995). This may lead via the GABAergic striatothalamic projection to a thalamic disinhibition and glutamatergic overflow to the cortex or hippocampus (Carlsson et al. 2001). Accordingly, absolute concentrations of glutamate have been shown to be increased in the prefrontal cortex and hippocampus of first-episode schizophrenia patients (Olbrich et al. 2008). In the hippocampus and medial prefrontal cortex, increased glutamate levels were associated with poor cognitive performance (Rüsch et al. 2008; Shirayama et al. 2010).

There is strong support for the assumption of disturbed glutamatergic and GABAergic microcircuitry on the molecular level as well, mainly affecting the expression of presynaptic vesicle proteins including the synaptosome-associated protein 25 (SNAP-25) and syntaxin, which together with vesicle-associated membrane protein (VAMP) form the soluble N-ethylmaleimide-sensitive factor attachment protein receptor (SNARE) complex. Synaptophysin, complexin 1, and complexin 2 operate in synaptic vesicle exocytosis by stabilizing the SNARE complex (Brose 2008). In the hippocampus it was shown that neuron populations had fewer dendritic spines and reduced dendritic arborizations (Kolomeets et al. 2005). A more consistent finding within the hippocampus and dentate gyrus is abnormal synaptic connectivity, as shown by decreased expression of the synaptic proteins synapsin, synaptophysin, spinophilin, and SNAP-25 (Young et al. 1998; Eastwood and Harrison 2001; Fatemi et al. 2001; Knable et al. 2004; Sawada et al. 2005). In addition, disturbances of complexins in the hippocampus were associated with the severity of cognitive impairment in lifetime (Sawada et al. 2005). Expression studies on whole tissue blocks containing different cell populations are common, but investigation of cell-type specific alterations may contribute to elucidate pathophysiological mechanisms in schizophrenia. In a laser capture microdissection (LCM) study in isolated dentate granular neurons, decreased expression of cluster of genes related to synaptic plasticity, cytoskeletal proteins, neurite outgrowth, energy metabolism, and protein turnover have been detected (Altar et al. 2005) supporting the notions of disturbed regenerative mechanisms in schizophrenia.

A decreased nerve pulse propagation of glutamatergic axonal projections may also be based on the dysfunction of surrounding oligodendrocytes. The main function of oligodendrocytes is axon myelination and maintenance of myelin in the central nervous system. Reduction or malformation of the myelin sheath may lead to increased ion leakage, resulting in reduced propagation of nerve impulses and dysconnectivity (Du and Dreyfus 2002). Other oligodendrocyte functions are trophic signalling to nearby neurons, synthesis of growth factors, neuronal survival and development, neurotransmission, and synaptic function (Du and Dreyfus 2002). Oligodendrocyte dysfunction has been an important focus of contemporary schizophrenia research (Segal et al. 2007). Evidence for disordered myelination of oligodendrocytes in the hippocampus of schizophrenia patients has been provided by transcriptomics (Katsel et al. 2005; Dracheva et al. 2006; Haroutunian et al. 2006, 2007). Additionally, proteomic studies reported oligodendrocyte dysfunction and related disturbances of myelination in schizophrenia. Myelin basic protein (MBP, the major constituent of the myelin sheath produced by oligodendrocytes and Schwann cells) and myelin oligodendrocyte glycoprotein (MOG) were found to be downregulated in the hippocampus and entorhinal cortex (Chambers and Perrone-Bizzozero 2004; Parlapani et al. 2010). Additionally the oligodendrocyte marker 2′,3′ cyclic nucleotide, 3′-phosphodiesterase (CNP) has been found downregulated. Other commonly altered myelin-related genes are quaking (QKI), transferrin (TF), myelin-associated gylcoprotein (MAG), claudin 11, ErbB3, myelin and leucocytes protein (MAL), proteolipid protein (PLP), gelsolin (GSN), Plasmolipin (PLLP, TM4SF11), sex determining region Y-box 10 (SOX10), claudin 5, and peripheral myelin protein 22 (PMP22) (Katsel et al. 2005; Dracheva et al. 2006; Haroutunian et al. 2006, 2007). They play a role in development, growth, and maintenance of myelin and axons (McInnes and Lauriat 2006) as well as synaptogenesis in the grey matter (Kaifu et al. 2003). These proteins might be potential biomarkers for the disease (Martins-de-Souza 2010).

The most pronounced decrease in expression of oligodendrocyte-related proteins has been reported in the hippocampus of schizophrenia patients (Katsel et al. 2005; Dracheva et al. 2006). Quaking is an oligodendrocyte-related gene influencing splicing and stabilization of RNA transcripts and supporting maintenance of myelin and oligodendrocyte differentiation. Quaking isoforms have been shown to be decreased in schizophrenia and there is genetic evidence for the association with schizophrenia by a large Swedish pedigree (Aberg et al. 2006; Harountunian et al. 2007; Lauriat et al. 2008). These microarray and our light microscopic findings of reduced oligodendrocyte number in the hippocampus have been supported by *in vivo* DTI studies of white matter tracts of the fornix body and posterior hippocampus in schizophrenia, showing decreased fractional anisotropy and thus supporting the hypothesis of functional dysconnectivity in schizophrenia (Kalus et al. 2004; Kuroki et al. 2006; White et al. 2007; Takei et al. 2008; Zhou et al. 2008). Moreover, cognitive function such as verbal declarative memory is correlated with the fractional anisotropy of the hippocampus in schizophrenia (Lim et al. 2006). In addition, fMRI studies showed reduced functional connectivity and dysfunction of the hippocampus (Achim et al. 2007; Reske et al. 2007; Zhou et al. 2008; Henseler et al. 2009, 2010; Kircher et al. 2009).

Functional role of schizophrenia risk genes in the hippocampus

Susceptibility genes are assumed to play a specific role in the pathophysiology of schizophrenia via abnormal synaptic connectivity. While the involvement of a single risk gene is considered rather small, the coherent influence of multiple susceptibility genes contributes significantly to the risk of developing schizophrenia in either an epistatic or a polygenic mode (Harrison and Weinberger 2005). Genome-wide scanning and mapping of regions associated with risk for schizophrenia resulted in the detection of several potential susceptibility genes. These genes encode proteins involved in processes starting from brain development to the maintenance of glutamatergic transmission in the adult brain. The D-amino-acid oxidase (DAAO) risk gene has been shown to be associated with schizophrenia (Chumakov 2002; Liu 2004; Schumacher et al. 2004; Corvin et al. 2007; Wood et al. 2007). In a recent gene expression study we could show increased expression of DAAO in the left and right hippocampal CA3 of patients with chronic schizophrenia (Habl et al. 2009). DAAO is involved in the NMDA receptor hypofunction that is hypothesized to play a major role in the pathophysiology of schizophrenia (Coyle et al. 1996). D-serine, an allosteric activator of the NMDA-type glutamate receptor in the brain (Schell et al. 1995), is degraded by DAAO. Accordingly, increased DAAO levels might lead to NMDAR dysfunction via decreased serine levels (Kapoor 2006; Bendikov 2007).

Neuregulin-1 maps to the short arm of chromosome 8, which has been shown to be associated with schizophrenia in genome-wide linkage studies and meta-analyses of whole-genome linkage scans (Pulver et al. 1995; Kendler et al. 1996, 2000; Blouin et al. 1998; Stefansson et al. 2002; Lewis et al. 2003). The neuregulin-1 family specifically activates receptor tyrosine kinases of the ErB receptors ErB2, ErB3, and ErB4 and encodes at least 15 different splicing variants, classified into three primary isoform groups (Buonnano and Fischbach 2001). Different nomenclature was given to these isoforms, based on their functions. Type I are known as neu differentiation factor (NDF), heregulin (HRG), or acetylcholine receptor-inducing activity (ARIA), type II as glial growth factors (GGF-I, -II, and -III), while type III has been called as sensory and motor neuron-derived factor (SMDF). Of late there is evidence of a type IV isoform (Law et al. 2006).

The neuregulin-1 isoforms are synthesized by neurons and play an important role in cell differentiation, apoptosis, migration, myelination, and proliferation of neurons and oligodendrocytes (Syroid et al. 1996; Anton et al. 1997; Falls et al. 2003; Corfas et al. 2004). Dysfunctions in these processes may be related to neurodevelopmental disturbances in schizophrenia (Weinberger 1996). Additionally, neuregulin-1 regulates the expression of the NMDA receptor subunit NR2C, of the ß2 subunit of the $GABA_A$ receptor and of subunits of the nicotinic acetylcholine receptor, most of them involved in the pathophysiology of schizophrenia (Ozaki et al. 1997; Yang et al. 2003; Rieff et al. 1999; Liu et al. 2001). However, the mechanisms underlying the contribution of neuregulin-1 risk haplotypes to disease pathophysiology remain elusive.

First insights came from a structural MRI study showing decreased hippocampal volumes in schizophrenia patients and relatives carrying the neuregulin-1 haplotype HAP(ICE) (Gruber et al. 2008). During a working memory task, patients with the risk allele (SNP8NRG221533) from the HAP(ICE) haplotype exhibited lower activations in the left hippocampus and cingulate cortex (Kircher et al. 2009). A gene expression study in a large group showed an increased gene expression of the isoform type I in the hippocampus of schizophrenia patients and interaction with the genotype SNP8NRG221132, whereas the type IV isoform was influenced by SNP8NRG243177 (Law et al. 2006). However, a second study found no differences in this region (Parlapani et al. 2010), which partially may be due to the smaller sample size. Interestingly, a transgenic animal model showing overexpression of neuregulin-type I isoform had deficits in PPI of the acoustic startle response, which points to consequences on schizophrenia-related behaviour (Deakin et al. 2009).

Genetic variation in DTNBP1, which encodes dysbindin-1, is associated with schizophrenia and lower cognitive ability (Strohmaier et al. 2010; Zhang et al. 2010). Risk SNPs and haplotypes are known to influence dysbindin-1 gene expression (Bray et al. 2005; Weickert et al. 2008; Balu and Coyle 2011). A reduced expression of dysbindin-1 has been reported in hippocampal dentate granule and polymorph cells as well as CA3. Furthermore, there was a correlation between dysbindin-1 expression and mRNA levels of the synaptic genes synaptophysin and spinophilin, suggesting the contribution of dysbindin to synaptic deficits in schizophrenia (Weickert et al. 2008). In accordance with these molecular neurobiological findings, a recent genomic imaging study provided first direct *in vivo* evidence of dysbindin-1 SNP rs2619522 being associated with variation of grey matter volumes bilaterally in the human hippocampus (Gruber et al. 2010a).

Chromosome 1q42 is among the genomic regions showing replicated linkage with schizophrenia. A segment of 1q42 is spanning the DISC1 gene. It is expressed in neurons and glial cells and translated to a protein which is involved in neuritic outgrowth, neuronal migration, synaptogenesis and glutamatergic neurotransmission (Brandon et al. 2009). Inherited variations in DISC1 have been shown to be associated with schizophrenia and reduced grey matter density in prefrontal cortex, as well as with decreased hippocampus volumes (Cannon et al. 2005; Gruber et al. 2010b). In summary, MRI and postmortem studies of the hippocampal formation in schizophrenia may contribute to elucidate functional consequences of risk factors in schizophrenia.

Influence of antipsychotics on the hippocampus

Antipsychotic drug treatment does not appear to correct the low hippocampal volume in schizophrenia, although only a few longitudinal studies exist on this subject (Wood et al. 2001; Chakos et al. 2005; Velakoulis et al. 2006). There is some indication that changes in medication associated with increases of hippocampal volume over time are associated with a slight symptom improvement (Panenka et al. 2007). Since most of the patients involved in postmortem studies or chronic patients in MRI studies had been treated with antipsychotics over a long period, medication effects may have influenced the results.

In most animal studies investigating the effects of antipsychotic medication, treatment lasted for only 1 month or less with low dosages. This does not permit comparisons between animal and human treatment conditions. To achieve comparative treatment effects, animal studies with extended medication times and doses equivalent to the therapeutic range in humans are required. To address this issue, we closely investigated the effects of a typical (haloperidol) and an atypical (clozapine) antipsychotic medication on NMDA receptor binding and gene expression of NMDA subunits in different rat brain regions after drug administration for up to 6 months by *in situ* hybridization. Analysis of gene expression revealed a significant NR2A subunit mRNA suppression in the ventral and posterior hippocampal subregions CA1, CA2, CA3 and in the dentate gyrus by both antipsychotic drugs when given in clinical doses compared to control animals (Schmitt et al. 2003). We did not detect alterations of gene expression in other subunits and did not confirm NR2A and NR2B upregulations in the hippocampus after acute haloperidol treatment. Our results of decreased NR2A mRNA by both antipsychotics may be due to a long-term regulation of gene expression and may be a compensatory mechanism to protect neurons against glutamate neurotoxicity. Mutant mice, deficient in the NR2A subunit, revealed significantly smaller injury sites after brain ischaemia, suggesting a protective effect of lower NR2A subunits expression to glutamate neurotoxicity (Morikawa et al. 1998).

Clozapine significantly downregulated the glutamate transporter (EAAT) 2 in hippocampal CA1 and downregulated EAAT3 in hippocampal CA1 and CA2 compared to controls. Haloperidol downregulated EAAT3 in CA2. We hypothesize that glutamate transporter downregulation by neuroleptics increases glutamatergic action at the postsynaptic neuron and thereby may be related to beneficial antipsychotic effects and side effects (Schmitt et al. 2003).

To date, the effects of antipsychotics on hippocampal volume are under debate (Panenka et al. 2007). A prospective study over 5 years showed decreased volumes in the early course but not in the chronic state of the disease. Patients treated with higher doses of antipsychotics revealed smaller volume loss of the hippocampus than those with lower doses, which points to neuroprotective effects of antipsychotics (Koolschijn et al. 2010). Accordingly, chronic haloperidol treatment increased total hippocampal volume suggesting that haloperidol alters neuroplastic processes (Schmitt et al. 2004). In animal studies of neurogenesis in the dentate gyrus, cell proliferation has been reported to be increased after treatment with atypical antipsychotics (Newton and Duman 2007). However, results are contradictory (Schmitt et al. 2004) and other brain regions as the dentate gyrus came into focus by researching the influence of antipsychotics on neurogenesis (Nasrallah et al. 2010).

Expression of the synaptic protein SNAP-25 may be increased by some antipsychotics (Barr et al. 2006). In contrast, others like complexins seem not to be influenced in animal models (Eastwood et al. 2000; Sawada et al. 2005). Therefore, in schizophrenia, new neuroprotective and regenerative therapy strategies are needed. Add-on therapy such as physical exercise (cycling) has been shown to increase total hippocampus volumes in schizophrenia patients and healthy probands. Accordingly, the patients' short-term memory ameliorated. Moreover, in the schizophrenia exercise group, the ratio NAA/creatine was shown to be increased, suggesting improvement of neuronal function (Pajonk et al. 2009). In animal models, exercise entailed increased synaptic plasticity and cognitive skills (Hu et al. 2009). New treatment strategies in schizophrenia should focus on pathophysiological mechanisms such as neuroplastic processes in the hippocampus.

References

Aberg, K., Saetre P., Lindhom E., *et al.* (2006). Human QKI, a new candidate gene for schizophrenia involved in myelination. *Am J Med Genet B Neuropsychiatr Genet*, 141B(1), 84–90.

Abbott, C., and Bustillo, J. (2006). What have we learned from proton magnetic resonance spectroscopy about schizophrenia? A critical update. *Curr Opin Psychiatry*, 19(2), 135–9.

Achim, A. M., Bertrand, M. C., Sutton, H., *et al.* (2007). Selective abnormal modulation of hippocampal activity during memory formation in first-episode psychosis. *Arch Gen Psychiatry*, 64(9), 999–1014.

Altar, C. A., Jurata, L. W., Charles, V., *et al.* (2005). Deficient hippocampal neuron expression of proteasome, ubiquitin, and mitochondrial genes in multiple schizophrenia cohorts. *Biol Psychiatry*, 58(2), 85–96.

Amaral, D., and Lavanex, P. (2007). Hippocampal neuroanatomy. In Andersen, P., Morris, R., Amaral, D., Bliss, T., and O'Keefe, J. (Eds.) *The Hippocampus Book*, pp. 37–114. Oxford: Oxford University Press.

Andlin-Sobocki, P., and Rossler, W. (2005). Cost of psychotic disorders in Europe. *Eur J Neurol*, 12(Suppl 1), 74–7.

Andreasen, N. C., Carpenter, W. T. Jr, Kane, J. M., *et al.* (2005). Remission in schizophrenia: proposed criteria and rationale for consensus. *Am J Psychiatry*, 162(3), 441–9.

Anton, E. S., Marchionni, M. A., Lee, K. F., and Rakic, P. (1997). Role of GGF/neuregulin signaling in interactions between migrating neurons and radial glia in the developing cerebral cortex. *Development*, 124(18), 3501–10.

Arnold, S. E., Franz, B. R., Gur, R. C., *et al.* (1995). Smaller neuron size in schizophrenia in hippocampal subfields that mediate cortical-hippocampal interactions. *Am J Psychiatry*, 152, 738–48.

Balu, D. T. and Coyle, J. T. (2011). Neuroplasticity signalling pathways linked to the pathophysiology of schizophrenia. *Neurosci Biobehav Rev*, 35(3), 848–70.

Barr, A. M., Young, C. E., Phillips, A. G., and Honer, W. G. (2006). Selective effects of typical antipsychotic drugs on SNAP-25 and synaptophysin in the hippocampal trisynaptic pathway. *Int J Neuropsychopharmacol*, 9(4), 457–63.

Bartzokis, G., Nuechterlein, K. H., Lu, P. H., *et al.* (2003). Dysregulated brain development in adult men with schizophrenia: a magnetic resonance imaging study. *Biol Psychiatry*, 53, 412–21.

Bendikov, I., Nadri, C., Amar, S., *et al.* (2007). A CSF and postmortem brain study of D-serine metabolic parameters in schizophrenia. *Schizophr Res*, 90, 41–51.

Benes, F. M. (1989). Myelination of cortical-hippocampal relays during late adolescence. *Schizophr Bull*, 15(4), 585–93.

Benes, F. M. (2010). Amygdalocortical circuitry in schizophrenia: from circuits to molecules. *Neuropsychopharmacology*, 35(1), 239–57.

Benes, F. M., Kwok, E. W., Vincent, S. L., and Todtenkopf, M. S. (1998). A reduction of nonpyramidal cells in sector CA2 of schizophrenics and manic depressives. *Biol Psychiatry*, 44, 88–97.

Blouin, J. L., Dombroski, B. A., Nath, S. K., *et al.* (1998). Schizophrenia susceptibility loci on chromosomes 13q32 and 8p21. *Nat Genet*, 20(1), 70–3.

Bogerts, B. (1997). The temporolimbic system theory of positive schizophrenic symptoms. *Schizophr Bull*, 23, 423–35.

Bogerts, B., Falkai, P., Haupts, M., *et al.* (1990). Post-mortem volume measurements of limbic system and basal ganglia structures in chronic schizophrenics. Initial results from a new brain collection. *Schizophr Res*, 3, 295–301.

Bogerts, B., Lieberman, J. A., Ashtari, M., *et al.* (1993). Hippocampus-amygdala volumes and psychopathology in chronic schizophrenia. *Biol Psychiatry*, 33, 236–46.

Braff, D. L., Geyer, M. A., Light, G. A., *et al.* (2001). Impact of prepulse characteristics on the detection of sensorimotor gating deficits in schizophrenia. *Schizophr Res*, 49, 171–8.

Brandon, N. J., Millar, J. K., Korth, C., Sive, H., Singh, K. K., and Sawa, A. (2009). Understanding the role of DISC1 in psychiatric disease and during normal development. *J Neurosci*, 29(41), 12768–75.

Bray, N. J., Preece, A., Williams, N. M., *et al.* (2005). Haplotypes at the dystrobrevib binding protein (DTNBP1) gene locus mediate risk for schizophrenia through reduced DTNBP1 expression. *Hum Mol Genet*, 14(14), 1947–54.

Breese, C. R., Freedman, R., and Leonard, S. S. (1995). Glutamate receptor subtype expression in human postmortem brain tissue from schizophrenics and alcohol abusers. *Brain Res*, 674, 82–90.

Brose, N. (2008). Altered complexin expression in psychiatric and neurological disorders: cause or consequence? *Mol Cells*, 25(1), 7–19.

Buonanno, A., and Fischbach, G. D. (2001). Neuregulin and ErbB receptor signaling pathways in the nervous system. *Curr Opin Neurobiol*, 11(3), 287–96.

Cannon, T. D., Hennah, W., van Erp, T. G., *et al.* (2005). Association of DISC1/TRAX haplotypes with schizophrenia, reduced prefrontal gray matter, and impaired short- and long-term memory. *Arch Gen Psychiatry*, 62(11), 1205–13.

Cannon, M., Jones, P. B., and Murray, R. M. (2002). Obstetric complications and schizophrenia: historical and meta-analytic review. *Am J Psychiatry*, 159(7), 1080–92.

Carlsson, A. (1995). The dopamine theory revisited. In Hirsch, S. R., and Weinberger, D. R. (Eds.) *Schizophrenia*, pp. 379–400. Oxford: Blackwell Science.

Carlsson, A., Waters, N., Holm-Waters, S., Tedroff, J., Nilsson, M., and Carlsson, M. L. (2001). Interactions between monoamines, glutamate, and GABA in schizophrenia: new evidence. *Ann Rev Pharmacol Toxikol*, 41, 237–60.

Casanova, M. F., Stevens, J. R., and Kleinman, J. E. (1990). Astrocytosis in the molecular layer of the dentate gyrus: a study in Alzheimer's disease and schizophrenia. *Psychiatry Res*, 35, 149–66.

Chakos, M. H., Schobel, S. A., Gu, H., *et al.* (2005). Duration of illness and treatment effects on hippocampal volume in male patients with schizophrenia. *Br J Psychiatry*, 186, 26–31.

Chambers, J. S., and Perrone-Bizzozero, N. I. (2004). Altered myelination of the hippocampal formation in subjects with schizophrenia and bipolar disorder. *Neurochem Res*, 29, 2293–302.

Choi, D. W. (1992). Excitotoxic cell death. *J Neurobiol*, 23, 1261–76.

Chumakov, I., Blumenfeld, M., Guerassimenko, O., *et al.* (2002). Genetic and physiological data implicating the new human gene G72 and the gene for D-amino acid ocidase in schizophrenia. *Proc Natl Acad Sci USA*, 99, 13675–80.

Cirillo, M. A., and Seidman, L. J. (2003). Verbal declarative memory dysfunction in schizophrenia: from clinical assessment to genetics and brain mechanisms. *Neuropsychol Rev*, 13(2), 43–77.

Conover, J. C., and Allen, R. L. (2002). The subventricular zone: new molecular and cellular developments. *Cell Mol Life Sci*, 59(12), 2128–35.

Corfas, G., Roy, K., and Buxbaum, J. D. (2004). Neuregulin 1-ErbB signaling and the molecular/cellular basis of schizophrenia. *Nat Neurosci*, 7(6), 575–80.

Corvin, A., Donohoe, G., McGhee, K., *et al.* (2007). D-amino acid oxidase (DAO) genotype and mood symptomatology in schizophrenia. *Neurosci Lett*, 426(2), 97–100.

Coyle, J. T. (1996). The glutamatergic dysfunction hypothesis for schizophrenia. *Harv Rev Psychiatry*, 3(5), 241–53.

Deakin, I. H., Law, A. J., Oliver, P. L., *et al.* (2009). Behavioural characterization of neuregulin 1 type I overexpressing transgenic mice. *Neuroreport*, 20(17), 1523–8.

Dean, B., Scarr, E., Bradbury, R., and Copolov, D. (1999). Decreased hippocampal (CA3) NMDA receptors in schizophrenia. *Synapse*, 32, 67–9.

Dracheva, S., Davis, K. L., Chin, B., Woo, D. A., Schmeidler, J., and Haroutunian, V. (2006). Myelin-associated mRNA and protein expression deficits in the anterior cingulate cortex and hippocampus in elderly schizophrenia patients. *Neurobiol Dis*, 21, 531–40.

Du, Y., and Dreyfus, C. F. (2002). Oligodendrocytes as providers of growth factors. *J Neurosci Res*, 68, 647–54.

Dwork, A. J. (1997). Postmortem studies of the hippocampal formation in schizophrenia. *Schizophr Bull*, 23, 385–402.

Eastwood, S. L., Brunet, P. W. J., and Harrison, P. J. (2000). Expression of complexin I and II mRNAs and their regulation by antipsychotic drugs in the rat forebrain. *Synapse*, 36, 167–77.

Eastwood, S. L., and Harrison, P. J. (2001). Synaptic pathology in the anterior cingulate cortex in schizophrenia and mood disorders. A review and a Western blot study of synaptophysin, GAP-43 and the complexins. *Brain Res Bull*, 55(5), 569–78.

Eastwood, S. L., Kerwin, R. W., and Harrison, P. J. (1997). Immunoautoradiographic evidence for a loss of alpha-amino-3-hydroxy-5-methyl-4-isoxazole propionate-preferring non-N-methyl-D-aspartate glutamate receptors within the medial temporal lobe in schizophrenia. *Biol Psychiatry*, 41, 636–43.

Ebner, F., Tepest, R., Dani, I., *et al.* (2008). The hippocampus in families with schizophrenia in relation to obstetric complications. *Schizophr Res*, 104(1–3), 71–8.

Falls, D. L. (2003). Neuregulins: functions, forms, and signaling strategies. *Exp Cell Res*, 284(1), 14–30.

Falkai, P., and Bogerts, B. (1986). Cell loss in the hippocampus of schizophrenics. *Eur Arch Psychiatry Neurol Sci*, 236, 154–61.

Falkai, P., Honer, W. G., David, S., *et al.* (1999). No evidence for astrogliosis in brains of schizophrenic patients. A post-mortem study. *Neuropathol Appl Neurobiol*, 25, 48–53.

Farber, N. B., Newcomer, J. W., and Olney, J. W. (1999). Glycine agonists: what can they teach us about schizophrenia? *Arch Gen Psychiatry*, 56, 13–17.

Fatemi, S. H., Earle, J. A., Stary, J. M., Lee, S., and Sedgewick, J. (2001). Altered levels of the synaptosomal associated protein SNAP-25 in hippocampus of subjects with mood disorders and schizophrenia. *Neuroreport*, 12(15), 3257–62.

Fatemi, S. H., Folsom, T. D., Reutiman, T. J., Huang, H., Oishi, K., and Mori, S. (2009). Prenatal viral infection of mice at E16 causes changes in gene expression in hippocampi of the offspring. *Eur Neuropsychopharmacol*, 19(9), 648–53.

Fendt, M., Lex, A., Falkai, P., Henn, F. A., and Schmitt, A. (2008). Behavioural alterations after postnatal hypoxia and effects of clozapine: implications for schizophrenia. *Pharmacopsychiatry*, 41 (4), 138–45.

Gao, X., Sakai, K., Roberts, R. C., Conley, R. R., Dean, B., and Tamminga, C. A. (2000). Ionotropic glutamate receptors and expression of N-methyl-D-aspartate receptor subunits in subregions of human hippocampus: effects of schizophrenia. *Am J Psychiatry*, 157, 1141–9.

Goff, D. C., and Coyle, J. T. (2001). The emerging role of glutamate in the pathophysiology and treatment of schizophrenia. *Am J Psychiatry*, 158(9), 1367–77.

Gruber, O., Falkai, P., Schneider-Axmann, T., Schwab, S. G., Wagner, M., and Maier, W. (2008). Neuregulin-1 haplotype HAP(ICE) is associated with lower hippocampal volumes in schizophrenic patients and in non-affected family members. *J Psychiatr Res*, 43(1), 1–6.

Gruber, O., Trost, S., Wobrock, T., *et al.* (2010a). Association of DISC1 genetic variation with intermediate phenotypes in neuroimaging: replication and further extension. *Biol Psychiatry*, 67(9), 163S.

Gruber, O., Trost, S., Wobrock, T., *et al.* (2010b). Association of schizophrenia susceptibility genes with intermediate phenotypes: new findings from genomic imaging. *Schizophr Res*, 117(2–3), S335–S336.

Gur, R. E., Keshavan, M. S., and Lawrie, S. M. (2007). Deconstructing psychosis with human brain imaging. *Schizophr Bull*, 33(4), 921–31.

Habl, G., Zink, M., Petroianu, G., *et al.* (2009). Increased D-amino acid oxidase (DAAO) expression in the bilateral hippocampus of schizophrenia patients: a post-mortem study. *J Neural Transm*, 116, 1657–65.

Haroutunian, V., Katsel, P., Dracheva, S., and Davis, K. L. (2006). The human homolog of the QKI gene affected in the severe dysmaleination 'quaking' mouse phenotype: downregulated in multiple brain regions in schizophrenia. *Am J Psychiatry*, 163(10), 1834–7.

Haroutunian, V., Katsel, P., Dracheva, S., Stewart, D. G., and Davis, K. L. (2007). Variations in oligodendrocyte-related gene expression across multiple cortical regions: implications for the pathophysiology of schizophrenia. *Int J Neuropsychopharmacol*, 10, 565–73.

Harrison, P. J. (2004). The hippocampus in schizophrenia: a review of the neuropathological evidence and ist pathophysiological implications. *Psychopharmacology*, 174(1), 151–62.

Harrison, P. J., and Eastwood, S. L. (2001). Neuropathological studies of synaptic connectivity in the hippocampal formation in schizophrenia. *Hippocampus*, 11, 508–19.

Harrison, P. J., and Weinberger, D. R. (2005). Schizophrenia genes, gene expression, and neuropathology: on the matter of their convergence. *Mol Psychiatry*, 10(1), 40–68.

Heckers, S. (2001). Neuroimaging studies of the hippocampus in schizophrenia. *Hippocampus,* 11, 520–8.

Heckers, S., Heinsen, H., Geiger, B., *et al.* (1991). Hippocampal neuron number in schizophrenia. A stereological study. *Arch Gen Psychiatry*, 48, 1002–8.

Henseler, I., Falkai, P., and Gruber, O. (2009). A systematic fMRI investigation of the brain systems subserving different working memory components in schizophrenia. *Eur J Neurosci*, 2009, 30(4), 693–702.

Henseler, I., Falkai, P., and Gruber, O. (2010). Disturbed functional connectivity within networks subserving domain-specific subcomponents of working memory in schizophrenia: relation to performance and clinical symptoms. *J Psychiatr Res*, 44(6), 364–72.

Hsu, J. C., Zhang, Y., Takagi, N., *et al.* (1998). Decreased expression and functionality of NMDA receptor complexes persist in the CA1, but not in the dentate gyrus after transient cerebral ischemia. *J Cereb Blood Flow Metab*, 18, 768–75.

Hu, S., Ying, Z., Gomez-Pinilla, F., and Frautschy, S. A. (2009). Exercise can increase small heat shock proteins (sHSP) and pre- and post-synaptic proteins in the hippocampus. *Brain Research,* 1249, 191–201.

Hurlemann, R., Tepest, R., Maier, W., Falkai, P., and Vogeley, K. (2005). Intact hippocampal gray matter in schizophrenia as revealed by automatized image analysis postmortem. *Anat Embryol (Berl)*, 210, 513–17.

Ioannidis, J. P., Ntzani, E. E., and Trikalinos, T. A. (2001). Contopoulos-Ioannidis DG. Replication validity of genetic association studies. *Nat Genet*, 29(3), 306–9.

International Schizophrenia Consortium, Purcell, S. M., Wray, N. R., *et al.* (2009). Common polygenic variation contributes to risk of schizophrenia and bipolar disorder. *Nature*, 460(7256), 748–52.

Jessen, F., Scheef, L., Germershausen, L., *et al.* (2003). Reduced hippocampal activation during encoding and recognition of words in schizophrenia patients. *Am J Psychiatry*, 160, 1305–12.

Jones-Gotman, M., and Milner, B. (1978). Right temporal-lobe contribution to image mediated verbal learning. *Neuropsychologia*, 16, 61–71.

Kaifu, T., Nakahara, J., Inui, M., *et al.* (2003). Osteopetrosis and thalamic hypomyelinosis with synaptic degeneration in DAP12-deficient mice. *J Clin Invest*, 111(3), 323–32.

Kalus, P., Buri, C., Slotboom, J., *et al* (2004). Volumetry and diffusion tensor imaging of hippocampal subregions in schizophrenia. *Neuroreport*, 15, 867–71.

Kapoor, R., Lim, K. S., Cheng, A., Garrick, T., and Kapoor, V. (2006). Preliminary evidence for a link between schizophrenia and NMDA-glycine site receptor ligand metabolic enzymes, d-amino acid oxidase (DAAO) and kynurenine aminotransferase-1 (KAT-1). *Brain Res*, 1106(1), 205–10.

Katsel, P., Davis, K. L., and Haroutunian, V. (2005). Variations in myelin and oligodendrocyte-related gene expression across multiple brain regions in schizophrenia: a gene ontology study. *Schizophr Res*, 79, 157–73.

Kendler, K. S., MacLean, C. J., O'Neill, F. A., *et al.* (1996). Evidence for a schizophrenia vulnerability locus on chromosome 8p in the Irish study of high-density schizophrenia families. *Am J Psychiatry*, 153(12), 1534–40.

Kerwin, R., Patel, S., and Meldrum, B. (1990). Quantitative autoradiographic analysis of glutamate binding sites in the hippocampal formation in normal and schizophrenic brain post mortem. *Neuroscience*, 39, 25–32.

Kircher, T., Thienel, R., Wagner, M., *et al.* (2009). Neuregulin 1 ICE-single nucleotide polymorphism in first episode schizophrenia correlates with cerebral activation in fronto-temporal areas. *Eur Arch Psychiatry Clin Neurosci*, 259(2), 72–9.

Knable, M. B., Barci, B. M., Webster, M. J., *et al.* (2004). Molecular abnormalities of the hippocampus in severe psychiatric illness: postmortem findings from the Stanley Neuropathology Consortium. *Mol Psychiatry*, 9(6), 609–20.

Kolomeets, N. S., Orlocskaya, D. D., Rachmanova, V. I., and Uranova, N. A. (2005). Ultrastructural alterations in hippocampal mossy fiber synapses in schizophrenia: a postmortem morphometric study. *Synapse*, 57(1), 47–55.

Koolschijn, P. C., van Haren, N. E., Cahn, W., *et al.* (2010). Hippocampal volume change in schizophrenia. *J Clin Psychiatry*, 71(6), 737–44.

Kraepelin, E. (1919). *Dementia praecox and paraphrenia*. Edinburgh: Livingstone.

Kuroki, N., Kubicki, M., Nestor, P. G., *et al* (2006). Fornix integrity and hippocampal volume in male schizophrenic patients. *Biol Psychiatry*, 60, 22–31.

Lauriat, T. L., Shiue, K., Haroutunian, V., *et al.* (2008). Developmental expression profile of quaking, a candidate gene for schizophrenia, and its target genes in human prefrontal cortex and hippocampus shows regional specificity. *J Neurosci Res*, 86(4), 785–96.

Law, A. J., Lipska, B. K., Weickert, C. S., *et al.* (2006). Neuregulin 1 transcripts are differentially expressed in schizophrenia and regulated by 5' SNPs associated with the disease. *Proc Natl Acad Sci U S A*, 103(17), 6747–52.

Lewis, C. M., Levinson, D. F., Wise, L. H., *et al.* (2003). Genome scan meta-analysis of schizophrenia and bipolar disorder, part II: Schizophrenia. *Am J Hum Genet*, 73(1), 34–48.

Lim, K. O., Ardekani, B. A., Nierenberg, J., *et al.* (2006). Voxelwise correlational analyses of white matter integrity in multiple cognitive domains in schizophrenia. *Am J Psychiatry*, 163, 2008–10.

Lipska, B. K., Al-Amin, H. A., and Weinberger, D. R. (1998). Excitotoxic lesions of the rat medial prefrontal cortex. Effects on abnormal behaviors associated with neonatal hippocampal damage. *Neuropsychopharmacology*, 19, 451–64.

Lipska, B. K., Aultman, J. M., Verma, A., Weinberger, D. R., and Moghaddam, B. (2002). Neonatal damage of the ventral hippocampus impairs working memory in the rat. *Neuropsychopharmacology*, 27, 47–54.

Liu, X., Hea, G., Wanga, X., *et al.* (2004). Association of *DAAO* with schizophrenia in the Chinese population. *Neuroscience Letters*, 369, 228–33.

Liu, Y., Ford, B., Mann, M. A., and Fischbach, G. D. (2001). Neuregulins increase alpha7 nicotinic acetylcholine receptors and enhance excitatory synaptic transmission in GABAergic interneurons of the hippocampus. *J Neurosci*, 21(15), 5660–9.

Martins-de-Souza, D., Dias-Neto, E., Schmitt, A., *et al.* (2010). Proteome analysis of schizophrenia brain tissue. *World J Biol Psychiatry*, 11(2), 110–20.

McInnes, L. A., and Lauriat, T. L. (2006). RNA metabolism and dysmyelination in schizophrenia. *Neurosci Biobehav Rev*, 30(4), 551–61.

Milner, B., Corkin, C., and Teuber, H. L. (1968). Further analysis of the hippocampal manestic syndrome: 14-year follow-up study of HM. *Neuropsychologia*, 6, 215–34.

Monyer, H., Burnashev, N., Laurie, DJ., *et al.* (1994). Developmental and regional expression in the rat brain and functional properties of four NMDA receptors. *Neuron*, 12(3), 529–40.

Morikawa, E., Mori, H., Kiyama, Y., Mishina, M., Asano, T., and Kirino, T. (1998). Attenuation of focal ischemic brain injury in mice deficient in the epsilon1 (NR2A) subunit of NMDA receptor. *J Neurosci*, 18, 9727–32.

Mouton, P. R., Long, J. M., Lei, D. L., *et al.* (2002). Age and gender effects on microglia and astrocyte numbers in brains of mice. *Brain Res*, 956, 30–5.

Nasrallah, H. A., Hopkins, T., and Pixley, S. K. (2010). Differential effects of antipsychotic and antidepressant drugs on neurogenic regions in rats. *Brain Res*, 1354, 23–9.

Newton, S. S., and Duman, R. S. (2007). Neurogenic actions of atypical antipsychotic drugs and therapeutic implications. *CNS Drugs*, 21(9), 715–25.

O'Donovan, M. C., Craddock, N., Norton, N., *et al.* (2008). Identification of loci associated with schizophrenia by genome-wide association and follow-up. *Nat Genet*, 40(9), 1053–5.

Olbrich, H. M., Valerius, G., Rüsch, N., *et al.* (2008). Frontolimbic glutamate alterations in first episode schizophrenia: evidence from a magnetic resonance spectroscopy study. *World J Biol Psychiatry*, 9(1), 59–63.

Olney, J. W., and Farber, N. B. (1995). Glutamate receptor dysfunction and schizophrenia. *Arch Gen Psychiatry*, 52, 998–1007.

Ozaki, M., Sasner, M., Yano, R., Lu, H. S., and Buonanno, A. (1997). Neuregulin-beta induces expression of an NMDA-receptor subunit. *Nature*, 390(6661), 691–4.

Pajonk, F., Wobrock, T., Gruber, O., *et al.* (2010). Hippocampal plasticity in response to exercise in schizophrenia. *Arch Gen Psychiatry*, 67(2), 133–43.

Panenka, W. J., Khorram, B., Barr, A. M., *et al.* (2007). A longitudinal study on effects of typical versus atypical antipsychotic drugs on hippocampal volume in schizophrenia. *Schizophr Res*, 94(1–3), 288–92.

Parlapani, E., Schmitt, A., Bergmann, A., *et al.* (2009). Association between myelin basic protein expression and left entorhinal cortex pre-alpha cell layer disorganization in schizophrenia. *Brain Res*, 1301, 126–34.

Parlapani, E., Schmitt, A., Wirths, O., *et al.* (2010). Gene expression of neuregulin-1 isoforms in different brain regions of elderly schizophrenia patients. *World Journal of Biol Psychiatry*, 11, 243–50.

Pulver, A. E., Lasseter, V. K., Kasch, L., *et al.* (1995). Schizophrenia: a genome scan targets chromosomes 3p and 8p as potential sites of susceptibility genes. *Am J Med Genet*, 60(3), 252–60.

Rametti, G., Junque, C., Vendreil, P., *et al.* (2009). Hippocampal underactivation in an fMRI study of word and face memory recognition in schizophrenia. *Eur Arch Psychiatry Clin Neurosci*, 259(4), 203–11.

Reif, A., Fritzen, S., Finger, M., *et al.* (2006). Neural stem cell proliferation is decreased in schizophrenia, but not in depression. *Mol Psychiatry*, 11, 514–22.

Reske, M., Kellermann, T., Habel, U., *et al.* (2007). Stability of emotional dysfunctionsß A long-term fMRI study in first-episode schizophrenia. *J Psychiatr Res*, 41(11), 918–27.

Rieff, H. I., Raetzman, L.T., Sapp, D. W., Yeh, H. H., Siegel, R. E., and Corfas, G. (1999). Neuregulin induces GABA(A) receptor subunit expression and neurite outgrowth in cerebellar granule cells. *J Neurosci*, 19(24), 10757–66.

Robinson, D. G., Woerner, M. G., Alvir, J. M., *et al.* (2002). Predictors of medication discontinuation by patients with first-episode schizophrenia and schizoaffective disorder. *Schizophr Res*, 57(2–3), 209–19.

Roth, T. L., Lubin, F. D., Sodhi, M., *et al.* (2009). Epigenetic mechanisms in schizophrenia. *Biochim Biophys Acta*, 1790(9), 869–77.

Rüsch, N., Taebartz van Elst, L., Valerius, G., *et al.* (2008). Neurochemical and structural correlates of executive dysfunction in schizophrenia. *Schizophr Res*, 99(1–3), 155–63.

Saha, S., Chant, D., and McGrath, J. (2007). A systematic review of mortality in schizophrenia: is the differential mortality gap worsening over time? *Arch Gen Psychiatry* 2007, 64(10), 1123–31.

Sawada, K., Barr, A. M., Nakamura, M., *et al.* (2005). Hippocampal complexin proteins and cognitive dysfunction in schizophrenia. *Arch Gen Psychiatry*, 62(3), 263–72.

Schell, M. J., Molliver, M. E., and Snyder, S. H. (1995). D-serine, an endogenous synaptic modulator: localization to astrocytes and glutamate-stimulated release. *Proc Natl Acad Sci USA*, 92, 3948–52.

Schmidt-Kastner, R., and Freund, T. F. (1991). Selective vulnerability of the hippocampus in brain ischemia. *Neuroscience*, 40, 599–636.

Schmitt, A., Zink, M., May, B., *et al.* (2003). Decreased gene expression of glial and neuronal glutamate transporters after antipsychotic treatment in rat brain. *Neurosci Lett*, 347(2), 81–4.

Schmitt, A., Zink, M., Müller, B., *et al.* (2003). Effects of long-term antipsychotic treatment on NMDA receptor binding and gene expression of subunits. *Neurochem Res*, 28(2), 235–41.

Schmitt, A., Weber, S., Jatzko, A., Braus, D. F., and Henn, F. A. (2004). Hippocampal volume and cell proliferation after acute and chronic clozapine or haloperidol treatment. *J Neural Transm*, 111 (1), 91–100.

Schmitt, A., Fendt, M., Zink, M., *et al.* (2007). Altered NMDA receptor expression and behaviour following postnatal hypoxia and its potential relevance to schizophrenia. *J Neural Transm*, 114 (2), 239–48.

Schmitt, A., Parlapani, E., Gruber, O., Wobrock, T., and Falkai, P. (2008). Impact of neuregulin-1 on the pathophysiology of schizophrenia in human post-mortem studies. *Eur Arch Psychiatry Clin Neurosci*, 258(Suppl 5), 35–9.

Schmitt, A., Steyskal, C., Bernstein, H. G., *et al.* (2009). Stereologic investigation of the posterior part of the hippocampus in schizophrenia. *Acta Neuropathol*, 117(4), 395–407.

Schumacher, J., Abon Jamra, R., Freudenberg, J., *et al.* (2004). Examination of G72 and D-amino-acid oxidase as genetic risk factors for schizophrenia and bipolar affective disorder. *Mol Psychiatry*, 9, 203–7.

Segal, D., Koschnick, J. R., Slegers, L. H., and Hof, P. R. (2007). Oligodendrocyte pathophysiology: a new view of schizophrenia. *Int J Neuropsychopharmacol,* 10(4), 503–11.

Shenton, M. E., Dickey, C., Frumin, C., and McCarley, R. W. (2001). A review of MRI findings in schizophrenia. *Schizophr Res,* 49, 1–52.

Shih, R. A., Belmonte, P. L., and Zandi, P. P. (2004). A review of the evidence from family, twin and adoption studies for a genetic contribution to adult psychiatric disorders. *Int Rev Psychiatry*, 16(4), 260–83.

Smith, M. L., and Milner, B. (1981). The role of the right hippocampus in the recall of spatial location. *Neuropsychologica*, 19, 781–93.

Shirayama, Y., Obala, T., Matsuzawa, D., *et al.* (2010). Specific metabolites in the medial prefrontal cortex are associated with the neurocognitive deficits in schizophrenia: a preliminary study. *Neuroimage*, 49(3), 2783–90.

Shukitt-Hale, B., Kadar, T., Marlowe, B. E., *et al.* (1996). Morphological alterations in the hippocampus following hypobaric hypoxia. *Hum Exp Toxicol*, 15, 312–19.

Smith, G. N., Lang, D. J., Kopala, L. C., Lapointe, J. S., Falkai, P., and Honer, W. G. (2003). Developmental abnormailities of the hippocampus in first-episode schizophrenia. *Biol Psychiatry*, 53(7), 555–61.

Sokolov, B. P. (1998). Expression of NMDAR1, GluR1, GluR7, and KA1 glutamate receptor mRNAs is decreased in frontal cortex of 'neuroleptic-free' schizophrenics: evidence on reversible up-regulation by typical neuroleptics. *J Neurochem*, 71, 2454–64.

Steen, R. G., Hamer, R. M., and Liebermen, J. A. (2005). Measurement of brain metabolites by 1H magnetic resonance spectroscopy in patients with schizophrenia: a systematic review and meta-analysis. *Neuropsychopharmacology*, 30(11), 1949–62.

Steen, R. G., Mull, C., McClure, R., Hamer, R. M., and Lieberman, J. A. (2006). Brain volume in first-episode schizophrenia: systematic review and meta-analysis of magnetic resonance imaging studies. *Br J Psychiatry*, 188, 510–18.

Stefansson, H., Ophoff, R. A., Steinberg, S., *et al.* (2009). Common variants conferring risk of schizophrenia. *Nature*, 460(7256), 744–7.

Stefansson, H., Sigurdsson, E., Steinthorsdottir, V., *et al.* (2002). Neuregulin 1 and susceptibility to schizophrenia. *Am J Hum Genet*, 71(4), 877–92.

Steiner, J., Bernstein, H. G., Bielau, H., *et al.* (2008). S100B-immunopositive glia is elevated in paranoid as compared to residual schizophrenia: A morphometric study. *J Psychiatr Res*, 42, 868–76.

Strohmaier, J., Frank, J., Wendland, J. R., *et al.* (2010). A reappraisal of the association between dysbindin (DTNBP1) and schizophrenia in a large combined case-control and family-based sample of german ancestry. *Schizophr Res*, 118(1–3), 98–105.

Swerdlow, N. R., Hanlon, F. M., Henning, L., Kim, Y. K., Gaudet, I., and Halim, N. D. (2001). Regulation of sensorimotor gating in rats by hippocampal NMDA: anatomical localization. *Brain Res*, 898, 195–203.

Syroid, D. E., Maycox, P. R., Burrola, P. G., *et al.* (1996). Cell death in the Schwann cell lineage and its regulation by neuregulin. *Proc Natl Acad Sci U S A*, 93(17), 9229–34.

Takei, K., Yamasue, H., Abe, O., *et al.* (2008). Disrupted integrity of the fornix is associated with impaired memory organization in schizophrenia. *Schizophr Res*, 103(1–3), 52–61.

Tkachev, D., Mimmack, M. L., Ryan, M. M., *et al.* (2003). Oligodendrocyte dysfunction in schizophrenia and bipolar disorder. *Lancet*, 362(9386), 798–805.

Tudesco Die, S., Vaz, L. J., Mantoan, M. A., *et al.* (2010). Assessment of working memory in patients with mesial temporal lobe epilepsy associated with unilateral hippocampal sclerosis. *Epilepsy Behav*, 18(3), 223–8.

Uranova, N. A., Vostrikov, V. M., Vikhreva, O. V., *et al.* (2007). The role of oligodendrocyte pathology in schizophrenia. *Int J Neuropsychopharmacol*, 10, 537–45.

van den Buuse, M. (2003). Deficient prepulse inhibition of acoustic startle in Hooded-Wistar rats compared with Sprague-Dawley rats. *Clin Exp Pharmacol Physiol*, 30, 254–61.

Velakoulis, D., Stuart, G. W., Wood, S. J., *et al.* (2001). Selective bilateral hippocampal volume loss in chronic schizophrenia. *Biol Psychiatry*, 50, 531–9.

Velakoulis, D., Wood, S.J., Wong, M. T., *et al.* (2006). Hippocampal and amygdale volumes according to psychosis stage and diagnosis. A magnetic resonance imaging study of chronic schizophrenia, first-episode psychosis, and ultra-high-risk individuals. *Arch Gen Psychiatry*, 63(2), 139–49.

Walker, M. A., Highley, J. R., Esiri, M. M., *et al.* (2002). Estimated neuronal populations and volumes of the hippocampus and its subfields in schizophrenia. *Am J Psychiatry*, 159, 821–8.

Weickert, C. S., Rothmond, D. A., Hyde, T. M., Kleinman, J. E., and Straub, R. E. (2008). Reduced DTNBP1 (dysbindin-1) mRNA in the hippocampal formation of schizophrenia patients. *Schizophr Res*, 98(1–3), 105–10.

Weinberger, D. R. (1996). On the plausibility of 'the neuro-developmental hypothesis' of schizophrenia. *Neuropsychopharmacology*, 14(3 Suppl), 1S–11S.

Weinberger, D. R. (1999). Cell biology of the hippocampal formation in schizophrenia. *Biol Psychiatry*, 45, 395–402.

Weiss, A. P., and Heckers, S. (2001). Neuroimaging of declarative memory in schizophrenia. *Scand J Psychol*, 42, 239–50.

White, T., Kendi, A. T., Lehericy, S., *et al.* (2007). Disruption of hippocampal connectivity in children and adolescents with schizophrenia – a voxel-based diffusion tensor imaging study. *Schizophr Res*, 90, 302–7.

Williams, R. W., and Rakic, P. (1988). Three-dimensional counting: an accurate and direct method to estimate numbers of cells in sectioned material. *J Comp Neurol*, 278, 344–52.

Witter, M. P. (1993). Orgainzation of the entorhinal-hippocampal system: A review of the current anatomical data. *Hippocampus*, 3, 33–44.

Wood, L. S., Pickering, E. H., and Dechaio, B. M. (2007). Significant support for DAO as a schizophrenia susceptibility locus: examination of five genes putatively associated with schizophrenia. *Biol Psychiatry*, 61(10), 1195–9.

Wood, S. J., Velakoulis, D., Smith, D. J., *et al.* (2001). A longitudinal study of hippocampal volume in first episode psychosis and chronic schizophrenia. *Schizophr Res*, 52(1–2), 37–46.

Wright, I. C., Rabe-Hesketh, S., Woodruff, P. W., David, A. S., Murray, R. M., and Bullmore, E. T. (2000). Meta-analysis of regional brain volumes in schizophrenia. *Am J Psychiatry*, 157, 16–25.

Yakovlev, P. I., and Lecours, A. R. (1967). *The myelogenetic cycles of regional maturation of the brain*, pp. 3–70. Oxford: Blackwell Scientific Publications.

Yang, J. Z., Si, T. M., Ruan, Y., *et al.* (2003). Association study of neuregulin 1 gene with schizophrenia. *Mol Psychiatry*, 8(7), 706–9.

Young, C. E., Arima, K., Xie, J., *et al.* (1998). SNAP-25 deficit and hippocampal connectivity in schizophrenia. *Cereb Cortex*, 8(3), 261–8.

Zhang, J. P., Burdick, K. E., Lencz, T., and Malhotra, A. K. (2010). Meta-analysis of genetic variation in DTNBP1 and general cognitive ability. *Biol Psychiatry*, 688(12), 1126–33.

Zhang, L., Hsu, J. C., Takagi, N., Gurd, J. W., Wallace, M. C., and Eubanks, J. H. (1997). Transient global ischemia alters NMDA receptor expression in rat hippocampus: correlation with decreased immunoreactive protein levels of the NR2A/2B subunits, and an altered NMDA receptor functionality. *J Neurochem*, 69, 1983–94.

Zhang, Z. J., and Reynolds, G. P. (2002). A selective decrease in the relative density of parvalbumin-immunoreactive neurons in the hippocampus in schizophrenia. *Schizophr Res*, 55, 1–10.

Zhou, Y., Shu, N., Liu, Y., *et al.* (2008). Altered resting-state functional connectivity and anatomical connectivity of hippocampus in schizophrenia. *Schizophr Res*, 100, 120–32.

Zola-Morgan, S., and Squire, L. R. (1993). Neuroanatomy of memory. *Annu Rev Neurosci*, 16, 547–64.

Author Index

Aimone, J. B. 108, 110,113
Andreasen, N. C. 175

Bailey, A. 182
Bauman, M. L. 182
Bayley, P. J. 46
Benson, D. F. 203
Bird, C. M. 49
Blatt, G. J. 182
Blümcke, I. 247
Borrell, J. 86
Braak, E. 58
Braak, H. 58, 153
Brierley, J. 206
Brinks, V. 84–5
Burgess, N. 49
Burke, S. N. 166

Charney, D. S. 263
Clelland, C. D. 112
Cohen, N. J. 40, 43, 45, 47
Conturo, T. E. 186
Cork, L. C. 153

Dager, S. R. 184–5
Dalton, K. M. 189
de Kloet, E. R. 89
Di, S. 90
Döbrössy, M. D. 112
Duff, N. J. 45
Dutta, R. 216

Eichenbaum, H. 47, 49
Endo, T. 185

Frodl, T. 281
Fujioka, M. 205

Garthe, A. 116
Giovanello, K. S. 49
Glisky, E. L. 46

Gould, E. 112, 113
Graf, P. 43, 154
Graham, K. S., 51
Guptill, J. T. 182

Hämäläinen, A. 236–7
Hannula, D. E. 51
Härtig, W. 153
Hassabis, D. 51
Huang, Y. Y. 162

Isaacs, E. B. 177

Jabès, A. 28
Joëls, M. 89, 97

Kandel, E. R. 162
Kaymak, S. U. 282
Kazsniak, A. W. 154
Kemper, T. L. 182
Kempermann, G. 115
Knowlton, B. J. 43
Kronmuller, K. T. 281–2
Kumaran, D. 48

Levine, S. 93–4
Liu, Y. 229

Maguire, E. A. 134
Manns, J. R. 49
McKinnon, M. C. 276
Moscovitch, M. 44, 46

Newman, M. C. 154
Nicolson, R. 184

O'Brien, J. L. 235
Oitzl, M. 85

Page, L. A. 185
Palmen, S. J. 184

Penner, M. R. 163

Rapp, P. R. 154
Reed, J. M. 50
Revest, J.-M. 117
Ridder, S. 85
Ryan, J. D. 48, 51

Saitoh, O. 32, 183
Santarelli, L. 117
Sarabdjitsingh, R. A. 81
Saxe, M. D. 113
Schacter, D. L. 45, 46
Schumann, C. M. 175, 184
Schwabe, L. 85
Selye, H. 77
Shors, T. J. 113
Sparks, B. F. 184
Spielmeyer, W. 55
Squire, L. R. 40, 43, 46, 50
Steinvorth, S. 46
Szabo, K. 203

Tashiro, A. 116
Teng, E. 50
Tronche, F. 85
Tulving, E. 46

Uttl, B. 154

Vogt, O. 55
Vythilingam, M. 278

Wiskott, V. 110

Zhao, Z. 274

Subject Index

11β-HSD type 1/2 79–80

Aβ plaques 58, 225, 226
N-acetylaspartate (NAA) 134, 185, 212, 253, 264–5, 294
activity-regulated cytoskeletal gene (Arc) 163–4
adaptation 5
adrenaline 86
adrenocorticotropin (ACTH) 78, 80, 81
aerobic glycolysis 60
affective behaviour 117–18
after-hyperpolarizing potential (AHP) 159, 166
aggression 85
ageing 152–68
  after-hyperpolarizing potential 159, 166
  Ca^{2+} regulation 159, 161
  cAMP/CREB 161–2
  cross-species cognitive ageing 153
  dopamine system 162
  dynamics of neural ensembles 164–8
  electrophysiological changes 159–60
  excitatory postsynaptic potential 159–60
  fMRI 130–1, 233–4
  free radical hypothesis 61
  gene expression 60, 162–3
  hippocampal atrophy 228
  hippocampal neuroanatomy 58, 154, 156–8
  hippocampal perfusion 132
  hippocampal volume 57–8
  HPA axis 81
  long-term potentiation 160–2, 163, 166
  NMDA receptors 161
  place fields 166
  plasticity 160–2
  population spikes 160
  presynaptic fibre potential 159
  selective neuronal vulnerability 57–8
  spatial learning 116
  spatial memory 154
  synapses 58, 158
alcohol, prenatal exposure 180
aldosterone 79–80
allostasis and allostatic load 78, 263
alveus 20
Alzheimer's disease (AD) 223–38
  Aβ plaques 58, 225, 226
  apolipoprotein E 225, 228–9
  biomarkers 224, 225, 230–1
  CA1 58, 226, 227, 229
  clinical diagnosis 225
  clinical features 224–5
  cognitive decline prediction 235–6
  combined MRI studies 236–7
  cross-species pathology 153
  early-onset 225
  fMRI 130–1, 224, 231–3, 234, 235–6
  frontal lobes 232
  genetic risk 131, 225
  hippocampal atrophy 224, 226–7, 230–1
  hippocampal deformation 227–8
  hippocampal perfusion 132
  hippocampal volume 224, 226–7
  major depression association 282
  medial temporal lobe atrophy 123, 224, 226–7
  memory loss 225, 232
  mitochondrial function 60–1
  neurofibrillary tangles 225–6
  neuropathological diagnosis 225–6
  NIA-R consensus criteria 226
  NINCDS-ADRDA criteria 225
  PET 230
  pharmacological fMRI after cholinergic enhancement 131
  posteromedial cortical function 230
  predictive role of MRI 229
  repetition suppression 234–5
  selective neuronal vulnerability 55, 58
  sporadic/late-onset 225
  staging 226
  stress effects 97
  structural MRI 224, 226–8
Ammon's horn 2
  sclerosis 246
amnesia 13, 25, 39–51
  developmental 28
  H.M. 4, 39–43, 46–7, 49–50
  hyperspecificity of preserved memories 46
  imagining the future 51
  short-term relational memory 51
  spared learning ability 40, 43–4
  sufficient conditions for intact memory 44–5
  temporal gradient 49–50
  see also transient global amnesia
amnestic stroke 203
AMPA receptors 90, 293–4
AMT-PET 255
amygdala 23, 78, 79, 82–3, 263
β-amyloid plaques 58, 225, 226
amyotrophic lateral sclerosis 55–6
Angelman syndrome 186–7
anhedonia 276
anoxic encephalopathy 204–5
anoxic long-term potentiation 202
anterior temporal lobe surgery 4, 256–7
antidepressants 89, 282
antipsychotic therapy 296–7
anxiety disorders
  hypocortisolaemia 82
  neurogenesis 117
apolipoprotein E (APOE) 225
  carriers 132, 228–9, 234
APP 187
apparent diffusion coefficient (ADC) 203
  hippocampal sclerosis 253
  limbic encephalitis 205, 207
  seizures 208–9
  transient global amnesia 212
Arc 163–4
area 29: 13
area 30: 13
area 35: 12, 123
area 36: 12, 27, 123
area TE 12
area TEO 12
area TF 12, 27, 123
area TF1 12
area TFm 12
area TH 12, 123
Aristotle 1
arterial spin labelling 132
artificial grammar learning 43
astrocytes
  ischaemia 202
  normal coverage and processes 30–2, 34
  schizophrenia 292
ATP synthesis 60
attention-deficit hyperactivity disorder (ADHD) 188–90
autism spectrum disorder 181–6
  N-acetylaspartate 185
  choline-containing compounds 185, 186
  creatine and phosphocreatine 185, 186
  diffusion tensor imaging and tracking 186
  GABAergic system 182
  glutamine/glutamate 185
  hippocampal neuropathology 32–3
  hippocampal shape 184–5
  hippocampal volume 32, 182–3, 184
  memory 186
  MRI 32, 182–6
  MRS 185
  neuronal density 182
  postmortem studies 182
  seizures 185, 186
autoimmune limbic encephalitis 206–7

basket cells 15, 18, 19, 33
borderline personality disorder 262, 265
bottleneck 108
brain-derived neurotrophic factor (BDNF) 134, 277, 280–1
Bratz sector 5
burst firing 246

c-fos 162
Ca^{2+} 64–5, 159, 161, 201
CA1 21–4
  ageing 58, 156, 158, 160, 161, 163, 164, 166
  Alzheimer's disease 58, 226, 227, 229
  as a comparator 109
  cortical spreading depression 211
  decoding network 109
  gene expression 66–7
  hypoglycaemic encephalopathy 210
  ischaemic sensitivity 56–7, 201, 202
  oxidative stress sensitivity 61, 66–7
  postnatal maturation 29
  projections from CA3 21–2
  projections from entorhinal cortex 22–3

CA1 (*cont.*)
  projections to entorhinal cortex 23–4
  projections to subiculum 23
  pyramidal cells 20, 24
  Schaffer collaterals 21, 158, 159, 201
  schizophrenia 59
  subcortical connections 23
  transient global amnesia 212
  trauma 209
CA2 20, 21
CA3 20, 21
  ageing 156, 158, 160, 161, 166
  burst firing 246
  gene expression 66, 67
  ischaemic sensitivity 56–7
  memory storage and retrieval 109
  pattern completion 109
  postnatal maturation 29
  projections from dentate gyrus 18–20
  projections from entorhinal cortex 21
  projections to CA1 21–2
  PTSD 265
  pyramidal cells 18–20, 22, 29
  schizophrenia 291
  stress 263
  subcortical inputs 21
  temporary autoassociative memory 108
calcium/calmodulin-dependent protein kinase II (CaMKII) 187
cAMP/CREB 161–2
cannabinoid receptors 191
cardiorespiratory arrest 57, 204–5
caspase-3 201
catastrophic interference 108, 116
catFISH 163
choline 180, 185, 186, 253
cholinergic inputs 18
chromosome 1q42 296
circadian rhythm 80
classical conditioning 43–4
clozapine 296–7
cocaine 181
cognitive maps 47, 107, 114
cognitive reserve 168
cognitive training 106
collateral sulcus 123
common ground 45
contextual fear conditioning 113
cornu ammonis 2; *see also* CA1, CA2, CA3
cortical efferents 25–7
cortical spreading depression 211
corticotrophin-releasing hormone 78
corticotropin-releasing factor 266
cortisol/corticosterone (CORT) 78, 79, 80
  network function 89–90
  pulsed secretion 80–2
  stress-induced activity 82
CPP 166
creatine 185, 186, 253
Cushing's disease 97
cyclopiazonic acid 162
cytochrome c 201

D-amino-acid oxidase gene 295
declarative memory 40, 51, 107
  fundamental features 44–9
  PTSD 264
  sleep 146
deep brain stimulation 5, 257
default mode network 133
degu 153
delay-conditioning 44
delayed neuronal death 201
dendrites 29
  ageing 58, 154, 156, 158
  Down syndrome 34
  stress-induced reorganization 88
dentate gyrus 15–20
  ageing 156, 158, 159–60, 161, 164
  basket cells 15, 18, 19, 33
  competitive learning network 108
  differences between rodents and primates 16–17
  Down syndrome 187
  granule cell layer 15; *see also* granule cells
  hypoglycaemic encephalopathy 210
  laminar organization 15–16
  long-term potentiation 161
  molecular layer 15
  mossy cells 16–17, 19, 28–9, 33–4, 108, 247
  neurogenesis 28, 78, 106, 116–17
  pattern separation 20, 106, 108
  polymorphic layer 15–16, 17, 18, 29, 33
  postnatal maturation 28–9
  projections to CA3 18–20
  retired neurons 110
  stress-induced gene expression 91
  subcortical inputs 18
  thorny excrescences 16
depression
  circadian rhythm 80, 81
  cortisol/corticosterone (CORT) 80, 82
  hippocampal volume 88, 265
  mineralocorticoid receptors (MR) 98
  trauma spectrum disorder 262
  see also major depression
depth electrodes 251–2
developmental amnesia 28
dexamethasone suppression test 82
diabetes 86, 97, 210
diathesis–stress model 78
diencephalic amnesia 25
diffusion tensor imaging (DTI) 203
  autism 186
  herpes simplex limbic encephalitis 205–6
  schizophrenia 290–1
diffusion tensor tracking 186
diffusion-weighted imaging 203
  epilepsy 208–9
  ex vivo 135–6
  herpes simplex limbic encephalitis 205
  hypoglycaemic encephalopathy 210
  status epilepticus 209
  temporal lobe epilepsy 253
  transient global amnesia 212
dihydroepiandrosterone 268
Dilantin® 267–8
DISC1 296
dissociative identity disorder 262, 265
distress 78
DNA methylation 163–4
dopaminergic system 14, 91, 162, 190
dot pattern completion 43
Down syndrome 34, 187–8
  Alzheimer's-like neuropathology 187, 188
  astrocytic coverage and processes 34
  dendritic spines 34
  dentate gyrus 187
  glial fibrillary acidic protein expression 34
  hippocampal volume 34, 187
  memory 187, 188
  S-100-beta expression 34
  structural MRI 34
DTNBP1 296
duroquinone 67
dysbindin-1 296

E6-AP 187
early life experiences 93–5, 278–9
Egr-1 91
electroconvulsive therapy 50, 283
electroencephalography 250–2
electron transport chain enzymes 60
emetine 162
emotions 84–5, 189
end bulb 20
endocrine challenge test 82
enriched environment 111–12, 176
entorhinal cortex 13–15, 123
  projections from CA1 23–4
  projections from presubiculum 25
  projections to and from subiculum 24
  projections to CA1 22–3
  projections to CA3 21
environmental enrichment 111–12, 176
epilepsy
  autism 185, 186
  diffusion-weighted imaging 208–9
  hippocampal sclerosis 57
  medial temporal sclerosis 246–8
  neurogenesis 118
  postictal stage 208–9
  stress effects 97
  syndromes 245
  temporal lobe resection 4
  vasomotor disturbances 55
  see also temporal lobe epilepsy
epinephrine 90
episodic memory 46–7, 107, 130
eustress 78
event-related fMRI 129
ex vivo imaging 135–7
excitatory postsynaptic potential 159–60
excitotoxicity 62–5
experimental autoimmune encephalomyelitis 216
explicit memory 45
extinction 83

familiarity 47
FDG-PET 230, 254–5
fear 84–5, 249
febrile seizures 32, 179, 247
fetal alcohol spectrum disorders 180
fetal cocaine exposure 181
fibre potential 159
FMR1 188–9
FMZ-PET 255
food caching 176, 186
forced extinction 84
fractional anisotropy 206
fragile X syndrome 188–9
free radical hypothesis of ageing 61

functional connectivity
neuroimaging 132–3
functional magnetic resonance
imaging (fMRI) 128
ADHD 190
ageing 130–1, 233–4
Alzheimer's disease 130–1, 224, 231–3, 234, 235–6
APOE carriers 234
challenges 128
cocaine users 181
combined techniques 133–4
event-related 129
fragile X syndrome 189
memory 128–30
mild cognitive impairment 130–1, 224, 232–3
pharmacological 131
preterm adults 178–9
PTSD 267
schizophrenia 290–1, 295
temporal lobe epilepsy 253
transient global amnesia 212

GABAergic system 18, 59, 182, 293, 294
gender, hippocampal development 177, 183–4
gene expression
ageing 60, 162–3
hippocampal development 30–2
selective neuronal vulnerability 65–70
stress-induce changes 91–2
general adaptation syndrome 77
Gilles de la Tourette syndrome 190–1
glial cells 66, 202, 209, 278, 291
glial fibrillary acidic protein 34
glucocorticoid receptors (GR) 79, 81, 82–7, 90
MR:GR balance hypothesis 95–6
polymorphisms 92–3
PTSD 267
glucocorticoids 78, 88, 278
Glud1 transgenic mice 62, 69
GluR2 90, 201
glutamate 62–5, 185, 201, 278, 293, 294
glycogen synthase kinase-3 277
Gollins partial pictures test 40
granule cells 15
ageing 160, 161, 163
major depression 278
postnatal development 28
projections to CA3 18–20
rodents 16
seizure induction 246
Greek mythology 1, 2
GSK3 277

haloperidol 296–7
handling 94
HAP(ICE) haplotype 296
herpes simplex limbic encephalitis 205–6
hierarchical organization 13
hippocampal formation 11, 123
cortical efferents 25–7
overall organization 11–27
postnatal development 27–32
hippocampal inversion 177
hippocampal perfusion 132
hippocampal sclerosis 57
dual pathology 253
febrile seizures 179
MRI 252–3
temporal lobe epilepsy 33, 179, 245–8, 252–3
hippocampal shape 127–8
Alzheimer's disease and MCI 227–8
autism 184–5
hippocampal volume 124–7
ADHD 189–90
ageing 57–8
Alzheimer's disease 224, 226–7
antipsychotics 297
autism 32, 182–3, 184
borderline personality disorder 265
cardiorespiratory arrest 205
depression 88, 265, 274, 276–7, 281–3
diabetes 210
dissociative identity disorder 265
Down syndrome 34, 187
early childhood stress 278–9
electroconvulsive therapy 283
fetal alcohol exposure 180
fragile X syndrome 189
genetic factors 277
glucocorticoid therapy 88
intelligence 175–6
mild cognitive impairment 227
multiple sclerosis 215–16
preterm births 176–8
PTSD 264–5
schizophrenics and their relatives 33, 290, 291
stress-induced changes 88
stress x gene interactions 279–81
subregional analysis 128
surrogate marker in clinical trials 125–6
temporal lobe epilepsy 33
Tourette syndrome 191
trauma spectrum disorders 265
traumatic brain injury 209
hippocampus
blood supply 203
coining of term 2, 10–11
early anatomical studies 3–4
fields 20
functions attributed to 6, 107–12
history of hippocampal research 2–6
interneurons 21, 59, 202
laminae 20
Hippocampus debate 2–3
hippocampus proper 11, 123
histone deacetylase 164
H.M. 4, 39–43, 46–7, 49–50, 137
hole board task 84
Homer 1a 163
5HT 90–1
5-HTTLPR 277, 279
human herpesvirus-6 limbic encephalitis 207
Huntington's disease 55, 58–9
Huxley–Owen debate 2–3
hydrogen peroxide 65
hypercortisolaemia 80, 82
hyperspecific memories 46
hypocortisolaemia 82
hypoglycaemia 201
hypoglycaemic encephalopathy 210
hypothalamic–pituitary–adrenal (HPA) axis 78, 87–8
major depression 276
pulsatility 80–2
stress-induced activity 82–3
hypoxia 30–1, 56–7, 293

imagining 51
immediate-early genes 162–3
implicit memory 45
indomethacin 178
inferences 130
inflammatory disorders 80–1
inflammatory response 66
information processing 78, 107, 108–9
intelligence 175–6
interneurons 21, 59, 202
ischaemia 30–1, 56–7, 67, 201, 202, 203–4, 293

K.C. 46
ketogenic diet 257

L-type Ca^{2+} channels 159, 161
lactate 212, 253
laser capture microdissection 66, 294
latent period 247
learning 5, 6, 31, 34, 108
neurogenesis 106, 112
sparing in amnesia 40, 43–4
limbic encephalitis 205–7
limbic system 6
lithium 282
long-term depression (LTD) 89, 278
long-term potentiation (LTP) 89, 94, 202, 278
ageing 160–2, 163, 166
Ube3a mutants 187

magnetic resonance imaging (MRI) 122–37
ADC, *see* apparent diffusion coefficient
ageing 228
Alzheimer's disease 224, 226–8
APOE carriers 228–9
arterial spin labelling 132
autism 32, 182–6
autoimmune limbic encephalitis 206, 207
automated segmentation techniques 125
correlates of hippocampal damage 203
Down syndrome 34
ex vivo 135–7
febrile seizures 179
fragile X syndrome 189
functional connectivity MRI 133
herpes simplex limbic encephalitis 205–6
HHV6 encephalitis 207
hippocampal sclerosis 252–3
historical background 123–4
mild cognitive impairment 226–8
multiple sclerosis 215
PCA-stroke 203
post-transplant acute limbic encephalitis 207
prediction of Alzheimer's disease 229

magnetic resonance imaging (MRI) (*cont.*)
  quantification methods 124–8; *see also* hippocampal shape; hippocampal volume
  schizophrenia 33
  temporal lobe epilepsy 33, 252–3
  transient global amnesia 212
  voxel-based morphometry 126–7, 229
  see also functional magnetic resonance imaging
magnetic resonance spectroscopy (MRS) 134
  autism 185
  PTSD 265
  schizophrenia 294
  temporal lobe epilepsy 253
  transient global amnesia 212
major depression 273–83
  Alzheimer's disease association 282
  anhedonia 276
  brain-derived neurotrophic factor 277
  cognitive impairment 282
  episodic nature 273
  glycogen synthase kinase-3 277
  hippocampal dysregulation 275–6
  hippocampal volume 265, 274, 276–7, 281–3
  HPA axis 276
  5-HTTLPR 277, 279
  memory 274
  neurogenesis 117
  postmortem hippocampal studies 278
  regional morphological variation in hippocampus 274–5
  serotonin system 276
  stress sensitivity 274
  treatment 282–3
MAPK 91
maternal care 94
medial prefrontal cortex 263
medial temporal lobe atrophy 123, 224, 226–7
medial temporal lobe memory system 123
memantine 166–8
memory
  ADHD 190
  ageing 154
  Alzheimer's disease 225, 232
  autism 186
  cocaine users 181
  consolidation of 13, 27, 49–50, 83, 109; *see also* sleep-related memory consolidation
  Down syndrome 187, 188
  fetal alcohol exposure 180
  fMRI 128–30
  fragile X syndrome 189
  glucocorticoid and mineralocorticoid receptors 83
  major depression 274
  PTSD 264
  retrieval 83
  schizophrenia 290
  switching between memory systems 85–6
  Tourette syndrome 191
  traumatic brain injury 209
  see also specific types of memory
mesial temporal sclerosis 246–8
metabolic stress 86
methylphenidate 190
metyrapone 85, 86
microglia 66–7, 209
mifepristone 86, 87, 97
migraine 211
mild cognitive impairment (MCI) 223
  Alzheimer's disease risk 225, 229
  combined MRI studies 236–7
  diagnostic criteria 225
  fMRI 130–1, 224, 232–3
  hippocampal shape 227–8
  hippocampal volume 227
  medial temporal lobe atrophy 224
  neuropathology 226
  structural MRI 226–8
  synaptic loss 58
mineralocorticoid receptors (MR) 79–80, 81, 82–7
  depression 98
  MR:GR balance hypothesis 95–6
  polymorphisms 92–3
mirror drawing 40
mirror-reversed word reading 40
mitochondria 60–1
Molaison, H., *see* H.M.
Morris water maze 83, 106, 111, 114
mossy cells 16–17, 19, 20, 28–9, 33–4, 108, 247
MR:GR balance hypothesis 95–6
multiple sclerosis 215–16
multiple single-unit recordings 164
myelination 216, 291–2
  preterm infants 177
  schizophrenia 295

N-acetylaspartate (NAA) 134, 185, 212, 253, 264–5, 294
Narp 163
navigation 6, 107, 109, 114–16
network theories 108–9
neuregulin-1 295–6
neurofibrillary tangles 225–6
neurogenesis 105–18
  ablation of 110–11
  affective behaviour 117–18
  antidepressants 89
  anxiety disorders 117
  capacity of dentate gyrus 116–17
  cognitive training 106
  dentate gyrus 28, 78, 106, 116–17
  depression 117
  dual role 117
  environmental enrichment 111–12
  epilepsy 118
  imaging as a surrogate marker 135
  learning 106, 112
  maturation process 110
  navigation 114–16
  negative effects of 110
  neurogenic reserve hypothesis 111
  pattern integration 112–13
  pattern separation 112–13
  physical exercise 106, 111–12, 114
  schizophrenia 117, 292
  stability–plasticity dilemma 109
  stress 88–9, 278
  theoretical models 109–10
neuromodulatory stimulation 257
neuronal encoding of environment 164–6
neuroplasticity, *see* plasticity
nitric oxide synthase 64
NMDA receptor antagonists 166–7
NMDA receptors 161, 289, 293–4
non-paraneoplastic limbic encephalitis 206–7
non-rapid eye movement sleep 145
noradrenergic inputs 18, 21, 23
norepinephrine 91

olfaction 6, 14
oligodendrocytes 294–5
oxidative phosphorylation 60
oxidative stress 60–1, 66–7, 201

paired-pulse facilitation 278
Papez circuit 6
paradoxical sleep 145
parahippocampal cortex 12–13, 123
parahippocampal gyrus 123, 156
paraneoplastic limbic encephalitis 206
parasubiculum 25, 30
paraventricular nucleus (PVN) 78, 83, 87
Parkinson's disease 55
paroxetine 268, 282
paroxysmal depolarizing shift 246
pathoclisis 55
pattern completion 107, 109
pattern integration 112–13
pattern separation 20, 106, 107, 108, 110, 112–13, 130
Pavlovian conditioning 43–4
pentraxin 163
perceptual processing 51
perforant path 158
perfusion imaging 132
Period 1 (Per1) 81
perirhinal cortex 12–13, 25, 123
pharmacological fMRI 131
phenytoin 267–8
phosphocreatine 185, 186
physical exercise
  hippocampal plasticity 134
  neurogenesis 106, 111–12, 114
  schizophrenia 297
Pick's disease 59
pimozide 191
Pittsburgh Compound B (PIB-PET) 230
place cells and place fields 6, 109, 164–8
plasticity 5
  ageing 160–2
  exercise-induced 134
  imaging 134–5
  stress 278
polymorphic layer 15–16, 17, 18, 29, 33
population spikes 159, 160
positron emission tomography (PET) 132
  Alzheimer's disease 230
  Down syndrome 188
  post-ischaemic neuronal death 203
  PTSD 267
  schizophrenia 290
  temporal lobe epilepsy 254–5
  Tourette syndrome 191
  transient global amnesia 211–12
post-transplant acute limbic encephalitis 207

post-traumatic stress disorder (PTSD)
CA3 265
corticotropin-releasing factor 266
diagnosis 263
functional imaging 267
genetic risk 266
glucocorticoid receptor expression 267
hippocampal activation 265–6
hippocampal function 264–6
hippocampal volume 264–5
hypocortisolaemia 82
memory 264
MRS 265
pharmacotherapy 267–8
psychobiology 263
risk factors 262–3
posterior cerebral artery insults 203
postictal switch 254
postnatal development 27–32
postrhinal cortex 12, 13, 25
potassium channel complex antibody-associated encephalitis 207
prefrontal cortex 82–3
prepulse inhibition of acoustic startle response 293
presubiculum 25
presynaptic fibre potential 159
preterm infants 176–9
priming 43
procedural memory 40, 44–9
propiomelanocortin peptides 78
prostaglandins 32
psychotic states
circadian rhythm 81
hypercortisolaemia 82
mifepristone treatment 97
pyramidal cells
ageing 158, 159, 160, 161, 166
autism 32–3
CA1 20, 24
CA2 21
CA3 18–20, 22, 29
schizophrenia 33
subiculum 24
temporal lobe epilepsy 33
vulnerability to ischaemia and hypoxia 56, 57

quaking 295

rapid eye movement sleep 145
reactive oxygen species 60–1
reading in amnesia 44
recollection 47
relational memory 47–9, 51
REM sleep 145
repetition suppression 234–5
resting-state functional connectivity MRI 133
retrieval 83
retrosplenial cortex 13
RU38486, *see* mifepristone

S100B 34, 187
Schaffer collaterals 21, 158, 159, 201
schemas 47
schizophrenia 288–97
aetiology 289
AMPA receptors 293–4
antipsychotic therapy 296–7
astrocytosis 292
CA1 59
CA3 291
cognitive dysfunction 290
diffusion tensor imaging 290–1
epidemiology 288–9
fMRI 290–1, 295
GABAergic system 59, 293, 294
genetic factors 33, 289, 295–6
glutamate dysfunction 293, 294
hippocampal involvement 33, 289
hippocampal volume 33, 290, 291
memory deficits 290
MRI 33
MRS 294
myelination 295
neurodevelopmental hypothesis 33, 289, 293
neurogenesis 117, 292
neurotransmitter hypothesis 289
NMDA receptors 289, 293–4
oligodendrocytes 294–5
PET 290
physical exercise 297
postmortem studies 291–2
selective neuronal vulnerability 59
sensory information processing 293
SPECT 290
synaptic connectivity 294
seizures
autism 185, 186
febrile 32, 179, 247
selective neuronal damage 57
see also epilepsy
selective neuronal vulnerability 54–71
ageing 57–8
concept 55–6
gene expression 65–70
glutamate 62–5, 201
hippocampus 56–9
mechanisms 59–70
neurological disease 55–6, 58–9, 200–2
reactive oxygen species 60–1
vascular hypothesis 55, 59
selective serotonin reuptake inhibitors 268
semantic memory 46–7, 107
septal projections 18, 21, 23
sequence learning 44
serial reaction time 44
serotonin system 14, 18, 21, 23, 90–1, 276
sertraline 282
shape analysis, *see* hippocampal shape
silent period 247
single nucleotide polymorphisms 92
single photon emission computed tomography (SPECT)
Down syndrome 188
PTSD 267
schizophrenia 290
temporal lobe epilepsy 254
Tourette syndrome 191
transient global amnesia 211–12
skill learning 40, 43
sleep-related memory consolidation 144–50
behavioural evidence 146
hippocampal-neocortical dialogue 147–8
paradoxical sleep 145
post-training wakefulness 149–50
regular/NREM sleep 145
REM sleep 145
sharp waves and ripples 145
sleep spindles 145
slow oscillations 145
synaptic downscaling hypothesis 146–7
SNARE complex 294
sodium valproate 164
Sommer's sector 5
source memory 50
spatial learning 112, 114–16, 154
Spielmeyer sector 5
stability–plasticity dilemma 109
status epilepticus 209, 246
steroids, postnatal 178–9
stratum lacunosum-moleculare 20
stratum lucidum 20
stratum oriens 20
stratum radiatum 20
stress 77–98
alarm phase 77
Alzheimer's disease 97
CA3 263
chronic stress 78
components 77
cortisol/corticosterone (CORT) 81
dendritic reorganization 88
diabetes 97
early life experiences 93–5
electrical properties of hippocampal cells 89–90
epilepsy 97
exhaustion phase 77–8
gene expression 91–2
general adaptation syndrome 77
hippocampal effects 263, 278
hippocampal volume 88
HPA axis 82–3
information processing 78
long-term potentiation 89, 94
major depression 274
molecular effects 90–2
network function 89–90
neuroendocrine substrate 78–9
neurogenesis 88–9, 278
resistance phase 77
stress hyporesponsive period 94
therapy directed at stress system 97
transient global amnesia 214–15
structural magnetic resonance imaging, *see* magnetic resonance imaging
subcortical inputs 14, 18, 21, 23, 25
subiculum 23, 24–5
ageing 156, 158, 161
postnatal maturation 29–30
subventricular zone 106
superoxide 65
sustained repetitive firing 246
sympathetic nervous system 78, 82
synapses
ageing 58, 158
mild cognitive impairment 58
schizophrenia 294
synapsin-Ia/Ib 91
synaptic downscaling hypothesis 146–7

taxi drivers 112, 134, 176, 186
temporal lobe amnesia 25

temporal lobe epilepsy (TLE) 245–57
  abdominal aura 249
  aetiology 245
  anterior temporal lobectomy 256–7
  anti-epileptic drug therapy 256
  automatisms 249
  autonomic phenomena 249
  autoscopy 249
  deep brain stimulation 5, 257
  dentate gyrus 33
  depth electrode recordings 251–2
  diffusion-weighted imaging 253
  dissociation 249
  electroencephalography 250–2
  feeling of fear 249
  first acknowledgement of hippocampal involvement 5
  fMRI 253
  focal seizures with generalized tonic–clonic convulsions 248, 249
  focal seizures with impaired awareness 249
  focal seizures with retained awareness 248–9
  frontobasal-cingulate seizures 250
  hallucinations and illusions 249
  hippocampal-amygdala seizures 250
  hippocampal neuropathology 33–4
  hippocampal perfusion 132
  hippocampal sclerosis 33, 179, 245–8, 252–3
  hippocampal volume 33
  ketogenic diet 257
  lateral temporal neocortical seizures 250
  lateral TLE 245, 249, 250–1
  mesial temporal sclerosis 246–8
  mesial TLE 245, 249, 250, 251
  motor phenomena 249
  MRS 253
  neurogenesis 118
  neuromodulatory stimulation 257
  opercular-insular seizures 250
  PET 254–5
  postictal state 249
  postictal switch 254
  pre-surgical assessment 253
  psychic phenomenon 249
  SPECT 254
  structural MRI 33, 252–3
  subclassification 249–50
  temporal polar seizures 250
  treatment 256–7
  vagus nerve stimulators 257
  vertigo 249
theta rhythm 109
thorny excrescences 16
tianeptine 267–8
time travelling 46
Tourette syndrome (TS) 190–1
trace conditioning 44, 112, 113
transgenic mice 62, 69, 85
transient global amnesia 210–15
  CA1 212
  diagnostic criteria 210–11
  hippocampal dysfunction 212, 214–15
  imaging 211–12
  jugular venous flow 211
  migraine link 211
  stress 214–15
transient ischaemic attacks 211
transient receptor potential melastatin 7 (TRPM7) 205
transitive inferences 130
trauma spectrum disorders 262–3, 265
traumatic brain injury 209
trisynaptic circuit 108–9

Ube3a 187
ultraradian rhythm 80
ultrasound scans 177

vagus nerve stimulators 257
Valsalva manoeuvre 211
vasopressin 78
ventral tegmental area (VTA) 108, 276
virtual maze 116, 154
visceral brain 6
visual areas 12
volumetry 124–7; *see also* hippocampal volume
voxel-based morphometry 126–7, 229
vulnerability of hippocampus 1, 5; *see also* selective neuronal vulnerability

Wada test 5, 256
water maze 83, 106, 111, 114, 116
word stem completion 43
Wyler score 247–8

zif268 162–3
zinc 64, 201